INTERNATIONAL SERIES OF MONOGRAPHS IN
NATURAL PHILOSOPHY
GENERAL EDITOR: D. TER HAAR

VOLUME 25

RADIO EMISSION OF THE SUN
AND PLANETS

*OTHER TITLES IN THE SERIES
IN NATURAL PHILOSOPHY*

- Vol. 1. DAVYDOV—Quantum Mechanics
- Vol. 2. FOKKER—Time and Space, Weight and Inertia
- Vol. 3. KAPLAN—Interstellar Gas Dynamics
- Vol. 4. ABRIKOSOV, GOR'KOV and DZYALOSHINSKII—Quantum Field Theoretical Methods in Statistical Physics
- Vol. 5. OKUN'—Weak Interaction of Elementary Particles
- Vol. 6. SHKLOVSKII—Physics of the Solar Corona
- Vol. 7. AKHIEZER et al.—Collective Oscillations in a Plasma
- Vol. 8. KIRZHNITS—Field Theoretical Methods in Many-body Systems
- Vol. 9. KLIMONTOVICH—The Statistical Theory of Non-equilibrium Processes in a Plasma
- Vol. 10. KURTH—Introduction to Stellar Statistics
- Vol. 11. CHALMERS—Atmospheric Electricity, 2nd edition
- Vol. 12. RENNER—Current Algebras and their Applications
- Vol. 13. FAIN and KHANIN—Quantum Electronics Vol. 1—Basic Theory
- Vol. 14. FAIN and KHANIN—Quantum Electronics Vol. 2—Maser Amplifiers and Oscillators
- Vol. 15. MARCH—Liquid Metals
- Vol. 16. HORI—Spectral Properties of Disordered Chains and Lattices
- Vol. 17. SAINT JAMES, SARMA and THOMAS—Type II Superconductivity
- Vol. 18. MARGENAU + KESTNER—Theory of Intermolecular Forces
- Vol. 19. JANCEL—Foundations of Classical and Quantum Statistical Mechanics
- Vol. 20. TAKAHASHI—An Introduction to field Quantization
- Vol. 21. YVON—Correlations and Entropy in Classical Statistical Mechanics
- Vol. 22. PENROSE—Foundations of Statistical Mechanics
- Vol. 23. VISCONTI—Quantum Field Theory—Vol. 1
- Vol. 24. FURTH—Fundamental Principles of Theoretical Physics

RADIO EMISSION OF THE SUN AND PLANETS

BY
V. V. ZHELEZNYAKOV

Professor at the Radio-Physical Institute
Gor'kii State University

TRANSLATED BY
H. S. H. MASSEY

EDITED BY
J. S. HEY

PERGAMON PRESS
OXFORD · LONDON · EDINBURGH · NEW YORK
TORONTO · SYDNEY · PARIS · BRAUNSCHWEIG

Pergamon Press Ltd., Headington Hill Hall, Oxford
4 & 5 Fitzroy Square, London W.1
Pergamon Press (Scotland) Ltd., 2 & 3 Teviot Place, Edinburgh 1
Pergamon Press Inc., Maxwell House, Fairview Park, Elmsford, New York 10523
Pergamon of Canada Ltd., 207 Queen's Quay West, Toronto 1
Pergamon Press (Aust.) Pty. Ltd., 19a Boundary Street, Rushcutters Bay, N.S.W. 2011, Australia
Pergamon Press S.A.R.L., 24 rue des Écoles, Paris 5ᵉ
Vieweg & Sohn GmbH, Burgplatz 1, Braunschweig

Copyright © 1970
Pergamon Press Ltd.

All Rights Reserved. No part of this publication may be reproduced, stored in a retrieval system, or transmitted, in any form or by any means, electronic, mechanical, photocopying, recording or otherwise, without the prior permission of Pergamon Press Limited.

First edition 1970

(This book is translation of Радиоизлучение Солнца и Планет 'Radioizlucheniye Solntsa i Planet') published by Izdatel'stvo 'Nauka", Moscow, 1964

Library of Congress Catalog Card No. 75-76797

08 013061 5

Printed in Hungary

Contents

FOREWORD	ix
FOREWORD TO THE ENGLISH EDITION	xiii
CHAPTER I. PHYSICAL CONDITIONS OF THE SUN, MOON AND PLANETS	1
1. The Sun's Atmosphere	1
The chromosphere (2). The corona (3)	
2. Solar Activity	8
Plages and flocculi (8). Sunspots (9). Flares (12). Coronal condensations (14)	
3. The Moon and Planets	15
Mercury, Venus and Mars (15). Jupiter and Saturn (17). The Moon (18)	
CHAPTER II. BASIC CHARACTERISTICS OF EXTRATERRESTRIAL RADIO EMISSION AND METHODS FOR STUDYING THEM	20
4. Frequency Spectrum	21
Aerial temperature and effective temperature of radio emission (23). Studying the radio-emission frequency spectrum. Multichannel receiving devices and radio spectrographs (27)	
5. Angular Spectrum	30
Aerial system requirements in radio astronomy. Parabolic aerials (30). The two-element interferometer (31). Modifications of the two-element interferometer (33). The problem of studying the radio brightness distribution over a source. Variable-baseline interferometer (36). The multiple-element interferometer (39). The Mills Cross (42). Eclipse observations (44)	
6. Polarization of Radio Emission	46
Polarization parameters (46). Methods of polarization measurements in the metric waveband (51). Polarization measurements in the centimetric band (59)	
7. Effect of the Earth's Atmosphere on the Observed Radio Emission	63
Absorption of radio waves in the troposphere (63). Absorption of radio waves in the ionosphere (64). Effects connected with refraction of radio waves in the atmosphere (65). Polarization change of the radio emission as it passes through the ionosphere (68)	
CHAPTER III. RESULTS FROM OBSERVATIONS OF THE RADIO EMISSION OF THE "QUIET" SUN	73
8. Frequency Spectrum of the "Quiet" Sun's Radio Emission	74
Determining the level of the "quiet" Sun's radio emission (74). Observed dependence of T_{eff} on wavelength (77)	
9. Distribution of Radio Brightness over the Sun's Disk	83
Remarks on methods of investigation. Some preliminary data (83). Features of the T_{eff} distribution over the disk of the "quiet" Sun in the radio-frequency band (86)	

Contents

CHAPTER IV. RESULTS OF OBSERVATIONS OF THE SUN'S SPORADIC
RADIO EMISSION ... 99

10. The Slowly Varying Component ... 101
    General characteristics. Correlation of the radio-emission flux with sunspots (101). Position, form and size of local sources (102). Radio-emission frequency spectrum (110). Directional properties and polarization (113). Altitude of local sources above the photosphere. Connection with optical features of the solar corona (116)
11. Microwave Bursts ... 120
    General characteristics. Basic types of microwave bursts (120). Frequency spectrum of bursts (125). Polarization of radio emission (128). Microwave bursts and chromospheric flares (130)
12. Noise Storms (enhanced radio emission and type I bursts) ... 135
    Time characteristics of noise storms (136). Frequency spectrum (137). Connection with optical features (139). Directional features of the radio emission (143). Size and position of radio-emission sources in the corona (144). Polarization of noise storms (148)
13. Type II Bursts ... 154
    General characteristics (154). Harmonics of type II bursts (157). Fine structure of type II bursts (165). Frequency drift and its interpretation (167)
14. Type III Bursts ... 176
    General characteristics (176). Polarization of bursts (180). Connection with optical phenomena (181). Position and movement of an emitting region in the corona. Frequency drift of bursts (185). U-bursts (191)
15. Types IV and V Radio Emission ... 194
    Basic characteristics of type IV radio emission (194). Type V bursts (201)
16. Other Forms of Burst ... 202
    Decimetric continuum (202). Rapidly drifting decimetric bursts (205). Continuum storms (208). The event of 4 November 1957 (210). Wide-band bursts of short duration (212). Reverse-drift pairs (212)
17. Sporadic Radio Emission and Geophysical Phenomena ... 215
    Preliminary remarks (215). Radio emission of the Sun and sudden ionospheric disturbances. Connection between microwave bursts and hard solar radiation (217). Solar radio emission and magnetic storms with a sudden beginning. Properties of geoeffective corpuscular streams (223). Radio emission of the Sun and polar blackouts. Connection between continuum-type radio emission and the appearance of energetic particles (230). General picture of the Sun's sporadic radio emission (237)

CHAPTER V. RESULTS OF OBSERVATIONS OF RADIO EMISSION OF THE
PLANETS AND THE MOON ... 244

18. First Investigations into the Radio Emission of the Moon, Planets and Comets ... 244
    First study of the Moon and planets in the radio-frequency band (244). Radio emission of comets (249)
19. Sporadic Radio Emission of Jupiter ... 250
    Radio emission flux and its time dependence (250). Frequency spectrum (253). Polarization (257). Local sources of sporadic

Contents

radio emission, their period of rotation and position on Jupiter's disk (258). Directional features of radio emission and size of local sources (264). Connection with solar activity (266)

20. Continuous Radio Emission of the Planets   269
    Radio emission of Saturn (269). Radio emission of Jupiter (270). Radio emission of Mars (277). Radio emission of Venus (277). Radio emission of Mercury (283)

21. Radio Emission of the Moon   283
    Preliminary remarks (283). Frequency spectrum and phase dependence of the Moon's radio emission (285). Radio brightness distribution over the lunar disk (292)

CHAPTER VI. PROPAGATION OF ELECTROMAGNETIC WAVES IN THE SOLAR CORONA   297

22. Propagation of Electromagnetic Waves in an Isotropic Coronal Plasma (approximation of geometrical optics)   298
    Quasi-hydrodynamic method and approximation of geometrical optics (298). Waves in an isotropic plasma (305).

23. Propagation of Electromagnetic Waves in a Magnetoactive Coronal Plasma (approximation of geometrical optics)   318
    Electromagnetic waves in a homogeneous plasma in the presence of a constant magnetic field (318). Waves in a non-uniform magnetoactive plasma (325). Faraday effect in the solar corona (329). Depolarizing factors and the question of elliptical polarization of certain bursts of solar radio emission (334)

24. Coupling of Electromagnetic Waves in a Plasma and Polarization of Solar Radio Emission   342
    Limiting polarization of emission leaving the coronal plasma (344). Preliminary remarks on the effect of the coupling of waves in the region of a quasi-transverse magnetic field (350). Calculations of coupling by the phase integral method (353). Certain features of solar radio emission polarization and their interpretation on the basis of wave coupling in the region of a quasi-transverse magnetic field in the corona (365)

25. Coupling of Electromagnetic Waves and the Problem of the Escape of Radio Emission from the Corona   373
    Preliminary remarks (373). Conversion of plasma waves into electromagnetic waves in a smoothly non-uniform isotropic plasma (376). Wave coupling in a smoothly non-uniform magnetoactive plasma (385). Conversion of plasma waves into electromagnetic waves because of scattering on electron density fluctuations (391)

CHAPTER VII. GENERATION AND ABSORPTION OF ELECTROMAGNETIC WAVES IN THE SOLAR CORONA   408

26. Emission and Absorption of Electromagnetic Waves in an Equilibrium Plasma   408
    Emission transfer equation (408). Electromagnetic wave emission by individual particles (413). Absorption of electromagnetic waves in an isotropic plasma (430). Absorption of electromagnetic waves in a magnetoactive plasma (440). Gyro-resonance absorption in the solar corona (452)

27. Emission, Absorption and Amplification of Electromagnetic Waves in a Non-equilibrium Plasma   459
    The kinetic equation method and the Einstein coefficients

Contents

method. The problem of wave amplification and instability in a plasma (459). Reabsorption and amplification of plasma waves in a non-equilibrium plasma with $H_0 = 0$ (quantum treatment) (471). Amplification and instability of plasma waves in a non-equilibrium plasma with $H_0 = 0$ (classical treatment) (478). Maximum amplitude and harmonics of amplified plasma waves (482). Reabsorption and amplification of electromagnetic waves in a non-equilibrium magnetoactive plasma (487). The appearance of plasma waves in shock wave fronts (501)

CHAPTER VIII. THEORY OF THE SUN'S THERMAL RADIO EMISSION ... 508

28. Theory of the "Quiet" Sun's Radio Emission ... 511
   Radio emission mechanism (511). Theory of the B-component in the simplest model of the chromosphere and corona (513). Interpretation of certain features in the distribution of the radio brightness over the Sun's disk on the basis of more complex models of the corona and chromosphere (521). Construction of a model of the solar atmosphere from radio data (531)

29. Origin of the Slowly Varying Component of the Sun's Radio Emission ... 538
   Thermal nature of the S-component of the sporadic radio emission (538). Bremsstrahlung mechanism of the local S-component sources above spots (542). Magnetic-bremsstrahlung mechanism of slowly varying emission (551). Origin of radio emission of haloes and local sources above flocculi free of spots (563)

CHAPTER IX. THEORY OF THE SUN'S NON-THERMAL RADIO EMISSION ... 568

30. Generation of Continuum-type Sporadic Radio Emission ... 568
   Origin of microwave bursts and certain phenomena accompanying them (569). Origin of the enhanced radio emission connected with sunspots (579). Mechanism of type IV radio emission (583)

31. Generation of Types I, II and III Bursts ... 589
   Theory of type III bursts (590). Mechanism of type II bursts (602). Generation of type I bursts (606)

CHAPTER X. ORIGIN OF RADIO EMISSION OF THE PLANETS AND THE MOON ... 610

32. Hypotheses on the Mechanism of Jupiter's Sporadic Radio Emission ... 610
   The "thunderstorm" hypothesis (610). Mechanism of plasma oscillations (612). Plasma hypothesis of the origin of Jupiter's radio emission when the planet's magnetic field is taken into account (616)

33. Origin of the Continuous Radio Emission of Jupiter and Saturn ... 624
   Radiation belts as the source of Jupiter's decimetric radio emission (629). Conditions of generation of Saturn's radio emission (631)

34. Sources of Venus's Radio Emission ... 638
   The "ionospheric" model (639). The "hot" surface model (645)

35. Theory of the Moon's Radio Emission ... 651
   Basic relations (651). Interpretation of the results of observations of the Moon's radio emission and the physical characteristics of its surface (657)

REFERENCES ... 669
INDEX ... 693

Foreword

THE task of present-day radio astronomy is to study extraterrestrial objects by means of the nature of the radio emission coming from them. Radio astronomy research is valuable for the significance of the results obtained which by no means duplicate the data of optical astronomy. It has therefore now become the basic source of information on the regions of space and objects in space which, whilst they play a basic part in the generation, reflection or scattering of radio waves, make no significant contribution to the optical part of the spectrum.

Radio astronomy divides into two fields from the point of view of the radio methods used. In one extraterrestrial objects are studied by radio signals transmitted from the Earth and reflected from them (radar astronomy); in the other information on the nature of extraterrestrial objects is obtained by studying the natural radio emission of these objects (galactic radio astronomy and radio astronomy of the solar system). As the names themselves indicate, galactic radio astronomy covers investigations into the radio emission coming to us from outside the solar system, whilst radio astronomy of the solar system covers investigations into the radio emission of the Sun, the Moon and the planets. On the borderline between radio astronomy and geophysics lie investigations of the radio emission generated in the Earth's upper atmosphere.

Shklovskii (1956) deals with galactic radio astronomy in his monograph; the problem of the generation of cosmic radio emission, which is closely linked with the question of the origin of cosmic rays, is discussed in detail in the book by Ginzburg and Syrovatskii (1964).[†] In the world's scientific literature there is, however, at present no publication summarizing the results of the extensive observations of the radio emission of the Sun, the Moon and the planets and discussing in detail from a single point of view its mechanisms. The appropriate sections of Steinberg and Lequeux (1960), Pawsey and Bracewell (1955), Pawsey and Smerd (1953), Kaidanovskii (1960), Shklovskii (1962) and the surveys by Zheleznyakov (1958a)

---

† And also in the monograph by S. A. Kaplan and S. B. Pikel'ner *The Interstellar Medium* (in Russian), Fizmatgiz, 1963.

Foreword

and Coutrez (1960) are either incomplete or considerably out of date.†
The present monograph is an attempt to fill this gap.

Extensive research into the Sun's radio emission started after the Second World War in 1945–6. Solar radio astronomy is thus not very old: it is only just about two decades old (planetary radio astronomy is even younger: its beginning was the unexpected discovery of Jupiter's sporadic radio emission in 1955).

Nevertheless, in this very short time a large amount of information has been obtained on the features of the Sun's radio emission, its connections with the dynamic processes in the Sun's atmosphere and geophysical phenomena of various kinds. The rapid progress of experimental radio astronomy and plasma physics has laid the foundations for solving the extremely difficult and complex problem of the origin of solar radio emission.

It is clear that without a theory for the solar radio emission it is impossible to understand the reasons for its connections with the complex of phenomena on the Sun and the Earth or to be able to judge on the basis of radio observations the physical conditions in the upper layers of the solar atmosphere and the processes occurring there. It should be mentioned that considerable success has been achieved recently in this field as well. Whilst several years ago one could argue about the generation mechanisms of all the basic components of the Sun's sporadic radio emission without exception, now the relative role of the various emission mechanisms in the coronal and chromospheric plasma has become clearer and better defined. At the same time as developing qualitative ideas on the generation conditions a preliminary quantitative theory has been worked out for the majority of the components; in the remaining cases there have been clear expressions of alternative points of view on the origin of the radio emission and experiments have been suggested which will finally enable us to select the generation mechanism.

Research into the Moon's radio emission is developing rapidly. Information on the composition and structure of its surface layers given by the analysis of results of observations of the Moon's natural radio emission and the data from optical and radar astronomy is now very much to the point because of the scientific apparatus to be sent to the Moon and Man's flight there.

It is likewise difficult to overestimate the importance of planetary radio astronomy which is at present taking its first steps. The relatively low level

---

† Good surveys on the radio emission of the Sun, the planets and the Moon have recently been produced by Wild, Smerd and Weiss (1963 and 1964), Roberts (1963, 1964) and Krotikov and Troitskii (1963).

Foreword

of our knowledge of the physics of planetary atmospheres (when compared with the state of the fields of astrophysics in which the atmospheres of the Sun and stars are studied) is due to the limitations of the methods of optical astronomy which permits study of the planets in reflected light only. Only the extension of the part of the spectrum studied into the long-wave region and in particular the transition into the radio-frequency band have made it possible to study the natural emission of the planets. As a result extremely important discoveries have been made here, the most important being the magnetic field and radiation belt of Jupiter and the temperature conditions of the surface of Venus.

The ideas expressed in the monograph on the origin of the radio emission of the planets cannot be considered final. It may be assumed, however, that these ideas will serve as a basis for further development of the theory of planetary radio emission.

When working on the manuscript we tried to reflect as fully as possible the outstanding successes achieved by radio astronomy of the solar system. At the same time stress has been laid also on as yet unsolved theoretical and experimental questions.

The list of references at the end of the book is not exhaustive but it does, in our opinion, cover all the basic publications on the question under discussion up to the middle of 1963.

The author is extremely grateful to V. L. Ginzburg, who suggested the writing of this book, for his valuable discussions of various aspects of solar radio emission theory and plasma physics and to G. A. Semenova for her great help in compiling the manuscript and making the numerical calculations.

V. V. ZHELEZNYAKOV

Foreword to the English Edition

INVESTIGATIONS of the radio emission of the Sun and planets do not yield such outstanding discoveries as those in which the galactic and metagalactic radio astronomy is so rich. Nevertheless, the study of the radio emission coming from the Sun and planets gave much rather significant and very often unexpected information on the processes in the upper layers of the solar and planetary atmospheres, as well as on the physical characteristics of the surface of the Moon and planets. This information is set forth in the English edition of my book which is now presented to the reader. I think that it will be of interest for the radio astronomy specialists who are carrying out experimental and theoretical investigations of the radio emission of the Sun and planets and for those who wish to make the acquaintance with the state of this field of radio astronomy.

At the same time the book will be valuable for radio astronomers and astrophysicists who make investigations or are interested in the state of investigations in galactic radio astronomy. The fact is that the radio emission from the objects such as pulsars, flare stars and interstellar OH emission sources can be interpreted on the basis of coherent mechanisms of radiation; the mechanisms of such kind were developed and used long ago and successfully for explaining many components of sporadic radio emission of the Sun. In this respect Chapters VI and VII, which are devoted to the general problems on the generation and propagation of radio waves in a cosmic plasma, are of special interest.

Soon after the Russian variant of this book had appeared, an excellent monograph by M. R. Kundu, *Radio Emission of the Sun*, was published. In this monograph the basic consideration is concentrated upon the analysis and preliminary interpretation of the results of observations of the solar radio emission. The present book supplements well the monograph by Kundu as to the problems on the generation and propagation of electromagnetic waves in a cosmic plasma and the problem of the origin of the radio emission from the Sun and planets. It is these problems which are mainly accentuated in this book. At the same time the subject of this book is broader than the monograph by Kundu in which the problems concerned with the radio emission of the Moon and planets are not discussed.

Foreword to the English Edition

Five years lying between the Russian and English editions is a large period in such a rapidly developing field as radio astronomy. Though the main content of the book retained conserves its importance up to the present, much new material was obtained from the theory and observations during that period. Among new results, we note here first the investigations of decameter radio emission from Jupiter (the effect of the Io satellite and the interplanetary medium on the features of the radio emission, etc.). In the theory of the generation and propagation of radio waves in a cosmic plasma of great importance are the discovery of the effect of synchrotron instability (it made possible to realize the coherent synchrotron mechanism of radiation), the investigations of non-linear processes of evolution of excited plasma waves in a stream-plasma system, and the allowance for the process of induced scattering in the conversion of plasma waves into electromagnetic ones. Considerable progress has been achieved also in the origin of the solar radio emission during recent years (here we mean the papers on the generation of type V solar bursts, the development of a more correct theory of type III bursts, taking into account a possible part of the coherent synchrotron mechanism in the occurrence of type IV radio emission, etc.).

A detail consideration of all these problems would require an essential reprocessing and increasing of the volume of the book which is already very large. I confined myself, therefore, in preparing this edition with some supplements and changes and also with references to the sources where one can make the acquaintance of the basic data and ideas which became familiar after 1964 and are concerned with the subject of this book.

All noticed inaccuracies and misprints of the Russian edition have been eliminated in the monograph. In this connection the remarks made by my colleagues V. V. Zaitsev and E. Ya. Zlotnik played a great part and I am sincerely thankful to them. I express many thanks to E. Ya. Zlotnik and E. V. Suvorov for their great help in reading the proofs of the book. I am grateful to D. ter Haar, J. S. Hey and H. S. H. Massey, thanks to whom it became possible to publish the English edition of this monograph.

V. V. ZHELEZNYAKOV

CHAPTER I

Physical Conditions of the Sun, Moon and Planets

THIS chapter provides the necessary information on the physical conditions in the upper layers of the Sun, the Moon and the planets for the discussion that follows. The reader can obtain more detailed information from monographs, surveys and original papers.†

1. The Sun's Atmosphere

Three basic layers are generally recognized in the solar atmosphere: the photosphere, the chromosphere and the corona. The photosphere is the name given to the thin layer (only about 300 km thick altogether) which is the main source of the Sun's optical radiation. The external radius of the photosphere is therefore taken as the radius of the visible solar disk $R_\odot \approx 6.95 \times 10^5$ km. From the Earth, i.e. at a distance of one astronomical unit (1.5×10^8 km), the solar disk subtends an angle of 32'.

The temperature of the lower photosphere, which is responsible for the production of the optical radiation in the continuous spectrum, is about 6×10^3 °K.

In the upper photosphere (the reversing layer), which is largely responsible for the formation of the Fraunhofer absorption lines, the temperature is lower. The decrease in temperature in the photosphere with height makes it possible to understand the observed "darkening" of the Sun at the edge of its disk: the reason is that near the Sun's limb the radiation comes from higher and colder layers of the photosphere than in the centre of the disk.

Above the photosphere there is the more rarefied chromosphere, which at heights of $h \sim 2 \times 10^4$ km above the level of the photosphere changes into an extremely extensive formation—the corona. The latter can be

---

† See, e.g., Shklovskii (1962), Kuiper and Middlehurst (1961), Kuiper (1947, 1953), Ambartsumyan et al. (1952), Sharonov (1958), Moore (1959), Vaucouleurs (1951), Peck (1958), Markov (1960), Flügge (1959) and Allen (1955).

1

observed by radio astronomy methods (from the scatter of the radio emission of discrete sources) right out to a distance of the order of $(40–55)R_\odot$ from the Sun, and sometimes as far as about $120R_\odot$, where the corona gradually changes into the interplanetary medium (Slee, 1961). The outermost layers of the corona are sometimes called the "supercorona".

THE CHROMOSPHERE

In photographs of the chromosphere in monochromatic light, e.g. in H_α light (the first line of the Balmer series), structural elements of the chromosphere (spicules) can be clearly distinguished near the Sun's limb. These spicules take the form of numerous jets about 10^3 km thick rising from an altitude of $h \sim 5\times 10^3$ km (i.e. from the accepted boundary between the lower and upper chromosphere) to an altitude of the order of 10^4 km and taking several minutes to fall back. The spicules are colder than the ambient medium and have a temperature of $T \sim (2–4)\times 10^4$ °K. In the deeper layers of the chromosphere cold and hot elements also coexist at the same level; the first are responsible for the metal lines with a temperature of $T \sim 4\times 10^3$ °K, and the latter for the hydrogen emission with $T \sim 10^4$ °K. At present it is not clear whether the spicules are a continuation of the cold or the hot elements of the lower chromosphere, although the latter is more probable.

For what is to follow the distribution of the electron concentration N and the kinetic temperature T in the chromosphere is a major point of interest. N and T can be determined as functions of the altitude from an analysis of the emission line intensity; Thomas and Athay (1957) discuss methods of solving this problem which is by no means simple. There are big differences between the results of different authors (Flügge, 1959). The values of N and T are most uncertain in the upper chromosphere, i.e. in the transition region between the lower chromosphere and the corona. More certain results have been obtained for the lower chromosphere although even these, to judge from the data of radio measurements, need considerable corrections.

In general it follows from the optical observations that the electron concentration decreases and the kinetic temperature increases with altitude, starting from values of $N \sim 10^{12}$ electrons/cm^3 and $T \sim 4400$°K at the boundary with the photosphere. Since in the layers of the photosphere which are responsible for the optical radiation in the continuous spectrum the temperature is about 6000°K, it is clear that at the boundary between the photosphere and the chromosphere there is a kinetic temperature minimum. At the same time the degree of ionization of the chromosphere material, which is largely hydrogen, increases rapidly with the rise in temperature as we move away from the photosphere, starting at $\sim 10^{-4}$ at the $h = 0$ level and approaching unity at altitudes of $h \gtrsim 6\times 10^3$ km. The comparatively slow rise in temperature in the lower chromosphere changes to a very rapid rise in its upper layers. At altitudes of $h \sim 2\times 10^4$ km the temperature becomes comparable with 10^6 °K and the electron concentration with 5×10^8 electrons/cm³, i.e. to the values characteristic for the lower corona.

TABLE 1

h, km	$\log N_a$	$\log N$	T, °K	$\log N$	T, °K	A
	Cold elements			Hot elements		
0	15.6		4400		4400	
1000	13.5	11.3	5000		9000	
2000	12.8	10.8	5500	11.3	12,000	0.15
3000	12.3	10.5	6000	10.9	15,000	0.1
4000	11.9	10.0		10.6		0.05
	Materials between spicules			Spicules		
6000	9.4	9.4		10.9	25,000	0.02
8000	9.0	9.0	$> 10^5$			0.01
10,000	8.8	8.8			40,000	0.001

As a guide Table 1 shows a two-component model of the chromosphere —the distribution at an altitude h of the kinetic temperature T, the electron concentration N and the concentration N_a of all the atoms (both the ionized ones and the neutral ones). The table also gives approximate values of the relative area of the solar surface A occupied by the "hot" elements of the chromosphere ($h < 5\times 10^3$ km) and the spicules ($h > 5\times 10^3$ km).

For a detailed treatment of models of the chromosphere based on radio astronomy data see section 28.

THE CORONA

Optical investigations of the corona in the continuous spectrum and in the emission lines provide data on the temperature and density. The continuous spectrum emission of the corona consists of two components— polarized and unpolarized. The polarized emission (or the K-corona) arises as the result of scattering on the coronal electrons of light passing from the photosphere. The second component (the F-corona) is due to scattering of light on dust particles reaching the corona from interplanetary space. Photometry of the K-corona makes it possible to determine the distribution of the electron concentration in the solar corona from the upper boundary of the chromosphere out to distances R from the centre of the Sun at which the concentration N is not less than 10^5 electrons/cm^3. At greater distances investigations of the K-corona lead to less reliable estimates of N.

Since the values of the electron concentration depend on the phase of the solar activity, and are different at different latitudes, and are determined to a considerable extent by the corona's structural features during the observations, it is not surprising that the values of N given by different authors do not agree (Shklovskii, 1962; Kuiper, 1953; Allen, 1955; Thomas and Athay, 1957; Newkirk, 1959; Bogorodskii and Khinkulova 1950). As a whole, however, closer results are obtained than for the chromosphere and they have a greater degree of reliability.

The corona, just like the chromosphere, is largely inhomogeneous. Amongst its structural features the most noticeable are the so-called coronal rays, which are bright extensive formations with an enhanced concentration of electrons sometimes several solar radii long.

The dimensions and form of the corona change considerably over the 11-year cycle of solar activity. At the cycle maximum when the Sun's activity, largely expressed in the appearance of spots, flares and the phenomena accompanying them, reaches a peak the corona takes on a more or less spherical shape: the extent of the corona at all heliographic latitudes becomes approximately the same (although it is slightly less in the polar regions). At the time of a minimum this "maximum" shape of the corona gives way to the "minimum" shape, which differs in its smaller extent, particularly in the polar regions, and as a result the coronal isophots assume a characteristic oblate form.

We note that the maximum shape reaches complete development a little before the solar activity maximum. Soon after it the corona assumes its minimum form, which thus remains for the majority of the 11-year cycle.

The distributions of the electron concentration N and the kinetic temperature T averaged over the corona's structural features are important for solar radio astronomy. One such model of the corona is given in Table 2 which gives values of the electron concentration $N(R)$ in electrons per cubic centimetre for the maximum and minimum forms of the "quiet" Sun's corona. In both cases the distributions $N(R)$ are given separately for the equatorial and polar directions. Table 2 was compiled by De Jager (1959b) on the basis of the following measurements of the K-corona: for the minimum form in the range $1\cdot03 < R/R_\odot < 3$ (Van de Hulst, 1953) and for the maximum form in the range $1\cdot03 < R/R_\odot < 2$ (Newkirk, 1959); the

§ 1] The Sun's Atmosphere

values of N for $R > 3R_\odot$ are given in accordance with Blackwell's (1956) and Elsässer's (1957) data. The table also gives the distribution $N(R)$ in the corona above a centre of activity (Newkirk, 1959) and the averaged values of N in the upper chromosphere (De Jager, 1959a).

In radio astronomical calculations and estimates of N in the corona we often use the so-called Baumbach–Allen formula:

$$N(R) = 10^8[1\cdot55(R/R_\odot)^{-6} + 2\cdot99(R/R_\odot)^{-16}]. \tag{1.1}$$

The values of $N(R)$ defined by (1.1) take up an intermediate position between the distributions of the electron concentration in the equatorial zone near the maximum and minimum of solar activity. At the same time they are closest to the values of N given in Table 2 for the maximum form of the corona in the polar latitudes.

As well as the values of the electron concentration we shall be needing the values of the characteristic (Langmuir) frequency of the coronal plasma oscillations $f_L = 2\pi\omega_L$, where $\omega_L = (4\pi e^2 N/m)^{1/2}$ (e is the charge of an electron with a mass m). The distribution $f_L(R)$ in the upper chromosphere and the corona is also given in Table 2 (in c/s).

It follows from the data given that at the lower latitudes the mean electron concentration changes by a factor of about 4 in the change from the minimum to the maximum form of the corona; this change is half as much according to other estimates (Kuiper, 1953). The discrepancy evidently occurs because the values of $N(R)$ for the maximum corona given in Table 2 relate to the last, exceptionally strong solar activity maximum.

All the available astrophysical data point to a very high kinetic temperature in the corona—of the order of 10^6 °K. Only the existence of this sort of temperature can explain the presence in the coronal spectrum of the emission lines of multi-ionized atoms such as the red line $\lambda = 6374$ Å belonging to Fe X, the green line $\lambda = 5303$ Å of the Fe XIV ion and, lastly, the yellow line $\lambda = 5694$ Å, which evidently is due to the appearance of Ca XV in the active regions of the corona. The observed width of the emission lines connected with the Doppler frequency shift of ions moving at thermal velocities leads to a kinetic temperature of the same order. The high temperature explains the considerable extent of the corona and the high intensity of its thermal radio emission (the radio emission of the "quiet" Sun, see section 28).

The kinetic temperature in the corona varies slightly with altitude, remaining close to 10^6 °K in the internal regions of the corona and apparently rises by several factors in layers located above the centres of activity. In the external layers of the corona the temperature should decrease,

TABLE 2

$R/R_\odot$	Minimum form				Maximum form	
	equator		pole		equator	
	N	f_L	N	f_L	N	f_L
1·011	$1·1\times10^9$	$3·1\times10^8$	$8·9\times10^8$	$2·7\times10^8$		
1·014	$4·0\times10^8$	$1·8\times10^8$	$2·8\times10^8$	$1·5\times10^8$	$8·0\times10^8$	$2·5\times10^8$
1·022	$2·0\times10^8$	$1·3\times10^8$	$1·3\times10^8$	$1·0\times10^8$		
1·03	$1·6\times10^8$	$1·1\times10^8$	$1·0\times10^8$	$9·0\times10^7$	$6·3\times10^8$	$2·2\times10^8$
1·06	$1·3\times10^8$	$1·0\times10^8$	$7·1\times10^7$	$7·6\times10^7$	$5·0\times10^8$	$2·0\times10^8$
1·1	$7·9\times10^7$	$8·0\times10^7$	$4·0\times10^7$	$5·7\times10^7$	$3·5\times10^8$	$1·7\times10^8$
1·2	$4·0\times10^7$	$5·7\times10^7$	$1·3\times10^7$	$3·2\times10^7$	$1·6\times10^8$	$1·1\times10^8$
1·4	$1·3\times10^7$	$3·2\times10^7$	$2·5\times10^7$	$1·4\times10^7$	$5·0\times10^7$	$6·3\times10^7$
1·6	$5·6\times10^6$	$2·1\times10^7$	$7·9\times10^5$	$8·0\times10^6$	$2·0\times10^7$	$4·0\times10^7$
2·0	$1·8\times10^6$	$1·2\times10^7$	$2·0\times10^5$	$4·0\times10^6$	$6·3\times10^6$	$2·2\times10^7$
3·0	$4·0\times10^5$	$5·7\times10^6$	$1·6\times10^4$	$1·1\times10^6$	$4·0\times10^5$	$5·7\times10^6$
5·0	$6·3\times10^4$	$2·2\times10^6$	$2·5\times10^3$	$4·5\times10^5$	$4·0\times10^4$	$1·8\times10^6$
10	$1·0\times10^4$	$9·0\times10^5$				
20	$2·5\times10^3$	$4·5\times10^5$				
60	$\sim 10^3$	$2·8\times10^5$				

$R/R_\odot$	Maximum form				Baumbach–Allen distribution	
	pole		active region			
	N	f_L	N	f_L	N	f_L
1·011			$1·5\times10^9$	$3·5\times10^8$		
1·014	$4·0\times10^8$	$1·8\times10^8$	$1·4\times10^9$	$3·4\times10^8$		
1·022			$1·4\times10^9$	$3·3\times10^8$		
1·03	$3·2\times10^8$	$1·6\times10^8$	$1·3\times10^9$	$3·2\times10^8$	$3·2\times10^8$	$1·6\times10^8$
1·06	$2·5\times10^8$	$1·4\times10^8$	$1·0\times10^9$	$2·8\times10^8$	$2·3\times10^8$	$1·4\times10^8$
1·1	$1·8\times10^8$	$1·2\times10^8$	$7·1\times10^8$	$2·4\times10^8$	$1·5\times10^8$	$1·1\times10^8$
1·2	$7·9\times10^7$	$8·0\times10^7$	$3·4\times10^8$	$1·7\times10^8$	$6·8\times10^7$	$7·4\times10^7$
1·4	$2·6\times10^7$	$4·6\times10^7$	$1·0\times10^8$	$1·0\times10^8$	$2·2\times10^7$	$4·2\times10^7$
1·6	$1·0\times10^7$	$2·4\times10^7$	$4·0\times10^7$	$5·7\times10^7$	$9·4\times10^6$	$2·7\times10^7$
2·0	$2·9\times10^6$	$1·5\times10^7$	$9·0\times10^6$	$2·7\times10^7$	$2·4\times10^6$	$1·4\times10^7$
3·0					$2·2\times10^5$	$4·2\times10^6$
5·0					$9·9\times10^3$	$8·9\times10^5$
10					$1·5\times10$	$3·5\times10^4$
20					$2·4$	$1·4\times10^4$
60						

§ 1] The Sun's Atmosphere

gradually approaching the temperature of the interplanetary medium ($\sim 10^4$ °K).

The high kinetic temperature leads to the coronal material (largely hydrogen) being in the state of an almost completely ionized plasma: only a 10^{-7} part of all the atoms of hydrogen remains neutral.

The Zeeman effect in the solar spectrum, the characteristic configuration of the coronal rays in the polar regions and the features of the motion of the material in the corona above groups of spots (see section 2) indicate the presence of magnetic fields on the Sun. In high heliographic latitudes where there are no centres of activity the general magnetic field of the Sun is probably close to a dipole. At the level of the photosphere its strength, according to the latest data, does not exceed 1 oersted (oe). At low latitudes this field is masked by the stronger local fields due to the centres of activity; we shall deal with them in the next section. The polarity of the Sun's total field does not remain the same. During the last solar activity maximum (1957–8) the direction of the magnetic dipole switched round, so that at present the orientation of the Sun's total field is the same as the direction of the Earth's magnetic field (Babcock, 1959).

Very valuable information on the upper layers of the corona (the "supercorona") have been obtained by observations of the radio eclipses of discrete sources (chiefly the Crab Nebula) by the solar corona (Vitkevich, 1959, 1960a, 1960b and 1961; Machin and Smith, 1951a and 1951b; Hewish, 1958 and 1959; Slee, 1961). The minimum angular distance by which the Crab Nebula approaches the Sun is about 70′; its radio emission then passes through the corona at a distance of $4 \cdot 5 R_\odot$ from the centre of the Sun. The observed angular size of this discrete source (or, what is the same thing, the width of the angular radio emission spectrum), which a long way from the Sun is close to 6·5′ (Vitkevich and Udal'tsov, 1958), increases sharply as one approaches the Sun.† This effect is caused by the scatter of the received radio emission on the coronal inhomogeneities of the electron concentration; it may also prove to be significant when estimating the true extent of the sources of solar radio emission from their observed angular size.

Investigations of the passage of radio emission from the Crab Nebula through the supercorona have shown (Vitkevich, 1960; Hewish, 1959) that the scatter of the radio waves in a radial direction is much less than the scattering at right angles to it, although noticeable deviations in the orientation of the axis of maximum scatter have also been found. This anisotropic scattering effect can be explained by the "anisotropy" of the

† Jupiter can be used as a "test" source of decametric radio emission (see Vitkevich, 1957a).

inhomogeneities in the supercorona, i.e. their elongated form and their orientation largely in directions that are close to radial. It is not excluded that similar inhomogeneities of the supercorona are an extension of the radial structure of the internal parts of the corona. The form and orientation of the inhomogeneities, in all probability, are caused by the effect of the Sun's quasi-radial magnetic field which renders difficult the diffusion of charged particles across the lines of force.

These investigations therefore provide important indications of the existence of magnetic fields in the supercorona and their difference from the magnetic dipole field in low heliographic latitudes (in the polar regions the dipole field is clearly quasi-radial in nature). It is possible that the observed orientation of the magnetic field at low latitudes is established by the action of corpuscular streams leaving centres of activity (in this connection see section 17).

Valuable information can be obtained on the magnitude of the magnetic field in the supercorona by measuring the polarization of the Crab Nebula's radio emission during its eclipse by the solar corona (Ginzburg, 1960a). See Gol'nev, Pariiskii and Soboleva (1963) on the first observations of this kind.

2. Solar Activity

In the previous section we have spoken about the physical conditions in the relatively stationary atmosphere of the "quiet" Sun. It is rare that the Sun can be observed in this state, however: in its atmosphere there are generally local disturbances—plages, spots, flares and prominences which are closely interconnected and are grouped into so-called centres of activity.

PLAGES AND FLOCCULI

The development of a centre of activity starts with the appearance on the solar disk of a plage—a bright formation visible in white light near the limb and in the light of the intense chromosphere lines (chiefly the H_α hydrogen line and the K-line of doubly ionized calcium) over the whole disk. The areas observed in white light are called photospheric plages and their upper part recorded in monochromatic light chromospheric plages. The more intense chromospheric plages are sometimes called flocculi.

The plages may be very extensive, forming plage fields which cover a considerable part of the solar surface in the low latitudes. The plages which are not connected with sunspots are small, less bright and, as a rule, do not last longer than a month. On the other hand, the plages among which groups of spots develop evolve rapidly, forming in the course of a

§ 2] Solar Activity

few days bright compact groups of flocculi which last for several months. They can also be observed after the decay of a group of spots in a time that considerably exceeds the life of the latter. The extent of these plages is comparable with the linear size of a group of spots or several times greater.

In regions of the solar disk occupied by plage fields (groups of flocculi) and active prominences, local magnetic fields are found whose strength in the photosphere is as much as 100–200 oe (generally several times less) (Zirin and Severny, 1961; Zirin, 1961). Apparently the lines of force run largely along the normal to the solar surface. In cases when there is a group of spots in the plage field the distribution of the magnetic fields outside the spots matches up badly with the structure of this group. On the other hand the general form of the extensive regions occupied by local magnetic fields repeats the contours of the plages defined by the distribution of the emission in the Ca II K-line (Leighton, 1959).

SUNSPOTS

At almost the same time as the flocculi (generally a few hours or days later) sunspots begin to develop in them. The sunspots can be observed by eye on the Sun's disk as dark formations due to the sharp contrast between the photosphere heated to 6×10^3 °K and the "colder" region of the spot: the temperature in the central part (in the umbra) of a large, well-developed spot is about $4 \cdot 3 \times 10^3$ °K; on the periphery of a spot (in the penumbra) it is about $5 \cdot 5 \times 10^3$ °K.

Spots appear in two zones 15–20 heliographic degrees wide at a distance of not more than 45° from the Sun's equator.

In the process of its development a group of spots passes through various stages each of which is distinguished by a type or class of spots. The accepted classification includes nine types of spot denoted by the letters from A to J; it is based on features of the spots such as size, the presence of penumbra and the nature of the magnetic field.

The development of a very large group of spots is achieved by the successive transition of the group from one class to another. The life of such a group may be more than two months. Groups of spots reach their greatest development in classes E and F. Class E includes large (not less than 10° in longitude) bipolar groups† with a complex structure; both the main spots

---

† A bipolar group extends in a direction approximately parallel to the equator. It consists of spots concentrated chiefly in the eastern and western parts of the group. In these regions are localized two large spots of opposite magnetic polarity (see below). Since the group moves from east to west because of the Sun's rotation the spot in the western part of the group is called the leader and the one in the eastern part the follower. The area of the leading spot is generally equal to or greater than the area of the following one, although it is not unusual for the opposite to be the case.

in the group have penumbrae; there are numerous small spots between the main spots. Very large (not less than 15° in size), well-developed bipolar or multipolar groups of spots form class F. At late stages in their lives, after the decay of one of the main spots, the groups become unipolar. The smaller groups miss out the more developed stages of E and F; their lives are accordingly shorter and on the average are no more than a month.

The total area σ of the spots visible on the surface of the Sun is a convenient index of solar activity. As well as this another quantity is used as a characteristic of the level of solar activity: the sunspot number (the Wolf number, W), which characterizes not only the number of spots but also their degree of concentration into groups (Kuiper, 1953).

Many years of observations have shown up characteristic changes in the values of σ and W with a period of about 11·1 years, which reflect the presence of a basic cycle of solar activity.

With the sunspots are connected strong magnetic fields whose strength increases with the size of the spot, reaching maxima of 3500 and even 4000 oe for large spots $\sim 5 \times 10^4$ km in diameter. The reduction in area of a spot during its decay has little effect on the strength of the magnetic field which remains at the maximum level for most of the spot's life. Sometimes a noticeable magnetic field remains even when the group has disappeared; strong magnetic fields (up to $1 \cdot 5 \times 10^3$ oe) sometimes appear outside the spots in regions which are not occupied by any active formation (Babcock, 1959).

The field strength distribution at the photosphere level has a maximum in the centre of a spot and drops at the outer edge of the penumbra (up to values characteristic for the plage magnetic fields of $H_0 \lesssim 100\text{--}200$ oe). The orientation of the field also changes with the transition from the centre to the periphery: the angle ϑ between the lines of force and the normal to the Sun's surface approaches 0 at the centre and $\pi/2$ at the edge of the spot.

The majority of well-developed groups of sunspots, particularly those with clearly defined leader and follower spots, are bipolar in nature: the magnetic fields in the preceding and following parts of the group run in opposite directions. The flux through the leading spot is generally several times higher than the flux through the following spot (four times as high on the average). It follows from this that most of the flux from the leading spot is scattered in a region not occupied by spots (largely on the plages).

The polarities of the leader spots in groups located in the northern and southern solar hemispheres are different. In this case the sign of the magnetic field remains constant for the whole of the 11-year cycle, including the activity maximum, changing to the opposite at the cycle minimum.

§ 2] Solar Activity

It follows from this that the complete magnetic cycle of the spots is 22 years.

The position and nature of the motion of sunspot-type prominences in the corona† and the circular polarization of the radio emission leaving the regions above centres of activity (see Chapter IV) indicate that the magnetic fields of the spots penetrate the corona. However the greater width of the coronal emission lines and the lower values of the field do not here permit direct measurements of the strength from the Zeeman effect.‡

For practical calculations it is sometimes assumed that the field above a spot can be approximated by the field of a circular bar magnet $2b$ in diameter and $l \gg b$ long, one pole of which is located at the level of the photosphere (Fig. 1a). Here the field strength on the axis of the spot

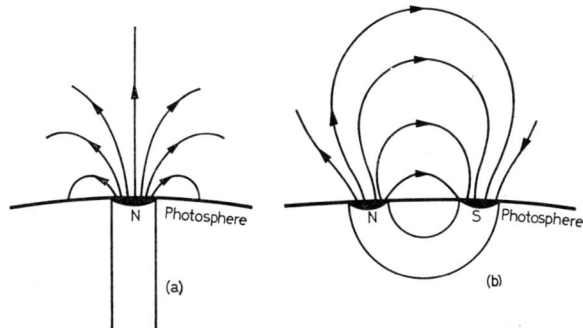

FIG. 1. Models of the magnetic field of sunspots: (a) unipolar spot; (b) bipolar spot

(i.e. on the normal to the Sun's surface passing through the centre of the spot)

$$H_0 = H_{0b}\left(1 - \frac{h}{\sqrt{h^2+b^2}}\right), \qquad (2.1)$$

where H_{0b} is the field at the base of a spot of radius b, h is the altitude above the photosphere. The values of the magnetic field H_0 in the chromosphere and the corona defined by the expression (2.1) are given in Table 3

---

† Sunspot-type prominences appear in the corona above spots because of condensation and cooling of the coronal material. The matter making up a prominence is drawn into the spot along constant trajectories. The form of these trajectories is determined in all probability by the configuration of the local magnetic field's lines of force along which the ionized matter moves in the corona.

‡ Zirin (1961) and Severny (1958) give data on the existence of magnetic fields with a strength of $H_0 \sim 10^3$ oe in the lower chromosphere above a spot.

for $H_{0b} = 2 \cdot 5 \times 10^3$ oe and $b^2 = 10^{19}$ cm². The table also gives the values of the electron gyro-frequencies $f_H = \omega_H/2\pi$, where $\omega_H = eH_0/mc$ (c is the velocity of light in a vacuum). This kind of model obviously best corresponds to a unipolar spot group in which only one magnetic pole is clearly defined. As a guide we can use the unipolar model in a bipolar group as well for the field characteristic in the immediate vicinity of the spots, at altitudes of $h \leqslant 2b$, since in these layers of the chromosphere and the inner corona the magnetic field is determined chiefly by one spot. On the other hand the unipolar model is also applicable to a certain degree at great altitudes in the corona (much greater than the linear dimensions of the group) in the case when in the bipolar group the magnetic flux through the follower spot is several times less than the flux through the leader spot. In the case of a bipolar group with magnetic fluxes of the same strength through the leading and following spots the field of the spots can be modelled by the field of a horseshoe magnet with poles located at the level of the photosphere (Fig. 1b).†

TABLE 3

$R/R_\odot$	h, km	H_0, oe	f_H, c/s	$R/R_\odot$	h, km	H_0, oe	f_H, c/s
1·00	0	2·5×10³	7·0×10⁹	1·1	70×10³	2·2×10²	6·3×10⁸
	2×10³	2·3×10³	6·6×10⁹	1·2	14×10⁴	6·1×10	1·7×10⁸
	5×10³	2·1×10³	5·9×10⁹	1·4	28×10⁴	1·5×10	4·3×10⁷
	10×10³	1·7×10³	4·9×10⁹	1·6	42×10⁴	7·3	2·0×10⁷
	16×10³	1·4×10³	3·8×10⁹	2·0	70×10⁴	2·5	7·0×10⁶
1·03	21×10³	1·1×10³	3·1×10⁹	3·0	14×10⁵	6·5×10⁻¹	1·8×10⁶
1·06	42×10³	5·0×10²	1·4×10⁹	5·0	28×10⁵	1·5×10⁻¹	4·2×10⁵

FLARES

At times in a centre of activity there are sudden increases in the brightness of the flocculi in the monochromatic spectrum—chromospheric flares. They are particularly noticeable in the H_α and K Ca II lines. In very rare cases when the flare becomes very strong it may also become visible in integrated light. The time-dependence of the emission in different parts of the flare (particularly a large one) may be different, but on the whole

† It is often convenient to use another model of the magnetic field above spots in which the field is made up of elementary magnetic dipoles located below the photosphere. In the case of a unipolar spot one vertical dipole is taken, for a bipolar one one horizontal or inclined dipole or two vertical dipoles pointing in opposite directions.

§ 2] Solar Activity

it can be taken that the emission rises quite rapidly (in a few minutes) to a maximum and then gradually decreases for 10–100 min.

As a rule flares appear around sunspots, chiefly near actively developing groups, in the first 10–15 days of their existence and only in rare cases on plages without spots. Flares appear quite often: during the life of a large well-developed group of spots several tens of flares may occur.

All flares are divided into several classes according to their scales. The basis of the classification is the area taken up by the flare on the Sun; in addition, the different classes (or degrees) of flares are distinguished by duration, width and intensity of the H_α line. All these quantities rise with the degree of the flare. The weakest flares (microflares) belong to class 1−. Their area expressed in millionth parts of the Sun's hemisphere is no more than 100 units. Class 1, 2 and 3 flares are localized on the Sun in regions with areas of 100–250, 250–600 and 600–1200 units respectively. Exceptionally powerful and rare events occupying an area greater than 1200 units belong to class 3+. Flares arise in the chromosphere and sometimes also embrace the transition region between the chromosphere and the corona. The height of the flare—the distance from its top to the photosphere—shows considerable variation from case to case, being on the average about 8×10^3 km although a considerable number of flares (about 15%) have a height of more than 18×10^3 km (Warwick and Wood, 1959).

The temperature in the region of a flare is of the order of $(1-1 \cdot 5) \times 10^4 \,°K$. According to Severnyi's (1957) estimates the particle concentration is about 10^{13} atoms/cm^3; the degree of ionization is of the order of unity, i.e. the value of the electron concentration is also close to 10^{13} electrons/cm^3. According to different data (Allen, 1955) the density of the matter is several orders higher.

Cinematographic and spectrographic observations show that the phenomenon of a flare proceeds in an extremely non-stationary manner. During its existence the form of a flare undergoes considerable changes. Material is thrown out of the flare region into the corona (prominences of the surge type) and this is clearly visible in the H_α line (see Fig. 54). The probability of an eruption rises with the class of the flare, reaching 80% for class 3 events. The ejected material often returns to the chromosphere along one and the same trajectory. The latter is evidently determined to a considerable extent by the orientation of the magnetic field in the corona. The average velocity is not more than 150 km/sec, but sometimes it is more than 300 km/sec.

Higher velocities are observed when flares during their development pass through a characteristic "explosive" phase (during which the edges of the flare suddenly start to expand with a velocity of several hundreds of kilometres per second†) and cause rapid changes and decay of the dark filaments (prominences) located at great distances in other centres of activity. From the time lag of the moments of activation relative to the period of the sudden expansion it may be concluded that the velocity of the propagation of disturbances from the flare is of the order of 10^3 km/sec. For three events of this kind the presence of high velocities has been confirmed by direct observations of the Doppler shift of the H_α line in the region of the flares. In one case the velocity found in this way for a large cloud formation along the line of sight was not less than 2500 km/sec, whilst the transverse velocity component of the disturbances was 550–2500 km/sec. It is quite possible that we are dealing here with direct observations of the corpuscular fluxes from the flare (or its immediate vicinity) which on passing through the corona cause type II and IV radio emission, and near the Earth magnetic storms with a sudden beginning (see sections 13, 15 and 17 for greater detail) (Moreton, 1960; Athay and Moreton, 1961).

It is important to stress that these fluxes with a velocity of $V_s \gtrsim 10^3$ km/sec are not the same as the eruptions of material from the flares which can be seen in ordinary observations in H_α light, since the velocity of the latter is as a rule less than parabolic (< 620 km/sec) and they cannot go far beyond the bounds of the corona. At the same time the injection of the corpuscular fluxes is apparently somehow connected with the eruption from the flare and takes place at the same time as or after the latter.

CORONAL CONDENSATIONS

The electron concentration N in the corona above a centre of activity (groups of flocculi and spots) is generally high. Table 2 (see p. 6), which gives the results of measurements of N near a solar activity maximum in the equatorial region of the "quiet" Sun and in active regions of the corona‡ (Newkirk, 1959), gives some idea of its magnitude. The values of N in the active regions is about two times higher than the average values of N at the same altitude.

In the lower layers of the corona ($R \lesssim 1 \cdot 1 R_\odot$) above a centre of activity we can observe coronal condensations which are extensive formations with a diameter of the order of 2×10^5 km with an elevated electron density (the mean value of N in a condensation is $4 \cdot 5 \times 10^9$ electrons/cm^3 and the

---

† Motions with a velocity of 80–250 km/sec in the flares are also recorded spectroscopically in observations in the wings of the H_α line (Severnyi, 1958).

‡ The concentration N in the active regions is obtained on the assumption that the density of the material is high when compared with the normal density in a cone with an angle of spread of 25° and an axis coinciding with the radius of the Sun.

§ 3] The Moon and Planets

maximum is $8\cdot 2\times 10^9$ electrons/cm³ (Waldmeier, 1956)). The kinetic temperature in these regions reaches $(2-4)\times 10^6$ °K, and their life is as much as several days. In long-lasting condensations of this kind sporadic condensations often occur as the result of a change in temperature or density; the latter phenomena are closely connected with the formation of sunspot-type prominences.

3. The Moon and the Planets

MERCURY, VENUS AND MARS

Mercury's period of rotation is the same as its period of revolution round the Sun. Therefore the same side of Mercury is always turned towards the Sun. This fact and its closeness to the Sun lead to strong heating of the illuminated hemisphere and to intense cooling of the hemisphere of Mercury that is in shadow. The temperature of the planet's surface at a point beneath the sun, measured from the intensity of the infrared emission, is 610°K. It is close to the equilibrium temperature (633°K), which is defined as the temperature of an absolutely black body placed in an orbit normal to the Sun's rays, provided that the only source of cooling is thermal radiation from the illuminated surface. No measurements have yet been taken of the unilluminated part of Mercury.

There are indications that Mercury retains a very rarefied atmosphere, whose reduced height† is of the order of 25 m.

The period of rotation of Venus around its axis is unknown. The dense uniform cloud cover that conceals its surface from us prevents us from determining the period by visual observations. Attempts to find the Doppler shift caused by the planet's rotation for the monochromatic lines in the sunlight reflected from Venus have also proved unsuccessful. It can be expected that in the near future the period of rotation will be found by radar sounding of Venus from the Doppler broadening of the spectrum of the signal when it is reflected from the planet's surface (the data available at present (Kotelnikov, 1961; Malling and Golomb, 1961) on this question are contradictory). The direction of rotation of Venus is judged from the results of analysing the phase curve of this planet's radio emission (see section 20).

---

† The thickness of planetary atmospheres is generally described by the reduced height for the atmospheres as a whole and the gaseous components comprising them. The reduced height is the thickness of a uniform layer which will be formed by the atmosphere under normal conditions (a temperature of 273°K and a pressure of 760 mm Hg).

Measurements of the emission from Venus in the far infrared spectrum have shown that the temperatures of the day and night sides of the planet are very close to each other and to the value of 234°K. This information relates to the upper layer of clouds. In the underlying layers of the atmosphere the temperature rises (according to observations of the vibrational–rotational CO_2 bands it is 285°K). Optical astronomy is unable to provide information on the temperature conditions of the layers of the atmosphere under the clouds or of the surface of Venus, which are completely unreachable for observations in the optical and infrared spectra. Radio astronomy is playing an important part in the solution of this problem (sections 20, 34).

Strong CO_2 absorption bands in the infrared spectrum of the solar radiation reflected from Venus indicate the presence of carbon dioxide in its atmosphere. The CO_2 content above the cloud layer corresponds to a reduced height of 100–1000 m. Allen (1955) indicates that the oxygen's reduced height cannot exceed 2 m and that of water 1 m, i.e. on Venus the amount of O_2 and H_2O above the clouds is less than $1\cdot 5 \times 10^{-3}$ and 5×10^{-2} of the corresponding values in the Earth's atmosphere. According to Strong's preliminary data (see Kellogg and Sagan, 1962) the abundance of water vapour in Venus's atmosphere above the clouds is about 19 microns of precipitated water. Nor is it excluded that the atmosphere of Venus contains a considerable quantity of nitrogen (up to 80%) even though it does not give any absorption bands in the infrared spectrum. Traces of nitrogen (N_2 and N_2^+) at great altitudes have been found in studies of the noctilucence spectrum of Venus (Kozyrev, 1954).

The composition of the clouds that envelop Venus is unknown, although probably the most widely held opinions are that they are formed by water droplets or ice crystals, or dust raised by wind from the planet's surface.

The measurements made by the "Mariner II" space probe did not discover Venus's magnetic field. It was possible to estimate the upper limit of the field strength (5–10% of the corresponding value on Earth) (*Sky and Telescope*, 1963).

The mean temperature of the illuminated disk of Mars is about 245°K. In the tropical zone in the middle of the southern summer the temperature reaches a maximum (~ 300°K) soon after midday; in the morning and evening it may decrease by about 80°. The temperature also varies noticeably with latitude.

The reduced height of the atmosphere of Mars is estimated as 1750 m from measurements of the scattering of light by the planet's gas shell. Only carbon dioxide CO_2 can be found spectroscopically of all the atmospheric components; its reduced height on Mars is 13 times greater than on Earth and is about 36 m (Grandjean and Goody, 1955). The other components of the Martian atmosphere are unknown but in all probability the basic component is nitrogen. It is possible that argon is present evolved in radioactive processes in the planet's crust; its content is estimated at approximately 20 m. The upper limit for the amount of oxygen in the Martian atmosphere is 2·5 m, whilst that for water is 0·2 m and possibly even 0·03–0·05 m.

JUPITER AND SATURN

Just like Venus these planets are covered by a dense layer of clouds (consisting, apparently, of ammonia crystals) which conceal the lower atmosphere and the surface of the planets from us. Light and dark bands running parallel to the equator can be seen in photographs of Jupiter and Saturn. The first are apparently higher and thicker cloud formations than the second. The detailed structure of these bands and the nature of the transition between them show noticeable changes in the course of a few days. A particular feature of the visible surface of Jupiter are dark and white spots which indicate motions of the cyclone and anticyclone type in its atmosphere. As a whole the changeability and complexity of the visible surface of Jupiter (and to a lesser degree of Saturn) are connected with intense processes of formation, rebuilding and breakdown of cloud layers at different altitudes. These processes in their turn may be caused by intense horizontal currents, convective currents and temperature variations in the atmospheres of these planets. The most stable element of Jupiter's visible surface is the so-called Red Spot, which is located near the equator (to the south of it).

The period of Jupiter and Saturn can be determined from the rate of movement of the surface's visible details over the disk. The rotation of Jupiter and Saturn is differential in nature: the angular velocity of the cloud layers depends on the latitude. For these planets the equatorial periods of rotation (or, as is generally said, the periods in coordinate system I) are $9^h50^m30^s$ and 10^h14^m respectively. The time for a complete revolution of Jupiter II, which is used for latitudes $\varphi > 12°$, is $9^h55^m24^s$; for the middle latitudes on Saturn the corresponding figure is 10^h38^m.

In the infrared part of the solar spectrum reflected from Jupiter and Saturn there are powerful absorption bands which indicate the presence of large quantities of methane and ammonia in the atmospheres of these planets. The reduced heights of CH_4 and NH_3 above Jupiter's cloud layer are approximately 150 m and 7 m. The content of these gases on Saturn is given by values of 350 m and 2 m. As well as the methane and ammonia the atmospheres of the large planets contains H_2,† probably helium and in smaller amounts nitrogen and argon. The infra-red temperature of Jupiter which characterizes its cloud cover is about 130°K; for Saturn it is probably slightly less (120°K).

The atmospheres of Uranus and Neptune differ by a higher content of methane than the atmospheres of Jupiter and Saturn (2200 m and 3700 m); the quantity of ammonia, however, is too small to be found by spectroscopic methods. This is due to the "freezing out" of the NH_3 from the atmosphere because of the low surface temperature of Uranus and Neptune ($\sim$ 100°K). At the same time a spectroscopically significant amount of hydrogen has been found in the atmospheres of these planets; its reduced height for Uranus is estimated as 5×10^4–10^5 m.

THE MOON

The Moon, seen from the Earth, subtends an angle of 31′, which is close to the angular diameter of the Sun. The Moon's period of rotation is exactly the same as the time it takes to revolve round the Earth. The same side of the Moon is therefore always turned towards us, unless one considers certain "rockings" of the Moon—librations connected with the ellipticity of its orbit and the inclination of the latter to the plane of the lunar equator.

The area of the lunar disk illuminated by the Sun varies depending upon the positions of the Moon, the Earth and the Sun in relation to each other. The degree of illumination is described by the Moon's phase ψ, which is defined as the angular distance between the Sun and the Moon in the heavens. The period of a complete change of phases (from 0° to 360°) is called lunation; it is 29·53 days.

On the surface of the Moon we can see dark plains—the so-called "maria"—and light mountainous areas—"continents"—surrounding them.

According to measurements of the Moon's own emission in the infrared spectrum the temperature of its surface varies over a wide range depending on the degree of illumination by the Sun's rays. The maximum temperature at the centre of the disk occurs at full moon and the minimum at the new moon; in the intermediate phases the greatest temperature is reached at the point lying beneath the Sun. Therefore no noticeable shift is observed in the phase between the illumination of the Sun and the temperature of the Moon.

---

† The presence of hydrogen in Jupiter's atmosphere has been recently confirmed by spectroscopic observations (Kiess, Corliss and Kiess, 1960); their processing has made it possible to estimate the hydrogen content above the cloud layer at 5500 m of atmosphere (Report of Princeton University Observatory, 1960).

§ 3] The Moon and Planets

The temperature at the centre of the disk varies between 407° and 120°K. However, measurements of the absolute values of T are very uncertain. Therefore a value of 375°K obtained from theoretical considerations is often taken for the temperature at the point beneath the Sun. There is a considerable change in the temperature not only when there are the comparatively slow variations in the conditions of illumination during the lunation but also during lunar eclipses. In the latter case a temperature drop of the order of 150–200° takes place very rapidly, taking about 1 hour. This fact, as well as the absence of any noticeable lag in the temperature of the Moon behind the change in phase, obviously indicates a very low thermal conductivity for the lunar surface, which ensures that the side of the Moon in shadow cools rapidly by radiation and the illuminated side heats up rapidly.

There is practically no atmosphere on the Moon: its reduced height is no more than 10^{-3} m. It has been possible to obtain the upper limit for the electron concentration N in the Moon's ionosphere (10^3–10^4 electrons/cm^3) when measuring the refraction of the radio emission of discrete sources (the Crab Nebula) at metric wavelengths when they pass close to the Moon's limb (Elsmore, 1959).

CHAPTER II

Basic Characteristics of Extraterrestrial Radio Emission and Methods for Studying Them

COSMIC sources of radio emission create on Earth an electromagnetic field $E(t)$, $H(t)$, the total energy flux in which is

$$S = \iint I(\omega, \Omega) \, d\omega \, d\Omega.$$

Here I is the spectral intensity of the radio emission, i.e. the energy flux in a unit range of angular frequencies $\omega = 2\pi f$ and solid angles Ω. The dependence of I on the frequency ω determines the frequency spectrum and the dependence on the direction in space Ω determines the angular spectrum of the radio emission.

In a vacuum the quantity I can always be represented in the form of the sum of the intensities of two elliptically polarized waves 1 and 2:

$$I = I_1 + I_2,$$

the field vectors in which rotate in opposite directions describing in a plane orthogonal to the direction of propagation similar ellipses with their long axes at right angles to each other. The ratio of the axes in each ellipse, in particular, may be equal to zero (two linearly polarized components) and unity (two circularly polarized components). The relationship between the intensities of the waves 1 and 2, the degree of coherency and the phase difference between them determine the nature of the polarization of the radio emission.

Investigation of the three characteristics of radio emission—the frequency spectrum, the angular spectrum and the polarization—by themselves and in connection with different kinds of optical phenomena is the basic source of information in radio astronomy about the nature of cosmic objects. We obtain information on these characteristics by radio telescopes,

§ 4] Frequency Spectrum

which are devices consisting of an aerial system and a power-measuring device (radiometer). The magnitude of the signal at the output of the radiometer (the radio telescope's response) Ξ when receiving the radio emission is proportional to the expression

$$\iint [K_1(\omega, \Omega) I_1(\omega, \Omega) + K_2(\omega, \Omega) I_2(\omega, \Omega)] \, d\omega \, d\Omega.$$

The aerial system as a rule receives waves of one polarization preferentially (circular, elliptical or linear). Therefore $K_1 \ll K_2$, or on the other hand $K_2 \ll K_1$, and in the expression for Ξ only I_1 and I_2 actually figure. The dependence of K_1 or K_2 on the frequency ω determine the frequency response of the radio telescope and the dependence on the direction Ω determines its directional characteristics. From the magnitude of the response Ξ to the radio emission received by devices with appropriately selected frequency characteristics and directional characteristics we can judge the frequency and angular spectra and also the polarization of this emission.

The corresponding methods of investigating extraterrestrial radio emission will be described briefly in this chapter. The reservation must be made here that we shall leave aside any discussion of the operating principles of radiometers (the design of systems with given frequency responses, methods of distinguishing a weak signal at the level of the intense natural noise of the receiving apparatus, etc.).

When studying the features of extraterrestrial radio emission from its characteristics on Earth it must be remembered that the properties of the radio emission may change noticeably in the process of propagation from the source to the receiver. The present chapter therefore provides information on the degree to which the Earth's atmosphere affects the propagation of radio waves necessary to draw conclusions on the properties of the sources of radio emission from their observed characteristics, and conditions are explained when this effect can be neglected.

4. Frequency Spectrum

In all cases of practical interest the receiving apparatus is designed so that the function $K_j(\omega, \Omega)$ $(j = 1, 2)$, which has a maximum at the "working frequency" $\omega = \omega_0$, drops fairly rapidly as ω moves away from ω_0. If we define the concept of the frequency bandwidth of the receiving device $\Delta\omega_b$ at the half intensity level by the relation $\Delta\omega_b = |\omega_1 - \omega_2|$, in which ω_1 and ω_2 satisfy the equations $K_j(\omega_1) = K_j(\omega_2) = \frac{1}{2} K_j(\omega_0)$, then the expression "fairly rapidly" will mean that $\Delta\omega_b$ is small when compared with the frequency range $\Delta\omega$ in which there is a noticeable change in the value of the radio emission's spectral intensity I_j. Then in the expression

$$\Xi = A \iint K_j(\omega, \Omega) I_j(\omega, \Omega) \, d\omega \, d\Omega \tag{4.1}$$

we can take away the slowly varying function I_j from the integral sign in ω with a value at the point $\omega = \omega_0$:

$$\Xi \approx A \int I_j(\omega_0, \Omega) F_j(\Omega) \, d\Omega. \tag{4.2}$$

Here A denotes a certain constant coefficient characterizing the amplification of the whole receiving section and $F_j(\Omega)$ the integral

$$\int K_j(\omega, \Omega) \, d\omega,$$

which we shall call the aerial polar diagram in the frequency band $\Delta\omega_b$.

For making relative measurements of the spectral intensity of the radio emission it is not necessary to know the value of A in (4.1)–(4.2): it is sufficient to be certain that it remains constant in the process of observation. However, it is impossible to make absolute measurements of I_j (to be more precise $S_{j\omega} = \int I_j F_j \, d\Omega$) without determining the value of A, i.e. without carrying out an absolute calibration of the apparatus. The latter consists of bringing to the input of the receiver (to the aerial or the radiometer) the emission of a source whose intensity is considered to be known.

In the ordinary method of absolute measurements, when the signal received is compared with the emission of a thermal standard (a "black" body heated to a definite temperature and connected up to the radiometer input instead of the aerial) an exact knowledge of the aerial polar diagram is obviously necessary. In practice, however, determination of the diagram (in particular allowing for scattering in the side lobes) is a very difficult problem, which is a source of systematic errors in measurements. Therefore the accuracy of the majority of absolute radio-astronomical measurements of radio emission intensity is no better than 10%.

At the same time this accuracy is clearly insufficient for a wide range of problems and in particular for measuring the radio emission flux of the Sun, the Moon and the planets. Therefore Krotikov, Porfir'yev and Troitskii (1961) suggested a new method for absolute measurements which has begun to be used in studies of the Moon's radio emission (see section 21). This method consists of comparing the response to the Moon's radio emission with the quantity Ξ when receiving the emission of a standard; for this purpose an "artificial Moon" is used (a "black" body with angular dimensions the same as the visible diameter of the Moon) placed in the wave zone of the aerial system. Since the radio emission of the Moon and

§ 4] Frequency Spectrum

the standard is received under identical conditions with respect to the aerial system there is no need with this method to determine the aerial parameters (see equation (4.2)). This makes it possible to eliminate the main source of errors and increase the accuracy of measurements up to 2–3%.

AERIAL TEMPERATURE AND EFFECTIVE TEMPERATURE OF RADIO EMISSION

In radio astronomy the spectral intensity $I_j(\omega, \Omega)$ of the radio emission is generally described by the value of the effective (or brightness) temperature T_{eff}. It is defined as the temperature of an equilibrium emission whose intensity

$$I_j^{(0)} = \frac{\omega^2 \varkappa T}{8\pi^3 c^2} \tag{4.3}$$

is equal† to $I_j(\omega, \Omega)$ (T is the equilibrium emission temperature, $\varkappa$ is the Boltzmann constant, c is the velocity of light in a vacuum). In accordance with this definition T_{eff} is in this case a function of the frequency ω and the direction Ω connected with $I_j(\omega, \Omega)$ by the relation

$$T_{\text{eff}} = \frac{8\pi^3 c^2 I_j}{\omega^2 \varkappa}. \tag{4.4}$$

The point of introducing the effective temperature is that in the case of thermal emission T_{eff} is connected with the kinetic temperature T of the source by a very simple relation, and is equal to T when the emitting region is thick enough (see section 26).

On the other hand, the concept of the aerial temperature T_A is extensively used for the characteristics of the radio emission flux received by the aerial. It is defined as the temperature of an opaque cavity in which the response Ξ when the aerial system is placed in it will be equal to the response to the emission received. In other words, the aerial temperature is the temperature of an equilibrium (isotropic) emission whose action on the receiving device‡

$$\Xi = AI_j^{(0)}(\omega_0) \int F \, d\Omega \tag{4.5}$$

is equivalent to the action of the investigated (non-isotropic) radio emission (see equation (4.2)).

---

† Planck's law becomes the Rayleigh–Jeans formula (4.3) provided that $\hbar\omega \ll \varkappa T$, where $\hbar$ is Planck's constant. In the radio band at frequencies $\omega \lesssim 2 \times 10^{11}$ sec^{-1} (i.e. at wavelengths $\lambda \gtrsim 1$ cm) this inequality is valid for emission temperatures $T \gg 1°$K.

‡ Here and in future we shall omit the suffix j in F_j for brevity of notation.

Equating the expressions (4.2) and (4.5) and taking (4.3) and (4.4) into consideration, we obtain a simple relation connecting the aerial temperature with the distribution of the effective radio emission temperature over the sky:

$$T_A = \frac{\int_{4\pi} T_{\text{eff}} F \, d\Omega}{\int_{4\pi} F \, d\Omega}. \qquad (4.6)$$

The aerial temperature T_A is also the quantity which is measured when receiving the radio emission from the response of the radio telescope Ξ after corresponding calibration of the apparatus.

It is clear from equation (4.6) that in the general case T_A depends not only on the distribution of the effective temperature over the sky $T_{\text{eff}}(\Omega)$ but also on the nature of the polar diagram $F(\Omega)$ of the aerial system, determining the mean diagram-weighted effective temperature of the observed radio emission.

Let us examine the case when the diagram $F(\Omega)$ has one high maximum in the direction Ω_0 and determine for it the width of the diagram at the half-intensity level (the width of the main lobe) as the range of angles $\Delta\theta_A$ around Ω_0 in which $F(\Omega) \geq \frac{1}{2}F(\Omega_0)$.

If the width of the aerial diagram $\Delta\theta_A$ pointed at a radio emission source is small when compared with the angular size $\Delta\theta_s$[†] of the source, then the effective temperature T_{eff} in equation (4.6) can be taken out of the integral with a value of Ω_0 in the direction of the main maximum of the diagram. Then the aerial temperature is obviously the same as the effective temperature T_{eff} of the source in the direction Ω_0:

$$T_A \approx T_{\text{eff}}(\Omega_0). \qquad (4.7)$$

In accordance with equation (4.2) the signal at the receiver output will here be directly proportional to the intensity of the radio emission in the direction Ω_0.

The angular sizes of the sources of radio emission in the solar system are very small: for the Sun and the Moon they are no more than fractions of a degree and no more than a few minutes for the planets. Therefore the problem of designing aerial devices with highly directional properties suitable for satisfying the inequality $\Delta\theta_A \ll \Delta\theta_s$ (when we can study the detailed distribution of T_{eff} over the source and judge its exact size) is very complex. This explains the fact that in many cases radio astronomy observations are made with aerials with comparatively poor directional

---

† To be more precise, with characteristic angular distances at which there is a noticeable change in the magnitude of the brightness temperature.

§ 4] Frequency Spectrum

properties: $\Delta\theta_A \gg \Delta\theta_s$. In this extreme case

$$T_A \approx F(\Omega_s) \frac{\int T_{\text{eff}} \, d\Omega}{\int F \, d\Omega}, \tag{4.8}$$

where Ω_s is the direction of the centre of the source. If the latter is the same as Ω_0, then

$$T_A \approx F(\Omega_0) \frac{\int T_{\text{eff}} \, d\Omega}{\int F \, d\Omega}. \tag{4.9}$$

The latter relation can be written slightly differently if we define the solid angle subtended by the source, for example, by the expression

$$\Omega_s = \int d\Omega \tag{4.10}$$

in which the integral is taken over the region of the source where $T_{\text{eff}} > \frac{1}{2} T_{\text{eff max}}$, and introduce the mean effective temperature of the emitting region

$$\bar{T}_{\text{eff}} = \frac{\int T_{\text{eff}} \, d\Omega}{\Omega_s}. \tag{4.11}$$

Then obviously

$$T_A \approx \frac{F(\Omega_0) \bar{T}_{\text{eff}} \Omega_s}{\int F \, d\Omega} \sim \frac{\bar{T}_{\text{eff}} \Omega_s}{\Omega_A}. \tag{4.12}$$

In changing to the last expression we have said approximately

$$\int F \, d\Omega \sim F(\Omega_0) \Omega_A,$$

where Ω_A is the solid angle occupied by the main lobe of the aerial. In accordance with equation (4.2) in the extreme case in question the signal Ξ at the output and the aerial temperature T_A will be directly proportional to the integral

$$S_{j\omega} = \int I_j \, d\Omega, \tag{4.13}$$

whose significance is the spectral flux of the radio emission from the whole source.

Therefore the use of aerials with poor directional properties does not permit us to determine the true size of the source and therefore does not make it possible to find its effective temperature T_{eff} from the known aerial temperature T_A (or the flux $S_{j\omega}$). This is also one of the reasons[†] why instead of the effective temperature T_{eff} use is often made of the so-called reduced effective temperature, relating it to the optical size of the source

---

[†] Another reason is that effective temperatures of sources that are different in size reduced to the same area make possible a direct comparison of the magnitude of the spectral emission fluxes from these sources (see equation (4.14)).

(in particular to the visible disks of the Sun, the Moon and the planets), and it is denoted by $T_{\text{eff}\odot}$, $T_{\text{eff}\mathbb{C}}$ etc.

By definition the reduced effective temperature of the solar radio emission related to the Sun's optical disk is the name given to the temperature which the Sun's surface should have when emitting as an absolutely black body to produce near the Earth a radio emission flux $S_{j\omega}^{(0)}$ at a frequency ω equal to the observed $S_{j\omega}$. Since the spectral flux of the emission of a black body under the above conditions is

$$S_{j\omega}^{(0)} = I_j^{(0)} \Omega_\odot,$$

where $I_j^{(0)}$ is the quantity from the formula (4.3) and $\Omega_\odot$ is the solid angle subtended by the optical disk and is equal to $\pi R_\odot^2 / R_{S-E}^2$ ($R_\odot$ is the radius of the Sun, R_{S-E} is the distance from the Sun to the Earth), it becomes clear that $T_{\text{eff}\odot}$ is determined by the relation†

$$S_{j\omega} = \frac{\omega^2 \varkappa T_{\text{eff}\odot}}{8\pi^3 c^2} \Omega_\odot \qquad (4.14)$$

or

$$S_{jf} = \frac{\varkappa T_{\text{eff}\odot}}{\lambda^2} \Omega_\odot \approx \frac{\varkappa T_{\text{eff}\odot}}{\lambda^2} \frac{\pi R_\odot^2}{R_{S-E}^2} \approx 9.4 \times 10^{-24} \frac{T_{\text{eff}\odot}}{\lambda^2} \text{ W m}^{-2} \text{ c/s}^{-1}. \quad (4.15)$$

Here $\omega = 2\pi f = 2\pi c/\lambda$, λ is in centimetres, $T_{\text{eff}\odot}$ is in °K.

For comparison it should be recalled that $T_{\text{eff}\odot} \sim 10^6$ °K in the metric band (at $\lambda \sim 3$ m) corresponds to a flux of $S_{jf} \sim 10^{-22}$ W m^{-2} c/s^{-1}.

By comparing the expressions (4.13) and (4.14) for $S_{j\omega}$ (remembering the definition of T_{eff} (4.4)) it is easy to find the connection between the "true" effective temperature and the "reduced" temperature of the source:

$$\int T_{\text{eff}} \, d\Omega = T_{\text{eff}\odot} \Omega_\odot. \qquad (4.16)$$

Since at the same time

$$\int T_{\text{eff}} \, d\Omega = \bar{T}_{\text{eff}} \Omega_s$$

(see (4.11)) it is clear that

$$\frac{\bar{T}_{\text{eff}}}{T_{\text{eff}\odot}} = \frac{\Omega_\odot}{\Omega_s}. \qquad (4.17)$$

The definition of the reduced effective temperature given above as applied to the solar radio emission is easy to rephrase for the case of the planets. In this case in the formulae (4.14)–(4.17) we must replace $T_{\text{eff}\odot}$, $\Omega_\odot$, $R_\odot$ and R_{S-E} by new notations and the coefficient 9.4×10^{-24} (unless great accuracy is required this coefficient can be left the same for the Moon

---

† We would stress once again that here it is the flux $S_{j\omega}$ for one polarization that figures. In the case of unpolarized (naturally polarized) radio emission the flux $S_{j\omega}$ is equal to half the total spectral flux S_ω.

§ 4] Frequency Spectrum

since $\Omega_c \approx \Omega_\odot$).

STUDYING THE RADIO-EMISSION FREQUENCY SPECTRUM. MULTI-CHANNEL RECEIVING DEVICES AND RADIO SPECTROGRAPHS

When the generation and propagation conditions of the radio waves change the radio emission does not remain constant. Although the spectral intensity I_j, which characterizes the energy flux in the Fourier component of an expansion of the electromagnetic field in the frequencies ω and the directions Ω, is by its very nature a strictly constant quantity, a radio telescope can find these changes since the signal at the output is in fact determined by the intensity in the whole frequency band $\Delta\omega_b$ and not at one frequency. In this case the response Ξ will reflect the quasi-stationary variations in the radio emission flux $S(t)$ in the range $\Delta\omega_b$ (with the $\Delta\omega_b$ band-averaged quantity $S_\omega(t) = S(t)/\Delta\omega_b)^\dagger$ if the characteristic time of these variations is $t \gg 1/\Delta\omega_b$. For the radio emission of the Sun and the planets t is known to be not less than 10^{-3} sec, so the inequality above is easily satisfied with a band wider than 1 kc/s. In practice the band $\Delta\omega_b$ is generally many times greater than this value and the value of t in fact has its lower limit set by the radio telescope's response, which is characterized by the time constant t_0.

We can obviously use a series of separate receivers with different working frequencies in the band of interest to us to study the frequency spectrum of the radio emission (i.e. the frequency dependence of the intensity or flux). Sometimes these devices are designed to combine into a single installation consisting of a common aerial and a multi-channel receiver (see, e.g., Vitkevich, Kameneva and Kovalevskii, 1956). The number of these devices (the number of channels) should be fairly large. This is necessary for the differences $\Delta\omega_0$ between the frequencies of adjacent channels to be less than the characteristic frequency range $\Delta\omega$ in which there is a significant change in the value of the intensity of the radio emission being received.

This requirement is comparatively easy to satisfy when studying smooth spectra for which the range $\Delta\omega$ is generally comparable with ω; then the number of channels does not as a rule exceed ten. As examples of these spectra we can take the radio emission spectrum of the "quiet" Sun, of some components of the sporadic solar radio emission (the slowly changing component, microwave bursts, enhanced radio emission and type IV events), and the spectra of the Moon's radio emission and the continuous radio emission of the planets (see Chapters III–V).

† Which we shall also call (slightly incorrectly) the spectral emission flux.

In the last case the basic problem is to obtain a sufficient ratio between the level of the useful signal at the output to the value of the fluctuation threshold of the receiver sensitivity, which is necessary for certain recording and measurement of the very weak radio emission flux. This is the reason for designing aerials with very large dimensions which will collect the radio emission over a large area and for using masers in radiometers to make a sharp reduction in the natural fluctuation level at the output of the receiver (see Alsop et al., 1959, and Giordmaine et al., 1959, on the use of masers in radio astronomy).

On the other hand the use of multi-channel receivers for a detailed analysis of the spectra of certain components of solar radio emission (chiefly types II and III bursts) becomes difficult. The point is that the frequency spectrum of these components of the radio emission at metric and decametric wavelengths changes sharply in a range of a few megacycles, whilst the entire event as a whole occupies a spectrum tens and hundreds of megacycles wide. It is clear from what has been said that obtaining satisfactory information on the spectrum of these components with a multi-channel receiver involves considerable complication of design since the number of channels must now be increased to many tens. Low-frequency resolution explains why multi-channel receivers have not been widely used to study the frequency spectra of sporadic radio emission.

At present the basic source of our knowledge about the complex spectra of the Sun's sporadic radio emission is a different type of receiver—the so-called radio spectrograph.† This is a combination of a wide-band aerial and a radiometer whose working frequency ω_0 changes in time, passing at metric wavelengths through a frequency range of several tens or hundreds of megacycles two or three times per second.‡ The signal at the output of the radio spectrograph modulates the brightness of a beam which moves linearly over the screen of an oscillograph in time with the change in frequency ω_0. The picture is photographed onto a continuously moving film strip; thus the latter records the "dynamic spectrum" of the

---

† Vitkevich, Kameneva and Kovalevskii (1956); Wild and McReady (1950); Wild, Murray and Rowe (1954); Goodman and Lebenbaum (1958); Thompson (1961); Markeyev (1961); Sheridan and Trent (1961); Sheridan and Attwood (1962); Boishot, Lee and Warwick (1960); Young et al. (1961); Kundu and Haddock (1961).

‡ The first radio spectrograph, which was designed in the fifties by Wild and his collaborators (Wild and McReady, 1950; Wild, Murray and Rowe, 1954), consisted of three similar devices operating in the 40–57, 75–140 and 140–240 Mc/s ranges respectively. Radio astronomers now have spectrographs with which they can study the radio-emission frequency spectrum in the range from 4000 to 2000 and from 950 to 5 Mc/s (see Goodman and Lebenbaum, 1958; Thompson, 1961; Markeyev, 1961; Sheridan and Trent, 1961; Sheridan and Attwood, 1962; Boishot, Lee and Warwick, 1960; Young et al., 1961; Kundu and Haddock, 1961).

§ 4] Frequency Spectrum

radio emission in which the spectral flux is shown by the degree of darkening (or lightening) on the "frequency–time" diagram. The dynamic spectrum (i.e. the frequency spectrum's change in time) is an important characteristic of the sporadic radio emission; the form of the dynamic spectrum basically determines which type a given event belongs to (see Chapter IV).

The frequency resolution of a radio spectrograph is determined by the bandwidth $\Delta\omega_b$ of the receiver at a fixed value of ω_0. In Wild's radio spectrograph the resolution is about 0·5 Mc/s, which is generally sufficient for studying all the components of the Sun's sporadic radio emission. The time resolution, which depends on the number of frequency scans in unit time, is fractions of a second as a rule. No more is required when studying almost all the components of the sporadic radio emission with a characteristic change time of a few seconds or more; this resolution is not sufficient, however, for a detailed investigation with type I bursts with a life of less than 1 sec. Because of the narrowness of the frequency range covering the spectrum of individual type I flares they are studied with a radio spectrograph with a higher flyback rate in a comparatively small range ($\sim$ 10 Mc/s (Elgarøy, 1959)) and generally on multi-channel devices with a low time constant (Groot, 1959). In the latter case sufficient frequency resolution is achieved by all the channels being concentrated in a narrow range of frequencies.

It should be pointed out that multi-channel devices or a set of installations with fixed working frequencies, on the one hand, and radio spectrographs, on the other, complement each other well in their properties. Radio spectrographs have high-frequency resolution and immediately provide a dynamic spectrum of the radio emission at the output. As a rule, however, they have lower sensitivity because of the use of wide-band aerials and the desire to obtain even amplification over the range. The problem of making a high-frequency (centimetric and millimetric band) radio spectrograph is very complicated because of difficulties of carrying out tuning within wide enough limits. On the other hand, multi-channel devices or a series of installations with fixed frequencies can work in any part of the radio band. Here it is easier to obtain high sensitivity and ensure good time resolution. The frequency resolution of this kind of system is not nearly so good as that of radio spectrographs, however. In this case the dynamic spectrum of the radio emission needs additional and very laborious (because of the great duration of the observations) processing of the output data, whilst in a radio spectrograph the dynamic spectrum is obtained directly in an easily comprehensible form.

5. Angular Spectrum

AERIAL SYSTEM REQUIREMENTS IN RADIO ASTRONOMY. PARABOLIC AERIALS

The angular spectrum $I_j(\Omega)$ of the radio emission characterizes the radio brightness distribution in the sky, i.e. the dependence of the effective temperature T_{eff} on the direction Ω. Since the polar diagrams of real aerials have a finite width, as the result of measurement we do not obtain the values of T_{eff} at each point in the sky but the value of the aerial temperature T_A which has the meaning of the diagram-averaged effective temperature (see equation (4.6)). Therefore when studying the radio brightness distribution with an aerial whose diagram width is $\Delta\theta_A$ it is impossible in practice to discover distribution details whose characteristic size is $\Delta\theta \ll \Delta\theta_A$. The narrower the aerial diagram the higher its angular resolution in the sense of the aerial temperature's reflecting finer and finer details of the real distribution of T_{eff} in the sky.

Therefore the ideal aerial system for the purposes of radio astronomy will obviously be a system the width of whose polar diagram is much less than the characteristic angular size of a significant change in the value of the radio brightness. This requirement is very rigid. In actual fact if as the aerial system we have an analogue of the optical telescope-reflector (a reflecting paraboloid with a receiving aerial set at the focus) the width of the polar diagram $\Delta\theta_A$ in all the planes passing through the axis of the paraboloid is the same as and close to λ/d, where d is the diameter of the mirror.† When studying the solar radio emission it is desirable to have a diagram with a width of up to 1' for resolving individual local sources on the disk, although in a number of cases even this is insufficient. At a wavelength of $\lambda \sim 1$ cm a 35-m diameter mirror has such a diagram; at wavelengths of $\lambda \sim 10$ cm and 100 cm this size is increased by a factor of ten and a hundred respectively. If at the same time we remember that when designing systems of this kind provision must be made for rotating the aerial in a given direction it becomes clear that parabolic aerials with good directional powers that are sufficient for studying the brightness distribution over the Sun's disk can be made only for the millimetric and centrimetric bands.

An easier problem than the design of a narrow ("pencil") beam is the obtaining of a so-called "knife-edge" beam with high azimuthal directional features and low angular position resolution. An example of an

---

† Here the width of the aerial diagram actually means the width of the main lobe at the half-intensity level.

§ 5] Angular Spectrum

aerial with a polar diagram of this type is the aerial of the Pulkovo radio telescope designed by Khaikin and Kaidanovskii (1959) (see also Khaikin *et al.*, 1960). It consists of a series of panels set on a curve whose shape ensures the reflection of a plane wave coming down at an angle into the receiving aerial located near the Earth's surface. Obviously the reflecting panels must be set on a curve formed by the intersection of a paraboloid of rotation (whose focus is at the point where the receiving aerial is) with a plane parallel to the Earth's surface in order to achieve this. This curve is an ellipse; its shape must obviously be altered slightly depending upon the angular position of the source of the radio emission being received (when the angular position of the source approaches zero the ellipse becomes a parabola). The design of the Pulkovo aerial provides for changing the shape of the curve by the ability of the reflecting panels to be moved within certain limits. With a total panel length of about 140 m, a height of 3 m and a special design of primary radiator the width of the polar diagram in azimuth at a wavelength of $\lambda = 3$ cm is of the order of $1'$ and of the order of $1°$ for angular position. The passage of a source of radio emission through the beam gives a one-dimensional distribution of the radio brightness over the disk[†]

$$G(\theta) = \int {}_{\text{eff}}(\theta, \xi) \, d\xi, \tag{5.1}$$

where θ and ξ are the angular coordinates in the sky.

THE TWO-ELEMENT INTERFEROMETER

A successful attempt to combine the requirement of high resolving power with a small aerial system area was the design of interferometers of different kinds. The latter are widely used in radio astronomical observations, particularly at wavelengths of $\lambda > 10$ cm where it becomes difficult to obtain high resolution with parabolic aerials.

The simplest two-element interferometer, which was used for studying the solar radio emission right at the beginning of the development of radio astronomy (Ryle and Vonberg, 1948), consists of two identical aerials a distance D apart and coupled by connecting lines with a single receiver. (In optics this kind of system is called a Rayleigh interferometer.)[‡]

If a plane monochromatic wave of frequency ω is incident on the interferometer at an angle θ' to the plane of the interferometer (i.e. the Fourier component of the frequency and angular spectra of the radio emission

---

[†] This is true provided that the angular diameter of the source is less than the beam width of this aerial with respect to the angular position but much greater than the azimuthal width.

[‡] This plane passes through the middle of the base (i.e. through the centre of the distance D between the two aerials) orthogonally to the latter.

with an intensity I_j), then the voltages E_1, E_2 produced at the junction of the cables coming from aerials 1 and 2 will be

$$\left. \begin{array}{l} E_1 = E_0 \sin(\omega t + \psi_1), \\ E_2 = E_0 \sin(\omega t + \psi_2 + kD \sin \theta'), \end{array} \right\} \quad (5.2)$$

where E_0^2 is proportional to I_j.[†] In the relations (5.2) $kD \sin \theta'$ is the phase lag due to the difference in travel $D \sin \theta'$ in the propagation of the wave to aerials 1 and 2 ($k = 2\pi/\lambda$ is the wave number), ψ_1 and ψ_2 are the phase shifts relative to the currents induced in the aerials which are made by the connecting cables. Since the voltages from the two aerials are added together at the cable junction point and the radiometer response $\varXi$ is proportional to the power of the signal arriving at the input we have

$$\varXi \propto \overline{(E_1 + E_2)^2} = E_0^2[1 + \cos(kD \sin \theta' + \varDelta\psi)], \quad (5.3)$$

where $\varDelta\psi = \psi_2 - \psi_1$, and the bar above the bracket denotes averaging for a time t.

It is clear from the relation (5.3) that a two-element interferometer with a baseline $D \gg 1/k$ has a multi-lobe polar diagram; for small θ' (when $\sin \theta' \approx \theta'$) the maxima (and minima) of the diagram are repeated at angular intervals of $\varDelta\theta' = 2\pi/kD = \lambda/D$. Since, as has already been pointed out, E_0^2 is proportional to I_j, i.e. $E_0^2 \propto T_{\text{eff}}(\theta, \xi)$, we finally obtain that the response of the radio interferometer to the radio emission received is

$$\varXi \propto \iint T_{\text{eff}}(\theta, \xi)[1 + \cos(kD \sin \theta' + \varDelta\psi)] \, d\theta \, d\xi. \quad (5.4)$$

In (5.4) θ and ξ are angular coordinates in the sky; the θ' coordinate is connected with the θ coordinate by the relation $\theta' = \theta - at$, where a is determined by the speed of rotation of the sphere of the heavens.

In future we shall limit ourselves to the case when a radio emission source with a distribution $T_{\text{eff}}(\theta, \xi)$ is not too far from the plane of the interferometer, so that in the integral (5.4) it is sufficiently accurate to put $\sin \theta' \approx \theta'$. Also let the connecting cables from the two aerials be completely identical: $\psi_1 = \psi_2$ and $\varDelta\psi = 0$. Then, remembering what has been said and bearing in mind the relation (5.1) for a one-dimensional brightness distribution, we write (5.4) in the following form:

$$\varXi \propto \int G(\theta) \, d\theta + \int G(\theta) \cos[kD(\theta - at)] \, d\theta. \quad (5.5)$$

The first term is clearly equal to the spectral flux S_ω of the radio emission from the whole source; the second term, as can easily be checked, is the

---

[†] We make no allowance for the directional properties of the separate aerials since they are considered to be great when compared with the size of the source of radio emission.

§ 5] Angular Spectrum

same as
$$g_{kD} \cos(kDat + \Psi_{kD}), \tag{5.6}$$

where the amplitude g_{kD} and the phase Ψ_{kD} can be determined from the relation

$$G_{kD} \equiv g_{kD}\, e^{i\Psi_{kD}} = \int_{-\infty}^{+\infty} G(\theta)\, e^{-ikD\theta}\, d\theta. \tag{5.7}$$

Therefore the signal at the output of the radio interferometer as the source passes through the beam changes as[†]

$$\Xi \propto S_\omega [1 + M \cos(kDat + \Psi_{kD})], \tag{5.8}$$

where the "modulation coefficient" is

$$M = \frac{g_{kD}}{S_\omega}. \tag{5.9}$$

In observations with an interferometer with a fixed baseline D the size of the radio emission source $\Delta\theta_s$ (see Vitkevich, 1952 and 1961) can be judged from the value of M. In accordance with equations (5.7) and (5.9) $M \approx 1$ if $\Delta\theta_s \ll 1/kD$; when this inequality is satisfied $\Delta\theta_s$ will also be much less than the width of the beam lobe $\Delta\theta_l = 2\pi/kD$.[‡] In the case of the opposite inequality $\Delta\theta_s \gg 1/kD$ the coefficient $M \ll 1$; in actual fact this coefficient becomes very small as soon as $\Delta\theta_s$ is comparable with $\Delta\theta_l$.

From the phase on the recording of the interference signal Ξ we can judge the position, and from the period of Ξ the velocity of the source relative to the interferometer beam; we can therefore judge the position and movement of a radio-emitting region in the sky. However, since it is not always possible to state which lobe the source is in from the nature of the recording, the θ coordinate (i.e. θ') of the source is given with the uncertainty $n\Delta\theta_l = n2\pi/kD$, where $n = 0, \pm 1, \pm 2, \ldots$

Modifications of the Two-element Interferometer

One of the other forms of the two-element interferometer is the so-called cliff interferometer (an aerial set on the seashore) (McCready, Pawsey and Payne-Scott, 1947). In this case the multi-lobe polar diagram is created by the interference of the direct beam with one reflected from the surface of the sea; the second element here is the mirror reflection of the receiving aerial relative to the conducting surface of the sea.

[†] The relations (5.4), (5.5) and (5.8) will be valid if the radiometer's frequency band $\Delta\omega_b$ is small enough for the non-chromaticity of the signal being received not to change significantly the phase relations in the interferometer that determine its polar diagram.

[‡] Here the lobe width is taken as the angular distance between adjacent nulls of the polar diagram of the radio interferometer.

The very simple two-element interferometer discussed in the preceding section has a number of disadvantages, amongst which there is in particular the possibility of receiving (as well as radio emission from sources of small angular size) emission with a broad angular spectrum—the galactic background and certain forms of terrestrial interference. Variations in the signal Ξ caused by the change in the interference level in time and the movement of the non-uniform background of galactic radio emission relative to the aerial beam, as well as fluctuations in the level of this radio emission due to a change in the amplification of the radiometer, badly distort the interference picture at the output, making it difficult to study weak local sources of small angular size.

The phase-switching interferometer (Ryle, 1952) is free from this disadvantage; here one of the connecting cables (running, let us say, from aerial 2) has a "half-wave" section switched in and out alternately to change the phase ψ_2 by π ($\psi_2 = \psi_1 + \pi$ and $\psi_2 = \psi_1$ respectively). Thanks to this operation the voltage at the radiometer input periodically becomes $E_1 + E_2$ and $E_1 - E_2$ (where E_1 and E_2 are taken from equation (5.2)). The radiometer is designed so that the signal at the output is determined by the difference of the mean squares of the voltages at the input, i.e.

$$\Xi \propto \overline{(E_1+E_2)^2} - \overline{(E_1-E_2)^2} = \overline{4E_1 E_2} = 2F_0^0 \cos(kD \sin\theta'). \quad (5.10)$$

and therefore when receiving radio emission from a source of finite size

$$\Xi \propto \int G(\theta) \cos[kD(\theta - at)] \, d\theta. \quad (5.11)$$

Comparing equations (5.10) and (5.11) with (5.3) and (5.5) we can see that switching the phase by π allows us to eliminate the constant component of the signal at the input and thus reduce the effect of the radio emission with the broad angular spectrum on the interference picture since the distributed radio emission's contribution to the value of (5.10) will be insignificant. Unfortunately the phase-switching interferometer cannot be used to measure the diameter of localized sources since the amplitude of the interference curve at the output depends not only on the size of the source but also on its intensity. Information about angular size can be obtained by repeating the observations on an interferometer with a different baseline or by using an aerial with low directional properties to measure the flux of the radio emission from the whole source.

Another serious disadvantage of the very simple interferometer is that it can be used to study only long-lived sources which have time to pass through several lobes of the beam (by taking part in the sky's rotation) and provide a clear interference picture before they disappear. However,

§ 5] Angular Spectrum

the life of such events as bursts of solar radio emission, which in some cases is fractions of a second, is clearly insufficient for this.

This difficulty can be got round by smoothly changing the electrical length of one arm of the interferometer so that the phase difference obtained in the coupling cables changes linearly in time: $\Delta\psi = \psi_2 - \psi_1 = kDa't$.

It is easy to see from the formulae (5.4)–(5.6) that for an interferometer of this kind with phase switching (or, as is often said, with a rocking polar diagram) the term $kDa't$ is contained in the expression (5.6) as well as $kDat$. Since $a' \gg a$ when the switching is rapid enough, the interference picture at the output $\Xi(t)$ will be chiefly conditioned by the movement of the beam lobes relative to the plane of the interferometer and not by the movement of the sky.

In the first experiments of this kind (Little and Payne-Scott, 1947) it took only a few hundredths of a second to obtain a unit interference picture; phase switching was repeated 25 times a second. This rapid scanning of the Sun made it possible to record the position and movement of short-lived local sources over the solar disk. The inevitable uncertainty of $n2\pi/kD$ in the position of the source θ was eliminated by simultaneous observations on interferometers with the two baselines D_1 and D_2. From the two series of the values of θ obtained here corresponding to the possible coordinates of the source one agreement—near the Sun—corresponds to the actual position of the source. The other coincidences are considerable distances away from the Sun, so can be rejected.

When studying the solar sporadic radio emission (in particular type II and III bursts) information on the frequency dependence of the time changes in the coordinates and the velocities of the local sources has an important part to play as well as data on these quantities. The necessary information can be obtained by combining a two-element interferometer with a radio spectrograph (Wild and Sheridan, 1958).

In an interferometer of this kind with frequency switching $\omega_0(t)$ the response Ξ to the radio emission of a point source with a broad frequency spectrum set at angle θ', as follows from (5.3), will take the form of interference with maxima at the frequencies

$$\omega_{max} = \frac{n \cdot 2\pi c}{D \sin \theta' + \Delta l}, \qquad (5.12)$$

for which

$$kD \sin \theta' + \Delta\psi = 2\pi n \qquad (n = 0, \pm 1, \pm 2, \ldots).$$

In the change from the last relation to (5.12) we took into consideration that $k = \omega/c$ and $\Delta\psi = k\Delta l$, where l is the difference in the values of the electrical length in the arms of the interferometer.

Unless the position of the source is frequency-dependent ($\theta' =$ const) the intervals between adjacent maxima will be the same; if, however, the coordinates θ' of the sources of radio emission at different frequencies are different these intervals will not be the same. In both cases the series of values of ω_{max} obtained in one passage of the radio spectrograph through the frequency band (i.e. in a fraction of a second) determines the "instantaneous" position θ' of the source of radio emission at these frequencies with the usual uncertainty, which can be eliminated to a certain extent by using a second interferometer with a different baseline. In the process of repeated passages of ω_0 over the band, therefore, the dependence of one of the source's coordinates θ' on frequency and time is determined. Just as before, the size of the observed source can be judged here from the degree of modulation of the interference picture at the receiver output.

In observations of Jupiter's sporadic radio emission in the decametric band with a phase-switching interferometer the signal is recorded just as in an ordinary radio spectrograph by modulating the brightness of the oscillograph's beam which moves in time with the change in frequency. Therefore a dynamic spectrum was recorded on a continuously moving tape. Due to the use of an interferometer, however, this spectrum had a characteristic feature: it was covered with a series of alternating dark and light bands corresponding to the maxima and minima of an interference picture created by a point source during the frequency switching (see Fig. 91 in section 19). Since disturbances, which have a broad angular spectrum, do not produce a similar interference picture, the problem of studying Jupiter's dynamic spectrum against a background of intense interference is thus considerably simplified.

THE PROBLEM OF STUDYING THE RADIO BRIGHTNESS DISTRIBUTION OVER A SOURCE. VARIABLE-BASELINE INTEROMETER.

As well as information on the position, effective size and nature of the movement of local sources an important part is played in radio astronomical observations by data on detailed study of the radio brightness distribution although, of course, it is a more difficult problem to obtain the latter than a simple estimate of the effective size of the source.

The one-dimensional distribution of the radio brightness over the source (5.1) can be found in observations on an interferometer with a variable base line. In actual fact, as follows from the relations (5.6)–(5.9), the variable component at the output of an interferometer with a fixed baseline D changes harmonically when the source passes through the multi-lobe polar diagram

§ 5] Angular Spectrum

with an amplitude and phase equal to the amplitude and phase of the component of a complex Fourier expansion of the function $G(\theta)$ in the "frequency" $kD = 2\pi D/\lambda$. In other words, when the source passes through the beam of the interferometer the latter separates from among all the components of the expansion of the one-dimensional radio brightness distribution $G(\theta)$ into a Fourier integral one component (with an angular period $2\pi/kD$ whose magnitude is the same as the width of an individual lobe of the aerial diagram). It can be seen from (5.11) that a two-aerial phase-switched interferometer can provide a similar operation.

In order to determine the spectrum as a whole we must carry out observations and obtain interference curves for all possible values of the baseline D without changing the orientation of the interferometer relative to the source. Knowing the whole set of the amplitudes $g_{kD}(kD)$ and phases $\psi_{kD}(kD)$ of the Fourier expansion the formula

$$G(\theta) = \int_{-\infty}^{+\infty} G_{kD}\, e^{ikD\theta}\, d(kD) \tag{5.13}$$

can be used to find the one-dimensional distribution of the effective temperature $G(\theta)$ easily.

For one-dimensional brightness distribution measurements the design of the interferometer must provide for moving one of the aerials (without changing the orientation of the installation) to obtain different values of the baseline D.

In practice interference pictures are taken not for a continuous series of D from 0 to ∞ but with a certain set of close enough individual values, the greatest of which is D_{max}. Therefore the spectrum G_{kD} obtained is limited—it contains only those Fourier components which correspond to an angular period of not less than $2\pi/kD_{max}$, and the distribution $G(\theta)$ found from the measurements differs from the actual one, being smoothed over angular intervals $\Delta\theta\ 2\pi/kD_{max}$. The latter means that the resolving power of an interferometer with a variable baseline determined by the value of D_{max} cannot be used to find radio brightness oscillations with an angular period less than $2\pi/kD_{max}$ in the actual distribution.

The two-dimensional distribution of the radio brightness $T_{eff}(\theta, \xi)$ cannot, of course, be determined from observations on an interferometer with a variable baseline which is orientated in a given way relative to the source. This is clear from the fact that any change in the dependence of $T_{eff}(\xi)$ on which preserves the value (5.1) of $G(\theta)$ is not reflected at all in the results of the observations on an interferometer whose baseline runs along θ.

A known function $G_{kD}(kD)$ can be used to judge the features of the two-dimensional distribution $T_{\text{eff}}(\theta, \xi)$ if we introduce an additional assumption about the form of the radio brightness distribution which in certain cases (for example, when studying the distribution of T_{eff} over the disk of the "quiet" Sun) can to a certain degree be justified by the visible configuration of the emitting object. Let us assume that the source of the radio emission has circular symmetry (Machin, 1951). Then $T_{\text{eff}}(\theta, \xi) = T_{\text{eff}}\left(\sqrt{\theta^2+\xi^2}\right)$ and $G(\theta)$ does not change when the sign of θ changes. This kind of assumption is obviously only possible if the observed value of the phase is $\Psi_{kD} = 2\pi n$ ($n = 0, \pm1, \pm2, \ldots$) for any kD. In this case

$$G_{kD} = g_{kD} = \int G(\theta) \cos(kD\theta) \, d\theta$$
$$= \iint_{-\infty}^{+\infty} T_{\text{eff}}\left(\sqrt{\theta^2+\xi^2}\right) \cos(kD\theta) \, d\theta \, d\xi \quad (5.14)$$

(see equations (5.7) and (5.1)). In the opposite case, in accordance with (5.7), $\int G(\theta) \sin(kD\theta) \, d\theta \neq 0$ and $G(\theta)$ will not be an even function of θ.

Introducing the new variables $r = \sqrt{\theta^2+\xi^2}$, $x = \theta/r$, we transform (5.14) to the form

$$G_{kD}(kD) = 4 \int_0^\infty r T_{\text{eff}}(r) \int_0^1 \frac{\cos(kDrx)}{\sqrt{1-x^2}} \, dx \, dr.$$

However,

$$\int_0^1 \frac{\cos(kDrx)}{\sqrt{1-x^2}} \, dx = \frac{\pi}{2} J_0(kDr),$$

where J_0 is a zero-order Bessel function. Therefore

$$G_{kD}(kD) = 2\pi \int_0^\infty r T_{\text{eff}}(r) J_0(kDr) \, dr \quad (5.15)$$

and so

$$T_{\text{eff}}(r) = \frac{1}{2\pi} \int_0^\infty kD G_{kD}(kD) J_0(kDr) \, d(kD), \quad (5.16)$$

since the relation (5.15) is a Hankel transform for the function $T_{\text{eff}}(r)$ (Tranter, 1956). The formula (5.16) solves the problem of finding the two-dimensional distribution of the radio brightness from the observed values of $G_{kD}(kD)$ in the case of circular symmetry of the source.

At the same time to obtain the two-dimensional distribution of the radio brightness $T_{\text{eff}}(\theta, \xi)$ without preliminary assumptions about its symmetry it is necessary to make a series of measurements which determine the function $G_{kD}(kD)$ for all possible orientations of the "plane" of the interferometer lobes relative to the source. Here one interferometer

§ 5] Angular Spectrum

with a variable baseline is no longer sufficient, (when the direction of the baseline is fixed) though even in this case the orientation of the source relative to the "plane" of the lobes changes within certain limits when the source moves through the sky.

The method of calculating $T_{\text{eff}}(\theta, \xi)$ from known values of the complex values $G_{kD}(k\mathbf{D})$ of the Fourier expansions of the one-dimensional distributions obtained in all possible orientations of the interferometer baseline $\mathbf{D}$ relative to the source is as follows (O'Brien, 1953a; see also Wild, 1960a, and Bracewell, 1960). First the $G_{kD}(k\mathbf{D})$ data are used to find the complex amplitudes of the two-dimensional Fourier expansion of $T_{\text{eff}}(\theta, \xi)$:

$$T_{kD}(k\mathbf{D}) \equiv T_{kD}(kD_\theta, kD_\xi)$$
$$= \int\!\!\int_{-\infty}^{+\infty} T_{\text{eff}}(\theta, \xi) \exp\left[-ik(D_\theta\theta + D_\xi\xi)\right] d\theta\, d\xi \qquad (5.17)$$

(D_θ and D_ξ are projections of the interferometer baseline onto the directions θ and ξ). In accordance with (5.1), (5.7) the Fourier transformation of the one-dimensional distribution $G(\theta)$ along the θ-axis is of the form[†]

$$G_{kD}(kD_\theta, 0) = \int\!\!\int_{-\infty}^{+\infty} T_{\text{eff}}(\theta, \xi) \exp\left[-ikD_\theta\theta\right] d\theta\, d\xi. \qquad (5.18)$$

By comparing (5.18) with (5.17) we can see that

$$G_{kD}(kD_\theta, 0) = T_{kD}(kD_\theta, 0).$$

We thus know the values of $T_{kD}(kD_\theta, kD_\xi)$ at all the points kD_θ along the θ axis. We can likewise obtain the amplitudes of the Fourier expansion of T_{kD} at all points along any other axis rotated relative to θ, if we take the values of G_{kD} measured with the corresponding orientation of the interferometer baseline $\mathbf{D}$. As a result, knowing $T_{kD}(kD_\theta, kD_\xi)$ at all points of the plane kD_θ, kD_ξ and by Fourier integration it is easy to find the unknown distribution $T_{\text{eff}}(\theta, \xi)$:

$$T_{\text{eff}}(\theta, \xi) = \int\!\!\int_{-\infty}^{+\infty} T_{kD}(kD_\theta, kD_\xi) \exp\left[ik(D_\theta\theta + D_\xi\xi)\right] d(kD_\theta)\, d(kD_\xi). \qquad (5.19)$$

THE MULTI-ELEMENT INTERFEROMETER

Measuring the spectrum G_{kD} on interferometers with a variable baseline requires lengthy and complex observations to be made. The results obtained can be noticeably distorted by a change in the distribution of the

[†] We previously denoted the quantity $G_{kD}(kD_\theta, 0)$ by $G_{kD}(k\mathbf{D})$.

radio brightness over the source during the observations. The method of determining the distribution $G(\theta)$ with aerial systems with high directivity for one coordinate (for θ) is free from the above disadvantages. An example of this kind of system is the Pulkovo radio telescope with a "knife-edge" beam mentioned above; another example is Christiansen's multi-element radio interferometer (Christiansen and Warburton, 1953). It consists of thirty-two parabolic aerials connected in phase synchronism and set along an east–west line at equal distances l from each other. The working principle of a multi-element interferometer is similar to that of an optical diffraction grating.

When receiving the emission of a point source at an angle θ' to the plane of the interferometer the difference in travel between adjacent aerials separated by a distance l is (just as in the case of a two-aerial interferometer) $l \sin \theta'$. It is clear from this that at the radiometer input the voltages from all N aerials are added with phases that form an arithmetic progression with a difference $kl \sin \theta'$. If at the same time we take into consideration the fact that the square of the amplitude of the oscillations from each aerial is proportional to $F_0(\theta', \xi')$ (the polar diagram of this aerial) it is easy to show that the signal at the radiometer output is

$$\Xi \propto F_0(\theta', \xi') \frac{\sin^2(N \cdot \tfrac{1}{2} kl \sin \theta')}{\sin^2(\tfrac{1}{2} kl \sin \theta')}. \tag{5.20}$$

The expression on the right-hand side of (5.20) clearly describes the polar diagram of a multi-element interferometer. Its principal maxima (whose value is $F_0 N^2$) correspond to angles $\theta'_{\max}$ for which

$$\tfrac{1}{2} kl \sin \theta'_{\max} = \pi n,$$

i.e.

$$\sin \theta'_{\max} = \frac{n\lambda}{l} \quad (n = 0;\ \pm 1;\ \pm 2;\ \ldots). \tag{5.21}$$

The width of the main lobe, which is determined by the distance $\Delta \theta_l$ between the polar diagram nulls θ'_0 and θ'_{00} nearest to the principal maximum, is given in accordance with (5.20) by the relations

$$N \cdot \tfrac{1}{2} kl \sin \theta'_0 = Nn\pi + \pi,$$

$$N \cdot \tfrac{1}{2} kl \sin \theta'_{00} = Nn\pi - \pi.$$

From this it follows that

$$\sin \theta'_0 - \sin \theta'_{00} = \frac{2\lambda}{D}$$

§ 5] Angular Spectrum

where $D = Nl$ is the length of the interferometer. If $\lambda \ll D$, then

$$\Delta\theta_l \approx \frac{2\lambda}{D \cos \theta'_{max}}. \tag{5.22}$$

The position of the principal maxima of a multi-element interferometer is thus determined by the distance between adjacent aerials (the "period of the grating") l and the width of the main lobes with respect to the angle θ' – the overall length D of the interferometer. For the central lobes (with $\theta'_{max} \ll 1$) the width $\Delta\theta_l$ with respect to the angle θ' is N times less than the distance between the lobes. At the same time the width of the beam in a plane of the interferometer orthogonal to the baseline D depends only on the directional properties of the individual aerials $F_0(\theta', \xi')$.

A good idea of the nature of the resultant polar diagram of a multi-element interferometer can be obtained from Fig. 2 which shows a cross-section of a Christiansen aerial diagram. When working on a wavelength of $\lambda = 21$ cm its aerial system (with $N = 32$, $l = 7.5$ m) receives the radio emission on a combination of narrow lobes about 3′ wide, 1·7° apart in a circle with a diameter of 11°. The last value is determined by the polar diagram $F_0(\theta', \xi')$.

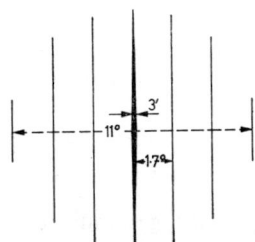

Fig. 2. Cross-section of polar diagram of thirty-two-element sky interferometer at half-intensity level

This kind of multi-element interferometer diagram makes it a very convenient instrument for studying the distribution of the radio brightness over a single powerful source such as the Sun. Since the angular diameter of the Sun ($\sim 32'$) is a third of the distance between the lobes the solar radio emission is received on only one lobe at a time; the signal Ξ at the receiver output is proportional to the intensity of a narrow strip of the Sun's disk about 3′ wide. As the Sun passes through each lobe on the recording of Ξ a curve is at once plotted of the one-dimensional distribution of the radio brightness $G(\theta)$, in which are resolved all non-uniformities of the effective temperature with a characteristic size greater than or of the order of 3′. Examples of recordings of one-dimensional radio-brightness distributions over the Sun are given in Fig. 12.

When studying the numerous weak discrete sources of cosmic radio emission the presence of several lobes and the great width of each lobe with respect to the angle ξ' make a further interpretation of the recordings obtained of Ξ more difficult or impossible. The number of lobes can be reduced, however, by increasing the ratio λ/l. In the case when the aerials are set so close together that $\lambda/l > 1$ only one main lobe with $\theta'_{max} = 0$ remains (see equation (5.21)).

In solar radio astronomy, thanks to the great intensity of the Sun's radio emission, the multi-lobe nature of the aerial system under discussion does not introduce any significant inconveniences since the effect of the comparatively weak discrete sources located at small angular distances from the Sun on the quantity Ξ is usually slight. On the other hand, the multiplicity of lobes is more of an advantage in this case, and not a disadvantage of the aerial system since during one passage of the Sun through the sky it permits the taking of several one-dimensional radio brightness distributions with different angles between the "plane" of the lobe and the source. When a further multi-element interferometer with a baseline set at right angles to the baseline of the first interferometer is added this makes it possible to obtain one-dimensional distributions when scanning the source at all possible angles. By expanding these distributions into Fourier integrals and using the method of calculation given above the two-dimensional distribution of the radio brightness over the Sun's disk can be found.

THE MILLS CROSS

Because of the above advantages multi-element interferometers have been widely used in radio astronomical observations of the Sun. The necessity for laborious processing of the results in order to obtain the two-dimensional distribution undoubtedly limits the scope for using aerial systems of this kind, however. On the other hand, this processing becomes superfluous if the aerial system has a narrow polar diagram lobe in the two dimensions. In the millimetric and partly in the centimetric bands this requirement is generally satisfied by a parabolic aerial of great enough diameter (see p. 30), and at longer wavelengths by using the system which has been given the name of the "Mills Cross" (Mills and Little, 1953).

The Mills Cross consists of two identical multi-element interferometers whose baselines are at right angles to each other. The voltages from the two interferometers are fed to the radiometer input first in phase and then in opposite phase by periodically introducing a "half-wave section" into the cable coupling one of the interferometers with the radiometer. The

§ 5] Angular Spectrum

signal $\varXi$ at the radiometer output is proportional to the difference of the signal powers at the input with in-phase and out-of-phase couplings of the two aerials.

Then, just as in the case of a two-element interferometer with phase-switching, during reception of radio emission on the Mills Cross $\varXi \propto \overline{E_1 E_2}$, where E_1 and E_2 are the voltages fed to the input from interferometers 1 and 2 when connected in phase. These voltages when receiving a point source characterized by a flux S_ω are respectively proportional to $\sqrt{S_\omega F_1(\theta', \xi')}$ and $\sqrt{S_\omega F_2(\theta', \xi')}$, where F_1 and F_2 are the polar diagrams of the individual interferometers 1 and 2. From what has been said it is clear that

$$\varXi \propto S_\omega \sqrt{F_1(\theta', \xi') F_2(\theta', \xi')},$$

and therefore the resultant polar diagram of the Mills Cross is determined by the product of the diagrams of the interferometers composing it:

$$F(\theta', \xi') = \sqrt{F_1(\theta', \xi') F_2(\theta', \xi')}. \tag{5.23}$$

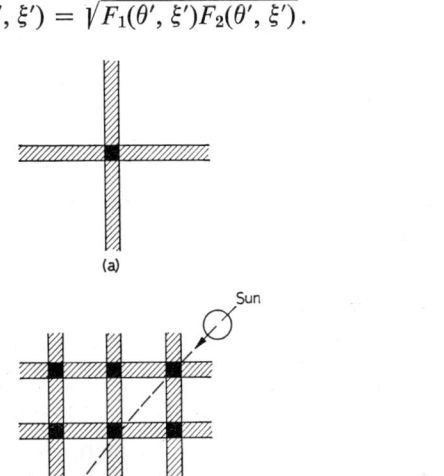

FIG. 3. Cross-sections of polar diagrams of a Mills Cross

Cross sections of the polar diagrams of the sky in two cases—when the individual interferometers have only one main lobe each and when the individual interferometers have a multi-lobe diagram—are shown diagrammatically in Figs. 3a and 3b respectively. In the first case the resultant polar diagram has one main lobe whose width in both measurements (along θ' and ξ') is defined by the relation (5.22). This kind of aerial system is widely used in studies of galactic radio emission. In the second case the resultant diagram is a combination of narrow ("pencil") lobes running along the intersection of the "knife-edge" lobes of the two interferometers. This kind of device, which combines the advantages of the Mills principle and of the multi-element interferometer (with a system of lobes a long way apart) is used in radio astronomy to study the radio brightness distribution over the Sun's disk (Christiansen and Mathewson, 1958).

In the latter case when the source passes through each "pencil" lobe, whose width is much less than the angular size of the source, the signal at the output is at any time equal to the flux of the radio emission created by a sector whose area is close to $(\Delta\theta_l)^2$. The width $\Delta\theta_l$ of the resultant lobes determines the angular resolution of the cross.[†] A two-dimensional distribution of the radio brightness over the disk is obtained by the Sun passing through the beam lobes several times. With a lobe width of the order of 3′ and distances between the lobes of the order of one degree the process of measuring $T_{\text{eff}}(\theta, \xi)$ takes about an hour. This time is short enough to be able to judge the real distribution of the radio brightness produced by the "quiet" Sun and the local sources of the slowly changing component, but it is too great to obtain the two-dimensional distribution $T_{\text{eff}}(\theta, \xi)$ at the time of bursts of radio emission whose life is generally not more than a few tens of minutes. On the other hand, it is quite possible to take the one-dimensional distribution $G(\theta)$ during a microwave burst with a multi-element interferometer since it takes only 2 minutes to scan the Sun's disk once.

ECLIPSE OBSERVATIONS

As well as the data on the nature of the distribution of the radio brightness over the Sun's disk provided by observations with complex aerial systems some information on the size, position and shape of radio emission sources can be obtained by a simpler method—measurements of the total radio emission flux of the Sun during solar eclipses. Measurements of this kind played a very important part in the first few years of the development of radio astronomy when the resolutions of the aerials was low; they could be used to estimate the size of the "radio Sun" at metric wavelengths, discover the existence of local sources (radio spots) on the Sun's disk, etc. (Khaikin and Chikhachev, 1947).

The resolution of eclipse observations is determined by the width of the diffraction picture produced by the radio emission of a point source localized on the Sun with diffraction on the edge of the lunar disk (Ginzburg and Getmantsev, 1950). This characteristic angular width $\Delta\theta$ depends on

[†] Often only a system of two interferometers with one main lobe is called a "Mills Cross".

§ 5] Angular Spectrum

the size of the first Fresnel zone $\sqrt{\lambda R_{E-M}}$ and is equal to $\sqrt{\lambda R_{E-M}}/R_{E-M}$, where R_{E-M} is the distance from the Earth to the Moon. If $\lambda = 100$ cm, $\Delta\theta \approx 0\cdot 2'$; if $\lambda = 1$ cm, $\Delta\theta \approx 0\cdot 02'$. It is clear from the estimates given that the resolution of eclipse measurements is far higher than the corresponding value for modern aerial systems: the latter is generally not better than 3', reaching 1–2' in the best case. Therefore eclipse observations are useful in the study of the size, position and mean effective temperature of small local sources on the Sun's disk. The duration of a "radio eclipse" and the magnitude of the residual flux of radio emission at the moment of total eclipse can be used to judge the effective size of the Sun in the radio beams; the magnitude and duration of the "jump" on the recording during the closing and opening of the local source on the Sun determine the emission flux and angular size respectively of the local source (generally speaking in two different directions). In certain cases the high resolution of eclipse observations can also be used for obtaining more detailed data on the nature of the radio brightness distribution over a small local source (see section 10).

It is easy to understand that in the sense of the volume of information obtained on the radio brightness distribution, observations during one eclipse are equivalent to two scans of the Sun (in the general case in two different directions) by an aerial with a "knife-edge" beam. The "one-dimensionality" noted for eclipse measurements does not let us find the two-dimensional distribution of radio brightness over the Sun during an eclipse: we cannot do without additional assumptions about the form of the radio-isophots† when drawing conclusions on the structure of the distribution.

Another disadvantage of eclipse observations is their episodic nature; they are therefore unsuitable when statistical processing of a series of experimental data is necessary to find some characteristic of the solar radio emission. For example, they are unsuitable for a reliable determination of the distribution of the radio brightness over the "quiet" Sun's disk— partly because of the "one-dimensionality" of the measurement results, partly because of the impossibility of distinguishing with certainty the distribution of the "quiet" Sun's radio emission from the sporadic component (radio emission of local sources), particularly in the metric and decimetric bands where the latter reaches a high intensity. In this respect eclipse observations are not nearly as good as investigations using modern aerial devices. It is true that there is a certain interest in finding the distribution of the radio brightness over the disk at shorter wavelengths during

† Radio-isophots are the geometrical position of points on the Sun's disk with a constant effective temperature.

eclipses to check data obtained by narrow-beam aerials. The point is that at centimetric and millimetric wavelengths the sporadic component is less intense and more closely linked with optical features on the Sun's disk. It is therefore easier to separate from the emission of the "quiet" Sun in subsequent processing of the experimental data.

6. Polarization of Radio Emission

POLARIZATION PARAMETERS[†]

Extraterrestrial radio emission consists, generally speaking, of two parts —a polarized one and an unpolarized one. Both components have a continuous frequency spectrum. In a polarized wave, however, the oscillations of the electric field components $E_x(t)$ and $E_y(t)$ along two axes at right angles to each other lying in the plane of the wave front are coherent in the sense that $\overline{E_x(t)E_y(t)} \neq 0$. In a non-polarized wave (also called naturally or chaotically polarized) these components are non-coherent, i.e. their time-averaged product is equal to zero.

In the general case the electrical vector $E(t)$ of a polarized wave describes an ellipse whose size fluctuates in such a way that the orientation of the ellipse and the ratio of its axes remain the same. Therefore in polarized emission received in the frequency range $\Delta\omega_b$ the field components E_x, E_y are of the form

$$E_x(t) = E_{0x} \sin(\omega_0 t - \psi_x), \quad E_y(t) = E_{0y} \sin(\omega_0 t - \psi_y), \quad (6.1)$$

where the amplitudes E_{0x}, E_{0y} and the phases ψ_x, ψ_y change in time[‡] although the ratio E_{0x}/E_{0y} and the phase difference $\psi_{xy} = \psi_x - \psi_y$ that define the orientation of the ellipse χ and the ratio of its axes p (Fig. 4) remain constant.

It follows from the above that the polarized part of the radio emission is characterized by three parameters: its intensity, orientation, and the ratio of the axes of the polarization ellipse (I_{pol}, χ, p); the direction of rotation of the vector $E(t)$ is allowed for here in the sign of p.[§] In a non-polarized wave both the direction and magnitude of the vector $E(t)$ vary chaotically. Therefore the non-polarized part of the radio emission is

---

[†] The content of this section and the next one is based to a considerable extent on an article by Cohen (1958).

[‡] The characteristic time for this change is of the order of $1/\Delta\omega_b$; with a small enough band $\Delta\omega_b \ll \omega_0$ the amplitudes and phases change slowly when compared with the period of the oscillations $2\pi/\omega_0$.

[§] Positive values of p correspond to clockwise rotation of the vector $E(t)$ (looking along the direction of propagation of the wave); negative ones correspond to anticlockwise rotation.

§ 6] Polarization of Radio Emission

characterized only by its intensity I_{nat}. As a whole the radio emission coming from a point in the sky lying in a direction Ω and received in the frequency band $\Delta\omega_b$ is defined by four parameters: the total intensity $I = I_{pol} + I_{nat}$, the degree of polarization $\varrho = I_{pol}/I$, the ratio of the axes p and the orientation χ of the polarization ellipse. If $\varrho = 0$ the emission is not polarized; when $\varrho = 1$ the emission is completely polarized; in the remaining cases it is partly polarized.

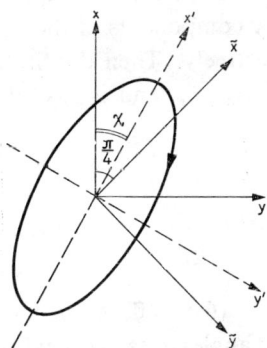

Fig. 4. Polarization ellipse

As the quantities describing the state of the radio emission's polarization we can also take the Stokes parameters I, Q, U, V, which are connected with I, ϱ, p, χ by the relations (Chandrasekar, 1953, section 15):

$$\left.\begin{aligned} I &= I, \\ Q &= \varrho I \cos(2\sigma) \cos(2\chi), \\ U &= \varrho I \cos(2\sigma) \sin(2\chi), \\ V &= \varrho I \sin(2\sigma), \end{aligned}\right\} \quad (6.2)$$

where $\sigma = \arctan p$. In accordance with equation (6.2)

$$\left.\begin{aligned} \varrho &= \frac{(Q^2+U^2+V^2)^{1/2}}{I}, \\ \sin 2\sigma &= \frac{V}{(Q^2+U^2+V^2)^{1/2}}, \\ \tan 2\chi &= \frac{U}{Q}. \end{aligned}\right\} \quad (6.3)$$

In the area of interest to us the value of the Stokes parameters is that they can be expressed very simply in terms of quantities that can be measured directly during polarization measurements.

In actual fact radio astronomical observations are as a rule made with aerials that receive the linearly polarized or circularly polarized component of the radio emission. For aerials orientated along the x- and y-axes the intensities of the corresponding linear components of the field will be

$$I_x = \overline{E_{0x}^2} + \frac{I_{\text{nat}}}{2}, \qquad I_y = \overline{E_{0y}^2} + \frac{I_{\text{nat}}}{2}. \tag{6.4}$$

In equation (6.4) $\overline{E}_{0x}^2$, $\overline{E}_{0y}^2$ are the time-averaged squares of the normalized amplitudes of the x and y components of the field of the polarized part of the radio emission respectively. Then the Stokes parameters can be expressed in terms of I_x, I_y, E_{0x}, E_{0y} and ψ_{xy} as follows:

$$\left.\begin{aligned}I &= I_x + I_y, \\ Q &= I_x - I_y, \\ U &= \overline{2E_{0x}E_{0y}} \cos \psi_{xy}, \\ V &= \overline{2E_{0x}E_{0y}} \sin \psi_{xy}.\end{aligned}\right\} \tag{6.5}$$

The first of the relations (6.5) reflects the fact that the two linear components which are at right angles to each other have opposite polarizations.† As a proof of the three remaining relations (6.5) we introduce the rectangular system of coordinates x', y' rotated by an angle χ relative to the x, y system. The new coordinate axes will be orientated along the axes of the polarization ellipse (Fig. 4). The components $E_{x'}$ and $E_{y'}$ of the electric field of the polarized part of the radio emission are here obviously of the form

$$\left.\begin{aligned}E_{x'} &= E_0 \cos \sigma \sin (\omega_0 t), \\ E_{y'} &= E_0 \sin \sigma \cos (\omega_0 t),\end{aligned}\right\} \tag{6.6}$$

where $\sigma = \arctan p$ and the quantity E_0 fluctuates with a characteristic time $1/\Delta\omega_b$. The oscillations of the field in the directions x and y are defined by the expressions

$$\left.\begin{aligned}E_x &= E_{x'} \cos \chi - E_{y'} \sin \chi \\ &= E_0[\cos \sigma \cos \chi \sin (\omega_0 t) - \sin \sigma \sin \chi \cos (\omega_0 t)], \\ E_y &= E_{x'} \sin \chi + E_{y'} \cos \chi \\ &= E_0[\cos \sigma \sin \chi \sin (\omega_0 t) + \sin \sigma \cos \chi \cos (\omega_0 t)],\end{aligned}\right\} \tag{6.7}$$

† Two waves are said to have opposite polarizations if the electrical vectors in these waves when rotating in opposite directions describe similar ellipses with axes at right angles to each other. In particular two circular polarizations with different signs of rotation will be opposite, as will two linear polarizations if the planes of polarization are orthogonal. The intensities of oppositely polarized waves are added no matter what the degree of coherency between them.

§ 6] Polarization of Radio Emission

which are the same as (6.1) if in the latter we put

$$\left.\begin{aligned}E_{0x} &= E_0(\cos^2\sigma\cos^2\chi+\sin^2\sigma\sin^2\chi)^{1/2},\\ E_{0y} &= E_0(\cos^2\sigma\sin^2\chi+\sin^2\sigma\cos^2\chi)^{1/2},\\ \tan\psi_x &= \tan\sigma\tan\chi,\\ \tan\psi_y &= -\tan\sigma\cot\chi.\end{aligned}\right\} \quad (6.8)$$

It follows from (6.8) that

$$\left.\begin{aligned}\overline{E_{0x}^2}+\overline{E_{0y}^2} &= \overline{E_0^2},\\ \overline{E_{0x}^2}-\overline{E_{0y}^2} &= \overline{E_0^2}\cos 2\sigma\cos\chi,\\ 2\overline{E_{0x}E_{0y}\cos\psi_{xy}} &= \overline{E_0^2}\cos 2\sigma\sin 2\chi,\\ 2\overline{E_{0x}E_{0y}\sin\psi_{xy}} &= \overline{E_0^2}\sin 2\sigma.\end{aligned}\right\} \quad (6.9)$$

Remembering (6.4) and the equalities $I_{\text{pol}} = \varrho I = \left(\overline{E_{0x}^2}+\overline{E_{0y}^2}\right)$ the relations obtained prove the validity of (6.5).

It is also easy to see that when the two linear aerials are turned by an angle $\pi/4$ the parameter Q becomes U and vice versa since the angle 2χ changes by $\pi/2$ (see equations (6.2)). Then the Stokes parameters in the new system $\tilde{x}, \tilde{y}$ become

$$\left.\begin{aligned}I &= I_{\tilde{x}}+I_{\tilde{y}},\\ Q &= 2\overline{E_{0\tilde{x}}E_{0\tilde{y}}}\cos\psi_{\tilde{x}\tilde{y}},\\ U &= I_{\tilde{x}}-I_{\tilde{y}},\\ V &= 2\overline{E_{0\tilde{x}}E_{0\tilde{y}}}\sin\psi_{\tilde{x}\tilde{y}},\end{aligned}\right\} \quad (6.10)$$

if the angle χ is as before read from the x-axis.

Further, for aerials receiving radio emission components with clockwise and anticlockwise polarization the intensities of the latter will be respectively

$$I_r = \overline{E_{0r}^2}+\frac{I_{\text{nat}}}{2}, \quad I_l = \overline{E_{0l}^2}+\frac{I_{\text{nat}}}{2}. \quad (6.11)$$

Here E_{0r}, E_{0l} denote the field amplitudes of the clockwise (right-handed) and anticlockwise (left-handed) polarized components respectively of the polarized part of the radio emission; the constant phase difference between them is ψ_{rl}. In this case the Stokes parameters become

$$\left.\begin{aligned}I &= I_r+I_l,\\ Q &= 2\overline{E_{0r}E_{0l}}\cos\psi_{rl},\\ U &= 2\overline{E_{0r}E_{0l}}\sin\psi_{rl},\\ V &= I_l-I_r.\end{aligned}\right\} \quad (6.12)$$

The validity of the relations (6.12) can easily be checked in the same way as for the formulae (6.5); the corresponding proof is given in (Cohen, 1958).

Measuring the intensities of two opposite polarizations I_x, I_y (or I_r, I_l) gives us two Stokes parameters; the phase difference ψ_{xy} between the field components E_x, E_y (or ψ_{rl} between E_r, E_l) and the product $\overline{E_{0x} E_{0y}}$ (or $\overline{E_{0r} E_{0l}}$) allow us to find the other two parameters. To measure $\overline{E_{0x} E_{0y}}$ and ψ_{xy}, or $\overline{E_{0r} E_{0l}}$ and ψ_{rl}, we use the fact that they are closely connected with the correlation function of the two linearly polarized or two circularly polarized components of the emission field. If we use $E'_x(t)$ and $E'_y(t)$ to denote the field components of the non-polarized part of the emission, then the correlation function of the two linear components can be written as follows:

$$\mathcal{G}(t_0) = \frac{\overline{[E_x(t)+E'_x(t)][E_y(t+t_0)+E'_y(t+t_0)]}}{\{\overline{[E_x(t)+E'_x(t)]^2}\,\overline{[E_y(t)+E'_y(t)]^2}\}^{1/2}}. \qquad (6.13)$$

Bearing in mind that with a narrow band $\Delta\omega_b$ the fields $E_x(t)$ and $E_y(t)$ are of the form (6.1), where the amplitudes and phases are slowly varying functions of t, we obtain

$$\mathcal{G}(t_0) = \tilde{\mathcal{G}} \cos(\omega_0 t_0 + \psi_{xy}), \qquad \tilde{\mathcal{G}} = \frac{\overline{E_{0x} E_{0y}}}{2(I_x I_y)^{1/2}}. \qquad (6.14)$$

Likewise for the circular components of the field

$$\mathcal{G}(t_0) = \tilde{\mathcal{G}} \cos(\omega_0 t_0 + \psi_{rl}), \qquad \tilde{\mathcal{G}} = \frac{\overline{E_{0r} E_{0l}}}{(I_r I_l)^{1/2}}. \qquad (6.15)$$

It follows from (6.14), (6.15) that the products $\overline{E_{0x} E_{0y}}$, $\overline{E_{0r} E_{0l}}$ and the phase differences ψ_{xy}, ψ_{rl} are determined by the amplitudes and phases of the corresponding correlation functions.

Polarization observations do not provide full information on the state of the polarization of extraterrestrial radio emission unless all the Stokes parameters are determined at the same time. For example, in investigations of solar radio emission it is typical to measure the intensities of the two circular components I_r, I_l, as a result of which the "degree of circular polarization" is found:

$$\varrho_c = \frac{I_l - I_r}{I_l + I_r} = \frac{V}{I}. \qquad (6.16)$$

This quantity is connected with the true degree of polarization ϱ by the relation

$$\varrho_c = \varrho \sin(2\sigma) \qquad (6.17)$$

(see equation (6.3)).

§ 6] Polarization of Radio Emission

The quantity ϱ_c is the same as ϱ only if $2\sigma = \pm\pi/2$.† Remembering the relation $\sigma = \arctan p$, we can see that the latter is valid only when the radio emission is completely or partly circularly polarized. In the general case of an elliptically polarized wave $|\varrho_c| < \varrho$; this is quite natural since the difference $I_l - I_r$ characterizes only the intensity of a circularly polarized wave without reflecting the presence of a linear component in the elliptically polarized emission in any way.

When measuring the intensities of the linear components of the field we introduce the definition of the "degree of linear polarization"

$$\varrho_l = \frac{I_x - I_y}{I_x + I_y} \frac{1}{\cos(2\chi)} = \frac{Q}{I \cos(2\chi)}. \tag{6.18}$$

In accordance with equation (6.3)

$$\varrho_l = \varrho \cos(2\sigma), \tag{6.19}$$

and in the general case $|\varrho_l| \leq \varrho$. They are equal when $2\sigma = 0, \pi$, i.e. when $p = 0, \infty$ and the radio emission is linearly polarized (fully or partly).

We notice that the field of an elliptically polarized wave can be represented as the superposition of circularly and linearly polarized components. However, the intensities of these components are not added since in the case in question the components themselves are coherent and their polarizations are not opposite. It follows from this that the sum $\varrho_c I + \varrho_l I$ is not equal to the intensity of the polarized part of the radio emission; this can also be seen, moreover, from the relations (6.17) and (6.19), according to which $\varrho^2 = \varrho_c^2 + \varrho_l^2$.

METHODS OF POLARIZATION MEASUREMENT IN THE METRIC WAVEBAND

The requirements imposed on the design of a polarimeter—a radio telescope specially designed for studying the polarization of extraterrestrial radio emission—are chiefly determined by the number of Stokes parameters to be measured and the specific features of the radio emission being studied (the frequency range, the nature of the angular spectrum, the time-dependence of the intensity, etc.).

In the majority of cases, in order not to complicate the design of the polarimeter, we limit ourselves in radio astronomy to obtaining information on the degree of circular or linear polarization. According to equation (6.16) the degree of circular polarization ϱ_c is defined by two Stokes parameters: I, V; it is sufficient to measure the intensities I_r, I_l of the circular components to obtain ϱ_c. At the same time measurements of the intensities I_x, I_y on two linear aerials with fixed axes are insufficient for

† Or, what is the same thing, $Q = U = 0$.

determining the degree of linear polarization ϱ_l since the orientation of the ellipse χ remains unknown in observations of this kind. The latter can be found by measuring $\overline{E_{0x}E_{0y}}$ and ψ_{xy} or making additional measurements of the intensities $I_{\tilde{x}}$, $I_{\tilde{y}}$ on a second pair of aerials at right angles to each other rotated relative to the first by a certain angle, let us say an angle $\pi/4$. Since the degree of linear polarization ϱ_l (6.18) does not depend on the choice of the axes x, y, the intensities $I_{\tilde{x}}$, $I_{\tilde{y}}$ will give us the value of $\varrho_l \cos(2\chi - \pi/2) = \varrho_l \sin(2\chi)$. Since we know at the same time the value of $\varrho_l \cos(2\chi)$ from the measurements of I_x, I_y, it is not hard to find ϱ_l and χ from this. We can also use measurements of the intensity made with a rotating linear aerial to find the value of χ: then the position of the aerial which corresponds to the maximum intensity value determines the unknown angle χ.

On the other hand there is insufficient information on the degree of linear or circular polarization to study the radio emission of the Sun and the planets. In recent years, therefore, a number of polarimeters have been made which can provide complete information on the nature of the polarization of the radio emission being observed.

It follows from what has been said earlier that the polarization of radio emission is fully described by four independent parameters which may be I, ϱ, p, χ or I, Q, U, V. To find the latter we must measure four independent quantities connected with them, e.g. the intensities of two field components with opposite polarizations and the amplitudes and phases of the correlation function of these components.

Two aerials are sufficient for measurements of this kind; however, if for some reason or other it is undesirable to determine the correlation parameters the number of aerials should be increased to four. In the latter case the knowledge of only the intensities of the four components provides full information on the nature of the polarization of the radio emission being studied. For example, after using two aerials with circular polarization to measure the intensities I_r, I_l we can find the Stokes parameters $I = I_r + I_l$, $V = I_l - I_r$ (see equation (6.12)). Information can be obtained about the other two parameters $Q = I_x - I_y$ (6.5) and $U = I_{\tilde{x}} - I_{\tilde{y}}$ (6.10) by using for the measurement of I_x, $I_{\tilde{x}}$ two linear aerials orientated at an angle of 45° to each other and remembering that $I_y = I - I_x$, $I_{\tilde{y}} = I - I_{\tilde{x}}$.

There may be different sets of four components in intensity measurements. However, the state of the polarization cannot be determined by measuring the intensities on four aerials with polarizations that are opposite in pairs (such as I_r, I_l, I_x, I_y or I_x, I_y, $I_{\tilde{x}}$, $I_{\tilde{y}}$) since in this case the results of the measurements are not independent: $I_r + I_l = I_x + I_y$; $I_x + I_y = I_{\tilde{x}} + I_{\tilde{y}}$.

§ 6] Polarization of Radio Emission

Having made these preliminary remarks let us examine in greater detail the working principles of polarimeters which are chiefly used in the metric waveband.

The block diagram of a typical two-aerial polarimeter, in which the Stokes parameters are found by measuring the intensities of two oppositely polarized components and by measuring the amplitude and phase of the correlation function between these components, is shown in Fig. 5. The

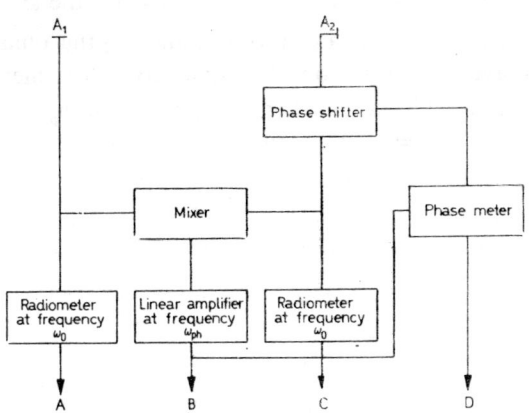

FIG. 5. Block diagram of a two-aerial polarimeter

radiometer outputs A and C give the values of the intensities of the two components with opposite polarizations (I_x, I_y or I_r, I_l, etc.; the nature of the components depends on the design of the aerials). A phase shifter is connected into one arm of the polarimeter for making the phase shift $\Delta\psi = \omega_{ph}t$ (i.e. the time lag) of the signal coming from aerial A_2 relative to the signal from aerial A_1.[†]

In a small enough frequency band $\Delta\omega_b$ the signal from aerial A_1 is proportional to

$$E_1(t) \cos [\omega_0 t + \psi_1(t)] + E_1'(t) \cos [\omega_0 t + \psi_1'(t)]; \quad (6.20)$$

after passing through the phase shifter the signal from aerial A_2 is proportional to

$$E_2(t) \cos [(\omega_0 + \omega_{ph})t + \psi_2(t)] + E_2'(t) \cos [(\omega_0 + \omega_{ph})t + \psi_2'(t)]. \quad (6.21)$$

The first term in equations (6.20) and (6.21) relates to the polarized part of the radio emission and the second to the non-polarized part. Therefore the amplitudes E_1', E_2' and the phases ψ_1', ψ_2' vary chaotically, whilst E_1, E_2 and

[†] In practice ω_{ph} is generally in the audio-frequency range so we always have $\omega_{ph} \ll \omega_0$.

ψ_1, ψ_2 fluctuate, keeping the ratio E_1/E_2 and the difference $\psi_1-\psi_2 \equiv \psi_{12}$ constant. Then both signals go to the mixer, the voltage at the output of which is proportional to the product of the voltages at the input, i.e. proportional to the product (6.20) and (6.21).

As a result of the subsequent filtration of the voltage from the mixer output by the linear amplifier, which passes only frequencies close to ω_{ph}, the signal at output B, as (6.20) and (6.21) can easily be used to show, is proportional to $\frac{1}{2}\overline{E_1 E_2} \cos(\omega_{ph}t+\psi_{12})$. Therefore the signal amplitude characterizes the amplitude $\overline{E_1 E_2}/2(I_1 I_2)^{1/2}$, and ψ_{12} the phase of the two components' correlation function. The quantity ψ_{12} is measured by the phase meter (output D) by comparing the signal phase $\omega_{ph}t+\psi_{12}$ at output B with the quantity $\Delta\psi = \omega_{ph}t$ introduced by the phase shifter.

If aerials A_1, A_2 are receiving linearly polarized components the outputs A, B, C, D give the values of I_x, $\overline{E_{0x}E_{0y}}$, I_y, ψ_{xy}. The Stokes parameters are then determined from the formulae (6.5). We notice that in accordance with (6.5) $I_x + I_y = I$, $I_x - I_y = Q$, $8(\overline{E_{0x}E_{0y}})^2 = U^2 + V^2$ and therefore the measurement of three quantities $(I_x, I_y, \overline{E_{0x}E_{0y}})$ enables us to find only the total intensity I and the degree of polarization ϱ of the radio emission (see equation (6.3)). The ellipticity p and the orientation χ in this case can be obtained only after phase measurements.

When A_1 and A_2 are aerials receiving circular components the outputs A, B, C, D give respectively the values of I_r, $\overline{E_{0r}E_{0l}}$, I_l, ψ_{rl} which enable us to determine the Stokes parameters from the formulae (6.12). If we take into consideration the fact that according to (6.12) $I_r + I_l = I$, $I_l - I_r = V$, $8(\overline{E_{0r}E_{0l}})^2 = Q^2 + U^2$ it becomes clear from (6.3) that in this case the three measured quantities I_r, I_l, $\overline{E_{0r}E_{0l}}$ define three characteristics of the radio emission: the intensity, the degree of polarization and the ellipticity. The fourth characteristic (the angle χ) can be obtained only as a result of phase measurements.†

From what has been said we can clearly see the advantages of polarization observations at metric wavelengths using aerials which receive circularly polarized emission over linear aerials: the former enable three polarization parameters from the value of the signals at the outputs A, B, C and the latter only two.

We shall now say a few words about the aerials for receiving the indi-

---

† In this connection we note that in the metric band the orientation of the polarization ellipse changes considerably (by several radians) as the radio emission passes through the ionosphere (see section 7). Since it is difficult to allow precisely for the rotation of the polarization ellipse in the ionosphere, measurement of the angle χ often loses any meaning.

§ 6] Polarization of Radio Emission

vidual polarized components. As linear aerials half-wave dipoles or arrays of these dipoles pointing in one direction are generally used. So-called "helical" aerials made in the form of sections of helices are sometimes used for the reception of circular polarization; here the sign of the polarization depends on which way the helix "twists". More often, however, an array of two dipole aerials crossed at right angles is used. The aerial outputs are coupled together, the voltages fed from the two dipoles being shifted in phase by $\pi/2$ at a common point. This is generally achieved by introducing a "quarter-wave line" between the common point and one of the dipoles.

The Cornell University polarimeter whose layout is as in Fig. 5 (Cohen, 1958) and consists of aerials in the form of crossed dipoles which receive circular components has been used by Cohen for studying type III solar bursts (see section 14). The polarimeter works at a frequency of $f \approx 200$ Mc/s; the band of the receiver is very narrow (about 10 kc/s). The possible error in determining the ellipticity and degree of polarization is $\pm 5\%$.

The principle of measuring the only intensities to obtain exhaustive information on the state of the polarization is used in the Tokyo Observatory polarimeter (Suzuki and Tsuchiya, 1958). This equipment is specially adapted for studying events of short duration in the composition of the Sun's sporadic radio emission (type I bursts; see section 12) at a frequency of 200 Mc/s: a complete cycle of measurements takes only 1/200 sec. The aerial array consists of four dipoles set at angles of 45° to each other. By appropriate switchings of the dipoles and introducing quarter-wave lines data can be obtained in the time indicated on the intensity of six components of the radio emission (two circular and four linear). The information supplied by the polarimeter is excessive since it is sufficient to know the value of the intensities of only four components to determine the Stokes parameters; the intensities of the other two components are needed only as a check.

When studying the polarization of sources of small angular size against a background of intense and, generally speaking, also polarized radio emission it is a good thing to set the receiving aerials far enough apart so as to use the advantages of interference reception (high resolution). The block diagram of an interference polarimeter is shown in Fig. 6. The aerials A_1, A_2 here may be any pair with opposite polarizations. Diagrams (a) and (b) in the figure are standard phase-shifted interferometers. The amplitudes of the interference pictures at the output A and C of these interferometers are proportional to the intensities I_1, I_2 of the oppositely polarized components of the radio emission. Output B in Fig. 6c provides the quantity $\overline{E_1 E_2}$, where E_1 and E_2 are the amplitudes of the opposite components of the polarized part of the radio emission being received, whilst output D gives the phase shift between the interference picture at output B and the phase shifter.

The way an interference polarimeter works (Fig. 6) is therefore similar to the way an ordinary two-aerial polarimeter works (Fig. 5), the only difference being that the amplitudes of the interference pictures at outputs A, B, C are determined not only by the quantities $I_1, \overline{E_1 E_2}$ and I_2 but also the angular size of the source of the polarized emission. All other things being equal, the scope of the interference picture is sharply reduced when the angular diameter of the source $\Delta\theta_s$ becomes less than $\Delta\theta_I \sim \lambda/D$—the width of one lobe of the interference picture (D is the interferometer's baseline; see section 5 in this connection).

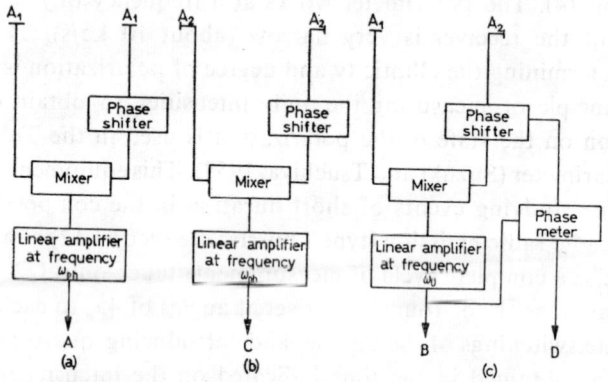

FIG. 6. Block diagram of an interference polarimeter

We notice that phase shifting is not necessary for polarization measurement since the corresponding time change in phase is ensured by the slow movement of the local source through the sky. However, just as in the case of studying the distribution of the radio brightness, phase switching by a special device ensures that an interference picture is obtained rapidly, which is of considerable importance when observing events of short duration.

In the type of interference polarimeter examined the data on the quadratic quantities (I_1, I_2, $\overline{E_1 E_2}$) are supplemented by phase measurements to obtain complete information on the state of the polarization. The latter can however be eliminated by adding a further interferometer of the type shown in Fig. 6a and b with aerials that will receive a further polarized component.

The first interference polarimeter was built in 1946. Ryle and Vonberg (1946) used it to find that at times the circularly polarized component is

§ 6] Polarization of Radio Emission

predominant in the composition of the solar radio emission at metric wavelengths. The phase is changed by the motion of the Sun through the sky. Phase shifting was used later in an interference polarimeter by Little and Payne–Scott (1951) when studying the polarization of bursts and enhanced radio emission connected with sunspots (see section 12). Basically the design of their polarimeter did not differ in any way from the installation whose block diagram is shown in Fig. 5.

Some information on the nature of the polarization of the radio emission can be obtained with a radio spectrograph (a frequency-shifted radiometer) whose input is coupled with two crossed linear aerials A_1 and A_2 (Komesaroff, 1958).

Let aerial A_1 receive the x-component and A_2 the y-component of the electric field of the radio emission. Then the voltages produced by the two aerials at the radiometer input (in the narrow frequency band $\Delta\omega_b$) will be proportional to the following expressions:

$$E_{0x} \sin(\omega_0 t - \psi_x - kl_1) + E'_{0x} \sin(\omega_0 t - \psi'_x - kl_1), \tag{6.22}$$

$$E_{0y} \sin(\omega_0 t - \psi_y - kl_2) + E'_{0y} \sin(\omega_0 t - \psi'_y - kl_2). \tag{6.23}$$

In (6.22), (6.23) E_{0x}, E_{0y} and ψ_x, ψ_y are the amplitudes and phases of the electric field of the polarized part of the radio emission near the aerials A_1 and A_2 (E_{0x}/E_{0y} and $\psi_x - \psi_y$ are not time-dependent); the primes denote the chaotically changing (in time) amplitudes and phases of the field of the non-polarized part of the emission; l_1 and l_2 are the respective electrical lengths of the lines coupling the aerials A_1, A_2 with the radiometer input. Remembering that the signal Ξ at the radiometer output is determined by the time-averaged square of the voltage at the input, it is not difficult to find from the expressions (6.22) and (6.23)

$$\Xi \propto \overline{E_{0x}^2} + \overline{E_{0y}^2} + \overline{E'^2_{0x}} + \overline{E'^2_{0y}} + 2\overline{E_{0x} E_{0y}} \cos(\psi_{xy} + k\Delta l), \tag{6.24}$$

where $\overline{E'^2_{0x}} + \overline{E'^2_{0y}} = I_{\text{nat}}$ is the intensity of the non-polarized part of the emission, $\Delta l = l_1 - l_2$ is the difference of the electrical lengths of the coupling lines. Since the wave number $k = \omega/c$, when the frequency of the radiometer is switched the signal Ξ at the output will change, reaching a maximum at frequencies

$$\omega_{\max} = \frac{c}{\Delta l}(2\pi n - \psi_{xy}) \tag{6.25}$$

and a minimum at frequencies

$$\omega_{\min} = \frac{c}{\Delta l}(2\pi n + \pi - \psi_{xy}) \tag{6.26}$$

($n = 0, 1, 2,\ldots$). The interval between adjacent maxima $\Delta\omega = 2\pi c/\Delta l$ can, by an appropriate choice of the value of Δl, be made sufficiently small when compared with the characteristic frequency range in which there is a noticeable change in the state of the radio emission's polarization (in particular the quantities ψ_{xy}, $\overline{E_{0x}E_{0y}}$ and the intensity $I = \overline{E_{0x}^2} + \overline{E_{0y}^2} + I_{\text{nat}}$).

Then, just as in the case of the combination of a radio spectrograph with an interferometer (section 5), the dynamic spectrum will be covered by a system of bands; however the position of these bands and their intensity will now determine not the coordinates and size of the source of the radio emission but the nature of the latter's polarization. The value of the data obtained here is chiefly that they give an idea of the change in polarization not only in time but also in frequency. In addition, when using this kind of installation to study the polarization from the nature of the dynamic spectrum the spectral type of the radio emission is determined at the same time, which is very important if we remember the complexity and variety of the phenomena studied on the Sun.

Unfortunately the information obtained here is insufficient to obtain a full idea on the nature of the polarization. In actual fact the constant component of the signal Ξ gives us the quantity $\overline{E_{0x}^2} + \overline{E_{0y}^2} + I_{\text{nat}}$, the amplitude of the variable component gives the value of $2\overline{E_{0x}E_{0y}}$ and the position of the maxima and minima on the frequency scale gives the phase ψ_{xy}. In accordance with expressions (6.4) and (6.5) these quantities can be used to find the three Stokes parameters I, U, V. From this we can determine only the degree of circular polarization $\varrho_c = V/I$ (6.16) and the sign of the rotation in the polarized wave.†

Some indirect data on the form of the polarization ellipse (the ratio of the axes p) can be obtained by starting with the values of the phase difference ψ_{xy} at different frequencies. Due to the presence of the geomagnetic field the polarization ellipse of the wave rotates as it passes through the ionosphere by a certain angle $\Delta\chi$ whose value in the metric band (at $\lambda \sim 3$ m) is as much as tens of radians (the Faraday effect; section 7). Since the angle of rotation is strongly dependent on the wavelength ($\Delta\chi \sim \lambda^2$), in the process of frequency tuning of the radio spectrograph the orientation of the polarization ellipse begins to change: when there is a wide enough frequency cover the rotation of the ellipse will enable it to take up all possible positions relative to the receiving aerials A_1 and A_2. In this

---

† If ψ_{xy} is in the range from 0 to π, then the angle σ lies in the range of 0 to $\pi/2$, which corresponds to a ratio of the axes of the polarization ellipse $p > 0$, i.e. to clockwise rotation. When ψ_{xy} is in the range from π to 2π, $p < 0$ and the electrical vector in the wave rotates anti-clockwise.

case the distribution of the phase difference ψ_{xy} will reflect the degree of ellipticity of the radio emission being studied. The quantity ψ_{xy} will be constant if the wave is linearly or circularly polarized. In the first case the polarization ellipse degenerates into a straight line and $\psi_{xy} = 0, 2\pi$; in the second case it becomes a circle for which $\psi_{xy} = \pm\pi/2$. For an elliptically polarized wave ψ_{xy} will take up different values with differing probability; the degree of ellipticity of the radio emission under study can be judged from the difference of the observed distribution of the phase difference ψ_{xy} from the above two extreme cases.

POLARIZATION MEASUREMENTS IN THE CENTIMETRIC BAND

Generally speaking apparatus similar to that described above can be used to make polarization measurements in the centimetric band. Due to the comparative shortness of the wavelength, however, methods which are either taken over directly from optics or are a further development of the methods of polarization measurements in the latter are preferable here for the transition to waveguide techniques. A system of two basic elements is used as a rule to study polarization in the centimetric band: (1) a phase-shifting device (the analogue of the optical "quarter-wave section") which introduces a phase difference of $\pi/2$ between the two linear components of the radio emission at right angles to each other; (2) an analyser placed after this device (the analogue of the Nicol prism or the polaroid lens in optics), which lets through only one linearly polarized component (Fig. 7).

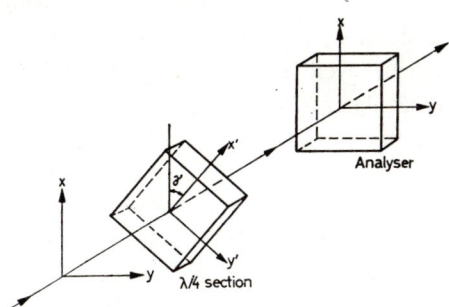

FIG. 7. Method of polarization measurements in the centimetric band

Let the phase-shifting device introduce a phase lag of $\pi/2$ into the component with an electrical vector running along x', whilst the analyser lets through only the x-component. When the circularly polarized emission passes through the "$\lambda/4$ section" it becomes linearly polarized in a direction subtending with the x'-axis an angle $\pm\pi/4$—depending on the sign of the rotation in the incident wave. Then the intensity of the wave at the output of the analyser in the two positions corresponding to the angles $\pm\pi/4$ between the axes x and x' will characterize the intensities of the clockwise and anti-clockwise polarized components I_r, I_l in the radio emission being studied. The data on the value of I_r, I_l enable us to find the degree of circular polarization ϱ_c.

This method of measuring ϱ_c was used by Covington (1951a) when studying microwave bursts of solar radio emission. The polarimeter he built is a parabolic mirror at whose focus there is a dipole (analyser). The phase-shifting device is a system of plane-parallel metal plates set in front of the paraboloid. As it passes through this device the phase velocity of the wave whose electrical vector is parallel to the plane of the plates rises: the distances between the plates and their width are selected so that the oppositely polarized component lags in phase by $\pi/2$. During the observations the dipole is rotated from one position subtending an angle of 45° with the plane of the plates into another where this angle is $-45°$. The process of measuring ϱ_c (determining I_r, I_l) once took 1 minute. If we remember that microwave bursts generally last for a few minutes it becomes clear that the operating speed of the polarimeter is sufficient for a rough investigation of the polarization of these bursts.

In order to obtain full information on the nature of the polarization the data on I_r, I_l must be supplemented by measurements of two further independent variables, let us say the intensities of the linear components orientated at an angle of 45° to each other. In the installation about which we have just spoken the cumbersome phase-shifting device must be moved away to do this; the observations become very inconvenient and each measurement cycle takes a long time. This obviously stops us from studying the polarization of events of like short duration bursts. On the other hand, the measurement process can be made more rapid and convenient and the whole design more compact if the phase-shifting device and analyser are made in the form of sections of waveguides placed after the aerial. A polarimeter of this kind for a wavelength of $\lambda = 3\cdot2$ cm has been designed by Kaidanovskii, Mirzabekyan and Khaikin (1956) specially for solar research. The receiving aerial used (which in this case should not distort the nature of the radio emission's polarization since it receives all the components with the same efficiency) is a circular horn at the focus of a parabolic mirror. The horn goes into a circular waveguide, a section of which can be rapidly replaced by a phase-shifting section. The analyser is a transition from the circular waveguide to a rectangular one; the size of the latter is selected so that only a wave of linear polarization can be propagated in it. The orientation of the analyser relative to the phase-shifting device is changed by introducing an additional section

§ 6] Polarization of Radio Emission

(a half-wave section) between them which is made to rotate around the axis of the waveguide.

Further developments of the type of polarimeter discussed above are the Pulkovo Observatory installations for measuring polarization at centimetric wavelengths (Korol'kov, 1962). In these installations the polarization is analysed as before in a waveguide section. There is no need for removable section here, however; all four Stokes parameters can be measured during the rotation of an additional section fitted between the phase-shifting device ("quarter-wave section") and the analyser provided the difference of travel introduced by this section is not a whole number of half-waves.

Moreover, for analysing the polarization the additional section is not in principle necessary and it can be eliminated as has been done in the Tokyo Observatory polarimeter which works at a wavelength of 3·3 cm (Akabane, 1958a). In the latter case the four Stokes parameters are measured while the phase-shifting device is rotated as follows. If polarized emission of the form (6.1) is fed to the input of the "quarter-wave section", then its components in the x', y' system of coordinates rotated by the angle γ relative to x, y can be written as follows:

$$E_{x'} = E_{0x} \cos \gamma \sin (\omega t - \psi_x) + E_{0y} \sin \gamma \sin (\omega t - \psi_y), \quad (6.27)$$

$$E_{y'} = -E_{0x} \sin \gamma \sin (\omega t - \psi_x) + E_{0y} \cos \gamma \sin (\omega t - \psi_y). \quad (6.28)$$

When it leaves the "quarter-wave section" the $E_{x'}$ component lags in phase behind $E_{y'}$ by $\pi/2$. It may be taken that $E_{y'}$ remains constant after passing through the phase-shifting device, whilst $E_{x'}$ becomes

$$E_{x'} = E_{0x} \cos \gamma \sin \left(\omega t - \psi_x - \frac{\pi}{2}\right) + E_{0y} \sin \gamma \sin \left(\omega t - \psi_y - \frac{\pi}{2}\right). \quad (6.29)$$

Further the analyser lets through only the x-component of the emission

$$\tilde{E}_x = E_{x'} \cos \gamma - E_{y'} \sin \gamma = \tilde{E}_{0x} \sin (\omega t - \psi), \quad (6.30)$$

where the amplitude of the x-component of the electric field of a wave which has passed through the analyser, in accordance with the relations (6.28)–(6.30), will be defined by the expression

$$\tilde{E}_{0x} = \left[\frac{3}{4} E_{0x}^2 + \frac{1}{4} E_{0y}^2 + E_{0x} E_{0y} \sin \psi_{xy} \sin (2\gamma) \right.$$
$$\left. + \frac{1}{2} E_{0x} E_{0y} \cos \psi_{xy} \sin (4\gamma) + \frac{1}{4} (E_{0x}^2 - E_{0y}^2) \cos (4\gamma) \right]^{1/2}.$$

As a result the signal at the output of the radiometer, which is connected to the analyser, will be

$$\Xi \propto \overline{E_{0x}^2} = \frac{3}{4}\overline{E_{0x}^2} + \frac{1}{4}\overline{E_{0y}^2} + \overline{E_{0x}E_{0y}}\sin\psi_{xy}\sin(2\gamma)$$
$$+ \frac{1}{2}\overline{E_{0x}E_{0y}}\cos\psi_{xy}\sin(4\gamma) + \frac{1}{4}(\overline{E_{0x}^2} - \overline{E_{0y}^2})\cos(4\gamma). \quad (6.31)$$

The formula (6.31) describes the response of the polarimeter to completely polarized radio emission. If in the emission being studied there is also a non-polarized component with an intensity $I_{nat}/2$ (calculated for one linear component) it must be added to $\overline{E_{0x}^2}$, so that in the general case $\Xi \propto \overline{E_{0x}^2} + I_{nat}/2$. Taking this into consideration and remembering the expressions (6.4) and (6.5) for the Stokes parameters, we finally find that the output signal in the polarimeter will be of the form

$$\Xi \propto \frac{1}{2}I + \frac{1}{4}Q + \frac{1}{2}V\sin(2\gamma) + U\sin(4\gamma) + \frac{1}{4}Q\cos(4\gamma). \quad (6.32)$$

When the phase-shifting device is rotated at a frequency ω_{ph} the angle γ changes in time as $\gamma = \omega_{ph}t$, whilst the signal Ξ has a constant component $\frac{1}{2}I + \frac{1}{4}Q$ and harmonic components corresponding to the frequencies $2\omega_{ph}$ and $4\omega_{ph}$.[†] The amplitudes of these components can be obtained by mixing Ξ with the oscillations $\sin(2\omega_{ph}t)$, $\sin(4\omega_{ph}t)$ and $\cos(4\omega_{ph}t)$ whose phase and frequency are given by the device that rotates the "quarter-wave section". After time-averaging this gives

$$\left.\begin{array}{c}\overline{\Xi\sin(2\omega_{ph}t)} \propto \dfrac{V}{4}, \\[4pt] \overline{\Xi\sin(4\omega_{ph}t)} \propto \dfrac{U}{2}, \\[4pt] \overline{\Xi\cos(4\omega_{ph}t)} \propto \dfrac{Q}{8}.\end{array}\right\} \quad (6.33)$$

Together with the constant component of Ξ the relations (6.33) fully describe the state of the polarization by determining (after appropriate calibration of the apparatus) all four Stokes parameters.

If an additional section is rotated in the polarimeter and the "quarter-wave section" is fixed the expression for Ξ looks slightly different; as before, however, the constant component and the amplitudes of the harmonic components give the full picture of the polarization of the radio emission being received.

[†] Only the constant component of the signal contains non-polarized emission and it makes no contribution to the variable part of Ξ since the rotation of the quarter-wave section makes no difference to the characteristics of this emission.

7. Effect of the Earth's Atmosphere on the Observed Radio Emission

ABSORPTION OF RADIO WAVES IN THE TROPOSPHERE

At present the radio emission of the Sun and planets is being studied over a wide range, starting at millimetric wavelengths and finishing at wavelengths of tens of metres. At millimetric wavelengths both the optical methods of infrared techniques (Sinton, 1952, 1955 and 1956) and electronic methods (Gordy et al., 1955; Coates, 1957; Kislyakov, 1961a and 1961b; Fedoseyev, 1963; Coates, 1961; Naumov, 1963) are used for indication and measurement of the emission of the Sun and the Moon; in this range, therefore, the two basic methods of studying cosmic objects actually meet. Reception at shorter wavelengths becomes difficult because of strong absorption of the radio waves by the molecules of oxygen and water in the Earth's atmosphere. Without going into any detailed discussion of this effect (see Al'pert, Ginzburg and Feinberg, 1953, sections 83, 84; Straiton and Tolbert, 1960) we shall content ourselves with remarking that the absorption in the oxygen is the result of the interaction of the electromagnetic field with the magnetic moment of the O_2 molecules and occurs in transitions between the rotational levels of these molecules; the absorption of the radio waves by water vapour is connected with the electrical moment of the H_2O molecules.

The optical thickness τ which characterizes the attenuation in the intensity of the radio emission as it passes through the atmosphere is defined by the relation

$$\tau = \int_0^\infty \mu(h) \sec \varphi \, dh, \tag{7.1}$$

where μ is the absorption coefficient, φ is the zenith angle of the source and h is the altitude above the Earth's surface.

During the normal passage of radio waves (when $\varphi = 0$) the absorption at $\lambda \gtrsim 2$ cm is determined by the oxygen; the value of τ_{O_2} in this range is small ($\lesssim 1 \cdot 5 \times 10^{-2}$), however, and in the majority of cases it can be neglected. As the wavelength decreases the absorption grows chiefly in the water vapour at first; at the maximum of the H_2O line $\lambda = 1 \cdot 35$ cm the optical thickness is $\tau \sim 0 \cdot 1$. The slight decrease in τ at wavelengths up to $\sim 0 \cdot 85$ cm is replaced by a sharp increase in the absorption because of the approach towards the oxygen absorption bands $\lambda \approx 0 \cdot 5$ cm and $0 \cdot 25$ cm. The absorption minimum between the lines comes at $\lambda \approx 0 \cdot 3$ cm, where $\tau \sim 0 \cdot 27$. At the short-wave end of the millimetric band the basic absorption is connected with the $\lambda = 0.163$ cm transition in water vapour. There is a small "window" at wavelengths of about 0·15 cm; the transition to the sub-millimetric band is accompanied by a sharp increase in the absorption by H_2O molecules.

It follows from the above that in most of the centimetric band (at $\lambda \gtrsim 2$ cm) the absorption of the radio waves is no more than $1\frac{1}{2}\%$. This absorption is insignificant in the majority of absolute radio astronomy measurements whose accuracy is not better than 10%. However, when high-accuracy measurements are made this absorption introduces a noticeable error for which allowance must be made. At shorter wavelengths, and in the millimetric band in particular, it becomes absolutely necessary to correct the observational data in accordance with the value of the molecular absorption.

ABSORPTION OF RADIO WAVES IN THE IONOSPHERE

In the long-wave part of the spectrum the band of frequencies studied is limited by the opacity of the ionosphere which increases as the wavelength grows. In practice extraterrestrial radio emission can pass through the ionosphere only at frequencies

$$f > f_{cr} = f_{L\,max} \sec \varphi_0, \qquad (7.2)$$

where f_{cr} is the critical frequency of the ionosphere with inclined incidence of the radiation on the layer at an angle of φ_0 to the vertical, $f_{L\,max}$ is the plasma frequency at the maximum of the ionospheric layer (see equation (22.39)). At frequencies of $f < f_{cr}$ the waves are reflected from the ionosphere.

The maximum electron concentration, as is well known, is reached in the F-layer ($N_{max} \sim 10^6$ electrons/cm³) which corresponds to $f_{L\,max} \sim 9$ Mc/s). In actual fact the value of N_{max} does not remain constant, so $f_{L\,max}$ may vary within certain limits without ever dropping, however, below 1·5–2 Mc/s, even at night in winter. The above values of $f_{L\,max}$ are the lower limit for the frequency of radio waves allowed through with normal incidence ($\varphi_0 = 0$); in the case of incidence at an angle the lower limit rises in accordance with the formula (7.2). Even at the higher frequencies, however, the reception of the radio emission (right up to 15–20 Mc/s) of the Sun and the planets is made difficult by the high level of atmospheric and industrial disturbance in this range.

The effect of the ionosphere on the level of the radio emission being received in the range $f > f_{cr}$ comes down to absorption and scattering of the latter. The ionosphere's optical thickness τ is defined by the relation (7.1) in which for this occasion the absorption coefficient μ is of the form

§ 7] Effect of the Earth's Atmosphere

(see section 26)

$$\mu = \frac{\omega_L^2}{\omega^2}\frac{\nu_{\text{eff}}}{cn} = \frac{4\pi e^2 N \nu_{\text{eff}}}{m\omega^2 cn} \qquad (7.3)$$

(n is the refractive index, ν_{eff} is the effective number of electron collisions in the ionospheric plasma). At frequencies far enough from $f_{L\,\text{max}}$ ($f^2 \gg f_{L\,\text{max}}^2$) we have $n \approx 1$, $\varphi \approx \varphi_0$, so

$$\tau = \sec\varphi_0 \frac{4\pi e^2}{mc\omega^2}\int_0^\infty N\nu_{\text{eff}}\,dh = Kf^{-2}\sec\varphi_0, \qquad (7.4)$$

where $K \sim 10^2$ if f is in Mc/s (see Benediktov and Mityakov, 1961).[†]

In accordance with equation (7.4) in the metric band (at a frequency of 100 Mc/s) $\tau \sim 10^{-2}$ for zenith angles that are not too great, i.e. the radio emission is only 1% attenuated in the ionosphere and this can be neglected. At the beginning of the decametric band (at 30 Mc/s) $\tau \sim 0\cdot1$, i.e. the absorption increases to 10% and it must be borne in mind, although studies of the radio emission of the Sun and the planets, which is largely sporadic in nature in this band, do not require high accuracy in absolute intensity measurements. At lower frequencies, particularly near a frequency $f_{L\,\text{max}} \sim 9$ Mc/s, the absorption increases so much that it is absolutely necessary to allow for it.

We also note that the degree of absorption in the ionosphere rises sharply when compared with that given above during the so-called fade-outs, which are distinguished by strong attenuation or cessation of radio communication and a decrease in the observed level of cosmic radio emission. These events, which are caused by the rise in the ionization of the lower layers of the ionosphere during chromospheric flares on the Sun, will be examined in section 17 when we discuss the connection of solar radio emission with geophysical phenomena.

Effects Connected with Refraction of Radio Waves in the Atmosphere

The observed position of a source of cosmic radio emission may differ noticeably from the actual position because of the refraction of the radio waves in the Earth's lower atmosphere and ionosphere.

Because of the absence of any noticeable dispersion in the refractive index n in dry air this difference is the same in the radio band as in the optical spectrum. The presence of water vapour in the air, however, in-

---

[†] The value of τ determines the midday absorption values; at night, when the electron concentration drops, τ is several times smaller.

creases the difference $n-1$ in the radio band, whilst having no significant effect on the refractive index in optics. Therefore the visible position of a source differs from that observed in the radio band; over a wide range of zenith angles (with the exception of values $\varphi_0 > 80°$), however, the refraction correction does not exceed the limits of error in radio astronomical observations whose angular accuracy is generally not better than $1'$. For zenith angles φ_0 which are only a few degrees away from $90°$ the refraction becomes significant. The sign of the refraction correction $\Delta\varphi$ for refraction in the lower atmosphere and ionosphere is negative; this makes it possible to observe the radio emission from sources which are in actual fact below the horizon (about $1°$ below the horizon).

As well as the regular refraction which is noticeable at high zenith angles irregular refraction caused by drifting inhomogeneities of the ionosphere have a significant effect on the radio emission coming through. The latter cause fluctuations in the angle of arrival of a ray relative to a certain mean value determined by the regular refraction. These fluctuations are closely connected with the fluctuations in the flux of the radio emission since they are of the same origin, and have an effect both at large and small zenith angles of the source. Studying the irregular refraction and the fluctuations in the flux of the radio emission from extraterrestrial sources is one of the basic methods of investigating the structure of the ionosphere. In the field of interest to us these phenomena only distort the information on the nature of the radio emission of the Sun and the planets obtained from observations on Earth.

The degree of irregular refraction, which is characterized by the mean square dispersion of the angle of arrival, is connected with λ and φ by the relation (Booker, 1958)

$$\sqrt{\overline{(\Delta\varphi)^2}} \propto \lambda^2 \sec \varphi. \tag{7.5}$$

With large zenith angles at $\lambda = 6.7$ m the value of $\sqrt{\overline{(\Delta\varphi)^2}}$ is of the order of $5'$; at $\lambda = 3.7$ m, in accordance with the formula (7.5), lower values—about 1–$2'$—are observed for the irregular refraction. Therefore the effect of the irregular refraction must be allowed for at metric wavelengths (chiefly at $\lambda \gtrsim 4$ m). In the decametric band it makes it quite difficult to obtain reliable information on the position of sources of radio emission, particularly short-lived ones, since in the latter case the observed coordinates cannot be averaged over the "period" of the fluctuations.

The effect of fluctuations in the flux of radio emission received on Earth is studied chiefly from the example of fluctuations of discrete sources of cosmic radio emission. The characteristic correlation time of the fluctuations is clearly determined by the drift velocity and size of the electron

§ 7] Effect of the Earth's Atmosphere

inhomogeneities of the ionosphere; it ranges between 8 sec and 2 min† and is apparently independent of the wavelength λ. The "amplitude" of the fluctuations noticeably varies during the day (with a maximum near midnight), has a positive correlation with the geomagnetic activity index and increases as λ rises (Booker, 1958).

The rise in the fluctuation level with the increase in wavelength makes us expect that the effect of the ionosphere on the time-dependence of the radio emission flux makes itself felt particularly strongly during investigations of the radio emission of the Sun and Jupiter in the decametric band. The degree of possible distortion can be judged from the magnitude of the fluctuations in the radio emission from discrete sources at these wavelengths. The presence of fluctuations caused by the ionosphere in the emission received from Jupiter is proved by comparing flux recordings made at different points. These experiments will be discussed in section 19 when dealing with the results of observations of this planet's sporadic radio emission.

Fokker (1957) has reported a particular type of fluctuation which has sometimes been observed at frequencies of 140, 200 and 545 Mc/s during reception of the radio emission both from the "quiet" Sun and during "disturbed" periods. The fluctuations differ from the oscillations in the flux of radio emission from discrete sources usually recorded since they largely take the form of fading with a depth of 20–40%. The fluctuation times are shown in the form of oscillations about a certain mean level, but increases in flux never predominate over decreases.‡ The intensity of the fluctuations reveals a close correlation at different frequencies, many oscillations being the same even in details. The probability of the appearance of fluctuations and the level of the latter rises in summertime, generally after local midday. In other respects (the length of the characteristic times of fading of ~ 1 min and the dependence of the level of the fluctuations on wavelength) the anomalous fluctuations of solar radio emission do not differ noticeably from the fluctuations in the radio emission from discrete sources. The characteristics of extraordinary fluctuations given by Fokker (1957) are largely confirmed by later observations (Tsuchiya and Morimoto, 1961). The fact that the intensity of the fluctuations decreases with the wavelength is proved because they have not been noted in the centimetric band even at times of strong oscillations at 200 Mc/s. Tsuchiya and

† Even faster fluctuations with a "period" of less than 6 sec are known (Warwick, 1963).

‡ Similar oscillations in the solar radio emission flux were observed during the event of 4 November 1957 which will be discussed in section 16. There are also some indications of the existence of fading phenomena in solar radio emission (Payne–Scott and Little, 1952; Owren, 1954).

Morimoto (1961) indicate more definitely the daily and seasonal variations shown by the anomalous fluctuations (with maxima before and after midday; as a result near 12 o'clock local time the probability of the appearance of fluctuations is minimal).

The source of fluctuations of this kind has not yet been established; nevertheless it must be of terrestrial origin. The point is that no definite connection has been found between the times of appearance, the intensity and the duration of the fluctuations in observations at two points 100 km apart. A certain similarity with the fluctuations of "radio stars" gives us a certain basis for assuming that some sort of specific phenomena in the ionosphere is the cause of the anomalous fluctuations. However, the absence of any noticeable correlation between the appearance of the fluctuations and the state of the ionosphere (in particular the appearance of the sporadic E layer), and estimates of the expected altitude of the objects causing the anomalous fluctuations made Tsuchiya and Morimoto inclined to believe in the tropospheric origin of the effect in question. It may be assumed that at the basis of the anomalous fluctuations are fading phenomena during interference at the point of reception of waves propagated in the troposphere along slightly different "trajectories" because of refraction on tropospheric inhomogeneities.

Since anomalous fluctuations are rarely observed and have their own particular features they may be connected only with some particular formations in the troposphere which appear under special conditions, let us say with masses of heated air which have broken away from the layer near the ground and formed closed volumes ("thermals"). These objects have apparently also been observed during radar sounding of the troposphere at $\lambda = 3\cdot 2$ cm (Gorelik and Kostarev, 1959). It is interesting to note that the appearance of reflected signals seems to have seasonal and 24-hour variations similar to those noted above for the anomalous fluctuations: the reflections are recorded during the warm part of the year, chiefly between 1 and 3 o'clock in the afternoon. It is clear from what has been said that detailed comparisons of the anomalous fluctuations with the results of simultaneous radar investigations of the troposphere may be of great help in solving the question of the origin of the anomalous fluctuations of the solar radio emission.

POLARIZATION CHANGE OF THE RADIO EMISSION AS IT PASSES THROUGH THE IONOSPHERE

The presence of the geomagnetic field in the ionosphere makes the medium magneto-active: the radio emission incident on the ionosphere "splits" there into two waves—an extraordinary and an ordinary one—

§ 7] Effect of the Earth's Atmosphere

with different polarization, different refractive indices (n_1 and n_2) and absorption coefficients (μ_1 and μ_2).

Since in the ionosphere at the frequencies of interest to us $\omega_H/\omega \ll 1$ ($\omega_H = eH_0/mc$ is the electron gyro-frequency) allowing for the magnetic field introduces only a relatively small correction into the estimates given above for the critical frequency f_{cr} and the optical thickness τ. On the other hand the part played by the geomagnetic field in changing the nature of the polarization of radio emission passing through the ionosphere, as we shall now see, is quite considerable.

Let us assume that an elliptically polarized wave of frequency ω, which can be looked upon as a superposition of two waves with clockwise and anticlockwise polarization, is incident on the ionosphere. In the ionosphere the latter will be propagated in the form of an extraordinary and an ordinary wave with different phase velocities determined by the value of the refractive index n_j ($j = 1, 2$).

The phase shift $\Delta\psi$ acquired in the medium between the circular components of the wave lead to a rotation of the plane of polarization by an angle†

$$\Delta\chi \approx \frac{1}{\omega^2 c} \int \frac{\omega_L^2 \omega_H \cos\alpha}{n_2 + n_1} dl = \frac{4 \cdot 7 \cdot 10^4}{f^2} \int \frac{NH_0 \cos\alpha}{n_2 + n_1} dl, \qquad (7.6)$$

where f is in c/s, N is in electrons/cm^3, H_0 is in oe, dl is in cm (the Faraday effect).

Using (7.6) it is easy to estimate $\Delta\chi$. Putting

$$2\int NH_0 \cos\alpha (n_2+n_1)^{-1} dl \sim NH_0 L,$$

where L is the thickness of the ionosphere and, considering that the values are $N \sim 10^6$ electrons/cm^3, $H_0 \sim 0.5$ oe, $L \sim 10^7$ cm we obtain: $\Delta\chi \sim 10^{17} f^{-2}$. It follows from the latter relation that $\Delta\chi \sim 1$ at frequencies $f \sim 3 \times 10^8$ c/s, i.e. at $\lambda \sim 1$ m; the change in the angle χ can be completely neglected at $\lambda \lesssim 30$ cm, where $\Delta\chi \lesssim 0.1$, but it becomes extremely significant in the metric band ($\Delta\chi \sim 10$ at 3 m) and reaches very high values on the transition into the decametric band ($\Delta\chi \sim 10^2$ at $\lambda \sim 10$ m). This is the reason why we cannot judge the orientation of the polarization ellipse of extraterrestrial radio emission at wavelengths $\lambda \gtrsim 1$ m without having any precise information on the value of $\int NH_0 \cos\alpha \, dl$ in the ionosphere at the time of observation; it is not easy to obtain the latter.

Polarization observations of the signals of orientated artificial Earth satellites and space vehicles, and also of reflected signals in radar studies of the Moon, may be of considerable help in this respect since in both

---

†The formula (7.6) is derived in section 23 (see expression (23.21) and above).

cases we know the initial direction of the preferential polarization of the radio emission.† On the other hand, it is difficult to determine the magnitude of the Faraday effect by comparing the values of the angle at the Earth's surface and beyond the ionosphere during the reception of extraterrestrial radio emission, since the original orientation of the polarization ellipse in this case is to a large extent unknown.

Since a polarimeter receives radio emission not at one frequency f_0 but in a certain band Δf_b, the radio emission being studied, because of the frequency dependence of $\Delta\chi$, will generally speaking consist of a combination of elliptical components orientated at different angles χ. The set of these angles contained in the range $\delta\chi$ whose value provided that $\Delta f_b \ll f_0$ is determined by the expression (Cohen, 1958) (see equation (7.6))

$$\delta\chi \approx \frac{d(\Delta\chi)}{df}\Delta f_b = 2\Delta\chi\frac{\Delta f_b}{f_0}. \tag{7.7}$$

The dispersion in the orientations of the polarization ellipse will obviously reduce the degree of linear polarization fixed by the polarimeter, the error increasing as the band of frequencies being received broadens (see section 23 for greater detail). In accordance with (7.7) the effect of the dispersion $\Delta\chi$ is slight if $|\delta\chi| \ll 1$. This condition is easily satisfied by narrowing the polarimeter's band down to values $\Delta f_b \ll f_0/2\Delta\chi$ (i.e. to $\Delta f_b \ll 5$ Mc/s at $f_0 \sim 100$ Mc/s and to $\Delta f_b \ll 0.15$ Mc/s at $f_0 \sim 30$ Mc/s).

The different optical thicknesses $\tau_j = \int \mu_j \, dl$ of the ionosphere for extraordinary and ordinary waves also causes changes in the state of the polarization of the radio emission passing through. In actual fact since $\tau_1 \neq \tau_2$ the observed intensity ratio of the circularly polarized components of the radio emission $(I_1/I_2)_{\text{obs}}$ will differ from the corresponding ratio outside the ionosphere:

$$\left(\frac{I_1}{I_2}\right)_{\text{obs}} = \frac{I_1}{I_2} e^{-\tau_1 + \tau_2}. \tag{7.8}$$

In accordance with equation (6.15) this circumstance leads to a change in the Stokes parameters characterizing the state of the polarization.

It follows from the relations (7.3) and (26.97) that the difference

$$\tau_1 - \tau_2 = \int (\mu_1 - \mu_2) \, dl \approx 4\tau \frac{\omega_H}{\omega} \cos\alpha, \tag{7.9}$$

---

†Radar studies of the Moon at $\lambda = 2.5$ m have shown (Evans, 1956) that the rotation of the plane of polarization when the radio emission passes once through the ionosphere is 10–13 radians, which is in close agreement with the estimate given for the value of $\Delta\chi$ at $\lambda \sim 3$ m.

§ 7] Effect of the Earth's Atmosphere

where τ is the optical thickness of the ionosphere without allowing for the magnetic field. The expression (7.9) is derived on the assumption that $\omega_H^2 \cos^2 \alpha/\omega^2 \ll 1$, $\omega_L^2/\omega^2 \ll 1$; these conditions are well satisfied in the ionosphere at frequencies higher than or of the order of 30 Mc/s (for $f_{L\,\text{max}} \sim$ 10 Mc/s); we can use them for estimates, however, up to $f \sim 15$ Mc/s.

Since $\tau \propto f^{-2}$, the difference $\tau_1 - \tau_2$ rapidly decreases as the frequency rises (by the law of f^{-3}). At $f \sim 30$ Mc/s $\tau_1 - \tau_2 \sim 2 \times 10^{-2}$†, so in practice over the whole range of frequencies being studied, starting at 15 Mc/s, the difference $\tau_1 - \tau_2$ is small in comparison with unity. Therefore, as can be seen from the ratio (7.8), the change in I_1/I_2 as the radio emission passes through the ionosphere is comparatively small:

$$\frac{I_1/I_2 - (I_1/I_2)_{\text{obs}}}{I_1/I_2} \approx \tau_1 - \tau_2 \ll 1. \tag{7.10}$$

It follows from this and from the expressions (6.12) for the Stokes parameters that the change in the quantities I, Q, U is insignificant. This cannot be said, however, of the parameter $V = I_r - I_l$ in the case when the degree of circular polarization $\varrho_c = (I_l - I_r)/(I_l + I_r)$ (6.15) is small. For example when non-polarized radio emission passes through the ionosphere (when $I_r = I_l$, i.e. in the notations used here $I_1 = I_2$) it becomes circularly polarized with the ordinary component predominant. The absolute magnitude of the observed degree of circular polarization, which is zero outside the ionosphere, is $(\tau_1 - \tau_2)/2$ in this case. The sign of ϱ_c is the same as the sign of the product of $\boldsymbol{H}_0 \boldsymbol{k}$ (where $\boldsymbol{k}$ is the wave vector).

From the estimates of the difference $\tau_1 - \tau_2$ we can draw the conclusion that the degree of polarization acquired in the ionosphere is about 1% at a frequency of 30 Mc/s, increasing rapidly as f decreases; at a frequency of 15 Mc/s it is as much as 10%. Therefore the part played by the ionosphere in changing the degree of polarization in the decametric band cannot be ignored at all. At higher frequencies the polarization introduced by the ionosphere is very small and in the majority of cases it can be neglected; however, even in the metric band this effect is significant when investigating very weak polarized signals against the background of intense non-polarized radio emission.

$$* \;\; *$$
$$*$$

It is clear from the contents of the present section that the lower layers of the atmosphere and the ionosphere have a noticeable effect on the level

---

†In actual fact from the estimates given above for the magnitude of the absorption in the ionosphere it follows that $\tau \sim 0.1$ for $\sec \varphi \sim 1$. Remembering at the same time that $\omega_H/\omega \approx 5 \times 10^{-2}$ and putting $\cos \alpha \sim 1$, we obtain this value of $\tau_1 - \tau_2$ from the relation (7.9).

of the radio emission received from the Sun and the planets, which is particularly significant in the millimetric band and at decametric wavelengths. The distorting effect of the atmosphere considerably complicates the investigation of the primary characteristics of extraterrestrial radio emission in these bands and makes it practically impossible to extend the frequency spectrum studied (beyond $\lambda \sim 1$ mm -30 m) in observations from the Earth's surface. A radical solution of this problem is the use of artificial Earth satellites and space vehicles for radio astronomy measurements (Benediktov, Getmantsev and Ginzburg, 1961; Getmantsev, Ginzburg and Shklovskii, 1958; Haddock, 1959a; Lovell, 1959; Tyas, Franklin and Molozzi, 1959). Reception of the radio emission from the Sun and the Moon in the millimetric band is quite possible with a parabolic aerial having a diameter of the order of 1 m; it is clearly quite permissible to fit such an aerial on a satellite. It should be pointed out that as the wavelength decreases the size of aerial required also decreases and, for example at a wavelength of $\lambda = 1$ mm the width of the polar diagram $\Delta \theta_A$ becomes close to the angular size of the Sun and the Moon ($\sim 30'$) for a paraboloid about 10 cm in diameter (section 5). Any further increase in the directional properties is obviously useless in this case since there will be no gain in the value of the aerial temperature (see equations (4.7) and (4.12)). It is difficult to use satellites to study the radio emission of the planets since because of the small angular size of the sources of radio emission the size of the aerials must be considerably increased (in order to obtain a noticeable value for the aerial temperature T_A). However, the exacting requirements for sensitivity and directivity of the receiving apparatus can be considerably reduced if space vehicles are used for the radio astronomy observations while they are passing near planets. The first experiment of this kind was the investigation of the radio brightness over the disk of Venus made by the space vehicle "Mariner II" (see section 20).

At wavelengths of tens of metres the reception of the sporadic radio emission of the Sun and Jupiter is possible with primitive aerials fitted in satellites. If here the trajectory of the satellite is outside the Earth's ionosphere the observed radio emission will be free from the distorting effect of the ionosphere and, at frequencies lower than critical, from atmospheric interference and the effect of terrestrial radio stations.

CHAPTER III

Results from Observations of the Radio Emission of the "Quiet" Sun

MEASUREMENTS of the intensity of the solar radio emission indicate the existence of a lower limit, which is generally taken as the emission level of the "quiet", unperturbed Sun. The concept of the radio emission level of the "quiet" Sun is in essence an idealization introduced to denote the amount of solar radio emission if there are no local sources of radio emission on the Sun. In actual fact the Sun is never entirely quiet: the stormy processes in the Sun's atmosphere and the corona lead to the appearance of local regions whose radio emission, being sporadic in nature, increases the observed intensity value when compared with the level of the "quiet" Sun. The sporadic radio emission in its turn is made up of relatively smooth and lengthy (of the order of hours and days) rises in intensity and comparatively sharp and short bursts (with a life of the order of seconds and minutes).

The story of the discovery and investigation of the Sun's radio emission starts with Oliver Lodge (see Haddock, 1958) who, as early as the end of the last century, at the dawn of development of radio, expressed a conviction that solar radio emission existed and even tried to find it. The equipment available to Lodge made the success of his experiment impossible; the Sun's radio emission was found only forty years later. The decisive role in this discovery was played by the development of radar which led to the production of sensitive receiving equipment.

The first documentary information on the reception of solar radio emission came from Hey and Southworth in 1942. Hey was studying the interference that makes the operation of radar in the metric band difficult and found that the Sun is a source of interference. At the time when this interference appears a large group of spots can be seen on the Sun's disk, and this gave reason for assuming that the observed radio emission was connected with the solar activity. Slightly later Southworth discovered the radio emission of the "quiet" Sun at centimetric wavelengths. During wartime, however, access to the reports on the investigations made was limited; the results of the observations did not see the light of day until the end of the Second World War (Hey, 1946; Southworth, 1945). In 1943 Reber rediscovered the Sun's radio emission at metric wavelengths and was the first to publish the data he obtained (Reber, 1944).

It must be pointed out that the sporadic radio emission had apparently also been picked up previously (see Arakawa, 1936; Heightman, 1938; Nagakami and Miya, 1939). Not one of the authors, however, looked upon the Sun as the direct source of the radio emission observed.

The present chapter discusses the characteristics of the background radiation (or B-component) of the "quiet" Sun; the features of the other, sporadic component are discussed in the next chapter.

8. Frequency Spectrum of the "Quiet" Sun's Radio Emission

DETERMINING THE LEVEL OF THE "QUIET" SUN'S RADIO EMISSION

As has already been pointed out in section 4, the flux of the Sun's radio emission is generally described by the value of the effective temperature related to the area of the Sun's disk $T_{\text{eff}\odot}$. At millimetric wavelengths there is no difficulty in determining the level of the "quiet" Sun's radio emission from the known dependence of $T_{\text{eff}\odot}$ on the time t, since bursts are a very rare phenomenon in this band and the enhanced radio emission that occurs when active regions appear on the Sun's disk raises $T_{\text{eff}\odot}$ by only fractions of a percent (Coates, 1960). However, measuring the absolute values of $T_{\text{eff}\odot}$ in this band (particularly in the short-wave part) is made more difficult by the considerable absorption of millimetric waves in the Earth's atmosphere (see section 7).

When we move into the region of the longer wavelengths the splitting of $T_{\text{eff}\odot}$ into two components—one relating to the sporadic component and another defining the level of the "quiet" Sun—is a more complex problem. This is because the sporadic radio emission in the centimetric, decimetric and metric wavebands is more intense than at millimetric wavelengths, particularly in years of high solar activity. The difficulties in separating the B-component from the total flux of the radio emission are caused not by bursts of sharply limited duration but by the lengthy smooth rises in the radio emission which occur when groups of spots and plages appear on the Sun's disk.

In the metric band the intensity of the enhanced radio emission connected with sunspots correlates well with the area of the spots located about the

§ 8] Frequency Spectrum

central meridian, and at shorter wavelengths with the area of the spots that can be seen on the Sun's disk (Piddington and Minnett, 1951). Therefore the effective temperature of the radio emission of the quiet Sun is generally found from a graph of $T_{\text{eff}\odot}(\sigma)$, where σ is the corresponding area of the spots, by extrapolating the area of the spots to zero (Fig. 8) (Pawsey and Yabsley, 1949).

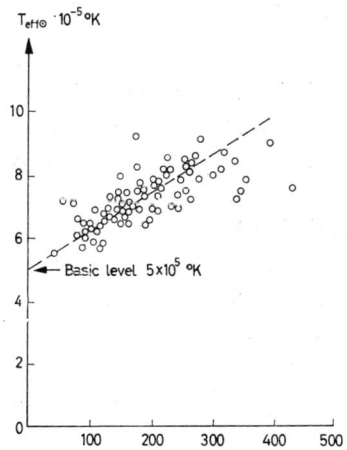

FIG. 8. Dependence of effective temperature $T_{\text{eff}\odot}$ of the Sun's radio emission at $\lambda = 50$ cm on the visible area of the spots σ. The points show the observed values of $T_{\text{eff}\odot}$ and the dotted line the averaged dependence of $T_{\text{eff}\odot}$ on σ

In this connection it should be pointed out that the radio-emitting phase of the spots, which at wavelengths of $\lambda \sim 1$–5 m is close to their lifetime, increases as the frequency rises: in the decimetric band the enhanced radio emission does not disappear at the same time as the group of spots decays and can often still be observed for one or two periods of the Sun's rotation.† During strong spot-formation activity of the Sun the basic level of the radio emission obtained by extrapolating the area of the spots visible on the Sun's disk to zero includes a considerable amount of enhanced emission from groups of spots that have already decayed. Therefore to obtain more accurate values of the $T_{\text{eff}\odot}$ of the basic level at wavelengths of $\lambda \sim 10$–50 cm we must compare the $T_{\text{eff}\odot}$ of the radio emission not with the area of the spots that can be seen on the Sun's disk but with the area of the flocculi which are connected, as observations show, with the enhanced emission in this band (see section 10 for more details).

†In the centimetric band the radio-emitting phase once more becomes close to the duration of the spot's life.

In order to allow for the influence of spots that have disappeared on the level of the solar radio emission we can also introduce the "complex area of the spots" σ_{comp} (Piddington and Davies, 1953), which contains as well as the area of the spots that can be seen on the disk during a given period of the Sun's rotation, the areas of the spots (in linear form) for the preceding periods of rotation σ_{-1}, σ_{-2}, etc:

$$\sigma_{comp} = \alpha_0\sigma_0 + \alpha_{-1}\sigma_{-1} + \alpha_{-2}\sigma_{-2} + \ldots \qquad (8.1)$$

The coefficients $\alpha_0, \alpha_{-1}, \alpha_{-2}, \ldots$ are chosen so that the correlation between the solar radio emission flux averaged for the period of rotation and the "complex area of the spots" σ_{comp} is as great as possible. After this a diagram of the effective temperature plotted against the "complex area of the spots", similar to the one shown in Fig. 8, is used to determine the level of the "quiet" Sun's emission by extrapolating σ_{comp} to zero. As is to be expected this method leads to a systematic reduction in the value of $T_{eff\odot}$ for the radio emission of the "quiet" Sun.

The above methods of processing the observational results unfortunately do not allow us to find the level of emission of the "quiet" Sun with sufficient accuracy and in actual fact provide only the correct order of magnitude at a period of high solar activity in the decimetric and metric bands. This is connected with the quite considerable difference from unity of the correlation coefficient between $T_{eff\odot}$ on the one hand and σ, σ_{comp} or the area of the plages on the other. Under real conditions a precise determination of the level of the "quiet" Sun by measurements of the total solar radio emission flux is, strictly speaking, possible only during low activity, at periods when there are no active regions on the disk, when the intensity of the emission approaches its lower limit.

We can count on a more accurate measurement of the "quiet" Sun's level by integrating the distribution of the radio brightness over the disk of the unperturbed Sun. This distribution can be obtained as a result of the appropriate processing of observations made with narrow-beam aerials, which will pick up the radio emission of only a part of the Sun's surface and not of the whole of the Sun (see section 9). This method is also valid in periods of high solar activity, when local radio-emitting regions can be seen all the time on the Sun's disk; all that is necessary is that these regions should not solidly cover the whole of the Sun's surface and should not make up a continuous band in longitude. In this case as the Sun rotates the brightness at each point of the solar disk will at certain times reach its lower limit, which is taken to be the radio brightness of the "quiet" Sun.

§ 8] Frequency Spectrum

OBSERVED DEPENDENCE OF $T_{\text{eff}\odot}$ ON WAVELENGTH

The values of the effective temperature $T_{\text{eff}\odot}$ of the "quiet" Sun at millimetric wavelengths obtained at different times from measurements at fixed frequencies are shown in Fig. 9. The numbers in square brackets refer to the articles from which the values of the effective temperature are taken.

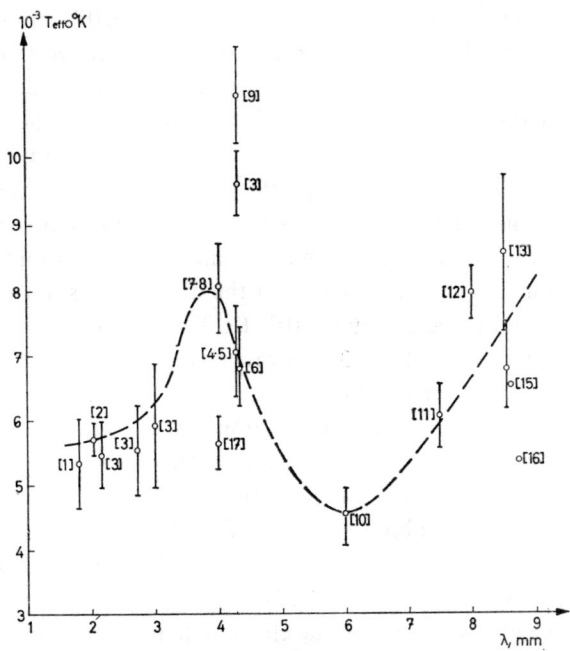

FIG. 9. Frequency spectrum of the "quiet" Sun's radio emission in the millimetric band. The vertical lines indicate the possible measurement errors given by the observers. The dotted line shows the most probable form of the frequency spectrum. The figures in square brackets indicate the sources as follows: 1—Gorokhov, Dryagin and Fedoseyev (1962); 2—Wort (1962); 3—Tolbert and Straiton (1961); 4—Coates (1958b); 5—Coates (1958a); 6—Edelson and Grant (1960); 7—Kislyakov (1961b); 8—Kislyakov (1961c); 9—Straiton, Tolbert and Britt (1958); 10—Whitehurst, Copeland and Mitchell (1957); 11—Whitehurst and Mitchell (1956a and 1956b); 12—Amenitskiy et al. (1958); 13—Hagen (1949); 14—Hagen (1951); 15—Weaver, Mitchell and Whitehurst (1958); 16—Aarons, Barrow and Castelli (1958), Coates (1957)

As was to be expected from theoretical considerations, according to which the millimetric band radio emission is generated in the lower layers of the chromosphere and its $T_{\text{eff}\odot} \sim T$, where T is the kinetic temperature of these layers, the observations lead to values $T_{\text{eff}\odot} \sim 5 \times 10^3 - 10^4 \,°\text{K}$. In order to get a clearer idea of the radio emission spectrum at millimetric wavelengths we must throw out the earlier (and, one must assume, less accurate) results of the same observers (Coates, 1957; Hagen, 1949) and also the values at $\lambda = 4\cdot3$ mm that are too high (Tolbert and Straiton, 1961; Straiton, Tolbert and Britt, 1958). On the other hand, the value of $T_{\text{eff}\odot}$ obtained by Coates, (1958b) at this wavelength should be increased by 5–10%, since in the processing of the measurements there a systematic error crept in when allowing for the molecular absorption because of an assumption about the isothermal nature of the Earth's atmosphere (see Kislyakov, 1961c). The results of measuring the effective temperature at $\lambda = 8\cdot7$ mm made by Aarons, Barrow and Castelli (1958) are also clearly far too low. It should be pointed out that the data available at present on $T_{\text{eff}\odot}$ at $\lambda \lesssim 3$ mm are still not entirely certain. An illustrative frequency spectrum of the radio emission $T_{\text{eff}\odot}(\lambda)$ in the millimetric band plotted to allow for the remarks made above is shown by the dotted line in Fig. 9. It shows a characteristic curve with a maximum at $\lambda \approx 4$ mm and a minimum at $\lambda \sim 6$ mm. The value of $T_{\text{eff}\odot}$ at the maximum is not very definite; it can be assumed to be about 8×10^3 °K. The minimum value of $T_{\text{eff}\odot}$ is close to $4\cdot5 \times 10^3$ °K. In the 2–3 mm range the effective temperature does not differ significantly from 6×10^3 °K (the temperature of the photosphere). As far as we know no information has been published about absolute measurements of $T_{\text{eff}\odot}$ at shorter wavelengths ($\lambda \sim 1$–$1\cdot5$ mm) corrected for absorption in the Earth's atmosphere apart from the data of Fedoseyev (1963), according to which at $\lambda = 1\cdot3$ mm $T_{\text{eff}\odot} = 6\cdot7 \times 10^3$ °K (for reception of the Sun's radio emission at $\lambda \sim 0\cdot3$–$1\cdot5$ mm see also Sinton, 1952 and 1955; Gebbie, 1957).

The presence of a minimum of the effective temperature in the $\lambda \sim 6$ mm region, which had been found earlier (Zheleznyakov, 1958a) was confirmed by a series of observations which Kislyakov (1961a) made with a wideband radiometer in the 3–7 mm band. In these experiments the frequency spectrum was analysed by placing waveguide filters with different critical frequencies between the aerial and the input of the radiometer. The preliminary results of the observations (the experimental values and the possible measurement error) are shown in Fig. 10, where the dotted curve indicates the most probable (from Kislyakov's point of view) form of the frequency spectrum of the "quiet" Sun's radio emission in the millimetric waveband (at $\lambda > 3$ mm). Since the $T_{\text{eff}\odot}$ of the radio emission is close to the temperature of the layers of the atmosphere with which this emission is connected, the minimum in the frequency spectrum at $\lambda \sim$ 5–6 mm indicates the existence of a minimum in the distribution of the kinetic temperature T in altitude in the lower layers of the chromosphere (for further details see section 28).

§ 8] Frequency Spectrum

With the transition from the millimetric to the centimetric and decimetric bands $T_{\text{eff}\odot}$ rises rapidly, reaching values of the order of 10^6 °K at metric wavelengths. In accordance with the theory of the Sun's thermal radio emission the value $T_{\text{eff}\odot} \sim 10^6$ °K for the radio emission coming from the corona points to its high kinetic temperature $T \sim 10^6$ °K.

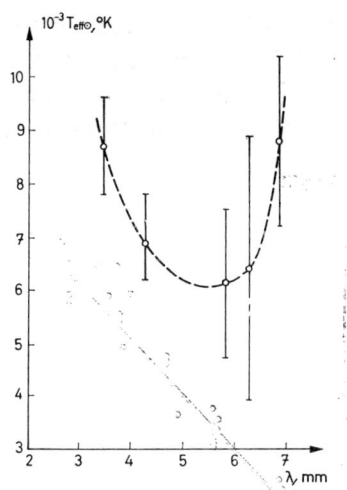

Fig. 10. Frequency spectrum of the "quiet" Sun's radio emission according to the data of Kislyakov (1961a). Markings as in Fig. 9

The values of $T_{\text{eff}\odot}$ obtained by a number of observers in the band $\lambda \sim 1$ cm–4 m are shown in Fig. 11 which also shows the averaged dependence of $T_{\text{eff}\odot}$ on the wavelength λ. The function $T_{\text{eff}\odot}(\lambda)$ is close to linear in the range of wavelengths from ~ 4 cm to $\sim 1\cdot 5$ m, where it can be quite well approximated by the relation (Zheleznyakov, 1958a)

$$T_{\text{eff}\odot} \text{ (in degrees)} = 5 \times 10^3 \lambda \text{ (in cm)}. \tag{8.2}$$

The nature of the function $T_{\text{eff}\odot}(\lambda)$ at wavelengths of $\lambda \sim 3$ m is still not clear at present.

There is a certain amount of interest in studying the variation in intensity of the "quiet" Sun's radio emission over a cycle of solar activity. This variation is possible because during a period of solar activity the state of the corona and the chromosphere (chiefly their density) alters and there is therefore a change in the intensity of the radio emission connected with them.

The available published information on this subject is extremely contradictory and incomplete. According to Covington and Medd (1954) the level of the "quiet" Sun's emission at a wavelength of 10·7 cm decreased by a factor of about 2 between 1947 and 1953. This period was distinguished by a general drop in solar activity which showed itself in a sharp reduction in the area of the sunspots. However, the method used by Covington and Medd (1954) for processing the results of the observations made no allowance for the emission connected with spots that had already decayed; therefore the value of the basic level of the radio emission for the period of high solar activity (1947–8) was about a factor of 2 too high (Piddington and Davies, 1953). This throws doubt on the correctness of the conclusions about the variation in the B-component of the Sun's radio emission.

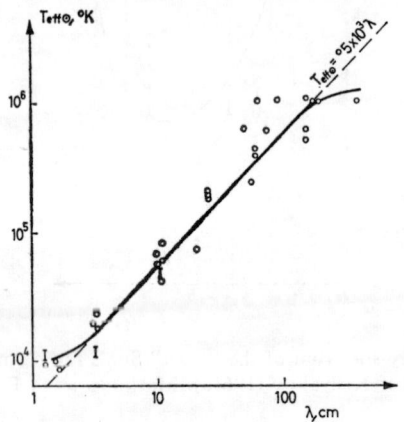

FIG. 11. Frequency spectrum of the "quiet" Sun's radio emission in the centimetric, decimetric and metric wavebands (from observational results in: Ryle and Vonberg, 1948; McReady, Pawsey and Payne-Scott, 1947; Machin, 1951; Pawsey and Yabsley, 1949; Piddington and Minnett, 1949; Dicke and Beringer, 1946; Tu Leng Yao et al., 1959; Minnett and Labrun, 1950; Sander, 1947; Troitskii et al., 1956; Piddington and Hindman, 1949; Hey and Hughes, 1958; Covington, 1948 and 1951; Christiansen and Warburton, 1955; Piddington, 1950; Lehany and Yabsley, 1949; Steinberg, 1953; Laffineur and Houtgast, 1949; Swarup and Parthasarathy, 1955; Reber, 1946; Seeger, Stumpers and van Hurck, 1959/60; Firor, 1959a; Pawsey, 1946; Priester and Dröge, 1955; Payne-Scott, 1946)

On the other hand the observations of Christiansen and Hindman (1951) at wavelengths of 10, 25 and 50 cm revealed no noticeable variation in the basic level of the radio emission for the same period with the exception of a sharp drop during 1950. In this year there was also a considerable reduction in the total area of spots. Since unsatisfactory processing of the measurements resulted in part of the emission connected with local regions on the Sun also being included in the B-component of the radio emission, the variation noted in the basic level was evidently caused to a

§ 8] Frequency Spectrum

considerable extent by the reduction in the intensity of radio emission from spots that had decayed.

The results of the more correct processing of the observational data carried out by Allen (1957) with allowances noted by Piddington and Davies (1953) indicate that at centimetric and decimetric wavelengths the $T_{\text{eff}\odot}$ values of the "quiet" Sun at maximum activity are systematically higher than the corresponding values at minimum activity. The greatest changes (by a factor of about 2) are found at wavelengths of about 30 cm. Allen's conclusions have been in fact confirmed (Christiansen and Warburton, 1955; Christiansen, Warburton and Davies, 1957; Labrum, 1960). Because of the complexity of separating the radio emission's B-component from its sporadic component in years of high solar activity and the resultant possible errors a certain amount of caution must be observed with the results given.

To conclude let us say a few words about the polarization of the radio emission and the $\lambda = 3\cdot 04$ cm hydrogen line in the spectrum of the "quiet" Sun.

According to modern ideas about the nature of the unperturbed solar radiation (see section 28) this polarization may be caused by the Sun's overall magnetic field, which leads to a difference in the optical thickness of ordinary and extraordinary waves in the coronal plasma. Therefore the intensity of both the normal components of the radio emission (ordinary and extraordinary) becomes different and, therefore, the emission coming from a certain part of a given region of the Sun's disk will be partly polarized.

It is easy to see, however, that provided there is sufficient symmetry in the magnetic field's lines of force the polarization effects from the separate parts of the Sun's disk cancel each other out. For example, in the most likely case, when the field is a dipole, the effects of circular polarization of the radio emission from areas on the Sun's disk located symmetrically with respect to the magnetic equator are cancelled out. At the same time neither of the normal components will be found to predominate in the overall radio emission of the Sun and circular polarization can be found only when receiving radio emission from part of the Sun's disk, e.g. during an eclipse.[†]

There are great difficulties involved in making measurements of this kind because of the smallness of the expected effect which, in addition, is masked by the strongly polarized radio emission connected with the spots and plages. The low degree of polarization of the "quiet" Sun's radio emission

[†] Since, when they leave the corona, the ordinary and extraordinary waves are circularly polarized (section 24), linear polarization is not to be expected when receiving radio emission from part of the Sun's disk.

is caused by the low strength of the Sun's overall magnetic field which in the photosphere is no more than about 1 oe (see section 1). At the same time it is quite important to measure the degree of polarization (particularly at metric wavelengths) since it will provide information on the magnitude of the magnetic field in the corona.

Calculations of the possible degree of polarization of the radio emission from the "quiet" Sun have been made by Smerd (1950b). The first eclipse observations of polarization of the solar radio emission at wavelengths of 10 and 50 cm did not reveal any noticeable predominance of one of the circularly polarized components (Piddington and Hindman, 1949; Christiansen, Yabsley and Mills, 1949). This led to the conclusion that the degree of polarization was not more than 1·5% and that the strength of the magnetic field at the poles was $H_0 < 11$ oe (Smerd, 1950b). No polarization component (with an accuracy of up to 0·2% of the total radio emission flux) was found at $\lambda = 8$ mm either (Amenitskiy et al. 1958).

During one of the eclipses when there were no active centres on the Sun before total eclipse and soon after it a circularly polarized component was found at a wavelength of $\lambda = 2$ m (Rydbeck, 1953); according to the data of Kaidanovskii, Mirzabekyan and Khaikin (1956) the content of this component at a wavelength of 3·7 cm is about 1%. Observations at $\lambda = 60$ cm (Conway, 1956) indicate that the degree of polarization corresponds to a magnetic field whose strength at the poles is not more than 2·5 oe. Therefore the radio results available do not contradict the results of the optical measurements of the total magnetic field at the level of the photosphere.

We have spoken above about the features of the continuous spectrum of the "quiet" Sun's radio emission, which, from the point of view of modern ideas about its origin, is the bremsstrahlung of electrons in collisions with ions in the corona and the chromosphere (section 28). As well as the continuous spectrum, however, the "quiet" Sun's radio emission strictly speaking also has a discrete spectrum caused by transitions of electrons between levels of atoms and ions. The most promising in this respect is the $\lambda = 3\cdot04$ cm line, which is connected with the $2^2S_{1/2} \rightleftarrows 2^2P_{3/2}$ transition between the levels of the fine structure of neutral hydrogen (Wild, 1952). The probability of a transition in this line is 2×10^8 times greater than the corresponding probability for a transition between the levels of the superfine structure of hydrogen accompanied by emission at $\lambda = 21$ cm (De Jager, 1959a) (study of the latter is playing an extremely important part in galactic radio astronomy).

It should be pointed out that great intensities are not to be expected on the Sun even in the 3·04-cm line; according to De Jager (1959a), if the population of the levels is determined by collisions under conditions of thermal equilibrium, the effective temperature at the frequency of the line will differ from T_{eff} in the continuous spectrum by only 5×10^{-3} °K and it is practically impossible to see the line. This value can increase significantly only in sharplynoon-equilibrium conditions, when the population of the levels is not determined by the kinetic temperature of the ambient medium. This explains the negative results of the observations by De Jager (1959a), Dravskikh (1960) of the "quiet" Sun, according to which the level of the radio emission at $\lambda = 3.04$ cm (in the 15 Mc/s band), even if it does differ from the level of emission in the continuous spectrum, does not differ by more than 1%. At the same time there are some indications of the existence of this line in the microwave bursts of radio emission (see section 11).

9. Distribution of Radio Brightness over the Sun's Disk

When investigating the "quiet" Sun's radio emission the most interesting and difficult problem is finding the angular distribution of the effective temperature T_{eff} over the source or, as is said, the distribution of the radio brightness over the Sun's disk.† The results obtained in the solution of this problem are of great importance for confirming the theory of the thermal radio emission of the "quiet" Sun and provide valuable information on the temperature distribution and electron concentration in the solar corona and chromosphere.

REMARKS ON METHODS OF INVESTIGATION. SOME PRELIMINARY DATA

The usual idea of the nature of the radio brightness distribution is given by the value of the Sun's effective radius in the radio-frequency band $R_{\text{eff}\odot}$ defined by the relation

$$\frac{\pi R^2_{\text{eff}\odot}}{\pi R_{\mathbb{C}}} = \frac{S_{\omega\,\text{max}}}{S_{\omega\,\text{max}} - S_{\omega\,\text{min}}} \tag{9.1}$$

from the change in the solar radio emission flux during an eclipse. In equation (9.1) $\pi R_{\mathbb{C}}$ is the area of the Sun's disk covered by the Moon ar totality; $S_{\omega\,\text{max}}$ and $S_{\omega\,\text{min}}$ are respectively the total emission flux (before the start of the radio eclipse) and the residual flux (at the period of the eclipse's maximum phase). This definition of $R_{\text{eff}\odot}$ has a clear-cut physical

---

† The latter expression is not completely accurate since a considerable part of the Sun's radio emission, in particular at decimetric and metric wavelengths, comes from regions lying a long way beyond the limits of the optical disk.

meaning: it is the radius of a uniform disk one astronomical unit (1 a.u.) from the Earth, the radio emission flux from which is equal to the observed values before the eclipse and at the moment of its totality.

The first eclipse observations of the Sun's radio emission, which were made by Khaikin and Chikhachev (1947) on 20 May 1947 at $\lambda = 1\cdot5$ m, showed that the effective radius of the Sun is 1·35 times greater than the optical radius $R_\odot$ and therefore the solar radio emission in this band owes its origin to the corona. Further measurements made it possible to determine that the altitude of the effectively emitting layer drops rapidly with the wavelength, reaching values of the order of $(4\cdot5-6)\times10^3$ km in the millimetric band (at $\lambda \approx 8$ mm) (Coates, Gibson and Hagen, 1958; Salomonovich, Pariiskii and Khangil'din, 1958).

Observations of solar eclipses at wavelengths of 1·5 m, 10 cm and 3·2 cm indicate that the radio diameter of the Sun decreases at a period of a drop in solar activity. This effect is more noticeable in the metric band; at centimetric wavelengths the change in $R_{\text{eff}\odot}$ is comparatively small (Troitskii *et al.*, 1956; Su Shih Weng *et al.*, 1962). Interferometer measurements (O'Brien, 1953a) also point to a change in the size of the radio Sun at $\lambda = 1\cdot4$ m: in 7 months of observations in 1951–2 its radio diameter decreased by 20%. This change is in close agreement with the variation in intensity of the green 5303 Å coronal line (O'Brien, 1953b); both the former and the latter phenomena are connected in all probability with the decrease in coronal density in the period of weakening of the solar activity. We notice, however, that at a wavelength of 7·9 cm in 1952 the radio brightness distribution was even slightly broader than in 1951 (O'Brien, 1953a), but careful observations in the decimetric band (at $\lambda = 21$ cm) did not reveal any changes in the form and extent of the radio sun at this time (Piddington, 1950).

Eclipse observations (Blum, Denisse and Steinberg, 1952a and 1952b) and interferometer measurements (Vitkevich, 1956a) indicate that the size of the radio sun at metric and decimetric wavelengths is considerably greater in the equatorial direction than the polar.

The data given above on the nature of the distribution of the radio brightness over the Sun's disk, which were chiefly obtained from eclipse observations of the change in the total radio emission flux, are highly schematic and incomplete. Further information on the distribution of T_{eff} on the basis of eclipse observations of the total solar radio emission flux is generally obtained by selecting a model for the distribution of the radio brightness over the Sun's disk which best corresponds to the observed curve of the flux variation. However, a whole number of factors—fluctuations in the amplification factor of the receiving apparatus, its inherent

§ 9] Radio Brightness over the Sun's Disk

noise and outside interference which distorts the progress of the radio eclipse, on the one hand, and the fact that the eclipse curves are not very critical in relation to the variations in the radio brightness distribution (particularly at metric wavelengths, when the size of the radio sun is far greater than the diameter of the optical disk), the difficulties in distinguishing the radio emission from the "quiet" Sun from the emission of local sources on the eclipse curves, and also the "uniformity" of the results obtained here noted in section 5, on the other hand—make it impossible to establish reliably the nature of the distribution of T_{eff} over the disk of the "quiet" Sun. At the same time the results of the eclipse observations, despite all their faults, are of definite importance in checking data obtained by other methods.

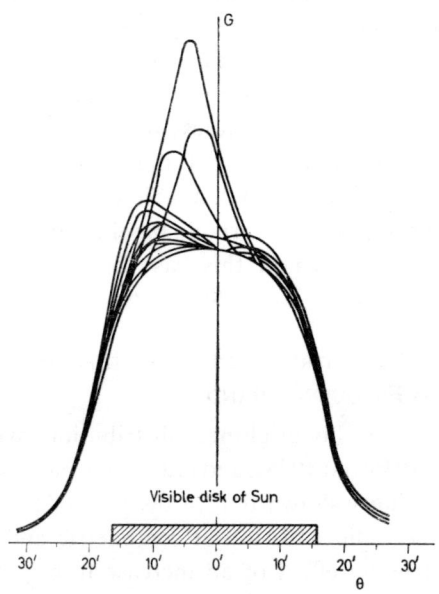

FIG. 12. Combined daily recordings of the one-dimensional distribution of radio brightness over the Sun's disk at $\lambda = 60$ cm

Narrow-beam parabolic aerials, multi-element interferometers and the combination of the latter into the Mills Cross (section 5) are of particular value in investigating the distribution of the radio brightness over the Sun's disk. The point is that these devices, which provide quite high resolution, give a radio brightness distribution curve at the output each time the Sun passes through a lobe. The distribution will be one-dimensional if the width of the lobe in one dimension is far greater than the angular size of the Sun. If the width of the lobe in both dimensions is much less than the angular size of the Sun the recording of the radio emission intensity as the Sun passes through the beam will be characteristic of the radio brightness along the scanning line. Moreover, the radio emission of the quiet Sun can be distinguished from the emission caused by active regions localized on the disk by combining a large number of daily recordings of the radio emission, as shown in Fig. 12: the one-dimensional distribution of the "quiet" Sun's radio brightness will in the latter case obviously be given by a curve fitting round the bottom of the combination of all the recordings.

A similar method of distinguishing the distribution of the "quiet" Sun's radio brightness from the emission of its sporadic component can in principle also be used when processing measurements made on interferometers with a variable baseline. However, the method of combining the radio brightness distributions is generally not used in the latter case since it is a very laborious task to obtain even one distribution in this way (see section 5). If at the same time we remember that when there are active regions on the Sun the radio brightness distribution may change in the time required to obtain one series of the Fourier components of the distribution, it becomes quite understandable why we try to make measurements on interferometers with a variable baseline at a period of low solar activity. It is true that even in this case one cannot be certain that the effect of active regions is not distorting the results obtained.

FEATURES OF THE T_{eff} DISTRIBUTION OVER THE DISK OF THE "QUIET" SUN IN THE RADIO-FREQUENCY BAND

Calculations of the radio brightness distribution over the Sun's disk made by Martyn (1946 and 1948), and later made more accurate by a number of other authors, have shown that in the case of thermal origin of the B-component of the radio emission at centimetric and decimetric wavelengths there must be an effect of an increase in brightness towards the edge of the Sun's disk. This effect is connected with the rise in temperature in the transition region from the chromosphere to the corona and its cause is that a considerable part of the radio emission from the central part of the disk is generated in the relatively "cold" layers of the chromosphere, whilst in the transition from the central part of the disk to the Sun's limb there is an increase in the contribution to the radio emission from the higher, hotter layers of the chromosphere and corona.

The increase in brightness at the edge of the disk should disappear at metric wavelengths, since in this band practically all the emission owes its origin to a corona with a kinetic temperature T which is constant or even slightly decreasing with altitude, the effective temperature of the radio

§ 9] Radio Brightness over the Sun's Disk

emission at the centre of the disk reaching a maximum possible value equal to T. The absence of brightening must also be expected in the millimetric band at wavelengths of $\lambda \sim 4$–5 mm, since in this case the radio emission is connected with the lower layers of the chromosphere in which, to judge from the observed spectrum of the radio emission (see Figs. 9 and 10), there is no temperature inversion (the temperature averaged for the fine structure of the chromosphere decreases as the altitude above the photosphere increases). Therefore in the transition from the centre to the limb T_{eff} at $\lambda \sim 4$–5 mm should also decrease since the emitting region in this case moves to the higher (and colder) layers of the chromosphere. At the same time brightening becomes possible again at $\lambda < 4$ mm, since here T_{eff} once more increases as we move away from the Sun's surface (see section 28 for more detail).

The first attempts to find an increase in the radio brightness at the Sun's limb were made during the eclipse of 1 November 1948. The existence of the brightening predicted by the theory would appear to follow from observations at a wavelength of 10 cm (Piddington and Hindman, 1949); however the results obtained were unconvincing since they related to a period of considerable solar activity which distorted the progress of the radio eclipse. Moreover the authors Christiansen, Yabsley and Mills, 1949), on the basis of measurements during the same eclipse, came to the opposite conclusion about the absence of brightening in the decimetric band. Further observations were made by Stanier (1950) on an interferometer with a variable baseline ($\lambda = 60$ cm) in order to find the brightening. It turned out, however, that the relation between the effective temperature and the distance from the centre of the disk obtained on the assumption that the radio brightness distribution has circular symmetry is a steady curve with gradual "darkening" towards the edge of the disk (see Fig. 19b).

The negative result of the observations of Stanier (1950) considerably increased interest in the solution of this problem. The point is that the absence of brightening cast doubt on either our knowledge of the physical conditions in the upper layers of the Sun's atmosphere, or the correctness of the interpretation of the "quiet" Sun's radio emission as thermal emission of the corona and chromosphere. The intensive research carried out after these observations have made it possible to obtain distributions of the effective emission temperature over the Sun's disk in a wide band of wavelengths from 4 mm to 8 m and definitely find the brightening effect towards the edge of the disk.

In the millimetric band the radio brightness distribution has been studied at wavelengths of 4·3 mm and 8·6 mm. At the first of these wavelengths the recording of the radio emission flux obtained by Coates (1958b) when scanning along the Sun's diameter with a parabolic aerial having a beam width of 6·7' is very satisfactory, however, both with respect to the model of the radio sun in the form of a uniform disk with a radius of $1·01R_\odot$ and to the model with a ring on the limb whose brightness is 3% greater than the brightness of the inner uniform part of the distribution (Coates, 1958d). (At the same time the agreement is not nearly so good if we put $R_{\text{eff}\odot} = R_\odot$.) Therefore, at present the important question of the detailed nature of the radio brightness distribution in the 4-mm band and particularly the existence of brightening at the Sun's limb remains open. It is in place to recall that the ideas given above, which are based on the theory of thermal radio emission and measurements of the Sun's spectrum at millimetric wavelengths, are not in favour of this effect.

More reliable data on the dependence of T_{eff} on the distance r to the centre of the disk have been obtained by Coates, Gibson and Hagen (1958) in observations during the eclipse of 30 June 1954 at a wavelength of $\lambda = 8·6$ mm. Their aerial system had low directivity along the line connecting the centres of the solar and lunar disks and high resolution in the direction at right angles. The position of the Sun and the Moon with respect to each other during the observations and the cross section of the polar diagram are shown in Fig. 13. It is clear from the figure that under these conditions

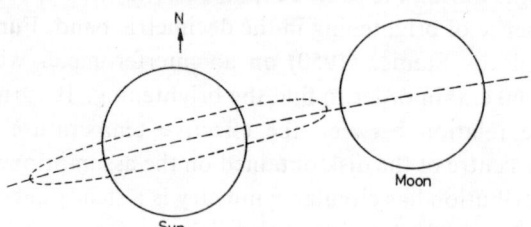

FIG. 13. Position of the Sun and the Moon in the sky and cross-section of the polar diagram of the sky during eclipse observations at $\lambda = 8·6$ mm on 30 June 1954

the distribution of the radio brightness along the line of the centres can be found, roughly speaking, by differentiating the curve of the variation in the solar radio emission flux. The results of processing the observational data are shown in Fig. 14, where we can see the features of the distribution $T_{\text{eff}}(r)$ quite well: the sharply bounded (at an altitude of about 5000 km above the photosphere) disk of the radio sun, the brightening at the limb (about 10% more than the average brightness), the constant brightness in the central region (for $r < 0·6R_\odot$) and the minimum of T_{eff} near $r = 0·9R_\odot$, the weak emission of the upper chromosphere.

§ 9] Radio Brightness over the Sun's Disk

A considerable rise in the effective temperature near the limb and a certain increase of T_{eff} in the central part of the disk are indicated (though without complete certainty) by measurements of the total radio emission flux of the Sun at a wavelength of about 8 mm made during the same eclipse (Salomonovich, Pariiskii and Khangil'din, 1958), and also by eclipse and interferometer observations at a wavelength of $\lambda = 3\cdot 2$ cm (Alon, Arsac and Steinberg, 1953 and 1955). The results of the latter are shown in Fig. 15 which gives two possible types of T_{eff} distribution over

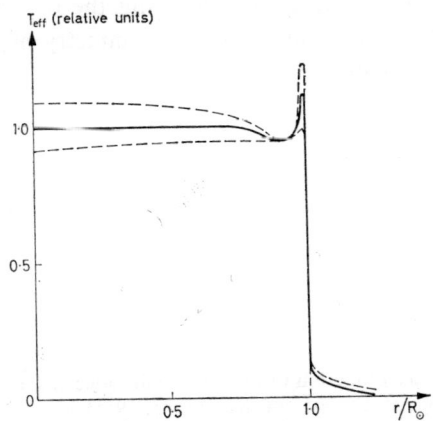

FIG. 14. Distribution of radio brightness over the "quiet" Sun's disk at $\lambda = 8\cdot 6$ mm (Coates, Gibson and Hagen, 1958) (solid curve; the dotted line indicates the limits of possible error)

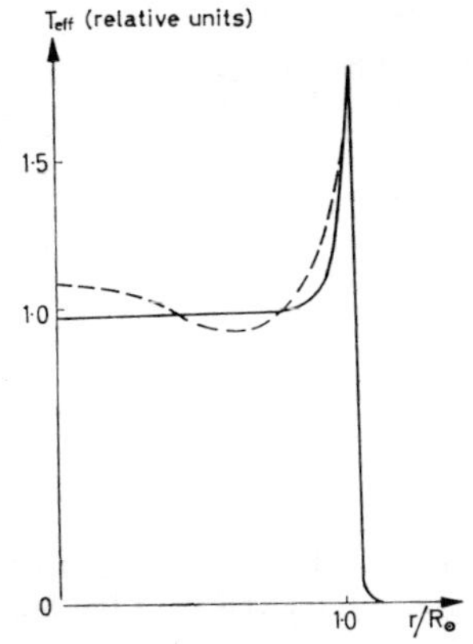

FIG. 15. Possible radio brightness distributions over the disk of the "quiet" Sun at $\lambda = 3\cdot 2$ cm (Alon, Arsac and Steinberg, 1953 and 1955)

the Sun's disk; unfortunately the accuracy of the results does not permit any definite choice between them. We notice, however, that the distribution obtained by Molchanov (1956) from eclipse observations at $\lambda = 3\cdot 2$ cm is close to that shown by the solid line in Fig. 15.

The increase in the brightness in the central part of the disk, which is indicated by the observations in the 8-mm band and which may still be there up to wavelengths of 3 cm, is apparently connected with the presence of spicules in the chromosphere. The interpretation of this effect from this point of view is given in section 28 when discussing the theory of the thermal radio emission of the "quiet" Sun. The existence of brightening near the Sun's limb in the centimetric band follows from the combined results of Tu Leng Yao et al. (1959); Blum, Denisse and Steinberg (1952a); Hagen, Haddock and Reber (1951); Bosson et al. (1951); Haddock (1957); Mayer, Sloanaker and Hagen (1957); Molchanov et al. (1959), although the doubts are not entirely non-existent (Troitskii et al., 1956).

It should be noted that the processing of the eclipse curves to obtain the radio brightness distribution was carried out on the assumption of circular symmetry of the distribution relative to the centre of the Sun's disk. This assumption is clearly quite permissible at wavelengths of $\lambda \lesssim 3$ cm since the deviations from circular symmetry are slight (Hachenberg, Fürstenberg and Prinzler, 1956; Molchanov, 1960). On the other hand, the radio brightness distribution at longer wavelengths (starting, for example, at 10 cm) display noticeable asymmetry, being elongated along the equator (see Hey, 1957, and the data given below). Therefore processing the observational results at $\lambda \gtrsim 10$ cm on the assumption of circular symmetry of the distribution leads to considerable errors.

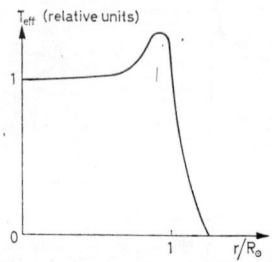

FIG. 16. Effective temperature T_{eff} as a function of the relative distance from the centre of the Sun $r/R_\odot$ at $\lambda = 7.5$ cm (Kakinuma, 1955)

Data on the radio brightness distribution have been obtained with multi-element interferometers, starting at $\lambda = 7.5$ cm. The distribution at this wavelength, which was found while scanning the Sun with an eight-element interferometer with a lobe width of 4·5′ (at the half-intensity points), is shown in Fig. 16 (Kakinuma, 1955). The results of the observations were processed on the assumption that the distribution of T_{eff} has circular symmetry. In the 10-cm band Covington and Broten (1954) when carrying out measurements on a multi-element interferometer made in the form of a waveguide with a large number of slot aerials found the function $T_{eff}(r)$ shown by the smooth curve in Fig. 17a. The same figure shows the later results which they made more accurately (Covington, *et al.*, 1955) (observations of the function $T_{eff}(r)$ in the equatorial and polar directions). During the last solar activity minimum Swarup (1961a and 1961b) also obtained a two-dimensional distribution of the radio brightness over the disk of the "quiet" Sun with a Mills Cross that had a beam consisting of a series of "pencil" lobes about 3′ wide (at half intensity) with a separation of 41′. This distribution is shown in Fig. 17b. The detail of the radio brightness distribution over the Sun's disk at $\lambda = 21$ cm has been established by Christiansen and Warburton (1955 and 1956) (see also Pawsey, 1957) on

§ 9] Radio Brightness over the Sun's Disk

the basis of measurements made first on one and then on two multi-element interferometers pointing at right angles to each other. The use of two interferometers made it possible to take one-dimensional distributions in practically every direction with respect to the solar equator and thus obtain (after the appropriate processing: see section 5) the two-dimensional distribution of the radio emission intensity of the "quiet" Sun without making any assumptions about the form of its radio iosphots (Fig. 18).

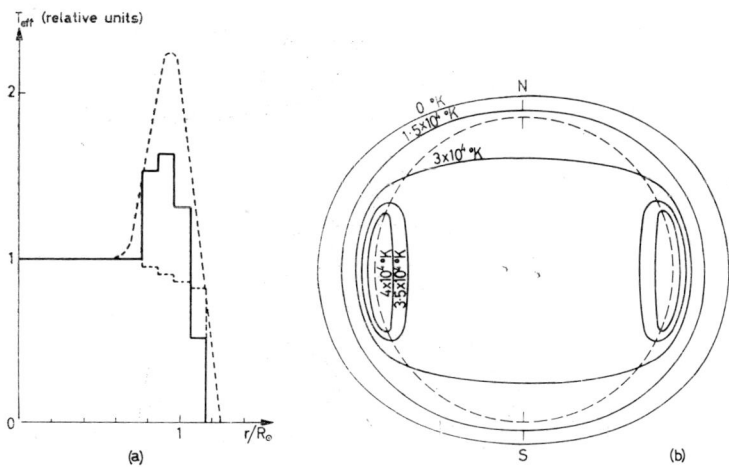

FIG. 17. (a) T_{eff} as a function of $r/R_\odot$ at $\lambda = 10\cdot3$ cm (smooth curve—preliminary data (Covington and Broten, 1954); the more accurate distributions in the equatorial and polar directions are shown by the solid and dotted histograms respectively (Covington et al., 1955)). (b) Radio isophots of the "quiet" Sun at $\lambda = 9\cdot1$ cm at the period of minimum solar activity. The effective temperature at the centre of the disk is 3×10^4 °K, at the point of intersection of the limb and the equator is $4\cdot5\times10^4$ °K and at the poles is 2×10^4 °K (Swarup, 1961a and 1961b)

At the longer wavelengths (in the 50–60 cm range) the radio brightness distribution has been studied by Ovsyankin and Panovkin (1956), Panovkin (1957) (see Fig. 19a), O'Brien and Tandberg–Hanssen (1955), Conway and O'Brien (1956) on interferometers with a variable baseline (Fig. 20) and also by Swarup and Parthasarathy (1956) on a multi-element interferometer with a lobe width of $8\cdot7'$ (Fig. 19b). The results of the observations of Swarup and Parthasarathy (1956) were processed on the assumption of circular symmetry of the radio isophots. As well as the function $T_{\text{eff}}(r)$ they obtained, Fig. 19b shows Stanier's distribution (Stanier, 1950); the latter differs sharply from the other distributions in the band under discussion by the absence of brightness towards the edge of the disk. This discrepancy can apparently not be explained by the difference in the state of the

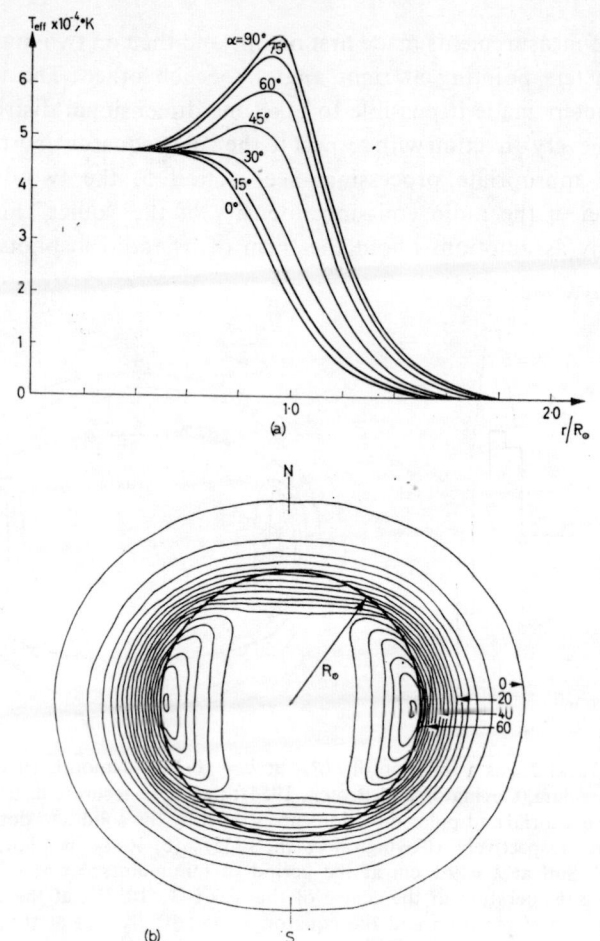

FIG. 18. Distribution of the radio brightness over the disk of the "quiet" Sun at a wavelength of 21 cm at a period of minimum solar activity. (a) Radial distributions of $T_{eff}(r)$ at different angles α to the central meridian; (b) radio isophots of the Sun (the contours are at intervals of 4×10^3 °K) (Christiansen and Warburton, 1955 and 1956)

corona at the time of the observations (Stanier, 1950; Swarup and Parthasarathy, 1956), since they were carried out at a period when the corona took its minimum form. It is more probable, therefore, that Stanier's curve does not reflect the true distribution of the radio brightness because of distortions introduced into it by the radio emission of local regions during the interference measurements.[†]

[†] Stanier's distribution agrees with the 1948 eclipse observations (Christiansen, Yabsley and Mills, 1949). The radio eclipse curves given in this reference agree closely, however, with distributions that differ markedly from Stanier's distribution.

§ 9] Radio Brightness over the Sun's Disk

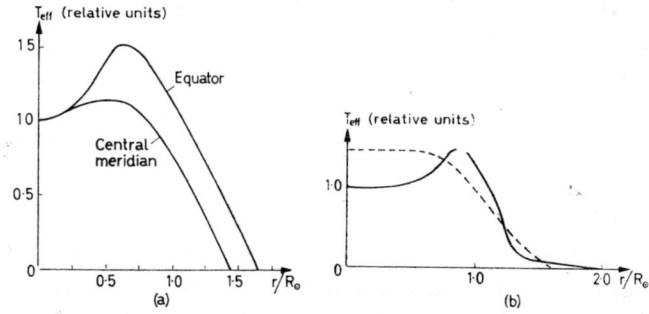

FIG. 19. Radio brightness distribution over the Sun's disk: (a) $\lambda = 50$ cm (Panovkin, 1957; Ovsyankin and Panovkin, 1956); (b) $\lambda = 60$ cm (Swarup and Parthasarathy, 1956 (solid line) and Stanier, 1950 (dotted line))

It is clear from the data given that the radio brightness distribution in the decimetric band (just as the distribution of the electron concentration in the corona: see section 1) deviates considerably from circular symmetry, revealing characteristic "flattening" towards the poles. If, for example, in Fig. 18 ($\lambda = 21$ cm) we take the contour at which the brightness is half that at the centre of the disk, then the distance from the contour to the centre along the equator is $1 \cdot 25 R_\odot$, whilst along the central meridian this distance will be only $0 \cdot 94 R_\odot$.

A very significant feature of the radio isophots at wavelengths of $\lambda \sim 10$–20 cm is that the degree of brightening decreases as one moves away from the equator; in the polar directions T_{eff} drops smoothly as the distance r from the centre of the disk increases. To judge from the data at a wave-

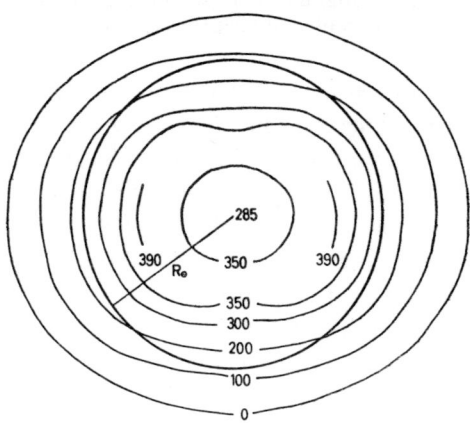

FIG. 20. Radio isophots of the quiet Sun at a wavelength of 60 cm in units of 10^3 °K (O'Brien and Tandberg–Hanssen, 1955; Conway and O'Brien, 1956)

length of 21 cm the brightening disappears at a latitude of about 60°, i.e. in the region where in the period of minimum solar activity there were considerable changes in the structure of the corona. At this latitude the fine polar rays ("brushes") characteristic of high latitudes appear, whilst in the region near the equator the corona look like a comparatively structureless, but more extensive formation.

Another feature of the radio brightness distribution in the band under discussion is that the T_{eff} maximum located at the equator does not reach to the edge of the optical disk (as at wavelengths $\lambda \lesssim 3$ cm) but is noticeably shifted towards the centre. This shift is small at $\lambda \sim 10\text{-}20$ cm

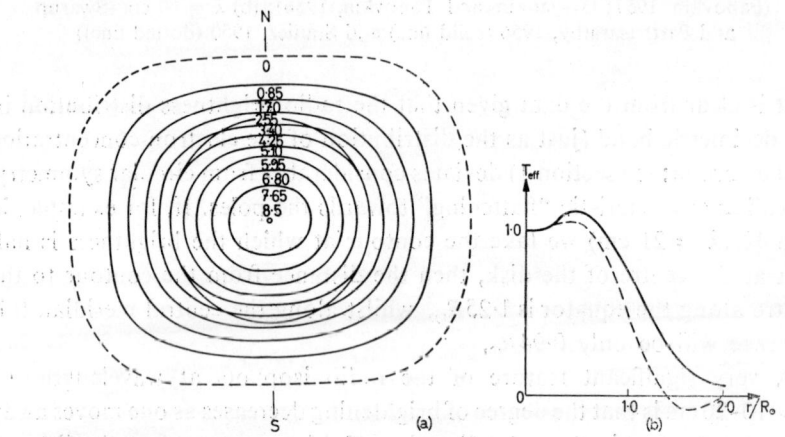

FIG. 21. (a) Radio isophots of the "quiet" Sun at $\lambda = 1\cdot 4$ m in units of 10^5 °K (O'Brien, 1953). (b) Radial distributions of the effective temperature at $\lambda = 1\cdot 45$ m in the equatorial direction (solid curve) and in the polar direction (dotted line) (Firor, 1957)

although it is noticeable at $\lambda = 7\cdot 5$ cm; the shift becomes considerable at wavelengths between 50 and 60 cm, where the T_{eff} maximum corresponds to a value $r \approx 0\cdot 9 R_\odot$ according to the data of Swarup and Parthasarathy (1955) and $r = (0\cdot 6\text{-}0\cdot 65)R_\odot$ to judge from the measurements by Panovkin (1957) and by O'Brien and Tandberg–Hanssen (1955). This feature has also been noted at the beginning of the metric band (at $\lambda = 1\cdot 45$ m), where the T_{eff} maximum is likewise shifted towards the centre of the disk, being at a distance $r \approx 0\cdot 6 R_\odot$ from it (see Fig. 21).

The radio brightness distribution over the Sun's disk at metric wavelengths was first found by Machin (1951) ($\lambda = 3\cdot 7$ m) and then by O'Brien (1953) ($\lambda = 1\cdot 4, 3\cdot 7, 7\cdot 9$ m) from observations on interferometers with a variable baseline. Unlike the distributions at shorter wavelengths the functions $T_{\text{eff}}(r)$ obtained on the assumption of circular symmetry of the

§ 9] Radio Brightness over the Sun's Disk

radio isophots and given in Fig. 155 (section 28) are monotonic with a gradual decrease in brightness as one moves away from the centre; for the three wavelengths given above the brightness drops to 0·1 of its value at the centre of the disk at distances r respectively to $1·6R_\odot$, $2·2R_\odot$ and $2·8R_\odot$.

During the observations, however, it was found that the radio sun did not extend the same distance in different directions relative to the Sun's equator. Therefore O'Brien made a series of measurements at $\lambda = 1·4$ m on variable-baseline interferometers pointing at different angles to the equator in order to obtain a two-dimensional distribution without further assumptions about circular symmetry. This distribution is shown at Fig. 21a; it agrees closely with the results of observations of the solar eclipse of 30 June 1954 at $\lambda = 1·5$ m (Priester and Dröge, 1955), whilst the interference measurements at $\lambda = 1·45$ m made by Firor (1955, 1956 and 1957) indicate (unlike the data just given) the existence of noticeable brightening towards the edge of the disk (Fig. 21b).†

The radio brightness distributions at wavelengths of 1·4, 3·7 and 7·9 m obtained by O'Brien (1953) were later improved by Conway and O'Brien (1956) (see also Hewish, 1957) on the basis of interference and eclipse observations. In the processing of the results of measurements on variable-baseline interferometers (the orientation of which was not changed during the observations) it was assumed that the Sun's radio isophots have the form of ellipses; the degree of ellipticity was found from the eclipse curves. The distributions of $T_{\text{eff}}(r)$ in the equatorial and polar directions plotted in this way are shown in Fig. 22. By way of comparison this figure also shows Machin's distribution ($\lambda = 3·7$ m). In general functions $T_{\text{eff}}(r)$ at $\lambda = 3·7$ and 7·9 m are similar to those obtained by O'Brien (1953); however in the radio brightness distribution at $\lambda = 1·4$ there is a slight increase in the brightness along the equator which confirms the results of Firor (1955, 1956 and 1957).

Data on the degree of brightening near the Sun's limb, i.e. the values of the ratio of the maximum effective temperature at the equator to the effective temperature at the centre of the disk $T_{\text{eff max}}/T_{\text{eff 0}}$ in the whole range under study from millimetric to metric wavelengths, are shown in the graph in Fig. 23. It can be seen from the figure that the models of the radio brightness distributions that best satisfy the results of the eclipse observations systematically have higher degrees of brightening than the distributions obtained by interferometer techniques. This circumstance is due to the different resolutions of these methods of measurement: unlike

† The negative values of T_{eff} in the distribution along the solar axis in Fig. 21b are connected with the uncertainty of the interference fringe amplitude in measurements on long baselines.

the eclipse observations which have high resolution, in interference measurements the actual radio brightness distribution (particularly the sharp maximum near the Sun's limb) is "smoothed out" over the beam which, as a rule, is not less than 3′ wide (Smerd and Wild, 1957).

The information given above on the distribution of T_{eff} over the Sun's disk relates basically to a period of low solar activity when the corona has its minimum form. It becomes difficult to obtain corresponding results for

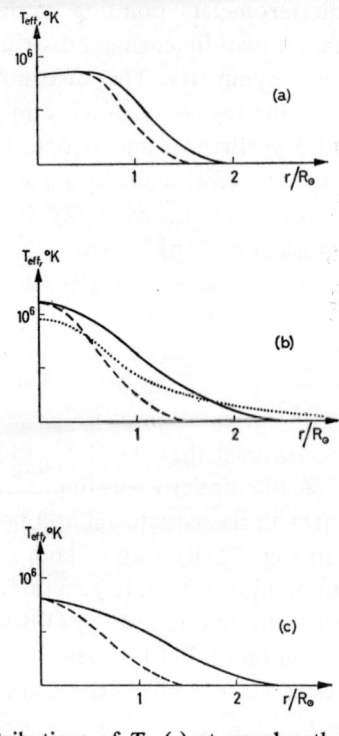

FIG. 22. Radial distributions of $T_{\text{eff}}(r)$ at wavelengths: (a) $\lambda = 1\cdot 4$ m, (b) $\lambda = 3\cdot 7$ m and (c) $\lambda = 7\cdot 9$ m (Conway and O'Brien, 1956) in the equatorial direction (solid curve) and the polar direction (dotted curve). The curve of small dots shows the distribution at $\lambda = 3\cdot 7$ m obtained by Machin (1951)

a cycle maximum because of the complexity of separating the B-component of the radio emission from the sporadic component. For example, during the last, very intense solar activity maximum (1957–8) the use of the method described above to determine the one-dimensional distribution of the "quiet" Sun as the envelope below all the one-dimensional distributions along the equator at $\lambda = 21$ cm turned out to be impossible: the local sources of the slowly changing component of the emission surrounded the Sun in a continuous band, so the envelope contained a considerable part of the sporadic component (Labrum, 1960). Scanning the Sun with a

§ 9] Radio Brightness over the Sun's Disk

multi-element interferometer along the solar axis made it possible, however, to find the one-dimensional radio brightness distribution in the high latitudes ($\pm 40°$) where there are no local sources. The T_{eff} distribution of the "quiet" Sun at low latitudes was then obtained as a result of extrapolation which was not entirely reliable. Nevertheless these observations give us a certain foundation for stating that the general nature of the T_{eff} distribution at $\lambda = 21$ cm did not change radically from the time of the observations of Christiansen and Warburton (1955) at the cycle minimum (see Fig. 18), at least with respect to the absence of brightening in the

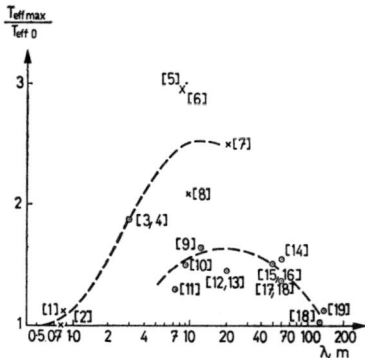

FIG. 23. Degree of increase in brightness $T_{\text{eff max}}/T_{\text{eff 0}}$ at the equator as a function of the wavelength λ: dots—results of measurements on interferometers; crosses—results of eclipse observations. The numbers in square brackets are the references from which the data are taken: (1) Coates, Gibson and Hagen (1958); (2) Salomonovich, Pariiskii and Khangil'din (1958); (3) Alon, Arsac and Steinberg (1953 and 1955); (4) Molchanov (1956); (5) Molchanov et al. (1959); (6) Haddock (1957); (7) Hachenberg, Fürstenberg and Prinzler (1956); (8) Haddock (1957); (9) Covington and Broten (1954); (10) Swarup (1961a and 1961b); (11) Kakinuma (1955); (12) Christiansen and Warburton (1953); (13) Christiansen and Warburton (1955); (14) Swarup and Parthasarathy (1955); (15) Panovkin (1957); (16) Ovsyankin and Panovkin (1956); (17) O'Brien and Tandberg-Hanssen (1955); (18) Conway and O'Brien (1956); (19) Firor (1955 and 1956)

polar directions. The conservation of the main features of the distribution in Fig. 18 for the whole of the solar cycle is also confirmed by eclipse data (Krishnan and Labrum, 1961) obtained at a falling-off in solar activity (8 April 1959).

* *
*

We can conclude from the results given above of observations of the "quiet" Sun's radio emission that this component has the following features.

The frequency spectrum of the "quiet" Sun's radio emission, which is characterized by the dependence of the effective temperature $T_{\text{eff}\odot}$ on the wavelength λ, varies within wide limits—from temperatures close to the temperature of the photosphere (6×10^3 °K) at the beginning of the millimetric band ($\lambda \sim$ 2–3 mm) to values of the order of the corona's kinetic temperature (10^6 °K) at metric wavelengths. The function $T_{\text{eff}\odot}(\lambda)$ is not monotonic: it has a maximum at $\lambda \sim$ 4 mm and a minimum at $\lambda \sim$ 6 mm; then $T_{\text{eff}\odot}$ again reaches a maximum in the metric band, dropping apparently with the transition to the decametric wavelengths.

The size of the radio sun at millimetric wavelengths is close to the size of the optical disk (being slightly larger than the latter); as the wavelength increases the extent of the radio sun rises and in the metric band the brightness is still of noticeable magnitude at distances $r \sim (1\cdot5\text{–}3)R_\odot$ from the centre of the disk.

The width of the radio brightness distribution over the Sun's disk is not the same for different directions in relation to the equator: as a rule it is greater in the equatorial direction than in the polar. This effect is found over the whole of the range studied, from $\lambda > 3$ cm, particularly at decimetric and metric wavelengths.

At metric wavelengths the radio brightness distribution is a monotonic curve decreasing gradually as one moves away from the centre of the disk. At wavelengths of $\lambda \sim 1\cdot5$ m brightening appears towards the edge of the disk; it is most clearly seen in the decimetric band at $\lambda \sim$ 10–20 cm. On a further increase in wavelength the degree of brightening once again drops. In the polar direction this brightening is less, as a rule, than in the equatorial direction; to judge from the data available in the 10–20 cm range there is no rise in radio brightening in the polar regions at all. The position of the $T_{\text{eff}\odot}$ maximum varies with the wavelength: at millimetric and centimetric wavelengths this maximum is at the limb but in the decimetric it is shifted towards the centre of the disk.

At wavelengths of $\lambda \sim 1$ cm as well as an increase in brightness at the limb there is a rise in T_{eff} in the central part of the disk.

CHAPTER IV

Results of Observations of the Sun's Sporadic Radio Emission

THE sporadic solar radio emission is one of the most complex of the phenomena included under the name of solar activity. The composition of the sporadic radio emission is extremely non-uniform: it is made up of several components differing in the magnitude of their intensity, the nature of the polarization, the directivity and frequency spectrum of the radio emission, features connecting them with other facets of solar activity, etc.

An initial acquaintanceship with the nature of the time-dependence of the solar radio emission intensity can be acquired to a certain extent by dividing the sporadic component into enhanced level radiation, which has a characteristic duration of the order of hours, days and months, and bursts with a "life" reckoned in minutes and seconds.

It is also possible to divide the bursts up in greater detail into separate types in accordance with the nature of the dependence of the flux on time, $S_\omega(t)$, if we take account of such features of the flux profile as the number of maxima, the "life" and the time the intensity takes to rise to the maximum value, the connection with other types of burst. This kind of classification of the sporadic radio emission is extensively used at present in the centimetric (and partly in the decimetric) waveband, where in the composition of the sporadic radio emission we distinguish the slowly varying component (or S-component), type A bursts (simple bursts), type B bursts (post-bursts) and type C (gradual rise and fall) (Covington, 1951). When processing the results of observations in the International Geophysical Year programme the classification of the sporadic radio emission was also extended to the metric wavelengths (*Solnechnyye Dannyye*, 1958 and 1960); a similar classification is also given by Coutrez (1960) and by Dodson, Hedeman and Owren (1953).

It should be noted, however, that the phenomenological separation of the sporadic radio emission into individual components in accordance with their $S_\omega(t)$ profile is incomplete and in many cases does not define the other features of the radio emission (polarization, frequency spectrum, etc.). It is clear that the basis of the classification must be a description of the sporadic radio emission which will also include the other significant features of the radio emission with sufficient accuracy. At metric and partly at decimetric wavelengths a suitable characteristic is the so-called dynamic spectrum of the radio emission, i.e. the time dependence of the frequency spectrum. As will be shown in Chapter IX the form of the frequency spectrum and its behaviour in time provide important information on the generation mechanism of the radio emission.

Wild and McCready (1950), who divided the bursts in the metric band into three types, started the classification of the sporadic radio emission by the nature of the dynamic spectra. Two further components—types IV and V radio emission—have now been added to the types I, II and III bursts. Together with the enhanced radio emission due to sunspots these five components comprise the basic group of phenomena observed in the metric band.†

We notice that the microwave radio emission of the "disturbed" Sun is not divided into its individual components in accordance with the nature of their dynamic spectrum at present largely because of a severe lack of spectral observations in this range. On the other hand, it is quite possible that at centimetric wavelengths (and in the short-wave part of the decimetric band) the dynamic spectrum as a whole is not a decisive characteristic for identifying the observed radio emission as it is at metric wavelengths. This is because the dynamic spectrum of the microwave radio emission does not have the richness and variety of details that distinguish it at the longer wavelengths.

The present chapter discusses the features of all the known components of the sporadic solar radio emission (sections 10–16) and the connection between these components and different kinds of geophysical phenomena (section 17). The information given is recapitulated at the end of section 17 and we give a general picture of the phenomenon of sporadic radio emission when a centre of activity appears on the Sun.

---

† Enhanced radio emission accompanied by type I bursts is called a "noise storm"; type II bursts are often spoken of as "outbursts".

10. The Slowly Varying Component

GENERAL CHARACTERISTICS. CORRELATION OF THE RADIO EMISSION FLUX WITH SUNSPOTS

On a recording of the flux of the Sun's radio emission the slowly varying component looks like smooth rises in the magnitude of the signal received above the emission level of the "quiet" Sun with a characteristic duration of the order of tens of days (Coutrez, 1960; Pawsey and Yabsley, 1949; Covington, 1948; Lehany and Yabsley, 1949). This component can be observed chiefly in the range of wavelengths from 3 to 50 cm and occupies a broad spectrum of frequencies of the order of several gigacycles. In the middle of this range the rise in the overall radio emission flux due to the slow component is not as a rule more than thrice the level of the "quiet" Sun's radio emission (Waldmeier and Müller, 1950). The relative contribution from the S-component outside the range $\lambda \sim$ 3–50 cm becomes very small, although it can be found at millimetric wavelengths where the rise in the flux is only fractions of a percent (Conway, 1956) and at the long-wave end of the decimetric band (Firor, 1959). We should mention that in the latter band of the spectrum observations of the S-component become far more complicated because of the masking action of the more intense enhanced radio emission connected with sunspots; for slow variations in the radio emission at metric wavelengths see Boishot and Simon (1959).

The time dependence of the S-component's flux shows a clear periodicity with a characteristic time of 27 days, equal to the Sun's period of rotation (Piddington and Minnett, 1951; Lehany and Yabsley, 1949). This indicates that the source of this component is not the Sun as a whole but individual local regions on its disk.

The variations in the enhanced level at centimetric and decimetric wavelengths correlate well with the relative number of sunspots W (the Wolf number); the corresponding correlation coefficient goes up to 0·8. In relative units the radio emission flux S_ω at $\lambda = 10$ cm is connected with W by the approximate relation (Waldmeier and Müller, 1950)

$$S_\omega \approx 1 + 7 \cdot 10^{-3} W. \tag{10.1}$$

For example, for a group consisting of two spots $W = 12$ (see Kuiper, 1953), and therefore the intensity of the radio emission connected with it is about one-tenth of the intensity of the "quiet" Sun's emission.

The intensity of the S-component also correlates closely with the total area occupied by spots on the Sun's disk (Pawsey and Yabsley, 1959; Minnett and Labrum, 1950; Covington, 1948; Lehany and Yabsley, 1949).

Roughly speaking the intensity S_ω of the S-component is proportional to the visible area σ of the spots, the coefficient of proportionality depending on wavelength (Pawsey and Yabsley, 1950). At the same time the correlation between $T_{\text{eff}\odot}$ and σ becomes worse when one moves into the metric band where the S-component gives way to enhanced radio emission; we shall speak about its features in section 12.

In section 9 we said that in the decimetric band the best index of the radio emission when compared with the magnitude of σ is the so-called "complex area of the spots" σ_{comp} which, as well as the area of the spots that can be seen on the disk σ, contains the area of spots that have already decayed 1 or 2 months ago. The enhanced correlation coefficient between S_ω and σ_{comp} indicates that radio emission continues from a centre of activity even after a group of spots has decayed.

The first indications that the regions which have previously been occupied by spots are sources of the slowly varying radio emission (as well as regions located near the spots) were obtained by Christiansen, Yabsley and Mills (1949) during the solar eclipse in November 1948 from the change in the radio emission flux while these regions were being covered and uncovered. The same conclusion was drawn from statistical analysis of lengthy measurements of the solar radio emission flux in the 10-cm band, according to which the emission flux is approximately proportional to the area of a spot in the first 15 days of its existence; when an active centre becomes older its emission does not decrease so rapidly as the area of the spot and remains after it has decayed (Vauquois, 1955 and 1959).

A similar conclusion is also drawn by Christiansen, Warburton and Davies (1957) from the data of observations at $\lambda = 21$ cm. A comparison made of the life of the radio emission sources studied and the life of sunspots showed that 2 months after the radio emission flux and the area of a group of spots reach a maximum the flux of the slowly varying radio emission is one-sixth of its greatest value, whilst the area of the spots is only about one-thirtieth. Therefore near the maximum there is a high correlation between the intensity of the radio emission and the area of the spots, whilst later the correlation deteriorates. Since, however, the radio emission from strong sources is generally dominant in the total flux of the slowly varying radio emission the degree of correlation between it and the area of the spots that can be seen on the disk remains quite high.

POSITION, FORM AND SIZE OF LOCAL SOURCES

Extensive observations were later made to study the connection of photospheric, chromospheric and coronal formations with local sources of radio emission and also to investigate the position, form and effective tempera-

§ 10] The Slowly Varying Component

ture of the sources of the S-component; these observations made it possible to obtain one-dimensional and two-dimensional distributions of the radio brightness over the disk of the "perturbed" Sun.† Typical examples of one-dimensional distributions of the radio brightness over the Sun's disk at centimetric, decimetric and metric wavelengths are shown in Figs. 24 and 25 (see also Fig. 12). At $\lambda = 3 \cdot 2$ cm the distribution was obtained with an aerial having a "knife-edge" beam about 1' wide. At the other wavelengths the observations were made with multi-element interfero-

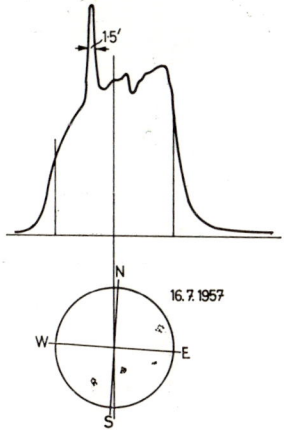

FIG. 24. One-dimensional distribution of radio brightness over the Sun's disk at $\lambda = 3 \cdot 2$ cm; the figure also shows the positions of sunspots on the disk during the observations (Ikhsanova, 1960a)

meters; their beams were series of widely separated lobes 3–5' wide. In this connection we must stress that the nature of the observed radio brightness distribution is sharply dependent on the resolution of the receiving aerials because of the averaging of T_{eff} over the beam (Covington, 1959).

Fuller data on the nature of the local sources of the S-component can be obtained if we have two-dimensional distributions of the radio brightness over the Sun's disk. Maps of the "disturbed" Sun's radio isophots of this kind have been made by Christiansen, Mathewson and Pawsey (1957), Christiansen *et al.* (1958, 1959, 1960) at $\lambda = 21$ cm and by Swarup (1960, 1961a, 1961b) at $\lambda = 9 \cdot 4$ cm with aerials of the "Mills Cross" type having

† Covington and Broten (1954); Christiansen and Mathewson (1958); Coates (1958a, 1958b and 1960); Christiansen, Warburton and Davies (1957); Firor (1959); Boishot and Simon (1959); Kislyakov and Salomonovich (1963); Salomonovich (1962a); Swarup and Parthasarathy (1958); Ikhsanova (1959); Hatanaka *et al.* (1956); Hatanaka (1957a); Christiansen *et al.* (1960); Christiansen, Mathewson and Pawsey (1957); Christiansen and Mathewson (1959); Vitkevich *et al.* (1958 and 1959); Swarup (1960).

a resolution of about 3′, and also by Vitkevich *et al.*, 1958 and 1959) at wavelengths of 10 cm and 3·2 cm by successive scanning of the Sun with a parabolic aerial with respective beam widths of 15′ and 6′. Some of these radio pictures of the Sun are shown in Fig. 26 (see also Figs. 31 and 32).

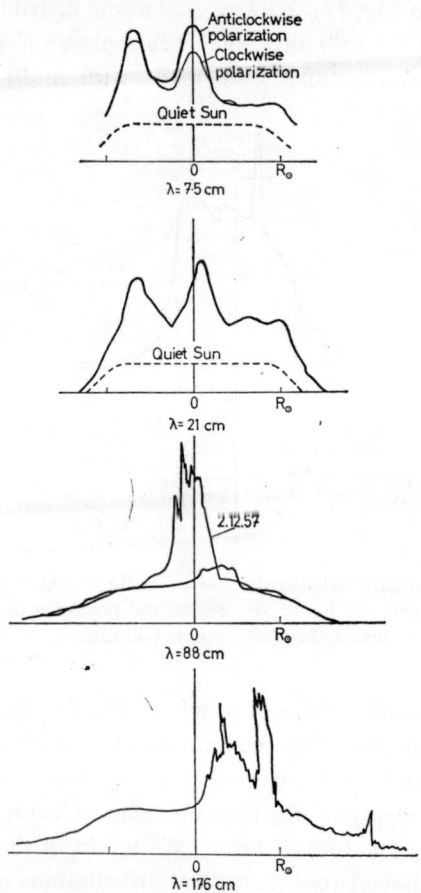

FIG. 25. One-dimensional distributions of the radio brightness over the disk of the Sun taken 3 December 1957 (Christiansen *et al.*, 1960)

On comparing the one-dimensional distributions over the disk of the "disturbed" Sun in Fig. 25 we see that the nature of the local sources changes sharply with the transition from wavelengths of 7·5–21 cm into the metric band where the radio emission connected with these sources generally has all the signs that are characteristic of a noise storm. At the intermediate wavelengths ($\lambda = 88$ cm) the radio emission (particularly

§ 10] The Slowly Varying Component

on 2 December) was also a typical noise storm, whilst at the shorter wavelengths the observed intensity varied only a little during the observations and the radio brightness distribution contours become smoother. It is true that in the long-wave part of the decimetric band sharp

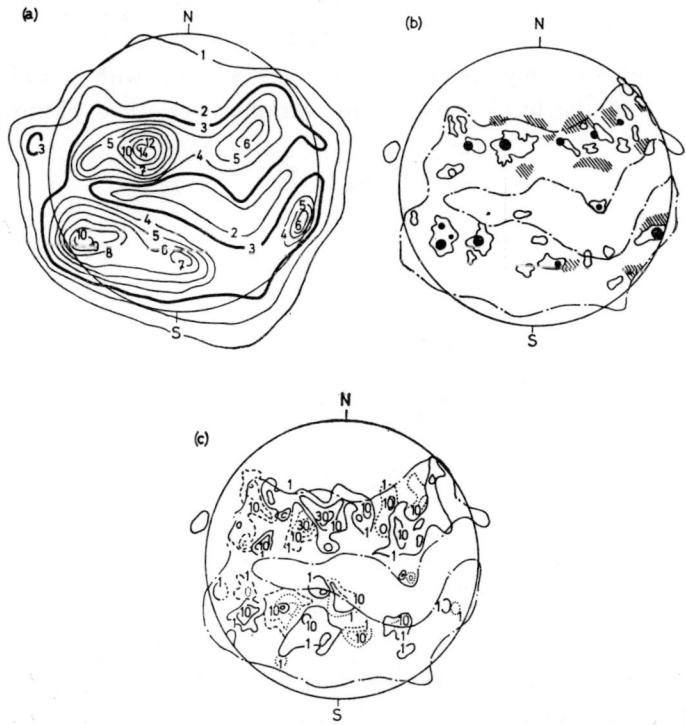

FIG. 26. Observations of the Sun on 11 November 1957: (a) radio isophots at $\lambda = 21$ cm (values of effective temperature in units of $7 \cdot 5 \times 10^4$ °K); (b) distribution of plages and position of groups of spots on disk; (c) magnetogram of Sun (contours correspond to 1,10 and 30 oe; southern polarity is shown by dotted lines, northern by solid lines). The lines of dashes and dots in the figures show the $T_{\text{eff}} = 2 \cdot 2 \times 10^5$ °K contours (Christiansen and Mathewson, 1959)

changes in intensity are not always observed; frequently the radio brightness distribution in this range is more like the distribution in the shorter wave ($\lambda \sim 3$–30 cm) range, where the S-component plays a major role. The general picture of the phenomena in the transition region between the decimetric and metric wavelengths is such that here (depending on the actual physical conditions in the region generating the radio emission) either the S-component is predominant or enhanced radio emission occurs, accompanied by intense bursts (Firor, 1959; Swarup and Parthasarathy, 1958).

Dodson (1954), when comparing the spectroheliograms in the K-line of twice ionized calcium with the one-dimensional radio brightness distributions of Covington and Broten (1954), came to the conclusion that the S-component at a wavelength of $\lambda = 10$ cm is very closely connected with the plages whose intensity in the K-line is almost 1·5 times higher than the corresponding intensity of the quiet Sun (i.e. with flocculi). The connection of the local sources with chromospheric and photospheric

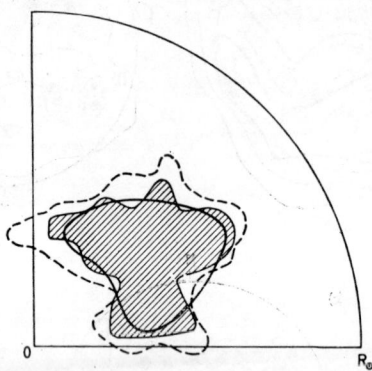

FIG. 27. Comparison of the form and size of a local source of radio emission at $\lambda = 21$ cm at half level of the maximum value of T_{eff} (solid line), contour of magnetic field with strength of 5 gauss (dotted line) and of plage (hatched) (Christiansen et al., 1960)

formations that can be seen in white light in the H_α and K Ca II lines was then studied in detail by other observers (Christiansen et al., 1960; Christiansen and Mathewson, 1959; Vitkevich and Mathewson, 1959).

It turned out that in the decimetric band the position of the local sources of the slowly changing radio emission on the Sun's disk is close to the position of bright plages and local magnetic fields which inevitably accompany the appearance of each source of radio emission. The form and size of the radio regions coincide to a considerable extent with the form and size of the plages, repeating moreover common contour lines of equal strength of the local magnetic fields in the photosphere. This is illustrated in Fig. 26, which shows the radio isophots at $\lambda = 21$ cm, a magnetogram and map of the Sun on which are shown the outlines of the plages and the position of the groups of spots. The close agreement between the lines of equal radio emission intensity, identical magnetic field strength and the limits of the plage in the H_α line for one of the centres of activity can also be clearly seen in Fig. 27.

§ 10] The Slowly Varying Component

The life of the local sources in the decimetric band is of the order of the duration of the existence of the plages and flocculi (on the average about 3 months). During this time a source has its own motion, changes in form and size but always remains connected with the same centre of activity (Christiansen and Mathewson, 1959).

The close connection of the local sources of radio emission with the plages that appear one or two days before the appearance of spots on them and are longer-lived formations than the spots explains the features noted above in the correlation of the S-component at $\lambda \gtrsim 10$ cm with the groups of spots, namely the increase in the correlation coefficient when we change from the area of the spots that can be seen on the disk to the "complex area", which includes spots that have already decayed as well.

According to Krishnan and Labrum (1961), Christiansen and Mathewson (1959) the radio brightness distribution at $\lambda = 20$ cm over the local source is very close to the distribution of the brightness of flocculi in the K-line of Ca II. This means that the effective size of the local sources at decimetric wavelengths is also close to the corresponding size of the flocculi. The typical extent of the local sources is about 5' (i.e. about $2 \cdot 5 \times 10^5$ km) although both smaller and larger sources are found: 3–10'' (Christiansen and Mathewson, 1959; see also Christiansen, Warburton and Davies, 1957; Christiansen, Yabsley and Mills, 1949; Swarup and Parthasarathy, 1958).

For intense local sources in the decimetric band a typical emission temperature is $T_{\text{eff}} \approx 10^6 – 2 \times 10^6$ °K (Christiansen, Warburton and Davies, 1959) although at wavelengths of $\lambda \gtrsim 50$ cm regions are found with a slightly higher temperature (Christiansen, Yabsley and Mills, 1949; Boishot and Simon, 1959; Swarup and Parthasarathy, 1958). According to the data given by Christiansen et al. (1960), Christiansen and Mathewson (1959) at $\lambda = 21$ cm T_{eff} in the brightest part of the source varies from small values to $1 \cdot 5 \times 10^6$ °K, the latter value playing the part of the upper limit for T_{eff}: greater brightness temperatures are generally not found; on the average $T_{\text{eff}} \approx 0 \cdot 6 \times 10^6$ °K. Therefore the observed values of T_{eff} are close to the generally accepted value of the corona's kinetic temperature (10^6 °K) or slightly higher than the latter.

Unlike the decimetric band, at centimetric wavelengths the local sources of radio emission are more closely connected with the spots than with the flocculi (Ikhsanova, 1960a; Ikhsanova, 1959; Vitkevich et al., 1959; Korol'kov and Soboleva, 1957; Molchanov et al., 1959; Korol'kov, Soboleva and Gelfreich, 1960; Gelfreich et al., 1959; Korol'kov and Soboleva, 1961; Ikhsanova, 1960b; Khaikin, 1960). Each group of spots larger than 50–100 millionths of the solar hemisphere has above it a region from which the S-component of the radio emission comes (see Fig. 24).† The flux of this radio emission rises as the area of the spots increases; the life of a local source with an accuracy up to days is the same as that for the group of spots connected with it.

The following is known about the size and distribution of the radio brightness over the sources in the centimetric band. At wavelengths of $\lambda \sim 8$–10 cm the size of the source and the value of the effective temperature hardly differ at all from those given above for 21 cm, although sometimes values of $T_{\text{eff}} > 3 \times 10^6$ °K are found (Swarup, 1961a and 1961b; Hatanaka et al., 1956; Khaikin, 1960; Kakinuma, 1956). A comparison of radio pictures of the Sun at $\lambda = 21$ cm and 3·2 cm also reveals quite close agreement between the radio isophots (Vitkevich and Mathewson, 1959). From this we could conclude that the sizes of the local sources at these wavelengths are close to each other. However, the considerable width of the aerial beams in the observations (Vitkevich and Mathewson, 1959; Salomonovich, 1962b; Vitkevich et al., 1958 and 1959) (about 6′) made it impossible to reveal the fine structure in the radio brightness distribution over a local source at centimetric wavelengths. This structure has been studied by means of eclipse observations and measurements by means of aerials with high resolution (Ikhsanova, 1960a; Korol'kov, Soboleva and Gelfreich, 1960; Gelfreich et al., 1959; Korol'kov and Soboleva, 1961; Ikhsanova, 1960b; Khaikin, 1960; Kundu, 1959a, 1958 and 1959b).

Kundu's observations (Kundu, 1958, 1959a and 1959b) on an interferometer with a variable baseline ($\lambda \sim 3$ cm) showed that the radio brightness distributions shown in Fig. 28 are typical for local sources of the S-component. It can be seen from the figure that a local source is usually a bright region about 1·5′ in diameter with an effective temperature of $T_{\text{eff}} \sim 5 \times 10^5$ °K, sometimes surrounded by a weak halo—an extensive source whose diameter reaches 10′ and whose effective temperature reaches 10^5 °K.‡ We note that Figs. 28b and c show radio brightness distributions for the same local source as Fig. 28a taken 3 and 8 days respectively after the first. In this time the bright region remains almost

---

† At the same time a comparison of the coordinates of the prominences with the position of the local sources of radio emission on the Sun's disk ($\lambda \sim 3$ cm: Ikhsanova, 1960a; Vitkevich et al., 1959) did not reveal the connection between them indicated earlier (Troitskii et al., 1956).

‡ Certain indications of this kind of radio brightness distribution over a local source have been obtained in Hatanaka et al. (1956); Hatanaka (1957a) from an analysis of interferometer and eclipse observations at $\lambda \sim 8$ cm (see also Tanaka and Kakinuma, 1958 and 1959). Salomonovich's investigations (Salomonovich, 1962b) with narrow-beam aerials indicate that in the millimetric band certain local sources also have a halo surrounding a bright region of small angular size (of the order of 1–2′).

§ 10] The Slowly Varying Component

constant, whilst the halo displays noticeable variations in diameter and effective temperature: the value of T_{eff} is reduced by a factor of more than 3 and the size of the course increases from 6' to 9', then decreases again to 5'.

Close results were obtained from eclipse observations and measurements of the radio brightness distribution with an aerial having a "knife-edge" beam up to 1' wide made by the Pulkovo group of radio astronomers.†

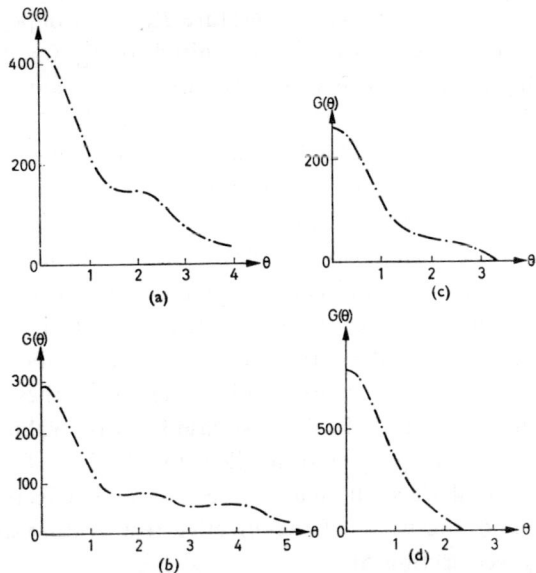

FIG. 28. One-dimensional distributions of radio brightness over local sources at $\lambda = 3$ cm (Kundu, 1959b) (θ in angular minutes)

According to the data they obtained the size of S-component sources at $\lambda \sim 2-5$ cm is not greater than the diameters of the spots with which the emission is connected and varies between values of about 1–3' (see Fig. 24). If the visible size of the spots is less than 1' the diameters of the emitting regions are not noticeably greater than this value. The effective temperature of these regions sometimes reaches 10^6 °K ($\lambda \sim 3$ cm) and is generally within the limits $(1-8) \times 10^5$ °K. In the regions above the flocculi surrounding the spots T_{eff} does not exceed $(3-5) \times 10^4$ °K according to the data in Khaikin (1960).

† Ikhsanova (1960a); Ikhsanova (1959); Korol'kov, Soboleva and Gelfreich (1960); Gelfreich et al., (1959); Korol'kov and Soboleva (1961); Ikhsanova (1960b); Khaikin (1960). See also Troitskii et al. (1956); Molchanov et al. (1959); Edelson, Grant and Corbett (1960).

Therefore the range of wavelengths around 10 cm is one of transition from the distribution with a halo to the more or less even distribution of T_{eff} over the source. In this range, therefore, depending on the actual conditions in the generation region a distribution of either one type or the other is realized (cf. Swarup, 1961a and 1961b; Khaikin, 1960).

RADIO-EMISSION FREQUENCY SPECTRUM

The frequency spectrum of the radio emission (the frequency dependence of the intensity I or the effective temperature T_{eff}) is of importance in the theory of the S-component. Orders of magnitude of T_{eff} for different local sources have been given above when discussing the results of measuring the effective sizes of emitting regions and the distribution of the radio brightness over the sources. This information is insufficient, however, for getting an idea about the S-component's frequency spectrum: reliable data on the frequency spectrum can obviously be obtained only as a result of simultaneous observations over a wide range of frequencies with narrow-beam aerial arrays. The requirements imposed on the aerials here are very rigid: since the characteristic size of local sources in the decimetric band is about 5', and at centimetric wavelengths about 1–2', it is clear that the width of the beams should be less than these values or at most comparable with them. In the first case from the value of the aerial temperature T_A we can judge the detailed distribution of T_{eff} over the source, in the second we shall be able to estimate the effective size of the source Ω_{source} and the value of the effective temperature averaged over the source $\bar{T}_{\text{eff}}$ (see section 5).

The absence at present of a large enough number of narrow-beam aerial arrays, particularly in the centimetric band where only the Pulkovo radio telescope has a resolution of up to 1', obliges us when investigating the frequency spectra of the S-component to turn to eclipse observations which provide high resolution with small aerials. Observations of this kind were carried out during the solar eclipses of 19 April 1958 and 15 February 1961 (see Molchanov and Korol'kov, 1961; Gol'nev *et al.*, 1961; Veisig and Borovik, 1961; Dravskikh and Dravskikh, 1961; Molchanov and Peterova, 1961; Kuznetsova, *et al.* 1961); the results obtained were collated by Molchanov (1961a, 1961b and 1962) and are shown in Fig. 29 in the form of a function of the wavelength λ of the overall radio emission flux from a local source S_ω (apart from the emission flux of the "quiet" Sun from the same region). The function $S_\omega(\lambda)$ in Fig. 29a is for a source located in the central part of the disk above a fairly large group of spots with an area of about 10^{-3} of the Sun's hemisphere, with a magnetic field $H_{0_b} \approx 2500$ oe. The graph of $S_\omega(\lambda)$ in Fig. 29b relates to a local region on the

§ 10] The Slowly Varying Component

limb connected with a small group of spots (area about 4×10^{-4} of the Sun's hemisphere, magnetic field strength 2200 oe). In addition Fig. 29 shows the function $S_\omega(\lambda)$ for a source located above flocculi in the centre of the disk.

It can be seen from the figures that for local sources connected with spots the dependence of the total flux on the wavelength has a characteristic curve with maxima in the region of $\lambda \sim 5\text{-}8$ cm; the quantity S_ω decreases both at the short- and long-wave ends.† For the source connected with

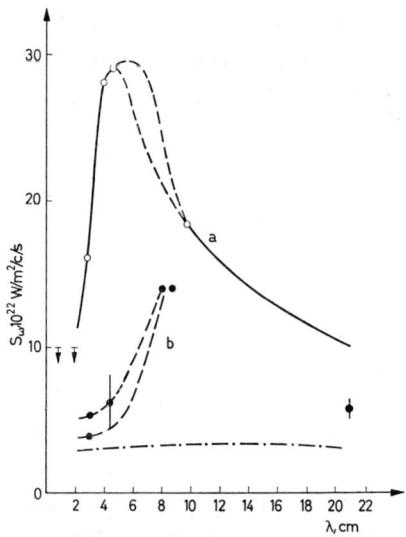

FIG. 29. Total radio emission flux from the bright part of local sources as a function of wavelength: (a) observations of 19 April 1958; (b) observations of 15 February 1961 (the line of dots and dashes indicates the spectrum of a local source located above a group of flocculi) (Molchanov, 1962)

flocculi (among which there were no spots) the curve of $S_\omega(\lambda)$ differs sharply from the one indicated: in the 3–21 cm range studied the value of the flux is comparatively small and does not depend on the wavelength.‡

In order to recover the frequency spectrum of the sources $\bar{T}_\text{eff}(\lambda)$ or

---

† A similar function $S_\omega(\lambda)$ was obtained considerably earlier by Piddington and Minnett (1951) on the basis of results from observations made by different researchers at different times. The existence of a maximum in the S-component's frequency spectrum (at wavelengths of $\lambda \sim 10$ cm) is also indicated by Tanaka and Kakinuma (1958), Tanaka (1964), Kakinuma and Swarup (1962a and 1962b) (see also Krüger, Krüger and Wallis, 1964).

‡ Radio emission of flocculi at a period when they are free of spots has also been found at millimetric wavelengths (Kislyakov and Salomonovich, 1963; Salomonovich, 1962b).

$I(\lambda)$ from the known function $S_\omega(\lambda)$ we need to know the sizes of the sources at different wavelengths. If we consider (in accordance with the information available) that at decimetric wavelengths the area of the emitting regions is the same as the area of the flocculi accompanying them and in the centimetric band is the same as the area of the spots and groups of spots[†] and does not change significantly with wavelength (within the limits of the bands in question) then the following conclusions can be drawn from the data given above. In the decimetric band $\bar{T}_{eff}$ for large sources reaches values of about 10^6 °K (or slightly higher) and they are of the same order of magnitude when the wavelength is reduced right down to 5–10 cm. In the centimetric band at wavelengths $\lambda < 5$–10 cm the effective temperature decreases together with the wavelength more rapidly than λ^2. This can easily be confirmed if we take into consideration that by definition (4.4) $T_{eff} \propto I\lambda^2$: as the intensity I decreases with the rise in wavelength in the $\lambda > 5$–10 cm region T_{eff} changes only slightly. On the other hand for the shorter wavelengths, where the intensity decreases, T_{eff} drops more rapidly than λ^2.

It must be stressed that the conclusion about the relative constancy of T_{eff} at wavelengths of $\lambda > 5$–10 cm and the sharp drop in I towards the short wavelengths at $\lambda < 5$–10 cm is of fundamental importance in the theory of the S-component (see section 29). Simultaneous investigations of local sources over the widest possible range, including the millimetric band, with accurate measurement of the emitting regions are absolutely necessary to find out the degree of reliability of the latter conclusion.

The data given on the spectra of the local sources relate only to wavelengths from 3 cm upwards; no definite results were obtained at shorter wavelengths during the eclipse observations. Some information on the value of the T_{eff} of local sources at millimetric wavelengths ($\lambda = 4$ mm and 8 mm) and on the results of comparing the values of $\bar{T}_{eff}$ at wavelengths of 8 mm and 3·2 cm is given by Kislyakov and Salomonovich (1963), Salomonovich (1962b) (see also Tolbert and Straiton, 1961; Coates, 1958a and 1958b; Hagen, 1951). For two sources studied (whose size was close to the size of the groups of spots connected with them $-1·5$–$2·0'$) the values of $\bar{T}_{eff}$ were about $1·5 \times 10^3$ °K and 4×10^3 °K respectively at $\lambda = 4$ mm and 8 mm. However, the low accuracy of the determination of the effective temperatures (up to 40%) made it impossible to draw reliable enough conclusions about the law of the change of T_{eff} in the millimetric band.

The comparison made by Salomonovich (1962b) of the radio emission fluxes for local sources at wavelengths of 8 mm and 3·2 cm reveals consider-

[†] The local sources whose spectra are shown in Fig. 29 had no noticeable haloes.

§ 10] The Slowly Varying Component

able scatter in the values, although on the average the ratio of the fluxes at these wavelengths is close to unity. The fluxes compared here relate to the source as a whole including the halo. The insufficient directivity of the aerials in these experiments made it impossible to determine the exact sizes of the local sources at $\lambda = 3\cdot 2$ cm. Therefore the nature of the change in T_{eff} in the transition from the centimetric band to the millimetric still remains unknown.

It is possible that the spectrum of $T_{\text{eff}}(\lambda)$ is not the same for the bright part of a source and the halo; for the latter it is probably close to the spectrum of local sources above flocculi which are free of spots.

DIRECTIONAL PROPERTIES AND POLARIZATION

At centimetric and decimetric wavelengths the S-component does not display highly directional features. This is clear from the circumstance mentioned above, according to which the radio emission appears earlier than a visible spot shows itself from behind the edge of the disk and disappears after the spot ceases to be visible because of the Sun's rotation.

We generally base ourselves when determining the S-component's angular spectrum on the daily values of the radio emission flux S_ω and the angle Θ between the Sun's radius passing through the source in the direction of the Earth. If the angular spectrum of the local source stays constant as one moves over the Sun's disk the observed dependence of S_ω on Θ would directly characterize the polar diagram of the radio emission. The quantity S_ω, however, depends not only on the position on the disk Θ but also on such source parameters as its age t, the area σ of the spots or flocculi connected with it, and other parameters. It is clear from this that in observations from the Earth it is impossible to determine the function $S_\omega(\lambda)$ with constant $t, \sigma, \ldots$ for each local source separately, since in the process of slowly moving over the disk not only the position of the source but also the values of $t, \sigma, \ldots$ change. In practice, therefore, it is reasonable to speak only of a certain directional characteristic averaged over many sources.

Examples of polar diagrams $f(\Theta)$ are shown in Fig. 30. The function $f(\Theta)$ at $\lambda = 25$ cm is obtained on the assumption that the flux is proportional to the area of the spots, the function $f(\Theta)$ at $\lambda = 50$ cm on the condition that the flux is proportional to the area of the flocculi. As can be seen from the figures there is very little difference between these curves. The polar diagram obtained by averaging the observed values of S_ω at $\lambda = 21$ cm for given Θ over many local sources with all possible values of the parameters σ and t is shown in Fig. 30b.

It can be seen from the figures that the flux of the emission from a local source varies as $S_\omega \propto \cos \Theta$ for values of Θ that are not too close to 90°. This dependence holds true over a wide range of wavelengths from 3 cm to 50 cm (Christiansen, Warburton and Davies, 1957; Vauquois, 1955 and 1959; Christiansen *et al.*, 1960; Christiansen and Mathewson, 1959; Gutmann and Steinberg, 1959; Kawabata, 1954; Waldmeier, 1953a and 1953b; Kawabata, 1960a).† However, the flux S_ω as a rule does not fall

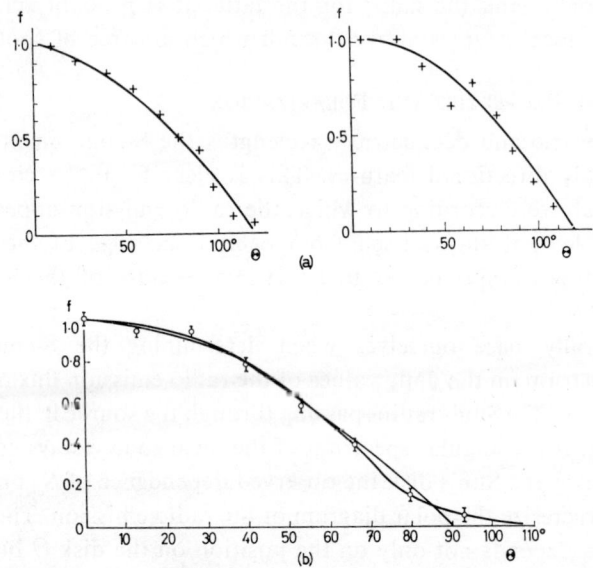

FIG. 30. Averaged polar diagrams of local sources $f(\Theta)$: (a) $\lambda = 25$ cm and $\lambda = 50$ cm (Vauquois, 1955 and 1959); (b) $\lambda = 21$ cm (Christiansen, Warburton and Davies, 1957)

to zero when $\Theta = 90°$; judging from Fig. 30 the flux values are then 0·06–0·4 of its maximum value at the centre of the disk at $\lambda \sim 20$–25 cm. According to the data of Swarup (1961a and 1961b), Waldmeier (1953b) and Vauquois (1959b) in the 10 cm band the decrease in the flux on the limb is about 0·2.

† The nature of the directivity at shorter wavelengths is unknown unless we consider the remark in Salomonovich (1962b) that this effect is absent at $\lambda = 8·6$ mm. On the other hand, in the band around 50 cm the function $S_\omega(\Theta)$ may differ noticeably from that given in Fig. 30, dropping more steeply as Θ rises (Swarup and Parthasarathy, 1958; Machin and O'Brien, 1954; Müller, 1956). At these wavelengths the effect of the enhanced radio emission connected with sunspots, which has higher directivity (section 12), makes itself felt.

§ 10] The Slowly Varying Component

The decrease in the total emission flux S_ω from a local source as it moves from the centre of the disk to the periphery is chiefly connected not with a decrease in the effective temperature T_{eff} but with a decrease in the angular size of the source in the direction from the centre of the disk to the limb (Christiansen and Mathewson, 1959). This obviously means that the extent of the S-component generation region in height is small in comparison with the size of this region along the solar surface.

Observations made in the centimetric band (Christiansen et al., 1960; Korol'kov, Soboleva and Gelfreich, 1960; Gelfreich et al., 1959; Korol'kov and Soboleva, 1961; Khaikin, 1960; Kundu, 1959a and 1959b; Tanaka and Kakinuma, 1958 and 1959; Covington, 1949) have shown that the S-component is partially polarized. The degree of circular polarization at $\lambda \sim 3\text{-}7 \cdot 5$ cm reaches 20–30% (Christiansen et al., 1960; Korol'kov, Soboleva and Gelfreich, 1960; Korol'kov and Soboleva, 1961)—40% according to Kakinuma and Swarup (1962a and 1962b)—and decreases as the wavelength rises: at $\lambda = 10$ cm it is less than or of the order of 10% (Piddington and Minnett, 1951; Kakinuma and Swarup, 1962a and 1962b), and at a wavelength of 20 cm the share of the circularly polarized component in the emission is known not to exceed 2% (Christiansen and Mathewson, 1959). As the wavelength further increases the degree of polarization of the local sources once again increases ($\lambda = 50$ cm (Vitkevich, 1956a)); however, it is apparently not connected with the S-component but is caused by enhanced radio emission whose effect becomes noticeable in this band. The degree of polarization apparently also decreases in the transition from the centimetric to the millimetric band. For example measurements (Amenitskii et al., 1958) at $\lambda = 8$ mm did not show up any difference between the intensities of the clockwise and anticlockwise polarized components; from this we may conclude (allowing for the sensitivity of the apparatus) that the amount of polarized emission in the total radio emission flux from a local source does not exceed 10%.

The polarization of the S-component is generally circular; no noticeable ellipticity can be observed as a rule. Korol'kov, Soboleva and Gelfreich (1960) mention several cases, however, when in the emission connected with large spots located near the central meridian linear polarization may have been observed whose degree ϱ_l was several percent of the total solar radio emission flux.

The results of eclipse observations made with aerials having a high resolution have made it possible to establish that the polarized emission owes its origin to a bright region of small angular size (of the order of 1–1·5′), whilst the extensive source (halo) surrounding this region makes no noticeable contribution to the polarized emission ($\lambda = 3\text{-}8$ cm (Kundu,

1959a and 1959b; Tanaka and Kakinuma, 1959 and 1958)). Observations by Korol'kov, Soboleva and Gelfreich (1960), Korol'kov and Soboleva (1961) at a wavelength of $\lambda \sim 3$ cm have shown that the circularly polarized component comes from regions located above spots at an altitude of the order of $0.06\,R_\odot$ from the photosphere; the size of the emitting region is of the order of the diameters of the spots and even of the order of the size of their nuclei. It was noted above that the effective diameter of the regions responsible for the total radio emission flux is the same as the sizes of spots and groups of spots, i.e. it is possibly slightly greater than the diameter of the sources of the polarized component, although it is not excluded that this conclusion is due to insufficient resolution obtained in the observations of the total radio emission flux.

It follows from a comparison of the direction of rotation of the polarization vector and the magnetic field in the spots (at the level of the photosphere) that the sign of the polarization corresponds to an extraordinary wave. Since the emission from regions located above spots in a bipolar group has clockwise and anticlockwise polarization, the resultant polarization when receiving the radio emission from the centre of activity as a whole will depend upon the relative contribution from the spots to the flux of the polarized emission. When this contribution changes, and also because the radio emission passes through the region of the transverse magnetic field in the corona, we may expect a change in the sign of the rotation in the total emission of a bipolar group (see section 24). This effect has actually been observed in certain cases when the central solar meridian cuts the active region (Piddington and Minnett, 1951; Covington and Minnett, 1949; Tanaka and Kakinuma, 1955).

According to Korol'kov, Soboleva and Gelfreich (1960) there are indications that the directional features of the polarized component are more sharply defined than those of the unpolarized component ($\lambda = 3$ cm); the polarized emission sometimes appears only if the local region is closer than 45° in longitude to the central meridian. This phenomenon has not been studied in detail, however.

ALTITUDE OF LOCAL SOURCES ABOVE THE PHOTOSPHERE. CONNECTION WITH OPTICAL FEATURES OF THE SOLAR CORONA

In maps of the radio isophots (see Fig. 26) it can be clearly seen that the sources of the S-component often extend beyond the optical disk of the Sun. This obviously means that they are located at quite a considerable altitude above the photosphere. Since at $\lambda \sim 3$ cm the angular size of a radio emission source is close to the size of the sunspots, this causes the radio emission to appear before the group of spots comes from

§ 10] The Slowly Varying Component

behind the Sun's disk and to disappear after the group is hidden behind the disk because of the Sun's rotation (Vitkevich, 1956a; Vauquois, 1955 and 1959a; Ikhsanova, 1960a; Covington, 1954; Takakura, 1953).

At $\lambda \sim 3$ cm the interval between the appearance of a local source and a group of spots is about a day. This corresponds to an altitude of the order of 2×10^9 cm above the level of the photosphere (i.e. about $0.03R_\odot$ where $R_\odot = 6.95 \times 10^{10}$ cm—the optical radius of the Sun). This interval increases with the wavelength, reaching two or three days at $\lambda = 10$ cm (Ikhsanova, 1960b). This dependence can be explained by the increase in the altitude of the local sources at longer wavelengths. It is possible, however, that the effect observed is caused chiefly by the extent of the sources increasing in heliographic longitude as the wavelength rises.

We said above that in the decimetric band the centre of a source of emission located near the central meridian coincides with the corresponding plage in the decimetric band and with the nuclei of spots at centimetric wavelengths. The difference in the altitudes at which the local sources and the optical centres of activity are located leads to a difference in the rates of motion over the Sun's disk; in addition the local sources lag behind the optical features connected with them right up to the moment they intersect the central meridian; after passing through the meridian the local sources are ahead of the spots and plages.

From an analysis of the rate of movement and the size of the shift of the radio sources relative to the plages and groups of spots we can also estimate the altitude of the layers of the solar atmosphere which are responsible for the creation of the S-component. It turns out here that the sources of radio emission at wavelengths of 3–20 cm are located on an average at an altitude of the order of $0.06R_\odot$ above the photosphere or less, with a considerable scatter about this value (from $0.03R_\odot$ to $0.15R_\odot$ at $\lambda \sim 20$ cm) (Christiansen, Warburton and Davies, 1957; Ikhsanova, 1960b; Christiansen 1960; Christiansen and Mathewson, 1959; Korol'kov, Soboleva and Gelfreich, 1960; Kaidanovskii, Molchanov and Peterova, 1960). Therefore the layers in which the slowly varying radio emission is generated are located as a rule slightly above the $0.03R_\odot$ level, which is often taken as the boundary between the chromosphere and the corona. The estimates do not reveal any tendency to a systematic variation with the wavelength. A study (Christiansen and Mathewson, 1959) of several sources undertaken to find the connection between the magnetic field strength and the altitude or "brightness" of the radio-emitting region did not lead to any definite results either.

At the same time the altitude found by Edelson, Grant and Corbett (1960) from the displacement of six sources of radio emission relative to the plages connected with them show a dependence on the wavelength, averaging $0\cdot04R_\odot$ and $0\cdot06R_\odot$ at $\lambda \sim 3\cdot15$ cm and $9\cdot4$ cm respectively. The presence of this dependence is at least not contradicted by the information given above on the lag of the optical centres relative to the appearance of the S-component when the active region comes from behind the edge of the disk. The increase in the altitude of the source together with the wavelength is also indicated by Kakinuma and Swarup (1962a and 1962b), but they give lower estimates of the average altitudes at wavelengths of 3 and 10 cm ($0\cdot018R_\odot$ and $0\cdot03R_\odot$ respectively). The latter altitude values are very favourable for the gyro resonance theory of the S-component developed in section 29.

It should be stressed that the estimates available for the altitudes of the local sources are not particularly reliable. Extensive observations with careful statistical processing of the results obtained must be made before a final conclusion can be drawn on the existence and nature of the dependence of the altitude on wavelength, magnetic field strength and other parameters characterizing the centre of activity.

Despite the fact that the local sources of the S-component are closely connected with spots and flocculi, measurements of source altitudes indicate that the generation regions, being located in the higher layers belonging to the inner corona, are not identical with these photospheric and chromospheric formations. From optical observations in continuous light and in the coronal lines we know that in the corona above the spots and flocculi there are regions of enhanced electron concentration which have a slightly higher kinetic temperature than the ambient medium (for further detail see section 2). The connection between the local sources of radio

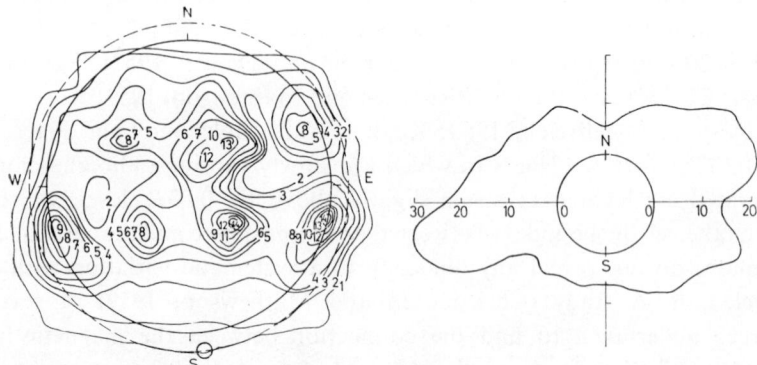

Fig. 31. Comparison of the Sun's radio isophots at $\lambda = 21$ cm with the polar diagram characterizing the intensity of the K-corona at a distance $1\cdot06R_\odot$ from the centre of the Sun's disk. The aperture of the coronograph is shown by the circle in the left-hand figure (Newkirk, 1959)

§ 10] The Slowly Varying Component

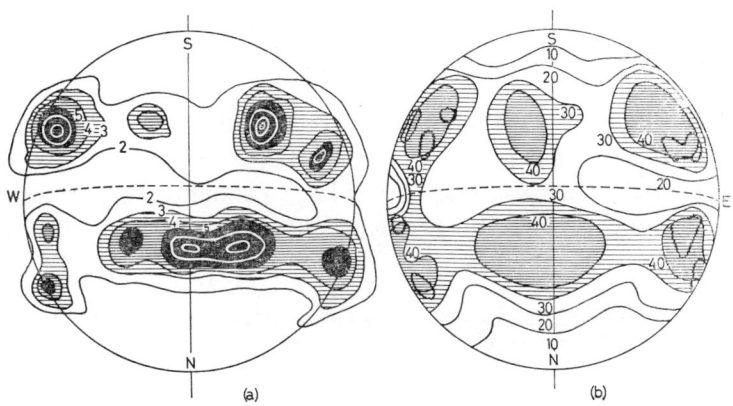

Fig. 32. Radio isophots of the Sun at $\lambda = 21$ cm (a) and isophots of the corona on the 5303 Å line (b) (Waldmeier, 1959). The radio brightness contours are drawn at intervals of 10^5 °K. The black areas in the right-hand figure are groups of spots; wavy lines—boundaries of plages

emission and the features of the solar corona in the $\lambda = 5303$ Å line and in continuous light has been studied by various authors (Waldmeier and Müller, 1950; Christiansen, 1960; Waldmeier, 1959; Laffineur *et al.*, 1954; Vitkevich and Sigal, 1956 and 1957; Newkirk, 1959).

Some results of observations which permit comparison of the features of the coronal emission in the optical region and the radio-frequency band are shown in Figs. 31 and 32. It can be seen from the first figure that the radio brightness distribution at a wavelength of 21 cm above the Sun's limb follows the variations in intensity of the K-corona's glow along the line which is shown by dashes. We pointed out in section 1 that the K-component of the continuous spectrum is the result of scattering of the photospheric emission by free electrons in the corona. Since the intensity of the scattered light is proportional to the number of electrons $\int Ndl$ along the line of sight, from what has been said the connection becomes clear between the effective temperature T_{eff} of the slowly varying radio emission at $\lambda = 20$ cm and the electron concentration in the region where this emission is generated. The second figure illustrates the presence of a general correspondence between the radio brightness distribution at $\lambda = 21$ cm and the coronal isophots in the 5303 Å line.

11. Microwave Bursts

GENERAL CHARACTERISTICS. BASIC TYPES OF MICROWAVE BURSTS

Despite the fact that the bursts in the centimetric band are less varied than at metric wavelengths much less is known about them. This is chiefly because the efforts of radio astronomers have been basically directed at

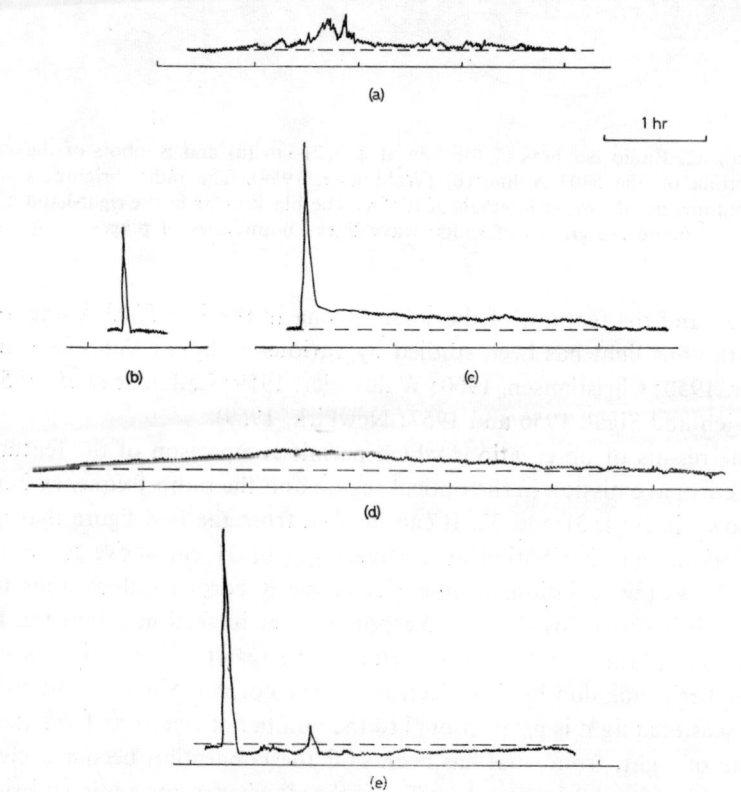

FIG. 33. Profiles of characteristic bursts of solar radio emission ($\lambda = 10$ cm (Covington, 1959))

studying metric bursts where the Sun's sporadic radio emission appears most brightly. Only in recent years has the importance of microwave bursts been recognized, particularly because of their connection with sudden ionospheric disturbances, geomagnetic storms and cosmic ray variations, bursts of gamma radiation, aurorae, etc. The limited nature of the information about this component of the solar radio emission is chiefly felt in the absence of spectrographic observations, which have become the

§ 11] Microwave Bursts

usual method of studying the sporadic radio emission at metric wavelengths. Only recently has this gap started to be filled.

The reservation must be made that the term "microwave (centimetric) bursts" must not be taken literally: it only stresses the most characteristic range in which these events occur. They are, however, also observed at millimetric and decimetric wavelengths; in a number of cases they are closely connected with sporadic phenomena in the metric band.

Profiles of typical microwave bursts are shown in Fig. 33. Although the bursts often appear in groups their component elements can apparently be discussed individually so that the grouping of the majority of centimetric bursts is not their characteristic sign. At rare periods of so-called "irregular activity" numerous relatively weak bursts of comparable intensity (Fig. 33a) appear (for several hours). The bursts which have only one maximum make up the greater majority of all events (Covington, 1951). Only a tenth have a more complex structure—usually with two, or more rarely three or more, maxima (Covington, 1951). We notice that in the clearly defined twin bursts the individual components, which are separated by intervals of the order of a minute, frequently show very similar characteristics (duration, intensity, size of the source of emission) (Kundu, 1959a).

Following Covington (1951 and 1959) (see also Kundu and Haddock, 1961; Dodson, Hedeman and Covington, 1954) we shall consider that the microwave bursts consist chiefly of elementary events of three types (A, B and C) which have the following features.

Type A bursts (simple bursts; see Fig. 33b) have the nature of a pulse, being distinguished by a rapid rise in intensity to a maximum value and a slower falling away to the pre-burst level. The ratio of the time elapsing from the beginning of the burst to the maximum to the overall duration of the burst is 0·1–0·5 with a mean value of 0·25 (Covington, 1959). The duration of a burst is 1–5 min.

Interference measurements have shown (Kundu and Haddock, 1961; Kundu, 1959a) that regions with a diameter of about 1′ are the source of weak type A bursts (with a radio emission flux less than 0·1 from the level of the "quiet" Sun); in the case of strong bursts the size of the generation region increases slightly (to 1·6′). The corresponding effective radio emission temperature is of the order of 10^6 °K for weak bursts and of the order of 10^7–10^9 °K for strong ones. The angular diameter of the source changes in the process of the development of the burst and reaches its minimum value at the time of the burst's maximum phase.

Type B bursts (post-bursts; see Fig. 33c) follow an individual simple burst or a group of such bursts. Their duration varies between a few minutes to a few hours. The generation region is fairly large in size (more than

2·5–3′ in diameter); the effective temperature of the source of emission is from 10^5 to 10^7 °K (Kundu and Haddock, 1961).

Type C bursts (gradual rise and fall; Fig. 33d) are distinguished by a slow rise in intensity to a maximum value and a slow fall to the preburst level. The ratio of the duration of the first period (from the beginning of the burst to its maximum phase) to the life of the burst averages 0·4 (Covington, 1959). The duration of these bursts varies between 10 minutes or so and several hours. The sources of the bursts are localized in regions of small diameter (less than 1′, on an average 0·8′) with an effective temperature of $T_{eff} < 10^6$ °K (Kundu and Haddock, 1961; Kundu, 1959a).

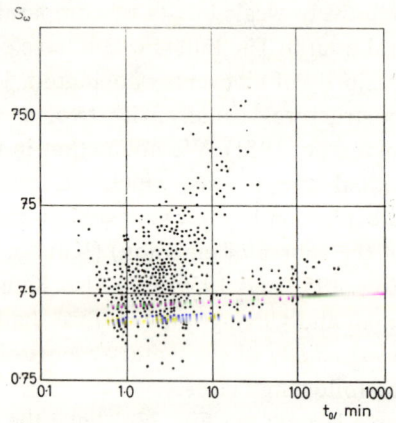

Fig. 34. Intensity–duration diagram for bursts having one maximum ($\lambda = 10$ cm) (Covington, 1959). (S_ω in relative units)

Type A, A–B and C bursts may be superimposed on each other or exist separately.

From the brief characteristics of microwave bursts given above it is clear that their most important feature, which allows us to draw fairly clear dividing lines between the basic types of burst, is (as well as the profile) the size of the radio-emission source.

The difference between the simple bursts (type A) and the phenomena of gradual rise and fall (type C) is easy to see in the diagram in Fig. 34, where each dot characterizes the intensity and duration of an individual burst. The diagram shows the tendency to an increase in the duration of the bursts as their intensity rises. All the points divide clearly into two branches: the top branch corresponds to the stronger and shorter type A bursts and the bottom one to lengthy and weaker type C phenomena. The two branches merge in the region of moderate intensity and life values.

§ 11] Microwave Bursts

The continuation of the arm corresponding to type C bursts in the intensity–duration diagram can obviously be the S-component which has much in common with the "gradual rise and fall" type of phenomena: close values of the effective temperature and position on the Sun's disk, a similar profile in observations at a fixed frequency (with the exception of the time scale) and similar polarization. This allows us to assume (Covington, 1959) that the type C bursts are the radio emission of short-lived coronal condensations, whilst the S-component is generated in regions of high density above centres of activity that last for a long time (for further detail see sections 29 and 30).

Unlike the type A–B phenomena, when the level of radio emission remains enhanced for some time after a simple burst, sometimes following the burst a temporary decrease is observed in the radio emission flux when compared with the pre-burst level (Covington, 1959). An example of this kind of phenomenon, which is connected in all probability with some sort of absorption process in the path of the radio wave propagation, is shown in Fig. 33e. In addition, on the radio emission intensity recordings of about a third of all bursts we find (Covington, 1951) weak fluctuations with a period from 1 sec (the time constant of the apparatus in the observations) up to several minutes. These fluctuations do not significantly alter the profile of a burst, sometimes appearing only in the period of rise or fall of the radio emission or grouping in the region of the maximum; they can therefore be ignored in the classification of the bursts.

The microwave outbursts, which have a very complex structure, are distinguished by a very high intensity of radio emission: the flux at the maximum of a burst at $\lambda = 3$ cm is generally higher than 10^{-20} W m^{-2} c/s^{-1}. The intense bursts appear more rarely than the weaker ones, so the frequency of the appearance of bursts at centimetric wavelengths drops as the radio emission flux S_ω rises approximately as $S_\omega^{-3/2}$. The outbursts generally consist of type A surges, having several maxima of comparable intensity; a post-burst of type B follows these peaks as a rule. Sometimes the appearance of an outburst is preceded by the appearance of one of type A or C bursts called a precursor (see in this connection Dodson, Hedeman and Covington, 1954). After the precursor the intensity rapidly (in 1–3 min) reaches its maximum value. The rapid rise and the short time during which the level of the radio emission is close to its maximum value indicates the pulse nature of the origin of the outbursts. On the other hand, the intensity during the fall after the burst decreases very slowly and reaches its normal level in a period varying between 10 min and several hours (Kundu and Haddock, 1961).

The sizes of the regions where the outbursts are generated are, generally speaking, characteristic of types A and B: at the time of maximum intensity (type A) the diameter remains between 1·5′ and 2′ (1·6′ on the average); after the maximum the size of the source increases to values of more than 3′ which is due to a transition to type B emission. Sometimes, however, complex bursts appear, the size of whose generation region does not alter despite sharp variations in intensity. The effective temperature of the outbursts is generally higher than 10^7 °K and may reach 10^9 °K at $\lambda \sim 10$ cm (Kundu and Haddock, 1961; Mullaly and Krishnan, 1963; Kundu, 1959a).

Sometimes the duration of the outbursts is very short (about 2–3 min in the 3-cm band); their intensity rises unusually rapidly to its maximum value, so the type A pulse period lasts for about a minute. The type B burst is either completely absent or is considerably shorter than in the case of ordinary outbursts. The rare short outbursts are generated by sources of a considerable size (more than 2·5′), whilst the angular diameter of ordinary outbursts is not more than 2·5′. The effective temperature of the rare short outbursts is of the same order as for the other outbursts.

The event of 20 March 1958 (Kundu and Haddock, 1961; Denisse, 1959b) can serve as an example of bursts of the type under discussion. The burst was exceptionally short: its intensity reached a maximum value in a time of less than 1 sec and then began to decrease exponentially with an attenuation constant of about 20 sec. The burst of radio emission was accompanied by a pulse of gamma-radiation. This indicates a very close connection between the phenomena under discussion.[†] The diameter of the radio emission source was very great ($\sim 5′$ at $\lambda = 3$ cm and $\sim 10′$ at $\lambda = 10$ cm). Since a source of such size appeared in a time of about 1 sec its appearance is connected with motions whose velocity is comparable with the velocity of light. The burst in question was one of the very strongest; the frequency spectrum was distinguished by a rapid decrease in intensity at the long-wave end: whilst at $\lambda = 3$ cm the emission flux exceeded 9×10^{-20} W m^{-2}c/s^{-1}, and at $\lambda = 10$ cm it exceeded $3·5 \times 10^{-20}$ W m^{-2} c/s^{-1}, at $\lambda = 21$ cm the flux was less than $2·5 \times 10^{-21}$ W m^{-2}c/s^{-1} and at $\lambda \gtrsim 60$ cm the outburst was not found at all.

It must be stressed that the essential difference between the ordinary and the short outbursts is their duration and the size of the sources. In addition the former occupy a broad frequency spectrum right up to metric wavelengths, whilst the spectrum of the latter is quite sharply limited to the low-frequencies.

---

† See section 17 for the connection between microwave outbursts and hard radiation and gamma-radiation in particular.

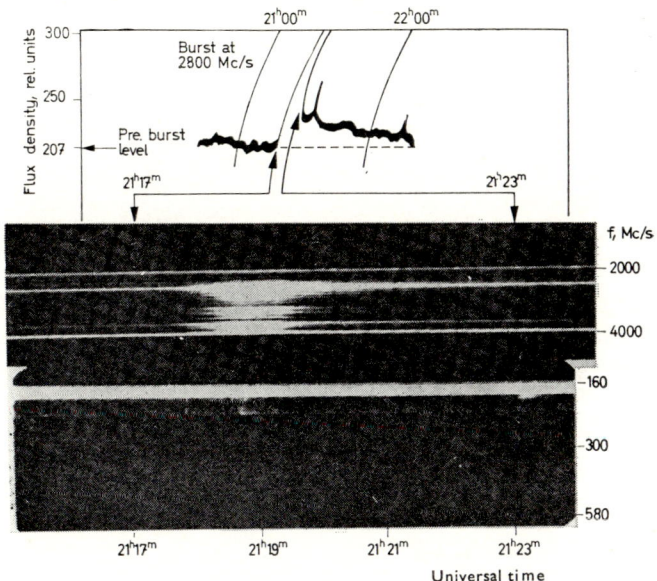

Fig. 35. Dynamic spectrum of the outburst of radio emission of 29 July 1959 in the 2000–4000 Mc/s band. At the top is the profile of the outburst at a frequency of 2800 Mc/s (Kundu and Haddock, 1961)

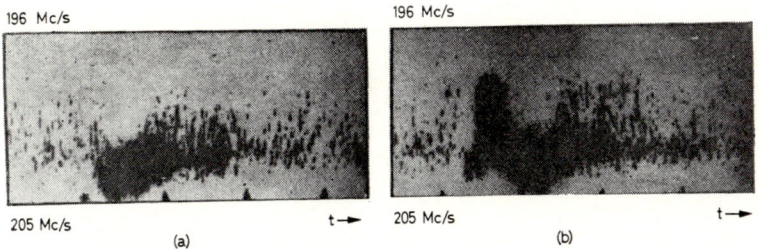

Fig. 43. Dynamic spectra of individual type I bursts: (a) drift towards low frequencies; (b) drift towards high frequencies (Elgarøy, 1959)

§ 11] Microwave Bursts

FREQUENCY SPECTRUM OF BURSTS

Figure 35 shows the first dynamic spectrum of an outburst obtained on a radio spectrograph in the 7·5–15 cm band (Kundu and Haddock, 1961). The same figure shows the recording of the burst's intensity while observing at a wavelength of 10·7 cm and the dynamic spectrum of type III bursts which occurred at the same time at metric wavelengths. The microwave outburst consisted of an intense type A pulse on which was superimposed another peak of low intensity. It was followed by a type B burst, during which one more type A simple burst appeared. The spectrum of the type B radio emission cannot be seen in the figure because of its low intensity; the spectrum of the type A burst indicates that this emission occupies a broad spectrum of frequencies. Type III bursts at metric wavelengths appeared at the time of the maximum phase of the microwave outburst; however, the nature of the dynamic spectra and the duration of these bursts are completely different. It is clear, therefore, that the type III bursts are not a simple continuation of the spectrum of the microwave outbursts into the metric waveband. The connection of the microwave outbursts with sporadic phenomena at metric wavelengths will be discussed in greater detail in sections 15 and 16.

Interesting data on the frequency spectrum of bursts of radio emission have been obtained by Hachenberg and Wallis (1960), Hachenberg (1960) at seven fixed frequencies in the 1·25–60 cm range. Ten bursts of different types according to Covington's classification were chosen from a large number of events that had been observed for the analysis. It turned out that the continuous spectrum of the microwave outbursts starts from about 30–60 cm, although in the case of large bursts the spectrum sometimes also includes the metric band.† The rise in intensity with the decrease in wavelength for the majority of events investigated by Hachenberg and Wallis (1960) can be described by the relation $S_\omega \sim \lambda^{-2}$. However, bursts with both steeper and shallower spectra than a quadratic one have been observed. Starting at 3–15 cm the rise in the intensity of the radio emission noted above often (but not always) changes by an interval in which the intensity is in practice independent of the frequency right up to the short-wave limit of the band under study (up to $\lambda = 1\cdot25$ cm). An example of a burst having a spectrum of this kind is shown in Fig. 36.

In certain cases the impression is created that on the radio emission

† The solar activity in the metric band generally speaking differs considerably from that observed at shorter wavelengths. According to Takakura (1959) the features of the bursts change in the 50–100 cm transition region. In this range the averaged frequency spectrum has a minimum: the radio emission intensity rises both at metric wavelengths and with the transition into the centimetric band.

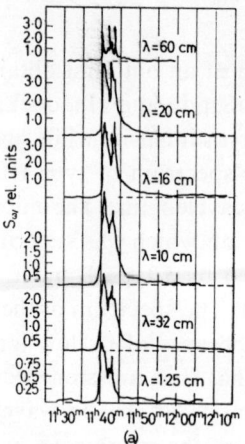
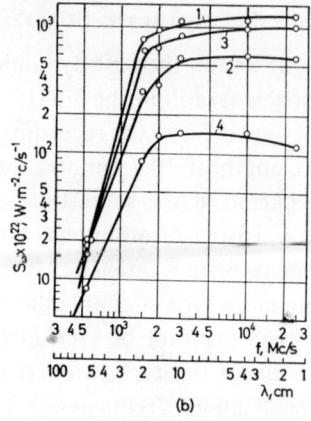

FIG. 36. Profiles (a) and frequency spectrum (b) of the burst of radio emission of 18 April 1959: 1—at the time of the maximum ($11^h 40^m$); 2—at the time of the minimum ($11^h 41\cdot 6^m$); 3—at the time of the maximum ($11^h 43\cdot 3^m$); 4—during the post-burst decay ($11^h 45^m$) (Hachenberg and Wallis, 1960)

with a spectrum of the type shown in Fig. 36 there is superimposed broadband emission with a clearly defined maximum of spectral intensity located at frequencies where the rise in emission alternates with a region of constant energy (Fig. 37). Sometimes the presence of the maximum of this broad-band emission affects the frequency spectrum in the interval where the intensity rises with the frequency.

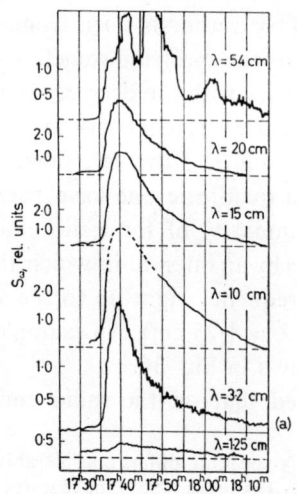
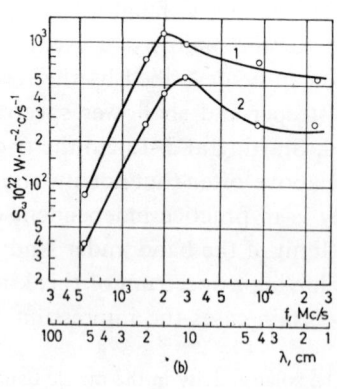

FIG. 37. The same as in Fig. 36 for the burst of 9 June 1959: 1—at the time of the maximum ($17^h 40^m$); 2—at the time of attenuation ($17^h 50^m$) (Hachenberg and Wallis, 1960)

§ 11] Microwave Bursts

It should be noted that the types of spectrum given do not always belong to different bursts: the nature of the frequency spectrum in the process of development of a complex burst may change. Unfortunately it is still not clear whether the features noted are simply variations of the frequency spectrum of a single sporadic phenomenon or whether we are dealing here with two different (although closely connected) forms of activity. The arbitrary position of the intensity maximum on the frequency scale in relation to a spectrum of the type in Fig. 36, and the appearance and disappearance of the maximum during the existence of certain complex bursts (with a fairly stable spectrum as a whole) are some argument in favour of the latter assumption.

In observations at three frequencies close to $\lambda \sim 3$ cm separated from each other by 500 or 170 Mc/s (Dravskikh, 1960a, 1960b) it turned out that during the development process of the majority of bursts the slope of the spectrum changes considerably. On the other hand, in measurements of the intensity of bursts in the 10-cm band at two frequencies 150 Mc/s apart no significant differences were found (Covington, 1958) in the form of the bursts and only in some cases were different intensities observed at the two wavelengths. The different results at 3 and 10 cm possibly reflect the actual differences in the nature of the frequency spectrum of the bursts at these wavelengths. It is not excluded, however, that the data of Covington (1958) can be explained by the lower accuracy of the method used for comparing the intensities at the different frequencies.

In the case when in the observations at the three frequencies in the 3 cm band the central frequency was 9850 Mc/s (i.e. corresponded to the $\lambda = 3\cdot04$ cm line connected with the transition between the fine structure levels of neutral hydrogen) Dravskikh (1960a) sometimes noticed both rises and falls in intensity at the central frequency when compared with the values at the side frequencies. The difference between these intensities as they varied during the development of the bursts at times was as much as 10%. At the same time, this effect was not noticed in the case when the central frequency differed from the line frequency by 1000 Mc/s or in the observations outside the bursts. The results given seem to indicate the existence of a hydrogen line in the spectrum of bursts of solar radio emission. More systematic observations must be made in this band, however; only after observations of this kind can the question of the existence of the hydrogen $\lambda = 3\cdot04$ cm radio line in the Sun's spectrum be finally solved.

Investigations at a number of fixed frequencies (Hachenberg, 1960) also permit definite conclusions to be drawn about the features of the dynamic spectra of microwave outbursts. Simple bursts or individual maxima in complex bursts whose frequency spectrum has an interval where the intensity is not frequency-dependent also preserve this feature during the decay of the bursts. If at the period of the maximum phase of the burst the frequency spectrum rises steadily with the increase in frequency, during the decay the spectrum becomes flatter (right up to the appearance in the short-wave part of the spectrum of the region where the intensity is not frequency-dependent). A burst whose spectrum has a maximum loses this feature in the decay period.

Above we have been speaking about the frequency spectrum of microwave bursts at centimetric and partly at decimetric wavelengths. In the millimetric band the study of bursts is still only just starting; the number of events observed here is literally one. Bursts at $\lambda < 1$ cm are recorded more rarely and their intensity is generally lower than at the longer wavelengths (Coates, 1960; Edelson and Grant, 1960; Edelson et al., Coates, 1960; Hagen and Hepburn, 1952).

The study of two bursts at a wavelength of 8 mm with "pencil-beam" aerials (beam width about 2′) permitted Salomonovich not only to measure the total radio emission flux but also to estimate the sizes of the sources (Salomonovich, 1960). It turned out that the extent of the generation region of one burst was apparently not more than 1′ and of the other was about 1·5′. The values of the radio emission flux correspond (allowing for the sizes given) to an effective temperature of about 5×10^5 and 10^6 °K.

POLARIZATION OF RADIO EMISSION

Some of the first observations during which polarization of microwave bursts was discovered were made by Covington (1951). According to data at $\lambda = 3$ cm (Kundu, 1959a) the emission of two-thirds of all bursts studied are partially circularly or elliptically polarized. The degree of polarization varies from a few percent to 30% and only in exceptional cases is as much as 50%. Sometimes even quasilinear polarization was noted; in another series of observations (Tanaka and Kakinuma, 1959), however, at $\lambda = 8$ cm with a polarimeter whose band was 10 Mc/s linear polarization was not found in a single burst among more than a hundred bursts studied.

The nature of the polarization of type B bursts is similar to the nature of the polarization of the preceding simple bursts. In the majority of cases type C bursts are partially circularly polarized similarly to the S-component (Kundu and Haddock, 1961).

The polarization of outbursts is typical for types A and B events (Covington, 1951a). A study of the radio brightness distribution over the sources of outbursts has shown (Kundu and Haddock, 1961; Kundu, 1959a) that

§ 11] Microwave Bursts

the generation area consists of two regions: a comparatively small sector with a low intensity and a long life responsible for the creation of polarized emission, and an extensive region with a high, strongly varying intensity and a short life which makes no noticeable contribution to the polarized part of the radio emission of the outburst.

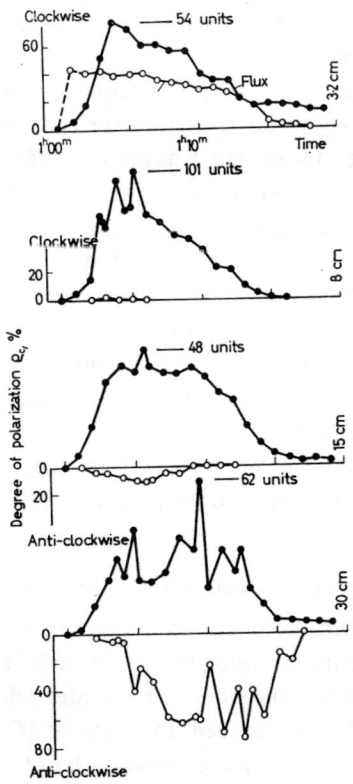

FIG. 38. Radio emission flux and degree of polarization of burst of 15 July 1957 at different frequencies (Tanaka and Kakinuma, 1959)

The simultaneous measurements of circular polarization at four wavelengths (3·2, 8, 15 and 30 cm) made by Tanaka and Kakinuma (1959) and Kakinuma (1958) showed that the number of polarized bursts is over 90%. At the wavelengths of 15 and 30 cm polarized bursts appear comparatively rarely but even at the shorter wavelengths the degree of circular polarization generally does not exceed 40%. In 70% of polarized bursts the magnitude and sign of the polarization vary with the frequency. This dependence is illustrated by Fig. 38, which shows the values of the radio emission flux and the degree of circular polarization ϱ_c at different wavelengths during the development of the burst. It can be seen from the figure that the degree of polarization at the wavelengths of 8 and 15 cm is small and the sign of the polarization is different, i.e. the direction of rotation changes in the 8–15 cm band. Any change of sign outside this range is rarely observed. Sometimes the polarization changes in the range 3–30 cm not once, but twice.

A change in the sign of the rotation with time is observed far more rarely than a change with frequency and occurs chiefly at the longer wavelengths (15 and 30 cm).

Bursts with the same sign of polarization reveal (more vaguely, it is true) the same tendency to distribution over the Sun's disk as the sources of the S-component (Tanaka and Kakinuma, 1959). If we remember that the polarization of the latter corresponds to an extraordinary wave (see section 10) and that the sources of the bursts are as a rule localized in the regions where the S-component is generated (Kundu, 1959b), it becomes clear that the sense of the polarization of bursts at high frequencies corresponds preferentially to an extraordinary wave. This is also confirmed by polarization observations of bursts and long-lived localized sources at $\lambda = 3$ cm (Gelfreich et al., 1959) at a period when a single active region predominated on the Sun. In this case the sign of the rotation was the same for the bursts and the S-component and corresponded (when compared with the resultant polarity of the magnetic field of the groups of spots) to an extraordinary wave.

MICROWAVE BURSTS AND CHROMOSPHERIC FLARES

The regions in which the majority of bursts are generated reveal no noticeable displacement; the velocity of motion of a source of one of the bursts at $\lambda = 3$ cm, when measured on a radio telescope with a "knife-edge" beam ($\sim 2'$), did not exceed 150 km/sec (Gelfreich et al., 1959). A similar position obtains in the decimetric band as well. The interferometer observations (Kundu, 1959b), however, indicate the existence in certain (very rare) cases of the systematic displacement of the source at higher velocities of the order of 1000 ± 300 km/sec. It is not excluded that the change noted above in the sense of the polarization during the development of a burst is connected with movements of this kind of the radio emission source in the magnetic fields of the active centres.

As has already been pointed out, the sources of the microwave bursts are localized in the regions of the Sun's disk from which the S-component of the radio emission originates.† A close connection is also known,

† According to data at $\lambda = 21$ cm (Mullaly and Krishnan, 1963) the centres of sources of a burst and the S-component do not generally differ by more than $0 \cdot 5'$ in magnitude. The sizes of sources of bursts vary between $1 \cdot 5'$ and $5'$; sometimes they reach the size of the local sources of the S-component but they are never larger.

§ 11] Microwave Bursts

between the intensity of this component and the probability of the appearance of centimetric bursts. At a time of high activity, however, very weak type A bursts also appear in regions which are not sources of the S-component (Kundu, 1959a and 1959c).

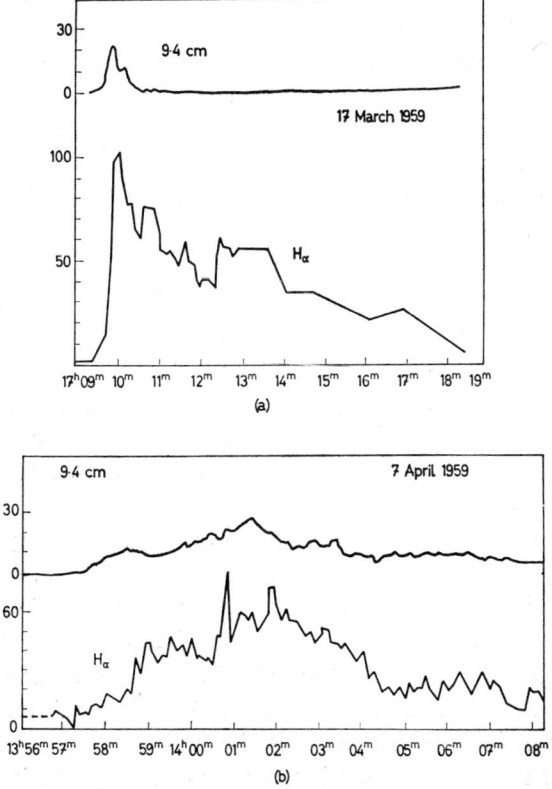

Fig. 39. Radio emission intensity at $\lambda = 9\cdot 4$ cm and brightness in the H_α line for two bursts and the flares connected with them (in percentages of the level of the radio emission and brightness existing before these events). The strong fluctuations in the H_α line in the second figure are probably caused by the effect of the Earth's atmosphere (Edelson *et al.*, 1960)

Microwave outbursts are closely connected with chromospheric flares (Appleton and Hey, 1946). It can be stated that all outbursts are accompanied by flares and that solar flares whose strength is greater than 2 (i.e. 3 and 3+) are accompanied by bursts in 75% of the cases (Kundu, 1959c). This connection becomes weaker for less intense bursts and flares (see Dodson, Hedeman and Covington, 1954; Hachenberg and Krüger, 1959).

The beginning of the flare that can be seen in the H_α line almost coincides with the beginning of the burst at $\lambda \sim 10$ cm, generally differing from it by not more than 1–2 min. Moreover the time of the flare's maximum brightness occurs slightly later than the maximum intensity of the burst at $\lambda \sim 10$ cm (Dodson, Hedeman and Covington, 1954; Edelson et al., 1960, 1959). For certain flares, in particular for those in which the brightness increases very rapidly, the times of the maximum phases almost coincide with the corresponding times for the bursts. It must be noted, however, that bursts of the "gradual rise and fall" type do not satisfy the above relation: these events can have a maximum either at the period of the greatest brightness of the H_α-burst or much later (Dodson, Hedeman and Covington, 1954). The events in the radio band whose maximum phase lags behind the period of a flare's maximum brightness are about 5% of all type C bursts (Edelson et al., 1960, 1959).

The time that type A bursts end comes at a period near the maximum of the H_α radiation. Remembering that the beginning of the bursts almost coincides with the beginning of the flares, it follows from what has been said that type A bursts occur at the time that the flares are increasing in brightness. Complex bursts frequently continue after the flare maximum as well. The duration of type B and C bursts is equal to or greater than the duration of the flares connected with them (Dodson, Hedeman and Covington, 1954).

The variations in the optical brightness of flares are very weakly connected with the variations in intensity of the type A bursts. A better agreement is sometimes observed between the curve of flare brightness and the profile of a type C burst. As an illustration Fig. 39 shows the functions of the radio emission intensity at $\lambda = 9.4$ cm and the brightness of the flares in the H_α line for two bursts: burst (a) consists of type A phenomena and burst (b) can be included among type C events.

The size of the flares ($\sim 0.5'$) is generally less than the diameter of outbursts and comparable with the size of regions where type C bursts are generated (Kundu, 1959a).

On the other hand, there is a very clear connection between the radio emission flux in a microwave burst and the width of the H_α emission line in the flare (Hachenberg, 1960). It must be pointed out that detailed agreement is hardly to be anticipated here (particularly because the measurements of the radio emission flux characterize all the generation region as a whole, whilst the spectrometric investigations in the monochromatic lines generally relate to individual regions of the flare in which the variation of the physical conditions with time may be different). Nevertheless, a close analogy between the profile of microwave bursts and the nature of the variation in the width exists for the great majority of events studied. The best examples of this agreement are shown in Fig. 40. It is clear from the figure

§ 11] Microwave Bursts

that there is agreement for the three elementary types of bursts and for bursts with a complex profile.

Although the statistics are still clearly insufficient the data given by Hachenberg (1960) indicate that the increase in the maximum width of the

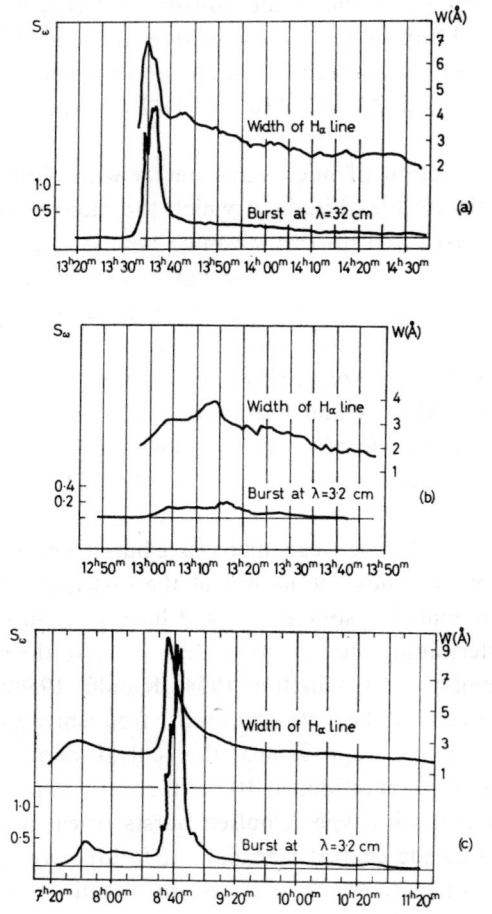

FIG. 40. Profiles of bursts at $\lambda = 3.2$ cm $S_\omega(t)$ and time-dependence of width W of H_α line in the chromospheric flares accompanying the bursts: (a) burst of 21 September 1957 type A–B (strength 2 flare); (b) type C burst of 7 July 1957 (strength 2+ flare); (c) outburst with complex profile of 3 July 1957 (strength 3 flare) (Hachenberg, 1960)

H_α line achieved during the flare occurs as the intensity rises at the burst maximum. In this case flares with a line width less than 3·5 Å have no noticeable effect in the centimetric band; starting at 3·5 Å the maximum intensity of the radio emission increases as the H_α line broadens. Further investigation is necessary for more definite conclusions.

There could be considerable interest in investigating the connection of microwave radio emission with the optical radiation of chromospheric flares in the continuous spectrum. Unfortunately in the continuous spectrum (against the background of the photosphere's intense emission) flares are observed very rarely; in one of the known cases (the flare of 31 August 1956) the spectral flux of the radio emission at 3 cm ($2 \cdot 3 \times 10^{-19}$ W m^{-2} c/s^{-1}) was one or two orders less than the corresponding value in the optical band at $\lambda = 5000$ Å ($2 \times 10^{-17} - 3 \times 10^{-18}$ W m^{-2} c/s^{-1}) (Hachenberg, 1960).

The close connection of microwave bursts with chromospheric flares allows us to estimate the altitude at which the radio emission generation regions are localized. To do this we must compare the positions of the source of the radio emission and the chromospheric flare on the Sun's disk, assuming that the source is located exactly above the flare. For two bursts at $\lambda = 3$ cm this led to values of the order of $(0 \cdot 1 - 0 \cdot 2)R_\odot$ above the photosphere (Gelfreich *et al.*, 1959) and for bursts at $\lambda = 21$ cm of not more than $0 \cdot 1 R_\odot$ (Mullaly and Krishnan, 1963). In the latter case we can also state that the altitude of the region in which the bursts are generated does not differ from the corresponding altitude for the S-component by a value greater than 2×10^4 km.

The connection noted between microwave bursts and optical flares also makes it possible to study the nature of the distribution over the Sun's disk of the radio emission sources without having recourse to using radio methods for determining their position. According to the preliminary data (Dodson, Hedeman and Covington, 1954; Kundu, 1959b) type C bursts and relatively large type B events occur more frequently and their intensity is higher when the flares appear near the central meridian. This effect is more noticeable in observations at 10 cm than at 3 cm. At the same time the flares connected with type A pulsed bursts reveal no noticeable concentration towards the central meridian. It follows from what has been said that types B and C bursts have certain directional properties which increase with the wavelength, whilst the type A radio emission has no such directional properties.

The directional nature of the bursts is quite noticeable in the decimetric band, being possibly caused not by a decrease in the "visible" area of the source as it moves away from the central meridian (as was the case with the S-component). In any case in observations (Mullaly and Krishnan, 1963) at $\lambda = 21$ cm the sources of bursts near the limb were of the same size as the sources of bursts in the central part of the disk.

Measurements of the radio emission flux and the coordinates of sources of bursts (Mullaly and Krishnan, 1963) indicate the clear predominance of bursts in the western part of the Sun's disk. The latter is obviously connected with the fact that the polar diagram of the radio emission is not symmetrical with respect to the vertical to the Sun's surface (possibly because of asymmetry of the magnetic field H_0 in the centre of activity which shows up particularly in the difference between the values H_0 of the leading and following spots; see section 2).

12. Noise Storms (enhanced radio emission and type I bursts)

With this section we start to discuss the sporadic phenomena characteristic of metric wavelengths, i.e. of the band in which the radio emission of the "disturbed" Sun appears in its most developed form (in the sense of intensity, variety of phenomena and their complexity).

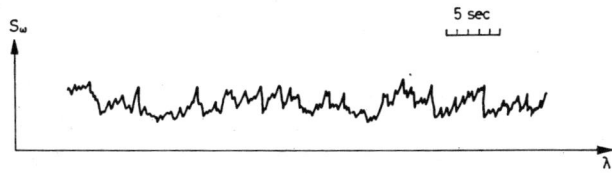

FIG. 41. High-speed recording of solar radio emission flux during a noise storm (Fokker, 1960b)

As was indicated in the introduction to this chapter, the term "noise storm" generally combines two closely interconnected events: slow rises in radio emission intensity with a characteristic time of the order of hours and days appearing over a wide range of frequencies of the order of hundreds of megacycles, and numerous brief bursts of a duration of the order of seconds and fractions of a second following rapidly upon one another. It should be pointed out that there are bursts of two different types (I and III) among surges of this kind. Noise storms containing type III bursts are rarer events than type I storms; they will be discussed in the section devoted to type III radio emission (section 14). Here we shall limit ourselves to type I storm although burst storm of types I and III are not always observed independently: they are sometimes superimposed on one another so that spectrographic studies are necessary for an exact classification of the individual bursts in this case (Wild, 1957).

Type I bursts are undoubtedly the most frequent sporadic phenomenon in the radio band (hundreds and thousands of bursts per hour during a noise storm). Their distinguishing feature is that each burst appears and stays for the whole of its "life" in a narrow frequency range whose width may be much less than the radio emission frequency.

Time Characteristics of Noise Storms

As a rule enhanced radio emission can be observed for several hours or days, its intensity revealing considerable changes in this time. The duration of individual bursts can be determined from recordings of the radio emission intensity obtained with high-speed devices having a small time constant. An example of one of these recordings is shown in Fig. 41. Type I surges show a noticeable tendency to collect in groups of two to three bursts; even more numerous groups are characteristic of this form of radio emission, however (De Jager and van't Veer, 1957). The interval between two successive bursts in a group is about 0·5 sec on the average (Elgarøy 1959, 1957). The profile of approximately half the bursts is symmetrical relative to the intensity maximum; in a third of the bursts the time that the intensity takes to rise is less than the time it takes to fall; in the remaining bursts the ratio between these time intervals is the opposite (Vitkevich and Gorelova, 1960). Sometimes bursts with two maxima appear (De Groot, 1960).†

According to the data of Vitkevich and Gorelova (1960) the duration of a type I burst varies approximately between 0·2 and 1·5 sec, averaging about 0·6–0·8 sec (from the zero level) at wavelengths of 2–4 m. No reliable dependence of the "life" on the wavelength in this range is observed.

Previously (Vitkevich, 1956a), however, larger values were given for the duration of bursts which increased as the wavelength rose (from 2 to 6 sec at wavelengths of 1·5–4 m). These data are clearly too high and relate to the duration of groups of bursts rather than individual surges; this is probably connected with the fact that the individual surges were not distinguished in the recording of the radio-emission's intensity.‡ What has been said is confirmed to a certain extent by earlier observations (Reber, 1955), according to which the duration of the burst in seconds is approximately equal to the wavelength in metres; the value for the "life" obtained from this dependence is, as can easily be seen, close to the values given by Vitkevich (1956a). In addition, in observations using receivers with a low time constant and a high recording speed at wavelengths from 0·6 to 6 m (Reber, 1955),

---

† At the same time attempts to find at $\lambda = 1\cdot5$ m a "radio echo" due to reflection in the solar corona (in the form of weak surges following strong type I bursts with a shift of a few seconds) proved unsuccessful (De Jager and van't Veer, 1957). Among the surges separated by an interval of 0–5 sec from several hundred strong type I bursts selected for analysis the number of surges preceding these bursts and following them was approximately the same. A "radio echo" could have been observed here if its intensity had been not less than 0·1 of the intensity of the type I bursts studied.

‡ The possibility is not excluded, moreover, that the results of Vitkevich (1956a) were affected by longer type III bursts which were not separated from the type I bursts in the analysis.

§ 12] Noise Storms

it was found that the individual "bursts" with a life of t_0 (sec) $\approx \lambda$ (m) consist of a series of surges whose duration is only tenths of a second. Similar results have been obtained (Blum, Denisse and Steinberg, 1951) at a wavelength of 1·8 m; surges fractions of a second long appear usually in groups whose duration is 2–3 sec.

Amongst the data on the "life" of type I bursts note should be taken of the results of Elgarøy (1957 and 1959) who established that the duration of the greater majority of bursts at $\lambda = 1·5$ m determined from the half-intensity level is between 0·1 and 0·5 sec; bursts lasting about 0·25 sec are most often found. In general outline this agrees with the earlier results of De Jager and van't Veer (1957), according to which the duration of the bursts at this wavelength (at the $1/e$ level) is 0·4–0·6 sec (sometimes up to 1 sec); the distribution of the number of bursts by their duration shows strong daily changes. At the shorter wavelengths the "life" of the bursts is shorter and is 0·15–0·25 sec (averaging about 0·2 sec) at $\lambda = 75$ cm and 90 cm. No bursts were noted that lasted less than 0·05 sec. It is significant that the duration of the surges does not apparently depend on their intensity.

FREQUENCY SPECTRUM

Noise storms are observed chiefly at metric wavelengths; in the decimetric band they are rarer and less intense (Firor, 1959b). At present the decimetric band is little studied: on the basis of the observations of Boishot, Lee and Warwick (1960), Blum, Denisse and Steinberg (1951), Erickson, (1961), Benediktov and Getmantsev (1961) we can say that noise storms are a rarer phenomenon there than in the metric band, although at times their intensity exceeds even the level of the radio emission in the latter. The intensity of the enhanced radio emission generally has a maximum at wavelengths of 2–4 m; the range of frequencies occupied rarely exceeds 250 Mc/s (Maxwell, Swarup and Thompson, 1958). The radio emission spectrum does not remain constant as a noise storm develops but changes noticeably in a few tens of minutes. The latter can be clearly seen in Fig. 42 which shows a dynamic spectrum of a noise storm (enhanced emission and type I bursts) in the 50–200 Mc/s band (Wild, 1951 and 1957).

When a noise storm starts the solar radio emission flux in the metric band generally increases by a factor of 10^2–10^3 or more. Since the effective temperature of the "quiet" Sun at these wavelengths is about 10^6 °K the $T_{\text{eff}\odot}$ for a noise storm at times reaches 10^8–10^9 °K and more (Wild and Sheridan, 1958; Goldstein, 1959; Maxwell, 1951). Such a high intensity, however, is characteristic only of certain noise storms at a period of high solar activity; near a solar activity minimum emitting centres appear more rarely on the Sun and the radio emission connected with them is generally comparable in magnitude of flux with the radio emission of the "quiet" Sun.

The relative intensity of the bursts and the enhanced radio emission, just like the frequency of appearance of bursts, vary within wide limits: the peak intensity of the bursts may be less than, comparable with or greater than the level of the enhanced radio emission. Type I bursts appear even in the absence of enhanced radio emission (Vitkevich, 1956a; Wild, 1951).

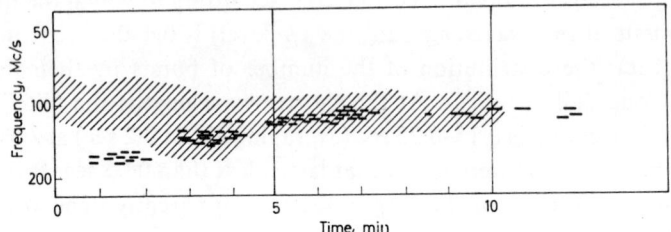

Fig. 42. Idealized dynamic spectrum of a noise storm—enhanced radio emission (diagonal strokes) and type I bursts (Wild, 1957)

Just like the enhanced radio emission the type I bursts appear most often at metric wavelengths; their intensity and frequency of appearance decrease as one moves into the decimetric band. Sometimes there is a sharp cut-off of the region where the bursts are generated at the long-wave end: cases have been noted (Vitkevich, 1956b) when intense bursts have appeared at $\lambda < 3$ m but have been absent at $\lambda > 3$ m. This feature is confirmed by spectrographic observations (Wild, 1957) according to which series of type I bursts often appear in a limited band of frequencies. It is very interesting that this band sometimes moves to lower frequencies as the burst storm develops and sometimes moves to higher frequencies. The frequency drift of a type I burst storm can be seen in Fig. 42.

The characteristics of the dynamic spectra of individual type I bursts have been studied by Elgarøy (1957, 1959), de Groot (1959, 1960), Vitkevich and Gorelova (1960).

As has already been pointed out, the distinguishing feature of type I bursts is the fact that each individual burst appears and stays for the whole of its "life" in a narrow spectral range. To judge from observations at several adjacent frequencies in the metric band (De Jager and van't Veer, 1957; Vitkevich, 1955, 1956b and 1957b) this range is generally 3–6 Mc/s; for the more intense bursts it increases to 12 Mc/s (Vitkevich,

§ 12] Noise Storms

1956b). According to later data (Vitkevich and Gorelova, 1960), however, the intensity and also the duration and profile asymmetry of wide-band bursts occupying a band from 6 to 26 Mc/s (on the average 13 Mc/s) are the same as for narrow-band bursts (<4 Mc/s).

The same band of frequencies is also occupied by type I bursts appearing at wavelengths shorter than 1 m: investigations with an eight-channel receiver in the 382–396 Mc/s and 324–336 Mc/s bands showed (De Groot, 1960; see also De Groot, 1959) that in 30% of the cases the frequency spectrum of an individual burst is gaussian in form with a mean width of 7 ± 2 Mc/s at the half-intensity level. The width of the spectrum does not show any radical variations during the "life" of a burst. The form of the frequency spectra of the remaining bursts is very irregular; the width of the spectrum is about 12 Mc/s. These bursts can clearly be looked upon as the result of the superposition of "simpler" bursts with a gaussian form of spectrum.

A comparison of the times of maximum intensity of bursts at two adjacent frequencies (199 and 200·5 Mc/s) showed (Elgarøy, 1957 and 1959) that about half of all the bursts develop first at the lower frequency and about a third at the higher one; the rest of the bursts appear simultaneously at both frequencies. The average time displacement is about 0·04–0·1 sec. The observed difference in the position of the bursts on the time scale depending on the frequency indicates the existence of frequency drift of individual type I bursts at a rate greater than or of the order of 15 Mc/sec^2. The frequency shift during the development of individual type I bursts can be found in dynamic spectra (Fig. 43) obtained with a radio spectrograph (Elgarøy, 1959; see also De Groot, 1959; Vitkevich and Gorelova, 1960). The ratio of the number of bursts appearing first at a lower and at higher frequencies varies during the development of a noise storm. This ratio, at least in certain cases, gradually increases, passing through unity at the period when the source of the noise storm is close to the central meridian.

CONNECTION WITH OPTICAL FEATURES

Noise storms are closely connected with sunspots; the position of the sources of radio emission, as interferometer observations show (Chikhachev, 1956), is close to the position of groups of spots on the Sun's disk, particularly in its central part. Enhanced radio emission and type I bursts generally appear in cases when large groups of spots containing well-developed large spots can be seen on the Sun's disk; the radio emission is connected, as a rule, with the largest spot in the group (Payne-Scott and Little, 1951). The spots accompanied by noise storms generally belong to classes E and F; G-spots, and D-spots in particular, show lower activity in the radio band (Fokker, 1960b; Simon, 1957). According to different data (Chikhachev, 1956), however, the intensity of the radio emission does not appear to depend on the class of the spot. It is also still unclear whether the rate of development of a group affects its activity at radio wavelengths; the published information on this subject (Fokker, 1960b; Chikhachev, 1956) is very contradictory.

The enhanced radio emission is connected with regions where there are sunspots with an area of $\sigma \gtrsim 130$ millionths of the solar hemisphere with a magnetic field at the base of the spot of $H_{ob} \gtrsim 1300$ oe (Dodson and Hedeman, 1957). Above the groups which are accompanied by enhanced radio emission and type I bursts there are prominences of the sunspot type (Wild and Zirin, 1956) and coronal emission is observed in the 5694 Å line (Simon, 1956) (which indicates the higher coronal temperature at a centre of activity $T \sim (2-4) \times 10^6$ °K).[†]

Since above large spots there are strong magnetic fields, with whose appearance the rise in the coronal temperature above the spots is also connected, and prominences of the above type are formed by condensation of the coronal plasma (probably by the action of the magnetic fields), it becomes clear that the signs given for the transition of spots into the radio-emitting phase are reduced in essence to the necessity for the appearance of strong ordered magnetic fields in the solar corona. The reservation must be made, however, that these signs were largely established from observations at a period of minimum solar activity; at the maximum of the 11-year cycle, when the number of spots on the Sun's disk is large, it is difficult to connect the local sources of radio emission with individual optical features (Zheleznyakov, 1959a).

Since a noise storm usually lasts for several hours or days and a group of spots exists for one or two months the question arises of which events in the active region give rise to the noise storm. In this connection it should be noted that certain noise storms develop gradually and relatively smoothly; it is difficult to say definitely when this kind of storm starts. Sometimes noise storms may appear suddenly (Fokker, 1960b); in this

---

[†] A comparison of the intensity at $\lambda = 5500$ Å in the umbra of sunspots with the mean daily flux of radio emission at $\lambda = 1 \cdot 5$ m has shown (Maltby, 1959) that sunspots progress to the radio-emitting phase only if the ratio of the intensity in the spot's umbra to the intensity in the surrounding photosphere is small enough ($\lesssim 0 \cdot 2$). If we take into consideration the fairly weak correlation of the radio emission with the area of the spots studied (each of which was sufficiently large: $\sigma \gtrsim 400$ millionths of the Sun's hemisphere) and the absence of a close connection between the radiation of the umbra and the size of the spots, it becomes clear that the stated criterion is not reduced to a requirement that the size of the spot be large.

§ 12] Noise Storms

case the change of the active region into the radio-emitting phase is characterized by the appearance of a sequence of type I bursts that become more and more intense and are superimposed on a more or less rapid rise in the enhanced radio emission. In the latter case, as Dodson (1958) found the start of noise storms shows a close correlation with chromospheric flares (or with sudden ionospheric disturbances which serve as a reliable indicator of flare-type phenomena on the Sun, even when they cannot be optically observed).† Characteristics of the flares connected with the sudden onset of noise storms are a complex profile of the time-variation of the radiation and a complex structure of the isophots in the H_α line. When this kind of flares develops, a successive rise in brightness of different parts of the disc often occurs, rather than the usual rise in brightness of the same region.

Sometimes a noise storm with a sudden onset appears in the wake of a type II outburst, so the noise storm is one of the forms of post-burst phenomena connected with chromospheric flares (as well as type IV radio emission; section 15) (Dodson, Hodeman and Owren, 1953; Dodson, 1958). However, whilst type IV radio emission chiefly follows strong force 2 or 3 flares, the noise storms (their sudden onsets and repeated sharp amplifications)‡ are largely connected with relatively weak chromospheric phenomena (strength 1 and 1+ flares) (Malinge, 1960).§

A further study of the connection between chromospheric flares and the sudden onset (and repeated strengthening of activity) of noise storms has been made by Malinge (1960) from the data of radio observations at $\lambda \approx 1\cdot 8$ m (including the localization of sources of radio emission on the Sun's disk in the east–west direction with a multi-element interferometer). The results obtained confirm the presence of a close correlation between the optical and radio events in question. The time interval separating the start of a flare and the start of a noise storm is generally not more than a few tens of minutes.

Interferometer measurements (Malinge, 1960) have made it possible to determine the angular distances θ_{source} of the sources of noise storms (R-centres) from the central meridian with an accuracy of up to 1'. The

---

† Bursts sometimes occur when eruptive prominences appear moving in the corona at a velocity of 150–300 km/sec (Das, Sethumadhavan and Davies, 1953) (for type I bursts connected with the development of prominences and filaments see also Wild and Zirin, 1956).

‡ Repeated sharp increases in activity may, generally speaking, be caused either by the formation of a new noise storm source or by the strengthening of an old one. Interferometer observations at $\lambda = 1\cdot 8$ m have shown (Malinge, 1960) that both cases are possible, the first occurring more rarely than the second.

§ No correlation can be found between noise storms and very weak 1-flares (microflares) (Lougheed, Roberts and McCabe, 1957).

distribution 1 obtained for the difference $\Delta\theta = \theta_{\text{source}} - \theta_{\text{flare}}$, where θ_{flare} is the angular distance from the central meridian for the accompanying flares shows (Fig. 44) that in 75% of the cases the value of $\Delta\theta$ is not more than 6'. The fact that the position of flares on the disk is really connected with the position of the R-centres is confirmed by comparing the distribution 2 in Fig. 44 of flares accompanying the beginnings of noise storms but separated from them by an interval of not more than 2 hours with the distribution 3 for flares which appeared in an interval of 4 hours to 1 hour before the onset of the noise storms. The latter distribution differs considerably from the former in its far lower concentration towards the small values of the difference $\Delta\theta$ which is natural for flares not connected with the events in question in the radio band. We notice that the distribution 1 (only for the flares preceding the start of storms)

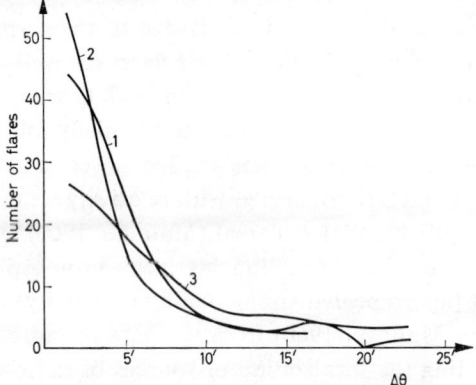

FIG. 44. Distribution of the number of chromospheric flares depending on the difference $\Delta\theta$ of the angular distances of the R-centres and optical centres from the Sun's central meridian: 1—distribution of flares preceding the start of noise storms; 2—distribution of flares both preceding and following the start of storms and separated from this time by an interval of not more than 2 hours; 3—distribution of flares from 4 hours to 1 hour before the start of noise storms (Malinge, 1960).

differs only a little from the distribution 2 which includes both the flares preceding and subsequent to the bursts. This result indicates that the accompanying events in the optical region may both precede the start of events in the radio band and lag behind it.

Data on the magnitude of $\Delta\theta$ for different values of θ_{flare} can be used to judge the position of sources of radio emission relative to chromospheric flares if one assumes that all the sources are located in the corona at the same altitude above the photosphere. It turned out that the observed position of the sources is then shifted on the average towards the central meridian in relation to the radius passing through the centre of the

§ 12] Noise Storms

accompanying flares. This effect increases for flares with large values of θ_{flare}; it is connected, in all probability, with the refraction of radio waves in the corona, as a result of which the observed position of a source of radio emission differs from the true one, being displaced towards the centre of the disk (for the path of rays in the corona see section 22).†

DIRECTIONAL FEATURES OF THE RADIO EMISSION

The radio emission intensity at the time of noise storms shows a weak correlation with the area of the spots on the Sun's disk. The connection becomes closer, however, if we take only those spots that are located near the central meridian (Hatanaka and Moriyama, 1952). The strong dependence of the intensity at $\lambda > 50$ cm on the heliographic longitude of the spots indicates the directional nature of the radio emission that is connected with it (Avignon, Boishot and Simon, 1959; Jorand, 1953; Vauquois, Cupiac and Laffineur, 1953; Tandberg-Hanssen, 1955; Simon, 1960a; Gnevyshev, 1960). According to the data of Machin and O'Brien (1954) the mean width of the polar diagram determined at the half-intensity level is 15°, 20° and 30° at wavelengths of 3·7 m, 1·7 m and 60 cm respectively. Processing results of observations in 1947–53 showed, however, that the angular separation does not vary significantly with the wavelength in the 1·5–5 m range, averaging about 34° at the half-intensity level (Müller, 1956).

The directional nature of the radio emission can also be judged by studying the longitude distribution of the number of radio-emitting centres (Avignon, Boishot and Simon, 1959). It turns out here that the number of R-centres is halved when one moves ±30° away from the central meridian. Therefore the noise storm type of radio emission connected with spots is far more directional than the S-component which is characterized by a cosine-wave variation of the flux with longitude (see section 10). In the latter case the flux is halved when the local source moves ±60° from the central meridian. It should also be noted that the analysis of the longitude distribution of the R-centres by Avignon, Boishot and Simon (1959) included two forms of radio-emitting centres. Some centres generate continuous radio emission with a small number of surges which change only slightly even in the course of several days. They can be observed both close to and a long way from the central meridian. The other centres are typical

† The connection between the sudden starts and intensifications of noise storms on the one hand and chromospheric flares on the other found in Dodson (1958); Lougheed, Roberts and McCabe (1957); (see also Fokker, 1960b) from observations at a period of minimum solar activity is not confirmed, however, by the data given by Swarup, Stone and Maxwell (1960). This is possibly because the latter observations were made near the maximum of the 11-year cycle at a time when the high level of activity in the optical region and in the radio band made it difficult to investigate this correlation.

centres of noise storms which generate numerous type I bursts as well as enhanced radio emission which changes noticeably from one day to the next. The directional nature of these centres is more sharply defined: they can be observed chiefly near the central meridian. It is clear from what we have said that the degree of directivity of the noise storm type of radio emission (without R-centres having no type I bursts) will actually be higher than the value given ($\pm 30°$). On the other hand, cases are well known when this directivity is relatively weak and in actual fact differs in no way from the directivity of the S-component (Fokker, 1960b; Avignon, Boishot and Simon, 1959; Morimoto and Kai, 1961).

Lastly information on the directional nature of noise storm type radio emission is provided by a study of the longitude distribution of chromospheric flares appearing near the sudden start of these storms (if we assume that the position of these flares on the disk defines the distance of the sources of radio emission from the central meridian). According to Avignon, Boishot and Simon (1959) flares of this kind show a noticeable concentration towards the central part of the disk when compared with the longitude distribution of all the chromospheric flares.

It is noted by Avignon, Boishot and Simon (1959) and Dodson (1957) that the ratio of the burst intensity to the intensity of the continuum apparently depends on the longitude of the source, increasing from values less than unity near the central meridian to a value greater than unity at a longitude greater than $\pm 50°$. It would seem to follow from this that the continuum radio emission is more directional than that of the bursts making up the noise storms with a sudden onset; in particular bursts are often observed even in the case when flares appear on the Sun's limb. It must be admitted, however, that quite inadequate study has been made of the question of the possible difference in degree of directivity of the two components of noise storms.

Therefore the noise-storm observations undoubtedly indicate the directional nature of their radio emission which, generally speaking, is more sharply defined than in the S-component although even the degree of directivity varies within very broad limits. The dependence of the nature of the directivity on the degree of polarization of noise storms will be discussed later.

SIZE AND POSITION OF RADIO-EMISSION SOURCES IN THE CORONA

Three methods are generally used to determine the altitude at which the sources of noise storms are located in the solar corona. The first of them is based on the fact that the sources of radio emission move over the Sun's disk at a higher velocity than the optical objects located at the level of the

§ 12] Noise Storms

photosphere. By finding the velocity of the R-centre from interferometer measurements and assuming that the angular velocity of rotation of the source around the Sun's axis is the same as the corresponding value for the equatorial regions of the photosphere we can determine the level at which the R-centre is located in the corona. This method is more suitable for sources near the central meridian when their relative velocity over the disk is greatest. The second method is based on measuring the difference in the positions of the source of the noise storm and the group of spots connected with it. In this case the altitude is found on the assumption that the source and the group of spots lie on the same line passing through the centre of the Sun. This method is obviously applicable for centres of activity a long way from the central meridian. The third method consists of direct measurement of the source's position at the time when it is furthest away from the central meridian (at longitude $\pm 90°$). In this case, unlike the preceding ones, no additional assumptions are required on the nature of the motion and the position of the R-centre in relation to the optical features of the photosphere. The last method (and the second to a slightly smaller degree), however, obviously gives only the lower limit for the altitude of the radio emission source since a long way from the central meridian the true position of the radio-emitting region may differ from the observed because of refraction in the solar corona (see section 22).

The distances from the photosphere to the sources of noise storms in the corona display considerable scatter and lie between $0.3 R_\odot$ and $1.0 R_\odot$ at metric wavelengths (Payne-Scott and Little, 1951; Avignon, Boishot and Simon, 1959). The mean altitude is about $0.6 R_\odot$ at $\lambda \approx 1.8$ m (Blum, Denisse and Steinberg, 1958), decreasing to $0.4 R_\odot$ at $\lambda \approx 1.2$ m (Fokker, 1960b) although observations (Chikhachev, 1956) at wavelengths of 2 and 1.5 m did not show any noticeable difference in the altitude position of the radio-emission sources (the mean altitude obtained by the second method is $0.4 R_\odot$). A close value for the mean altitude of noise storm sources at $\lambda = 1.5$ m was obtained by Suzuki (1961), although the observed scatter is very great (from $< 0.1 R_\odot$ to $0.65 R_\odot$). Here a certain tendency is noticed to a rise in altitude as the strength of the magnetic field in the accompanying groups of spots increases, although it is still too early to draw any definite conclusions on the existence of a dependence between these quantities. A slightly lower value for the mean altitude ($0.2 R_\odot$–$0.3 R_\odot$) of sources of type I bursts has been found at the same wavelength (Morimoto and Kai, 1961). In the decametric band the generation regions are located far higher: according to rather uncertain estimates (Erickson, 1959) at $\lambda \approx 11$ m the distance from the source of one of the noise storms to the photosphere was $(3-4) R_\odot$.

To judge from the data given the emitting region in the corona lies far higher than the level where the refractive index becomes zero: in the Baumbach–Allen model (see section 1) this layer is located in the transition region between the chromosphere and the corona at an altitude of about $0\cdot03 R_\odot$ for $\lambda = 1\cdot5$ m and at an altitude of about $1\cdot6 R_\odot$ for $\lambda = 10$ m.

Interference measurements have also made it possible to determine the size of noise storm sources. Early observations (Ryle and Vonberg, 1948; McCready, Pawsey and Payne-Scott, 1947) at wavelengths of $\lambda \approx 1\cdot5$ m showed that this size is of the order of the spots' diameter and in any case does not exceed $10'$. It follows from observations at a period of low solar activity (Chikhachev, 1956) from 1949 to 1950 that the average size of regions generating enhanced radio emission at $\lambda = 1\cdot5$ m is slightly greater than $6'$ and at $\lambda = 2$ m is about $8'$. Local sources with sizes of the order of 6–$8'$ have also been found at a wavelength of $\lambda = 3\cdot5$ m (Vitkevich, 1956c). Measurements by Avignon, Boishot and Simon (1959) and by Blum, Denisse and Steinberg (1958) of the diameter of "radio spots" at $\lambda \approx 1\cdot8$ m made with a multi-element interferometer have shown that the sizes of the sources are distributed evenly from values of $< 3'$ to $9'$.

The position and size of the sources of burst storms (according to the data of Erickson (1961) at $\lambda = 1\cdot5$ m and 2 m) are practically the same as those of the regions which generate enhanced radio emission. The established size of sources of noise storms at these wavelengths (6–$8'$) is several times greater than the size of the groups of spots with which the radio emission was connected ($\sim 2'$). During one type I burst storm the diameter of the source of the bursts averaged for an hour was about $10'$ at 45 Mc/s and about $6'$ at 65 Mc/s (Wild and Sheridan, 1958), i.e. it increased slightly as the wavelength rose. Values of the same order have also been obtained by other observers (Payne-Scott and Little, 1951; Högbom, 1959) in the metric waveband. In the decametric band the effective diameter of the source of one noise storm, according to Erickson (1961), did not exceed $15'$.

It must be remembered that the values given above characterize the angular size of the sources of the burst storms as a whole. These sizes, generally speaking, differ on the large side from the sizes of the regions where individual type I bursts are generated in a noise storm because of a certain difference in the size and position of the sources of the bursts on the Sun's disk. The results of preliminary measurements (Högbom, 1959) at $\lambda = 3\cdot7$ m show that the sizes of the sources of individual type I bursts during one noise storm were $6 \pm 1'$, whilst their positions during the noise storm differed by less than $3'$. A similar conclusion was also drawn in ob-

§ 12] Noise Storms

servations at $\lambda = 1\cdot 5$ m (Fokker, 1960b), according to which the size of the region in which an individual burst is generated (a few minutes of arc) is approximately equal to the diameter of the source of enhanced radio emission but is sometimes less. The scatter in the positions of the individual bursts is only 1–2′.

Therefore the local sources of the individual bursts are scattered over a region only slightly larger than the size of these sources. It is assumed by Högbom (1959) that this sort of effect may be caused by scattering of the radio emission on inhomogeneities of the solar corona. It should be noted that the effect of the scattering, if it does occur, by no means always makes itself felt in the observed size of the source, since in a number of cases at wavelengths of 2–3 m the diameters of sources of individual type I bursts did not exceed 1–1·6′ (Goldstein, 1959; Avignon, Boishot and Simon, 1959). It is possible that the degree of scattering in the solar corona has a connection with the strong variations in size of the R-centres (from <3′ to 9′) noted by Avignon, Boishot and Simon (1959) at $\lambda \approx 1\cdot 8$ m and the scatter in the sizes of the sources of type I bursts from the values of less than 1′ to 5′ observed at $\lambda = 1\cdot 5$ m (Cohen and Fokker, 1959).

As well as measurements of the effective sizes of sources of noise storms there is great interest in finding the radio brightness distribution in the region occupied by a source of this kind. Investigations in this direction have been made by Fokker (1960b) at $\lambda = 1\cdot 5$ m using three interferometers with baselines 390λ, 570λ and 960λ long.

It followed from observations on the first two baselines that the diameter of the sources found on the assumption that the latter are disks with a uniform radio brightness distribution is 2–6′. With this model it proved impossible, however, to match the high amplitude values of the interference recordings on the large baseline of 960λ, which can be explained by the presence of a sharp rise in the radio brightness in the centre of the noise storm source. The size of the "bright" region is generally one or two minutes of arc. In actual fact the results of the measurements on the three baselines agree well with model 1 of the distribution of the effective temperature T_{eff} over the local source shown in Fig. 45a. This is easy to see in Fig. 45b where the vertical lines indicate the results of the measurements on the three baselines, whilst curve 1 characterizes the modulation depth M of the interference curve as it passes through the polar diagram of a source with a radio brightness distribution corresponding to model 1. At the same time the observational results do not satisfy model 2 which has no sharp increase in brightness in the centre.

The observed radio brightness distribution is interpreted by Fokker (1960b) as the consequence of scattering of radio emission in the solar corona: the central part is the real source of the noise storms, whilst the peripheral region is the scattered radio emission of the central source.

The data given on the size of sources of enhanced radio emission and type I bursts make it possible to judge the magnitude of the effective temperature T_{eff} of these sources. It was pointed out above that in the metric waveband a noise storm sometimes increases the effective temperature reduced to the Sun's disk by a factor of 10^2–10^3, bringing it up to 10^8–10^9 °K. In this case the T_{eff} of a "radio spot" with an angular size of $d \sim$ 7–8′ is more than 10^9–10^{10} °K; for sources of the order of 2′ in diameter T_{eff} will be still higher ($\gtrsim 10^{10}$–10^{11} °K). We are of course giving here the

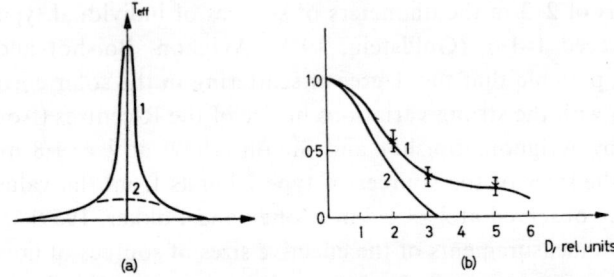

Fig. 45. (a) Models of radio brightness distribution over the local source of noise storms: 1—model with increase of brightness at centre; 2—model without increase of brightness. (b) Interference recording amplitude as a function of interferometer baseline D when passing through the source; curve 1 corresponds to model 1, curve 2 to model 2 (Fokker, 1960b).

record figures which are only reached during very strong noise storms. In the usual cases, when the level of solar radio emission rises by only a few factors when active regions appear on the Sun, the effective temperature of the sources will be far lower ($\sim 10^8$ °K for $d \sim 7$–8′ and 10^9 °K for $d \sim 2'$).

Polarization of Noise Storms

Polarization observations at metric wavelengths (Ryle and Vonberg, 1948; Payne-Scott and Little, 1951; Hatanaka and Moriyama, 1952; Cohen and Fokker, 1959; Fokker, 1960a) indicate that the enhanced radio emission generally consists of two components: a randomly polarized one and a circularly polarized one, the degree of polarization often being close to 100%. The sense of the polarization for an individual local region that is active in the radio band does not as a rule change during the whole existence of the source. Sometimes the degree of polarization of the enhanced radio emission which was at first weakly and irregularly polarized

§ 12] Noise Storms

increases as the local source develops. The opposite cases of a gradual decrease in the polarization of the "radio spots'" emission (to a state close to an unpolarized one) have not been observed. In certain cases the enhanced radio emission does not show any noticeable polarization during the whole existence of this component. When this is so, however, the radio emission flux connected with the spots is comparable with or less than the radio emission flux of the "quiet" Sun. On the other hand, intense enhanced emission whose level is more than four times higher than the basic level of the solar radio emission is almost always polarized (Cohen and Fokker, 1959).

Just like the enhanced emission, type I bursts connected with spots are generally strongly polarized. The observations of Hatanaka, Suzuki and Tsuchiya (1955a and 1955b), Hatanaka (1957b) and Suzuki (1961) at $\lambda = 1.5$ m with a polarimeter, which provides full information on the state of the polarization in 1/200 sec (see section 6), showed that the type I radio emission consists of two components—a randomly polarized one and an elliptically polarized one. The elliptical component is generally about 90% of all the emission (often about 100% (Cohen and Fokker, 1959)), but sometimes the degree of polarization decreases to 50% and even to 10%. For the majority of bursts the degree of polarization exceeds 50% (Komesaroff, 1958).

The degree of polarization of type I bursts reveals a tendency to increase as the observed size of the sources decreases. This can be explained by the fact that the magnetic field is more or less uniform within the small sources. This cannot be said of the large sources situated above bipolar or multipolar groups of spots. These sources occupy regions in which there are various magnetic field orientations, including directly opposite directions; the polarizations of the radio emission from these regions (with different signs of rotation) cancel each other out to a certain extent (Suzuki, 1961).

Moreover, the degree of polarization for strongly polarized storms depends in many cases on the position of the emitting region on the solar disk, decreasing as the source moves away from the central meridian (Suzuki, 1961). Weakly polarized and irregularly polarized noise storms do not display this dependence. This can be explained by the fact that the polarized component of the emission is more directional than the unpolarized (this effect has already been noted in section 10 for the S-component).

The ratio of the axes of the polarization ellipse varies from 1 to 0.1, being close to the ratio of the axes of the ellipse obtained by projecting onto the plane of the Sun's disk the circle lying in the plane of the Sun in the region of the actively emitting spot. The angle of inclination of the polarization ellipse is almost constant during the course of a day but is different on different days (Komesaroff, 1958). The change in orientation is evidently connected with the Faraday effect in the Earth's ionosphere and is caused by the daily variations in its electron concentration (see section 7 for further detail).

The nature of the polarization of bursts following one after the other is not repeated quite precisely, although the state of the polarization (degree of polarization, form of the ellipse and the sign of rotation) generally stays more or less constant for several days (and even during the passage of the noise storm source over the disk (Cohen and Fokker, 1959)). The change in the sign of rotation, if it occurs, generally takes place fairly rapidly—in a period of less than 12 hours (Alekseev and Vitkevich, 1959; see also Komesaroff, 1958).

It follows from the polarization observations by Komesaroff (1958) with a radio spectrograph (40–240 Mc/s) that the emission is polarized at all frequencies where there is a burst storm, and generally (but not always) has the same direction of rotation throughout the band. In one case polarization of both signs has been observed, but at different frequencies and different times.

The sense of the polarization of type I bursts is, as a rule, the same as the sense of the polarization of the enhanced radio emission which these bursts accompany; the degree of polarization of the surges is generally close, but not identical, to the corresponding value for the enhanced radio emission (Cohen and Fokker, 1959; Fokker, 1960a; Alekseev and Vitkevich, 1959). Certain bursts with an anomalous (higher or lower) degree of polarization may suddenly appear during a uniformly polarized noise storm.

There are exceptions to the rule given for the sameness of sense and closeness of degree of polarization even in cases when the source of the noise storms is connected only with one group of spots. Sometimes strongly polarized bursts are superimposed on weakly polarized enhanced radio emission and vice versa. At the same time there have been cases of "mixed" polarization of the radio emission in one centre of activity, when the variations in the degree of continuum polarization have been accompanied by changes in the sign of rotation in the bursts. When considerable differences appear between the polarizations of the enhanced radio emission and the bursts it may be thought that these two components arise in different parts of the local source† (with a magnetic field of opposite polarities) (Cohen and Fokker, 1959; Fokker, 1960a). However, no definite connection has been found between the sense of the polarization and the position of the source of the bursts relative to the centre of activity even in the cases when the

† This has been confirmed once by interferometer measurements.

§ 12] Noise Storms

bursts were generated above bipolar groups (Cohen and Fokker, 1959). This is not surprising, moreover, if one remembers the considerable size of the sources and their localization in the high layers of the corona at great distances from the centres of activity.

There are very serious difficulties in establishing the type of normal wave emitted by a "radio spot" (ordinary or extraordinary). The point is that the connection between the sense of the polarization and the direction of the magnetic field can be established by determining (simultaneously with the polarization observations) the position of the local source in the corona and the structure of the magnetic field in the region occupied by the source. However, the detailed structure of the magnetic fields in the corona is in actual fact unknown, and conclusions on the direction of the field in a radio-emitting region located at a great altitude are generally based on a very unreliable extrapolation of the results of observations of the structure of the magnetic field in the photosphere and the lower layers of the chromosphere.

This extrapolation generally consists of the following. The magnetic flux through one of the spots of a bipolar group (generally through the leading spot) is as a rule far greater than the corresponding value for the other spot. It can therefore be taken that at great distances in the corona above the group of spots the direction of the magnetic field is determined by the polarity of the spot with the highest flux value, i.e. in the majority of cases by the polarity of the head spot. During the 11-year solar cycle the polarities of the leader and follower spots in a group are determined unambiguously by which hemisphere—northern or southern—these spots appeared. It follows from this that if one type of normal wave is predominant in the generation of noise storms, then the observed sense of the rotation in the polarized radio emission will be determined by the hemisphere in which the local source of the radio emission and the bipolar group of spots connected with it are located. In the other cases, when the structure of the group of spots is more complex, the predominance of a magnetic field of one direction or another can be judged on the basis of magnetometer observations of the sunspots. Measurements of this kind, of course, are also useful for the bipolar groups since they make it possible to allow for the exceptions to the general rule of the predominance of the magnetic field of the leader spots.

Having made these preliminary remarks we shall give the results of polarization observations whose purpose was to establish the type of normal wave emitted by "radio spots".

The observations of Ryle and Vonberg (1948), Payne-Scott and Little (1951) Hatanaka and Moriyama (1952), Suzuki (1961) and Malinge (1960) indicate that the circularly polarized component corresponds largely to an ordinary wave. Among these observations notice should be taken first of all of the results of Payne-Scott and Little (1951) who in the process of their measurements also found the positions of the emitting region relative to the bipolar groups of spots. According to Payne-Scott and Little (1951) the enhanced radio emission is almost completely circularly polarized, the direction of rotation not changing during the spot's emission; in particular the sense of the polarization does not change when the spot passes through the central meridian. Any change in the direction of rotation, if it occurs, is connected with a change in the relative contribution to the radio emission from the two active regions. For groups of spots located in the northern hemisphere right-handed polarization occurred in 95% of the cases when the leader spot was equal in area to the follower, or larger. On the other hand, for groups to the south of the solar equator under the same conditions left-handed polarization was observed in 70% of the cases; once when the follower spot was larger in area than the leader the radio emission contained a right-handed polarized component. At the time of these observations the leading spots in the bipolar groups of the northern hemisphere were south magnetic poles, whilst in the southern hemisphere they had the opposite polarity. From this Payne-Scott and Little concluded that the enhanced radio emission connected with the spots corresponds chiefly to an ordinary wave. The connection of the sense of the polarization with the position of the radio emission source relative to the solar equator is confirmed by the observations of Malinge (1960) during the next cycle of activity (1959); the results she obtained also indicate the predominance of the ordinary component in the composition of the enhanced radio emission.

On the other hand, Stanier's measurements (see Ryle, 1958) indicated predominance of the extraordinary component; however, he was not sufficiently certain that the direction of the magnetic field had been correctly determined during the observations. The investigations of Cohen and Fokker (1959) during which, just as in the observations of Payne-Scott and Little (1951), both the polarization and the position of the emitting region on the Sun's disk were determined, indicated that both ordinary and extraordinary emission are of equal probability, since Cohen and Fokker (1959) found no definite correlation at all between the sense of the rotation and the position of the "radio spots" relative to the equator. Later, however, Fokker (1960a) re-examined his results and on the basis of polarization measurements during the International Geophysical Year came to conclusions, which agree in fact with Payne-Scott and Little's data: (1) the sense of the polarization of the radio emission coming from a single

§ 12] Noise Storms

source almost without exception remains the same during the whole existence of this source; (2) the sources of noise storms in the northern hemisphere mostly generated left-handed polarized radio emission, and those in the southern hemisphere right-handed polarized emission. If the direction of the magnetic field in the source is determined by the polarity of the leading spot it follows that the ordinary component is predominant in the radio emission of noise storms. This connection is, of course, far from unambiguous; it is sufficient to mention cases of the appearance of "mixed" polarization under conditions when radio emission with different directions of rotation is created by several sources located on one hemisphere, and even by one source.

Suzuki (1961) compared the sense of the polarization of the observed radio emission during noise storms at $\lambda = 1\cdot 5$ m directly with the polarity of the accompanying groups of spots (the position of the emitting region on the Sun's disk in the east–west direction was determined with an interferometer). It turned out that the sense of the polarization of type I bursts generated above unipolar groups of spots is given by the direction of the magnetic field and corresponds to an ordinary wave. The same tendency obtains in general in bipolar groups if the sense of the polarization is connected with the predominant polarity in these groups. For groups with a complex magnetic field structure the polarization of the bursts is mixed and changes frequently. Exceptions to these rules are not rare, however. In particular in bipolar groups with clear predominance of one polarity mixed, irregular polarization characteristic of groups of spots with a complex magnetic structure and stable, strong polarization with the sign of rotation characteristic of spots of opposite polarity have been observed.

Therefore the observations that have been made largely indicate the predominance of the ordinary component in the radio emission of noise storms (enhanced level and type I bursts). The effect of the exceptions to this rule, however, is to create the impression that noise-storm theory should first explain the possibility of the appearance of the ordinary and extraordinary components mixed in different proportions and then the preferential generation of ordinary waves in the majority of local sources.

It follows from the contents of this section that the enhanced radio emission and the type I bursts are two closely interconnected components of the Sun's sporadic radio emission. In published papers the opinion has been more than once stated (Vitkevich, 1956a; Brown, 1953) that the enhanced radio emission is the result of short type I bursts being superimposed on each other. The basis of this conclusion was the fact that a rise in the level is usually accompanied by more or less intense bursts, the latter also occurring in the absence of enhanced radio emission. The identical sense and the close degree of polarization of the continuum and its accompanying bursts are also in favour of this assumption.

Serious objections can be adduced against this idea, however. An increase in the level of radio emission is not always accompanied by bursts; even at the time of a noise storm intervals are found on the high-speed recordings of the solar radio emission when there are no bursts yet the flux of the enhanced radio emission remains at about the same level as during the bursts, without dropping to the value characteristic of the "quiet" Sun (Fokker, 1960b; see also the discussion in Vitkevich, 1957b). The appearance of bursts in a certain range of frequencies and their intensity are weakly connected (or generally unconnected) with the level of the enhanced radio emission in this range. In other words, the total frequency spectrum of type I bursts differs sharply from the spectrum of the enhanced radio emission (Wild, 1951; Billings, Pecker and Roberts, 1954) (see Fig. 42). If we also remember that sometimes significant differences can be observed in the relative content and even in the sign of the polarized component of the enhanced radio emission and the bursts, it becomes clear that the enhanced radio emission as a whole cannot be reduced to the sum of the type I bursts. In individual cases, however, the bursts observed to follow one another closely during a strong burst storm may, of course, cause a certain "continuous" rise in the level of the radio emission on the intensity recording, particularly in observations on equipment that has a large time constant ($\gtrsim 1$ sec).

13. Type II Bursts[†]

GENERAL CHARACTERISTICS

The second spectral type covers powerful bursts of solar radio emission at metric wavelengths with a "life" at a fixed frequency of the order of several minutes or tens of minutes; these bursts first appear at high frequencies and then in the process of the burst's development its spectral features (intensity maximum, etc.) shift towards the lower frequencies at a rate of up to 1 Mc/s per second (generally about $\frac{1}{4}$ Mc/s per second at $\lambda \sim 3$ m), so that as a whole an event of this type occupies a band up to hundreds of megacycles wide. Bursts of this kind with frequency drift were first found in simultaneous observations at several fixed frequencies (Payne-Scott, Yabsley and Bolton, 1947); later very valuable information

---

[†] Extensive use has been made in this section of the article by Roberts (1959a) which recapitulates the results of 65 type II events between 1952 and 1958. Basically these results agree with the data on the solar radio emission of 1956–60 in the form of type II bursts (138 events) given by Maxwell and Thompson (1962).

§ 13] Type II Bursts

on the dynamic spectra of type II bursts was obtained in the process of radio spectrographic observations by Wild and others (Wild and McCready 1950; Wild, Murray and Rowe, 1954; Wild, 1950; Wild, Murray and Rowe, 1953).

Type II bursts are some of the rarest events in the Sun's radio emission. Even at a period of maximum solar activity in every 50–100 hours of observations on the average only one burst of this type is found, whilst the mean frequency of the appearance of type I and III bursts is apparently one burst per minute (if we limit ourselves to periods of noise storms the latter value is even higher). The number of type II bursts observed shows considerable variations over a solar cycle, decreasing sharply near its minimum phase (Roberts, 1959a).

Type II bursts generally appear at the time of chromospheric flares (Wild, 1957; Swarup, Stone and Maxwell, 1960; Roberts, 1959a; Warwick, 1957; Giovanelli and Roberts, 1958, 1959) near the maximum of their radiation in the H_α line or slightly later, and generally stop earlier than the flare itself. In certain cases the appearance of bursts is not accompanied by any noticeable effects in the optical region. Flares of importance 1 or over have been observed in approximately 60–80% of the cases when type II bursts have appeared. At the same time only rare flares are connected with bursts of this type (only about 3% of the flares according to the data of 3 years of observations). However, the probability of the appearance of a type II burst when a flare arises on the Sun's disk rises rapidly with the class of flare—from 2% for class 1 flares to 30% for strong class 3 flares. The eruptive activity of the flares (ejection of material into the corona) is not apparently a sufficient condition for the generation of type II bursts since the ejection of prominences from the flares is a more frequent phenomenon than bursts with slow frequency drift (Roberts, 1959a).†

As has already been pointed out, type II bursts belong to the class of very intense phenomena in the radio band: during these bursts a rise in the radio emission flux at metric wavelengths to 10^{-19} W m^{-2} c/s^{-1} (and therefore an increase in the Sun's effective temperature $T_{\text{eff}\odot}$ to 10^9 °K) cannot be considered out of the way (Roberts, 1959a). Cases are known when the radio emission intensity was even higher. For example, on 8 April 1959 the flare-connected outburst, which lasted for about 4·5 min at $\lambda = 1\cdot5$ m, raised the level of the radio emission to a value of over 5×10^{-17}

---

† The close connection between the outbursts and chromospheric flares was also found by Dodson and her co-workers (Dodson, Hedeman and Owren, 1953; Dodson, 1958). It should be noted, however, that it is difficult to judge the spectral type of a burst from observations at one fixed frequency. Many outbursts connected with chromospheric flares are not type II but are large groups of very intense type III bursts, the latter event occurring even more often than type II bursts (Roberts, 1959a).

W m^{-2} c/s^{-1} ($T_{\text{eff}\odot} > 5\times 10^{11}$ °K) (Fokker, 1960a). The burst of 8 March 1947 was even more powerful; for 3·5 min the flux at $\lambda = 5$ m was greater than or of the order of 10^{-15} W m^{-2} c/s^{-1} ($T_{\text{eff}\odot} \gtrsim 10^{13}$ °K). This $T_{\text{eff}\odot}$ value, which is ten million times greater than the effective temperature of the "quiet" Sun, is a record for solar radio emission (Payne-Scott, Yabsley and Bolton, 1947).

Unfortunately the size of the sources of these bursts is still not known. If, however, we assume that the generation regions occupied 10^{-2} of the Sun's disk, then the brightness temperature T_{eff} of the local source of the bursts is (for $T_{\text{eff}\odot} \sim 10^9$ °K) about 10^{11} °K and its record value of 8 March 1947 will exceed 10^{15} °K!

Weiss and Sheridan (1962) have measured the diameters of sources of type II radio emission at frequencies of 40 and 60 Mc/s. According to the preliminary data the angular size of the central part of the sources is 5–7' (at the half-intensity level); this part is generally surrounded by a weak halo which is far larger (up to 40'). No noticeable difference can be seen between the size of the sources of the fundamental frequency and the second harmonic in observations at 40–60 Mc/s.

The data on the nature of the polarization of the type II bursts are very incomplete; we can say, however, that the type II bursts are either not polarized or are polarized very weakly. Komesaroff's observations (Komesaroff, 1958), which were made with a low-sensitivity polarimeter (which made it possible to determine the polarization only if the degree of polarization of the radio emission was not less than 25%), showed that if there is elliptical polarization of type II bursts it is very weak. For example, out of thirteen bursts investigated eight were not polarized (with the accuracy stated above);[†] in four cases it proved impossible to give a definite conclusion on the nature of the polarization, and only on one recording did a polarized component seem to appear for a short time. Moreover even in the last case the degree of polarization found (not more than 30%) was close to the threshold of the equipment sensitivity.

According to Fokker (1960a) the flux of the circularly polarized radio emission during the outburst of 8 April 1959 was not more than a few times greater than the value of 10^{-22} W m^{-2} c/s^{-1}. If we remember that the total flux of the radio emission (randomly polarized and circularly polarized) was more than 5×10^{-17} W m^{-2} c/s^{-1}, then we can conclude that the degree of polarization was less than 0·001%. Akabane and Cohen (1961) observing on a narrow-band polarimeter (with a 10 kc/s band), did

---

† In particular there was no polarization in type II bursts whose harmonic components showed their own peculiar splitting, which in certain papers is linked with the effect of the magnetic field (see below).

§ 13] Type II Bursts

not find any polarization of the radio emission in the three type II bursts investigated either.

On the other hand, the earlier observations of Payne-Scott and Little (1951) indicate that the bursts of radio emission (which from their intensity and duration can be included among type II)† have no polarization only at the initial stage of development: there is often a secondary rise in the intensity with elliptical polarization. It is possible, moreover, that the second period was not a type II burst but a post-burst phenomenon of the noise storm or type IV radio emission type. It is difficult to judge, however, from recordings at a single frequency.

HARMONICS OF TYPE II BURSTS

The most outstanding feature of type II bursts (and also type III) found by Wild, Murray and Rowe (1953, 1954) when investigating the dynamic spectra of bursts is the so-called second harmonic. In the dynamic spectra of the type II bursts shown in Figs. 46 and 47 it can be clearly seen that the radio emission occupies two bands several tens of megacycles wide whose contours, although not absolutely identical, repeat each other in general outline and in many details.‡ This does not allow us to doubt that the two bands are genetically connected and belong to radio emission generated in one source. Since the structure of one band at the frequency ω is repeated in the other band approximately at a frequency 2ω, this served as a basis for calling the first band of the radio emission the fundamental frequency and the second band the second harmonic.§

According to Roberts (1959a, 1959b) the second harmonic can be observed in 60% of the bursts that appear in the 40–240 Mc/s range. This figure rises if we limit ourselves to events of moderate intensity, with well-defined dynamic spectra. The percentage of bursts recorded with a second harmonic rises to 75–80% if we carry out spectroscopic observations in a broader range of frequencies (from 580 to 25 Mc/s) (Maxwell and Thompson, 1962; Wood, 1961). The simultaneous appearance of a first, second and third harmonic has not been observed; if a third harmonic does exist its intensity is less than one-fifth of the intensity of the second

---

† We note that the outbursts of 8 March 1947 and 8 April 1959 also belong to the second spectral type only according to these signs.

‡ The width of the band of frequencies occupied by a burst at any given point in time varies according to the band and averages $0 \cdot 3f$, where f is the frequency corresponding to the maximum spectral intensity of the emission (Maxwell and Thompson, 1962).

§ To avoid misconceptions we would stress once more that if the radio emission spectrum consists of several bands whose mean frequencies are in the ratio of 1 : 2 : 3, etc., then the first (low-frequency) band here and in future will be called the first harmonic or fundamental frequency, the second band the second harmonic and so on.

harmonic (even in the cases when the second harmonic is comparable with the first). In one case (25 April 1956) three bands were recorded in the dynamic spectrum in a ratio close to 2:3:4. It is possible that these

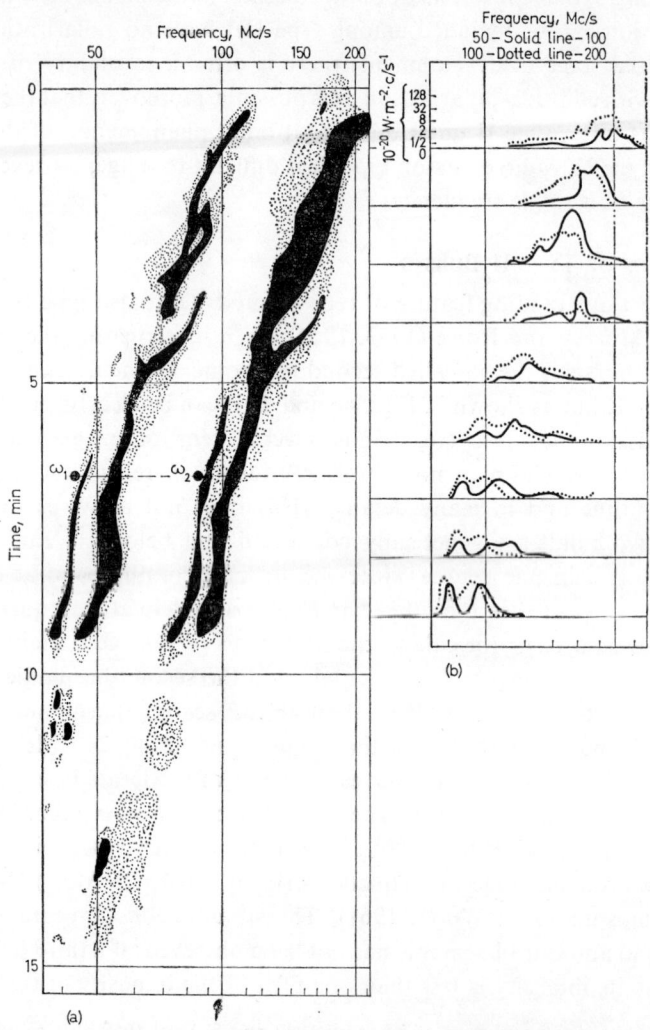

Fig. 46. Type II burst of 21 November 1952: (a) dynamic radio emission spectrum (intensity contours correspond to 5×10^{-21} W m^{-2} c/s^{-1} and 20×10^{-21} W m^{-2} c/s^{-1} levels); (b) frequency spectrum of radio emission (first harmonic—solid curve, second harmonic—dotted line) (Wild, Murray and Rowe, 1954).

bands are the second, third and fourth harmonics respectively of a type II burst; the range of frequencies corresponding here to the first harmonic were outside the range of the radio spectrograph. At the same time the

§ 13] Type II Bursts

absence of a third harmonic in the rest of the bursts studied makes us doubt the correctness of this interpretation of 26 April 1956; another possible explanation of this phenomenon is given in the footnote to p. 165.

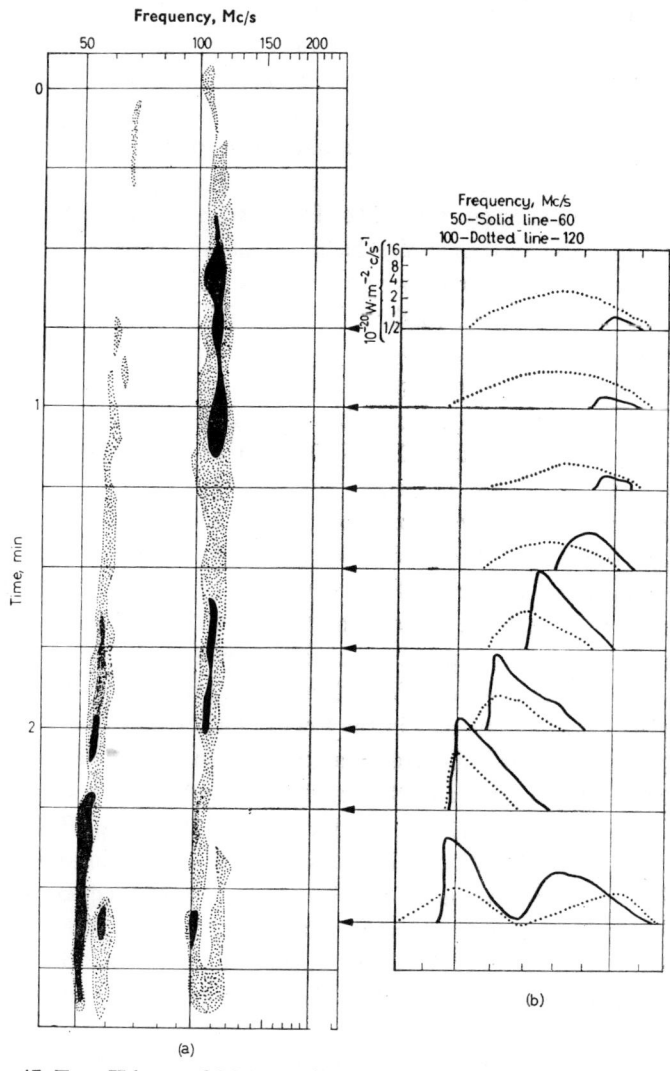

FIG. 47. Type II burst of 5 May 1953; (a) dynamic spectrum (levels the same as in Fig. 46); (b) frequency spectrum (the second harmonic is shown by the dotted line) (Wild, Murray and Rowe, 1954).

We notice that although the ratio of the mean frequencies of the second harmonic and the first harmonic is close to two it is less than two as a rule (Wild, Murray and Rowe, 1954; Roberts, 1959a). For the majority of bursts it lies in the range between 1·9 and 2·0, averaging 1·95. The intensities of the two harmonics are usually the same; sometimes one of the harmonics is more intense than the other. The ratio of the intensity of the first harmonic to that of the second does not apparently increase for the more powerful bursts; in many cases in quite weak type II events the second harmonics are more intense than the first (Roberts, 1959a); this, moreover, may also be connected with the difference in the conditions under which the radio emission at the first and second harmonics is propagated and leaves the corona.

To explain the observed ratio of the frequencies of the harmonic components we can assume that the first harmonic of type II bursts is generated in the layers of the solar corona where the characteristic frequency ω_L of the plasma oscillations is close to the radio emission frequencies of the first harmonic ω.† It is known, however, that radio waves whose frequencies are lower than ω_L do not go beyond the corona, and at frequencies close to ω_L they are subject to strong absorption, decreasing rapidly as the difference $\omega - \omega_L$ rises (sections 22, 26). The action of the above two causes ensures the "cut-off" of the spectrum of the radio emission leaving the corona at frequencies $\omega > \omega_L$ close to ω_L and at frequencies $\omega < \omega_L$. At the same time the radio emission corresponding to the second harmonic escapes completely. It is clear from what has been said that as a result the ratio of the mean frequencies of the harmonic components in the radio emission escaping beyond the corona will be different from two and will become less than two (Wild, Murray and Rowe, 1954).

The existence of a "cut-off" in the low-frequency part of the spectrum of the first harmonic is indicated by a comparison of the frequency spectrum of the first harmonic and that of the second harmonic of one of the bursts studied by Wild, Murray and Rowe (1954) (see Fig. 47). Weightier proof was obtained by Wood (1961) as a result of studying the dynamic spectra of slowly drifting bursts in the broad range of frequencies of 25–580 Mc/s. If low-frequency attenuation really does exist it obviously leads to the ratio of the frequencies ω_2/ω_1 corresponding to the low-frequency edges of the first and second harmonics becoming less than two, whilst the corresponding ratio of the frequencies of the high-frequency edges of the harmonic bands will be close to two.‡ The presence of this kind of

---

† The statement about the closeness of the frequency ω of the bursts of sporadic radio emission to the characteristic frequency of the plasma oscillations $\omega_L = (4\pi e^2 N/m)^{1/2}$, where e and m are the charge and mass of an electron and N is the electron concentration in the source, is the basic content of the so-called "plasma hypothesis". This hypothesis is extensively used for interpreting the radio emission of the "disturbed" Sun; details of it are given in section 31.

‡ The meaning of the frequencies ω_1 and ω_2 is clear from Fig. 46.

§ 13] Type II Bursts

effect is shown well in Fig. 48 where histograms are drawn of the number of type II bursts with a given value of the ratio ω_2/ω_1. The mean value of ω_2/ω_1 found from these histograms is 2·0 for the higher-frequency edge and 1·9 for the low-frequency edge of the harmonic components.

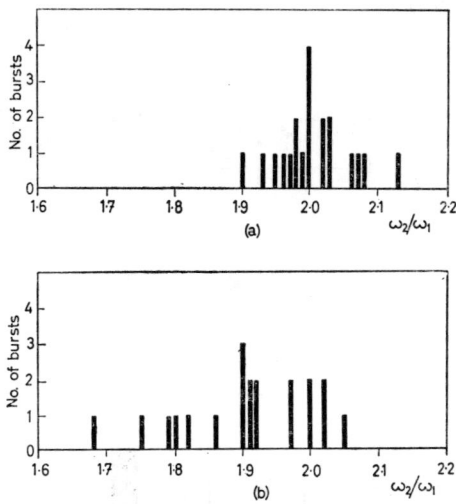

FIG. 48. Histograms of the number of type II bursts as a function of the value of the ratio ω_2/ω_1 at the high-frequency (a) and low-frequency (b) edges of the harmonic components (Wood, 1961).

It was assumed above that the source of the bursts is located in the centre of the Sun's disk: an analysis of the propagation of radio waves in a spherically symmetrical corona (section 22) shows that only in this case do waves of all frequencies satisfying the condition $\omega \geqslant \omega_L$ reach an observer from the source. Unless the angle θ between the straight lines connecting the centre of the Sun with the observer and the source is zero only radio emission of a higher frequency ($\omega \geqslant \omega_{min}$, where $\omega_{min} > \omega_L$) can leave the source in the direction of the observer because of refraction. In this case the ratio of the frequencies of the first and second harmonics should depend on the longitude of the source Θ, i.e. decrease as Θ rises. In addition, at large enough distances r of the source from the centre of the solar disk the first harmonic radio emission cannot generally be observed on Earth. Therefore in the latter case, which, according to Wild, Murray and Rowe (1954), is true approximately at a longitude of $\Theta > 30°$, i.e. at distances of $r > 0.5R_\odot$, type II bursts should have only one harmonic component.

Further observations (Roberts, 1959a; Giovanelli and Roberts, 1958; Roberts, 1959b; Wood, 1961), however, did not confirm this: it turned out that the ratio of the second harmonic frequencies and first harmonic frequencies does not depend on the distance of the source from the centre of the disk (Fig. 49a), whilst both harmonic components are observed at distances greater than $r = 0.5 R_\odot$ from the centre of the disk and the percentage of bursts having only one harmonic band rises only when $r > 0.9 R_\odot$ (Fig. 49b).

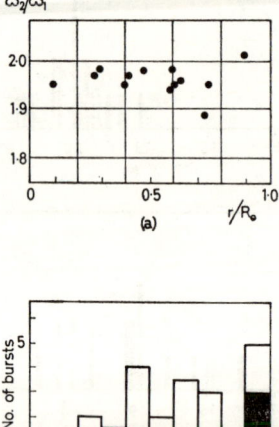

FIG. 49. (a) Ratio of second harmonic frequency to first harmonic frequency as a function of the distance of optical events connected with type II bursts from the centre of the Sun's disk. (b) Number of recorded cases of type II bursts with two harmonic components (unshaded part of histogram) and with one component (shaded part of histogram) for different distances from the centre of the disk of optical events connected with type II bursts (Roberts, 1959a).

The absence of any considerable directivity of the type II radio emission is also indicated by the nature of the distribution over the Sun's disk of the chromospheric flares accompanied by bursts (Roberts, 1959a; Wood, 1961; see also Maxwell and Thompson, 1962; Warwick and Warwick, 1959). Although the number of such flares in Roberts (1959a); Wood (1961) decreases noticeably as there is an increase in the longitude Θ read from the central meridian, the total number of flares recorded in the range between Θ and $\Theta + d\Theta$ also decreases as we move from the central part of the disk towards the limb because of deterioration of the conditions of observation. As a result the percentage of flares in the range between Θ and $\Theta + d\Theta$ accompanied by type II bursts varies very little from the

§ 13] Type II Bursts

total number of flares in this range as Θ changes, being not less than 0·6 of its value at the centre of the disk at the limb.

The non-directional nature of the type II bursts at the basic frequency and the second harmonic and the independence of the ratio of the frequencies of the harmonic components to the source's longitude indicate the weak effect of refraction on the radio emission. The latter circumstance may be connected with the fact that the fundamental frequency ω, not to mention the second harmonic 2ω, differs strongly from the plasma frequency ω_L of the unperturbed coronal plasma in the neighbourhood of the source, being much higher than ω_L. In this case the refractive index $n(\omega) = \sqrt{1 - \omega_L^2/\omega^2}$ in the surrounding corona will be comparable with unity and the effect of the refraction on the propagation of the radio waves after they have left the source will become insignificant; if the source itself produces non-directional emission it escapes into the space around the Sun over a wide range of solid angles and refraction can limit the reception of the radio emission only for bursts generated near the Sun's limb ($\Theta \approx \pm 90°$). Within the framework of the plasma hypothesis the difference between the frequency of the first harmonic and the plasma frequency ω_L of the corona surrounding the source can be explained by the enhanced electron concentration in the bunches of plasma (corpuscular streams) which are undoubtedly also the agent that causes the type II bursts (see p. 176).†

When studying the origin of the higher harmonic components in type II (and III) bursts the solution of the question whether the radio emission of the first and second harmonics is coherent (see section 31) is of importance. Preliminary results indicating the coherent nature of the radio emission during some bursts at frequencies whose values are in the ratio of 1:2 have been obtained by Jennison (1959). His equipment picked up the Sun's radio emission in two separate channels about 500 kc/s wide at frequencies of 127 and 254 Mc/s. The signals at the output of these channels were, however, at the same frequency since the frequency was doubled in the first channel. When outbursts appeared which, from their duration and intensity, could apparently be put into spectral type II, interference between the oscillations was observed at the output which indicated coherency of the radio emission in the bands near the harmonically connected frequencies (1:2). The reservation must be made that Jennison himself is far from being convinced of the existence of this

---

† It follows from section 31 that type II bursts are apparently generated not in the actual bundle but in the shock wave front travelling ahead of it, where the concentration is also enhanced when compared with the surrounding corona.

effect: the experiment was continued for too short a time and the type of burst was not checked with a radio spectrograph. Further investigation into the coherency of the first and second harmonics is therefore particularly desirable.

Smerd, Wild and Sheridan (1962) have obtained interesting results relating to the mutual position of sources of radio emission at the first and second harmonics (by measurements on interferometers with frequency tuning in the 40–70 Mc/s range; see section 5). Generally speaking, simultaneous measurements of the coordinates at the frequencies f and $2f$ should be made when studying the position of such sources. However, the data obtained here were noticeably distorted because of differing refraction in the ionosphere at different frequencies. Therefore in observations by Smerd, Wild and Sheridan (1962) the position of the sources of the first and second frequencies was measured at the same frequency f. In this case both measurements, of course, were separated in time: first the coordinate of the source of the fundamental frequency was determined and then (after a time necessary for the frequency of the second harmonic to change due to frequency drift from the value $2f$ to f) the coordinate was found of the source of the radio emission's second harmonic component. For four type II bursts studied it turned out unexpectedly that the source of the second harmonic "seen" in the radio emission was closer to the centre of the Sun's disk than the source of the first frequency. If we remember that the flares which accompanied these bursts were located even closer to the centre of the disk and accept the fact that the motion of the agent causing the type II bursts comes from a flare in a quasi-radial direction, the result obtained obviously indicates that the "visible" source of the second harmonic is located in deeper layers of the corona than the source of the fundamental frequency (despite the shift in time between the two directions during which the agent is moving away from the solar surface; at the first glance an inverse relation would be expected here between the observed positions of the two sources on the Sun's disk).

The anomalous position of the source of the second harmonic found by Smerd, Wild and Sheridan (1962) can be explained by the effect of reflection of this harmonic radio emission from the deep-lying layers of the solar corona, thanks to which we observe the harmonic emission chiefly in reflected light, whilst the emission of the fundamental frequency reaches the Earth directly from the generation region. When receiving the second harmonic of basically reflected radio emission recording, of course, is possible only if the radio emission of this harmonic escaping from the generation region towards the Sun's surface predominates over the emis-

§ 13] Type II Bursts

sion released away from the Sun.† The possible cause of this is discussed in section 31.

FINE STRUCTURE OF TYPE II BURSTS

Even in the first dynamic spectra of type II bursts obtained by Wild (1950a; see also Wild, Murray and Rowe, 1954) the individual harmonic bands showed their own "splitting" of frequencies into two sub-bands which can be seen clearly in Fig. 46. As well as the frequently observed "doubling", frequency "tripling" is found in certain cases. For the fundamental frequency the split bands are separated by an interval of the order of 10 Mc/s; the structure of the splitting is repeated in the second harmonic where the distance between the sub-bands is twice as great (about 20 Mc/s). In many cases both the split bands of each harmonic component are identical; sometimes, however, they differ considerably in the width of the frequency spectrum, intensity, etc. Although in each burst the variation in the magnitude of the splitting is slight the available data when taken altogether indicate that the degree of splitting increases with the frequency: if the splitting is about 5 Mc/s at 30 Mc/s it rises to 18 Mc/s at 80 Mc/s‡ (Roberts, 1959a; 1959b; see also Maxwell and Thompson, 1962).

The fine structure of the type II bursts is not limited to longitudinal splitting of the bands in the dynamic spectrum that belong to the individual harmonic components. Haddock (1958, 1959b) and Roberts (1959a, 1959b) (see also Maxwell and Thompson, 1962; Sheridan, Trent and Wild, 1959) have observed cases when during outbursts the dominant feature of the dynamic spectrum was a rapid sequence of short-lived, wide-band bursts with a rapid frequency drift. These bursts can be included in type III (see

---

† It must be pointed out that this "backwards" predominance of the radio emission for the second harmonic does not apparently occur for all type II bursts. This is indicated by the fact that for the bursts studied by Roberts (1959a) sharply defined details of the dynamic spectrum of the first harmonic appeared approximately 1 sec later than the corresponding features of the second harmonic. This lag is probably caused by the difference in the group velocities of the first and second harmonic radio emission; if the second harmonic here were observed in reflected beams the fundamental frequency would be expected to be ahead of the second harmonic instead of lagging behind (in this connection see section 14).

‡ It is not excluded that the appearance of three bands (whose frequencies were in the ratio of 2 : 3 : 4) during the event of 25 April 1956 may also be connected with the presence of splitting. From this point of view the two high-frequency bands are the second harmonic split into two sub-bands, whilst the low-frequency band belongs to the first harmonic, being one of its split components; the second component extends beyond the working limit of the radio spectrograph. The frequency ratio (3 : 4) in the split bands of the second harmonic must then be recognized to be random, although the magnitude of the splitting ($\sim$ 25 Mc/s) is slightly higher than that generally observed in type II bursts (Roberts, 1959a).

section 14). They appear first in a narrow spectral range drifting from high to low frequencies at a rate typical of type II bursts; then the type III bursts in the process of drifting in both directions (towards the low and the high frequencies) diverge rapidly, going away from this spectral range. On the whole the dynamic spectrum of this kind of event, which we shall call a type II burst with a type III fine structure,† is of the form shown in Fig. 50a. In the last stage of the development of this event ($2^h1\cdot5^m$ to $2^h2\cdot5^m$ universal time) the radio emission of the type II burst disappears and the narrow-band, slowly drifting part of the spectrum exists only as a "separator" between rapidly drifting elements diverging in two directions towards the low and the high frequencies.

These elements are apparently independent in time and in any case do not appear in pairs. More often only half the fine structure is observed, namely the part of the rapidly acting bursts which leaves the slowly drifting band in the dynamic spectrum towards the low frequencies.‡ Examples of dynamic spectra of events of this kind are shown in Fig. 50b. The dynamic spectrum of the event of 18 December 1958 with a type III fine structure at the last stage of development of the type II burst is also interesting in that it was received over a wide range of wavelengths including part of the decametric band.§ Type II phenomena appear very rarely here: generally a slowly drifting burst starts only at frequencies of 175 Mc/s or less (Maxwell and Thompson, 1962). The range of frequencies covered by each type III element is comparatively small and amounts to 10–50 Mc/s; the "life" at one frequency and the rate of drift may also differ slightly from the corresponding values characteristic of ordinary type III bursts.

The fine structure under discussion is the predominant feature of the dynamic spectrum only in very rare cases; however, in about 20% of the type II events this feature can be observed in part of the burst, although in a less clear-cut form than in Fig. 50. The type III fine structure is not excluded from being more widespread in type II events but it cannot be resolved on many recordings of the dynamic spectrum of the radio emission.

---

† In the papers by Roberts (1959a and 1959b) this type of burst feature is called "herring-bone structure"; generally speaking it is observed at the fundamental and second harmonics.

‡ Haddock (1959b) observed as well as the fine structure—type III bursts drifting from high to low frequencies—groups of U-bursts superimposed on the type II bursts (U-bursts are a variety of type III in which the drift to the low frequencies reverses during the burst's development; for further detail see section 14). The duration of each group is less than 30 sec, the return frequency being about the same for all the bursts in a group.

§ For observations of bursts in this range see Sheridan and Trent (1961), Sheridan and Attwood (1962), Maxwell and Thompson (1962), Warwick and Warwick (1959).

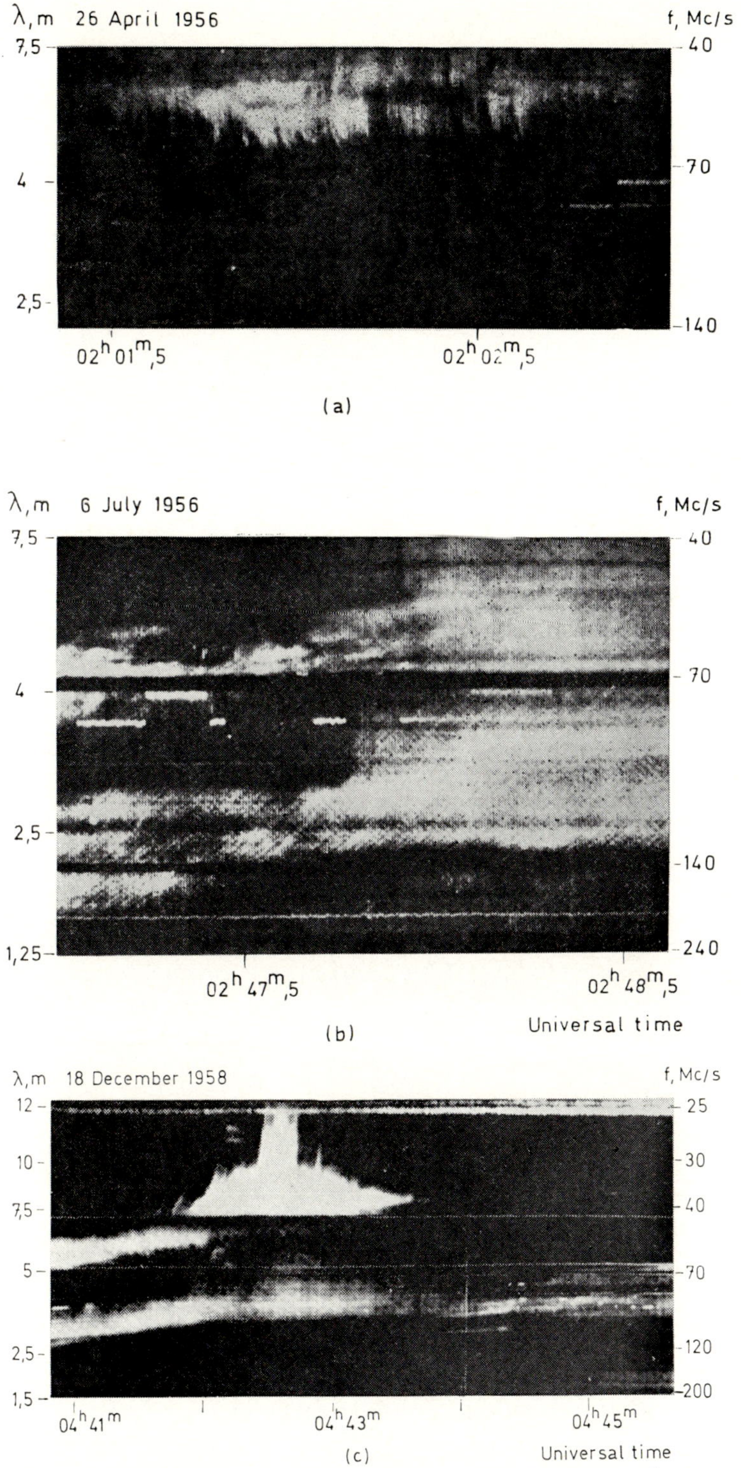

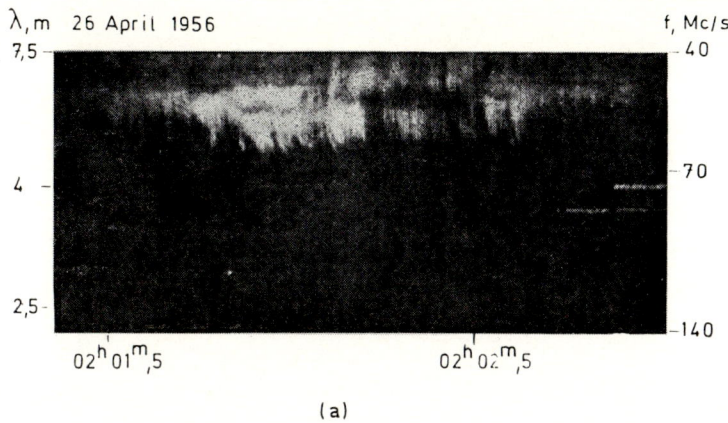

Fig. 50. Type III fine structure in type II bursts: (a) Event of 26 April 1956; (b) Event of 6 July 1956 (Roberts, 1959a); (c) Event of 18 December 1958 (Sheridan, Trent and Wild, 1959) (universal time)

§ 13] Type II Bursts

Haddock (1959b) has even assumed that rapidly drifting elements that are limited in frequency act as the basic component of the complex phenomenon which was previously fixed as a simple spectral type II event.

FREQUENCY DRIFT AND ITS INTERPRETATION

A study of the dynamic spectra of type II bursts has shown (Roberts, 1959a) that in the process of frequency drift these phenomena may cover a range of the order of hundreds of megacycles per second. Generally a type II burst starts almost simultaneously at the first and second harmonics, appearing suddenly at frequencies lying for the most part below 80 Mc/s (for the first harmonic). Sometimes the development of a type II burst suddenly and almost simultaneously ceases at a certain frequency below 40 Mc/s, sometimes reappearing at lower frequencies in the regions of the dynamic spectrum which are a continuation of the harmonic bands of the preceding radio emission.

The general impression is that generation conditions in the corona for type II bursts are realized only in certain frequency intervals connected by a ratio of 2 : 1 for the second and first harmonics at periods of time that are the same for both harmonic components. Since, as follows from what will be said later, the frequency drift of type II bursts can be explained by the motion of an agent from the region of a chromospheric flare through the corona, it follows from what has been said that the conditions for the generation of type II bursts are satisfied only at a high enough altitude in the corona. The appearance and disappearance of type II radio emission is in all probability connected with a change in the generation conditions and not in the the propagation of the radio waves from a source moving in the corona, since the difference in the propagation conditions of the first and second harmonics will then not allow us to explain the simultaneous appearance and disappearance of the radio emission at the harmonic frequencies.

The graph in Fig. 51 (Roberts, 1959a) shows data on the rate of frequency drift of twenty-four type II bursts. These bursts have a clear-cut second harmonic; however the values shown on the graph for the frequency drift rate df/dt relate only to the first harmonic. Lines mark the values of df/dt at different frequencies for one burst; if for a given burst the rate is measured only at one frequency the value obtained is marked with a plus sign. According to the figure the mean rate of frequency drift decreases (in absolute magnitude) from 0.35 Mc/s^2 at a frequency of 90 Mc/s and up to 0.04 Mc/s^2. The scatter of the values of df/dt also decreases with the frequency but the relative dispersion of the observed rates remains more or less constant.

Generally speaking there are several possible causes for the observed frequency drift (Payne-Scott, Yabsley and Bolton, 1947; Wild, Roberts and Murray, 1954; Wild, 1950b; Takakura, 1954); the question is which is the determining one.

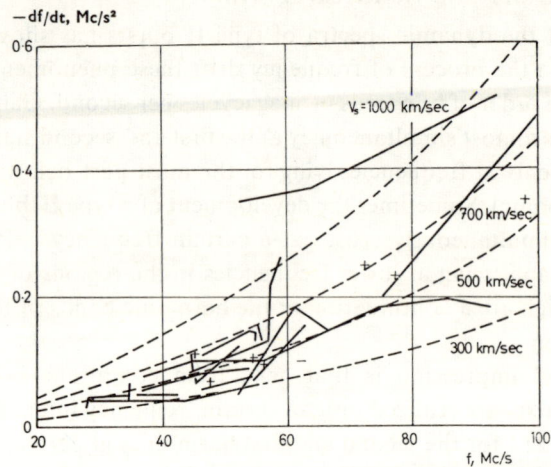

FIG. 51. Rate of frequency drift df/dt as a function of the first harmonic frequency f for type II bursts. The dotted curves show df/dt as functions of f for agents moving radially away from the Sun at fixed velocities V_s (Roberts, 1959a)

Above all the drift may be principally a consequence of the difference in the group velocity of the radio waves at different frequencies as the radio emission passes through the coronal plasma. At the same time the drift will occur if the frequency of the radio emission when a type II burst is generated is determined by the parameters of the coronal plasma (the magnetic field strength, the electron concentration, etc.) in the region occupied by the source. In the latter case we can imagine two models of the burst emission. In the first of them the agent that excites the type II bursts moves through the solar corona; in this case the frequency drift is explained by the change in the parameters of the corona in the layers through which the exciting agent passes. In the second the source of the burst during its development does not show any systematic displacement; the plasma parameters in the region of the source which determine the frequency of the radio emission vary; for example, because of the expansion of the dense ionized cloud in the corona.

In this connection it should be noted above all that a group lag cannot be used to explain the drift of type II bursts, since the difference in the times of arrival of type II bursts at frequencies the difference between

§ 13] Type II Bursts

which is $\Delta\omega \sim \omega$, is minutes and tens of minutes, whilst the difference of the lag times in this case (as the estimates given in section 22 show) is not more than seconds and fractions of a second. As for the possibility of explaining the drift based on the assumption of the radio emission frequency being connected with the plasma parameters, the first model (the agent moving in the corona) has from the very beginning (Payne-Scott, Yabsley and Bolton, 1947) seemed to be preferable to the second (the immobile source). In its favour was the connection of the type II bursts with the eruptions of material from the region of the flares and with the solar corpuscular streams which cause magnetic disturbances on the Earth 1 or 2 days later than strong type II bursts (see section 17). Finally the correctness of the first model was confirmed by the direct interferometer observations of Wild, Sheridan and Trent (1959) about which we shall speak a little later.

The absence of noticeable polarization in type II bursts renders highly improbable the assumption that there is a connection between the frequency of the radio emission and the frequency drift and the magnitude and variation of the magnetic field strength in the corona. Therefore currency has been given (Wild, Murray and Rowe, 1954; Payne-Scott, Yabsley and Bolton, 1947; Wild, 1950a; Shklovskii, 1946; Martyn, 1947) to another point of view according to which the frequency ω of the fundamental harmonic of the radio emission is determined by the value of the electron concentration N in the corona, ω being close to the characteristic parameter of the coronal plasma—the characteristic frequency of the oscillations $\omega_L = (4\pi e^2 N/m)^{1/2}$ in the region where the type II bursts are generated (plasma hypothesis). Since the electron concentration in the corona decreases as one moves away from the photosphere the frequency drift will be towards the lower frequencies if the agent is moving away from the Sun and, vice versa, the frequency starts to rise when the agent is moving in the opposite direction.

For type II bursts the first case is generally realized. However, according to data obtained by Roberts (1959a) during the development of the burst of 5 April 1957, the drift towards the low frequencies was replaced by the opposite; the frequency drift rates of both the first and second phase of the burst was typical for type II radio emission.† This event, which is similar to the often observed type III U-bursts (see section 14), can be explained in all probability by the fact that the agent causing the radio emission first moved into the external layers of the corona and

---

† There are certain indications that another one or two type II bursts have had this kind of dynamic spectrum.

then, having reached a certain maximum altitude, began once more to approach the photosphere.

If the frequency of the first harmonic radio emission of type II bursts is really close to the characteristic frequency of the coronal plasma, then, by giving a definite distribution of the electron concentration in the corona, it is not hard to estimate the altitude at which the source of the radio emission is at any time, and the known rate of frequency drift can also be used to find the velocity of the agent (to be more precise, its component along the gradient of the concentration in the corona).

In actual fact if the electron concentration in the generation region is the same as the concentration N in the corona in the absence of an agent causing type II bursts, then

$$\omega^2 \approx \omega_L^2 = \frac{4\pi e^2 N(\mathbf{R})}{m}, \qquad (13.1)$$

where the radius-vector $\mathbf{R}(t)$ characterizes the position of the source of the radio emission at the time t. Differentiating the relation (13.1) with respect to time we obtain:

$$\omega \frac{d\omega}{dt} \approx \frac{2\pi e^2}{m}\left(\operatorname{grad} N \frac{d\mathbf{R}}{dt}\right). \qquad (13.2)$$

Denoting the angle between the directions of the electron concentration gradient in the corona $\operatorname{grad} N$ and the agent's velocity $\mathbf{V}_s = d\mathbf{R}/dt$ by ϑ and remembering that

$$\operatorname{grad} N \frac{d\mathbf{R}}{dt} = |\operatorname{grad} N|\cdot|\mathbf{V}_s|\cos\vartheta,$$

we find from (13.2) that the projection of the agent's velocity onto the direction $\operatorname{grad} N$ is

$$|\mathbf{V}_s|\cos\vartheta \approx 2\frac{N}{f}\frac{\frac{df}{dt}}{|\operatorname{grad} N|}. \qquad (13.3)$$

Here df/dt is the frequency drift rate; the concentration N for a given frequency $f = \omega/2\pi$ can be determined from the relation (13.1).

If the distribution of the electron concentration in the corona is radially symmetrical, then $N(\mathbf{R}) = N(R)$, where R is the distance from a given point in the corona to the centre of the Sun. In this case the $\operatorname{grad} N$ runs along the radius towards the centre of the Sun and $|\mathbf{V}_s|\cos\vartheta$ is the radial component of the agent's velocity

$$V_{sR} \approx 2\frac{N}{f}\frac{\frac{df}{dt}}{|\operatorname{grad} N|}. \qquad (13.4)$$

§ 13] Type II Bursts

We notice that the V_{sR} obtained in this way is negative if the source of the radio emission is moving away from the Sun.

By using the formulae (13.1) and (13.4) it is easy to estimate the velocity of a source of radio emission V_{sR} and its distance from the Sun at the time a frequency f is generated at the first harmonic. For example, the frequency $f \sim 8 \times 10^7$ c/s corresponds to a concentration $N \sim 8 \times 10^7$ electrons/cm³, which is reached under the conditions of a stationary Baumbach–Allen corona with values of $R \approx 1 \cdot 2 R_\odot$ (see Table 2). Then by making the drift rate df/dt 2×10^5 c/s² in absolute magnitude and remembering that in the layers of the corona where N reaches these values $|\text{grad } N| \sim 8 \times 10^{-3}$ electrons/cm³, we find $V_{sR} \sim 5 \times 10^7$ cm/sec = 500 km/sec.[†] Velocities of the same order are given by the calculations of Roberts (1959a) for the type II bursts whose frequency drift rates are shown in Fig. 51. The majority of bursts have velocities V_{sR} between 300 and 700 km/sec and only a few bursts 800–1100 km/sec. A radial velocity of about 500 km/sec must be considered typical for type II bursts.

A clear idea of the magnitude of V_{sR} obtained on the basis of (13.4) can be obtained by comparing the observed dependences of the frequency drift rate df/dt on the frequency with the corresponding dependences obtained from the relation (13.4) for a series of values of the agent's velocity V_{sR}. Similar curves for the electron concentration given by the formula (1.1) and for values of V_{sR} from 300 to 1000 km/sec are shown in Fig. 51 by a dotted line.

Values of the altitude of sources above the level of the photosphere during the development of three type II bursts are shown by dots in Fig. 52. These dots relate to bright features of the second harmonic in the dynamic spectra of these bursts; they were found by means of the relation

$$\omega \approx 2\omega_L = 2 \left(\frac{4\pi e^2 N}{m} \right)^{1/2} \qquad (13.5)$$

(which is valid within the plasma hypothesis for the second harmonic) and the distribution $N(R)$ (1.1). Since the values of the altitudes of the sources above the photosphere are located approximately on straight lines in the figure it is clear that within the framework of the assumption (13.5) for the chosen type of electron concentration distribution in the corona there is little change in the radial component of the velocity during the burst's development. This is also true for the majority of the bursts in Fig. 51. Certain bursts, however, show a noticeable variation in

---

[†] A value of the same order is obtained if we allow for the lag time of the type II burst (in observations at a fixed frequency) relative to the start of the flare accompanying the burst.

the velocity V_{sR}; they sometimes may increase by a factor of more than 2 (Roberts, 1959a; Wild, 1950a).

In this connection it should be noted that (as the optical observations show) the material ejected from a flare in a number of cases is also accel-

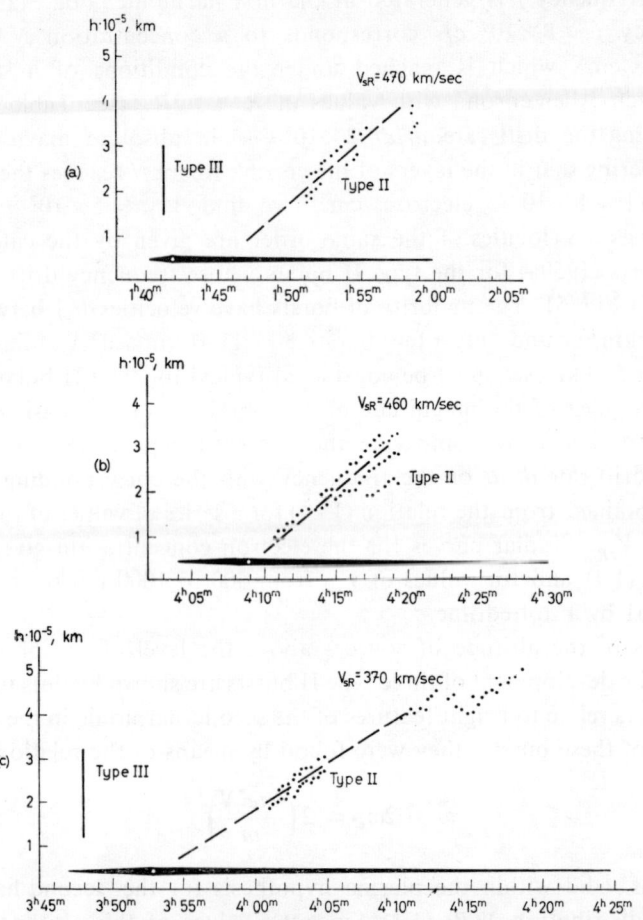

FIG. 52. Altitude h above photosphere of generation sources as a function of the time t during development of three type II bursts: (a) 11 September 1957; (b) 22 November 1957; (c) 6 December 1957 (universal time) (Roberts, 1959a)

erated when moving in the corona, although even if the relations (13.1) and (13.5) are valid the values of the velocity V_{sR} found from the frequency drift and the nature of its variation in the movement of the agent in the corona cannot have any great significance ascribed to them because of the large variations in the distribution of the electron concentration above the

§ 13] Type II Bursts

active regions. In addition the value of N in the generation region is apparently affected by the agent causing the bursts. Therefore calculations of the altitude of sources of radio emission and their velocities from radio spectrographic data are significant in the best case only as estimates of the order of magnitude.

At the same time the correctness of the general idea of the connection of the frequency drift of type II bursts with the motion of the radio emission's source in the corona is confirmed by interferometric determination of the position of the radio emission on the Sun's disk in the 40–70 Mc/s range of frequencies for two type II events (Wild, Sheridan and Trent, 1959).[†] In both the cases investigated the position of the local source of type II radio emission at a fixed frequency remained approximately constant for the whole of the burst's existence, whilst in measurements at successively decreasing frequencies the source showed a systematic displacement relative to the Sun's disk. Since the frequency f corresponding to the maximum of the radio emission's spectral intensity varies during the development of the burst it follows that the generation region of this frequency moves in the corona at a velocity whose component in the east–west direction is

$$V_{sE-W} = \frac{d\theta}{df}\frac{df}{dt} R_{S-E}. \tag{13.6}$$

Here the derivative $d\theta/df$ characterizes the angular variation in the position of the source on the disk in the above-mentioned direction when the frequency changes and df/dt defines the frequency drift rate (R_{S-E} is the distance of the Sun from the Earth).

In one case (7 July 1958) the type II burst, which appeared around the central meridian near a chromospheric flare, moved over the Sun's disk at a velocity of about 250 km/sec. In another case (26 June 1958) a class 2 flare localized near the limb was connected with a complex event whose first phase belonged to spectral type II; the radio emission in the second phase could be placed in type IV. The results of measuring the source position (with an accuracy of ±4′), the polarization of the radio emission and the dynamic spectrum of this event are shown in Fig. 53. The velocity component found from the data obtained for the source of a type II burst moving away from the Sun as the frequency decreases is about 2000 km/sec. This value is far greater than the radial velocity obtained from the data on the frequency drift rate by means of the relation (13.4)

[†] During these observations only one of the source coordinates (in the east–west direction) was determined. For the method of measuring the position of a source of radio emission using a variable-frequency interferometer see section 5.

and the Baumbach–Allen formula (about 500 km/sec for the burst of 26 June 1958). This difference is apparently a real one; if the relation (13.1) corresponds to the actual facts it can be explained by the fact that the gradient of the electron concentration in the corona is a few times less than that taken for the calculations (in this connection see Moiseyev, 1960 and 1961).

When investigating the nature of the agent whose motion in the corona stimulates the appearance of type II radio emission the first thought is to identify this agent with surge prominences which are injected into the corona from the region of a chromospheric flare. Phenomena of this kind can often be visually observed (generally in the H_α line) in a broad class of flares (see section 2).

According to the observational data of Maxwell, Howard and Garmire (1960), which are very incomplete and are by way of being preliminary, the appearance of prominences of the surge type is sometimes connected with type II bursts. In all cases when it was possible to establish the velocity of the material injected it did not exceed 250 km/sec.

To study the connection between optical events and type II bursts simultaneous observations were also made by Giovanelli and Roberts (1958, 1959) of the dynamic spectrum of the radio emission and phenomena near the limb of the Sun were photographed in the H_α line. As a result information was obtained on several type II bursts connected with prominences ejected at the edge of the disk; data on one of these events are given as an example in Fig. 54. Since all the bursts had only one harmonic band it is difficult to decide whether the radio emission received belonged to the first or second harmonic. Reception of only the second harmonic cannot be considered excluded since the generation sources were in the immediate vicinity of the limb where the effect of refraction in the corona above all limits the escape of the radio emission at the fundamental harmonic. If we assume that radio emission occurs at the first harmonic, then within the framework of the plasma hypothesis (13.1) for the Baumbach–Allen model of the corona the time-dependence of the altitude of the generation source will be of the form shown in Fig. 55a. For the second harmonic (13.5) this dependence will obviously be slightly different (see Fig. 55b).

By comparing the dependences of the altitude of the radio emission source on time in these figures with the corresponding data for the ejected prominence it is easy to check that, if the observed radio emission belongs to the second harmonic, the dependence obtained for the phenomenon in the radio band is a direct extension of the corresponding dependence in the optical range. In other cases, however, better agreement is obtained if the radio emission is ascribed to the fundamental harmonic.

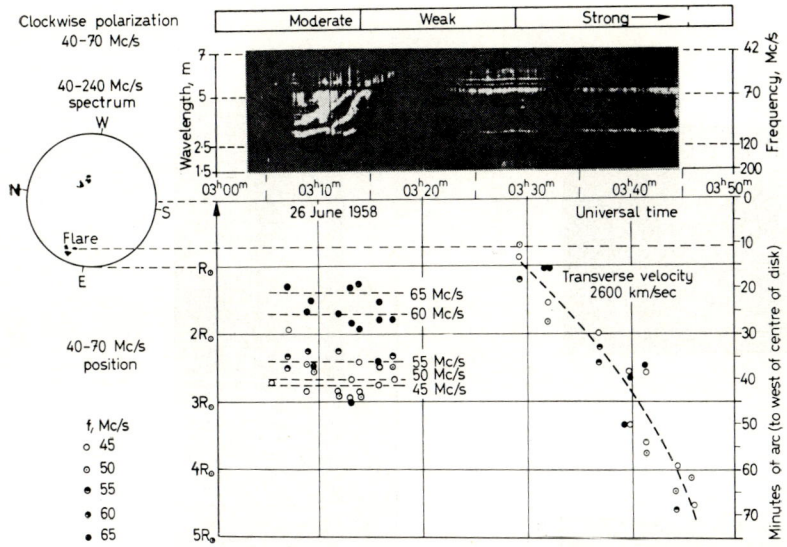

Fig. 53. Polarization, dynamic spectrum and coordinates of the complex type II–IV event of 26 June 1958 (Wild, Sheridan and Trent, 1959)

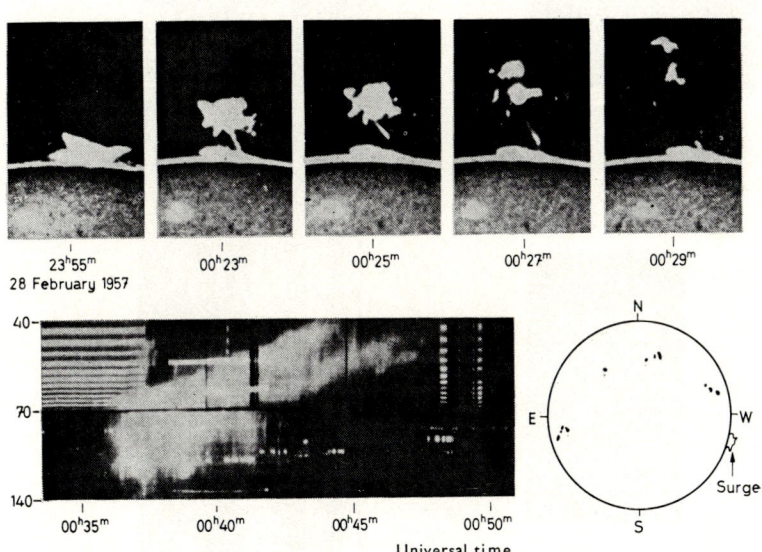

Fig. 54. Photographs of an ejected prominence on the Sun's limb in H_α line light and dynamic spectrum of the accompanying type II burst of 1 March 1957 (Giovanelli and Roberts, 1958)

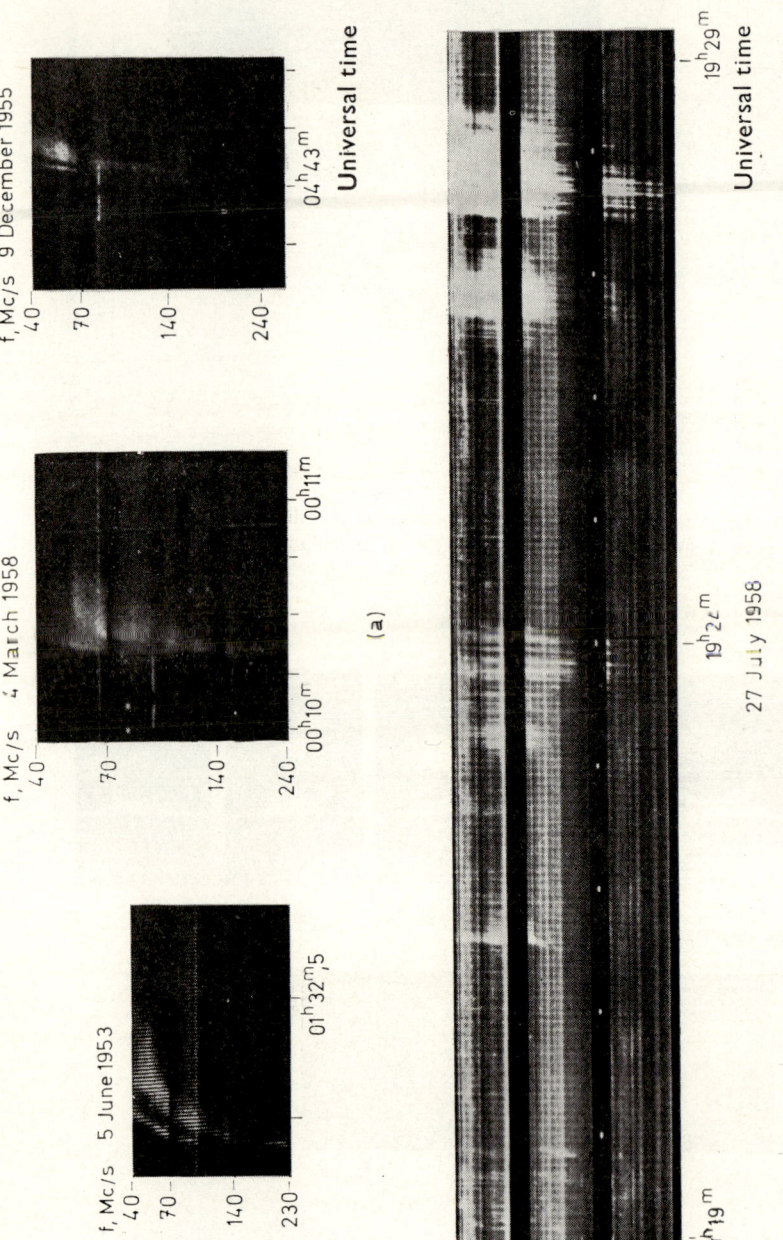

Fig. 56. Dynamic spectra of type III radio emission: (a) individual rapidly drifting bursts with a second harmonic (Wild, Sheridan and Neylan, 1959); (b) group of rapidly drifting bursts (Swarup, Stone and Maxwell, 1960)

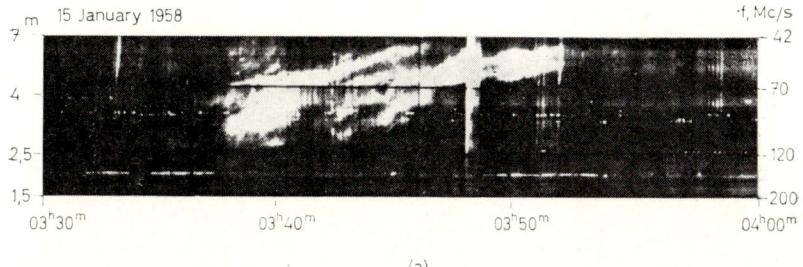

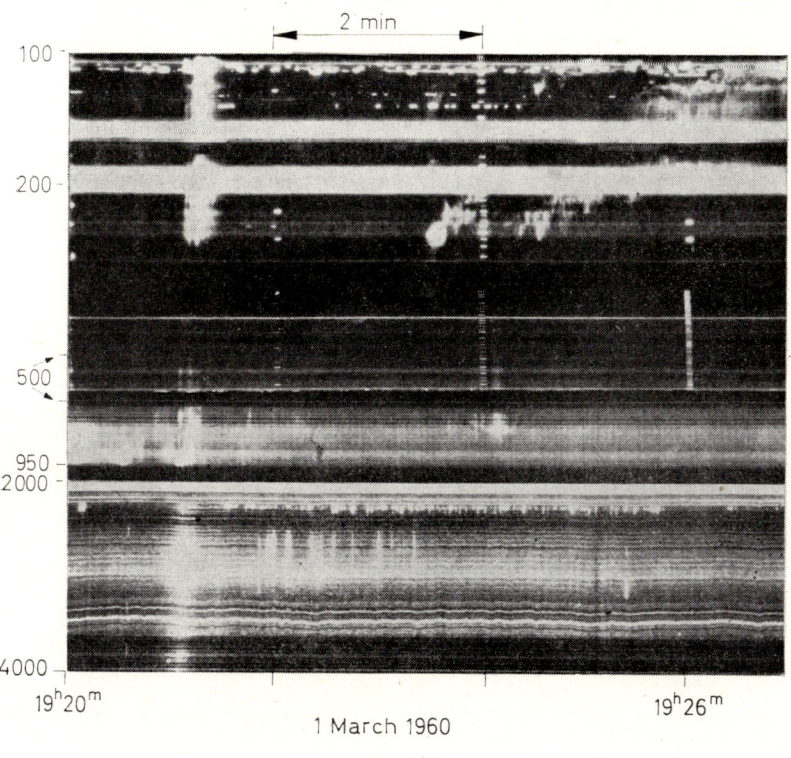

FIG. 58. Dynamic spectra of complex types III–II events: (a) event of 15 January 1958; type III radio emission precedes a type II burst by 5 min (Roberts, 1959a); (b) event of 1 March 1960; group of type III bursts appeared 2 min before type II radio emission (in the 2000–4000 Mc/s range only the smooth rise in the radio emission near 19^h21^m is of solar origin) (Kundu et al., 1961)

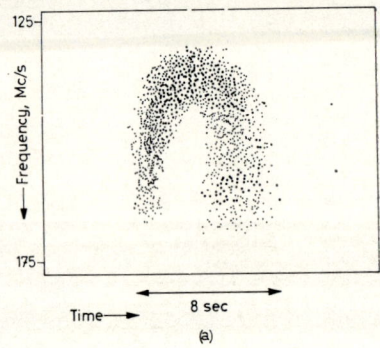

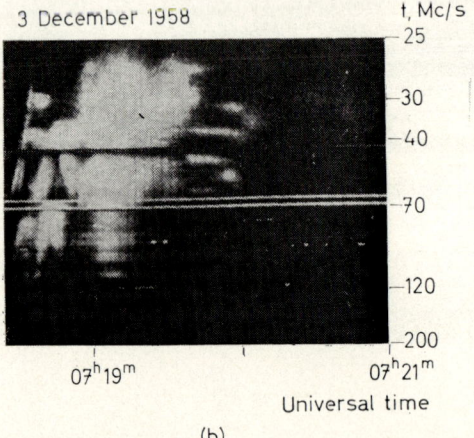

Fig. 61. (a) Dynamic spectrum of a U-burst (Maxwell and Swarup, 1958). (b) Dynamic spectrum of a complex event at the start of which can be seen a U-burst with a second harmonic (Sheridan, Trent and Wild, 1959)

§ 13] Type II Bursts

The data given would appear to confirm the idea that the agent causing type II bursts is prominences ejected from the region of flares. It should, however, be noted that the agreement between the optical and the radio observations (Giovanelli and Roberts, 1958 and 1959) looks unconvincing. The point is that the position of the region where type II bursts are generated was found by the Baumbach–Allen formula (1.1). It is known, however, that (see section 1) even in a stationary corona the distribution of the electron concentration differs noticeably from that accepted in the model

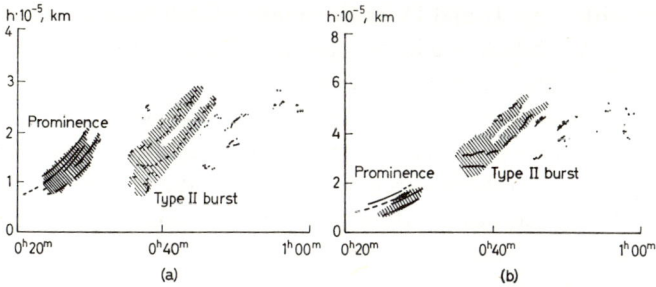

FIG. 55. Dependence of altitude of prominence and altitude of radio emission source above photosphere h on time t for the event of 1 March 1957: (a) if the radio emission belongs to the fundamental harmonic $\omega \approx \omega_L$; (b) if the radio emission comes from the second harmonic $\omega \approx 2\omega_L$ (Giovanelli and Roberts, 1958)

(1.1), not to mention the values of N above the centres of activity. In addition, in estimates of the velocity V_{sR} from the magnitude of the frequency drift no allowance is generally made for the variation in the concentration N in the generation region which is introduced by the agent causing the type II bursts. This variation may be quite significant (see section 31). In one way or another the absence of any noticeable directional properties, about which we spoke earlier, indicates that N deviates considerably from the values in a stationary corona. We also note that the ejection of prominences into the corona is a far more frequent event than the appearance of type II bursts, whilst the velocities of eruptions of this kind (Warwick, 1957; Giovanelli and Roberts, 1958 and 1959) are far lower than the velocities obtained by Wild, Sheridan and Trent (1959) by direct observations of the displacement of a source of type II radio emission connected with a flare near the Sun's limb.

A weighty argument against the type II bursts being connected with prominences of the surge type is the very fact that the motion of the latter is often reciprocating in nature: having risen to a certain altitude in the corona the ejected matter once again approaches the solar surface at approximately the same velocity. During this type of prominence motion the type II bursts should change the sign of the frequency drift rate but this is not usually observed.

At the same time surges of material from flares undergoing a stage of sudden expansion (an "explosive" phase, see section 2) have velocities of the order of 1000–2000 km/sec.† A comparable velocity is found in corpuscular streams which cause on Earth magnetic storms with a sudden onset and aurorae 1 to 3 days after the chromospheric flares. The features of these geoeffective streams and their connection with the solar activity (in particular with types II and IV radio emission) will be discussed in detail in section 17. As a whole the impression is created that there is a continuous chain of events starting with a chromospheric flare and finishing with geophysical phenomena like the magnetic storms and aurorae. The connecting link between these events is the corpuscular stream (plasma bunch) which is ejected from the flare region and moves at a velocity of thousands of km/sec. During its passage through the corona this stream excites radio emission of spectral types II and IV.‡

14. Type III Bursts

GENERAL CHARACTERISTICS

In the absence of type I noise storms the majority of sufficiently intense bursts of solar radio emission in the metric band belong to spectral type III. The dynamic spectra of this sporadic component given in Figs. 56 and 57 are similar in general features to the spectra of type II bursts: they also consist of one or two bands at frequencies in the ratio of 1 : 2, the maximum of the spectral intensity of the emission drifting in time (generally towards the low frequencies). The basic difference of type III bursts from type II bursts consists in that the former develop more rapidly by a factor of two orders. Therefore the basic characteristics of type III bursts can be obtained from the corresponding values for type II by changing

---

† In section 2 one of the arguments adduced in favour of ejection from the flare region of agents moving at a velocity of the order of 10^3 km/sec was the cases of reformation and destruction after a corresponding time of filaments located at great distances from the flare. A similar phenomenon also occurs in the radio band: after microwave bursts connected with chromospheric flares in a number of cases (4–8 min later) a rise has been observed in the radio brightness of regions on the disk at a distance of up to one solar radius away from them ($\lambda = 21$ cm). This lag corresponds to a velocity of 1000 or 2000 km/sec (Mullaly, 1961).

‡ The source of the type IV radio emission moves in the corona at a velocity which is approximately equal to the velocity of the displacement of the region in which type II bursts are generated (see section 15).

§ 14] Type III Bursts

the time scale—by reducing it by a factor of about 100. In actual fact the life of type II bursts is a few minutes or tens of minutes and the frequency drift rate is fractions of a Mc/s^2; for type III the duration at a fixed frequency is 3–15 sec and the drift is about 10–30 Mc/s^2. The "instantaneous" width of the frequency spectrum of rapidly drifting bursts of this kind is 10–50 Mc/s, but may be as much as 100 Mc/s (Wild and McCready, 1950; Wild, Murray and Rowe, 1954; Wild, 1960a; Bracewell, 1955 and 1956; Wild, Roberts and Murray, 1954; Wild, Sheridan and Neylan, 1959).

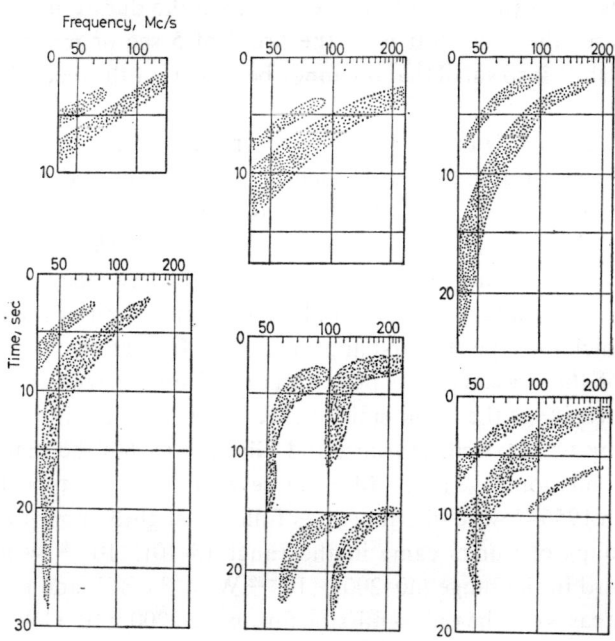

FIG. 57. Typical dynamic spectra of individual type III bursts (schematic) (Wild, Murray and Rowe, 1954)

Type III bursts appear in isolation or in groups containing up to ten elements; the duration of groups of this kind is of the order of a minute. At times type III bursts enter into the composition of type III burst storms lasting for hours which consist of intermittent, interrupted sequences of bursts covering a wide range of wavelengths and appearing at a frequency of 10–100 bursts per hour. Both classes of burst storm (types I and III) may appear independently; cases are known, however, when they have accompanied each other (Wild, 1957).

Bursts at the longer wavelengths (in the decametric band) are for the greater part a low-frequency extension of type III bursts observed at metric wavelengths. This is shown by a comparison of recordings of the solar radio emission intensity at $f = 18$ Mc/s with dynamic spectra in the 25–580 Mc/s range and by spectrographic observations at frequencies of 15–33 Mc/s (Boishot, Lee and Warwick, 1960). The low-frequency part of the bursts lasts longer than the high-frequency part: if the "life" of a type III event at metric wavelengths is up to 10 sec, in the decametric band it is of the order of half a minute and shows a tendency to increase with the wavelength.

The time lag in the arrival of bursts at low frequencies (16 Mc/s) relative to the higher ones (33 Mc/s) is always less than the duration of the burst. In some cases the time shift is of the order of 5 sec or more; for other bursts this shift is so small that it cannot be found on the recordings available.

The start of type III bursts generally occurs in the metric band, although they sometimes appear first at decimetric wavelengths and then drift towards the low frequencies (Malville, 1962). To judge from preliminary spectrographic observations (Rabben, 1960) in the 48–165 Mc/s range this limit depends on the intensity of the burst and the frequency drift rate: as the latter increase a tendency is found for the upper limit of the frequency range in which the type III burst appears to rise. Indications have also been obtained of the presence of a positive correlation between the drift rate and the intensity of the radio emission (Rabben, 1960).

According to Maxwell, Howard and Garmire (1960) the magnitude of the radio emission flux at 10 Mc/s of several thousand type III events observed in 1956–7 were distributed as follows: slightly more than half of all the groups of bursts came in the range $(5–40) \times 10^{-22}$ W m^{-2} c/s^{-1}, about a third in the range $(40–200) \times 10^{-22}$ W m^{-2} c/s^{-1}, and only in one-sixth of the cases did the radio emission flux exceed 200×10^{-22} W m^{-2} c/s^{-1}. The maximum value of the flux during type III events may reach 10^{-18} W m^{-2} c/s^{-1} (Goldstein, 1959; Wild, Sheridan and Neylan, 1959), which corresponds to an effective temperature of $T_{\text{eff}\odot} \sim 10^{10}$ °K when reduced to the Sun's disk. This value is $\sim 10^4$ times greater than the corresponding value of $T_{\text{eff}\odot}$ for the "quiet" Sun at metric wavelengths.

Interferometer measurements of the angular size of local sources of type III radio emission indicate that the sizes are greater than the diameter of the accompanying flares and average 8–9' at a frequency of 55 Mc/s. The diameter of the radio-emitting region does not remain constant: it decreases as the frequency rises approximately from 10' at $f = 45$ Mc/s to 6' at $f = 65$ Mc/s (Wild and Sheridan, 1958; Weiss and Sheridan, 1962). Still lower values of the angular size (down to 3') are obtained in the 105–140 Mc/s range (Goldstein, 1959).

§ 14] Type III Bursts

If we take account of the values given above for $T_{\text{eff}\odot}$ during type III bursts and the values of the angular sizes of the local sources, it turns out that the effective temperature T_{eff} of the sources reaches 10^{11} °K in certain cases.

Two harmonic bands with a frequency ratio of approximately 2:1[†] are observed only in a few percent of high-speed bursts, i.e. far more rarely than in type II events (Wild, 1960a; Bracewell, 1955 and 1956). This is apparently partly connected with the fact that in dynamic spectrum recordings it is difficult to resolve a type III burst into two bands because of the great width of the "instantaneous" frequency spectrum. In addition, it is not excluded that in some type III bursts only second harmonic (without the first) radio emission is observed, since the latter because of the directional nature of the radio emission cannot always be received on Earth (see below).

The intensity of the second harmonic in the cases when it is observed together with the first is often comparable with the intensity of the fundamental harmonic but is mostly less than it. In ordinary type III bursts a third harmonic is not found; all that can be said is that its intensity is less than 0·1 of the corresponding value for the second harmonic (Wild, Murray and Rowe, 1954; Wild, Roberts and Murray, 1954). Certain indications of the existence of higher harmonics have been obtained in observations of one of the varieties of type III bursts—the so-called U-bursts about which we shall speak later.

We notice that in type III bursts the same tendency is observed for mutual positioning on the Sun's disk of sources of the fundamental and second harmonics as for type II bursts (Smerd, Wild and Sheridan, 1962) (see section 13). Amongst six type III phenomena investigated in two bursts the sources of the second harmonic were closer to the centre than the sources of the fundamental harmonic, in three bursts the sources coincided, and in only one burst was the location of the sources the opposite (although it was this very positioning of the sources which was to be expected with the measurement method used[‡] if the positions of the sources of radio emission at the frequencies f and $2f$ were the same). Although the results for type III bursts proved to be less definite than for type II events, it may be noted here that in many cases it is not the radio-emission generation region that is apparently observed at the second harmonic but its image in radio rays reflected from the deeper lying layers of the solar corona.

[†] In type III bursts the frequency ratio shows more scatter than for type II events; in the majority of bursts, however, this ratio is close to 1·8–2·0 (Stewart, 1962).
[‡] For the method of observations of this kind see section 13.

A possible explanation of this effect will be given in section 31. All we shall say here is that this feature of the second harmonic radio emission agrees with the time lag of the second harmonic relative to the first (of 2–4 sec) observed for many bursts. It is true that for certain type III phenomena instead of a lag there is a time lead of up to 2 sec; the latter may be caused by a lower group velocity of the fundamental harmonic radio emission in the corona provided that it is the actual source of the second harmonic that is observed and not its reflection (Stewart, 1962).

POLARIZATION OF BURSTS

The combined polarization and spectrographic observations made by Komesaroff (1958)[†] showed that about half the type III bursts are not polarized (i.e. the degree of polarization is known to be less than 25%—the sensitivity threshold of the polarimeter used). The remaining type III bursts show quite strong polarization: the degree of polarization varies between 30% and 70%. In the general case the polarized component is elliptical; the form of the ellipse varies within wide limits. The appearance of circular or quasi-circular polarization is more probable, however. The type III bursts that appear in the course of a day generally have the same sign of rotation, which is the same as the sense of the rotation of the type I bursts observed at the same time. Some of the bursts that have a second harmonic are also polarized, but the degree of polarization at the fundamental frequency is higher than at the second harmonic. There is a basis for saying that the polarization effect observed is an objective characteristic of type III bursts and not a consequence of the differing absorption of ordinary and extraordinary waves in the Earth's ionosphere or the reception of radio emission reflected from the Earth's surface.

These results are basically confirmed by Cohen's data (Cohen and Fokker, 1959); he also observed weak elliptical and linear polarization in some type III bursts at $\lambda = 1\cdot 5$ m. His polarimeter operated in a narrow frequency band (10–18 Mc/s) and had a higher sensitivity than in the experiments of Komesaroff (1958). Just as in the latter case a radio spectrograph was used to check the type of burst. For the three groups of bursts studied in detail the degree of polarization ϱ was 5–30% and the ratio of the axes of the ellipse p varied from a value close to zero to 0·8.

Further investigations (Akabane and Cohen, 1961; Akabane and Cohen, 1960) on polarimeters with a broad band of 10 and 20 kc/s in one series of observations and of 10 and 300 kc/s in the other showed that a noticeable

[†] In these observations aerials orientated at right angles were coupled with a radio spectrograph covering the 40–140 Mc/s band. For the method of observations of this kind see section 6.

§ 14] Type III Bursts

part ($\sim 1/8$) of the strong type III bursts displays linear or elliptical polarization in the 10 kc/s band, although circular polarization is a more frequent phenomenon. A special study has been made (taking a few dozen bursts) (Akabane and Cohen, 1960) of the "degree of coherency" of the two circularly polarized components defined as the amplitude of the correlation function of these components (6.15):

$$\tilde{Q} = \frac{\overline{E_{0r}E_{0l}}}{(I_r I_l)^{1/2}}$$

(see section 6). Knowing the value of $\tilde{Q}$ from simultaneous observations in two frequency bands we can obtain valuable information on the propagation conditions in the corona of radio waves leaving the generation region and, in particular, the value of the Faraday rotation of the plane of polarization in the solar corona (see section 23).

The measured degree of coherency for the 10 kc/s band $\tilde{Q}_{10}$ does not exceed 0·3. The corresponding value $\tilde{Q}_{20}$ for the 20 kc/s band is always less than $\tilde{Q}_{10}$; the ratio $\tilde{Q}_{20}/\tilde{Q}_{10} \approx 0\cdot6$–$0\cdot8$. Lower ratios could not be obtained since to do this it is necessary to measure small values of $\tilde{Q}_{20}$; the latter was prevented by the low accuracy of the observations (not better than 10%).

A comparison of the results of simultaneous observations in bands 10 kc/s and 300 kc/s wide led to the conclusion that even in the cases when the type III bursts had a significant degree of coherency at 10 kc/s, over a wide band $\tilde{Q}$ did not differ from zero by a value exceeding the measurement error. The estimated upper limit of the ratio of the degrees of coherency was 0·2–0·5.

It may therefore be concluded from the above that for intense polarized type III bursts at $\lambda = 1\cdot5$ m the degree of coherency between the clockwise and anticlockwise components of the radio emission decreases as the bandwidth increases; in an interval of the order of hundreds of megacycles these bursts stand out as being unpolarized or weakly circularly polarized.

CONNECTION WITH OPTICAL PHENOMENA

Groups of type III bursts sometimes precede slowly drifting type II bursts (generally appearing during chromospheric flares), forming a complex type III–II event (Wild, Roberts and Murray, 1954). Examples of events of this kind can be seen in Fig. 52; their dynamic spectra are shown in Fig. 58. In relation to type II radio emission combined phenomena of this kind occur quite often: about 60% of the slowly drifting bursts are accompanied by type III bursts, preceding the type II by an average of

5 min. In this case the groups of type III bursts are "precursors" of type II radio emission. A group may include dozens of rapidly drifting elements which on the whole carry more energy than the slowly drifting burst following it. In the other cases the type III radio emission, on the other hand, is relatively weak and consists only of one or two individual bursts (Roberts, 1959a).

The connection between rapidly and slowly drifting bursts can be apparently explained (Roberts, 1959a; Wild, Roberts and Murray, 1954) by the fact that during the ejection from the chromospheric flare of a flux moving at a velocity of the order of 10^3 km/sec and responsible for the generation of type II radio emission a second agent is also injected into the corona which excites rapidly drifting type III bursts. Since the velocity of the latter is very high (of the order of 10^5 km/sec, as will be clear from what follows) this agent precedes the plasma bunch connected with the type II event and is the first to reach the high layers of the corona in which generation of metric band radio emission occurs.

It should be stressed, however, that since the type II radio emission is a very rare phenomenon the greater majority of type III bursts does not make part of type III–II events. Before the work of Lougheed, Roberts and McCabe (1957) this part of the rapidly drifting bursts was not connected with any optical events on the Sun's disk, thus giving a foundation for calling them "isolated" bursts. However, a comparison of the dynamic spectra in the 40–240 Mc/s band with ciné photography data of the Sun in H_α rays allows us to establish that type III bursts (both the individual surges and the groups) are closely connected with solar flares. Additional arguments in favour of this connection are provided by interferometer observations in the 40–70 Mc/s band (Lougheed, Roberts and McCabe, 1957), due to which the position of the source of the radio emission according to one coordinate is close to the position of the flare on the Sun's disk. The connection of type III with flares is confirmed by the investigations of Maxwell, Howard and Garmire (1960), Rabben (1960), Maxwell and Swarup (1958). According to Lougheed, Roberts and McCabe (1957), Maxwell, Howard and Garmire (1960) and Rabben (1960) up to 60–70% of the type III bursts and groups of bursts with an intensity of more than 5×10^{-21} W m^{-2} c/s^{-1} appear during solar flares (generally in the first stage of development between the start and the maximum of the flare) or slightly ahead of them. On the other hand, only about a quarter of all the flares correlate with type II emission. This very low percentage indicates that there are apparently some features or other whose presence determines the appearance of type III bursts.

The extent of the connection of type III emission with flares changes

§ 14] Type III Bursts

noticeably over the course of a few days. The flares appearing in some active regions show a high degree of correlation, whilst the flares from other centres of activity are not connected or are weakly connected with rapidly drifting bursts (Lougheed, Roberts and McCabe, 1957; Rabben, 1960). Apart from the dependence on the centres of activity which, moreover, is placed in doubt by Maxwell, Howard and Garmire (1960), the probability that a flare is accompanied by type III bursts depends on the importance of the flare: only 20% of microflares (i.e. flares of class 1−) are accompanied by rapidly drifting bursts, whilst for flares of class 1 and over this value increases sharply (up to 60%). However, since large chromospheric flares are a comparatively rare event the majority of bursts are connected with very weak flares (microflares) (Lougheed, Roberts and McCabe, 1957).

The correlation increases slightly for flares with ejections of material into the corona (Lougheed, Roberts and McCabe, 1957; Maxwell, Howard and Garmire, 1960): about a third of these flares are accompanied by type III radio emission, which is one and a half times greater than the corresponding value for flares without eruptions. It is known that eruptions of this kind are found in about a quarter of all flares.

Part of the flares that are accompanied by the ejection of diffuse material passes through a stage of sharp and rapid expansion for about 1 min, which is in the nature of an explosion (see section 2). It is this expansion, according to Giovanelli (1958 and 1959), that is the characteristic of flares with which the appearance of type III radio emission is closely connected: about 70% of the flares studied (mostly in the 1− class) when passing through the "explosive phase" are accompanied (with an accuracy of ±2 min with respect to the time of the explosion) by rapidly drifting bursts. Apparently, during this expansion as well as the slowly moving diffuse material (velocity 100 km/sec) a very fast agent responsible for creating the radio emission is injected into the corona.†

The close connection of type III bursts with flares allows us to obtain information on the degree of directivity in the radio emission from the nature of the distribution of the chromospheric flares accompanying the rapidly drifting bursts over the Sun's disk. The distribution obtained by Maxwell, Howard and Garmire (1960) and Rabben (1960) for the flares connected only with strong bursts shows a sharp decrease as one moves away from the central meridian indicating that the emission is directional. At the same time the flares connected with all the bursts that have been

---

† We recall that the flares passing through the "explosive phase" also eject streams with a velocity of the order of thousands of kilometres per second, which cause type II bursts as they move through the corona (see sections 2 and 13).

observed, both the strong and the weak ones, do not show any similar dependence on the distance from the central meridian (Lougheed, Roberts and McCabe, 1957; Morimoto and Kai, 1961; Maxwell, Howard and Garmire, 1960). The latter authors explain the quasi-uniform nature of the distribution in the latter case by the directional effect being masked in a large number of bursts and when the frequency of occurrence of bursts is high by an increase in the number of random coincidences of bursts and flares.

It is clear from what has been said that the question of the directional nature of the type III bursts still awaits elucidation. Directivity is quite possible although certain doubts about its existence are raised by the fact that the percentage of bursts having two harmonic components not only does not decrease with the distance of the radio emission source from the central meridian but, on the contrary, even increases. In the case of the directional radio emission chiefly occurring in the reception of the fundamental harmonic, this harmonic should be expected to be observed mostly in the bursts that appear in the central part of the Sun's disk; in this case the bursts a long way from the centre will start to contain only one component—the second harmonic.

At the same time the number of bursts connected with type III bursts is about one and a half times greater on the eastern than on the western side of the disk (Lougheed, Roberts and McCabe, 1957; Maxwell, Howard and Garmire, 1960; Wild, Sheridan and Neylan, 1959; Rabben, 1960). The east–west asymmetry for outbursts connected with chromospheric flares had been noted earlier from the data of observations at a fixed frequency (Hey, Parsons and Phillips, 1948; Hey and Hughes, 1955). It is possible that this effect is caused by the "non-symmetrical" nature of the refraction in the solar corona brought about by the east–west asymmetry in the electron distribution above the active regions of the corona (Hey and Hughes, 1955). From this point of view it becomes understandable why there is no east–west asymmetry for type III bursts that have been observed at higher frequencies (250–580 Mc/s) where the effect of refraction in the corona becomes insignificant (Maxwell, Howard and Garmire, 1960). We note that east–west asymmetry has also been observed for the sources of type I noise storms (Suzuki, 1961); in the latter case, however, the greater number of sources was not on the eastern but on the western half of the Sun's disk.

§ 14] Type III Bursts

Position and Movement of an Emitting Region in the Corona. Frequency Drift of Bursts

The observations of Wild, Sheridan and Trent (1959), Wild, Sheridan and Neylan (1959) on an interferometer with frequency tuning showed (see also Goldstein, 1959) that the position of the source of type III radio emission at a given frequency remains constant with an accuracy of up to 1′ for the whole life of an individual burst. However, the radio emission at the different frequencies comes from different regions; this effect is particularly noticeable in rapidly drifting bursts connected with flares that appear near the Sun's limb.

The above can be clearly seen in Fig. 59, where the positions of the sources of eight type III bursts at 45–60 Mc/s are marked by short lines. Since during the measurements the source of the bursts was localized in only the one coordinate θ (in the east–west direction), all that can be said is that the sources of the radio emission are located on lines which are extensions of these sections. The latter indicate the position of the radio emission source if the sources are located on an extension of a solar radius passing through the chromospheric flare connected with the burst. It is clear from the figure that the radio emission is generated at a considerable altitude in the corona since the source of the bursts is a long way outside the bounds of the Sun's disk.

From Fig. 59 and the nature of the distribution of the number of sources with respect to their distance from the centre of the disk at a frequency of 60 Mc/s it may be concluded that a radio emission source at this frequency

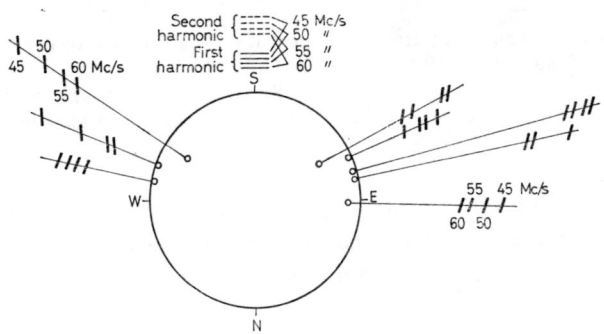

Fig. 59. Positions of eight sources of type III bursts at frequencies of 60, 55, 50 and 45 Mc/s. The flares connected with the bursts are marked by small circles on the Sun's disk. At the top of the drawing are given the levels in the corona responsible for emission at frequencies of 60–45 Mc/s in the Baumbach–Allen model of the corona and calculated on the assumption that the observed radio emission is either the fundamental or second harmonic (Wild, Sheridan and Neylan, 1959)

is at a distance of $\theta_{60}R_{S-E} < 1\cdot 66R_\odot$ from the centre. It follows from this that type III radio emission with a frequency of 60 Mc/s is generated at a level located in the corona at a distance $R_{60} \simeq 1\cdot 66R_\odot$ from the centre of the Sun. It is important to note that the radio emission at $f = 45$ Mc/s comes from a region located on an average $0\cdot 34R_\odot$ further from the centre of the disk than the emission at 60 Mc/s ($R_{45} \approx 2\cdot 00R_\odot$; see Fig. 59). The decrease in the generated frequency as the source gets further away from the solar surface is confirmed by interferometer observations of the coordinates of the regions in which type III emission is generated at frequencies of 19·7 Mc/s (Shain and Higgins, 1959) and 200 Mc/s (Morimoto and Kai, 1961). Here the distances from the sources to the centre of the Sun ($2\cdot 9R_\odot$ and $1\cdot 25 R_\odot$) were found by comparing the position of the flares on the Sun and that of the bursts connected with them (on the assumption that the sources on the average are located in the corona radially above the flares).†

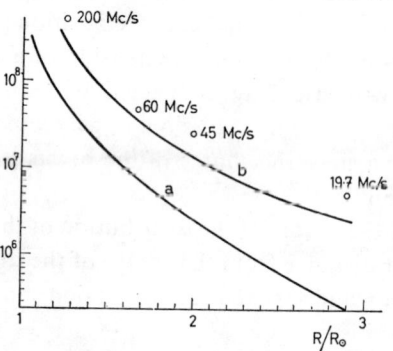

Fig. 60. Electron concentration in corona N_e as a function of distance from centre of Sun R: a—Baumbach–Allen model of stationary corona (1.1); b—Newkirk's model of active region. Small circles show values of N_e obtained by processing radio observation data on basis of "plasma hypothesis" (Wild, Sheridan and Neylan, 1957)

The distances given considerably exceed those calculated for the Baumbach–Allen model of the corona on the "plasma hypothesis"‡, not only on the assumption that the radio emission belongs to the fundamental

---

† We note that no allowance was made for refraction of the radio waves in the corona when finding these values; its effect does not apparently significantly alter the results obtained by Morimoto and Kai (1961), Wild, Sheridan and Neylan (1959), Shain and Higgins (1959) (in these connection see the remarks made in section 22).

‡ As we have already said in section 13, the basis of the "plasma hypothesis" is the statement that the frequency ω of radio emission generated in the corona is close to the characteristic frequency ω_L of the plasma oscillations in the generation region (13.1). From this standpoint the second harmonic will be twice ω_L (see (13.5)).

§ 14] Type III Bursts

harmonic (i.e. $\omega \approx \omega_L$) but also if it is the second harmonic ($\omega \approx 2\omega_L$; see Fig. 59). If the "plasma hypothesis" is true, this fact means that in the region where type III bursts are generated the electron concentration N that determines the characteristic frequency ω_L of the coronal plasma is far greater than that given by the Baumbach–Allen formula and agrees better with the model given in section 1 for the corona above active centres, but is slightly higher than the latter. This is easily checked by examining Fig. 60 where the small circles show the concentration values obtained by formula (13.1) for the frequencies 200, 60, 45 and 19·7 Mc/s which are emitted by sources located at the above-mentioned distances from the centre of the Sun.

Furthermore, it follows from the observations that have been made that the type III emission at the lower frequencies appears not only further from the photosphere but also later than at the high frequencies. This obviously means that during the development of a type III burst the generation region moves in the corona. From data on the time of a burst's appearance and the position of the emitting region at different frequencies we can find the agent's velocity component in the plane of the Sun's disk (from the observations of Wild, Sheridan and Trent (1959), Wild, Sheridan and Neylan (1959), it is the east–west component).

An analysis of the distribution of the difference of the coordinates of the bursts ($\theta_{45} - \theta_{60}$) in the east–west direction at frequencies of 45 and 60 Mc/s as a function of the position of the source of radio emission at 60 Mc/s (i.e. of θ_{60}) shows that there is considerable dispersion in this quantity. However, the value of $\theta_{45} - \theta_{60}$ averaged for many bursts varies almost linearly as θ_{60} changes, passing through zero when the θ_{60} coordinate coincides with the centre of the Sun's disk. In accordance with what has been said the east–west component of the agent's velocity in the plane of the Sun's disk calculated from the relation†

$$V_{s,E-W} = \frac{\theta_{45} - \theta_{60}}{t_{45} - t_{60}} R_{S-E} \qquad (14.1)$$

(t_{45} and t_{60} are respectively the times when the type III radio emission reaches its maximum intensity at the frequencies of 45 and 60 Mc/s) displays a systematic increase as one moves away from the centre of the disk, reaching maximum values of the order of 0·5c, where c is the velocity of light in a vacuum.

The features noted in the distribution of $\theta_{45} - \theta_{60}$ and $V_{s,\,E-W}$ with respect to θ_{60} indicate that the assumption of the radial nature of the motion of the

† Compare with the formula (13.6).

agent away from the Sun is on the average quite reasonably valid. Then the value of the agent's total velocity can be estimated from the relation

$$V_s \approx \frac{R_{45}-R_{60}}{t_{45}-t_{60}}, \tag{14.2}$$

where the distances R_{45} and R_{60} of the sources from the centre of the Sun have been found above.

No allowance has been made in the formulae (14.1) and (14.2) for the finite value of the difference in the propagation time of the radio emission from the sources of generation at the frequencies of 45 and 60 Mc/s to the observer; it is essential to allow for this difference if the velocity V_s is close to the velocity of light. In the latter case, as can easily be checked, instead of (14.2) we must use the formula

$$t_{45}-t_{60} = \left(\frac{1}{V_s}-\frac{1}{c}\cos\Theta\right)(R_{45}-R_{60}), \tag{14.3}$$

in the derivation of which it is assumed that the velocity of signal propagation in the corona is equal to the velocity of light[†] (Θ is the angle between the agent's velocity V_s and a line connecting the centre of the Sun with the observer). If we put[‡]

$$\theta_{45} \approx R_{45}\sin\Theta, \qquad \theta_{60} \approx R_{60}\sin\Theta,$$

then the formula (14.3) can be used to find the value of V_s from the measured values of θ_{60}, θ_{45} and $t_{45}-t_{60}$; in this case the value of Θ is determined from the position on the Sun's disk of the flare connected with the type III event. It turned out as a result that for the rapidly drifting bursts investigated by Wild, Sheridan and Neylan (1959) the velocities V_s lie in the range between $0.2c$ and $0.8c$ with a mean value of $V_s \approx 0.45c$. It is significant that the distribution of the bursts with respect to the velocities of their agents is sharply limited at the lower end: velocities of $V_s < 0.2c$ are not observed.

On the other hand the velocity of the agent causing type III bursts (to be more precise, its radial component V_{sR}) can be estimated on the basis of the "plasma hypothesis" by means of the formulae (13.1) and (13.4) if we know the frequency drift rate df/dt and state a definite rule for the variation of the electron concentration N in the source with altitude

[†] In actual fact this velocity in a plasma is less than c; the corresponding corrections to (14.3) are calculated by Wild, Sheridan and Neylan (1959) and taken into consideration in the estimates of the agent's velocity V_s which are given below.

[‡] These relations are valid if the source moves in an east–west direction over the disk; they can be accepted approximately for all type III events occurring in low heliographic latitudes.

§ 14] Type III Bursts

(Wild, 1950b). If we assume that this concentration is close to that determined by the Baumbach–Allen formula (1.1), then the values obtained in this case are in certain cases 2–5 times smaller than those found from interference observations. Generally estimates of this kind lead to values of $V_{sR} = 3 \times 10^4 - 10^5$ km/sec (Wild, Sheridan and Trent, 1959; Wild, Sheridan and Neylan, 1959); the same velocities are also given by Newkirk's estimates for the active region of the corona (Table 2 in section 1) which are based on data on the frequency drift in the 8–41 Mc/s range (Malville, 1962).

The situation here is just the same as for the type II bursts (section 13). It follows from the above discrepancy in the velocity estimates (within the framework of the concept of generation of the "fundamental frequency" of type III bursts at the $\omega \approx \omega_L$ level, where ω_L is the characteristic frequency of the oscillations of a plasma with a concentration N) that the magnitude of N in the generation region and its variation with altitude as the source moves in the corona differ from the values in the stationary corona, for instance, from the distribution (1.1).

The latter circumstance is confirmed by the anomalously high position of the source of type III radio emission in the corona noted above, which is apparently connected with the deviation of the electron concentration distribution above the centres of activity from those taken in the Baumbach–Allen model. The agent that excites the type III bursts, and is apparently corpuscular in nature, also has a certain influence on the magnitude of N in the generation region. Due to the latter $N = N_c + N_s$, where N_c and N_s are the electron concentrations in the corona and in the flux. It can be assumed, however, that the part played by the corpuscular flux in the variation of the magnitude of N is small in the majority of cases if the available and, it is true, very ambiguous data on the directional nature of the type III radio emission are confirmed. The point is that this kind of directivity cannot be explained if the flux density is greater than the density of the corona since the effect of refraction in the corona on the propagation of the radio waves from the emission source even at the fundamental harmonic (not to mention the second harmonic) will then be slight. When the emission is directional for the majority of bursts the agent's density N_s should be less than the density of the surrounding corona. In accordance with the formula (13.1) for the metric band ($f \sim 100$ Mc/s) the electron concentration in the generation region is $N \sim 10^8$ electrons/cm³, i.e. $N_s < 10^8$ electrons/cm³. If the directional feature is present right up to a frequency of $f \sim 16$ Mc/s (this is the minimum value of the frequency at which type III bursts have been observed (Boishot, Lee and Warwick, 1960)), then it follows from what has been said that the concentration in the corpuscular fluxes does not then exceed 3×10^6 electrons/cm³.

On the other hand, it is possible that in certain cases when the flux passes through the outer rarefied layers of the corona the case occurs when $N_s \gg N$. This assumption allows us to interpret one feature of the dynamic spectrum of type III bursts which can be seen in Fig. 57. The point is that sometimes during the development of a type III burst the rapid frequency drift (~ 20 Mc/s^2) that is characteristic of it slows down sharply to almost a dead stop at a certain frequency $f_{\min}$; for the fundamental harmonic radio emission $f_{\min} \lesssim 25$–50 Mc/s. The slowing down of the drift observed can be explained (Zheleznyakov, 1956) by assuming that in the high layers of the corona, where $N_s \gg N$, the radio emission frequency is determined by the parameters of the stream. From this it is clear that the limiting frequency achieved by a burst at the fundamental frequency is

$$\omega_{\min} = 2\pi f_{\min} \approx \omega_{L_s}, \qquad (14.4)$$

where $\omega_{L_s} = (4\pi e^2 N_s/m)^{1/2}$. The relation (14.4) allows us to estimate the density of the corpuscular flux if we know the value of $f_{\min}$: when $f_{\min} \lesssim (2 \cdot 5 - 5) \times 10^7$ c/s the concentration is $N_s \lesssim (1-3) \times 10^7$ electrons/cm^3. These values are very high. However, if the interpretation suggested for the reduction in the drift rate is true, then for bursts with this feature the emission of the fundamental harmonic in a rapidly drifting medium when $N_s \ll N$ should be directional in nature.

In the second phase, which is characterized by a sharp reduction in the frequency drift rate, $N_s \gg N$ and the radio emission becomes non-directional. It can therefore be assumed that the fundamental harmonic of these bursts a long way from the centre of the disk should be preferentially observed only during the last phase with a reduced drift rate, whilst the second harmonic can be recorded in both phases. This effect can be checked in two ways: first, by comparing the times of appearance of the fundamental and second harmonic in the dynamic spectrum of the type III bursts which appear in the centre of the Sun's disk and a long way from it and, secondly, by comparing the mean rate of frequency drift for the fundamental harmonic and second harmonic of the bursts that appear at great distances from the centre of the disk. Since, however, the second harmonic is a rather rare phenomenon in type III emission it is easier to investigate the dependence of the frequency drift rate of all type III bursts on the distance from the central meridian. Since these bursts also include events whose dynamic spectrum have the above-mentioned feature there should be a certain reduction in the frequency drift rate as the heliographic longitude rises if the suggested interpretation is correct. This effect appears to be confirmed by observations (Rabben, 1960).

§ 14] Type III Bursts

It is worth stressing that this explanation for the slowing down and stopping of the frequency drift seems to be hardly convincing. It is more likely that this effect, in principle, may be the consequence of a decrease in the velocity of the corpuscular flux because of energy losses in excitation of plasma waves by the latter. On the other hand, it is known that the reduction in the agent's velocity cannot be connected with collisions between the particles of the flux and the corona since the protons and electrons moving at a velocity of 10^5 km/sec fly through the corona with hardly any collisions. At the same time a reduction in the frequency drift rate because of energy losses of the flux in readjustment of the magnetic field "frozen" into the coronal plasma, or because of a change in the direction of the velocity of the flux causing the reduction in the radial component V_{sR}, cannot be excluded if the configuration of the lines of force is essentially not rectilinear. In the latter case bursts with a reduction in the drift rate should be looked upon as intermediate between ordinary type III bursts and U-bursts.

U-BURSTS

Above, we have been discussing the features of ordinary type III bursts, during the whole development of which the drift rate is towards the low frequencies (which corresponds to movement of the agent into the higher and more rarefied layers of the corona). In the type III fine structure of the type II bursts as well as the elements that drift rapidly towards the low frequencies there are surges with a frequency drift in the opposite direction (for further detail see section 13). Sometimes, however, in the composition of the Sun's sporadic radio emission bursts are also observed during whose development the drift towards the low frequencies first slows down, stops at a certain frequency f_{min} and is then replaced by a drift with an increase in frequency. The dynamic spectrum of bursts of this kind looks like an inverted letter U (Fig. 61); therefore this component of the radio emission (after its discovery by Maxwell and Swarup (1958)) has been given the name of a U-burst. Since the value of the frequency drift rate of U-bursts in both directions, the lifetime and the width of the frequency spectrum are characteristic of type III bursts. This component is not generally put into a separate spectral class but considered to be one of the variants of the rapidly drifting type III bursts.

According to Maxwell and Swarup (1958) the majority of U-bursts have a spectrum "turn" frequency f_{min} of about 100–150 Mc/s (Fig. 61a) although in certain cases the radio emission frequency does not drop below

250–570 Mc/s (Alsop *et al.*, 1959a and 1959b). According to the data reported by Goldberg (1958) the values of the "turn" frequency are distributed more or less evenly in the range of 100–580 Mc/s. Figure 61b shows an example of a U-burst in which f_{min} is only 35 Mc/s (Sheridan, Trent and Wild, 1959).

Just as in ordinary type III bursts the radio emission of the U-bursts sometimes consists of two bands whose frequencies are in the ratio of 2 : 1 (Fig. 61b). In one case (3 September 1957) the appearance was recorded (Haddock, 1959b; McMath 1958) of three U-bursts whose "turn" frequencies (130, 250 and 375 Mc/s) were reached simultaneously with an accuracy of up to the observational error ($\sim$ 1 sec). The duration of the radio emission in all three bursts, and likewise the variation in intensity during their development, was the same. If at the same time we remember that the ratio of the "turn" frequencies is close to 1 : 2 : 3 it is very probable that there is a close connection between the individual radio emission bands: in this case three harmonics of a single event seemed to be observed. This event had an exceptionally high intensity when compared with the other bursts; however, the low-frequency harmonic component (the fundamental harmonic), unlike the second and third harmonics, was very weak. It is possible that the directional nature of the fundamental harmonic radio emission was making itself felt here. If this is so it is not excluded that in the majority of U-bursts we are observing radio emission of the second and not the fundamental harmonic.

The frequency drift inversion characteristic of U-bursts can be explained by the fact that the agent causing the bursts first rises into the upper layers of the corona and then drops down. This kind of trajectory of the agent (if it is a corpuscular stream) cannot be caused by gravity since the velocity of the agents causing U-bursts estimated from the observed values of df/dt is far more than hyperbolic. On the other hand, it is quite possible that this kind of motion of the agent in the corona is caused by the magnetic field of bipolar groups of spots (see Fig. 1) along which the particles of the corpuscular stream are also moving. In this case the trajectory of the stream may be inclined to the surface of the Sun, i.e. the source of the radio emission as well as moving in a radial direction should show noticeable displacement along the solar surface. The latter is confirmed by interferometer observations (Wild, Sheridan and Trent, 1959), according to which in one case studied the positions on the Sun's disk of the regions responsible for the emission at a given frequency were different for the ascending and descending branches of the U-burst's dynamic spectrum. In favour of this interpretation there is also the high time correlation between the U-bursts and the chromospheric flares that generally appear near bipolar groups of

§ 14] Type III Bursts

spots (about 80% of U-bursts are accompanied by flares (Maxwell, Howard and Garmire, 1960; Maxwell and Swarup, 1958))†.

From this standpoint the difference between ordinary type III bursts and their U-burst variant is that the first are excited by agents moving along the magnetic field lines of force which stretch far out from the Sun's surface, whilst the U-bursts are generated by streams moving along arc-shaped lines of force whose apex is at comparatively low altitudes in the corona.‡

If this interpretation of the U-bursts is the true one, then the sign of the polarization of U-bursts should change to the opposite one when the bursts reach their minimum frequency. The latter is connected with the different direction of the magnetic field on the descending and ascending branches of the dynamic spectrum. Unfortunately nothing is known about the features of the polarization of U-bursts unless we consider the remarks made by Haddock (1959b) according to which in several cases investigated the degree of polarization of the U-bursts is close to the corresponding values for ordinary type III events observed at the same time.

In conclusion we should note one more interesting feature of U-bursts. According to Goldberg (1958) the "turn" frequency (and likewise the maximum altitude reached) is connected with the maximum drift rate of the burst: the "turn" frequency shows a tendency to decrease as $|df/dt|_{max}$ rises. Since the "turn" frequency defines the maximum altitude reached by the radio emission source and the frequency drift rate defines the agent's radial velocity, what has been said means that the stream of fast particles does not simply move along lines of force of a given configuration but partly deforms them, carrying the coronal matter with the magnetic field "frozen" into it to an altitude which is greater the higher the velocity (and

---

† Sometimes at the end of the development process of a U-burst an increase is observed in the width of the frequency spectrum of the radio emission (Fig. 61a). From the point of view of the "plasma hypothesis" this effect may be caused by an increase in the extent of the corpuscular stream generating the U-bursts because of dispersion of particle velocities (Maxwell and Swarup, 1958). In one case the rising branch of the dynamic spectrum characterizing the development of the U-burst split into three bands after reaching the "turn" frequency (see the discussion by Haddock, 1959b). This phenomenon is apparently connected with "splitting" of the particle flux caused either by the complex configuration of the magnetic fields above the group of spots, thanks to which the stream moved along three different trajectories towards the photosphere, or by the complex initial velocity distribution of the particles in the stream with three maxima, which then led to the stream being split into three bunches moving along a single trajectory.

‡ This altitude can be estimated (from the known "turn" frequency) from the relation (13.1) for a definite electron concentration distribution in the corona. For example, the "turn" frequency of 35 Mc/s, which was once observed at a period of maximum solar activity, corresponds to an altitude of about $0.7 R_\odot$ above the photosphere (if the distribution N is that given in Table 2 of section 1 for the active region of the corona). For ordinary U-bursts with a "turn" frequency of 150 Mc/s this altitude is far less ($\sim 0.2 R_\odot$).

energy) of this stream. The increase in the maximum altitude as the velocity of the agent rises, also to a certain extent accounts for the independence (with certain exceptions) noted by Haddock (1959b) and Goldberg (1958) of the overall duration of a U-burst with respect to the "turn" frequency.

It is quite possible that the motion of sources of ordinary type III bursts is also sometimes accompanied by considerable deformation of the magnetic field. The absence of frequency drift inversion can then be explained either by a lower magnetic field strength in those regions of the corona where the type III source is moving when compared with the U-burst generation regions, or by a greater energy of the corpuscular streams exciting the type III radio emission. In the first case, which is obviously realized when the chromospheric flares connected with the U-bursts and the ordinary type III bursts are not in the same position (relative to the sunspots), we should expect a different mean degree of polarization of these bursts. In the second case there should be (with more or less the same mean degree of polarization) a difference in the mean frequency drift rates and the intensities of the U-bursts and the ordinary type III bursts. However, the absence of sufficient experimental data at present does not permit any definite judgement to be given.[†]

15. Types IV and V Radio Emission

BASIC CHARACTERISTICS OF TYPE IV RADIO EMISSION

The bursts connected with large solar flares often have a very complex dynamic spectrum (Figs. 62 and 63; see also Fig. 53). In this case the development of a typical event in the metric band breaks down into three stages. The first stage, which coincides with the beginning of the chromospheric flare or is close to it, is characterized by the appearance of a group of short-lived type III bursts; in the second stage, which arrives a few minutes after the first, the level of the radio emission suddenly increases and, varying irregularly, stays enhanced for 10 min or so; in the third stage the level of the radio emission gradually rises and remains enhanced for a time from a few minutes to a few hours. The second stage in the dynamic spectrum corresponds to one or two slowly drifting bands which can be included in spectral type II. The wide band emission following them, which differs by a very smooth intensity curve (without any noticeable variations in the level of the emission in the form of short-lived bursts or with very weak fluctuations), is a particular form of activity. This kind of structureless continuum was first described by Boishot

---

[†] See also Hughes and Harkness (1963) on the properties of type III radio emission.

§ 15] Types IV and V Radio Emission

(1957, 1958, 1959) and was set apart as a separate sporadic component which was given the name of type IV radio emission.

To judge from the dynamic spectra obtained later (Maxwell, Swarup and Thompson, 1958; Haddock, 1959b; Wild, Sheridan and Trent, 1959; Thompson and Maxwell, 1962; McLean, 1959; Kundu, 1961a) the development of type IV radio emission generally proceeds as follows (see Fig. 62). A smoothed continuum, divested of any frequency or time details, appears after a type II burst with a shift of a few minutes. The intensity maximum is generally reached some dozens of minutes after the beginning of the event. The radio emission first appears at high frequencies, but as a rule not higher than 250 Mc/s (Kundu, 1961a),[†] and then gradually spreads to the lower frequencies. One gets the impression that the low-frequency edge of the spectrum, which sometimes drops more steeply than the high-frequency part, drifts towards a decrease in frequency at about the same rate as the preceding type II burst.

An example of this type of continuum can be seen in the dynamic spectrum in Fig. 63 accompanying a complex type III–II event. In this case the radio emission appeared almost simultaneously with the type II emission first in the decimetric band at frequencies of the order of 500 Mc/s, and then spread to the metric wavelengths. For the whole of its life (about 45 min), however, the intensity maximum of the continuum remained in the decimetric band.

Type IV radio emission in the metric waveband as a rule appears after the start of the outburst at centimetric wavelengths connected with the type II–IV event in the period when the latter reaches its maximum intensity, and can generally be observed for longer than the microwave burst (Thompson and Maxwell, 1962; Kundu, 1961a; Mogilevskii and Akin'yan, 1961). We can say that the type IV radio emission is not a low-frequency extension of the centimetric bursts and, vice versa, the latter are not the result of the spectrum of the type IV emission expanding into the higher frequencies. This can be clearly seen in an example of a dynamic spectrum (Fig. 64) restored from the data of simultaneous observations of solar radio emission at a number of fixed frequencies in the range between 67 and 9400 Mc/s (Takakura and Kai, 1961).

At the same time the two separate continua (one characteristic of metric and the other of centimetric wavelengths) do not always exist exclusively within the limits of the above bands. In certain cases the type IV radio emission starts at low frequencies (< 250 Mc/s) and then covers the higher ones (250–600 Mc/s) and can be observed at these frequencies

[†] Exceptions from this rule, when the upper limit is in the decimetric band, are by no means rare (Mogilevskii and Akin'yan, 1961).

until the low frequency part of the continuum disappears (Kundu, 1961a). The maximum intensity of the type IV is either in the metric or the decimetric band (Mogilevskii and Akin'yan, 1961). On the other hand, the centimetric bursts often embrace the decimetric band and at times even extend into the metric band. This continuum, appearing as a low-frequency continuation of the centimetric bursts, is called the first phase of type IV radio emission by Kundu (1961a), Mogilevskii and Akin'yan (1961), Pick-Gutmann (1961), and simply type IV emission by Haddock (1959b). In its properties, however, it differs from the structureless continuum following type II bursts discussed in the present section. Therefore we shall not use the term "type IV radio emission" for the decimetric continuum even in the cases when it extends to lower frequencies.

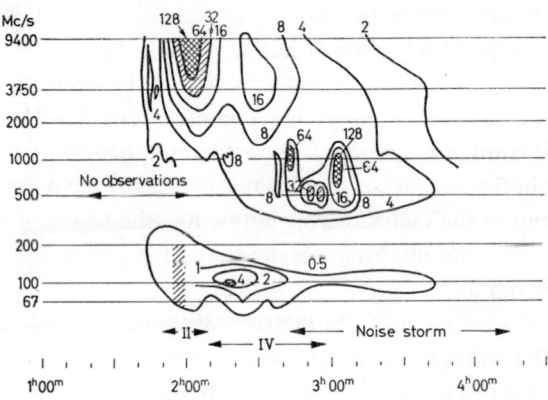

Fig. 64. Dynamic spectrum of radio emission in 67–9400 Mc/s band of 5 April 1960 plotted from observational data at fixed frequencies. The classification of the events according to the data of spectrographic observations at metric wavelengths is shown below (Takakura and Kai, 1961)

The features of the decimetric continuum—an extension of centimetric bursts into another frequency range—will be discussed in section 16. Here we shall content ourselves with saying that, to judge from Fig. 64 and other dynamic spectra obtained by Takakura and Kai (1961), it is quite possible in certain cases for the decimetric continuum to exist as a separate component and not as a simple extension of the centimetric bursts.

There is still one more wide-band continuum that we know of in the metric band—the enhanced radio emission in the composition of noise storms. However, the type IV radio emission differs from the latter in its smooth (or relatively smooth) profile, shorter duration and close connection with type II phenomena.

§ 15] Types IV and V Radio Emission

These features are in their turn closely interconnected. This is shown by a detailed study of the dynamic spectra in the 40–200 Mc/s range obtained in 1952–8 (McLean, 1959). From the available spectra twenty-two recordings were selected in which there were clear-cut rises in the level of radio emission for a time varying between 10 min and 5 hours starting from the time of the flares or after them. After dividing these recordings into two groups—one free of frequency and time details (thirteen events), and the other, in which brief narrow-band bursts prevailed (nine events)—it turned out that all the events of the first group (except for one) followed type II bursts, whilst the latter did not accompany events in the second group. Therefore the continuum-type radio emission that appears during the flares as a rule either has both signs of spectral type IV or does not have these features.

Nevertheless sometimes (as was remarked in section 12) enhanced radio emission is predominant in the composition of noise storms; type I bursts are rare and of low intensity. On the other hand, it is known that noise storms (in rare cases, it is true) are one of the forms of post-burst phenomena which accompany outbursts connected with chromospheric flares. Despite the results of McLean (1959) we cannot exclude the possibility that part of these bursts belong to spectral class II.

In connection with what we have said the problem arises of identifying the type IV radio emission since there may prove to be insufficient of the above-mentioned signs of this component based on the nature of the dynamic spectrum in certain cases for a clear solution of the question of whether an event belongs to type IV (particularly when the radio emission has only of one the two type IV features). In this respect a very valuable and, in essence, the determining feature of type IV radio emission is the circumstance that during the development of the continuum the position of the source of emission on the Sun's disk does not remain constant during observation at a fixed frequency.† This was discovered by Boishot (1958, 1959) during observations on a multi-element interferometer ($f = 169$ Mc/s): it turned out that in the initial stage the source of the type IV radio emission moves from its initial position on the Sun's disk close to a chromospheric flare to a distance of up to $(4-5)R_\odot$ from it far beyond the limits of the Sun's limb.

On the assumption of the radial nature of the motion of the source away from the chromospheric flare associated with it the displacement of

---

† It is appropriate to mention that the centres of noise storms and the regions where types II and III bursts are generated show no systematic displacements in observations at one frequency, remaining localized in the same part of the disk for the whole of their existence.

the generation region observed in the projection onto the plane of the Sun's disk can generally be interpreted as follows. On appearing in the lower corona the source of the type IV radio emission first rises rapidly upwards (at a velocity of the order of 1000 km/sec); having reached the upper layers of the corona it drops slowly down and after a few dozen minutes it stops at an altitude whose value varies considerably from event to event between $0·3R_\odot$ and $(4-5)R_\odot$.

However, the hypothesis of the radial nature of the motion is far from always satisfied. This is indicated, for example, by the case when the source of the type IV radio emission connected with the flare in the centre of the Sun's disk went far beyond the Sun's limb and did not remain within the limits of the latter (Kundu and Firor, 1961). Therefore the details of the picture given above of the motion (in particular the downward displacement of the source after it has reached its maximum altitude in the corona) are by no means without contention, and the values given for the velocities and altitude are only a guide.

The motion of a source of type IV radio emission in the corona was also later fixed in interferometer measurements at a frequency of 87 Mc/s (Kundu and Firor, 1961) and in observations on an interferometer with frequency tuning in the 40–70 Mc/s band (Wild, Sheridan and Trent, 1959) (Fig. 53). In the latter case it was established that, unlike the types II and III bursts, the position of the type IV activity source is not frequency-dependent, whilst the generation region of radio emission at a fixed frequency moves in the plane of the Sun's disk at a velocity of up to 5000 km/sec. For the type IV event in Fig. 53 this velocity was about 2600 km/sec, and the velocity of the agent causing the type II burst was close to the velocity of the type IV source.

According to the data of Kundu and Firor (1961) at $f = 87$ Mc/s, the displacement velocity of the type IV source is generally higher than or of the order of 10^3 km/sec. At higher frequencies ($f = 200$ Mc/s) no cases were noted when the source of the radio emission was further than a distance of $R_\odot$ from the photosphere, nor were rapid displacements at the high velocities characteristic of low frequencies (Takakura and Kai, 1961).[†]

The close connection of the type IV radio emission with type II bursts and the displacement of the generation regions of both events at approxi-

---

† Morimoto (1961) gives examples of several events which showed no systematic displacement in interferometer observations at a frequency of 200 Mc/s, remaining localized in the corona at an altitude of about $0·3R_\odot$ above the flare. Since these events had no characteristic sign distinguishing type IV radio emission they cannot be identified with this component (as was done by Morimoto (1961)). The phenomena observed should apparently be included among the decimetric continuum or continuum storms (see section 16).

§ 15] Types IV and V Radio Emission

mately the same velocity indicates that in the final count both phenomena are caused in all probability by a single agent moving in the corona from the region of a chromospheric flare.

The angular diameter of the region of generation of the type IV radio emission measured at 169 Mc/s (Boishot, 1958 and 1959) at the time of a rise in intensity is very great ($> 12'$). During the maximum phase and at the time of the decrease in the intensity of the radio emission the size of the source decreases slightly: to 7–12' at $f = 169$ Mc/s and to 5' at $f = 40$–60 Mc/s.[†] In all cases, however, the sources of type IV radio emission have a greater extent than the noise storm centres appearing at the same time. The data given agree with the results of Kundu and Firor (1961) at a frequency of 87 Mc/s (angular diameter of type IV generation regions greater than 10').

The effective temperature T_{eff} of the emitting region is also very great: in exceptional cases it reaches values of 10^{10}–10^{12} °K (Boishot, 1959) although generally $T_{eff} \sim 10^7$–10^{10} °K (Thompson and Maxwell, 1962; Kundu, 1961a).

Type IV radio emission is partly polarized: as well as the naturally polarized component it contains a circularly polarized component (Haddock, 1959b; Wild, Sheridan and Trent, 1959; Takakura and Kai, 1961).[‡] The sign of the polarization generally does not alter during the whole of a type IV event and is apparently not connected in any way with the sign of the polarization of noise storms preceding this event or following it. A degree of polarization of 30–70% is usual at a frequency of 200 Mc/s (Takakura and Kai, 1961). In particular the type IV radio emission whose dynamic spectrum is shown in Fig. 63 contained 30% of a circularly polarized component (see Haddock, 1959b).

We notice that the number of type IV events decreases as the distance of the chromospheric flares connected with them from the centre of the Sun's disk increases (Thompson and Maxwell, 1962; McLean, 1959; Kundu, 1961a). This circumstance indicates that the directional properties of the component in questions of sporadic emission cannot be considered to be finally solved.

An interesting feature of the type IV radio emission was discovered

---

† According to Weiss and Sheridan (1962) type IV sources have a halo which, it is true, is weaker than for type II events.

‡ Even in the early observations of Payne–Scott and Little (1952) it was noted that in large bursts connected with chromospheric flares there exist two phases: the first is not polarized (it can now be included in spectral type II) and the second is a partly circularly polarized component, which is now called type IV radio emission. It is possible that the motion of the sources of bursts that they observed at a frequency of 97 Mc/s belonged to this latter type of radio emission.

by Boishot and Simon (1960): in certain cases the increase in the level of the radio emission during a noise storm swamps the type I bursts, reducing their number and intensity (see Fig. 65). From the duration of the rise in radio emission relative to the "smooth" time-intensity curve and the connection with the chromospheric flares these phenomena can apparently be put among spectral type IV. This kind of "swamping" action of the continuum has been observed not only when the positions of the continuum source and the centre of the noise storm coincided but also when there

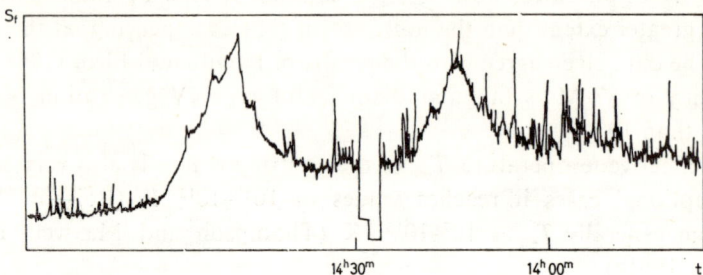

FIG. 65. Recording of solar radio emission intensity on 1 August 1957 at a frequency of 169 Mc/s. The intensity of the bursts decreases as the level of the enhanced radio emission rises (Boishot and Simon, 1961)

was some difference in their coordinates (up to 4'). However, when the region where the type IV radio emission is at far greater distance away from the centre of the noise storm this emission has no noticeable effect on the degree of type I activity.

It is possible that this effect is connected with the direct influence of the source of the type IV radio emission on the generation conditions of the type I bursts. However, unlike Boishot and Simon (1961), we feel it more probable that the reduction in the activity of the short bursts is caused by their absorption in the region where the type IV radio emission is generated located in higher layers of the corona. The greater size of this region ($\sim 10'$) explains the appearance of this effect even with a certain difference in the mean coordinates of the types IV and I sources on the Sun's disk. The attenuation is apparently caused by absorption of the emission by the particles generating the type IV component; it will considerably reduce the intensity of the type I bursts if the optical thickness of the system of emitting particles is $\tau > 1$.

The great interest at present in type IV radio emission is explained by the close connection of this component with the diverse geophysical phenomena (magnetic storms, increase in cosmic ray intensity, etc.). The different aspects of this connection will be specially discussed in section

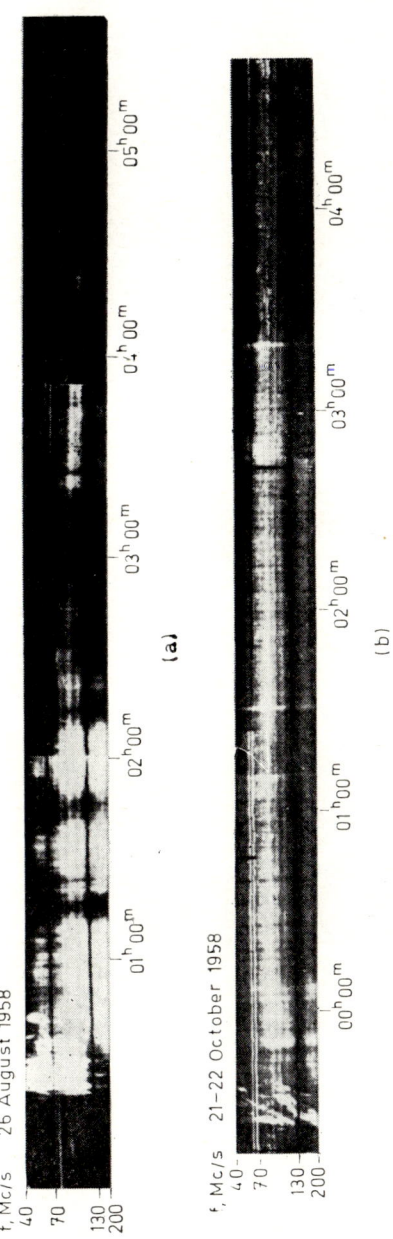

FIG. 62. Dynamic spectra of complex type II–IV events. In Fig. 62a type II–IV emission is preceded by a large group of intense type III bursts (McLean 1959)

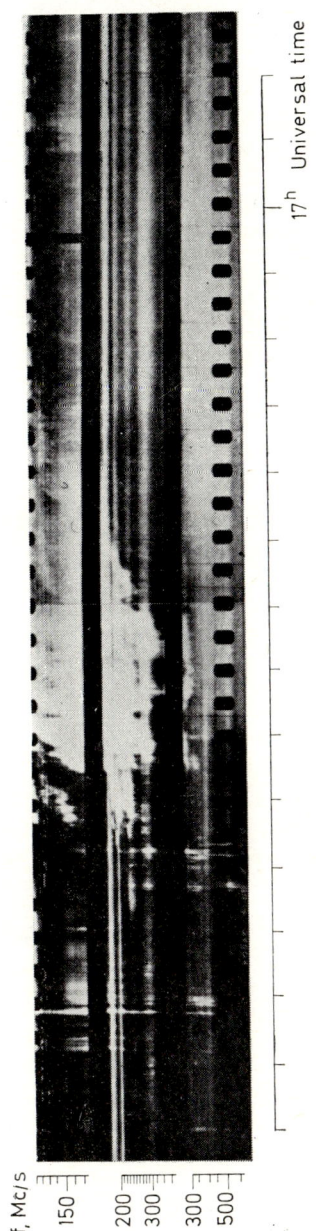

FIG. 63. Dynamic spectrum of a complex type III–II–IV event during a solar flare of 20 October 1957 (Haddock, 1959b). Time markers at 1 min intervals

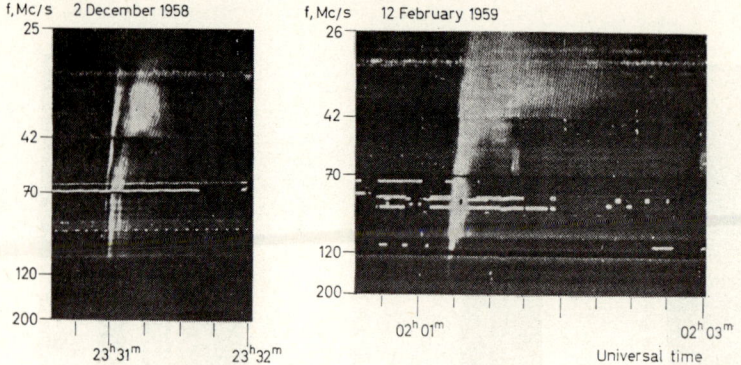

Fig. 66. Dynamic spectra of type V bursts following type III events (Wild, Sheridan and Neylan, 1959)

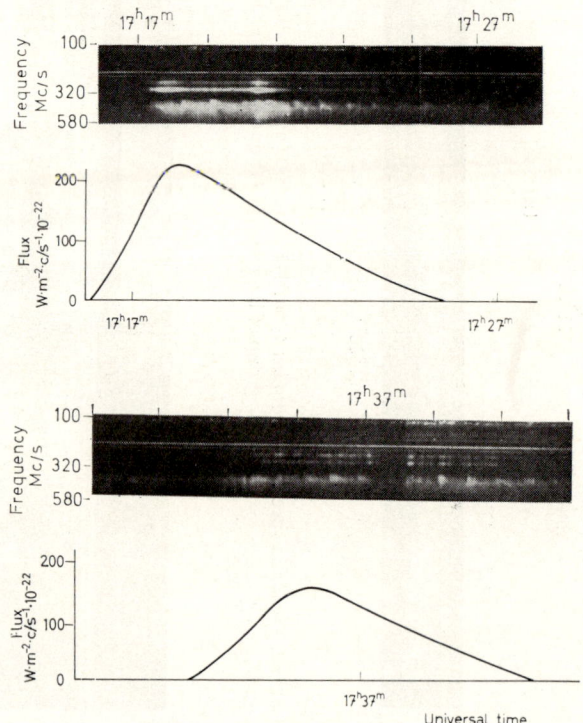

Fig. 67. Dynamic spectrum of decimetric continuum at frequencies of 100–580 Mc/s and profiles of centimetric bursts at 2800 Mc/s connected with it (22 December 1957) (Kundu, 1961a)

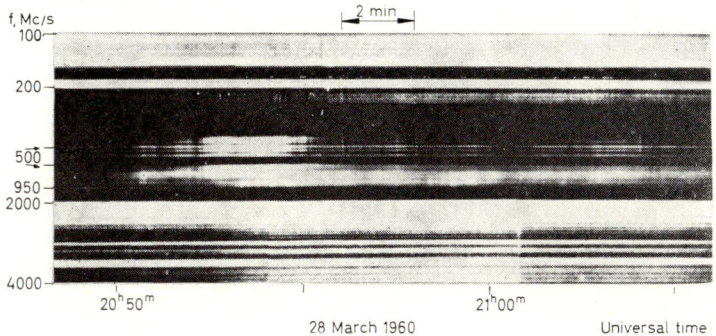

Fig. 69. Dynamic spectrum of a complex event and microwave burst, decimetric continuum + type II burst of 28 March 1960. The continuum appeared at $20^h 50^m$ world time, whilst the type II burst started at $20^h 56 \cdot 5^m$ and ended at $21^h 12^m$. The dynamic spectrum of the latter is hard to see in the figure, particularly at frequencies over 200 Mc/s (the light horizontal lines in the figure are not of solar origin) (Kundu, 1961a)

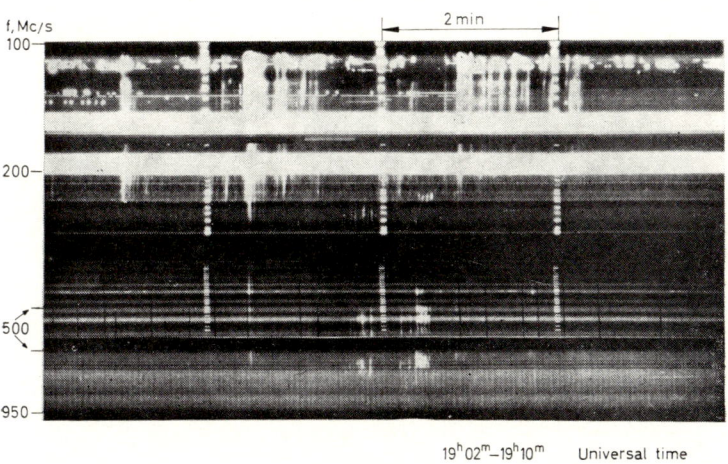

Fig. 70. Dynamic spectra of rapidly drifting metric and decimetric bursts (the light horizontal lines are an equipment effect or interference) (Kundu et al., 1961)

§ 15] Types IV and V Radio Emission

17; this will make it possible to get a more definite idea of the nature of the agent causing type IV events and the conditions obtaining in the region where the radio emission is generated.†

TYPE V BURSTS

In certain cases broad-band radio emission appears after type III bursts (see the dynamic spectra in Figs. 61b and 66). The frequency band occupies several tens (up to a hundred) of megacycles with a spectral intensity maximum generally located below 100 Mc/s. Often the spectrum drops sharply towards the high frequencies; there are no indications of the presence of frequency drift. The life of the component in question is short (0·5–3 min) and increases slightly at low frequencies (Thompson and Maxwell, 1962); the radio emission flux may be extremely small (as little as 10^{-18} W m^{-2} c/s^{-1}). In certain cases with observations at one frequency displacements of a type V source were recorded on the Sun's disk at a velocity of the order of 3×10^3 km/sec (i.e. a velocity close to that found for type IV activity) (Wild, Sheridan and Trent, 1959; Neylan, 1959).

This component, which was discovered by Wild, Sheridan and Trent (1959)‡ and called type V radio emission by them, is distinguished from the type IV continuum by its close connection with rapidly drifting bursts and a duration that is approximately two orders shorter.§

In a comparison of the radio data obtained with a radio spectrograph in the 40–240 Mc/s range and in observations at a number of fixed frequencies between 1000 and 9400 Mc/s a close correlation was found (Neylan, 1959) between type III–V events and centimetric bursts: of twenty-seven type III bursts accompanied by type V radio emission twenty appeared at the same time as radio emission in the centimetric band.§§ With few exceptions the duration of the events in the metric and centimetric bands was the same. At the same time among twenty-six selected type III bursts which were not accompanied by type V activity only one burst occurred at the same time as a rise in the level of emission in the centimetric band. It follows from this that the probability of a connection between type III bursts and centimetric bursts rises sharply if the former are accompanied by type V events.

† For features of type IV radio emission see also Ben'kova, Turbin and Fligel' (1961), Akin'yan and Mogilevskii (1961), Thompson (1962), Takakura (1963b).
‡ Apparently type V radio emission was observed at the same time by Maxwell on a radio spectrograph in the 100–570 Mc/s range (see remark by Boishot (1958)).
§ Here the ratio of the time scales is the same as in types II and III events.
§§ Judging from other data (Thompson and Maxwell, 1962) this correlation is only half as great.

Since each centimetric burst is almost certainly connected with some chromospheric flare it is clear that type V radio emission should also correlate closely with the flares. In actual fact, according to Neylan (1959) out of fifteen type V events fourteen appeared during flares, half of them starting (with an accuracy of up to a minute) in the period of the flare's "explosive" phase (a sharp sudden expansion), of which we have already spoken in section 14 when discussing the connection of flares with rapidly drifting bursts. No concentration of flares connected with type V phenomena was noted towards the centre of the disk; this obviously indicates that the radio emission is not directional (Thompson and Maxwell, 1962).

We notice that type V radio emission is not an extension of the frequency spectrum of a microwave burst into the metric band. This is indicated both by the drop in flux of the type V radio emission at the high-frequency end and by the difference in the coordinates of the sources: the region in which a type V burst is generated is in the upper corona whilst the sources of microwave bursts are located far lower down.

16. Other Forms of Bursts

In the preceding sections of this chapter we have discussed the basic components of the Sun's sporadic radio emission: the slowly varying component and microwave bursts, noise storms (enhanced radio emission + type I bursts), types II and III bursts and types IV and V continua. The majority of the events in the solar radio emission belong to one of these components; however the classification which we have expounded above does not exhaust the whole variety of observed phenomena. The point is that events are found in the composition of the solar radio emission ("broad-band bursts of short duration", "continuum storms", etc.) whose properties occupy an intermediate position between the individual classified components. It often remains uncertain whether phenomena of this kind are a special case of well-known components or whether they should be consigned to a new class of events. At the same time there are also phenomena in the solar radio emission which differ sharply from all the basic components; as an example we can take the event of 4 November 1957.

In the present section it is to these little-studied phenomena that we shall turn, starting the discussion with the decimetric band.

DECIMETRIC CONTINUUM

It was indicated in section 15 that centimetric bursts accompany certain groups of type III bursts (particularly if the latter are connected with type

§ 16] Other Forms of Bursts

V radio emission) and also type IV events that follow type II bursts. In this case the types II–IV event appears at the period when the microwave emission reaches its maximum intensity and generally lasts for longer than this emission. The features of the dynamic spectrum and the displacement of the region in which the type IV radio emission is generated defininitely indicate that the type IV and the centimetric bursts accompanying it are separate yet connected components of the Sun's sporadic radio emission. At the same time certain bursts in the centimetric band appear (with an accuracy of up to 2 min) at the same time as the broad-band radio emission at decimetric wavelengths, which we shall call the decimetric continuum. The time-dependence of the latter's level generally to a certain extent repeats the profile of the microwave burst; in particular their maxima and lives more or less agree (Fig. 67). Other features of the broadband radio emission at decimetric wavelengths (size of the generation region, its position on the Sun's disk, etc.) also enhance its likeness to the microwave bursts. We can therefore say that a decimetric continuum of this kind is a low-frequency continuation of microwave bursts and chiefly of type A outbursts (Kundu, 1961a; Pick–Gutmann, 1961).

In future we shall chiefly have in mind this variety of continuum, although in certain cases the situation is more complex: the decimetric continuum is of greater duration than the centimetric burst,† appears later and has a spectral intensity maximum at decimetric wavelengths (see the dynamic spectra in Figs. 64 and 68) (Takakura and Kai, 1961).

Generally the decimetric continuum exists at frequencies higher than 250 Mc/s. However, in certain cases, particularly when the intensity and duration of the microwave burst are sufficiently great, the radio emission also extends to frequencies below 250 Mc/s (Pick-Gutmann, 1961). This continuum at metric wavelengths, generally speaking, does not accompany a type II burst although when the intensity of the centimetric burst is very high (greater than 10^{-19} W m^{-2} c/s^{-1}) the appearance of a slowly drifting component is highly probable. It is characteristic that in this case, unlike a type II–IV event, the continuum appears earlier than the type II burst (Fig. 69).‡ This was to be expected since the microwave bursts with

---

† Moreover, the life of the decimetric continuum is as a rule shorter than that of the type IV radio emission and is rarely more than a quarter of an hour, being between 10 min and 2 hours; only in exceptional cases is it as much as 6 hours (Kundu, 1961a; Pick-Gutmann, 1961).

‡ This gave Kundu (1961a), who clearly separated the two phenomena of the decimetric continuum and type IV radio emission, a basis for calling them the first and second phases respectively of type IV radio emission. We shall not keep to this terminology since the features of the decimetric continuum distinguish it sharply from type IV radio emission and in no way boil down simply to precursors of type II bursts.

which the decimetric continuum is connected generally precede type II bursts.

The decimetric continuum is generated by sources of small angular size—in the majority of cases less than 4' at a frequency of 340 Mc/s (Kundu and Firor, 1961) and of the order of 2–5' at 1420 Mc/s (Krishnan and Mullaly, 1961 and 1962). This is close to the size of the sources of centimetric outbursts ($< 2.5'$) and differs strongly from the corresponding values for type IV. The effective temperature of the decimetric continuum radio emission is 10^6–10^9 °K (Kundu, 1961a; Krishnan and Mullaly, 1961 and 1962).

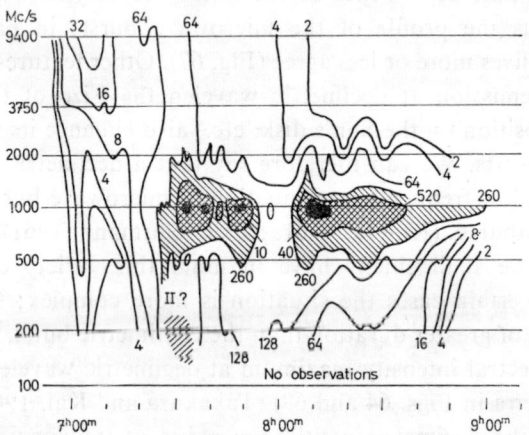

FIG. 68. Dynamic spectrum of decimetric continuum of 29 March 1960 according to simultaneous observations at frequencies in the 200–9400 Mc/s range (Takakura and Kai, 1961)

The generation regions are localized around flares generally in the same place as a source of the slowly varying component was before the appearance of the continuum. Unlike the type IV bursts the sources of the decimetric continuum (according to observations at a frequency of 340 Mc/s) do not show any significant displacements over the Sun's disk to a distance of more than 1–2' (Kundu and Firor, 1961; Boishot and Simon, 1960; Krishnan and Mullaly, 1961 and 1962). In this respect the decimetric continuum also recalls the microwave bursts.

In two cases when the active region was located on the Sun's limb it proved to be possible to estimate the altitude of the sources above the photosphere: it turned out that it does not exceed $(0.06$–$0.07)R_\odot$, i.e. the generation of the decimetric continuum occurs in the lowest layers of the corona (Kundu and Firor, 1961; Krishnan and Mullaly, 1961 and 1962). It follows from this that even in the cases when the decimetric continuum

§ 16] Other Forms of Bursts

is generated at the same time as type IV radio emission it is not a high-frequency continuation of the spectrum of this emission.

We notice that a continuum at metric wavelengths preceding a type II burst is very weakly polarized, whilst type IV radio emission following type II is far more strongly polarized. Sometimes the decimetric continuum at a frequency less than 250 Mc/s is weakly polarized at the beginning of an event and more strongly a few minutes later. Moreover, the degree of polarization of the continuum in question rises when we move into the decimetric and centimetric bands where the radio emission becomes noticeably polarized (Kundu, 1961a; Pick-Gutmann, 1961).

At metric wavelengths the decimetric continuum displays considerable directional features in its radio emission which show themselves in the concentration towards the centre of the Sun's disk of the flares connected with it in the cases when the continuum covers part of the metric band (Kundu, 1961a).

The weak decimetric continuum which does not accompany a type II event is sometimes connected with a type C microwave burst ("gradual rise and fall"). From observations in the 100–570 Mc/s range the continuum appears at almost the same time as a microwave burst (with an accuracy of up to 2–3 min) and reaches its maximum intensity at the same time as the latter. The generation region of this kind of continuum at 169 Mc/s does not show any significant displacements from its original position (unlike the type IV). Its angular size is of the same order as for a source of type IV radio emission ($\sim 10'$), the effective temperature is far lower ($T_{\text{eff}} \lesssim 10^7$ °K) (Kundu, 1961a).

RAPIDLY DRIFTING DECIMETRIC BURSTS

As well as the lengthy, structureless continuum in the decimetric band there are also bursts of short duration (with a life $\lesssim 1$ sec), which show frequency drift at a rate greater than or of the order of 100 Mc/s^2 (Alsop et al., 1959).

A comparison of the dynamic spectra of rapidly drifting bursts at metric and decimetric wavelengths obtained by spectrographic observations in the 100–950 and 2000–4000 Mc/s ranges (Kundu et al., 1961) showed that the majority of the bursts which appear in the decimetric band can be observed only at frequencies of not less than 400 Mc/s (usually in the 400–800 Mc/s range). Therefore the rapidly drifting decimetric outbursts are generally not a high-frequency continuation of metric band type III bursts (see Fig. 70).

Moreover if all the rapidly drifting bursts are characterized by the lowest frequency f_{min} which they reach, then histograms of the number of bursts for the values of f_{min} display two clear maxima—one at the lower limit of the frequency range studied (near 100 Mc/s) and the other in the 400–500 Mc/s range (Fig. 71). It follows from this that two separate classes of rapidly drifting bursts exist: one in the metric band (type III bursts) and another in the decimetric band (rapidly drifting decimetric bursts).

The properties of decimetric bursts differ to a considerable extent from the features of type III bursts discussed above. The absolute frequency drift rate of the decimetric bursts is higher ($\gtrsim$ 100 Mc/s²), the drift often being not towards the low frequencies (as in the majority of metric bursts)

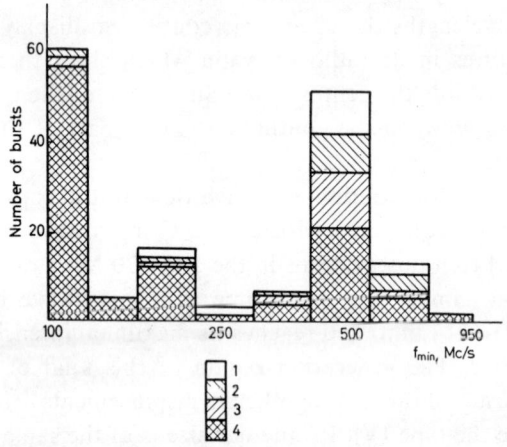

FIG. 71. Distribution of groups of rapidly drifting bursts with a differing kind of frequency drift according to the values of their minimum frequencies in the dynamic spectrum: 1—drift towards high frequencies; 2—simultaneous appearance at all frequencies ("infinitely rapid drift"); 3—"mixed drift" (towards low and high frequencies); 4—drift towards low frequencies (Kundu et al., 1961)

but towards the high ones.† In certain cases the direction of the drift is opposite at the high-frequency and low-frequency edges of the dynamic spectrum of the same decimetric burst. The number of bursts with a different type of frequency drift in the metric and decimetric wavebands can be judged from the histogram in Fig. 71.

† The observed drift obviously cannot be explained by the effect of group lag increasing with a decrease in frequency since both directions of drift occur in decimetric bursts. If we link the frequency drift with the advancing motion of a certain agent in the transition layers from the chromosphere to the corona, then the velocity of this agent, estimated on the assumption that the frequency of the radio emission is close to the plasma frequency, will be far less than the velocity of the agents causing type III bursts in the metric waveband (Alsop et al., 1959).

§ 16] Other Forms of Bursts

The bandwidth Δf covered by decimetric bursts while drifting is tens and hundreds of megacycles; generally these bursts exist in the 400–800 Mc/s range, although sometimes they drop right down to 200 Mc/s and are sometimes noted at frequencies of the order of 1000 Mc/s. However, the relative width of the range $\Delta f/f$ is as a rule less than for metric type III bursts.

Decimetric bursts form groups of three to ten elements; the duration of an individual burst is short—about 0·5 sec. Sometimes successive bursts in a group appear at frequency intervals displaced from each other, forming the characteristic discontinuous dynamic spectrum of the event. A discontinuous spectrum of a slightly different type appears if the sequence of bursts is observed simultaneously in different (and not intersecting) spectrum ranges (Kundu et al., 1961).

Groups of decimetric bursts may appear in isolation, precede or follow type III bursts at metric wavelengths (Fig. 70). Sometimes they appear at the same time. An example of this kind of event is shown in Fig. 58b, where a group of type III metric bursts preceding type II radio emission is accompanied by a group of decimetric bursts and one centimetric burst. It is not difficult to see that the duration of the latter is of the order of the duration of the whole group of type III bursts.

As well as the rapidly drifting bursts whose frequency drift rate exceeds 100 Mc/s^2, bursts with a moderate drift rate of 10–50 Mc/s^2 (intermediate-drift bursts) are found at decimetric wavelengths (Dicke and Beringer, 1956). The drift is generally towards the low frequencies. However, just as with the rapidly drifting decimetric bursts, the opposite direction of drift has also been noted here. The life of bursts at one frequency is 0·2–0·6 sec and the width of the frequency spectrum is of the order of 10 Mc/s, covering a frequency band of about 50–150 Mc/s during the drift. In certain cases the drift rate decreases in the low-frequency part of the band studied. The radio emission flux during the bursts may be about ten times higher than the level of the "quiet" Sun, amounting to 5×10^{-20} W m^{-2} c/s^{-1}. Intermediate-drift bursts at times accompany the appearance of the decimetric continuum.

To judge from dynamic spectra (Alsop et al., 1959) at frequencies of 500–950 Mc/s, in certain cases part of the decimetric continuum appears to show a fine structure: it breaks down into a combination of very short (of the order of 0·2 sec) broad-band bursts. However, this circumstance, which is obviously of importance in interpreting the decimetric continuum, needs further checking: it is quite possible that "bursts" of this kind, each of which was noted during one or two frequency scans of the radio spectrograph, are a consequence of equipment effects.

It must be pointed out that the characteristics of the Sun's sporadic radio emission in the decimetric band are still quite inadequately investigated; the papers on which the preceding discussion was based are actually the first attempts in this direction.

Let us now discuss special forms of solar radio emission in the metric waveband.

CONTINUUM STORMS

During her investigations into type IV radio emission Pick-Gutmann (1960, 1961) paid attention to the fact that sometimes after a type II burst a very lengthy and intense continuum with a low content of type I bursts appears.[†] The latter may at first be completely absent; only after a few hours or days do the bursts become more numerous and strong, since finally the radio emission changes into an ordinary noise storm. This long-lived continuum displays a number of features that distinguish it from noise storms and type IV radio emission. What has been said allowed Pick-Gutmann to treat it as a special variety of sporadic radio emission—the storm continuum.

Phenomena of this kind are connected with strong chromospheric flares of strength 2 or more; more than half the events come with strength 3 and 3+ flares. Continuum storms apparently do not always follow slowly drifting bursts. For example, the event in Fig. 72 was not preceded by a type II burst but by a group of type III bursts.

Continuum storms are not observed at frequencies over 500 Mc/s since they are an event that is characteristic only of the metric band; the radio emission flux in this range is generally less than or of the order of 10^{-19} W m^{-2} c/s^{-1}. We can get some sort of idea of their frequency spectrum from the simultaneous recordings of the radio emission flux at several frequencies in the range between 40 and 9400 Mc/s shown in Fig. 72. The flux increases at lower frequencies, reaching 10^{-18} W m^{-2} c/s^{-1} at a frequency of 18 Mc/s.[‡] The diameter of the sources of continuum storms, with a few exceptions, does not exceed 4' at a wavelength of $\lambda \approx 1.8$ m (for type IV this value is 8–12' and in the case of noise storms the angular size is evenly distributed from values of less than 3' to 9').

The sources of the form of radio emission under discussion, unlike the type IV, do not show any displacement over the Sun's disk in observations at $\lambda \approx 1.8$ m for several hours after the beginning of an event. If the

---

[†] One observation of this kind of continuum has already been reported by Boishot and Warwick (1959) (see Fig. 72).

[‡] The radio emission intensity profile corresponding to a frequency of 18 Mc/s is not shown in Fig. 72.

§ 16] Other Forms of Bursts

emission lasts for a few days a displacement of the source can be observed, which occurs at about the same velocity as the motion of the region on the Sun where the chromospheric flare accompanying the beginning of the storm continuum was located. A certain difference between the veloc-

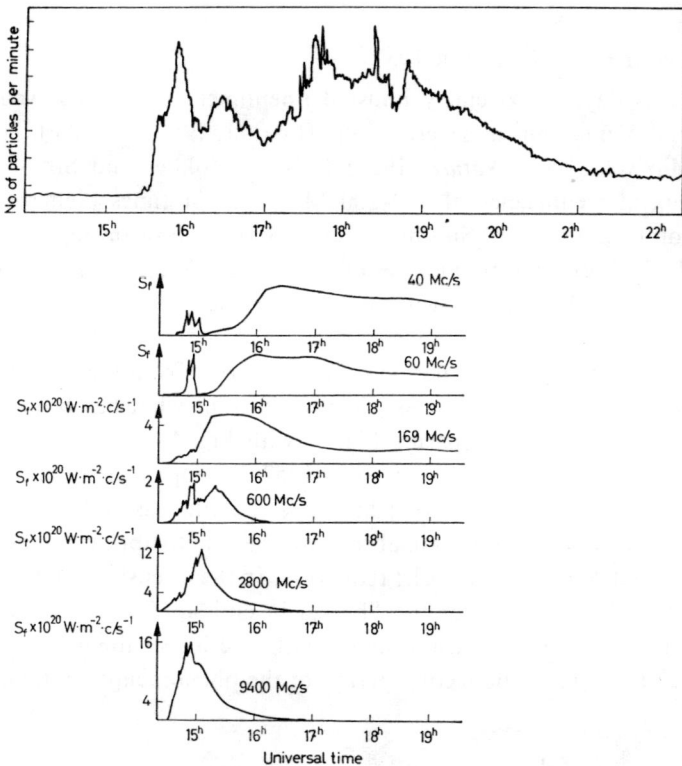

Fig. 72. Recordings of radio emission of 22 August 1958 at fixed frequencies in the 40–9400 Mc/s range. The storm continuum is the emission on the right-hand side in the metric and decimetric band recordings (40–169 Mc/s) (Pick-Gutmann, 1961; Boishot and Warwick, 1959). The top curve shows the variations of the cosmic rays (protons with an energy of about 170 MeV) at the same period (Anderson, 1958)

ities of the radio and optical objects can be explained by considering that the source of the radio emission is located on a straight line passing the centre of the Sun and the chromospheric flare at a distance of $0.3R_\odot$ from the photosphere. This is much less than the mean altitude of the sources of noise storms ($\sim 0.7R_\odot$) and type IV radio emission ($\sim R_\odot$) at these wavelengths.

The emission of storm continua is circularly polarized, the sign of the polarization showing the same dependence on the position of the source relative to the solar equator as for the noise storms. Hence, by analogy with section 12, we can conclude that the radio emission under discussion corresponds chiefly to an ordinary ray.

THE EVENT OF 4 NOVEMBER 1957

On this day an extremely unusual phenomenon was observed in the solar radio emission for several hours (Fokker, 1960b; Boishot, Haddock and Maxwell, 1960; *Nature*, 1958; Boishot, Fokker and Simon, 1959). The flux at frequencies of 150–200 Mc/s was at times higher than the emission of the "quiet" Sun by a factor of 10^3, the recordings of the first period of this event at frequencies of 169 and 200 Mc/s being distinguished by very rapid fluctuations with a characteristic time of 0·2–0·3 sec. The intensity variations were symmetrical about a mean level; as a whole the intensity recording at this period differed sharply from the corresponding recordings during noise storms.† The validity of what has been said can easily be checked by comparing Fig. 73 with Fig. 41.

As the phenomenon of 4 November 1957 developed the intensity variations became rarer and rarer; at the same time as well as the rapid "bursts in emission" "bursts in absorption" began to appear on the recordings. The latter looked like brief reductions in the intensity of the enhanced radio emission. Sometimes the "bursts in emission" and the "bursts in absorption" were observed alternately with a period of about a minute. The particular nature of the second period of the phenomenon of 4 November

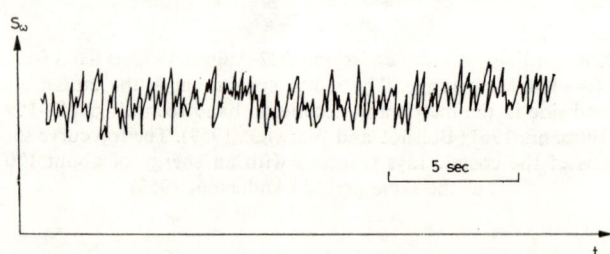

FIG. 73. High-speed recording of radio emission at 200 Mc/s during the first stage of the event of 4 November 1957 (Fokker, 1960b)

† The fluctuations during the first period of the event in question recalled the radio emission intensity oscillations of discrete sources which are affected by the Earth's ionosphere. However, on 4 November 1957 the ionosphere was quiet; it is possible that the fluctuations were of solar origin or were caused by non-uniformities in the Earth's troposphere (see section 7 in this connection).

§ 16] Other Forms of Bursts

1957 is also indicated by the dynamic spectra of the radio emission shown in Fig. 74 (Boishot, Haddock and Maxwell, 1960).

Above all it should be noted that spectra obtained at two points separated from each other by a distance of more than 2000 km are identical even in the smallest details. This proves that the features of the second phase of the event in question are of solar origin; in any case they are not connected with the effect of the Earth's atmosphere. In the dynamic spectra we can see characteristic bursts in the form of "spaghetti" which are unlike the types I–III events usually observed. Each burst takes up a very narrow frequency band of the order of 1 Mc/s (like type I) but exists for a far longer time (up to 1 min). The bursts display a certain slow drift both towards the low and the high frequencies with irregular variations in its magnitude and direction. In the dynamic spectra we can also clearly see the "bursts in absorption" in the form of dark strips edging the "bursts in emission" on the low-frequency side. This kind of mutual arrangement of the "bursts in emission" and "bursts in absorption" is a highly characteristic feature of this anomalous form of sporadic radio emission.†

To judge from recordings at fixed frequencies and from the dynamic spectra the phenomenon of 4 November 1957 observed in the 150–250 Mc/s range was absent at decimetric and centimetric wavelengths and wavelengths of about 6 m. At a frequency of 200 Mc/s the radio emission was completely polarized anti-clockwise during the whole of its development.

The position and size of the radio emission source were determined interferometrically at frequencies of 169 and 255 Mc/s. It turned out that the radio emission originated in a region with a diameter of the order of 9′ located approximately at a distance of $0 \cdot 1 R_\odot$ from the centre of the Sun's disk, no systematic displacements of the source being observed for 2 hours. In addition it can be stated that the instantaneous position of the centre of the emission did not deviate from the mean value by more than $\pm 1'$.

The event of 4 November 1957 was not connected with chromospheric flares and was not accompanied by its attendant phenomena (sudden ionospheric disturbances, geomagnetic effects and cosmic ray variations).

---

† According to Alsop *et al.* (1959 and 1960) the dynamic spectrum of the event of 31 May 1959 in the decimetric waveband (500–950 Mc/s) had the same structure. In the periods when bursts with an intermediate drift rate towards the high frequencies appeared on the background of the decimetric continuum they were edged on the low-frequency side with a "dark band" indicating that the intensity of the continuum decreases in this case. In several cases when intermediate-drift bursts drifting towards the low frequencies appeared on the background of the decimetric continuum they were also accompanied by "bursts in absorption", but this time on the high-frequency side.

WIDE-BAND BURSTS OF SHORT DURATION

Certain difficulties are also met when classifying the bursts of short duration (of the order of a second) and of relatively low intensity studied by Vitkevich (1956a, 1957c), Vitkevich and Gorelova (1960), Vitkevich, Gorelova and Lozinskaya (1959) on a multi-channel receiver operating in the metric waveband. The majority of bursts of this kind can undoubtedly be included among the solar radio emission components we know (among types I and III bursts to judge from the characteristic life).† However, Vitkevich chiefly investigated bursts whose intensity was comparable with the level of the "quiet" Sun or was two or three times greater than it. This circumstance allows us to assume that amongst these bursts there are components which are not observed or are observed extremely rarely in the dynamic spectra obtained by a radio spectrograph of relatively low sensitivity.

An example of this component is shown in Fig. 75a where the dynamic spectrum of the radio emission is plotted from the data of simultaneous observations at several fixed frequencies. This event is a group of surges ("small broad-band peaks") which in their properties occupy an intermediate position between types I and III bursts: the "life" of the surges (of the order of half a second) does not differ from the duration of type I bursts; the width of the frequency band occupied (6–26 Mc/s, averaging 13 Mc/s) is greater than for type I bursts but less than for type III bursts; the frequency drift rate is typical of type III bursts.

Activity of this kind was observed by Maxwell, Swarup and Thompson (1958) and Wild (see the remarks by Giordmaine *et al.*, 1959) as a variety of noise storm whose dynamic spectrum in this case had a "feather" structure (Fig. 75b).

REVERSE-DRIFT PAIRS

A fresh spectral feature of the solar radio emission in the metric wave band was found by Roberts (1958); it was given the name of reverse-drift pairs.

In the dynamic spectrum (Fig. 76) each such burst consists of two bands, the features of one band being repeated (at the same frequency) in the second band with a lag of 1·5–2 sec. Both elements generally appear at a fixed frequency but at different times. Therefore both elements of the

---

† In these papers type I bursts (according to the terminology used by Vitkevich and Gorelova (1960)) correspond to "small narrow-band peaks" ($\Delta f < 4$ Mc/s) and type III bursts to "small peaks of long duration" (several seconds) apparently.

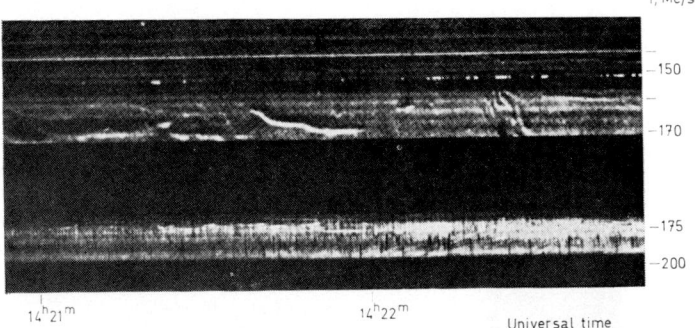

FIG. 74. Dynamic spectra of last stage of event of 4 November 1957 (Boishot, Haddock and Maxwell, 1960)

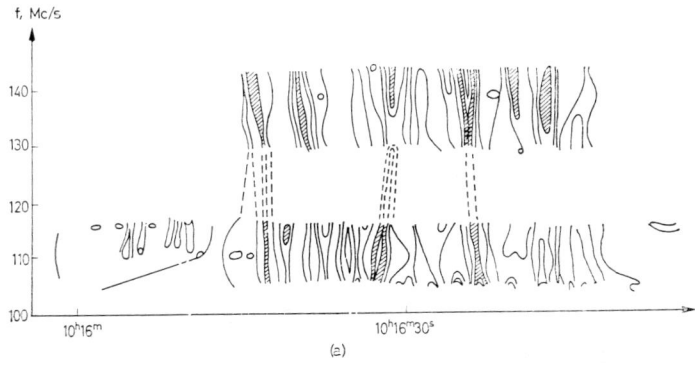

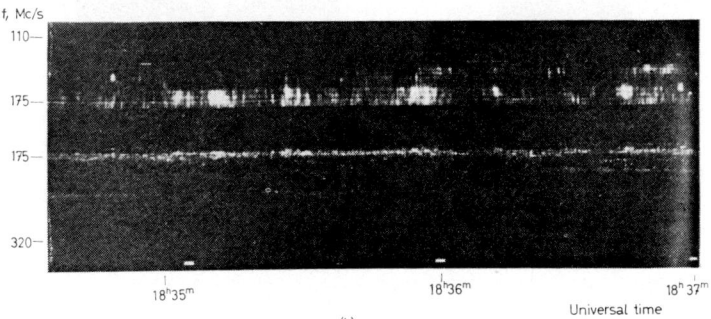

FIG. 75. Dynamic spectra of broad-band bursts of short duration: (a) group of bursts of 29 March 1958 (Vitkevich and Gorelova, 1960); (b) noise storm of 12 October 1957 showing the characteristic "feather" structure (Maxwell, Swarup and Thompson, 1958)

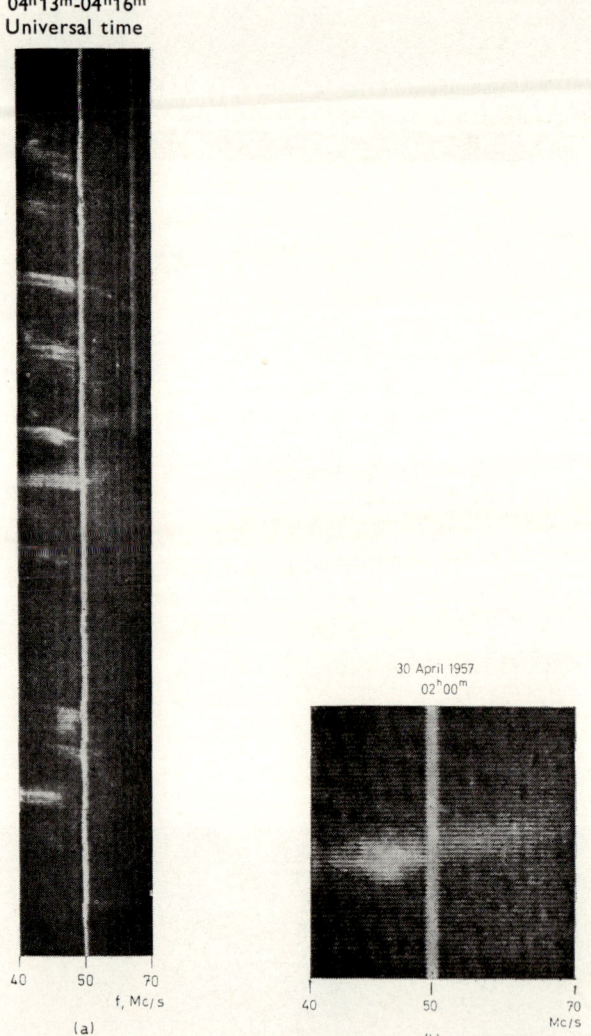

Fig. 76. Dynamic spectra of reverse-drift pairs (Roberts, 1958)

§ 16] Other Forms of Bursts

bursts in question are the result of a time "shift" and not frequency "splitting".†

In each element the frequency increases as the burst develops at a rate of 2–8 Mc/s².‡ The duration of an element at one frequency is very small (of the order of 0·5 sec); the total duration of the phenomena in the dynamic spectrum is from a few seconds to 10 sec. The "instantaneous" band-width of one element is between 1 and 10 Mc/s; the frequency band covered during the development of the burst is between a few megacycles and a few tens of megacycles.

About 10% of these bursts are located within the bands in the dynamic spectrum occupied by type III bursts (see Fig. 76b), forming the fine structure of rapidly drifting events. At the same time the reverse-drift pairs are more intense than type III events. However, the pairs are generally found as an independent phenomenon unaccompanied by noticeable type III radio emission. The pairs of reverse-drift bursts are of moderate intensity: the radio-emission flux during these bursts does not exceed 5×10^{-20} W m^{-2} c/s^{-1} in the 40–50 Mc/s range. The pairs appear very rarely at higher frequencies. Preliminary measurements indicate that their radio emission is not strongly polarized.

The solar origin of the pairs is confirmed by interferometric observations.

Since the features of one element of a twin burst are repeated with a certain lag at the same frequency in the other element, it is natural to assume, following Roberts (1958), that the second element is an "echo" from the first element caused by reflection of the radio waves from the deeper layers of the solar corona. When the radio waves are propagated along the electron concentration gradient the reflection originates from the level where the refractive index $n(\omega)$ is zero, i.e. where $\omega = \omega_L$ (without allowing for the magnetic field); when the propagation is at an angle to grad N the reflection is from layers slightly higher up (for further detail see section 22).

For this point of view to be argued it is necessary for the radio-emission

---

† We know that this is not so for types II and III harmonic bands: both bands appear at different frequencies at about the same time, and the details of the structure are repeated simultaneously at different frequencies. Moreover, dynamic spectra of five reverse-drift pairs were once recorded, whose elements were definitely (unlike ordinary twin bursts) the result of frequency "splitting" at 0·5–1 Mc/s and not of time lag. The low-frequency elements of the twin bursts also showed splitting into several tenths of a megacycle, which was also noticeable in the high-frequency elements where the splitting was even smaller (0·03 Mc/s). However, Roberts is not convinced that bursts of this kind are of solar origin.

‡ Sometimes (very rarely) bursts are found with a drift towards the low frequencies; in all other respects they do not differ in any way from reverse-drift (towards high frequencies) pairs.

frequency ω of the burst pairs to differ noticeably (be greater) from the plasma frequency ω_L in the layers of the corona which surround the region where the bursts were generated. Then the level $n(\omega) = 0$ will be far enough from the generation region, which provides the second element with the necessary lag. For example, if $\omega - \omega_L \approx \omega_L$ a lag of 1·5–2 sec can be explained by assuming that the gradient is close to that defined by the Baumbach–Allen formula (1.1).

Within the plasma hypothesis the relation $\omega - \omega_L \approx \omega_L$ for pairs will be satisfied if the observed radio emission is the second harmonic ($\omega \approx 2\omega_L$), whilst the reverse-drift bursts without splitting, generally speaking, may also correspond to radio emission at the fundamental harmonic $\omega \approx \omega_L$ (Roberts, 1958). This assumption would be confirmed if at the same time as the burst pairs reverse-drift bursts at half the frequency were recorded. Roberts did not find one event of this kind. In section 31 this is explained by the fact that the radio emission intensity of reverse-drift bursts at the fundamental frequency $\omega \approx \omega_L$ is low. It should also be borne in mind that since reverse-drift pairs are largely observed in the 40–50 Mc/s range a frequency of half that would be beyond the working limits of the radio spectrograph (40–240 Mc/s). The latter circumstance did not permit a final decision on whether the rapidly acting type III bands within whose limits certain burst pairs are located are the second harmonic.

In principle it is quite possible that the drift towards the high frequencies, i.e. in the direction opposite to that normally observed in types II and III bursts, is caused by motion of an agent at a velocity[†] of $(2-5) \times 10^9$ cm/sec in the regular corona towards the Sun's surface, i.e. into layers of the corona with a higher electron concentration. It is more probable, however, that this kind of frequency drift obtains with ordinary motion of the agent away from the Sun; in this case the bursts under discussion appear in the regions of the corona where the effect of local inhomogeneities leads (unlike the regular corona) to a rise in a certain region of the electron concentration as one moves away from the Sun.

From this point of view the connection between type III bursts and reverse-drift pairs is as follows. As it moves into the outer layers of the corona the agent generates an ordinary type III burst. On meeting a denser inhomogeneity than the surrounding corona the agent excites in the region of increasing concentration a short burst with reverse frequency drift, continuing at the same time to generate in the surrounding corona type III radio emission with a drift towards the low frequencies. In rare cases when the drift of the burst pairs (and of the type III phenomena accom-

† The velocity of the agent is estimated by the formulae (13.1) and (13.4) for the model (1.1) of the corona, proceeding from the value of the frequency drift.

§ 17] Sporadic Radio Emission

panying them) is towards the low frequencies, radio emission apparently appears in the second stage when the agent passes through an electron concentration maximum in a local inhomogeneity into a region with a concentration gradient running in the direction opposite to the motion.

17. Sporadic Radio Emission and Geophysical Phenomena

PRELIMINARY REMARKS

A number of geophysical phenomena in the upper atmosphere are undoubtedly connected with solar activity (chromospheric flares, bursts of radio emission, etc.). The nature of the agent which is the direct cause of some geophysical phenomenon becomes clear if we remember how the latter lags behind the corresponding event on the Sun.

All the comparatively brief geophysical phenomena can be divided into two groups in this way. The first group includes phenomena which are observed at practically the same time as the recording of the events occurring on the Sun (with an accuracy of up to a few minutes). Amongst these we must include above all the sharp drops in the level of the cosmic radio emission (blackouts) observed on Earth, the sudden attenuations and cessations in reception of short-wave transmissions (fade-outs) and the disappearances of signals reflected from the E- and F-layers of the ionosphere. The immediate cause of all the phenomena is the increase in the absorption of radio waves in the D-layer. To this group belong the phase anomalies in long-wave propagation and increases in the level of atmospherics caused by a change in the actual height and reflecting power of the D-layer respectively. In the final count all these effects, which are combined under the name of sudden atmospheric disturbances, and the variations in the geomagnetic field connected with the change in the ionospheric currents are caused by a rise in the ionization of the D-layer. If we remember that the electromagnetic radiation both in the radio band and at the higher frequencies travels the distance from the Sun to the Earth at the same velocity, whilst the sudden ionospheric disturbances are observed only during the day, it becomes clear from what we have been saying that the basis of these phenomena is the action of electromagnetic radiation or, to be more precise, of ionizing radiation from the X-ray (and far ultraviolet) part of the solar spectrum.

As well as the practically simultaneous phenomena there is a second group of events which occur with a lag of between a few tens of minutes and several days after the solar phenomena. They include cosmic-ray flares, the rise in the absorption of radio waves in the polar regions (polar black-outs), ionospheric storms, geomagnetic storms with a sudden beginning and aurorae. These events owe their origin to the action of solar corpuscular radiation.†

The characteristic time of the longer duration geophysical phenomena —slow variations in ionospheric parameters and the degree of the geomagnetic field's perturbation (magnetic storms with a gradual beginning)— is a few tens of days. These phenomena show 27-day variations whose presence indicates that on the Sun there are long-lived local centres which take part in the Sun's rotation about its axis and affect the upper layers of the Earth's atmosphere by means of ionizing radiation and corpuscular fluxes.

As well as the above-mentioned geophysical effects there are very slow long-period changes in the state of the upper atmosphere which depend on the general level of the solar activity over the whole 11-year cycle and are apparently determined by the combined action of ionizing radiation and corpuscular radiation.

It should be stressed that the correlation of the above-mentioned geophysical phenomena with the optical indices of the solar activity is by no means complete and neither is the correlation of the latter with the effects of the solar sporadic radio emission. The absence of an unambiguous agreement between the optical and the radio indices of the solar activity allows us to hope that certain components of the solar radio emission have a closer connection with geophysical phenomena than events observed in the optical part of the spectrum. In the latter case the nature of the geophysical phenomena allows us to judge the origin of the individual components of the Sun's radio emission. In addition this connection may be of considerable help in the matter of forecasting geophysical phenomena (Artem'yeva, Benediktov and Getmantsev, 1961).‡

---

† Since the velocity of the particles may be close to the velocity of light it can be taken *a priori* that the fast charged particles moving from the Sun to the Earth are also a basic cause of certain of the above-mentioned "simultaneous" events. However, because of the effect of the magnetic fields that deform the trajectories of the charged particles in the corona and in interplanetary space, the time taken to travel from the Sun to the Earth even by very fast particles making up cosmic rays of solar origin is usually not less than an hour.

‡ A large number of papers has been published on the Sun–Earth problem as a whole and on the question of the connection between solar radio emission and geophysical phenomena (see in particular Ellison, 1959; Eigenson *et al.*, 1948; *Proceedings of the Conference of the Commission on Solar Research*, 1957). However, some of the papers are rendered to a considerable extent valueless by the absence in them of any clear-cut classification according to the nature of the dynamic spectrum of the events

§ 17] Sporadic Radio Emission

RADIO EMISSION OF THE SUN AND SUDDEN IONOSPHERIC DISTURBANCES. CONNECTION BETWEEN MICROWAVE BURSTS AND HARD SOLAR RADIATION

When we come to discuss the connection between sporadic radio emission and geophysical phenomena of the first group caused by solar ionizing radiation, we must make the reservation that these phenomena provide only indirect information on the latter from the magnitude and nature of the variation in the electron concentration in the ionosphere. On the other hand, data on the radiation intensity cannot be obtained from terrestrial observations since the part of the solar spectrum of interest to us is absorbed in the atmosphere. Rocket investigations are not effective enough either when studying the correlation of the ionizing component with the solar activity phenomena because of the short duration of the experiment, although they have provided a number of valuable results on the spectrum of the ionizing radiation during chromospheric flares and its connection with microwave bursts (see below). Investigations in artificial Earth satellites may be of great help in this respect.

As for the sudden ionospheric disturbances, here the fact that the electron concentration N in the ionosphere increases as the flux S of the ionizing radiation at the edge of the Earth's atmosphere rises is qualitatively clear, although it is not so simple to establish quantitative relationships between N and S. In the majority of cases the dependence of N on S is determined from formulae that are valid for a simple ionized layer (the Chapman layer; see, e.g. Mitra, 1955 and Al'pert, 1960, Chapter V). It should not be forgotten, however, that this model is only a rather rough approximation to the truth.

For a simple layer the change in the electron concentration in time is described by the equation

$$\frac{dN}{dt} = J - \alpha_{\text{eff}} N^2, \qquad (17.1)$$

in the Sun's radio emission. This is particularly true of certain papers (Dodson, Hedeman and Owren, 1963; Avignon, Boishot and Simon, 1959; Sinno and Hakura, 1958a; Budějický and Švestka, 1958; Sinno and Hakura, 1958b; Mitra, 1955; Al'pert, 1960; Davies, 1954; Kawabata, 1960b; Hachenberg and Volland, 1959; Kundu, 1961b; Peterson and Winkler, 1959; Heitler, 1956; Denisse and Kundu, 1957; Kundu, 1960; Minnis and Bazzard, 1958; Taubenheim, 1958; Kerblai and Kovalevskaya, 1960; Obayashi, 1959; Yudovich and Fel'dshtein, 1958; Sinno, 1959; Zhigalov, 1960; also part of Budějický and Švestka, 1958; Sinno and Hakura, 1958b) in which the connection between geophysical phenomena and the Sun's radio emission was studied from observational data at fixed frequencies in the metric band. The division of sporadic events by the nature of their profile at one frequency used by Dodson, Hedeman and Owren (1963), Avignon, Boishot and Simon (1959), Yudovich and Fel'dshtein (1958) cannot provide enough information for definite identification of the observed events with the radio emission components discussed in this chapter.

where α_{eff} is the effective recombination coefficient, J is the ionization index (the number of atoms ionized in unit time in a unit volume). At the layer's maximum J is defined by the relation

$$J = \frac{S}{\varepsilon_i \, e h_{\text{eff}}} \cos \varphi. \qquad (17.2)$$

Here ε_i is the ionization potential, h_{eff} is the reduced height of the atmosphere, φ is the Sun's zenith angle, $e = 2 \cdot 71 \ldots$

With a quasi-static variation in the ionospheric parameters, when $|dN_{\max}/dt| \ll \alpha_{\text{eff}} N_{\max}^2$, it follows from (17.1) and (17.2) that

$$S \approx \varepsilon_i \, e h_{\text{eff}} \alpha_{\text{eff}} N_{\max}^2 \sec \varphi \sim f_{L \max}^4 \sec \varphi. \qquad (17.3)$$

The latter relation allows for the fact that the concentration at the layer maximum $N_{\max}$ is proportional to the square of the plasma frequency $f_{L \max}$ in this region (i.e. to the square of the layer's critical frequency with normal incidence; see section 7). Therefore when S (and φ) changes slowly enough the index of the level of the ionizing radiation can be the quantity $f_{L \max}^4 \sec \varphi$; for the E-layer $f_{L \max}$ is easy to find from the results of pulsed radio sounding of the ionosphere.

In the cases when $dN_{\max}/dt$ is a noticeable part of $\alpha_{\text{eff}} N_{\max}^2$ the quantity J, and therefore S, can be determined from the equation (17.1) taking $dN_{\max}/dt$ into account. The variations of S can be judged here from the change in $dN_{\max}/dt$ and $N_{\max}$ during ionospheric disturbances; for the D-layer the latter quantities can be obtained on the basis of investigations into the degree of absorption of the radio waves in the above period (see p. 334).

Sudden ionospheric disturbances caused by a rise in the ionization of the lower layers of the ionosphere caused by the Sun's soft X-radiation (mostly in the 2–10 Å range) show a clear correlation with chromospheric flares. The correlation increases as the strength of the flares and their altitude above the photosphere increase, becoming practically complete for 3 and 3+ flares (Warwick and Wood, 1959; Dodson, Hedeman and Owren, 1953; Hachenberg and Krüger, 1959).[†]

The ionizing radiation is generated by flares accompanying bursts at metric wavelengths and also by flares unconnected with noticeable effects in this range (Dodson, Hedeman and Owren, 1953; Budějický and Švestka, 1958). On the other hand, sudden ionospheric disturbances largely accompany only the chromospheric flares and bursts in the metric band which are connected with surges of radio emission at centimetric wave-

---

[†] We notice, by the way, that a similar connection with bursts also holds for microwave bursts of radio emission.

§ 17] Sporadic Radio Emission

lengths; if the spectrum of the sporadic radio emission during a flare is restricted to the metric waveband it cannot be the source of intense ionizing radiation (Warwick and Wood, 1959; Hachenberg, 1960). Therefore the correlation between the bursts of radio emission and the ionospheric disturbances increases as we move from the metric into the decimetric and then into the centimetric band (Artem'yeva, Benediktov and Getmantsev, 1961; Davies, 1954), where bursts accompany more than 80% of all sudden ionospheric disturbances (Hachenberg and Krüger, 1959) (according to different data (Kawabata, 1960b)—about 70% of all and 95% of the strong ionospheric disturbances). At the same time microwave bursts are more frequent phenomenon than sudden disturbances of the ionosphere: the latter accompany a little more than half (Dodson, Hedeman and Covington, 1954; Hachenberg and Krüger, 1959) (according to Kawabata, 1960b, about a third) of all the bursts.

The question naturally arises: how do the microwave bursts of radio emission accompanying these geophysical effects differ from bursts that are not connected with them? This difference is not determined unambiguously by the nature of the burst profile since all three types of burst classified in section 11 are found among the events in the radio band accompanying ionospheric disturbances: simple bursts (A), post-bursts (A–B) and gradual rises and falls (C) (Figs. 77–79). At the same time type B and C events are connected more often with sudden ionospheric disturbances, particularly if they reach a great strength (Kawabata, 1960b). Among the bursts without ionospheric disturbances we should note the group with a frequency spectrum that rises towards the decimetric wavelengths (Hachenberg and Krüger, 1959).

Interesting results are obtained when we compare the profiles of bursts of solar radio emission at $\lambda = 3\cdot 2$ cm with the variations in the intensity of the Sun's ionizing radiation during individual sharp increases in the absorption of short radio waves in the ionospheric D-layer (Hachenberg, 1960; Hachenberg and Volland, 1959).

Figures 77 and 78 show the results of observations of the variation in the intensity of the solar radio emission and the rise in the absorption in the D-layer when two chromospheric flares appeared on the Sun's disk. Both phenomena—bursts and fade-outs—were recorded almost simultaneously, although the nature of their time dependences has little in common.

The position changes if we move from the observed variation in the optical thickness τ in the D-layer to the time-dependence of the flux S of the ionizing radiation. The variation in τ during a flare, as can be easily checked by the example of Figs. 77 and 78, can be approximated with sufficient accuracy by a curve of the form

$$\Delta\tau = \tau - \tau_0 \propto \left(\frac{t}{t_0}\right) \exp\left(1 - \frac{t}{t_0}\right). \qquad (17.4)$$

Here τ_0 is the value of the optical thickness before the start of the fade-out and t_0 is a parameter describing the duration of the event. If we remember that in the D-layer the absorption coefficient μ and the optical thickness τ are proportional to the electron concentration N, then obviously

$$\Delta N \propto \Delta\tau \qquad (17.5)$$

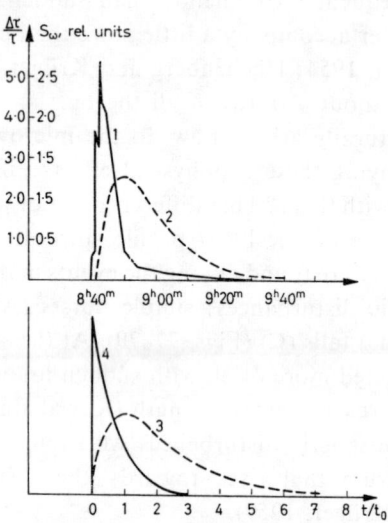

FIG. 77. Comparison of the profile of a microwave burst of solar radio emission during the flare of 5 June 1958 with the variation in the flux of the ionizing radiation calculated from data on the ionospheric absorption of signals from short-wave radio stations: 1—radio burst profile at $\lambda = 3.2$ cm; 2—relative increase in $\Delta\tau$, the optical thickness of the D-layer ($\Delta\tau/\tau$); 3—curve approximating the dependence of $\Delta\tau$ on the time t; 4—time-dependence of the flux of the ionizing radiation ΔS calculated for the Chapman layer (Hachenberg, 1960).

and the rise in the value of N during a flare can also be approximated by a curve like (17.4). A calculation of the variation in the flux of the ionizing radiation made by Hachenberg and Volland (1959) using relations of the (17.1) and (17.2) type (allowing for (17.4) and (17.5)) showed that the function $S(t)$ corresponds well to the profile of the microwave burst of radio emission accompanying the flare although it can be seen to be not completely identical (see Figs. 77 and 78).

§ 17] Sporadic Radio Emission

From the nature of their profiles the bursts of radio emission in the figures shown can be included in class A and A–B respectively. Figure 79 shows the variation in the absorption in the D-layer which accompanies a type C burst. Since the process takes place quite slowly it can be looked upon as being quasi-stationary by neglecting the term dN_{max}/dt in the equation (17.1). In this case, as can easily be followed from the formulae (17.1), (17.2) and (17.5), the small variations in the flux of the ionizing radiation simply repeat the variations in the degree of absorption. The latter, as is clear from the figure, are similar to the time-dependence of the radio emission flux at $\lambda = 3\cdot2$ cm.

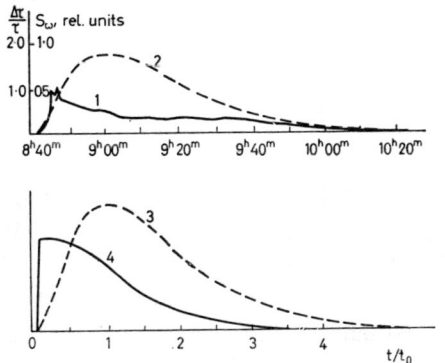
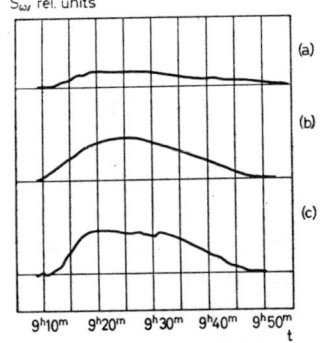

Fig. 78. The same as Fig. 77 for the flare of 14 December 1957 and an event of the "burst with gradual post-burst decay" type (Hachenberg, 1960).

Fig. 79. Profile of a "gradual rise and fall" type of burst at $\lambda = 3\cdot2$ cm (a), rise in the ionospheric absorption (b) and in the level of atmospherics during the flare of 2 November 1957 (c) (Hachenberg, 1960).

A close connection is found between microwave bursts and harder ionizing radiation when we compare the results of balloon and rocket measurements of the flux of radiation with an energy of 20–70 keV per quantum and the data from radio observations (Kundu, 1961b). All seven "bursts" of X-radiation (whose spectrum is actually a continuation into the shortwave band of the intense spectrum of the emission ionizing the D-layer during sudden ionospheric disturbances) investigated were accompanied by microwave bursts. Events in the metric band (type III) occurred only in three cases; in two of them the connection of these events with the X-radiation is doubtful, since the type III bursts were weak and were observed for a time far shorter than the period of intense radiation. On the other hand, the microwave bursts also show a detailed correlation with the variation in the X-radiation flux, as can easily be confirmed by looking at Fig. 80.

The microwave bursts accompanying hard X-radiation belong to the class of short-lived outbursts (see section 11). The size of the generation region may for them be quite large; for example, in one case a burst connected with hard radiation originated from a source about 4' and 8' in diameter at wavelengths of 3 and 21 cm, whilst in the rest of the cases investigated the size was not more than 2' and 4' respectively. The radio emission frequency spectrum $S_\omega(\lambda)$ falls rapidly as we move from the centimetric into the decimetric band and the brightness temperature at wavelengths of $\lambda > 3$ cm is 10^6–10^7 °K for bursts not accompanied by emission in the metric waveband; the spectrum appears to slope less and T_{eff} reaches 10^7–10^8 °K for two bursts accompanied by type III emission followed by type II and IV bursts.

The data given show that in the regions occupied by chromospheric flares or in the ones adjacent to them X-radiation is generated, the nature of its variation with time having much in common with the profile of microwave bursts.

We should point out that in exceptional cases flares are the sources of even harder radiation. This is shown by the event of 20 March 1958, when for about 20 sec a "burst" of γ-rays was observed on Earth with an energy of $\hbar\omega \approx 0.2$–0.5 MeV per quantum ($\hbar$ is Planck's constant) (Peterson and Winkler, 1959). The γ-quantum flux, which reached (1·6–2·8)$\times 10^{-5}$ erg/cm²/sec, was accompanied by a microwave outburst of radio emission. The duration of the radio burst was close to the duration of the gamma-radiation. More detailed information about this "burst" is given in section 11; ideas on the possible mechanisms of the gamma and radio emission are discussed in section 30.

It follows from section 30 that the γ-quanta in the final count owe their origin to relativistic particles as does the radio emission that accompanied the "burst" of γ-rays. An additional indication of the connection between the radio emission and relativistic velocities was the fact that a source at $\lambda = 21$ cm reached a size of more than 8' (3.5×10^5 km) for about 1 sec (Denisse, 1959b).

The comparison made of the nature of the time dependences of the solar emission flux at radio frequencies on the one hand and of X- and gamma rays on the other can be treated only as a first attempt in this direction. Even now, however, very valuable results have been obtained here that indicate the close connection of what appear to be such remote components of the solar spectrum.

Returning to the question of the correlation of the sporadic radio emission with the ionizing radiation we recall that above we had in mind the part of the radiation which causes the ionization of the ionospheric D-layer.

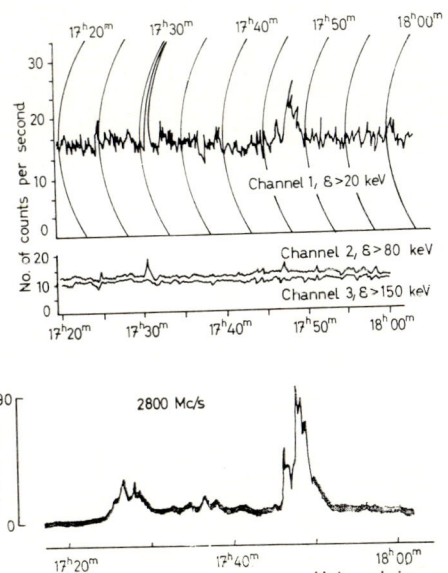
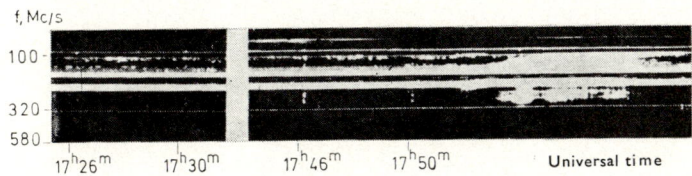

FIG. 80. Variation in fluxes of X-radiation with an energy $\mathscr{E} > 20$ keV and of radio emission at a frequency of 2800 Mc/s and in the 580–100 Mc/s range during the flare of 12 October 1960 (Kundu, 1961c)

§ 17] Sporadic Radio Emission

As for the E-layer, the slow variations in its state are closely connected with the level of the slowly varying component of the solar radio emission (Denisse and Kundu, 1957; Kundu, 1960; Minnis and Bazzard, 1958; Taubenheim 1958). If we take here as the initial data the values of the radio emission flux averaged for 5-day intervals and the index $f^4_{L\,\text{max}} \sec \varphi$[†] obtained from midday measurements of the critical frequency of the E-layer with normal incidence, then it turns out (Kundu, 1960) that the correlation coefficient, which is very high in the 3–30 cm range (about 0·8; see Fig. 81), drops sharply (to 0·2) as the wavelength increases.[‡] Since the

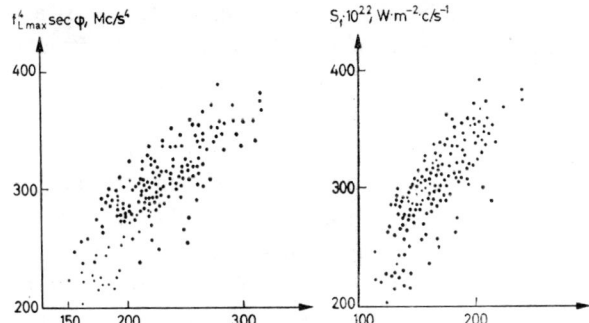

Fig. 81. Correlation diagram of combined five-day values of the parameter $f_{L\,\text{max}} \sec \varphi$ and the flux of solar radio emission S_f at $\lambda = 10\cdot 7$ cm (Kundu, 1960).

slowly varying component also reaches its maximum development in the $\lambda \sim 3\text{--}30$ cm range it becomes clear that the conditions under which the slowly varying component is generated in the inner layers of the corona also make possible an increase in ionizing radiation with a wavelength of a few tens of angstroms.§

SOLAR RADIO EMISSION AND MAGNETIC STORMS WITH A SUDDEN BEGINNING. PROPERTIES OF GEOEFFECTIVE CORPUSCULAR STREAMS

We spoke above of disturbances in the lower layers of the ionosphere whose source is ionizing radiation from the upper chromosphere or the

---

† The index $f^4_{L\,\text{max}} \sec \varphi$, according to what has been said above, characterizes the flux S of the ionizing radiation when there is a quasi-stationary change in the ionization conditions.

‡ For the correlation of the solar radio emission flux with the critical frequency see also Artem'yeva, Benediktov and Getmantsev (1961), Taubenheim (1958) Kerblai and Kovalevskaya (1960).

§ The reader can obtain fuller information on short-wave solar radiation from Shklovskii (1962), Nikol'skii (1962).

inner corona. We shall now discuss geophysical phenomena of the second group caused by the action of solar corpuscular fluxes, starting with ionospheric and magnetic storms and aurorae.

Ionospheric storms, unlike sudden ionospheric disturbances, are chiefly observed in the upper layers of the ionosphere (in the F_1 and F_2 layers). An ionospheric storm sometimes starts at night-time; it generally lasts for several days. An ionospheric storm is characterized by sharp changes in the critical frequency of the F_2 layer, a rise in the actual altitude of the F_1 layer and breakdowns in radio communication. In the polar regions, where the ionospheric storms are particularly strong, the regular structure of the upper ionosphere breaks down, the characteristic stratification disappears and the distribution of the electron concentration becomes extremely non-stationary.[†] These phenomena are accompanied by aurorae which are sometimes noticeable even at middle latitudes; ionospheric storms are also closely connected with strong disturbances of the Earth's magnetic field (geomagnetic storms with a sudden beginning); the state of the latter is a convenient index for quantitatively estimating the intensity and duration of the group of phenomena under discussion.

Geomagnetic disturbances generally follow chromospheric flares, lagging a few days behind them. This can easily be checked from the example of Fig. 82, where the values of the geomagnetic activity index (the K-index)[‡] were obtained by the method of "epoch superimposition" on the days preceding the flare and the days following it: the disturbances of the geomagnetic field increase after a flare, the increase in activity appearing about 2–3 days later (Avignon, Boishot and Simon, 1959). If at the same time we remember that the geophysical phenomena under discussion start not only during the day but also at night, it becomes clear that the cause of these phenomena may be only corpuscular streams moving away from the Sun at a velocity of the order of 10^3 km/sec. At the same time the sudden start of geomagnetic storms is connected with the fact that a rapidly moving ($V_s \sim 10^3$ km/sec) corpuscular stream meets a slowly moving Earth[§] chiefly from the side of its well formed front (Fig. 83a).

In the majority of cases it is hard to establish an unambiguous connection between the flare and the magnetic disturbance. However for certain particularly intense events (very strong flares accompanied by magnetic

---

[†] For the features of geomagnetic and ionospheric storms see, e.g., Obayashi (1959) and Al'pert (1960, section 18).

[‡] The K-index is a measure of the amplitude of the most disturbed element in the magnetograms during each 3-hour period. The K-index takes up values from 0 (magnetically quiet days) to 9 (strong magnetic storms).

[§] The orbital velocity of the Earth is far less—about 30 km/sec.

§ 17] Sporadic Radio Emission

storms of great strength)† this correspondence holds. The lag may be days or less; for example magnetic storms started only 19·5‡ and 26·5 hours after the strength 3+ flares observed on 28 February 1942 and 25 July 1946 respectively (Ellison, 1959). We notice that magnetic storms are largely connected with the flares that are located in the central part of the Sun's disk, which indicates the quasi-radial nature of the motion of the corpuscular streams away from the Sun (Kuiper, 1953, chapter 6).

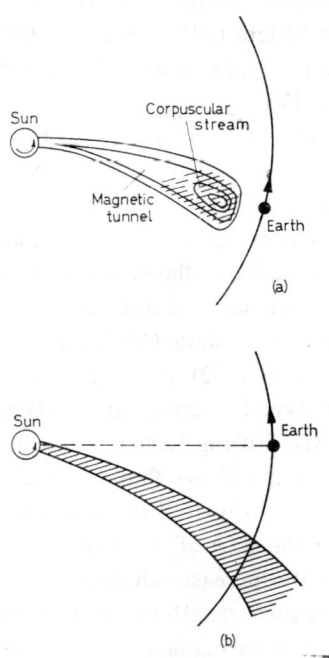

FIG. 82. Time-dependence of the K-index averaged over 115 events; the start of the reading was taken as the time the chromospheric flare appeared (Dodson, 1958).

FIG. 83. Motion of geoeffective corpuscular streams from the Sun to the Earth: (a) when magnetic storms with a sudden start appear; (b) when magnetic disturbances with a gradual start appear.

In Fig. 82 when investigating the connection of chromospheric flares with geomagnetic storms the only flares chosen were those that at the pre-maximum phase were accompanied by large bursts of radio emission at metric wavelengths.

† Strong magnetic storms include disturbances during which the change in the geomagnetic field strength exceeds 200–300γ ($\gamma = 10^{-5}$ oe). During particularly intense storms the variations reach 800–1000γ.

‡ This lag corresponds to a flux velocity of about 2×10^3 km/sec.

The close connection between magnetic disturbances and breakdowns in radio communication and powerful bursts having a complex $S_\omega(t)$ profile was established by Yudovich and Fel'dshtein (1958) (see also Zhigalov, 1960). According to the data of a statistical analysis of 10-year observations (Sinno, 1959) the probability of the appearance of magnetic storms rises noticeably as the strength of the flare increases and the intensity of the bursts accompanying them. Unfortunately observations at fixed frequencies did not make it possible to establish the spectral class of the bursts investigated by Avignon, Boishot and Simon (1959), Yudovich and Fel'dshtein (1958), Sinno (1959) and Zhigalov (1960) although it is quite possible that some of them belong to type II (in Zhigalov (1960) to type IV).

Correlation of geomagnetic disturbances with type II bursts was noted by Roberts (1959a, 1959b), Maxwell, Thompson and Garmire (1959) and Thompson (1959). The disturbance of the magnetic field increased 1·5–2 days after bursts of this kind; at the same time there was no correlation for strength 2 or 3 flares or for powerful type III bursts in the cases when these phenomena were not accompanied by type II events. However, to judge from other data (McLean, 1959; Sinno, 1959; Simon, 1960b; Kamiya, 1961 and 1962), the magnetic disturbances are more closely connected not with type II events, but with type IV radio emission following certain of the slowly drifting bursts. What has been said is confirmed by Fig. 84, from which it follows that geomagnetic storms show no correlation with the group of slowly drifting bursts not followed by a type IV continuum (or with the start of noise storms). At the same time the level of geomagnetic activity increases sharply 1–3 days after the recording of type IV events. Here the strength of the magnetic storm rises if the type IV radio emission appears during large flares (strength 2+ or 3) (Simon, 1960b), particularly if they are located near the central meridian (Kamiya, 1961 and 1962). In the figure we can also see a certain (but less definite than for type IV) rise in the disturbance of the Earth's magnetic field after strength 3 flares not accompanied by a type IV continuum.

In the light of what has been said the connection between type II bursts and geomagnetic disturbances can be explained by the fact that many of the slowly drifting events analysed by Roberts (1959a and 1959b), Thompson (1959) were accompanied by type IV radio emission. However, as a whole the question of whether magnetic storms are connected with all type II bursts or only with some of the bursts preceding type IV radio emission remains unclarified (in this connection, see Wood, 1961). It is quite possible that geomagnetic disturbances are connected with both types of event and the insufficient number of observations affects the results of

§ 17] Sporadic Radio Emission

analysing this connection which is generally carried out by the "superimposing of epochs" method.

In one way or another a close correlation of magnetic storms with types II and IV phenomena exists and is quite natural since these components of the radio emission are caused by agents moving in the corona at velocities which are the same in order of magnitude as the estimates of the velocity of the geoeffective corpuscular streams. In sections 13 and 15 it was found that the velocity of the radio emission agents is about 10^3 km/sec

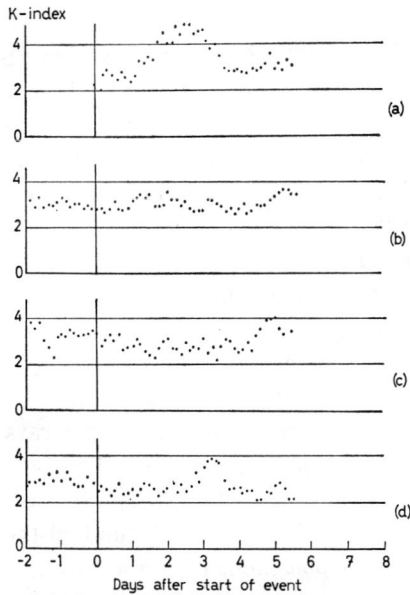

FIG. 84. Correlation of geomagnetic disturbances: (a) with type IV radio emission; (b) with type II bursts without type IV; (c) with noise storms whose start is connected with chromospheric flares; (d) with strong flares unconnected with noticeable type IV radio emission. The mean values of the K-index are plotted at 3-hour intervals (McLean, 1959).

or slightly more. The velocities of the geoeffective streams obtained from the lag time of the magnetic storms on Earth relative to the flares on the Sun also amount to 1000 or 2000 kilometres per second. On the other hand, they agree with the velocities of the most rapid ejections of material from a region of chromospheric flares (section 2). The close connection between all these phenomena and the agreement of the orders of the velocities convince us that we are dealing with a combination of events which in the final count are produced by a common cause and develop sequentially during the motion of a corpuscular stream from the flare towards the

Earth.† This circumstance is very important since it allows the observed features of the magnetic storms and geoeffective corpuscular streams to be used to get an idea of the nature and properties of these streams during the generation of the radio emission in the solar corona.

The concentration $N_\oplus$ of particles in the solar streams near the Earth can be estimated if we know the change in the geomagnetic field ΔH_0 in the initial phase of the magnetic storm (Martyn, 1951):

$$N_\oplus \sim \frac{8(\Delta H_0)^2}{\pi m_i V_s^2}. \tag{17.6}$$

Putting $\Delta H_0 \geqslant 200\gamma = 2\times 10^{-3}$ oe, the mass of one particle (a proton) in the stream $m_i = 1.7\times 10^{-24}$ g and the velocity $V_s \sim 10^8$ cm/sec, we find that the particle concentration in the geoeffective stream near the Earth is $N_\oplus \gtrsim 5\times 10^2$ protons/cm³. The question of the particle concentration in the geoeffective flux is discussed in detail in *Proc. of the Conf. of the Comm. on Solar Research* (1957).

The corpuscular bunches that cause magnetic storms with a sudden start carry magnetic fields with them. The screening action of these fields can be used to explain the decrease of several percent in the intensity of cosmic rays of galactic origin during very strong storms (the Forbush effect). According to Dorman (1957) a bunch with an extent of 0 1 a.u. (1·5 × 10⁷ km) will cause the observed decrease in the cosmic rays on Earth if the magnetic field in the stream is $H \gtrsim 10^{-4}$ oe. Direct measurements of the magnetic field by the artificial satellite "Explorer X", whose maximum distance away from the Earth was several hundred thousand kilometres, made it possible to find (*Geomagnetizm i Aeronomiya*, 1961) that at periods of strong disturbances that are probably connected with the passage of corpuscular streams from chromospheric flares the field strength changed from 5 to 40γ (i.e. from 0.5×10^{-4} to 4×10^{-4} oe). This agrees with the estimate given above for the field in the streams according to the Forbush effect.

---

† The velocity of plasma bunches—the sources of type IV radio emission and geomagnetic storms with a sudden beginning— is far greater than parabolic (on the Sun's surface the latter is 6×10^2 km/sec). In certain cases the plasma bunches definitely move in the corona at a velocity far less than parabolic; the reciprocating motions of the agent that occur in this case explain the appearance of the rare type II U-bursts (see section 13). It is not out of the question that sometimes a radio-emitting plasma bunch ejected from a flare at a velocity that is close to parabolic but does not reach it can be considered a natural satellite of the Sun until the coronal material's resistance to its motion significantly reduces its velocity (Vitkevich, 1962b). However, cases of this kind of periodic motion of a source of radio emission around the Sun with a period that differs significantly from 27 days (i.e. from the period of the Sun's rotation about its axis) have not yet been observed by anybody.

§ 17] Sporadic Radio Emission

Therefore a solar corpuscular stream that causes magnetic storms with a sudden beginning on Earth and the phenomena accompanying them is ejected from the region of a chromospheric flare at a velocity of the order of 10^3 km. When it passes through the corona it causes radio emission of spectral types II and IV. Since this stream carries a magnetic field it would be incorrect to look upon it as a system of separate particles penetrating the solar corona and interplanetary space. The stream is more a bunch of plasma with a magnetic field "frozen" into it.† The configuration of the lines of force to a certain extent ensures "magnetic insulation" of the stream from the penetration into it of charged particles from outside (Forbush effect).‡ Below we shall be able to confirm that this insulation permits fast charged particles accelerated in the solar atmosphere to be retained in the region occupied by the stream and to be transported toward the Earth's orbit.

Apart from magnetic storms with a sudden beginning which develop completely in a time of the order of an hour, there are also storms with a gradual beginning in which the geomagnetic field disturbances are weaker and more regular and increase in the course of a day. If the phenomena with a sudden beginning are caused by plasma bunches, whose sharply limited front flows towards the Earth (Fig. 83a), then the disturbances with a gradual beginning owe their appearance to lengthy corpuscular streams emitted from active regions of the Sun's disk occupied by spots and flocculi. Streams of this kind, taking part in the rotation of the Sun around its axis, catch up the Earth in its orbit so that the Earth largely enters the stream from the side where the particle density (and accordingly the extent of the magnetic disturbances) rises gradually (see Fig. 83b).

The duration of the emission of streams from active regions can be judged from the clear tendency to a 27-day repetition of the magnetic field's K-index value. The absence of the Forbush effect and the low amplitude of the magnetic disturbances as the Earth passes through the long-lived streams indicate that the latter do not bear with them "frozen in" magnetic fields and their intensity is much less than the corresponding value for sources of magnetic storms with a sudden beginning.

Magnetic disturbances with a gradual beginning show a clear correlation

---

† We note that geoeffective streams in the form of plasma bunches with a magnetic field ejected from chromospheric flares are, of course, quasi-neutral in nature: the electrical charge of the positive particles (mostly protons) is compensated for with great accuracy by the charge of the negative particles (electrons). We can therefore say that the electron concentration in the stream is close to the proton concentration.

‡ According to the data of Kamiya (1961 and 1962) there is a very close connection between the Forbush effect and type IV events: almost all the decreases in the cosmic ray intensity on Earth follow flares accompanied by type IV radio emission.

with the appearance in the central part of the Sun's disk of very large spots (with an area of not less than 10^{-3} of the Sun's hemisphere), the maximum of the magnetic storm appearing about 2 days after the group of spots intersects the central meridian and 2–4 days after the radio emission maximum at metric wavelengths (Gnevyshev, 1960; Van Sabben, 1953; Kuiper, 1953, chapter V). This correlation is apparently unconnected with activity of the type of chromospheric flares near spots (Kuiper, 1953; Simon, 1955a and 1955b).

It is not difficult to estimate the velocity of the particles making up the corpuscular stream from the value of the above lag connected with the distortion of the corpuscular streams due to the rotation of the Sun (Fig. 83b); it is equal to approximately 10^3 km/sec or slightly less (for further detail, see Mustel', 1957).

Of undoubted interest is the fact that disturbances of the geomagnetic field and the ionosphere with a gradual beginning largely accompany the spots which display noticeable activity in the radio band. On the other hand, when the central meridian is intersected by a group of spots with which radio emission sources of the noise storm type are not connected, in certain cases it is not an increase but even a decrease in geomagnetic activity that is observed (Simon, 1955a; Denisse, 1953; Simon, 1955c). The latter may possibly be caused by the fact that the inactive region itself does not emit geoeffective streams, creating at the same time, due to the magnetic field of the group of spots, a so-called "cone of avoidance" with its apex in this group. Since it is difficult for streams from other centres of activity to penetrate inside the cone the passage of the Earth through the "cone of avoidance" is accompanied by a weakening of the degree of the geomagnetic disturbances (Roberts, 1955). It follows from this that the radio emission connected with the spots is one of the most important characteristics which to a considerable extent determines the effect of the solar centres of activity on the Earth's magnetic field and ionosphere.[†]

RADIO EMISSION OF THE SUN AND POLAR BLACKOUTS. CONNECTION BETWEEN CONTINUUM-TYPE RADIO EMISSION AND THE APPEARANCE OF ENERGETIC PARTICLES

At a period of magnetic storms with a sudden beginning in polar regions there is often a strong rise in the absorption in the ionosphere at altitudes

---

[†] According to Mustel' (1957) corpuscular streams of long duration are actually connected not with spots but with flocculi. It is therefore of interest to study the correlation between geomagnetic storms with a sudden beginning and the slowly varying component of the solar radio emission in the decimetric band whose close correspondence with flocculi is solidly established.

§ 17] Sporadic Radio Emission

below the E-layer. As a result reflected signals disappear during pulsed ionospheric sounding (or the value of the minimum frequency at which reflected pulses can still be observed); in addition there is a sharp drop in the level of the cosmic radio emission received on Earth. The combination of these phenomena has been given the name of polar blackouts. As well as the blackouts developing together with magnetic storms and the Forbush effect (Freier, Ney and Winckler, 1959), a rise in the absorption in the lower ionosphere near the geomagnetic poles may also occur long before the start of a magnetic storm, preceding the latter by an average of 20 hours. In these cases the polar blackout is recorded an hour or more after the large chromospheric flares.

A blackout develops as follows. It is first localized in two comparatively small regions near the geomagnetic poles; gradually the area of these regions increases, and at the time of arrival of the magnetic storm the enhanced absorption completely covers the polar regions. In the process of a magnetic storm's development the blackout moves into lower latitudes, so that in the main phase of the magnetic storm the absorption becomes weaker in the polar regions and becomes stronger in the auroral zone (Hakura and Goh, 1959; Hakura, Takenoshita and Otsuki, 1958).

The short time lag of many polar blackouts relative to events on the Sun indicates that these phenomena are caused by particles whose velocity is much greater than the velocity of the corpuscular streams responsible for the appearance of magnetic storms with a sudden beginning.

According to Bailey (1959), Reid and Leinbach (1959), the ionizing agent may be protons with an energy of the order of several tens of mega electron-volts injected by the Sun during chromospheric flares. Charged particles with this kind of energy can penetrate to altitudes of about 100 km only in the polar regions of the Earth, this being prevented by the magnetic field in the low geomagnetic latitudes; from this point of view it becomes understandable why the phenomenon of polar blackouts is observed only in high latitudes. At the same time the protons that cause the blackouts cannot be recorded at the Earth's surface since their energies are insufficient for them to pass right through the Earth's atmosphere. Therefore the appearance of protons with an energy between several tens and several hundreds of mega electron-volts in the neighbourhood of the Earth after flares and the close connection of polar blackouts with similar particles have been established only by observations at great altitudes by sounding balloons and artificial Earth satellites (Anderson, 1958; Freier *et al.*, 1959; Winckler *et al.*, 1959a and 1959b; Rothwell and McIlwan, 1959; Brown and D'Arcy, 1959; Charakhch'yan *et al.*, 1961).

If the differential energy spectrum of protons of solar origin can be characterized by a power function of the form

$$P(\mathcal{E}_{kin})\, d\mathcal{E}_{kin} = F\mathcal{E}_{kin}^{-\gamma}\, d\mathcal{E}_{kin}, \tag{17.7}$$

then the parameters F and γ are different for different flares. In the energy range from several tens to several hundreds of MeV the values of γ are generally 4–5, although they may be more or less than this. The coefficient F characterizing the flux density of the energetic protons in the Earth's vicinity varies within very broad limits; as a guide we can take it that F reaches 2×10^{10} if $\mathcal{E}_{kin}$ is expressed in MeV and P in cm^{-2} sec^{-1} (Anderson and Enemark, 1960).

Polar blackouts show a close connection with type IV radio emission (Hakura and Goh, 1959; Anderson and Enemark, 1960; Thompson and Maxwell, 1960a and 1960b; Besprozvannaya and Driatskii, 1960). According to data obtained during the last solar activity maximum, blackouts preceding magnetic storms were not observed if the magnetic disturbances were weak ($\Delta H_0 < 100\, \gamma$) and did not follow type IV events; among the stronger magnetic storms ($\Delta H_0 > 100\, \gamma$) only a fifth followed polar blackouts. The position changes sharply for magnetic storms connected with type IV radio emission: in this case more than half the weak storms and about 85% of the strong magnetic storms follow the polar blackouts (Hakura and Goh, 1959).

A comparison of the results of radio observations at a number of fixed frequencies in the 9400–1000 Mc/s range and of spectral observations in the 600–40 Mc/s range with data on the absorption of cosmic radio emission in the polar regions (unrelated to their connection with geomagnetic disturbances) has shown (Kundu and Haddock, 1960) that polar blackouts show an even closer correlation with microwave bursts than with type IV phenomena. Practically all blackouts follow microwave bursts and the majority of blackouts follow type IV phenomena.[†]

It follows from what has been said that in the active regions conditions obtain in which it becomes possible for charged particles (protons) to be accelerated up to energies of the order of 100 MeV. It is very probable that as well as the energetic protons—the sources of polar blackouts in the regions indicated—electrons are also accelerated to comparable energies,

---

[†] Polar blackouts also accompany other types of radio emission with a broad frequency spectrum. For example a polar blackout and the stream of protons with an energy of several hundreds of MeV that caused it were recorded after the flare of 22 August 1958 which was accompanied by a storm continuum (section 16). In this case the variation in the flux of the energetic particles in the vicinity of the Earth to a certain extent repeated the intensity curve of the radio emission at metric wavelengths (particularly at $\lambda = 7{\cdot}5$ m; see Fig. 72) (Boishot and Warwick, 1959).

§ 17] Sporadic Radio Emission

although a feature of the energetic particles approaching the Earth is the absence of any noticeable quantity of fast electrons among them. Possible causes for this are indicated in section 30.

We notice that in exceptional cases when a chromospheric flare and the phenomena connected with it reach enormous intensity particles are generated in the flare region with an even higher energy (2–10 GeV). About an hour after the flare these particles (cosmic rays of solar origin) reach the vicinity of the Earth, raising the level of the corpuscular radiation recorded at the Earth's surface by tens and hundreds of percent. Outside the Earth's atmosphere the spectrum of the solar particles (protons) in the $\mathscr{E} \sim 2\text{--}10$ GeV apparently differs strongly for different flares and drops sharply as $\mathscr{E}$ rises approximately in accordance with (17.7) with a γ of a few units (according to Dorman (1957) $\gamma \sim 4\text{--}7$). The more usual phenomena, although they are more difficult to observe are the small effects of solar flares in cosmic rays which raise the level of the cosmic rays on Earth by only a few tenths of a percent. These effects can be seen better at a period of minimum solar activity after large flares. The rises in the intensity of the cosmic rays show a correlation with type IV bursts. Here in the region of the flare or in the region where the type IV radio emission is generated particles are also accelerated to energies of several GeV but, of course, there are far fewer of them (Dorman, 1957; Dorman and Kolomeyets, 1961; Kolomeyets, 1961).

It follows from the picture of the development of polar blackouts given above that in certain cases energetic charged particles reach the vicinity of the Earth long before the start of a magnetic storm; being deflected towards the poles by the geomagnetic field they penetrate into the lower ionosphere where they carry out additional ionization of the latter.† In other cases the arrival of energetic protons heralding the start of a polar blackout occurs only in the company of a solar stream that is the source of magnetic storms with a sudden beginning.

The possibility of the appearance of a polar blackout before the start of a magnetic storm and independently of it is apparently determined by the geometry of the magnetic fields in the solar envelope and in interplanetary space. Under favourable conditions the charged particles accelerated on the Sun approach the Earth and cause the blackout an hour or more after the chromospheric flare.‡ When the particles move on a radial trajectory this lag corresponds to a velocity of $V \lesssim 0\cdot 15\, c$. At the same time the

---

† For the nature of the motion of energetic particles in the Earth's magnetic field during a polar blackout see Sakurai (1961).

‡ The minimum lag of a blackout after type IV radio emission is about 40 min (Thompson and Maxwell, 1960a and 1960b).

velocities of the protons with $\mathscr{E} \sim$ 30–300 MeV that are responsible for the rise in the absorption in the polar regions lie in the range between 0·25 and 0·65 c, i.e. on the average they are several times greater than the values found from the time lag of the blackouts relative to optical or radio events on the Sun. This discrepancy is connected with the finite time taken by the energetic particles to leave the magnetic "trap" in the solar corona where they were accelerated, and with the complex trajectory of these particles in interplanetary space whose nature is determined by the configuration of the magnetic fields in existence there.

Since as the energy of the charged particles increases they leave the confines of the magnetic "trap" more efficiently and their velocity away from the Sun rises accordingly, it is to be expected that the lag of the polar blackouts after events on the Sun will be less than for the blackouts which are caused by more energetic protons. This dependence has not yet been investigated, unfortunately. In its favour, however, there is the close connection between the lag of the blackout and the intensity of the preceding microwave bursts (Kundu and Haddock, 1960; Kamiya, 1961 and 1962). This time does not exceed 5 hours for large bursts (with a radio emission flux of $(2-5) \times 10^{-20}$ W m^{-2} c/s^{-1} or more), increasing for the weaker bursts. All this becomes comprehensible on the natural assumption that the intensity of the radio emission increases with a rise in the energy of the emitting particles, i.e. electrons, if we remember that apparently the conditions under which there is efficient acceleration of protons in the region of a chromospheric flare will also be favourable for the acceleration of electrons.

It has already been pointed out above that in certain cases the arrival of energetic protons in the vicinity of the Earth and the appearance of a polar blackout occur only with the beginning of a magnetic storm. In certain cases the depth of a polar blackout preceding a magnetic storm increases when the latter starts. These facts indicate the very important circumstance that there are also fast particles in the composition of a slow geoeffective flux (a plasma bunch with a magnetic field "frozen" into it). It must be pointed out that in certain cases the stream bears with it not only energetic protons with $\mathscr{E}_{\text{kin}} \sim 10^2$ MeV but also soft cosmic rays of solar origin with an energy of the order of 10 GeV (Charakhch'yan, Tulinov and Charakhch'yan, 1961; Steljes, Carmichael and McCracken, 1961; Roederer *et al.*, 1961). In the latter case the increase in cosmic rays lags several tens of hours behind the chromospheric flare.

It is known, furthermore, that the duration of polar blackouts is often several days, i.e. is far greater than the time the geoeffective stream is passing near the Earth, not to mention the duration of the chro-

§ 17] Sporadic Radio Emission

mospheric flare. It may be assumed that a corpuscular stream with a magnetic field "frozen" into it changes the geometry of the lines of force in circumsolar and interplanetary space, "stretching" them from the active region on the Sun into the vicinity of the Earth (see Fig. 83a). This permits rapid motion of energetic particles from the Sun to the Earth, along the "magnetic tunnel" that is formed, after the geoeffective corpuscular flux has left the Earth's orbit, and is localized within the "tunnel".

It is interesting to note that the duration of this phase of the polar blackouts is comparable with the life of noise storm sources on the Sun (R-centres). This does not apparently occur by chance (Kundu and Haddock, 1960). The point is that only a quarter of all polar blackouts occur on days free of noise storms; the remaining three quarters come in periods when there are active centres on the Sun generating enhanced radio emission and type I bursts. Moreover the duration of polar blackouts rises with an increase in the duration of the noise storms, starting from the time of the appearance of a microwave burst or type IV event connected with the beginning of the blackouts; here the absorption effect in the polar regions mostly disappears sooner than the noise storm. On the other hand, there is no correlation with the time that the noise storm is in existence from its start to the appearance of the events on the Sun after which the blackouts start.

The preliminary data given, according to which the time energetic protons are present near the Earth is determined to a considerable extent by the length of the noise storms after the corresponding flares on the Sun, provide a very good basis for the assumption that the regions in which the noise storms are generated are a kind of reservoir that holds the energetic particles in a "trap" formed by the bipolar magnetic field of the sunspots. The energetic particles appear here either because of acceleration in the sunspots' fields or as a result of the injection of particles from the chromospheric flare region. The formation of the "magnetic tunnel" and the rapid movement towards the Earth of the energetic particles that have left the "trap" above the spots allow a polar blackout to be maintained for a long time.

The data given above allow no room for doubt that the regions where the microwave bursts are generated near chromospheric flares and the plasma bunches (geoeffective corpuscular fluxes) ejected from the flares exciting type II and IV radio emission as they move in the corona contain protons with energies of 10–100 MeV, and in certain cases particles with an energy of several GeV. It is possible that the regions above the spots where noise storms are generated also retain energetic protons. The possibility cannot be excluded that in these regions of the corona there are also electrons accelerated to comparable energies, although the latter have not been found near the Earth during polar blackouts. In section 30 we shall show that the emission of fast electrons in the magnetic fields of sunspots, plasma bunches ejected from flares, and the flares themselves may explain the observed features of the enhanced radio emission making up noise storms and type IV radio emission, and certain properties of microwave bursts respectively.†

Energetic particles can be localized in regions occupied by the streams near chromospheric flares and above sunspots provided that the magnetic fields there are strong enough and their configuration satisfies the requirements of a magnetic "trap". Without dwelling in greater detail on the question of the retention of charged particles, which by the way is very important for microwave burst generation mechanisms, noise storm generation and type IV radio emission generation, we would point out that, since for a number of reasons the magnetic insulation of these regions cannot be absolute, one would expect some energetic particles to get out of them. We have spoken above about the possibility of polar blackouts being maintained for a long time because of particles leaving the bipolar fields of sunspots. It is highly probable that it is this effect of energetic particles leaving "magnetic traps" that is also the cause of the appearance of "precursors" (type III bursts preceding type II bursts) and the appearance of fine type III structure in type II bursts. In the first case energetic particles, on leaving the region of a chromospheric flare, move in the corona at a velocity that is a significant fraction of the velocity of light and generate rapidly drifting type III bursts. In the second case energetic particles leave a plasma bunch with a quasi-closed magnetic field moving at a velocity of $\sim 10^3$ km/sec and excite in the surrounding corona rapidly drifting surges of radio emission, giving the characteristic structure discussed in detail in section 13 (see also section 14) in the dynamic spectrum.

In conclusion we should like to make some remarks about the fore-

---

† Since type III bursts are, in all probability, caused by streams of particles with a velocity of $V_s \sim 0.5$ c (which corresponds to a proton energy of $\mathscr{E}_{kin} \sim 100$ MeV), attempts have also been made (Thompson, 1959; Thompson and Maxwell, 1960a and 1960b) to find a possible correlation between these bursts (without a type II–IV phenomenon) on the one hand and polar blackouts and a rise in cosmic ray intensity on the other. To date, however, these attempts have proved unsuccessful; this is not surprising since the blackout statistics are clearly inadequate, and the level of the cosmic rays at the Earth's surface is not connected with particles having energies of the order of 100 MeV, since the latter are unable to pass right through the Earth's atmosphere. Therefore observations must be made in the polar regions at great altitudes to establish the connection between the streams of energetic particles and type III bursts, using sounding balloons and artificial Earth satellites to do this.

§ 17] Sporadic Radio Emission

casting of geophysical phenomena from observations of solar sporadic radio emission.

Firstly, it becomes clear from what has been said that it is impossible to forecast individual phenomena on Earth connected with a rise in the absorption in the D-layer and caused by the action of X-radiation, since these phenomena appear almost simultaneously with the recording of the corresponding events on the Sun (chromospheric flares, microwave bursts). Therefore we can only make probability statements about the expected frequency and intensity of these phenomena, judging from the extent of the flare activity, the frequency of appearance and intensity of microwave bursts. There are no particular advantages either in comparing the critical frequencies of the E-layer with the level of the slowly varying component (when compared with the area of the plages), or in comparing the 11-year variations in the state of the ionosphere and the Earth's magnetic field with the radio indices of solar activity, and not with the Wolf numbers, sunspot area, etc. (Artem'yeva, Benediktov and Getmantsev, 1961).

There is more future in forecasting phenomena like magnetic and ionospheric storms, aurorae and blackouts which are caused by streams of particles, closely connected in their turn with such events on the Sun as generation of type IV radio emission and microwave bursts. In this case the correlation with radio emission is undoubtedly closer than with optical phenomena of the chromospheric flare type. The lag (not less than an hour for polar blackouts or tens of hours for the rest of the geophysical phenomena) is also sufficient in many cases for timely forecasting. Certain possibilities for forecasting magnetic storms with a gradual beginning and the ionospheric disturbances accompanying them are opened up by correlation of the latter with R-centres (radio-emitting groups of spots).

GENERAL PICTURE OF THE SUN'S SPORADIC RADIO EMISSION

In the earlier sections of this chapter we have discussed in detail the features of the different components of the sporadic solar radio emission and their connections with a number of phenomena on the Sun and the Earth. However, because of the complexity and wealth of experimental data it is difficult to get any general idea about the Sun's sporadic radio emission, its quantitative characteristics and the sequence of development of the individual components. In this respect the reader will be helped by the present section which briefly recapitulates the content of Chapter IV.

The dynamic spectra of the basic components of sporadic radio emission arising when a centre of activity appears on the Sun are shown diagrammatically in Fig. 85 (Wild, 1962, 1960b). The data on these components and

TABLE 4

Type of radio emission	Source of radio emission	Nature of source		Nature of radio emission			Relative time for which given type of activity can be observed (%)	Generation mechanism†
		Life	Size of generation region	Effective temperature	Band	Polarization		
"Quiet" Sun radio emission (B-component)	Undisturbed chromosphere and corona	Always present	>32′; increases with wavelengths	~10^6 °K at metric wavelengths; decreases with wavelength to 6×10^3 at $\lambda \sim 2$ mm	mm, cm, dm, m	Very weakly polarized	100	Bremsstrahlung
Slowly varying radio emission (S-component)	Regions of high density, magnetic field and temperature in lower corona above centres of activity	Few months	1–10′; increases on moving from centimetric to decimetric band	$\lesssim 2 \times 10^5$ °K at centimetric and decimetric wavelengths	mm, cm, dm	Partly polarized (with extraordinary emission predominant), degree of polarization decreases with rise in wavelength		Bremsstrahlung combined with magnetobremsstrahlung non-coherent

238

§17] Sporadic Radio Emission

	Type	Generation region	Duration	Size	Temperature	Wavelength	Polarization		Mechanism
Microwave bursts	Type A (plain bursts), type B (postburst fall), type C (gradual rise and fall)	Generation region usually in same place as source of S-component	Few minutes. Few minutes to few hours. Tens of minutes to few hours.	∼1′ More than 2–3′ ∼1′	Up to 10^9 °K Up to 10^7 °K <10^6 °K	mm, cm, dm	Partly polarized (extraordinary emission apparently predominant)		Bremsstrahlung combined with magnetobremsstrahlung non-coherent
Type I noise storms	Enhanced radio emission connected with sunspots, type I bursts	Region of corona above groups of spots	From few hours to few days	Up to 10′	Up to 10^9–10^{10} °K or more	m	Strongly polarized, apparently with ordinary waves predominant	13	Magnetobremsstrahlung non-coherent. Magnetobremsstrahlung coherent or non coherent; Cherenkov coherent
			Fractions of a second	Less than or close to size of source of enhanced radio emission					
	Type II bursts	Region in corona moving away from chromospheric flare at a velocity of ∼10^3 km/sec	Few minutes	5–7′	Up to 10^{15} °K, generally 10^{11} °K	m	Not polarized	0·05	Cherenkov coherent

† From the results of the theoretical treatment in Chapters VIII–IX.

(*contd.*)

TABLE 4 (*continued*)

Type of radio emission	Nature of source			Nature of radio emission				Relative time for which given type of activity can be observed (%)	Generation mechanism
	Source of radio emission	Life	Size of generation region	Effective temperature	Band	Polarization			
Type III bursts	Region in corona moving away from chromospheric flare at a velocity of $\sim 10^5$ km/sec	Few seconds	3–10'; increases with wavelength	Up to 10^{11} °K	m, dm	Sometimes partly polarized		0·25	Cherenkov coherent
Type IV radio emission (follows type II burst)	Region in corona moving away from chromospheric flare at a velocity of $\sim 10^3$ km/sec	From few minutes to few hours	Up to 10' or more	Up to 10^{12} °K, usually 10^7–10^{10} °K	dm, m	Partly polarized		0·28	Magnetobremsstrahlung non-coherent and coherent
Type V radio emission (follows type III bursts)	Region in corona	Minutes	?	?	m	?			Cherenkov coherent

† From the results of the theoretical treatment in Chapters VIII–IX.

§ 17] Sporadic Radio Emission

the characteristics of the "quiet" Sun's radio emission are summarized in Table 4, which differs comparatively little from the summary of basic solar radiation characteristics given by Denisse (1959a). The relative contribution from the individual components of the Sun's sporadic radio emission can be judged from the data in the last column but one of the table, which shows the percentage of time for which each type of solar burst can be observed at a frequency of 125 Mc/s (with $S_\omega > 10^{-21}$ Wm^{-2} c/s^{-1}) at a period of maximum solar activity (Maxwell, Howard and Garmire, 1960).†

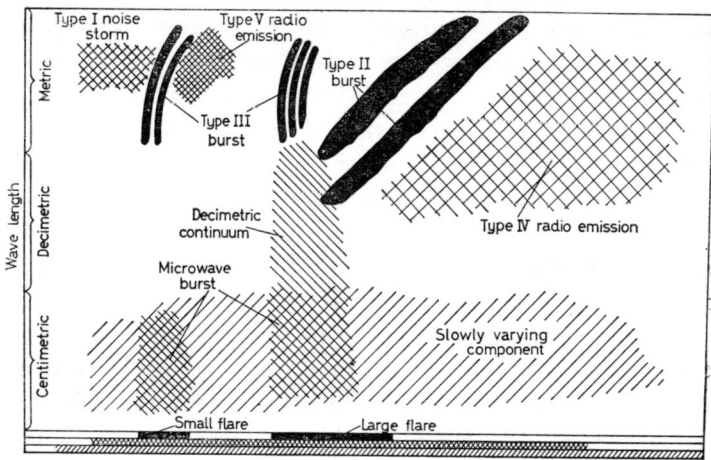

FIG. 85. Idealized dynamic spectrum of sporadic radio emission from a centre of activity.

From observations of the solar radio emission and the phenomena accompanying them in the solar and terrestrial atmospheres we can conclude that the sporadic radio emission (in its fullest form) largely develops as follows (see Fig. 85).

The appearance of a centre of activity—flocculi and groups of spots with which local magnetic fields are connected—on the Sun stimulates the formation in the lower corona of regions of enhanced density which act as the source of the slowly varying component of the sporadic radio emission. The latter can chiefly be found at wavelengths of 3–30 cm.

In the higher layers of the corona where the strong magnetic fields of sunspots penetrate, conditions are realized under which it becomes possible in the metric band for strongly polarized type I noise storms with a gradual

† A catalogue of type II and IV events recorded in 1957–62 is given by Kislyakov, Losovskii and Salomonovich (1960); a list of types II, III, IV and V phenomena observed in the period 1952–60 is given by Boorman et al. (1961).

beginning to be generated (enhanced radio emission with a wide frequency spectrum and short-lived narrow-band type I bursts lasting less than 1 sec; T_{eff} is comparable with the corresponding value for the enhanced radio emission). As a result of chromospheric flares near active groups of sunspots noise storms with a sudden beginning appear and the sources of noise storms already in existence are strengthened. The noise storm generation region apparently contains energetic particles (protons with an energy of the order of 100 MeV) which maintain lengthy polar blackouts on Earth after strong chromospheric flares.

Agents (apparently streams of charged particles) leave the weak chromospheric flares; when they move into the denser layers of the corona at a velocity of about half the velocity of light, non-polarized or moderately polarized type III bursts are generated. The type III bursts are distinguished by a rapid frequency drift towards the low frequencies at a rate of about 20 Mc/s^2. This effect can be explained by the decrease in the electron concentration in the region where the radio emission is generated as the agent moves in the corona.

Sometimes a fine structure—reverse-drift pairs (drifting towards the high frequencies)—appear in the type III bursts. The change in the sign of the drift is probably connected with the fact that the variety of radio emission in question appears at the time when the agent meets a local inhomogeneity of enhanced density in its path. The reverse-drift bursts are often not accompanied by any noticeable type III radio emission.

The frequency drift towards the low frequencies in type III bursts sometimes changes to the opposite in the process of the development of the bursts. This is interpreted as the consequence of the special nature of the motion of the corpuscular agent in the magnetic field of a bipolar group of spots when the stream first goes away from the Sun's surface and then approaches it. In such a case the dynamic spectrum of the type III bursts acquires features characteristic of their variety—the so-called U-bursts.

Type III surges accompanied by bursts of radio emission at centimetric wavelengths precede the appearance of an intense broad-band continuum—type V radio emission lasting for several minutes.

Large flares act firstly as the sources of microwave bursts whose spectrum also spreads at times into the decimetric band, creating radio emission that varies smoothly in frequency and in time—the decimetric continuum. Sometimes the spectrum of the radio emission is so broad that this continuum can also be observed at metric wavelengths.

The above-mentioned components connected with the appearance of large flares on the Sun's disk are generated in the lower layers of the corona

§ 17] Sporadic Radio Emission

immediately adjacent to the region of the chromospheric flares. Besides, at the time of the sudden expansion and ejection of material from the flare into the corona rapidly moving agents are injected which excite type III bursts. These bursts and the decimetric continuum precede the most powerful of all the components of solar radio emission—non-polarized type II bursts at metric wavelengths having a slow frequency drift at a rate of the order of $\frac{1}{4}$ Mc/sec^2. The type II bursts are caused by agents moving in the corona at a velocity of the order of 10^3 km/sec; these agents are probably plasma bundles with magnetic fields "frozen" into them ejected into the corona from the flare region.

Faster particles also make up these plasma bunches as well as relatively slow particles; having broken out of the "trap" formed by the magnetic field of the plasma bundle, they move in the corona at a velocity of the order of 10^5 km/sec, generating type III bursts which in the present case are one of the elements of the fine structure of the type II event.

After the type II radio emission from the region of a bunch moving in the corona there appears a type IV continuum—broad-band structureless radio emission of very high intensity lasting from tens of minutes to several hours. The basic difference of the type IV radio emission from the remaining components of the continuum group is the circumstance that they are generated in a region moving in the corona at a distance of up to five solar radii.

On approaching the Earth, the plasma bunches with their "frozen" in magnetic field, now called geoeffective corpuscular fluxes, cause magnetic and ionospheric storms with a sudden beginning and aurorae. The fast particles (chiefly protons with an energy of the order of 100 MeV) leave the region of the chromospheric flare and the plasma bunch, even while it is moving in the corona, precede the geoeffective flux and appear in the vicinity of the Earth an hour or more after the beginning of the events on the Sun, which are connected with the chromospheric flare. These fast particles are the cause of polar blackouts; the energetic particles that do not have time to leave the geoeffective flux earlier are carried along with the latter into the vicinity of the Earth and deepen the blackout.

CHAPTER V

Results of Observations of Radio Emission of the Planets and the Moon

18. First Investigations into the Radio Emission of the Moon, Planets and Comets

FIRST STUDY OF THE MOON AND PLANETS IN THE RADIO-FREQUENCY BAND

Among the planets of the solar system the methods of radio astronomy are being used at present to study Mercury, Venus, Mars, Jupiter and Saturn; attempts to find the radio emission of Uranus and Neptune have not yet been crowned with success (Goldberg, 1960; Smith and Carr, 1959; Smith, H. G., 1959; Brissenden and Erickson, 1962). Among the planetary satellites only the nearby Moon creates on Earth a radio emission flux that can be measured by present-day receiving equipment.

The Moon was first studied in the radio-frequency band by Dicke and Beringer (1946), who in 1945 measured its effective temperature at a wavelength of $\lambda = 1\cdot 25$ cm. The numerous measurements made later have made it possible to obtain a more precise value for $T_{\text{eff}\mathbb{C}}$, to elucidate its dependence on frequency and the phase of the Moon, to obtain information on the radio brightness distribution over the Moon's disk, etc. (see section 21).

At the beginning of 1955 Burke and Franklin (1955) unexpectedly found at a frequency of 22·2 Mc/s a very interesting radio object which differed sharply from previously studied discrete sources of cosmic radio emission. The recordings of the emission flux from the new source obtained as it passed through the beam of a fixed aerial (a "Mills Cross" with a lobe width of about 2·5°) are reproduced in Fig. 86. It is clear from the figure that, unlike a typical discrete source—the Crab Nebula (see the recording on the right)—the radio emission found is very irregular in nature (it consists of a series of short bursts very similar to the interference from lightning discharges in the Earth's atmosphere), and the time of appearance of the bursts varies in the course of time, i.e. the source moves in the sky. It

§ 18] First Investigations

followed from this that the new source was located far closer to the Earth than discrete sources of galactic and metagalactic origin; it could be assumed that it was part of the solar system.

Burke and Franklin obtained more definite results when they compared the coordinates of the new discrete source with the coordinates of Jupiter,

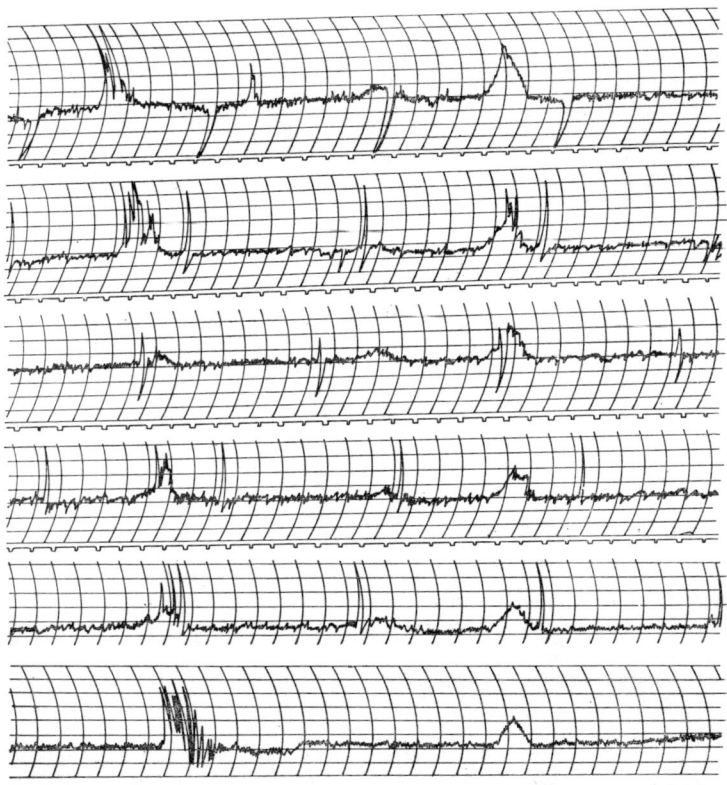

FIG. 86. Samples of recordings of the radio emission flux of Jupiter (series of bursts on the left) and the Crab Nebula (right) at a wavelength of $\lambda = 13\cdot 5$ as the sources passed through the beam of a fixed aerial (Burke and Franklin, 1955)

Uranus and the two galactic objects the nebulae NGC 2420 and NGC 2392 (at the time of the observations the visible position of these planets and nebulae was close to the position of the source of the bursts in the sky). This comparison of one coordinate (right ascension) is shown in Fig. 87. The periods when the bursts appeared are marked by wavy lines; the points show the position of Jupiter during the observations. The position of the radio emission source on this graph obviously corresponds to the middle of the band bounded by dotted lines. It is easy to see that the spherical coordinates of the new radio object agree (for the given observational accuracy) with the coordinates of Jupiter. The source found repeats the planet's motion through the sky and can therefore be definitely identified with the planet Jupiter.

A study made very soon afterwards by Shain (1955, 1956) of old recordings of cosmic radio emission at a frequency of 18·3 Mc/s showed that as early as 1950–1 radio astronomers had been picking up Jupiter's emission but had not attached any particular significance to it, assuming that the

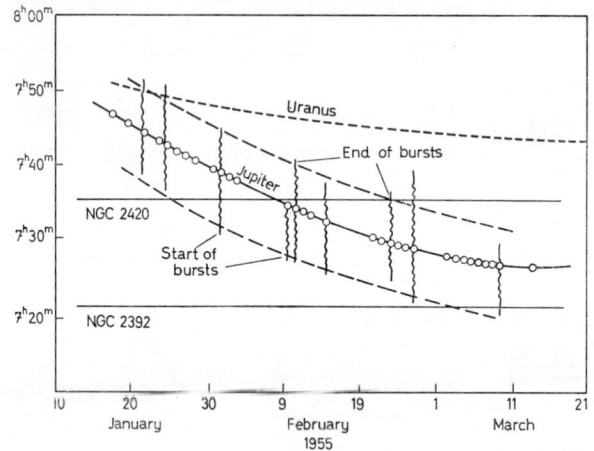

FIG. 87. Comparison of coordinates of the new radio object with the coordiates f Jupiter, Uranu and two galactic nebulae: NGC 2420 and NGC 2392 (from right ascension) (Burke and Franklin, 1955)

series of bursts from the new source were caused only by random atmospheric interference.†

After Burke and Franklin's work the radio emission of the planets was intensively studied. Mayer, McCullogh and Sloanaker (1958a, 1958b and 1958c) in 1956 measured the radio emission flux of Venus, Mars and Jupiter at centimetric wavelengths; radio emission from Saturn has been measured in this same band (Drake and Ewen, 1958). In recent years the band studied has been considerably expanded both towards the millimetric and decimetric wavelengths. A negative result was obtained at first in the radio observations of Mercury (Goldberg, 1960; H. J. Smith, 1959). Then Howard, Barret and Haddock (1961, 1962) in 1960–1 found the radio emission of this planet and measured its effective temperature in the 3-centimetre band.

The features of the planetary radio emission are discussed in detail in

† Further details of the history of the discovery of Jupiter's radio emission are given by Haddock (1958) and Franklin (1959).

§ 18] First Investigations

sections 19 and 20. Here we shall content ourselves with saying that attempts continue to be made to find radio bursts from other members of the solar system as well as the sporadic radio emission of the Sun and Jupiter. It is highly improbable, of course, that sporadic radio emission is generated by objects which have hardly any gaseous shell (Mercury, asteroids and the greater part of planetary satellites, including the Moon). Pluto is also an unsuitable object in this respect because of its distance away from the Sun, its small size and its low temperature. No sporadic emission has been found coming from Neptune, Uranus or Mars; if it does exist, then its flux on Earth is definitely much less than the magnitude characteristic of the decametric radio emission of Jupiter (Smith and Carr, 1959; Smith, H. J., 1959; Brissenden and Erickson, 1962). Three possibilities therefore remain for the source of the sporadic radio emission: Venus, Saturn and comets.

In 1956 Kraus (1956a, 1956b, 1957a and 1957b) reported reception of sporadic radio emission from Venus at decametric wavelengths. According to the information he obtained the surges of radio emission of short duration from Venus are similar to the sporadic bursts coming from Jupiter; the maximum emission flux from Venus at $\lambda = 11$ m reaches 10^{-21} W m^{-2} c/s^{-1}, which is several times greater than the corresponding value for the discrete source Cygnus-A at the same wavelength.

Kraus's results are not confirmed, however, by other researchers. Burke and Erickson (Burke, 1960) in 1956 found no traces of radio emission from Venus at $\lambda = 11$ and 13·6 m with an accuracy of up to 10^{-23} W m^{-2} c/s^{-1}. Smith and Carr (1959) and H. J. Smith (1959) carried out studies of Venus at a wavelength of 11 m for several months in 1958 without observing a single event which could be attributed to sporadic radio emission of this planet. (The sensitivity of their apparatus was quite sufficient to record radio emission at the intensity indicated in Kraus, (1956a).) In the next 2 years observations were continued at wavelengths of 11 and 16·5 m (Carr et al., 1961); they too were unsuccessful as were observations of Brissenden and Erickson (1962).

It must be stressed that when studying the decametric radio emission of the planets, errors are very probable, due to the similarity between the planetary sporadic radio emission, terrestrial atmospherics and bursts of solar origin. According to Roberts (1958) the bursts described by Kraus are similar to those reverse-drift pairs observed at the same time at frequencies of 40–50 Mc/s (section 16). The latter are known to have been connected with the Sun since the source of the bursts was located on a line passing through the solar disk; the distance from Venus to this line was quite large (several degrees).

Further observations and an analysis of the results obtained forced Kraus (1960) to recognize that some of the bursts he found were definitely not connected with Venus and were interference of terrestrial origin. As for the rest of the bursts he reported that the available data were insufficient to conclude that they were generated in the region of Venus, although this possibility cannot be completely denied. As a whole the question of the existence of sporadic radio emission from Venus still remains open.

In 1957 Smith and Douglas (1957, 1959), Smith (1959) observed at a frequency of 21·1 Mc/s about ten events which were possibly connected with the planet Saturn. The times the activity appeared (with one exception) agree closely with the period of $10^h 22^m$, which is close to Saturn's period of rotation ($10^h 14^m$). Only two events which can be attributed to this planet were recorded in several months of the next year at $f = 22 \cdot 2$ and 23 Mc/s. In all cases definite identification proved to be impossible since the interference picture in the radio emission recordings made by interferometers was very weak and the level of terrestrial disturbances was high.

Another group of radio astronomers (Smith and Carr, 1959) observed Saturn in 1957 at 18 Mc/s and in the next year at 18 and 22·2 Mc/s. The perturbations that presumably proceeded from Saturn were obtained during six out of the fifty-four nights favourable for observations at 18 Mc/s and for six out of thirty-three nights at 22·2 Mc/s. The activity state generally lasted for 1 to 5 minutes with a maximum radio emission flux of up to 6×10^{-21} W m^{-2} c/s^{-1}.

Unfortunately it proved impossible to compare the recordings obtained by the two groups: in the periods when one of the groups was recording the radio emission of Saturn, the apparatus of the other group was not operating.

In 1960 during observations at 18 Mc/s Carr *et al.* (1961) noted seven very weak and for the most part brief events; not one of them, however, can definitely be said to have been caused by Saturn. All we can say is that all the events occurred at one and the same position of the planet, if we assume that the period of rotation of the radio emission source on Saturn is $11^h 57 \cdot 8^m$. The latter value differs considerably from the rotational period of the planet's cloud layer determined by optical observations.

Therefore the available data cannot be used as a basis for definitely judging the existence of noticeable sporadic radio emission from Saturn, although it is quite possible that such emission does occur.

§ 18] First Investigations

RADIO EMISSION OF COMETS

Among the comets to have been studied by radio astronomy methods are Comet Arend-Roland (1956h) in the period of its closest approach to the Sun (April 1957), Burnham's Comet (1959k) and Wilson's Comet (1961d).

From observations at a wavelength of $\lambda = 11$ m Kraus (1958a) concluded that Comet Arend-Roland generates radio emission whose intensity is subject to strong oscillations. On the average the radio emission flux on Earth was about 5×10^{-22} W m^{-2} c/s^{-1}. However, Shain and Slee's measurements (Shain and Slee, 1957), which were made at the same time at wavelengths of 15 and 3·5 m with a far better aerial, did not confirm Kraus's data. The receiver sensitivity at $\lambda = 15$ m proved to be insufficient when scanning the aerial through the region occupied by the comet (its core and tail). It followed from this that the comet's radio emission flux was therefore at least 25–50 times less than the value indicated by Kraus (1958a). No absorption of cosmic radio emission as it passed through the comet was observed either. At $\lambda = 3·5$ m the upper limit of the radio emission flux from the core was about 10^{-25} W m^{-2} c/s^{-1}. A number of recordings was made during observations with an aerial having a "knife-edge" beam in order to find any possible emission from the comet's tail. No radio emission was found in this case either above the sensitivity threshold of the apparatus (10^{-24} W m^{-2} c/s^{-1}). A negative result (with an accuracy up to values of the same order) was obtained by Whitfield and Högbom (1957, 1959) (see also Seeger, Westerhout and Conway, 1957) in interferometer observations of the core of Comet Arend-Roland at $\lambda = 3·7$ and 7·9 m. The emission from the extended tail at these wavelengths was definitely less than 5×10^{-23} W m^{-2} c/s^{-1}.

At the same time two groups of observers have reported (Coutrez, Hunaerts and Koeckelenbergh, 1958; Müller, Priester and Fischer, 1957) picking up radio emission from Comet Arend-Roland at shorter wavelengths (50 and 21 cm). The emission flux at $\lambda = 50$ cm was $(3-8) \times 10^{-23}$ W m^{-2} c/s^{-1}, the radio emission originated from the head of the comet or from a certain "halo" slightly ahead of the comet as it moved towards the Sun.

The existence of this "halo" was confirmed by visual observations. At a wavelength of 21 cm the emission of the comet was in the nature of bursts with a life of 20–60 sec.

It is difficult now to judge the reliability of these results. All we shall say is that the attempts to find the radio emission of another comet (Burnham's) at wavelengths of 20–125 cm made at Jodrell Bank using the 250-foot radio telescope were unsuccessful (Conway, Chuter and Wild, 1961).

As a result the upper limit for the radio emission flux was estimated at $(1-3) \times 10^{-26}$ W m^{-2} c/s^{-1}. No effect was found in the $\lambda = 21$ cm neutral hydrogen line. Observations of Wilson's Comet at a frequency of 26·3 Mc/s also produced a negative result (Erickson and Brissenden, 1962). Therefore the existence of any noticeable radio emission from comets still remains unproved.†

19. Sporadic Radio Emission of Jupiter

RADIO EMISSION FLUX AND ITS TIME DEPENDENCE

In the decametric band Jupiter's radio emission is sporadic in nature and consists of a series of bursts appearing from time to time. As can be seen from Fig. 88 (see also Fig. 86) the recording of Jupiter's radio emission flux is similar to corresponding recordings of solar radio emission at metric wavelengths or of radio interference of atmospheric origin.

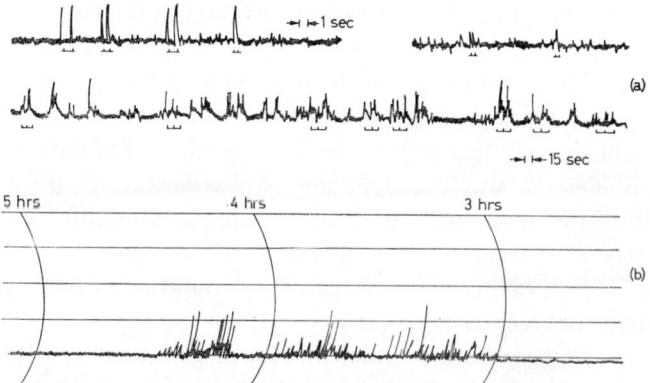

FIG. 88. Recordings of Jupiter's radio emission flux made with high-speed (a) and low-speed (b) recording devices during reception at a frequency of 18 Mc/s ($\lambda = 16 \cdot 7$ m) (Carr et al., 1958).

The intensity of Jupiter's sporadic radio emission is quite considerable. For example, at a frequency of 18 Mc/s the radio emission flux for typical bursts is of the order of 10^{-20} W m^{-2} c/s^{-1} (Shain, 1956, Carr et al., 1958), and at a frequency of 22·2 Mc/s is comparable with the flux of the radio emission from the Crab Nebula (5×10^{-23} W m^{-2} c/s^{-1}) or is a few times greater (Burke and Franklin, 1955). Radio observations of Jupiter at a frequency of 26·6 Mc/s show that the intensity of the bursts may be higher than the level of the radio emission from the discrete source Cassiopeia-A (Kraus, 1956c).

† The radio emission of comets has been discussed from the theoretical standpoint by Poloskov (1953), Khanin and Yudin (1955) and Dobrovol'skii (1961).

§ 19] Sporadic Radio Emission of Jupiter

According to the data of Gardner and Shain (1958), Barrow and Carr (1958) Jupiter's radio emission flux in the 18–20 Mc/s range sometimes reaches values of about 10^{-19} W m^{-2} c/s^{-1}, which corresponds to a radio emission source power of 5×10^5 W c/s^{-1}. This maximum flux value is of the same order as for outbursts connected with chromospheric flares and is more than a hundred times higher than the level of emission from the "brightest" discrete sources at these frequencies. Therefore in the decametric band Jupiter is the next most intense source of radio emission after the active Sun.

According to Kraus (1956c and 1958b), Gallet and Bowles (1956), and Gallet (1961) the "life" of individual Jovian bursts is a few hundredths or thousandths of a second. As well as these short impulses longer surges with a duration of a few seconds are observed. It is shown by Carr *et al.* (1958) that the duration of the bursts of radio emission varies from a value less than 0·1 sec up to 1 sec; sometimes, however, even longer surges with a life of 10–15 sec are recorded. Bursts of this kind, following one after the other and superimposed on each other can keep Jupiter's radio emission level high for several minutes.

In a case when Jupiter's activity is not very great it can be clearly seen on a recording of the radio emission flux that the bursts sometimes appear in doublets and triplets with intervals between the individual bursts of between a quarter and one second. It is noted by Kraus (1956c, 1958b) that the second burst is weaker than the first and the third weaker than the second, the life of each successive burst being longer than the preceding one. This tendency was not found, however, by Carr *et al.* (1958) in high-speed recordings of Jupiter's radio emission: the relative intensity of the individual components of the doublets and triplets, generally speaking, is close to unity, a weak burst often preceding the appearance of a stronger burst. This is illustrated by the recordings of Jovian bursts given in Fig. 88.

Later observations of Jupiter for the most part do not confirm the existence of very short bursts (with a duration of a few milliseconds).† Gardner and Shain (1958) generally deny the presence in Jupiter's sporadic radio emission of noticeable bursts with a life of much less than a second, indicating that the characteristic intensity rise time in individual bursts does not drop below 0·2 sec. The shorter surges noted on the recordings were due to atmospheric interference of the nature of lightning, since their envelope (unlike the envelope of the lengthy bursts) displayed no alternating maxima and minima characterictic of interferometer reception of the radio emission of point sources. These data are confirmed by the ob-

† The work of a Florida group of astronomers (see Carr, 1959) is apparently an exception in this respect.

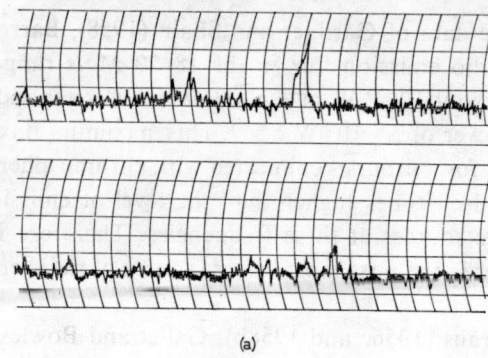

FIG. 89. Case of poor correlation between simultaneous recordings of the intensity of Jupiter's radio emission at a frequency of 18 Mc/s at two points separated by a distance of 7000 km (Smith et al., 1960)

servations of Smith, Lasker and Douglas (1960), according to which the duration of the bursts averages out at 0·7 sec; no short-lived bursts (of a few microseconds) were found in Jupiter's radio emission in this case either.

It follows from what has been said that at present the question of the duration of Jupiter's bursts of radio emission, which is of very great importance in the theory of their origin, has clearly been insufficiently studied. It will become clear from the contents of section 32 that millisecond bursts, whose existence has been called in doubt by many observers, are of particular importance in this respect.†

When solving the question of the life of Jovian bursts it is necessary to be sure to separate Jupiter's radio emission from interference of terrestrial origin and solar bursts; in addition our ionosphere has a considerable

† Smith, Carr, Mock, Six and Bollhagen (1963) have once again reported the presence of short bursts in the Jovian radio emission: according to their data there are three types of bursts with mean durations of 8 msec, 0·4 sec and 8 sec.

§ 19] Sporadic Radio Emission of Jupiter

effect on the characteristics of the decametric radio emission received on Earth.

A comparison of recordings of Jupiter's radio emission obtained simultaneously on receivers separated by several tens (Gardner and Shain, 1958) and several thousands (Carr *et al.*, 1960; Smith *et al.*, 1960) of kilometres points to a considerable difference in the variations in intensity at the reception points. Some groups of bursts obtained on one recording are absent at the other; time shifts can be observed between the times of the appearance of the corresponding bursts; the detailed structure of the groups of bursts is often very different (Fig. 89a). A detailed study of the simultaneous recordings showed that the strongest differences may be explained by fading or oscillations in intensity at the two stations occurring at different times. The characteristic period of these oscillations is of the order of 30–40 sec (Fig. 89b).

We pointed out in section 7 that fluctuations in the radio emission of discrete sources with a characteristic time of between 2 min and 8 sec or less are caused by the effect of the Earth's atmosphere, i.e. by scattering of the radio emission passing through on moving electron inhomogeneities. This effect (whose influence increases with the wavelength) can, in all probability, be used to explain the breakdown in correlation on the recordings of Jupiter's radio emission at points some distance apart.

What has been said obliges us to be cautious when drawing conclusions on the actual oscillations of Jupiter's radio emission from the results of terrestrial observations, although the periods of good correlation noted by Smith and Douglas (1959), Smith *et al.* (1960), provide a basis for assuming that the fine structure of the radio emission received with a characteristic time of the order of a second is to a considerable extent a feature of Jupiter's radio emission and does not entirely owe its origin to the Earth's ionosphere.† However, the second fluctuations of Jupiter's radio emission were later shown (Slee and Higgins, 1966; Douglas and Smith, 1967) to be due to the effect of scattering by the inhomogeneities of the interplanetary medium.

Frequency Spectrum

Simultaneous observations (Warwick, 1963a; Smith and Carr, 1959; Gallet and Bowles, 1956; Franklin and Burke, 1956 and 1958) of Jupiter at several frequencies 2–4 Mc/s apart showed that there is only a general

---

† We note that in this respect studies of Jupiter's decametric radio emission by artificial satellites and space vehicles may be of great help. The latter would make it possible to eliminate the distorting action of the ionosphere and attenuate interference of terrestrial origin. Rockets and satellites would also be useful in finding the sporadic radio emission of Venus and Saturn.

correspondence of the planet's activity at these frequencies, whilst there is generally no detailed correlation between the appearances of individual bursts. However, at the closer frequencies of 22·2 and 23 Mc/s about 80% of the bursts appear simultaneously (Smith and Douglas, 1959). It follows from this that the width of the frequency spectrum of an individual burst is between 1 and 2 Mc/s, the frequency corresponding to the spectral intensity maximum staying essentially the same during the life of the burst. In general these results agree with the observational results (Smith, Lasker and Douglas, 1960) from a multi-channel receiver (operating in the band around 22 Mc/s), from which it follows that the mean width of the frequency spectrum of an individual burst is about 2 Mc/s at the half-intensity level; according to other information, however, it does not exceed 0·5 Mc/s (Carr, 1959; Carr et al.,1960).

The spectral characteristics of the individual bursts have also been studied with a radio spectrograph whose panoramic scan rate was up to 60 times per second; a ciné camera recorded the frequency spectrum of the radio emission received from the oscillograph screen twice per second (Carr et al., 1961). Sequences of spectrograms for two bursts of radio emission obtained in this way are shown in Fig. 90. The width of the frequency spectrum of the bursts measured at half-intensity level is about 0·4 (Fig. 90a) and 0·2 Mc/s (Fig. 90b). These values are typical of the bursts observed in 1960, although many surges recorded in the following year had a broader frequency spectrum (of the order of 1 Mc/s). Despite the fact that the interval occupied by an individual burst shows no significant displacement along the frequency scale, changes are often observed in the detailed stucture of the spectrum during the burst's existence (see Fig. 90b). A kind of multiplet feature can even be seen amongst them.

Warwick's observations (Warwick, 1961, 1961–2 and 1963a) on a radio spectrograph for 15–34 Mc/s (later from 8 to 40 Mc/s) made it possible to find an interesting feature of Jupiter's sporadic radio emission. Jupiter's high-activity state, which sometimes lasts for an hour or more (noise storm), often appears over a broad range of frequencies (over 5 Mc/s). Often this interval occupied by narrow-band bursts moves slowly over the frequency scale at a rate of up to Mc/s per minute (generally about 0·5 Mc/s per minute). The drift is both towards the low and the high frequencies. It turned out that its direction is closely connected with the longitude of Jupiter's central meridian during the observations (see Fig. 92).

The presence of drift has also been noted in observations using multichannel equipment (Carr et al., 1961, 1960; Smith, Lasker and Douglas, 1960). According to Carr et al. (1961) the drift rate varies within broad

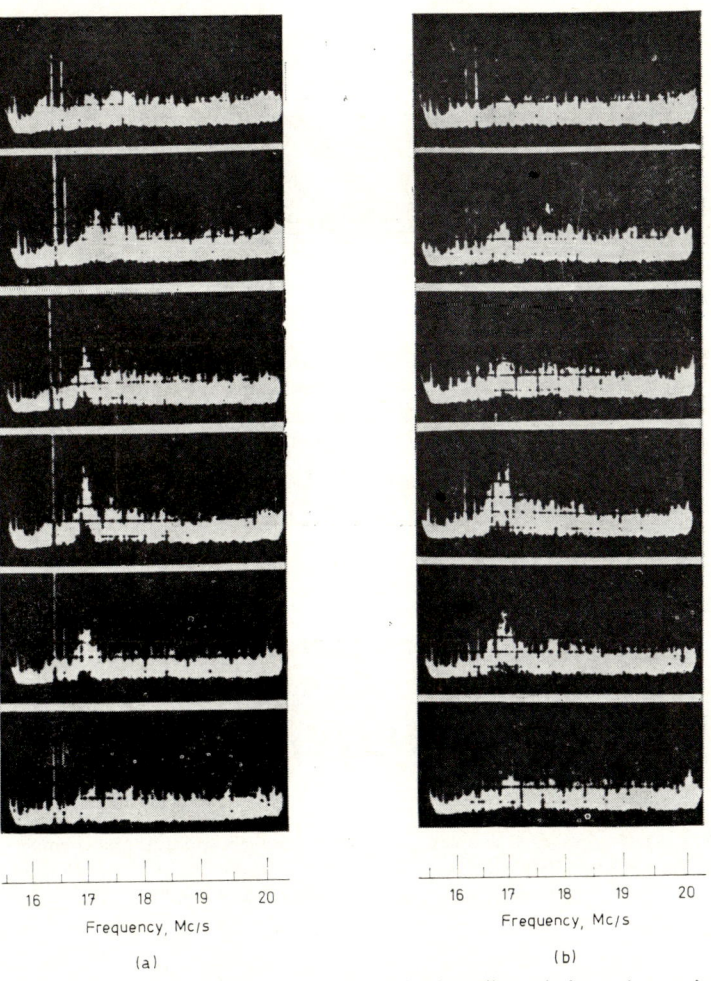

Fig. 90. Spectrograms of two bursts of Jupiter's radio emission taken at intervals of 0·5 sec on 29 March 1960. The narrow peaks on the left are interference from radio stations (Carr *et al.*, 1961)

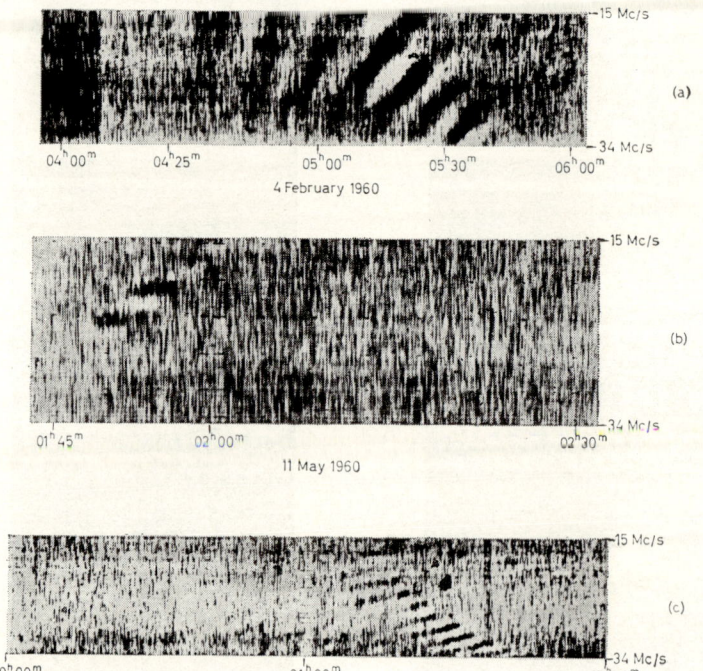

FIG. 91. Examples of dynamic spectra of Jupiter's sporadic radio emission. The alternating light and dark bands are due to the use of an interferometer as a receiving aerial (Warwick, 1963a)

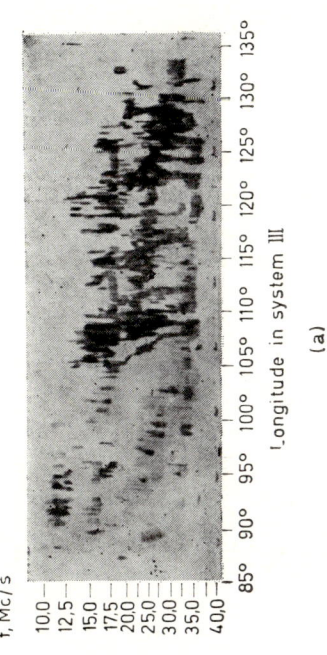
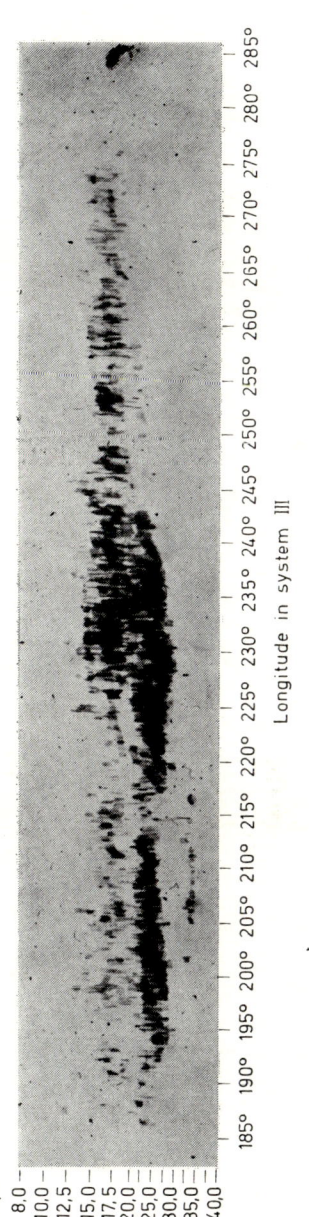

Fig. 92. Number of recordings of Jupiter's radio emission on a chart of the radio emission frequency against the longitude of the central meridian in system III (Warwick, 1961 and 1961–2)

§ 19] Sporadic Radio Emission of Jupiter

limits, generally being several Mc/s per minute. Drift towards the low frequencies occurs more often than drift in the opposite direction. In certain cases the direction of the drift is reversed during a single burst storm.† It is highly significant that on one occasion the drift towards the low frequencies (at about the same rate) was recorded simultaneously on a radio spectrograph (Warwick, 1961 and 1961–2) and a multi-channel receiver (Smith, Lasker and Douglas, 1960) at points several thousands of kilometres apart. The latter makes it certain that the observed effect was extraterrestrial in origin.

This feature of Jupiter's radio emission can be seen clearly in Fig. 91 which shows dynamic spectra obtained by Warwick. Here the burst storms occupy regions on the "frequency–time" diagrams in which we can see alternating light and dark bands of the interference type. Their appearance is because the aerial system of the radio spectrograph was an interferometer (see section 5); this made it possible to identify the radio emission being received with greater certainty. In Fig. 91a, c the frequency spectrum of the burst storm shifts in the course of time towards the high frequencies and in Fig. 91b towards the low ones.

The dependence of the direction of the frequency drift on the longitude of Jupiter's central meridian is clearly seen in Fig. 92 which shows a chart of radio emission frequency plotted against longitude of the central meridian in system III during the observations,‡ plotted by Warwick on the basis of dynamic spectra. In the figure the degree of darkening characterizes the number of recordings of radio emission at the given frequency and longitude. Drift towards the high frequencies occurs in the range of longitudes between 75° and 140°; drift towards the low frequencies occurs in the 185–280° region; there is no significant drift in the 140–185° transition region. It is significant that the negative drift (towards the low frequencies) is connected with a region where the main local source of the bursts is located (with its centre at a longitude of about 178°; see below), or follows after it, whilst positive drift is characteristic of a weaker source (longitude 76°). Warwick (1963b) shows that a dynamic spectrum of the type shown in Fig. 92 is a stable characteristic of Jupiter's decametric radio emission which lasts for a long time. Despite the fact that the dynamic spectra of Jupiter's individual periods of activity may differ strongly from that shown in Fig. 92, a well-developed burst storm usually has a spectrum of just this type.

A careful study of the dynamic spectrum in Fig. 91c reveals yet another

---

† It must be stressed that the individual bursts making up the noise storm show none of the frequency drift characteristic of the phenomenon as a whole.

‡ Here the period of rotation of system III is taken as $9^h\ 55^m\ 29 \cdot 5^s$.

feature of the bursts of radio emission. To the left of the broad, easily distinguished "corrugated" band characterizing the dynamic spectrum of the burst storm that started at low frequencies at about 1^h10^m and finished at high frequencies at 1^h40^m we can see a further band at frequencies of about 30 Mc/s in the period near 1^h10^m. Moreover, in the same frequency interval at about 0^h50^m in Fig. 91c the presence of a third band can be guessed at. If the latter two bands are extended parallel to the first it turns out that their frequencies for the same point in time are approximately in a harmonic ratio to the frequency of the first band of 3 : 2 : 1. We should note that the second band is not an exceptional event at the time of burst storms: out of the ten phenomena analysed by Warwick (1961 and 1961–2) this effect was found in three cases and possibly in a further case.

On the other hand, simultaneous observations (Carr *et al.*, 1960) at frequencies of 10 and 20 Mc/s point to the absence of any correlation between bursts at frequencies in the ratio of 1 : 2. This can be made to fit in with what has been said above if we assume that the "harmonics"† of Jupiter's bursts of radio emission are not in such a simple ratio to the "fundamental harmonic" or that bursts at frequencies of ω and 2ω do not appear simultaneously (i.e. the harmonic bands appear only for the burst storm as a whole). It is also possible that the bands in the dynamic spectra are not genetically connected and are, for example, radio emission generated in different layers of Jupiter's ionosphere.

The following should be noted in regard to Jupiter's degree of activity at different frequencies. Simultaneous observations at a number of fixed frequencies in the 10–27 Mc/s range indicate that the intensity, duration and "frequency of appearance" of the sporadic radio emission are maximum at frequencies of around 18 Mc/s and drop by several factors as we move towards the ends of this range (Smith, 1963; Smith and Douglas, 1959; Gardner and Shain, 1958; Carr, 1959; Franklin and Burke, 1958; Gardner, 1957). The extent to which the activity decreases as one moves away from the optimum frequency does not apparently remain constant, varying from year to year.

Investigations into Jupiter's radio emission at frequencies below 10 Mc/s are considerably prevented by the rise in the level of interference and the screening effect of the ionosphere. Recently, however, some data have been obtained on Jovian bursts at $f = 4.8$ Mc/s (see Ellis, 1962; Ellis and McCulloch, 1963); the maximum intensity here reached 2×10^{-20}

---

† By the use of inverted commas we wish to stress the conventional nature of this term as applied to Jupiter's radio emission.

§ 19] Sporadic Radio Emission of Jupiter

W m^{-2} c/s^{-1}, although the mean level of the radio emission in the periods of the planet's activity was two orders lower.

Bursts are found very rarely at frequencies over 27 Mc/s, although there are data (Warwick, 1963a; 1961 and 1961–2) on observations of Jupiter's activity right up to 37 Mc/s (see Figs. 91, 92). Some results from recording Jupiter's sporadic radio emission in the 30–40 Mc/s range are given in (Kraus, 1958b). The bursts at these frequencies, however, are weak and small in number; they have not been found at all at higher frequencies (in this connection see Burke and Franklin, 1955; F. G. Smith, 1955).

POLARIZATION

In 1956 Franklin and Burke (Franklin and Burke, 1956 and 1958) established that Jupiter's radio emission is polarized. In certain cases the degree of polarization is close to 100%; the sense of rotation is usually right-handed. Only right-hand polarization has been found in the radio emission from the main local source; mixed polarization, and likewise left-hand polarization, have been recorded in two cases when receiving radio emission from one of the weak sources.

Further observations have confirmed that the radio emission is circularly or elliptically polarized, this polarization being always or almost always right-handed (in 94% or more of the cases; see Carr *et al.*, 1961). According to Carr *et al.*, (1961) the ratio of the axes of the ellipse for sharply defined bursts varies noticeably within the course of a few hours, displays very considerable scatter and lies between 1 (circular polarization) and a value close to 0 (almost linear polarization). The mean ratio of the axes (judging from several hundred analysed bursts observed in 1960) is 0·34. An attempt in these experiments to establish a dependence between the ratio of the axes of the polarization ellipse and the longitude of the central meridian (in particular the difference between the values of this ratio for three local sources of Jupiter's radio emission) led to no definite results.

The polarization observations discussed above were made only at frequencies close to 20 Mc/s. Sherrill and Castles (1963) have reported on the results of studying the polarization of Jupiter's sporadic radio emission in the 15–24 Mc/s range, making it possible to establish the frequency dependence of the polarization and its features for different local sources. It turned out firstly that at frequencies below 20 Mc/s the polarization characteristics of the bursts become more complex: there is a noticeable rise in the probability of appearance of left-handed polarization and also of chaotic or mixed polarization in all three local sources. The latter, however, differ in the ellipticity of their emission: the mean ratio of the axes of the ellipse for the main local source is 0·83; this ratio is less (about 0·5) for the two weaker sources.

We notice that the polarization investigations (Smith and Carr, 1959; Franklin and Burke, 1956 and 1958) were carried out at stations located in the northern hemisphere; one observation (Gardner and Shain, 1958) (at a different time) has been made in the southern hemisphere. Lastly, a series of simultaneous polarization measurements has been made at points with geomagnetic latitudes of +41° and −22° (located on opposite sides of the geomagnetic equator) (Carr et al., 1960, 1961; Smith et al., 1960; Douglas, 1960). The fact that the sign of the rotation in these experiments was the same, despite the opposite direction of the Earth's magnetic field component along the line of sight, allows us to say that the polarization observed does not owe its origin to the Earth's ionosphere and is a characteristic feature of Jupiter's sporadic radio emission.† The presence of such a feature is decisive proof of the existence of a Jovian magnetic field, whose discovery can rightly be looked upon as an outstanding success of planetary radio astronomy.

LOCAL SOURCES OF SPORADIC RADIO EMISSION, THEIR PERIOD OF ROTATION AND POSITION ON JUPITER'S DISK

A study by Shain (1955, 1956) of old recordings of cosmic radio emission of 1950–1 ($f = 18\cdot3$ Mc/s), in which bursts of radio emission from Jupiter were recorded, made it possible to find a close connection between the appearance of activity in the radio band and the planet's period of rotation, indicating in particular the localization of the sources of the bursts on Jupiter's disk.

Jupiter's radio emission in the form of groups of bursts was recorded almost every day for 1 or 2 hours, whilst the aerial array of the radio telescope made it possible to receive this planet's emission for almost the whole of Jupiter's period of rotation.‡ It follows from this that the angle Θ by which Jupiter turns in the period of intense emission is 35–70°.

A comparison of the time of appearance of the sporadic radio emission with the longitude of the central meridian in system I showed no definite periodicity; this periodicity, however, became clearly noticeable when we changed to system II (Fig. 93a). The figure shows that the position of the source of bursts on the planet in this system does not remain fixed, i.e.

---

† For the extent of the effect of the Earth's atmosphere on the nature of the polarization of radio emission received see section 7.

‡ We recall that Jupiter's equatorial period of rotation (system I) is $9^{h}50^{m}30\cdot003^{s}$, and the period of rotation of the moderate belts and the polar regions (system II) is $9^{h}55^{m}40^{s}$ (Akabane, 1958b).

§ 19] Sporadic Radio Emission of Jupiter

its period of rotation differs slightly from the period of system II and is obviously less than the latter. For the period in the new system III established on the basis of radio observations Shain gave a value of $9^h55^m13^s \pm 5$ sec. The values of the central meridian longitude in system III are shown in Fig. 93b. Later the period of rotation in system III was determined more precisely by a number of other observers (Smith and Carr, 1959;

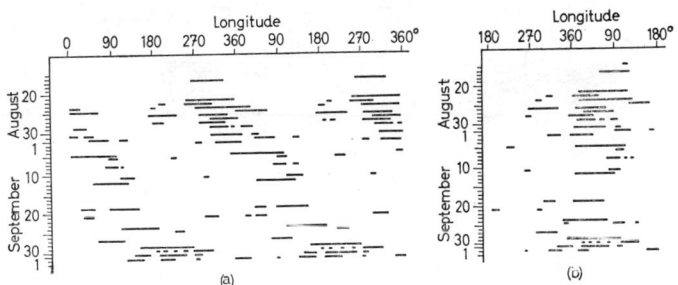

FIG. 93. Value of the longitude of Jupiter's central meridian during a period of the planet's activity at a frequency of 18·3 Mc/s (1951): (a) in system II; (b) in system III. The short horizontal lines are the intervals during which bursts were observed; the position of these lines indicates the longitude of Jupiter's central meridian during the observations (Shain, 1956)

Carr et al., 1958, 1961; Gardner and Shain, 1958; Franklin and Burke, 1958; Gallet, 1958) who had more extensive observational data available. When processing the results of radio observations at present use is often made of the period of rotation of the local sources determined by Smith and Carr (1959)

$$T^{III}_{2\!\!\downarrow} = 9^h55^m28\cdot8^s. \tag{19.1}$$

Here the new longitude system III, by definition, coincides with system II at 0^h universal time on 1 January 1957 and rotates with a period $T^{III}_{2\!\!\downarrow}$. The longitude of Jupiter's central meridian Θ^{III} in system III at 0^h universal time for the Jovian data j can be calculated from the published data for system II by the formula

$$\Theta^{III} = \Theta^{II} + 0\cdot28845(j - 2{,}435{,}839\cdot5), \tag{19.2}$$

However, the coordinates of the local sources, strictly speaking, do not remain constant in a system with the period (19.1). According to Douglas's data (Douglas, 1960) obtained by homogeneous statistical analysis of the results of all known observations of Jupiter's decametric radio emission for 10 years (1950–60) the period of system III is†

$$T^{III}_{2\!\!\downarrow} = 9^h55^m29\cdot37^s \pm 0\cdot16^s. \tag{19.3}$$

† This very high accuracy for the quantity $T^{III}_{2\!\!\downarrow}$ is not surprising: it is the result of the fact that the observations cover thousands of periods of Jupiter's rotation over many years.

In that case:

$$\Theta^{III} = \Theta^{II} + 0.2742 \times (j - 2{,}433{,}339 \cdot 5). \tag{19.4}$$

A value $T_{2\!\!\!\perp}^{III} = 9^h55^m29 \cdot 35^s$, which is very close to (19.3), is given by Carr *et al.* (1961).

Figure 94 shows histograms covering a 7-year period for the number of observations of Jupiter's activity in the decametric band at a frequency of 18–20 Mc/s as a function of the longitude of the central meridian in

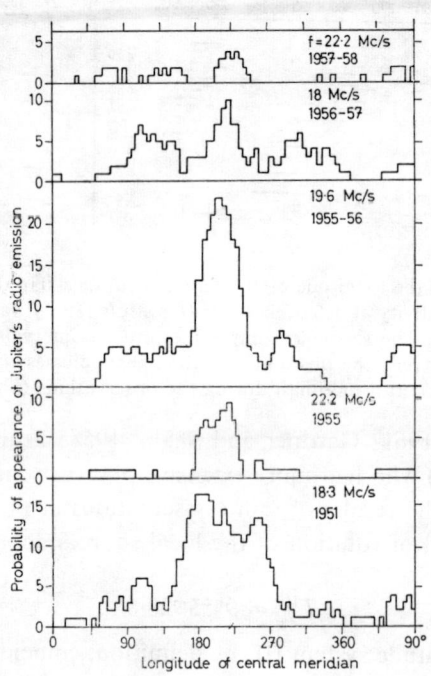

FIG. 94. Histograms of a number of observations of Jupiter's sporadic radio emission as a function of the longitude of the central meridian in system III (period $9^h55^m28 \cdot 8^s$) at frequencies of 18–22 Mc/s in 1951–8 (Smith and Carr, 1959)

system III. It is clear from these that there are three local sources of radio emission on Jupiter (Carr *et al.*, 1958; Gardner and Shain, 1958; Franklin and Burke, 1958; Barrow, Carr and Smith, 1957). The basic source (with a longitude around 200°) stands out in all the histograms. Two weaker sources (with a longitude of about 110 and 300°)† were not found in Burke and Franklin's observations (1955); during Shain's observations

† Here the longitude is given in the system with $T_{2\!\!\!\perp}^{III} = 9^h55^m28 \cdot 8^s$. In the more exact system III with $T_{2\!\!\!\perp}^{III} = 9^h55^m23 \cdot 37^s$ the values of Θ^{III} will be slightly different: $178° \pm 7°$ for the basic source and $76° \pm 13°$ and $254° \pm 15°$ for the weaker ones. Therefore the difference in the longitudes is close to 180° for the latter two sources.

§ 19] Sporadic Radio Emission of Jupiter

(1951) the latter sources were in fact merged with the basic one. It should also be noted that the three sources of bursts are observed only in the 16–22 Mc/s range. At higher frequencies only one (or more rarely two) centre of activity is generally to be found. At lower frequencies (5–10 Mc/s) the connection between the degree of activity and the longitude of the central meridian of the planet becomes very vague (Fig. 95) (Carr *et al.*,

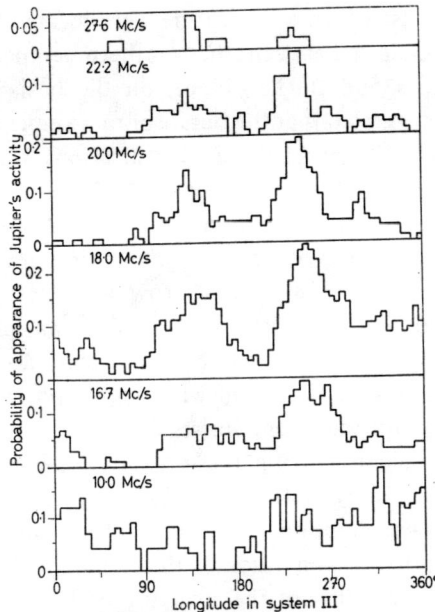

FIG. 95. Histograms of the probability of appearance of Jupiter's activity as a function of the central meridian's longitude in 1960 at frequencies of 10–27·6 Mc/s (Carr *et al.*, 1961)

1961; Gardner and Shain, 1958). This connection, however, becomes quite clear if the degree of activity is characterized not by the number of bursts recorded but by their mean intensity. Then, according to Ellis (1962), Ellis and McCulloch (1963), at $f = 4·8$ Mc/s there are two sources—at longitudes of about 20 and 200° (in the system with $T_{2\!\!\!\;4}^{III}$ (19.1)). The latter obviously coincides with the main centre of radio emission in the 20 Mc/s band.

To judge from the data of 10 years of observations, the radio-emitting centres on Jupiter (and above all the main local source) are very stable in nature: both the period and the relative longitude position of the local sources remain fixed within the limits of the observational accuracy (Douglas, 1960). A comparison of the coordinates of the main source at different frequencies shows that the longitude corresponding to the radio emission maximum appears to be slightly less at high frequencies, the difference being about 15° in the interval between 17 and 22 Mc/s (see Fig. 95) (Carr *et al.*, 1961; Gardner and Shain, 1958).

No definite connection has been found between the emitting centres and the structural features of Jupiter's visible surface. Shain (1955, 1956) from observations in 1951–2 identified the sources of radio emission with particular white spots, which at that time were located in a zone of high activity on the boundary between the southern temperate belt and the polar zone. Franklin and Burke (1956), on the basis of observational results for 1955–6, suggested that the bursts of radio emission are connected with the Red Spot. However, these identifications were recognized as unfounded (Carr *et al.*, 1958; Gardner and Shain, 1958; Franklin and Burke, 1956) because of the clear difference between the periods of rotation of the white spots and the Red Spot and the period of system III. At the same time (Carr *et al.* (1958) and Barrow and Carr (1958) suggested that in 1957 the second most active source of bursts was possibly connected with a southern tropical disturbance. No confirmation of this connection was obtained. Scanning of Jupiter in white light and in the H_α line with photoelectric devices during strong bursts of radio emission did not reveal any optical effects accompanying these phenomena either (Morris and Berge, 1962; see also Jelley and Petford, 1961).

If we take into consideration the remarkable constancy of the period of rotation and the Jovian longitude of the radio-emitting centres on the one hand, and the variability of the structure of Jupiter's visible surface, the difference in the period of rotation of the local centres from the periods of systems I and II and the lack of success in attempts to connect the local sources with features of the cloud layer on this planet on the other, it becomes clear that the local sources of sporadic radio emission (their period of rotation and position on the disk) are connected with deep-lying layers of the planet rotating as a solid body. This cannot, of course, be understood in the sense that the regions where the sporadic radio emission is generated are located in Jupiter's sub-cloud layers; they may even be located in the upper layers of the planetary atmosphere. In the latter case the connecting link between the deep-lying layers and the radio-emission sources determining the period of rotation of the latter (and possibly even their localization on the planetary disk (Warwick, 1963a; Carr, Smith and Bollhagen, 1960)) may be Jupiter's magnetic field.

As well as the 10-hour cycle determined by the length of the Jovian days, in 1957 a well-defined cycle of the planet's activity was found with a period of about 8 days (Carr *et al.*, 1958). This can be checked in Fig. 96,

§ 19] Sporadic Radio Emission of Jupiter

which shows the index of Jupiter's daily activity at a frequency of 18 Mc/s.†
(In this figure we can also see a 32-day cycle with maxima near 16 January 1957 and 17 February 1957.) The 8-day cycle, however, was not noticeable in the observations at the end of 1957 and beginning of 1958 at frequencies of 18 and 22·2 Mc/s (Smith and Carr, 1959). This may be connected with the very considerable weakening of Jupiter's activity at decametric wavelengths (rarer recordings of radio emission when compared with the preceding observations).

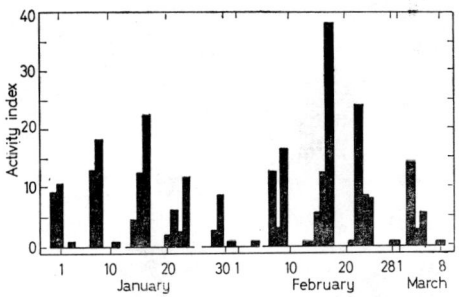

FIG. 96. Index of Jupiter's activity at a frequency of 18 Mc/s at the beginning of 1957 (Carr et al., 1958)

A study of the correlation of the 8-day cycle with the number of sunspots, the index of geomagnetic activity and with the level of cosmic radio emission (which characterizes the absorption in the Earth's atmosphere, this depending in its turn on the quantity of solar ionizing radiation) did not lead to any positive results (Carr et al., 1958). The suggestion that this cycle is connected with tidal phenomena on Jupiter that alter the transparency of the Jovian ionosphere and caused by the most massive of this planet's satellites, Ganymede (period of rotation about 7 days), is most improbable: the tidal forces from Ganymede are far less than from the nearer satellites. It is more probable that this cycle is a consequence of the stroboscopic effect occurring during the rotation of the Earth and Jupiter: every 5 and 7 days the emitting region on Jupiter passes through the planet's central meridian at the same terrestrial time (with an accuracy of up to an hour). Since the 3-hour sessions of night observations were also made at a constant time it may be assumed that this circumstance also caused the short-period variations in the activity index (5 and 7 days), whilst the throbbing between these periods led to long-period variations with a characteristic time of 35 days (Carr et al., 1958; Carr, 1959).‡

---

† The activity index is characterized here in a certain arbitrary scale based on an estimate of the intensity and duration of the sporadic radio emission.
‡ 32 days, to judge from preliminary observations (see Fig. 96).

DIRECTIONAL FEATURES OF RADIO EMISSION AND SIZE OF LOCAL SOURCES

When processing the observations of 1950–1 Shain noticed the short duration of Jupiter's radio-emitting phase when compared with the planet's period of rotation: the angle $\Delta\Theta$ which Jupiter turned during the recording of intense emission was only 35–70° (see Fig. 93). These data are confirmed by later investigations (Carr *et al.*, 1958; Franklin and Burke, 1956), according to which $\Delta\Theta \approx 40$–50° at frequencies of 18–22 Mc/s.

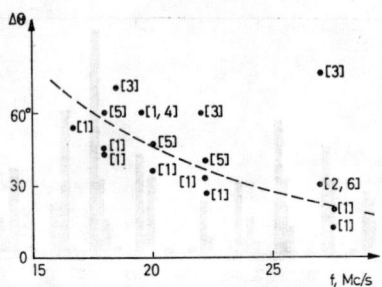

FIG. 97. Dependence of width of basic source of radio emission $\Delta\Theta$ on the frequency f. The value of $\Delta\Theta$ is determined at half level from the histograms characterizing the probability of appearance of Jupiter's activity as a function of central meridian longitude. Sources: 1—Carr *et al.*, 1961; 2—Gardner and Shain, 1958; 3—Franklin and Burke, 1958; 4—Gardner, 1957; 5—Douglas, 1960; 6—Morris and Berge, 1962

The value of $\Delta\Theta$ can also be judged from the histograms in Figs. 94 and 95. Here, however, the width of the maxima tends to provide values of $\Delta\Theta$ that are too high because of averaging of the times of appearance of activity that show noticeable dispersion. This is easily checked by an example from Figs. 93 and 94: according to the first of them $\Delta\Theta \sim 35$–70° and according to the second $\Delta\Theta \approx 135°$ (at 18·3 Mc/s).

In simultaneous observations at two frequencies Gardner and Shain (Gardner and Shain, 1958; Gardner, 1957) found that $\Delta\Theta$ decreases as the frequency rises. Figure 97 shows the results of these observations together with data (Carr *et al.*, 1961; Douglas, 1960) that confirm this dependence. The exceptions are the values of $\Delta\Theta$ found by Franklin and Burke (1958) and also shown in the figure: in general they do not vary with the frequency.

The last results are the least convincing since they are based on far less statistical material. Barrow (1960), on the other hand, has cast doubt on the reliability of the frequency dependence of the angle $\Delta\Theta$ on the basis that the small range of angles $\Delta\Theta$ occupied by the local source on the high-frequency histograms (27 Mc/s) may be caused by the rarer recording of Jupiter's radio emission because of its lower intensity at these frequencies

§ 19] Sporadic Radio Emission of Jupiter

when compared with the 18–22 Mc/s range. This objection is obviously valid provided that in the observations at high frequencies it was mostly those cases of Jupiter's activity that were recorded when the level of the radio emission was close to the minimum detection threshold even when the source intersected the central meridian. Since Jupiter's activity index, as has already been pointed out above, drops sharply as we move from 18 Mc/s to the higher frequencies this argument is not without foundation. It should, however, be noted that the conditions are more favourable for observation at high frequencies because of the reduction in the noise level, which also determines the detection threshold for the Jovian radio emission. At the same time the dependence of $\Delta\Theta$ on the frequency f is also observed in the 16–22 Mc/s range where the activity of Jupiter is greatest. Here Barrow's arguments lose their force.

It is clear from what has been said that the nature of the dependence of $\Delta\Theta$ on f needs further analysis, although the combined available data quite definitely indicates a rise in the interval $\Delta\Theta$ as the frequency drops.

The fact that the radio emission ceases or is strongly attenuated when the local sources are a comparatively small distance away from the planet's central meridian is due to the combined action of two causes: the small extent of the local sources in longitude $\Delta\Theta_s$ and the directional nature of their radio emission characterized by the beam width $\Delta\Theta_{\text{beam}}$.[†] It is easy to see that the range of longitudes $\Delta\Theta$ in which the activity of Jupiter can be observed is connected with $\Delta\Theta_s$ and $\Delta\Theta_{\text{beam}}$ by the relation

$$(\Delta\Theta)^2 \approx (\Delta\Theta_s)^2 + (\Delta\Theta_{\text{beam}})^2, \tag{19.5}$$

from which it follows that the quantity $\Delta\Theta$ plays the part of the upper limit for the intervals $\Delta\Theta_s$ and $\Delta\Theta_{\text{beam}}$.

The maximum size of the regions in which the Jovian bursts are generated has also recently been determined from interferometer observations (Slee and Higgins, 1963) ($f = 19\cdot7$ Mc/s). Despite the considerable separation of the interferometer aerials (32·3 km, which is equivalent to 1940 λ), the effect of ionospheric fluctuations did not prevent the success of the experiment. It was found that the diameter of the source of a burst is less than one-third of the planet's diameter, all the bursts during a noise storm coming from one region with a size of less than half the planet's diameter. These results obviously do not contradict the data given above on the upper limit of $\Delta\Theta_s$.

[†] There is less probability of being able to explain the smallness of the angle $\Delta\Theta$ not by the directivity of the emission but by the sharp dependence of the latter on the degree of illumination of the local source by the Sun's rays (the visible positions of the Earth and the Sun in the sky are very close: the angular distance between them is not more than ten degrees or so) (Shain, 1956).

CONNECTION WITH SOLAR ACTIVITY

The presence of a close connection of the various geophysical phenomena with solar activity allows us to assume that there is a similar connection for the other planets of the solar system as well, and that it manifests itself in the level of Jupiter's sporadic radio emission. Indeed, Warwick (1960) obtained a number of data indicating a correlation between Jupiter's decametric radio emission and solar radio emission in the same waveband (see also Kraus, 1958b).

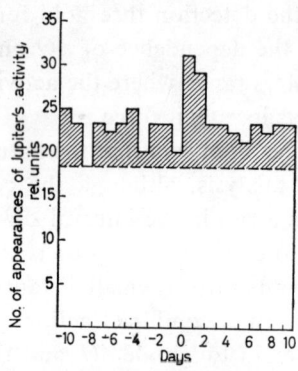

FIG. 98. Number of cases of recording of Jupiter's radio emission on days of an increase in the continuous solar radio emission, and also on the preceding ("minus" sign) and succeeding ("plus" sign) days. The dotted line shows the mean number of appearances of Jupiter's radio emission in a period of 5 months (Warwick, 1960)

A comparison of recordings of Jovian and solar radio emission made on radio spectrographs operating in the 15–34 Mc/s range showed a general agreement between the dates of the appearance of Jovian activity and the level of the Sun's radio emission. In particular the considerable rise in the level of the solar radio emission on 28 March to 4 April 1960 and 21 April to 14 May 1960 was accompanied by several outstanding events in Jupiter's radio emission. The connection between the activities of Jupiter and the Sun can be seen more definitely from the diagram in Fig. 98, which was plotted by the method of superimposing epochs from the results of observations during the spring of 1960. The diagram shows the number of cases of the appearance of Jovian radio emission on days when a rise was observed in the level of the solar radio emission, and also on the preceding and succeeding days. It is clear from the figure that the Jovian's activity rises noticeably 1 or 2 days after intense solar radio emission has been recorded.† In addition, the index of Jupiter's activity in periods of strong

† According to different data (Douglas, 1960) this lag is slightly greater (3–4 days).

§ 19] Sporadic Radio Emission of Jupiter

solar activity (to which the data given in Fig. 98 relate) is systematically higher than the mean index of activity in the spring of 1960. Less definite results were obtained by Warwick (1960) in a comparison of the data of Burke and Franklin's observations in 1955 with the number of sunspots.

The lag noted corresponds to motion of an agent from the Sun towards Jupiter at a velocity of about $0 \cdot 1c$ (c is the velocity of light in a vacuum). This agent may be energetic protons ejected from the region of chromospheric flares on the Sun which cause the polar blackouts on Earth (see section 17).

In this connection it would be of interest to make a direct comparison of Jupiter's activity index with the absorption of radio waves in the Earth's polar regions.

Since corpuscular streams of solar origin have a significant effect on Jupiter's activity, during opposition we may expect the appearance of a correlation between the sporadic radio emission of this planet and disturbances of the Earth's magnetic field by the action of corpuscular streams moving at velocities of the order of 10^3 km/sec. Certain indications of the existence of a close connection have been obtained by Carr, Smith and Bollhagen (1960) (see also Carr et al., 1961) from radio observations at 10 and 18 Mc/s. It turned out that there is correspondence between events on Earth and on Jupiter if we allow for the shift in time (about 8 days) required for the stream of particles to move from the Earth to Jupiter. The magnitude of this shift was determined from the velocity of the streams, which in its turn was found from the lag of the geomagnetic disturbances behind the chromospheric flares.

We spoke above about the positive correlation of Jupiter's sporadic radio emission with solar activity manifested in the increase in Jupiter's activity index at decametric wavelengths following sporadic phenomena on the Sun. It is quite possible, however, that there is also a negative correlation between Jupiter's radio emission and solar activity covering the 11-year cycle of solar activity (Carr, 1959; Warwick, 1960). In actual fact, Jupiter's radio emission was observed almost every day and was very intense in 1951 (Shain, 1956) a few years after the maximum of the solar cycle in 1947. In 1955–6 bursts were recorded in the best case once every 3 days for half an hour in 5 hours of daily observations (Gardner and Shain, 1958). In other words, at this period Jupiter was active for about 3% of the observational time. There was a particularly sharp decrease in the planet's activity in 1957–8, i.e. in the years of the next solar activity maximum (Smith and Carr, 1959). And only 2 years later, at a period of considerable weakening in solar activity, Jupiter's radio emission showed a tendency towards a restoration of the previous high level (Warwick, 1960).

The variation in the extent of Jupiter's activity in the radio-frequency band near a sunspot activity maximum can be clearly seen in Fig. 99, which shows the mean values of the number of sunspots (Wolf number) and the probabilities of the appearance of Jupiter's activity in 1957–60 (Carr et al., 1961). (The latter quantity was determined as the ratio of the time during which Jupiter's radio emission was recorded to the whole time of the observations.)

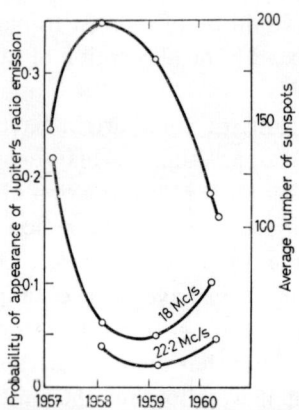

FIG. 99. Comparison of mean probability of appearance of Jupiter's activity in the radio-frequency band with the relative number of sunspots in 1957–60 (Carr et al., 1961).

The small amount of data given on the connection of phenomena on the Sun and on the Earth with Jupiter's sporadic radio emission is by way of being a preliminary. The negative correlation has been followed for only one solar cycle; data on the positive correlation are based on insufficiently extensive statistical material. More definite information on the nature of the connection of Jupiter's sporadic radio emission with solar activity can be obtained only as the result of further lengthy observations.

In conclusion we would remark that the characteristics of Jupiter's sporadic radio emission show a very close analogy with type I solar bursts (see section 12). This similarity is manifested in features of the dynamic spectra both of individual bursts (including the life, the relative width $\Delta f/f$ of the frequency spectrum, the absence of any significant drift) and of burst storms as a whole (in the limited width of the spectrum and the slow frequency drift). The nature of the polarization and the directivities of the two phenomena also have much in common. This analogy is apparently not a random one: it is probably connected with the common features of the radio emission mechanisms and with a certain similarity in the conditions of the generation of this radio emission on the Sun and on Jupiter.†

20. Continuous Radio Emission of the Planets

We shall call the radio emission of the planets in the millimetric, centimetric and decimetric bands continuous since (despite certain variations connected in particular with the motion of the planets) it can always be observed and does not have a discontinuous nature like the radio emission of Jupiter that appears sporadically (from time to time) at decametric wavelengths.

Radio Emission of Saturn

The original observations of the planet with an 8·5 m paraboloid and a wide-band radiometer operating in the $\lambda = 4$–4.5 cm band gave a value for the spectral radio emission flux of about 4×10^{-26} W m^{-2} c/s^{-1} (Drake and Ewen, 1958). The later, more accurate measurements of Cook et al. (1960) with better equipment (an aerial 26 m in diameter and a maser) showed that this value was far too high. In actual fact at $\lambda = 3.45$ cm the effective temperature of the planet related to its optical disk (without allowing for Saturn's rings) is $T_{\text{eff}\hbar} = 106 \pm 21°$K, which corresponds to a spectral radio emission flux of $S_f \approx 0.7 \times 10^{-26}$ W m^{-2} c/s^{-1} (for one polarization).‡

The value obtained for $T_{\text{eff}\hbar}$ agrees with the measurements of Saturn's temperature in infrared rays (120°K). It can be expected, therefore, that all the observed radio emission comes from regions of the atmosphere adjacent to the planetary cloud layer. Saturn's brightness temperature at $\lambda \sim 10$ cm is noticeably higher than at $\lambda \sim 3$ cm and equals 180–200°K (Davis, Beard and Cooper, 1964; Rose, Bologna and Sloanaker, 1963a and 1963b; Drake, 1962b).

The data available on the degree of polarization of the radio emission are contradictory. Rose, Bologna and Sloanaker (1963a and 1963b) have reported a high degree of polarization at $\lambda = 9.4$ cm ($\varrho_l > 20\% \pm 8\%$); after subtracting the non-polarized thermal component with $T_{\text{eff}\hbar} = 106°$K (as at $\lambda = 3.45$ cm) the degree of polarization ϱ_l becomes not less than

† For the features of Jupiter's radio emission see also articles by Roberts (1963, 1964) and by Douglas and Smith (1963). Data obtained later on the radio emission of Jupiter including the I_0 relation with the decametric burst level and the nature of the dynamic spectrum are reported in a review by Warwick (1967).

‡ It is easy to change from the value of S_f to the quantity $T_{\text{eff}\hbar}$ by a formula of the (4.15) type, remembering that the radius of Saturn is $R_{\hbar} \approx 6 \times 10^9$ cm and assuming that the distance from the Earth to Saturn in the observation period was about 9·5 a.u. $\approx 1.5 \times 10^{14}$ cm.

$51\pm22\%$. Davis, Beard and Cooper (1964), however, did not find such polarization at $\lambda = 11\cdot3$ cm: during their observations ϱ_l did not exceed 6%.

RADIO EMISSION OF JUPITER

All the values known at present for Jupiter's effective temperature $T_{\text{eff}2\!\!\!\!\!\text{\tiny{2}}}$ obtained as the result of measurements[†] in the centimetric and decimetric wavebands are shown in the graph in Fig. 100a. This also shows the upper limits for the value of $T_{\text{eff}2\!\!\!\!\!\text{\tiny{2}}}$ at $\lambda = 8\cdot6$ mm (McClain, 1960), $\lambda = 73\cdot5$ cm (Long and Elsmore, 1960) and $\lambda = 3\cdot5$ m (Mills et al., 1958).

The values of $T_{\text{eff}2\!\!\!\!\!\text{\tiny{2}}}$, as a rule, are obtained as the result of averaging tens and hundreds of recordings made as the planet passed through the beams of the receiving aerials. It must be pointed out that this operation was used not only in the cases when the level of the signal being received was less than the fluctuation threshold of the sensitivity and the averaging became completely necessary to pick out the signal on the background of intense equipment noise: this method was used to raise the measurement accuracy even when the useful signal could be clearly seen even on a single recording.

The values of Jupiter's spectral radio emission flux S_f given in Fig. 100b are calculated from the $T_{\text{eff}2\!\!\!\!\!\text{\tiny{2}}}$ in Fig. 100a by a formula of the (4.15) type, in which the distance from the Earth to Jupiter was taken as $5\cdot2$ a.u. for all the observations.

The first series of observations of Jupiter's continuous radio emission at $\lambda = 3\cdot15$ cm with an ordinary radiometer (Mayer, McCullough and Sloanaker, 1958) (see also Mayer, 1959) led first to a value for $T_{\text{eff}2\!\!\!\!\!\text{\tiny{2}}}$ of $140\pm38°$K, then $145\pm26°$K. These values agree within the limits of the measurement accuracy achieved with the temperature of the upper layer of clouds on the planet (130°K).

The effective temperature in the 3 centimetre band was determined more exactly later as the result of observations (Alsop et al., 1959a and 1959b) using masers which increased the receiver sensitivity by an order of magnitude. It turned out that at $\lambda \sim 3$ cm $T_{\text{eff}2\!\!\!\!\!\text{\tiny{2}}}$ is slightly higher than Jupiter's temperature in infrared rays and rises slightly with wavelength: $T_{\text{eff}2\!\!\!\!\!\text{\tiny{2}}} = 171\pm20°$K at $\lambda = 3\cdot03$ cm and $T_{\text{eff}2\!\!\!\!\!\text{\tiny{2}}} = 189\pm20°$K at $\lambda = 3\cdot36$ cm.[‡]

[†] Alsop et al. (1959a and 1959b); Mayer, McCullough and Sloanaker (1958); Drake and Ewen (1958); Rose, Bologna and Sloanaker (1963a and 1963b); Gibson and Corbett (1963); Sloanaker and Boland (1961); McClain and Sloanaker (1959); McClain, Nichols and Waak (1960 and 1962); Epstein (1959); Drake and Hvatum (1959); Roberts and Stanley (1959); Bibinova et al. (1962) (see also Mayer, 1959; Field, 1959; Sloanaker, 1959).

[‡] In this connection it should be noted that the relative measurement accuracy achieved by Rose, Bologna and Sloanaker (1963a and 1963b) was higher than the ab-

§ 20] Continuous Radio Emission of the Planets

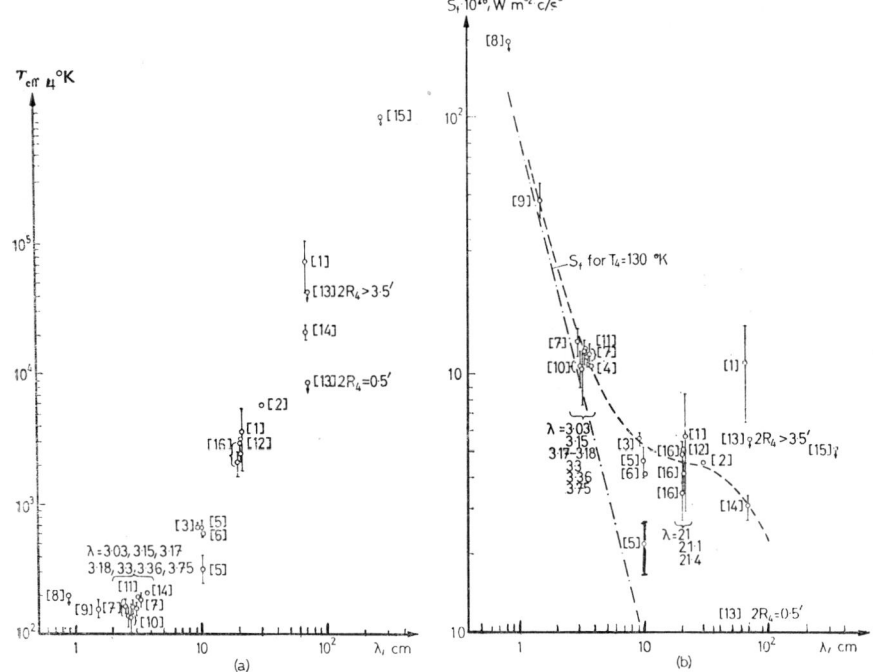

Fig. 100. Values of Jupiter's effective temperature $T_{eff\, 2\!\!\!\downarrow}$ related to the planet's optical disk and the spectral radio emission flux S_f at different wavelengths. The vertical lines show the possible measurement errors. The dotted curves show the most probable frequency spectrum of Jupiter's continuous radio emission. The wavelengths λ in the figure are in cm. Sources: 1—Drake and Hvatum (1959); 2—Roberts and Stanley (1959); 3—Rose, Bologna and Sloanaker (1963a and 1963b); 4—Drake and Ewen (1958); 5—Sloanaker and Boland (1961); 6—McClain and Sloanaker (1959); 7—Alsop et al. (1959a and 1959b); 8—McClain (1960); 9—Vetukhnovskaya et al. (1963); 10—Seeger, Westerhout and Conway (1957); 11—Bibinova et al. (1962); 12—Epstein (1959); 13—Long and Elsmore (1960); 14—Zakharov et al. (1964); 15—Mills et al. (1958); 16—McClain, Nichols and Waak (1962).

The closeness of the effective radio temperature to the temperature in infra-red rays gave a reason for assuming that the radio emission at $\lambda = 3$ cm is generated in regions of Jupiter's atmosphere adjacent to the visible layer of clouds. Here it first seemed possible (Alsop et al., 1959a and 1959b; Ellis, 1962; Ellis and McCulloch, 1963; Mayer, 1959) that the higher values of $T_{eff\, 2\!\!\!\downarrow}$ at $\lambda \sim 10$ cm (about 600°K (Sloanaker and Boland,

solute accuracy indicated here. The result of the observations (Drake and Ewen, 1958; Field, 1959) in the 3-cm band, according to which $T_{eff\, 2\!\!\!\downarrow}$ is about 210°K, is unreliable and apparently contains an error (see Mayer, 1959).

1961; McClain and Sloanaker, 1959)) were connected with the planet's hotter layers beneath the clouds, from which the radio emission in this band also comes. However, later measurements of Jupiter's effective temperature at decimetric wavelengths cast doubt on the correctness of this interpretation. It is improbable that the radio emission in this band, which has a higher effective temperature (about 3×10^3 °K at 20 cm (McClain, Nichols and Waak, 1960 and 1962; Epstein, 1959; Drake and Hvatum, 1959) and 5.5×10^3 °K at 30 cm (Roberts and Stanley, 1959)), is generated by layers of the planet beneath the clouds, since in this case their kinetic temperature should be close to the above values for the radio temperature $T_{\text{eff} 2\!\!\!\!\!/}$, i.e. approach the temperature of the solar photosphere, which is completely unlikely. Moreover, if Drake and Hvatum's data (1959) ($T_{\text{eff} 2\!\!\!\!\!/} = (7 \pm 3) \times 10^4$ °K at $\lambda \approx 68$ cm) correspond to the truth, regions with an even higher temperature should exist in the layers under Jupiter's clouds.

It is true that the latter value of $T_{\text{eff} 2\!\!\!\!\!/}$ is apparently somewhat too high. This is indicated by Long and Elsmore's interferometer observations (Long and Elsmore, 1960) who, having failed to find radio emission from Jupiter at $\lambda = 73.5$ cm, set the upper limit for the spectral radio emission flux S_f at $1 \cdot 1 \times 10^{-26}$ W m^{-2} c/s^{-1} provided that the diameter of the emitting region is close to the planet's optical diameter (about 0·6′). As we increase the assumed size of the source, which it was not possible to establish in Long and Elsmore (1960), the upper estimate for S_f rises, reaching 5.4×10^{-26} W m^{-2} c/s^{-1} (for a source with a diameter of more than 3·5′). These two values of S_f correspond to the following values of the effective temperature related to Jupiter's optical disk: $T_{\text{eff} 2\!\!\!\!\!/} < 8 \times 10^3$ °K and $T_{\text{eff} 2\!\!\!\!\!/} < 4 \times 10^4$ °K respectively, which is less than the value given by Drake and Hvatum (1959). The latter estimate agrees with the data of recent measurements (Zakharov, Krotikov, Troitskii and Tseitlin, 1964) at $\lambda = 70$ cm, according to which $T_{\text{eff} 2\!\!\!\!\!/} = 2 \times 10^4$ °K with an accuracy of $\pm 10\%$.

It is as well to point out that the question of Jupiter's radio emission at wavelengths of $\lambda > 30$ cm plays an important part in determining the nature of the frequency spectrum. In actual fact, it follows from all the known data on the magnitude of S_f (including the result of Drake and Hvatum (1959) at $\lambda \sim 70$ cm) that the radio emission flux increases with λ in the range of wavelengths from 30 to 70 cm. If, in accordance with Roberts and Stanley (1959), Zakharov et al. (1964) we take the smaller values of S_f at wavelengths of 70 cm, it turns out that the spectral radio emission flux decreases here as the wavelength rises (see Fig. 100).

On the basis of what we have been saying above the suggestion was put

§ 20] Continuous Radio Emission of the Planets

forward by Drake and Hvatum (1959), Roberts and Stanley (1959) and Field (1959) that the decimetric radio emission is not generated in the layers beneath Jupiter's clouds but in the regions surrounding the planet that on Earth are occupied by the Van Allen belts. It might be expected in the latter case that the angular size of the radio emission source would noticeably exceed the visible diameter of Jupiter. Since the emission in this case is obviously connected with charged particles moving in Jupiter's magnetic field, the presence of polarization in the observed radio emission from the planet is possible from this point of view.

In order to check this hypothesis Radhakrishnan and Roberts (1960a and 1960b) carried out some very important work, the results of which throw light on the problem of the origin of Jupiter's continuous radio emission: they measured the angular size of the source of the radio emission and investigated the nature of the latter's polarization. Their radio telescope at $\lambda = 31\cdot3$ cm was an interferometer with a scanning pattern. The distance between the adjacent lobes of the beam could be altered during the observations by changing the interferometer baseline; it took up values about 18′, 9′, 4·5′ and 2′. Provision was made in the interferometer to change the orientation of the aerials in both arms so that it could work as an interference polarimeter (see sections 5 and 6).

In their observations of Jupiter Radhakrishnan and Roberts found no circularly polarized emission; if it does exist its relative contribution does not exceed 6%. At the same time they noted linear polarization in Jupiter's continuous radio emission, whose value ϱ_l (according to the data of measurements made on different days) varied from 25% to 43% with a mean value of 33%.[†] The electrical vector of the polarized component is located almost parallel to Jupiter's equatorial plane, not deviating from it by more than $\pm 12°$.

Subsequent observations (Morris and Berge, 1962) showed that during the rotation of Jupiter the plane of polarization is subject to systematic oscillations (with the same period); their amplitude is $9\pm3°$. This effect is obviously caused by the inclination of the magnetic dipole's axis to the planet's axis of rotation by an angle of $9\pm3°$. Judging from the nature of the deviation of the plane of polarization from the equatorial plane in system III, the magnetic poles are located at longitudes of 200 and 20° with an accuracy of up to 10°. Within the limits of observational error these data agree with those obtained by Rose, Bologna and Sloanaker (1963a and

---

[†] The use of a slightly different method of measuring the degree of polarization gave a lower mean value (22%). Close values of the degree of polarization were then obtained at even shorter wavelengths ($\varrho_1 \approx$ 25–45% at $\lambda = 21$ cm (Miller and Gary, 1962) and $\varrho_1 \approx 21\%$ at $\lambda = 9\cdot4$ cm (Rose, Bologna and Sloanaker, 1963a and 1963b)).

1963b) at $\lambda = 9.4$ cm. Therefore the longitude position of one of the magnetic poles is close to the position of the main local source of decametric radio emission (at a longitude of 200°; see section 19).

Recordings of Jupiter's radio emission at different distances from the plane of the interferometer (and therefore at different distances between the lobes of the beam) have shown (Radhakrishnan and Roberts, 1960a and 1960b) that the modulation depth of the interference recording varies when the mean distance between the lobes reaches 2'. It follows from this that the extent of the radio emission source is likewise about 2', i.e. it is approximately three times greater than the visible diameter of Jupiter.

Therefore the hypothesis of the generation of the decimetric radio emission in the outermost regions of Jupiter has been brilliantly confirmed.

The observations made by Radhakrishnan and Roberts (1960a and 1960b) are insufficient for us to judge the details of the radio brightness distribution over Jupiter. It was noticed, however, that the modulation depth of the polarized component decreases as the width of the lobes gets less more rapidly than the modulation depth of the radio emission's non-polarized component. This effect is, in all probability, caused by the fact that the outer regions of the source are more strongly polarized than its central part.

Later Morris and Berge (1962) have once more measured the linear polarization and investigated the radio brightness distribution over the source of Jupiter's continuous radio emission at $\lambda = 31.3$ and 21.6 cm. For the degree of linear polarization ϱ_l they obtained values of $33\pm7\%$ and $28\pm6\%$, which are close to those indicated earlier. The distribution of T_{eff} over the source can be approximated with sufficient accuracy by a Gaussian curve; the diameter of the local source determined at a level of $1/e$ of the maximum T_{eff} value exceeds the optical diameter by a factor of 3.3 at $\lambda = 31.3$ cm and a factor of 2.9 at $\lambda = 21.6$ cm. These values of the effective radio diameter characterize the size of the emitting region in the equatorial direction; in the polar direction the radio diameter is far less (~ 3 and 2.4 times respectively at the above wavelengths), i.e. it is only slightly larger than the optical diameter.

Let us now examine the data on the variations in the flux of the Jovian radio emission.

The level of Jupiter's radio emission in the 3-cm band is relatively constant. The observations (with an accuracy of up to 10% of the total flux value) do not show any rapid variations with a characteristic time of the order of tens of hours or several days, in particular variations connected with the planet's rotation (the only exception is an indication of an anomalous increase in $T_{\text{eff}\;\gtrless}$ to 270°K in the period between 30 April and

§ 20] Continuous Radio Emission of the Planets

1 May 1958). At the same time there is a noticeable difference in the effective temperature values obtained in 1956–7 (140–145°K (Mayer, McCullough and Sloanaker, 1958c; Gibson and Corbett, 1963)) and in 1958–9 (165–190°K (Alsop *et al.* 1959a and 1959b)). This slow rise in the level of the radio emission is approximately one and a half times higher than the possible observational errors. Slow variations in the radio emission are also found at wavelengths of 10 cm (Sloanaker and Boland, 1961); for example the mean value of $T_{\text{eff }\mathcal{2}\!\!\!\mid}$, which according to the measurements of 1958 was $640\pm85°K$, dropped in the following year to $315\pm65°K$.[†]

In the decimetric band the strong time dependence of the level of radio emission has been noted by many observers (see, e.g., McClain and Sloanaker, 1959; Epstein, 1959; Drake and Hvatum, 1959; Roberts and Stanley, 1959). However, Roberts (1963 and 1964) now considers it almost certain that in many cases the variations were caused by subsidiary things: low equipment stability, difficulties of picking out the source on a background of intense cosmic radio emission, polarization of the Jovian emission, etc.

Attempts to find any correlation of Jupiter's effective temperature in the decimetric band with the chromospheric flare index, the level of the solar radio emission at $\lambda = 10$ cm and geomagnetic disturbances (i.e. with solar corpuscular streams) are still unsuccessful (McClain, Nichols and Waak, 1960, 1962; Roberts and Stanley, 1959). There are only a few indications of an increase in the intensity of Jupiter's emission at $\lambda = 21$ cm after a strength 3 solar flare which occurred on 10 May 1959 and the intense aurora that was observed a day later (McClain, Nichols and Waak, 1960 and 1962).

Sloanaker and Boland (1961) have investigated the connection of the level of the radio emission observed at $\lambda \approx 10$ cm with Jupiter's rotation. (This connection is obviously possible if the source of the radio emission is not symmetrical relative to the planet's axis of rotation.) They found no definite correlation of $T_{\text{eff }\mathcal{2}\!\!\!\mid}$ with the longitude of Jupiter's central meridian in systems I, II or in a system with a period of $T^{\text{II}}_{\mathcal{2}\!\!\!\mid} +41$ sec. However,

---

[†] The graphs for the decimetric band in Fig. 100 show the values of $T_{\text{eff }\mathcal{2}\!\!\!\mid}$ and S averaged from observations made for many days. The different values of $T_{\text{eff }\mathcal{2}\!\!\!\mid}$ and S at $\lambda \sim 10$ cm (Sloanaker and Boland, 1961) and $\lambda \sim 20$ cm (McClain, Nichols and Waak, 1960, 1962) shown in Fig. 100 therefore reflect only slow variations in the radio emission level with a characteristic time of the order of a month or more. In principle it is not impossible that the different results of the measurements (Drake and Hvatum, 1959; Long and Elsmore, 1960) at $\lambda \sim 70$ cm can be explained by variations in the Jovian radio emission flux. The absence of any significant variations in the value of $T_{\text{eff }\mathcal{2}\!\!\!\mid}$ in the experiments of Drake and Hvatum (1959) cannot be a convincing argument against this since the observations were made for a total of only 2 days.

in a system with a period of $T_{2\!\!\!\!/}^{II} + 82$ sec the effective temperature of one side of the planet turned out to be 30% greater than the $T_{\text{eff }2\!\!\!\!/}$ of the other side. This correlation seems also to appear in a coordinate system rotating 123 sec more slowly than system II.

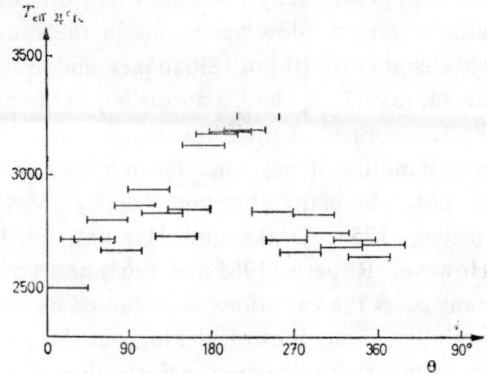

FIG. 101. Dependence of Jupiter's effective radio emission temperature $T_{\text{eff }2\!\!\!\!/}$ at $\lambda = 21$ cm on the longitude of the central meridian Θ in system II (McClain, Nichols and Waak, 1960 and 1962)

The comparison of the longitude of the central meridian with the variations in $T_{\text{eff }2\!\!\!\!/}$ at a wavelength of 21 cm made by McClain, Nichols and Waak (1960, 1962) (see also Miller and Gary, 1962) confirms the presence of this effect. Just as at 10 cm no correlation was found in system I, although there is a clear connection between the planet's effective temperature and the longitude of the central meridian in system II. It shows up in the 20% increase in $T_{\text{eff }2\!\!\!\!/}$ between 175 and 225° of longitude. This correlation can be clearly seen in Fig. 101, where the longitude of the central meridian in system II is plotted on the abscissa and the effective emission temperature averaged over the time Jupiter takes to rotate an angle of 30° is plotted on the ordinate; this averaging was carried out at successive intervals of 45°.†

The results given for the comparison of the level of radio emission with the longitude of Jupiter's central meridian are interesting, although not very convincing because of the limited number of observations. As experimental material is accumulated it will presumably become possible to

† No significant differences of the correlation in system II and in system III (introduced for the characteristic curve of Jupiter's local sources of sporadic radio emission) were found during the observation time because of the closeness of these systems' periods. We note that the position of Jupiter corresponding to the radio emission maximum at $\lambda = 21$ cm (longitude 110° in system III) does not correspond to intersection of the central meridian by the principal source of the bursts since the longitude of the latter is about 200° in this system.

§ 20] Continuous Radio Emission of the Planets

connect the variation in $T_{\text{eff}\,\mathrm{2L}}$ more certainly with the planet's rotation and obtain more definite information on the extent of the asymmetry of Jupiter's source of continuous radio emission (in this connection see also Roberts, 1963 and 1964; Morris and Berge, 1962; Gary, 1963). A review of later results of decimetric observations of the Jupiter radio emission is given by Warwick (1967).

RADIO EMISSION OF MARS

There are very few radio observational data on this planet. The radio emission of Mars has been studied only in the 3-cm band. The first observations (Mayer, McCullough and Sloanaker, 1958c) made with a 15 m parabolic aerial and a radiometer of comparatively low sensitivity showed that at the period of opposition the planet's effective temperature at $\lambda = 3\cdot 15$ cm was $T_{\text{eff}\,\delta} = 218\pm 50°\text{K}$. Further measurements of the effective temperature of Mars were made at the same wavelength 6 weeks after opposition (Rose, Bologna and Sloanaker, 1963a and 1963b). The more accurate value of $T_{\text{eff}\,\delta}$ of $211\pm 20°\text{K}$ obtained with a maser is about 35° less than the temperature of the illuminated part of Mars in infrared rays (see section 3). However, the measurement accuracies in the radio and infrared parts of the spectrum are such that this difference cannot for the time being considered the actual one since it is not outside the possible limits of observational error.

RADIO EMISSION OF VENUS

Results of measurements[†] of the effective temperature $T_{\text{eff}\,\female}$ related to the optical disk of Venus in the range of wavelengths from 4 mm to 40 cm are shown in Fig. 102a.[‡] We note that the figure shows the values of $T_{\text{eff}\,\female}$ during observations made at different phase angles of Venus. Therefore the possible variations in $T_{\text{eff}\,\female}$ depending on the planet's phase (i.e. on the extent that it is illuminated by the Sun's rays) are not reflected here. Moreover, the observations of Venus are generally carried out not far from lower conjunction, so the possible differences in the value of the effective temperature caused by the latter's phase dependence are comparatively small (see below).

The values of the spectral radio emission flux S_f corresponding to these

[†] Alsop, Giordmaine, Mayer and Townes (1959a and 1959b); Mayer, McCullough and Sloanaker (1958a and 1958b); Gibson and Corbett (1963); Kislyakov, Kuz'min and Salomonovich (1961 and 1962); Grant and Corbett (1962); Kuz'min and Salomonovich (1960 and 1961); Gibson and McEwan (1959); Mayer, McCullough and Sloanaker (1960); Lilley (1961); Drake (1962a); Vetukhnovskaya et al. (1963); Drake (1962c); Bibinova (1962).

[‡] For observations of the radio emission of Venus see also Roberts (1963, 1964) and Lynn, Meeks and Sohigian (1963).

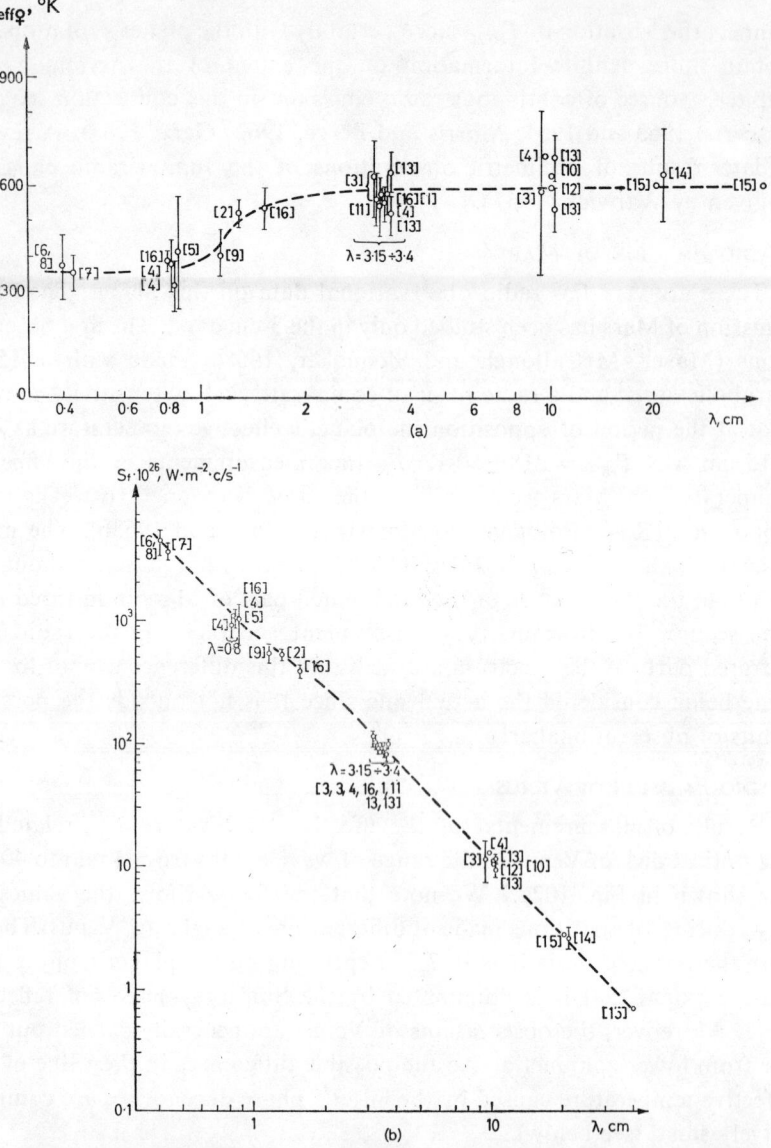

Fig. 102. Values of effective temperature $T_{\text{eff}\,\venus}$, °K (a) and of radio emission flux $S_{f\,\venus}$ (b) of Venus in the millimetric and centimetric bands. The dotted line shows the most probable curve of $T_{\text{eff}\,\venus}(\lambda)$ and the most probable function $S_{f\,\venus}(\lambda)$. The notations are the same as in Fig. 100. The wavelengths λ are given in centimetres. Sources: 1—Alsop et al. (1959a and 1959b); Gibson and Corbett (1963); Mayer, McCullough and Sloanaker (1958a and 1958b); Kuz'min and Salomonovich (1960 and 1961); Gibson and McEwan (1959); Kislyakov, Kuz'min and Salomonovich (1961 and 1962); Grant and Corbett (1962); Kislyakov and Plechkov (1963); Staelin, Barrett and Kusse (1963); Drake (1962b); Bibinova et al. (1962); Drake (1962a); Mayer, McCullough and Sloanaker (1960); Lilley (1961); Drake (1963); Vetukhnovskaya et al. (1963).

§ 20] Continuous Radio Emission of the Planets

$T_{\text{eff}♀}$ are shown in Fig. 102b. These values are calculated by a formula of the (4.15) type for the distance from the Earth to Venus at the period of lower conjunction (0·28 a.u.).

It is clear from Fig. 102 that the values of the effective temperature of Venus in the radio band are systematically higher than the values in the infra-red band (about 233°K, see section 3), $T_{\text{eff}♀}$ rising as one moves from the millimetric towards the centimetric wavelengths. In the $\lambda \approx$ 1·5–21 cm range this value remains constant with an accuracy of up to the possible measurement errors: $T_{\text{eff}♀} \approx 600°K$. This unexpected result, according to which the radio temperature is far higher than the temperature of the upper layer of clouds on Venus, was obtained in the very first observations of the planet at centimetric wavelengths (Mayer, Sloanaker and McCullough, 1958a and 1958b).

This difference in temperatures indicates that the upper part of the planet's cloud cover cannot be responsible for the creation of the observed radio emission: generally speaking, it comes either from the deeper-lying layers of Venus beneath the clouds or is generated in its outermost regions.

In this connection it should be noted that, unlike Jupiter's radio emission, there is no noticeable linearly polarized component in the radio emission of Venus (Mayer, McCullough and Sloanaker, 1958a and 1958b), whilst the spectral flux of the emission is characterized by considerable constancy, unless we consider the slow variations in S_f caused by the phase variation or connected with the change in the distance from the Earth to Venus.† The data given indicate that a radiation belt on Venus does not play such an important part in the creation of its radio emission as on Jupiter. A weighty argument for or against the generation of radio emission from Venus at great distances from its surface is provided by the data on the effective diameter of Venus at a wavelength of 3 cm obtained by Korol'kov et al. (1963) using the Pulkovo telescope. It follows from recordings during the planet's passage through the aerial's 1·2′ knife-edge beam that there is actually no radio emission at a distance of $1·07R_♀$ from the centre of Venus's disk. This means that the observed radio emission is generated in layers not more than about 420 km above the clouds, i.e. it is definitely not connected with the planet's radiation belts. On the other hand, the radio emission of Venus cannot be connected with its ionosphere either (see section 34) and, in all probability, comes from the layers of the planet under the clouds which are not accessible to study by the methods of optical astronomy. In this case the high values for $T_{\text{eff}♀}$ in the centimetric

---

† The presence of variations in $T_{\text{eff}♀}$ at 9·6 cm, which was reported by Kuz'min and Salomonovich (1960, 1961), is not confirmed (see Kellog and Sagan, 1962, p. 101).

band point to the high kinetic temperature of the lower layers of the planet's atmosphere and surface.

Experiments at $\lambda \approx 3$ cm (Korol'kov et al., 1963), about which we have just spoken, also indicate the absence on Venus of brightening at the limb; moreover, they point (very uncertainly, it is true) to a "darkening" effect towards the edge of the planet's disk. Apparently more reliable results on the radio brightness distribution over the disk were obtained in measurements made by Barrett and Lilley (1963) with the space probe "Mariner II" at a wavelength of 1·9 cm. The observations were made at the period when the distance from the rocket to Venus was a little more than 4×10^4 km. Scanning of the disk of Venus with an aerial having a beam width of one-eighth of the planet's diameter made it possible to state that T_{eff}, which reaches 570°K $\pm 15\%$ at the centre of the disk, decreases noticeably at the periphery. This is of great importance for the theory of Venus's radio emission (see section 34).

Whilst in the case of Jupiter it is largely the daytime hemisphere that faces the Earth, Venus passes successively through all phases from eclipsed to completely illuminated. Since we may expect, generally speaking, a noticeable difference in the effective radio temperatures of the night and day sides of Venus,[†] we are faced with the problem of investigating the phase dependence of $T_{\text{eff}\venus}$. The presence of phase dependence can be concluded from all the radio observations that have been made. However, the insufficient accuracy and duration of the measurements prevent entirely definite quantitative conclusions from being drawn on the amplitude of the oscillations of $T_{\text{eff}\venus}$, not to mention the phase shift of these oscillations. This is because Venus's radio emission can, as a rule, be reliably received only in a limited range of phase angles ψ: a long way from inferior conjunction recording and measurement of the radio emission are made difficult because of the reduction in the flux as Venus moves further from the Earth. In the majority of cases, however, even this range has been far from completely investigated.

Preliminary data on the existence of the dependence of $T_{\text{eff}\venus}$ on ψ at centimetric wavelengths (on the increase in the effective temperature as we move away from inferior conjunction) are given by Mayer, McCullough and Sloanaker (1958a, 1958b, 1960), Gibson and Corbett (1963), Bibinova et al. (1962); however, the range of ψ investigated was insufficient for an absolutely definite conclusion about the presence and nature of phase

---

[†] It should be pointed out, moreover, that, judging from observations in infrared rays, no essential difference can be found between the temperatures of the illuminated and eclipsed parts of the disk of Venus and therefore there is no phase dependence of the temperature of the planet's cloud cover averaged over the disk.

§ 20] Continuous Radio Emission of the Planets

dependence. Mayer, McCullough and Sloanaker (McClain, 1962) made measurements of $T_{\text{eff}\,♀}$ at $\lambda = 3\cdot15$ cm in a very broad range of phase angles $\psi = 62$–$234°$,[†] obtaining a clear-cut phase dependence of the planet's radio temperature. If we approximate the latter by a curve of the form

$$T_{\text{eff}\,♀}(\psi) = T_{♀=} + T_{♀\sim} \cos(\psi - \psi_0), \tag{20.1}$$

then it turns out that the constant component of the temperature $T_{♀=}$ is approximately $630°K$, whilst the amplitude of the variable part is $T_{♀\sim} = 70$–$90°K$; the angle ψ_0 is about $12°$. Therefore the radio temperature of the daytime hemisphere of Venus exceeds by 140–$180°$ the corresponding value for the planet's night-time hemisphere.

Drake's measurements (Drake, 1962a) at $\lambda = 10$ cm at a period near inferior conjunction led to the conclusion that the phase dependence at these wavelengths is half as strong as at $\lambda \sim 3$ cm; $T_{♀=} \approx 629°K$, $T_{♀\sim} \approx 39°K$, $\psi_0 \approx 17°$. The phase dependence obtained agrees within the limits of error with the results of measurements (Drake, 1962b) of the temperature of Venus near exterior conjunction (with $\psi \approx 25°$ the radio temperature was $610 \pm 55°K$).

A very important result of the observations at 3 and 10 cm which, of course, needs careful checking is the presence of the phase shift $\psi_0 \approx 12$–$17°$. This shift means that the minimum effective temperature is reached not precisely at inferior conjunction, when the eclipsed side of Venus can be seen from the Earth, but slightly later.

A rise in the effective temperature as one moves away from inferior conjunction, which is outside the limits of possible measurement errors, has also been observed in the millimetric band at $\lambda \approx 8$ mm (Kuz'min and Salomonovich, 1960 and 1961) and $\lambda \approx 4$ mm (Kislyakov, Kuz'min and Salomonovich, 1961 and 1962; Grant and Corbett, 1962)). The results of studying the phase dependence of radio temperature of Venus at 4 mm are shown in Fig. 103. The width of the shaded band characterizes the possible relative errors between the individual measurements at the time of the observations. Because of the absolute error of the observations, which is about 30%, this band may shift upwards or downwards on the graph by this value. The curve of the phase dependence $T_{\text{eff}\,♀}(\psi)$ is located within the shaded band, the most probable dependence of $T_{\text{eff}\,♀}$ on the phase of Venus being in the middle of this band.[‡]

[†] The angle ψ is here taken as $0°$ at upper and $180°$ at lower conjunction.
[‡] The results of Copeland and Tyler's measurements (1962) are also in favour of phase dependence at 8 mm. However, judging from these experiments, the phase dependence at 8 mm is far more weakly defined than was indicated by Kuz'min and Salomonovich (1960 and 1961).

The presence of the phase dependence of $T_{\text{eff}\,\venus}$ is an argument against a small period of rotation for Venus (if the radio emission is connected with the layers of the planet beneath the clouds). With rapid rotation the temperature conditions of the lower atmosphere and the surface cannot change in the transition from day to night so much as to cause a noticeable change in the value of $T_{\text{eff}\,\venus}$. On the other hand, the period of rotation of Venus may not be the same as the orbital period, since in this case there would be no phase shift ψ_0 in the phase dependence of the radio temperature at centimetric wavelengths. This shift can obviously be connected with

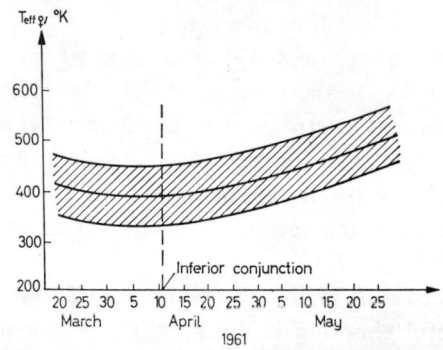

FIG. 103. Phase dependence of effective temperature of Venus from observations at a wavelength of $\lambda \approx 4$ mm (Grant and Corbett, 1962).

the thermal inertia of the layers of Venus which are responsible for the observed radio emission; this inertia consists of slow heating of the effectively emitting layers during the day and slow cooling at night. (The situation here is analogous to that obtaining on the Moon where the phase shift of the radio emission is due to the same reasons.)

Furthermore, it is easy to see that the sign of the angle ψ_0 can be used to judge direction of the planet's rotation. For example, the lag of the minimum $T_{\text{eff}\,\venus}$ in the experiments (McClain, 1962; Drake, 1962a), of which we have spoken above, indicates reverse rotation of Venus, i.e. rotation about its axis in the opposite direction to the rotation of the other planets. If this result is confirmed it will obviously be of great importance in solving the question of the origin of the solar system. In this case we note that the conclusion about the reverse rotation at least does not contradict the data on radar sounding of Venus (W. B. Smith, 1963). For radio astronomy methods of determining the period and direction of Venus's rotation see also Kuz'min and Salomonovich (1963).

RADIO EMISSION OF MERCURY

At present there is only one series of measurements of Mercury's radio emission made at $\lambda = 3\cdot45$ cm and $\lambda = 3\cdot75$ cm with a 28-m diameter aerial (using a maser at the first wavelength) (Howard, Barrett and Haddock, 1961 and 1962). The effective temperature reduced to the planet's disk averaged 400°K. During the measurements Mercury was in elongation so that its disk was not completely illuminated by the Sun. If we neglect the contribution to the radio emission from the unilluminated side of the planet and assume that the distribution of the effective temperature T_{eff} over the illuminated side is subject to the law $T_{\text{eff}} = T_0 \cos^{1/4} \varphi$ (where T_0 is the effective temperature at the point beneath the Sun, φ is the zenith angle of the Sun at a given point on the surface of Mercury), then the value of T_0 is 1100 ± 300°K. In the case of even distribution of T_{eff} over the daylight side of the planet this value should be reduced by about 30%. Allowing for the low accuracy of the radio observations, we can take into account that the value obtained for T_{eff} does not diverge from the results of measurements of Mercury's temperature in infra-red rays (610°; see section 3), although it is possible (Roberts, 1963 and 1964) that it is systematically higher than the infrared temperature.

In conclusion we shall say that in the sections dealing with the continuous radio emission of Jupiter, Mars and Venus we discussed emission having a broad frequency spectrum. At the same time in the composition of the emission from these planets there are undoubtedly also monochromatic lines which owe their appearance to electron transitions in molecules of the gases making up the planetary atmospheres. The question of the expected line intensities on Venus and the possibilities of finding them are discussed in section 34.

21. Radio Emission of the Moon

PRELIMINARY REMARKS

Investigations of the Moon's radio emission are being carried on at present over a broad range of wavelengths from 1·5 mm to 1·5 m. The data obtained here together with the results of measuring the lunar radiation in the infrared part of the spectrum are playing an important part in the study of the properties of the Moon's upper layers. According to present-day ideas, the infrared emission is connected with a very thin layer on the surface of the Moon, whilst the longer radio wavelengths come from a far thicker layer of lunar rock. Thanks to this study of the nature of the Moon's radio emission we can judge the temperature conditions and also the thermal conductivity and electrical conductivity of the upper layers of the Moon; a comparison of these characteristics with the corresponding values for terrestrial rock makes it possible to draw certain conclusions on the composition and structure of the lunar surface (see section 35). Apart from being of general importance to astrophysics, the information obtained is of particular value in connection with the problem of delivering scientific apparatus onto the Moon and the lunar landing of space vehicles: here the question of the properties of the Moon's surface layers becomes a major one.

Both infrared techniques (Sinton, 1955 and 1956) ($\lambda \sim 1\cdot5$ mm) and electronic methods of investigation (Fedoseyev (1963) at $\lambda = 1\cdot3$ mm, and Naumov (1963) at $\lambda = 1\cdot8$ mm) are used for picking up the Moon's radio emission at the lower end of the millimetric band. At longer wavelengths infrared techniques give way to electronic methods of investigation.

Since the effective temperature T_{eff} of the Moon's radio emission is non-zero only within the limits of the lunar disk, the aerial temperature T_A (4.6) in our case will be of the form

$$T_A = \frac{\int_{\Omega_{\mathbb{C}}} T_{\text{eff}} F \, d\Omega}{\int_{4\pi} F \, d\Omega}, \tag{21.1}$$

where $\Omega_{\mathbb{C}}$ is the solid angle subtended by the Moon's disk. Having measured the value of T_A we can find as a result the value of the Moon's radio temperature averaged over the aerial beam:

$$T_F = \frac{\int_{\Omega_{\mathbb{C}}} T_{\text{eff}} F \, d\Omega}{\int_{\Omega_{\mathbb{C}}} F \, d\Omega}. \tag{21.2}$$

If we compare the expressions (21.1) and (21.2) it is easy to see that T_F is connected with T_A by the relation

$$T_F = T_A \left(1 + \frac{\int_{4\pi - \Omega_{\mathbb{C}}} F \, d\Omega}{\int_{\Omega_{\mathbb{C}}} F \, d\Omega} \right). \tag{21.3}$$

To change from T_A to T_F we must precisely know the polar diagram $F(\Omega)$ not only within the main lobe but also in the region of the rear and side

§ 21] Radio Emission of the Moon

lobes, since the latter's contribution to the value of the integral $\int_{4\pi} F\,d\Omega$ is tens of percent for parabolic aerials. This circumstance has been fully recognized and only in recent years has begun to be used when processing observations. It was not taken into consideration in the majority of early papers, which was a source of considerable systematic errors, leading to values of T_F that were too low.

The radio temperature will obviously depend not only on the nature of the distribution of T_{eff} over the lunar disk but also on the nature of the beam and the position of the aerial axis relative to the disk. If, however, the width of the main lobe of the beam is such that the emission is chiefly received in the solid angle $\Omega_A \ll \Omega_\mathrm{C}$, then in accordance with (21.2) $T_F \approx T_{\text{eff}}(\Omega_0)$, where Ω_0 is the direction of the aerial axis. In particular the temperature T_F averaged over the beam is the same as the effective temperature of the centre of the lunar disk $T_{\text{eff}}(0)$ if the axis of the aerial is pointing towards it. It should be noted that the equality $T_F \approx T_{\text{eff}}(0)$ is satisfied even with the less rigid condition $\Omega_A < \Omega_\mathrm{C}$, since (as follows from what we shall be saying later) the distribution of the radio brightness T_{eff} over the Moon's disk is fairly uniform: T_{eff} noticeably decreases only in the immediate vicinity of the limb, at a distance of 2–3′ from it (with an angular diameter of the Moon of 31′). In the other extreme case, when $\Omega_A \gg \Omega_\mathrm{C}$,† T_F is equal to the effective temperature related to the Moon's optical disk

$$T_{\text{eff}\,\mathrm{C}} = \Omega_\mathrm{C}^{-1} \int_{\Omega_\mathrm{C}} T_{\text{eff}}\,d\Omega \quad \text{(see (21.2))}.$$

Moreover, owing to the quasiuniform distribution of T_{eff} over the lunar disk the values of $T_{\text{eff}\,\mathrm{C}}$ differ but little from $T_{\text{eff}}(0)$, being slightly less than the latter.

FREQUENCY SPECTRUM AND PHASE DEPENDENCE OF THE MOON'S RADIO EMISSION

The first observations of the Moon in the radio band ($\lambda = 1.25$ cm) made by Dicke and Beringer (1946) gave a value for T_F of 270°K (allowing for a correction (Piddington and Minnett, 1949b)). Later the value of T_F was determined more accurately by Piddington and Minnett (1948, 1949b), who observed the Moon's radio emission at the same wavelength with an aerial whose beam width (45′) considerably exceeded the angular size of the lunar disk. According to Piddington and Minnett (1949b) the mean temperature during lunation was $T_{\text{eff}\,\mathrm{C}} \approx T_F = 215\pm10$°K. In the

---

† It is this situation that obtains in observations from the Earth of the continuous radio emission of planets because of the very small angular sizes of the latter.

course of the measurements they also found a clear-cut phase dependence of the lunar radio temperature, which can be approximated well by a function of the form

$$T_F = T_{F=} - T_{F\sim} \cos(\psi - \psi_0), \qquad (21.4)$$

where $\psi = \omega_\mathbb{C} t$ is the lunar phase. The amplitude of the variation in the radio temperature $T_{F\sim}$ is 36°K and the phase shift is 45°. The value of the constant component $T_{F=}$ is given above.

Therefore, judging from the observations at $\lambda = 1\cdot 25$ cm, the variation of the radio temperature of the Moon lags noticeably in phase behind the similar dependence in the infra-red part of the spectrum, where $\psi_0 = 0$ for the Moon's mean temperature over the disk. From the point of view of modern ideas on the thermal nature of the Moon's radio emission, according to which the effective temperature T_{eff} is determined by the temperature of the layers of the Moon from which the observed radio emission comes (section 35), this is quite natural since a definite time is required for heating the sub-surface layers of the Moon by the Sun's rays and cooling by radiation. For a phase shift of $\psi_0 = 45°$ it is one-eighth of a lunar day. It will become clear from what follows that Piddington and Minnett's results are in close agreement with the data of present-day observations of the Moon's radio emission.

Figures 104–6 show the results of all the measurements we know of the quantities $T_{F=}$, $T_{F\sim}$ and ψ_0. The light circles in the figures denote data obtained with narrow-beam aerials ($\Omega_A < \Omega_\mathbb{C}$); here the values of $T_{F=}$, $T_{F\sim}$ and ψ_0 characterize the magnitude and variation in time of the effective temperature of the central part of the lunar disk. The dark circles, on the other hand, correspond to measurements with relatively broad beams ($\Omega_A > \Omega_\mathbb{C}$) providing information on the effective temperature $T_{\text{eff}\mathbb{C}}$ averaged over the disk. Figure 104 shows the observed values of T_F time-averaged for the lunation (i.e. the quantity $T_{F=}$) in the range of wavelengths from 1·3 mm to 1·68 m.

The vertical lines characterize the possible absolute measurement errors indicated by the observers themselves. The ratios of the amplitude of the phase dependence of the radio temperature $T_{F\sim}$ to its constant component $T_{F=}$ are plotted in Fig. 105. The measurement errors noted here are determined by the magnitude of the relative error characterizing the dispersion of the experimental values of T_F throughout the whole cycle of observations. Lastly, Fig. 106 gives an idea on the values of the phase shift ψ_0 and the possible error in its measurements.

The reservation made above about the errors of measurement of $T_{F=}$ indicated by the observers themselves is not a chance one. The point is that

§ 21] Radio Emission of the Moon

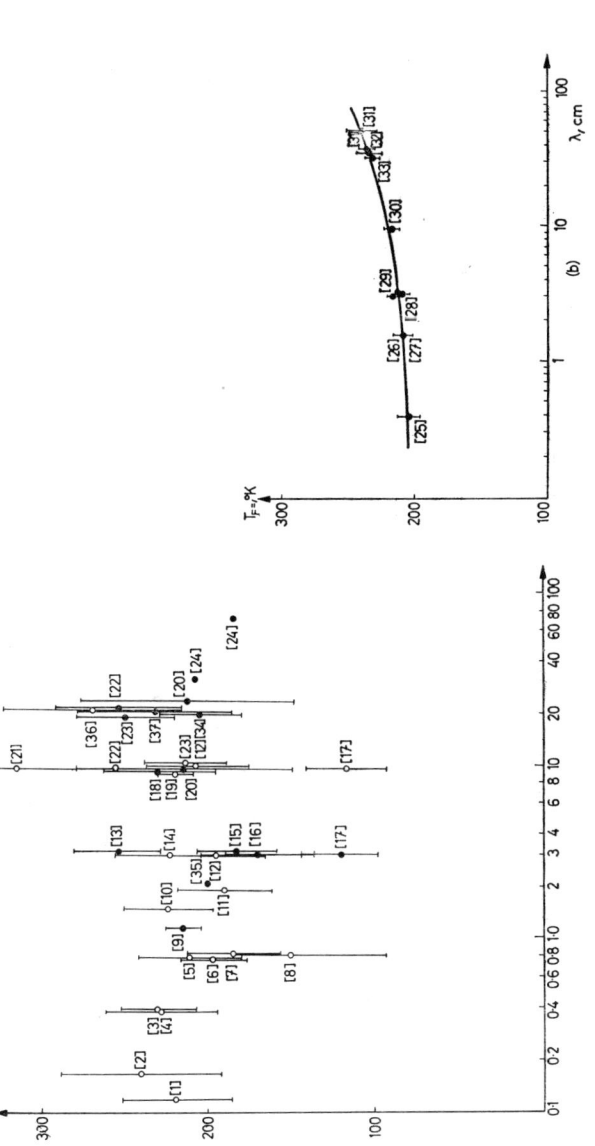

FIG. 104. Values of the lunar temperatures $T_{F=}$ in the radio band averaged throughout lunation. Graph (b) shows the results of Troitskii *et al.* whose accuracy is higher. Sources: 1—Fedoseyev (1963); 2—Naumov (1963); 3—Kislyakov (1961); 4—Kislyakov and Salomonovich (1963); 5—Salomonovich and Losovskii (1962); 6—Salomonovich (1958); 7—Gibson (1958); 8—Hagen (1949); 9—Piddington and Minnett (1949b); 10—Zelinskaya, Troitskii and Fedoseyev (1959a and 1959b); 11—Salomonovich and Koshchenko (1961); 12—Strezhneva and Troitskii (1961); 13—Koshchenko, Losovskii and Salomonovich (1961); 14—Zelinskaya and Troitskii (1956); 15—Troitskii and Zelinskaya (1955); 16—Kaidanovskii, Turusbekov and Khaikin (1956); 17—Koshchenko, Kuz'min and Salomonovich (1961); 18—Medd and Broten (1961); 19—Piddington and Minnett (1951); 20—Akabane (1955); 21—Castelli, Ferioli and Aarons (1960); 22—Mezger and Strassl (1960); 23—Seeger, Westerhout and Conway (1957); 24—Kislyakov and Plechkov (1963); 25—Kamenskaya, Semenov, Troitskii and Plechkov (1962); 26—Dmitrenko, Kamenskaya and Rakhlin (1964); 27—Krotikov, Porfir'yev and Troitskii (1961); 28—Bondar' *et al.* (1962); 29—Krotikov (1962) 30—Krotikov (1963a); 31—Krotikov and Porfir'yev (1963); 32—Razin and Fedorov (1963); 33—Waak (1961); 34—Grebenkemper (1958); 35—Davies and Jennison (1960); 36—Westerhout (1958); Mezger (1958).

in a number of cases the accuracy of the experiment is far lower than that shown in Fig. 104a because of the systematic errors occurring in the observations. This is chiefly true of the low values for the constant component of the radio temperature of $T_{F=} < 200°K$ obtained (Hagen (1949); *Report U.S. Commission V URSI* (1960); Salomonovich (1958); Gibson (1958); Salomonovich and Koshchenko (1961; Troitskii and Zelinskaya (1955);

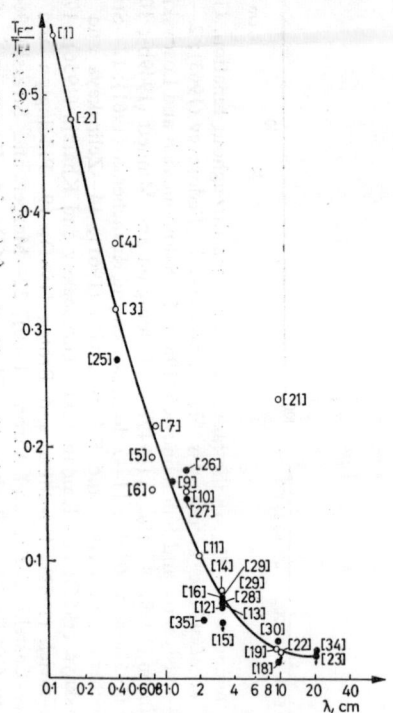

FIG. 105. Ratio of amplitude of phase dependence $T_{F\sim}$ of the Moon to its constant component $T_{F=}$ in the radio band. Sources: see Fig. 104.

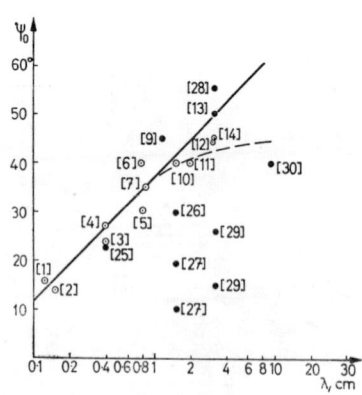

FIG. 106. Phase shift ψ_0 between variations in the Moon's illumination and its radio temperature. The accuracy of the measurements is generally ± 5–$10°$. Sources: see Fig. 104.

Kaidanovskii, Turusbekov and Khaikin (1956); Zelinskaya and Troitskii, (1956); Seeger, Westerhout and Conway (1957) apparently because of neglect or insufficiently accurate allowance being made for the effect of the side and rear lobes of the polar diagrams when changing from the aerial temperature T_A to the Moon's radio temperature T_F; see (21.3)). If we eliminate these data and also the clearly erroneous result of Akabane (1955) at $\lambda = 10$ cm ($T_{F=} \approx 315°K$), then it follows from all the rest (the later) of the measurements shown in Fig. 104a that the constant component

§ 21] Radio Emission of the Moon

of the Moon's radio temperature over the whole range of wavelengths investigated is between 200° and 250°K, mostly between 215° and 230°K both for the central part of the Moon and the effective temperature averaged over the disk.

Particular note should be taken of the measurements made by Troitskii and his collaborators at wavelengths of 0·4–50 cm with an absolute error of ±2–3% (whilst usually the accuracy of absolute radio astronomy measurements is no better than 10%[†]); the values of $T_{F=}$ they obtained are plotted in a separate graph (Fig. 104b). The increase in accuracy here was achieved thanks to a new method of calibrating the apparatus consisting of using an "artificial Moon" as a radio-emission standard. Since with this method it is not necessary to determine the parameters of the aerial array (the nature of its polar diagram), the main source of errors in the measurements connected with allowing for the effect of the rear and side lobes (see section 4) is eliminated.

In Fig. 104b the curve shows the most probable frequency spectrum of the constant component of the lunar radio emission plotted from the data of higher-accuracy measurements. This spectrum characterizes the effective temperature $T_{(=}$ averaged over the disk as a function of the wavelength λ, since all these measurements were made with beams whose width was greater than the angular size of the Moon. It follows from the data shown in the figure that in the range of wavelengths from 4 mm to 50 cm the value of $T_{F=} \approx T_{\text{eff}(}$ does not remain constant: it increases with λ, the temperature difference at the edges of this range being about 37°. The latter, as will be shown in section 35, makes it possible to estimate the heat flux from the Moon's core and from its value to estimate the thermal conditions of the internal layers and the thermal history of the Moon.

The following can be said about the phase dependence of the Moon's radio temperature.

Generally the observed function $T_F(\psi)$ can be approximated quite accurately by a function of the form of (21.4). This can be checked by the example in Fig. 107 which shows the function $T_F(\psi)$ in the millimetric band (Kislyakov, 1961b).[‡] It is difficult to give a definite answer here to the

---

[†] This can easily be checked by examining Fig. 104 (and also Figs. 100 and 102 in section 20). We would also point out that the measurement accuracy indicated by Baldwin (1961) for $T_{\text{eff}(}$ at $\lambda = 1·5$ m ($±3·5\%$) is too high, since the error in the magnitude of the background galactic radio emission, with which the Moon's radio emission was compared, is of the order of 10% (Troitskii, 1962c).

[‡] The reliability of the observations in Kislyakov (1961b) can be judged from the excellent agreement of the results of the latter with the data of three single measurements made independently of the lunar radio temperature (Coates, 1961) (see Fig. 107).

question of the relative content of the higher harmonic components in the curve of $T_F(\psi)$ because of the considerable dispersion of the experimental points used to plot the averaged function $T_F(\psi)$; however, the deviations of the phase dependence T_F from the law (21.4) have been noted in a number of papers (see, e.g., Kislyakov, 1961c; *Report U.S. Commission V URSI*, 1960; Salomonovich, 1958; Gibson, 1958). The amplitudes of the second and third harmonics in the function $T_F(\psi)$ shown in Fig. 107

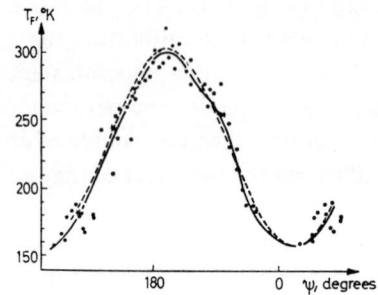

FIG. 107. Radio temperature T_F as a function of the phases of the Moon at $\lambda = 4$ mm (Kislyakov, 1961b). The asterisks indicate the values obtained by Coates (1961).

do not exceed 10% of the amplitude of the fundamental oscillation. Later observations (Tyler and Copeland, 1961) made it possible to establish that the content of the second and third harmonics in the phase dependence of the T_{eff} of the central part of the lunar disk reaches 7% and 4% respectively of the value of the constant component T_{eff}. As we move away from the centre of the disk (along the equator) the amplitudes of the higher harmonics gradually decrease. At a wavelength of 8 mm (to judge from the data of *Report U.S. Commission V URSI* (1960); and Gibson (1958)) the asymmetry is greater: the maximum T_F is reached 4 days after full moon, whilst the minimum occurs near the new moon. At the same time other observers (Salomonovich and Losovskii, 1962) have not found such a distinct asymmetry of $T_F(\psi)$ in the 8-mm band.

The most probable curve for the ratio $T_{F\sim}/T_{F=}$ as a function of the wavelength λ is shown in Fig. 105. The amplitude of the change in the radio temperature in the course of a lunation steadily decreases as λ rises. At $\lambda \sim 4$ mm the ratio $T_{F\sim}/T_{F=}$ is about 30%; in the 3-cm band it decreases to 7% and at wavelengths of $\lambda \sim 10$ cm to 3%. In the last case the low amplitude values of $T_{F\sim}$ make it very difficult to study the phase dependence of the radio temperature, imposing rigid requirements on the operating stability and calibration accuracy of the receiving equipment.

§ 21] Radio Emission of the Moon

At $\lambda \sim 10$ cm the features of the phase dependence have been studied by Medd and Broten (1961) and Krotikov (1962); a variable component $T_F(\psi)$ with an amplitude of the same order as in Krotikov, (1962), Medd and Broten (1961) is noted by Waak (1961).[†] There is no information yet on the magnitude of the phase dependence at longer wavelengths, unless we consider the preliminary report of Waak (1961), according to which the ratio $T_{F\sim}/T_{F=}$ at $\lambda \approx 20$ cm is 2·5% (i.e. comparable with the corresponding ratio at $\lambda \approx 10$ cm). This result requires further checking.

The observed decrease in the amplitude of the phase dependence of the Moon's radio temperature is quite natural from the point of view of the theory of the Moon's radio emission (section 35). This decrease is connected with the low thermal conductivity of the lunar rocks and is due to the fact that the deeper lying layers that are responsible for the emission of the longer wavelengths cool down less during the lunar night and do not have time to heat up strongly during the lunar day.

The slow heating of the sub-surface layers of the Moon by the Sun's rays and their cooling after sunset also explains both the considerable phase shift ψ_0 between the illumination of the lunar surface and the infrared temperature of the latter on the one hand, and the radio temperature on the other. This shift, as follows from Fig. 106, increases as we move away from the infrared part of the spectrum (move from millimetric to centimetric wavelengths): if at $\lambda \sim 1\cdot3$–$1\cdot8$ mm ψ_0 is about 14–16°, then by $\lambda \sim 0\cdot8$–10 cm it goes up to 40–50°.[‡] It is possible, moreover, that at the beginning of this range of wavelengths the shift is slightly less: 30–35°.

As well as investigating the oscillations in the Moon's radio temperature as there is a slow change in the extent of its illumination during a lunation, there is also interest in studying the variations of the quantity T_F with a more rapid (over a few hours) variation in the illumination during lunar eclipses. It might be expected that these variations will be far less than $T_{F\sim}$, since during the rapid change in illumination the thermal inertia of the Moon's upper layers will have a stronger effect. This obviously explains the lack of success in the majority of attempts to find and measure eclipse variations of T_F at wavelengths of 0·75–23 cm (Gibson, 1958; Kaidanovskii, Turusbekov and Khaikin, 1956; Castelli, Ferioli and Aarons, 1960; Mezger and Strassl, 1959; Mitchell and Whitehurst, 1958; Gibson, 1961).

[†] The result of Akabane (1955), according to which at wavelengths of $\lambda \sim 10$ cm the ratio $T_{F\sim}/T_{F=}$ is 0·24, is obviously erroneous: it would be easy to find such a strong phase dependence in this band.

[‡] Exceptions in this respect are the observations of Dmitrenko et al. (1964) at $\lambda = 1\cdot63$ cm and Bondar' et al. (1962) at $\lambda = 3\cdot2$ cm, according to which ψ_0 is 8–18° and 15–26° respectively. Although the accuracy of the phase shift determination is low these data obviously disagree with the results of other observations.

All that can be said now is that the drop in the Moon's radio temperature during an eclipse is not more than 2·5% at wavelengths of 10 and 23 cm (Castelli, Ferioli and Aarons, 1960) and 1% at a wavelength of 8·6 mm (Gibson, 1961); the latter, however, does not agree with the data of Tyler and Copeland (1961), according to which the "cooling" of the Moon at $\lambda = 8\cdot6$ mm is about 10%. At shorter wavelengths there are only the observations of Sinton (Sinton, 1956) in a broad band near $\lambda = 1\cdot5$ mm, according to which an hour after totality the radio temperature $T_{\text{eff}}(0)$ of the central part of the lunar disk falls from $\sim 300°$K to a minimum value of about 160°K. We notice that the lunar temperature measured in the infrared part of the spectrum during an eclipse also changes by a factor of about 2.

The rapid drop in the temperature of the Moon at wavelengths less than or of the order of a millimetre can be explained by the low thermal conductivity of the surface layer which is responsible for the emission in this range. Therefore the intense cooling of the eclipsed surface is not compensated by a flow of heat from the deeper layers of the Moon. At the same time the low thermal conductivity prevents cooling of the deep-lying layers from which the radio emission at the longer wavelengths comes; this is the cause of the decrease in the eclipse variations of T_F as λ rises.

It is clear from what we have been saying above that the millimetric wavelengths are the most favourable for carrying out investigations to find eclipse variations in the radio temperature; it is apparently quite possible to measure these variations at wavelengths of $\lambda \sim 4$ mm.

Radio Brightness Distribution over the Lunar Disk

The unequal illumination of the day and night sides of the Moon disturbs the central symmetry of the radio brightness distribution over the lunar disk, leading in particular to periodic displacements of the effective centre of the radio emission in the process of lunations. The coordinates of this centre, by definition, are:

$$\theta_{\mathbb{C}} = \frac{\int_{\Omega_{\mathbb{C}}} T_{\text{eff}}(\theta, \xi)\, \theta\, d\theta\, d\xi}{\int_{\Omega_{\mathbb{C}}} T_{\text{eff}}(\theta, \xi)\, d\theta\, d\xi}, \qquad (21.5)$$

$$\xi_{\mathbb{C}} = \frac{\int_{\Omega_{\mathbb{C}}} T_{\text{eff}}(\theta, \xi)\, \xi\, d\xi\, d\theta}{\int_{\Omega_{\mathbb{C}}} T_{\text{eff}}(\theta, \xi)\, d\xi\, d\theta}. \qquad (21.5a)$$

§ 21] Radio Emission of the Moon

Here θ and ξ are an orthogonal system of angular coordinates in the sky with its origin at the centre of the lunar disk. In practice $\theta_{\mathbb{C}}$ and $\xi_{\mathbb{C}}$ can be found, for example by scanning the Moon with an aerial with a "knife-edge" beam that gives a one-dimensional distribution of the radio brightness over the disk. For example, when scanning along θ with an aerial whose beam width in θ is much less and in ξ is much greater than the angular size of the lunar disk the aerial temperature T_A will be proportional to $\int T_{\text{eff}}(\theta, \xi)\, d\xi$. Having obtained the function $T_A(\theta)$ we use the formula (21.5) to find the coordinate $\theta_{\mathbb{C}}$. The centre of radio emission can also be found with a narrow-beam aerial ($\Omega_A \ll \Omega_{\mathbb{C}}$), for which $T_A \sim T_{\text{eff}}(\theta, \xi)$. We note that in the latter case the radio brightness distribution along the scanning line is obtained at the radiometer output in one scan over the disk; this makes it possible to determine the centre of radio emission along such a line.

It must be pointed out that when studying the variation of T_{eff} during a lunation measurement of the shift of the effective centre of radio emission has definite advantages over measurement of the Moon's radio temperature T_F averaged over the beam. The first needs no absolute calibration of the equipment, which is the main source of errors in measurements of T_F; equipment stability requirements are also far lower in the first case since it is only necessary for the amplification factor to be constant for one or a few scans to measure the displacement of $\theta_{\mathbb{C}}$. At the same time measurements of $\theta_{\mathbb{C}}$, $\xi_{\mathbb{C}}$ are possible only if we have highly directional aerials (much less than the angular diameter of the Moon if only in one direction), whilst such high directivity is not necessary for studying the function $T_F(\psi)$.

The first attempt (Kaidanovskii, Turusbekov and Khaikin, 1956) to find the displacement of the effective centre of the Moon's radio emission at a wavelength of 3·2 cm proved unsuccessful because of the low directivity of the receiving aerial (about 30′), which made it impossible to record any shift $\theta_{\mathbb{C}} < 0\cdot 5'$. Noticeable asymmetry in the equatorial distribution of the radio brightness was found by Gibson (1958) at $\lambda = 8\cdot 6$ mm when scanning the Moon with an aerial having a narrower beam (about 12′). The success of the observations was also helped by the fact that the variations in T_{eff} during a lunation are far greater at millimetric wavelengths than in the centimetric band. In the centimetric band (at a wavelength of 2·3 cm) the displacement of the effective radio emission centre has been investigated by Kaidanovskii et al. (1961) in observations with an aerial having a "knife-edge" beam about 2′ wide in the horizontal and 20–60′ in the vertical direction. The function they obtained for the displacement of the effective centre in the direction at right angles to the line of the horns can be approximated by the sine curve

$$\theta_{\mathbb{C}} \approx 0',17 \sin(\psi - 35°), \qquad (21.6)$$

where the value of the maximum displacement is determined with an accuracy of ±30%.

The development of aerial techniques, and in particular the design of parabolic aerials with high resolution, has made it possible to study in greater detail the nature of the distribution of the effective temperature T_{eff} over the lunar disk and its variation during the lunar day.

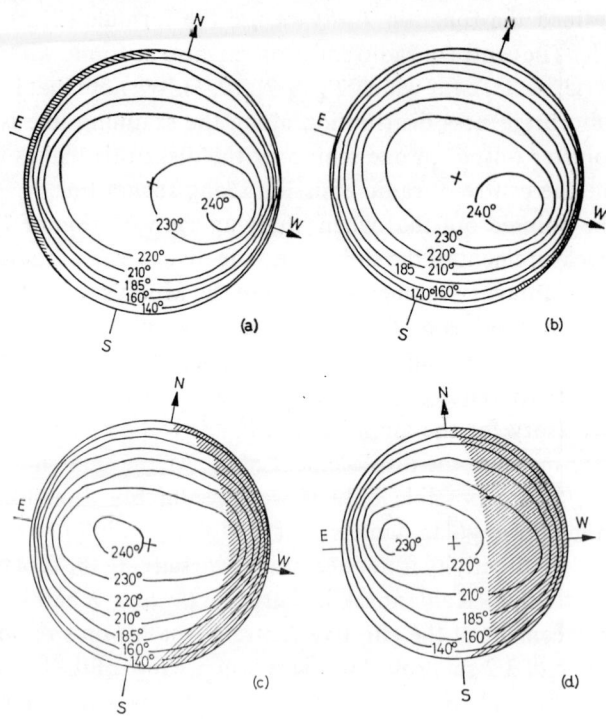

Fig. 109. Radio isophots of the Moon at $\lambda = 3 \cdot 2$ cm (Koshchenko, Losovskii and Salomonovich, 1961): (a) $\psi = 168°$, (b) $\psi = 194°$, (c) $\psi = 230°$, (d) $\psi = 262°$. Values of T_{eff} are in °K

Two-dimensional distributions of T_{eff} in the centimetric band were obtained at $\lambda = 2$ cm (Salomonovich and Koshchenko, 1961) and $\lambda = 3 \cdot 2$ cm (Koshchenko, Losovskii and Salomonovich, 1961) by successive azimuth scans of the Moon at different angles of aerial elevation. The measurements at a wavelength of 2 cm were made with a narrow-beam aerial (beam width about 4′); the resolution of the aerials at 3·2 cm was lower (about 6–7′). Similar distributions were obtained at millimetric wavelengths by Coates (1961) ($\lambda = 4 \cdot 3$ mm, 6′ beam) and Salomonovich and Losovskii (1962) ($\lambda = 8$ mm, 2′ beam). The Moon's radio isophots are shown in

§ 21] Radio Emission of the Moon

Figs. 108 and 109; the shaded areas in the figures correspond to the parts of the disk that were not illuminated by the Sun during the observations.

The form of the radio isophots for different values of the Moon's phase indicates that the area of maximum T_{eff} is in the equatorial belt and moves over the disk behind the point beneath the Sun, lagging systematically behind it. The radio brightness distribution (in accordance with the data given above on the displacement of the effective centre of radio emission) becomes more asymmetrical soon after the first and last quarters of the Moon. At this time the radio brightness maximum displays its maximum displacement relative to the centre of the disk in the east–west direction.

The elliptical nature of the isophots at 3·2 cm (Fig. 109) indicates that the Moon's radio temperature decreases as we move from the equatorial to the polar regions. This feature also obtains at the shorter wavelengths (Fig. 108) although in the latter case deviations of the radio isophots from the right form become noticeable.

A comparison of the isophots at $\lambda = 4·3$ mm (Fig. 108a) with a photograph of the lunar disk gives certain indications that the dark plains ("maria") have a slightly higher radio temperature than the light-coloured hilly areas ("continents"). This difference becomes less noticeable with large phases corresponding to longer illumination of the "maria" and "continents" by the Sun (Fig. 108b). In the case of Fig. 108c the Mare Tranquillitatis in the part of the disk in shadow has a lower radio temperature than the surrounding regions, whils the "maria" illuminated by the Sun preserve enhanced T_{eff} values. It may be concluded from what has been said that the sub-surface layers of the lunar "maria" are heated by the Sun during the day and are cooled at night more rapidly than the "continents". This effect is apparently connected with the difference in the absorptive and emissive capacities of the dark and light-coloured regions or with the difference in their thermal conductivity.

The results of Coates (1961), according to which the distribution of the radio brightness over the lunar disk is determined not only by the conditions of its illumination by the Sun's rays but also by the visible features of the lunar surface, are not unexpected. There is little probability that the electrical and thermal properties of the upper layers of the lunar surface are the same everywhere; this is obviously also reflected in the form of the radio isophots in the millimetric band (and also at a wavelength of $\lambda = 2$ cm). The more regular nature of the radio isophots at $\lambda = 3·2$ cm indicates that the distribution of the parameters of the lunar rocks in the layers from which the 3-cm band radio emission comes is more uniform than in the layers above that are responsible for the shorter wavelength emission.

The reservation must be made, however, that the two-dimensional distributions at the wavelengths of 4·3 mm and 3·2 cm are a very rough approximation of the truth since they were obtained with aerials of comparatively poor directivity. Therefore the conclusions drawn above about the connection of the radio emission in the millimetric band with the structural features of the lunar surface and the absence of any connection at wavelengths around 3 cm are uncertain and need further checking with aerials having higher resolution.

Such a check in the millimetric band (at 4 and 8 mm) has been made by Kislyakov, Losovskii and Salomonovich (1963). An analysis of meridian recordings of the Moon's radio emission obtained with a resolution of 2–3' made it possible to find a systematic difference in the effective temperatures of the "maria" and the "continents", which can also be found in infra-red observations (Geoffrion, Korner and Sinton, 1961). At the same time recordings of the T_{eff} distributions along the lunar equator showed no connection between the radio emission and optical details, probably because the equatorial part of the lunar disk is poor in optical contrasts (the greater part of the equatorial belt is filled with "maria") (Kislyakov, Losovskii and Salomonovich, 1963; Kislyakov and Salomonovich, 1963).

In conclusion we note that, as follows from the theory of the Moon's radio emission, the latter should be linearly polarized with a degree of polarization of up to several per cent. This polarization has actually been observed recently at a wavelength of $\lambda = 3·2$ cm in the process of scanning the lunar disk with an aerial having a knife-edge beam (Soboleva, 1962).

CHAPTER VI

Propagation of Electromagnetic Waves in the Solar Corona

WITH this chapter we shall start a systematic exposition of the theory of the radio emission of the Sun and the planets. The central problem here is the question of the origin of the radio emission, i.e. the question of the actual generation "mechanisms" and the explanation of their action. At the same time a very significant part is played by the problem of the propagation of electromagnetic waves in the atmospheres of the Sun and the planets and the variation in the radio emission characteristics connected with this process.

Since matter is strongly ionized in the emission sources (the solar corona, Jupiter's ionosphere, etc.), i.e. is in a plasma state, it is quite natural that present-day ideas on the origin and propagation of the radio emission of the Sun and planets is essentially based on plasma physics and above all on the theory of the generation, propagation and absorption of electromagnetic waves in this medium. The very extensive range of problems in this sphere have not yet been reflected in monographs and are scattered through numerous articles in journals. Here we have in mind chiefly such questions as the escape of the radio emission from the plasma, the scattering of plasma waves, the generation of electromagnetic waves by individual particles and by the particles collectively in an equilibrium and a non-equilibrium plasma. It must be pointed out that under cosmic conditions all these problems (just like the propagation of waves in a uniform and a regularly non-uniform equilibrium plasma as treated by Ginzburg (1960)) have their own specific features determined by the way the problems are stated and the magnitude of the characteristic parameters. This is why it is best to precede Chapters VIII–X, which discuss the possible mechanisms and explain the properties of the different components of the radio emission of the Sun and the planets, by two chapters devoted to a detailed study of the generation, propagation, amplification and absorption of electromagnetic waves in the cosmic plasma. Most attention is paid here to the corona as being the main source of the solar radio emission (which justifies the title of these chapters); however, many results are of more general interest and will be used when discussing the theory of planetary radio emission.

22. Propagation of Electromagnetic Waves in an Isotropic Coronal Plasma (Approximation of Geometrical Optics)

QUASI-HYDRODYNAMIC METHOD AND APPROXIMATION OF GEOMETRICAL OPTICS

When studying electromagnetic waves in a plasma we proceed from the Maxwell equations for the electrical (E) and magnetic (H) fields:

$$\left.\begin{array}{l} \text{curl } H = \dfrac{4\pi}{c}j + \dfrac{1}{c}\dfrac{\partial E}{\partial t}, \quad \text{curl } E = -\dfrac{1}{c}\dfrac{\partial H}{\partial t}, \\ \text{div } E = 4\pi\varrho, \quad\quad\quad\quad\quad \text{div } H = 0. \end{array}\right\} \quad (22.1)$$

(c is the velocity of light in a vacuum). The electromagnetic properties of a plasma depend on the nature of the connection of the density of the total current j and the density of the electric charge ϱ with the fields E and H. The completeness and accuracy of the description of the electrical processes in the plasma depend to a considerable extent on how accurately this connection is determined. If we limit ourselves for the meantime to studying an equilibrium[†] plasma, then, with a few exceptions which will be specially noted in sections 26 and 27, the connection between the current j and the charge ϱ, on the one hand, and the fields E and H, on the other, can be obtained with sufficient accuracy for our purposes by means of the so-called "quasi-hydrodynamic method", which is based on the concept of the electron and ion components of the plasma in the form of "fluids" that penetrate each other and interact with each other via the electromagnetic fields E and H.[‡]

Within the framework of this method the electron and ion components are separately subject to the equations of hydrodynamics—the Euler equation and the continuity equation. For the electron "fluid" they can be written as follows:

$$\left.\begin{array}{l} \dfrac{\partial V}{\partial t} + (V\nabla)V = -\dfrac{\nabla p}{mN} - \dfrac{e}{m}\left(E + \dfrac{1}{c}[V, H]\right), \\ \dfrac{\partial N}{\partial t} + \text{div}(NV) = 0, \end{array}\right\} \quad (22.2)$$

[†] In the sense of the velocity distribution of the plasma particles in the absense of electromagnetic waves.
[‡] For this method see also Ginzburg (1960b, section 13).

§ 22] Waves in an Isotropic Plasma

where V is the velocity of the electron component, N is the concentration of the electrons in the plasma, e and m are the absolute values of the charge and mass of the electron, $p = N\varkappa T$ is the electron pressure ($\varkappa$ is the Boltzmann constant and T the kinetic temperature).[†] The quantities $\boldsymbol{j}$ and ϱ which figure in the Maxwell equations (22.1) can be expressed in terms of the concentrations N, N_i and velocities $\boldsymbol{V}$, $\boldsymbol{V}_i$ of the electrons and ions as follows:[‡]

$$\boldsymbol{j} = -eN\boldsymbol{V} + eN_i\boldsymbol{V}_i, \quad \varrho = -eN + eN_i. \tag{22.3}$$

We note that in the right-hand side of the equations (22.2) another term $\nu_{\text{eff}}(\boldsymbol{V}_i - \boldsymbol{V})$ is generally added, where ν_{eff} is the effective frequency of collisions of electrons with ions. The introduction of the electron pressure p and the number of collisions ν_{eff} allows for the presence of thermal motion in the plasma, although not entirely logically. In sections 22–25 the effect of collisions on the process of electromagnetic wave propagation is for the most part neglected, with the exception of a few cases for which special reservations are made; the part played by collisions in the absorption and emission of electromagnetic waves will be discussed in section 26.

Equations similar to (22.2) also hold for the ions. We shall not start to write them out since the motion of heavy ions under the action of rapidly varying (high-frequency) fields is generally insignificant when compared with the motion of the light electrons. It follows from this that when studying the propagation of high-frequency electromagnetic waves in a plasma we can assume that the motion of the ions is given and independent of the fields of these waves:

$$N_i = N_i(\boldsymbol{R}), \quad \boldsymbol{V}_i = \boldsymbol{V}_i(\boldsymbol{R})$$

(here $\boldsymbol{R}$ is a radius vector). The necessity of equations for the ions is thus eliminated although they are, of course, necessary for finding the stationary distribution of the ion concentrations $N_i(\boldsymbol{R})$ and the velocities $\boldsymbol{V}_i(\boldsymbol{R})$. It is generally also assumed that the kinetic temperature has no rapidly varying

---

[†] The relation $p = N\varkappa T$ is valid only for ideal gases, among which we can include the electron component, and all the plasma as a whole if the mean energy of interaction between the particles $e^2/r \sim e^2 N^{1/3}$ is small when compared with the mean kinetic energy of these particles $\mathscr{E}_{\text{kin}} \sim \varkappa T$ (here r is the average distance between the closest particles): $N^{-1/3}T \gg e^2/\varkappa \sim 10^{-3}$ degree cm. This inequality is well satisfied under cosmic conditions.

At the same time the equations of motion written in the form (22.2) are valid only in the case of non-relativistic velocities, when $V^2/c^2 \ll 1$, $V_{\text{th}}^2/c^2 \ll 1$ ($V_{\text{th}} = \sqrt{\varkappa T/m}$ is the thermal velocity of the plasma electrons).

[‡] It is assumed in (22.3) that the ions have a charge $+e$. The effect of multiply ionized atoms and negative ions under cosmic conditions, as a rule, can be neglected because of the low concentration of the latter when compared with particles that have been ionized once.

part, i.e. the propagation of electromagnetic waves in a plasma is an isothermal process.†

In the absence of rapidly varying processes, when the equations (22.1), (22.2) describe only stationary or quasi-stationary phenomena, considerable deviations of the electron concentration N from the corresponding value of N_i for the ions are impossible. The point is that thanks to the high mobility of the electrons the appearance in the cosmic plasma of Coulomb fields connected with the charge $\varrho = -e(N-N_i)$ causes displacement of the electrons which reduces the difference $N-N_i$ and compensates for the charge which appears. This kind of compensation occurs at distances of the order of the so-called Debye radius $D = (\varkappa T/4\pi e^2 N)^{1/2}$; therefore in the case that the ion concentration $N_i(R)$ varies only a little over the range D the "quasi-neutrality" condition $N \approx N_i(R)$ is found in the plasma. In the cosmic plasma the latter condition is satisfied with great accuracy since the value of D is very small here: for example in the solar corona it is of the order of only 1 cm. As well as the "quasi-neutrality" condition, which means that the charge density $\varrho \ll eN_i$, in the cosmic plasma because of its high self-induction and mobility the condition of equality of the electron and ion fluxes $NV \approx N_i V_i$ is also well satisfied. We notice, by the way, that the approximate relation $V \approx V_i$ follows from this and from the condition $N \approx N_i$.

Rapidly varying wave processes disturb these relations; however, the time- or space-averaged values of N and V (we shall denote them by N_0 and V_0) remain as before close to the values of N_i and V_i:

$$N_0 \approx N_i, \quad V_0 \approx V_i. \tag{22.4}$$

Here the precise values of N_0, V_0 and the stationary values of the electric (E_0) and magnetic (H_0) fields can be found from the Maxwell equations (22.1) and the hydrodynamic equations (22.2), which take the following form:

$$\left.\begin{aligned}(V_0 \nabla) V_0 &= -\frac{\varkappa \nabla (TN_0)}{mN_0} - \frac{e}{m}\left(E_0 + \frac{1}{c}[V_0, H_0]\right) - \nu_{\text{eff}}(V_0 - V_i),\\ \text{div}\,(N_0 V_0) &= 0.\end{aligned}\right\} \tag{22.5}$$

† The thermal motion in the plasma can be properly allowed for only within the framework of the kinetic equation, which, unlike (22.2), allows for the existence of a whole spectrum of particle velocities in each element of the volume. The effects that appear with this approach to problems of wave propagation in a plasma will be discussed in sections 26 and 27. Here all we shall say is that in the cases when the quasi-hydrodynamic method is adequate for the problem under study qualitatively it does not lead to strict quantitative results, allowing for the effect of thermal motion with an accuracy up to coefficients of the order of unity. At the same time this approach is quite legitimate where thermal motion can generally be neglected. The extensive application of the quasi-hydrodynamic method is justified by its simplicity when compared with the kinetic equation method.

§ 22] Waves in an Isotropic Plasma

The stationary magnetic field H_0 that figures in (22.5) plays a very important part in the propagation of waves in a plasma, since in the presence of this field the latter sharply alters its properties, becoming a magnetoactive medium (see sections 23–27).

The equations (22.2) and the first of the relations (22.3) are non-linear in the variables H, N and V. This circumstance considerably complicates the study of electromagnetic processes in a plasma. In practice (with a few exceptions which will be discussed in sections 25 and 27), however, the high-frequency fields $E' = E - E_0$, $H' = H - H_0$ are sufficiently small, as are the deviations V', N' connected with them of the velocity V and the concentration N from their stationary values V_0, N_0. Thanks to this relations (22.2), (22.3) can be linearized by neglecting in them the terms that are quadratic in H', N' and V'. Remembering what has been said and provided that (22.1)–(22.3) are satisfied in the absence of high-frequency perturbations, we obtain for the latter:

$$\operatorname{curl} H' = \frac{4\pi}{c} j' + \frac{1}{c}\frac{\partial E'}{\partial t}, \quad \operatorname{curl} E' = -\frac{1}{c}\frac{\partial H'}{\partial t}, \\ \operatorname{div} E' = 4\pi\varrho', \quad \operatorname{div} H' = 0. \tag{22.6}$$

$$\frac{\partial V'}{\partial t} + (V'\nabla)V_0 + (V_0\nabla)V' = -\frac{\varkappa}{m}\left(\frac{\nabla(TN')}{N_0} - \frac{\nabla(TN_0)}{N_0^2}N'\right) \\ -\frac{e}{m}\left(E' + \frac{1}{c}[V' H_0] + \frac{1}{c}[V_0 H']\right) - \nu_{\text{eff}} V'. \tag{22.7}$$

$$\left.\begin{array}{l}\dfrac{\partial N'}{\partial t} + \operatorname{div}(N' V_0 + N_0 V') = 0, \\ j' = -eN_0 V' - eN' V_0, \quad \varrho' = -eN'.\end{array}\right\} \tag{22.8}$$

The system of linear equations (22.6)–(22.8) defines the distribution of the weak high-frequency electromagnetic perturbations in a non-uniform moving plasma. It holds until the quasi-hydrodynamic approach is applicable to a plasma at each point in which there is local thermodynamic equilibrium when there are no perturbations (as regards the velocity distribution of the particles).

Sections 22–26 discuss the propagation of waves in a motionless plasma ($V_0 = V_i = 0$) with a uniform temperature distribution ($T = \text{const}$). This significantly simplifies the equations (22.7), (22.8) which become:†

---

† The term $(\nabla N_0/N_0^2)N'$ is sometimes not taken into consideration in the first of the equations (22.9); generally speaking, this is not legitimate (for this see Zheleznyakov and Zlotnik (1963).

$$\frac{\partial V'}{\partial t} = -\frac{\varkappa T}{m}\left(\frac{\nabla N'}{N_0} - \frac{\nabla N_0}{N_0^2}N'\right) - \frac{e}{m}\left(E' + \frac{1}{c}[V'\,H_0]\right) - v_{\text{eff}}V',$$

$$\frac{\partial N'}{\partial t} + \text{div}\,(N_0\,V') = 0,$$

$$j' = -eN_0 V', \quad \varrho' = -eN'. \qquad (22.9)$$

In future we shall omit the primes when writing the high-frequency fields E', H' and the currents j' and charge densities ϱ' connected with them.

In a uniform plasma, where $N_0 = \text{const}$, $T = \text{const}$, $H_0 = \text{const}$, $V_0 = \text{const}$, the solution of the system (22.6)–(22.8) can be found in the form of plane monochromatic waves in which the electromagnetic field varies as

$$E = E_\alpha e^{i\omega t - ikr}, \quad H = H_\alpha e^{i\omega t - ikr} \qquad (22.10)$$

(E_a and H_a are amplitudes that are constant in time and space, ω and k are the frequency and wave vector). Substituting these expressions for E, H (and also the similar functions for V', N') in (22.6)–(22.8) we obtain a homogeneous system of linear algebraic equations for the amplitudes E_a, H_a, V'_a, N'_a. The ratios of the components of the vector E_a along the coordinate axes defined by this system characterize the polarization properties of the wave. Further, from the requirement of non-triviality for the solution of the homogeneous equations we can find the so-called dispersion equation which connects the frequency of the wave with its wave vector:

$$\varphi(\omega, k) = 0. \qquad (22.11)$$

The dispersion equation (22.11) can be written in a more concrete form if we introduce the dielectric permittivity tensor $\varepsilon_{il}(\omega, k)$, which characterizes the linear connection of the electric induction $D \equiv E - i4\pi j/\omega$ with the electric field E in a monochromatic wave:

$$D_i = \varepsilon_{il}(\omega, k) E_l \qquad (22.12)$$

(D_i and E_l are components of the vectors D and E along the axes of a cartesian system of coordinates). The tensor ε_{il} can be obtained by finding the connection between the current density j and the field E by means of the equations (22.7), (22.8). Then the system of homogeneous algebraic equations mentioned above becomes

$$\left(k^2 \delta_{il} - k^2 \cos \alpha_i \cos \alpha_l - \frac{\varepsilon_{il}\omega^2}{c^2}\right) E_{al} = 0 \qquad (i, l = 1, 2, 3) \quad (22.13)$$

(α_i, α_l are the angles subtended by the vector k with the coordinate axes i and l; E_{al} is the component of the amplitude E_a along the l axis). For the

§ 22] Waves in an Isotropic Plasma

system (22.13) to have a non-trivial solution (i.e. for E_{al} to be non-zero even with a single value of l) it is necessary to require that the determinant

$$\det \left(k^2 \delta_{il} - k^2 \cos \alpha_i \cos \alpha_l - \frac{\varepsilon_{il} \omega^2}{c^2} \right) \tag{22.14}$$

should become zero. Here det denotes a third-order determinant whose element standing at the intersection of line i and column l is equal to the expression contained in the brackets; δ_{il} is Kroneker's symbol ($\delta_{il} = 1$ if $i = l$, and $\delta_{il} = 0$ if $i \neq l$) (Silin and Rukhadze, 1961, section 6).

The dispersion equation gives the wave number k or the refractive index $n = ck/\omega$ as a function of the frequency ω for a wave being propagated in a direction k. If the functions $k(\omega)$ (or $n(\omega)$) are ambiguous (i.e. with a given frequency ω the dispersion equation has several roots k_j), several waves may be propagated in the plasma, each with its own refractive index $n_j(\omega)$ (the index of j is the "number" of this wave). It should be noted that since the linear system (22.6)–(22.8) is satisfied not only by the functions (22.10) with appropriate values of $k_j(\omega)$ but also by any linear combinations of these functions, in a uniform plasma each weak wave is propagated independently of the other waves. An arbitrary electromagnetic field of a frequency ω in a plasma (with certain limitations on its behaviour at infinity) can always be represented in the form of a superposition of waves (22.10); therefore each wave (22.10) corresponding to one of the solutions of the dispersion equation is called a "normal" wave.

Unless the medium is uniform the functions (22.10) cannot satisfy the system of equations (22.6)–(22.8) and the problem of finding its precise solutions is considerably complicated. However in the case when the properties of the plasma change only a little in a wavelength the system (22.6)–(22.8) can be solved in the so-called geometrical optics approximation. This approximation is based on the circumstance that with a slow variation in space of the properties of the medium the propagation of waves in each sufficiently small region occurs in just about the same way as in a uniform medium with parameters close to the parameters of this region. Therefore the solution in the approximation of geometrical optics should basically recall (22.10); however it must be borne in mind that when R changes the value of k_j alters, so the phase shift will be $\int k_j \, dR$ and not simply $k_j R$:

$$E = E_a(R) \, e^{i\omega t - i \int k_j(R) \, dR}, \quad H = H_a(R) \, e^{i\omega t - i \int k_j(R) \, dR}. \tag{22.15}$$

In addition, in the approximation in question the amplitude of the wave ceases to be constant; this is also taken into consideration when writing (22.15).

It should be stressed that the vector k_j and the ratios of the amplitude components E_a (and H_a) in (22.15) can be found respectively from the dispersion equation (22.14) and the system of algebraic equations (22.13) for a uniform plasma whose properties coincide with those of a non-uniform medium at a given point R. The necessity for special treatment of the equations (22.6)–(22.8) in a non-uniform medium arises here only when finding the dependence of the magnitude of the wave amplitude on the radius vector R. It is clear from what has been said that in the approximation of geometrical optics the values of the refractive index, the number of possible waves in the plasma and the nature of their polarization remain the same as in a uniform medium.

Unless there are regions in a non-uniform plasma in which the geometrical optics approximation would be inapplicable, or if there are such regions but the waves do not reach them, then the solutions of (22.15) with different k_j will be independent. The arbitrary electromagnetic field can then be approximately represented in the form of a combination of solutions of (22.15) with different values of k_j and amplitudes E_a, H_a. This lets us conventionally call the geometrical optics solutions in a smoothly non-uniform medium "normal" waves.

On the other hand, in the case when the waves in the course of propagation pass through a region where geometrical optics are invalid (where the expressions for the fields E, H differ essentially from (22.15) and from any linear combination of these functions with identical or different values of k_j), the geometrical optics solutions of the (22.15) type no longer remain independent outside this region: their amplitudes will be connected by definite relations. This phenomenon bears the name of the coupling of "normal" waves. This kind of coupling is obviously not caused by a violation of the principle of superposition for fields in a plasma, i.e. by the non-linear nature of the electromagnetic waves: this effect is caused by the fact that in certain regions of the plasma an expansion of the field into "normal" waves is inapplicable, if by them we understand solutions written in the approximation of geometrical optics. Due to the phenomenon of interaction an arbitrary field can be made up only of precise solutions of the equations (22.6)–(22.8) which describe the electromagnetic waves in the whole plasma including the region where geometrical optics are inapplicable;[†] a long way from these regions the accurate solutions can be represented asymptotically in the form of linear combinations of the functions (22.15) with a definite connection between their amplitudes, i.e. in the form of "coupled" waves of the geometric–optical approximation.

[†] Only these accurate solutions now play the part of the true normal waves.

§ 22] Waves in an Isotropic Plasma

A well-known case of interaction is the effect of electromagnetic wave reflection: here the connection between the "normal" waves (direct and reflected) is provided by simultaneous violation of geometrical optics for both waves in the layer called the reflection region. The more complex effects of the coupling of different types of wave, which are of great importance to the theory of solar radio emission, will be discussed later (see sections 24, 25). In sections 22 and 23 we shall limit ourselves to discussing the features of wave propagation, taking the approximation of geometrical optics as a basis. In a smoothly non-uniform medium this has very extensive applicability, being valid everywhere with the exception only of certain limited regions of comparatively small extent.

The present section goes on to study wave propagation in an isotropic plasma, i.e. under conditions when H_0 can be taken as equal to zero. The constant magnetic field is allowed for in section 23.

WAVES IN AN ISOTROPIC PLASMA

If the effect of the constant magnetic field H_0 can be neglected,† then in a uniform motionless plasma the dispersion equation (22.13) splits into two:

$$\omega^2 = \omega_L^2 + c^2 k^2, \quad \omega^2 = \omega_L^2 + V_{th}^2 k^2. \tag{22.16}$$

Here $\omega_L = (4\pi e^2 N_0/m)^{1/2}$ is the so-called characteristic (Langmuir) frequency of the plasma oscillations, $V_{th} = \sqrt{\varkappa T/m}$ is the thermal velocity of the electrons. The connection between the frequency ω and the wave number $k = |k|$ defined by these equations does not depend on the direction of the wave propagation (the orientation of the vector k), i.e. a motionless plasma with $H_0 = 0$ is an isotropic medium.

The first of the relations (22.16) describes the refractive index of transverse electromagnetic waves:

$$n_{1,2}^2 = \varepsilon \equiv 1-v, \tag{22.17}$$

where ε is a quantity the same as the dielectric constant of an isotropic plasma in the absence of thermal motion, and the parameter v, which we shall be making use of often, is equal to ω_L^2/ω^2. The phase velocity of the transverse waves is $V_{ph} \equiv \omega/k = c/\sqrt{\varepsilon} \geqslant c$; the group velocity is $V_{gr} \equiv d\omega/dk = c\sqrt{\varepsilon} \leqslant c$; the wavelength is $\lambda = 2\pi c/\omega\sqrt{\varepsilon}$.

In these waves the vectors of the electrical (E) and magnetic (H) fields are located in a plane orthogonal to k. In the rest, their polarization

---

† In the majority of cases the effect of the field H_0 on the propagation of the waves can be neglected if the frequency ω is much greater than the electron gyro-frequency $\omega_H = eH_0/mc$. However, for certain effects, for example the Faraday effect, the condition $\omega \gg \omega_H$ becomes insufficient.

remains arbitrary, so that in the expansion of the given electromagnetic field any two waves of the (22.10) type with opposite polarizations[†] can be taken as the normal ones. If the case of an isotropic plasma obtains as the result of a transition to the limit as the magnetic field H_0 applied to the plasma approaches zero, the polarization of the transverse waves will be quite definite (see (23.10)).

The mean energy flux in a transverse electromagnetic field with an electrical field amplitude E_a is given by the expression

$$S_{el} = \frac{c\sqrt{\varepsilon}}{8\pi}|E_a|^2. \tag{22.18}$$

The second of the relations (22.16) gives the refractive index of a ongitudinal plasma wave

$$n_3^2 = \frac{\varepsilon}{\beta_{th}^2} = \frac{1-v}{\beta_{th}^2}, \quad \beta_{th} = \frac{V_{th}}{c}. \tag{22.19a}$$

In this wave the electrical field E runs along the vector k, i.e. curl $E = 0$. In accordance with (22.6) this implies that the magnetic field H is equal to zero. The phase velocity of the plasma waves is $V_{ph} = V_{th}/\sqrt{\varepsilon} \gg V_{th}$, the group velocity is $V_{gr} = V_{th}\sqrt{\varepsilon} \leqslant V_{th}$ and the wavelength is $\lambda = 2\pi V_{th}/\omega\sqrt{\varepsilon}$. Unlike the transverse waves, a longitudinal wave with a finite value of the refractive index and a non-zero group velocity appears only if the thermal motion is allowed for (when $\beta_{th} \neq 0$).

We notice that the incompleteness of the quasi-hydrodynamic approach affects the expression for n_3^2 and the limits of its applicability. Using the more accurate kinetic method gives (Ginzburg, 1960b, section 8)

$$n_3^2 \approx \frac{1-v}{3\beta_{th}^2} \tag{22.19b}$$

for waves with a phase velocity $V_{ph} \gg V_{th}$ (i.e. with a wavelength $\lambda \gg 2\pi D$, where $D = V_{th}/\omega_L$ is the Debye radius). For plasma waves that do not satisfy the condition $\lambda \gg 2\pi D$ the expression (22.19b) becomes untrue; in addition, these waves are strongly damped (at a distance of the order of λ) even without allowing for collisions of electrons with ions (see section 26). The latter circumstance is not found with the quasi-hydrodynamic treatment.

The mean energy density in a plasma wave (in the quasi-hydrodynamic approximation) is

$$W_{pl} = \frac{E_a^2}{8\pi(1-\varepsilon)} \approx \frac{E_a^2}{8\pi}, \tag{22.20}$$

[†] These waves are also noted in (22.17) by the suffices $j = 1, 2$. For the concept of waves with opposite polarization see section 6.

§ 22] Waves in an Isotropic Plasma

and the mean energy flux is

$$S_{pl} = W_{pl} V_{gr} = \frac{V_{th}\sqrt{\varepsilon}}{8\pi(1-\varepsilon)} E_a^2 \approx \frac{V_{th}\sqrt{\varepsilon}}{8\pi} E_a^2. \tag{22.21}$$

The last equalities in (22.20), (22.21) are valid if $\omega \approx \omega_L$, i.e. if $\lambda \gg 2\pi D$; only in this case can one actually use (with the reservations made above) the quasi-hydrodynamic method.

According to (22.17), (22.19) the refractive index of a plasma wave, all other things being equal, is $\sim \beta_{th}^{-1}$ times greater than the refractive index of transverse waves; accordingly the length of the plasma wave is as many times less than the length of the transverse waves. Under the conditions of

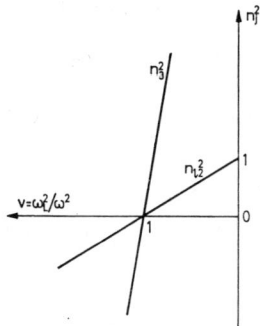

FIG. 110. Square of refractive index n_j^2 as a function of the parameter $v = \omega_L^2/\omega^2$ in an isotropic plasma

the solar corona, where $T \sim 10^6$ °K, $V_{th} \sim 4\times 10^8$ cm/sec and $\beta_{th} \sim 10^{-2}$, the length of a plasma wave is two orders less than the corresponding value for an electromagnetic wave.

The dependence of the refractive indices n_j on v in an isotropic plasma is shown in Fig. 110. If the plasma is non-uniform and the electron concentration depends linearly only on the single coordinate z,† then the graphs shown in the figure reflect (with a corresponding change of scale along the abscissa) the function $n_j^2(z)$.

In a smoothly changing plane-layered medium geometrical optics are applicable everywhere for waves propagated along the z-axis, with the exception of the vicinity of the point z where $v = 1$ (i.e. $\varepsilon = 0$ and $n_j^2 = 0$). The latter does not appear unexpected since as we approach the $n_j^2 = 0$ level the wavelength steadily rises, so the relative change in the properties of the medium at a distance of the order of $\lambda/2\pi$ becomes greater and greater; the relative smallness of this kind of change is a condition of the

† A medium whose properties depend only on a single cartesian coordinate is called plane-layered.

applicability of the geometrical optics approximation

$$\frac{d\varepsilon}{dz}\frac{\lambda}{2\pi} \ll \varepsilon. \qquad (22.22)$$

By approximating the function $\varepsilon(z)$ in the $\varepsilon \sim 0$ layer by the linear function $z \, \text{grad } \varepsilon$ we can reduce this inequality to the form

$$|z| \gg \left(\frac{c}{\omega\sqrt{\text{grad } \varepsilon}}\right)^{2/3} \qquad (22.23)$$

for transverse and

$$|z| \gg \left(\frac{V_{\text{th}}}{\omega\sqrt{\text{grad } \varepsilon}}\right)^{2/3} \qquad (22.24)$$

for longitudinal waves. It follows from this that in the solar corona, where grad $\varepsilon \sim 10^{-10}$ cm^{-1}, geometrical optics at a frequency of $\omega \sim 2\pi \times 10^8$ sec^{-1} is completely violated only in the narrow layer $|z| \lesssim 2 \times 10^4$ cm for electromagnetic and $|z| \lesssim 10^3$ cm for plasma waves.[†]

In layers where (22.23) is satisfied for electromagnetic waves ($j = 1, 2$) the approximation of geometrical optics is of the form (Ginzburg, 1960b, section 16)

$$E = c_1 \frac{a_j}{\sqrt{n_j}} e^{i\omega t - i\frac{\omega}{c}\int n_j \, dz} + c_2 \frac{a_j}{\sqrt{n_j}} e^{i\omega t + i\frac{\omega}{c}\int n_j \, dz}. \qquad (22.25)$$

The factor $1/\sqrt{n_j}$ determines the dependence of the electric field amplitude on the z coordinate and the constants c_1 and c_2 determine the absolute magnitude of the amplitude; the vector a_j characterizes the polarization of the jth wave. For plasma waves ($j = 3$) the expression for the electric field is similar to that given, but with the difference that instead of the factor $1/\sqrt{n_j}$ the factor $\omega_L/\omega\sqrt{n_j}$ figures in it (Zheleznyakov and Zlotnik, 1963).

In the $z < 0$ region, where $n_j^2(z) > 0$ and $n_j(z)$ is the real value, the first term in (22.25) defines a wave travelling in the direction of positive z, and the second term a wave travelling in the opposite direction. In $z > 0$ layers, where $n_j^2(z) < 0$ and the quantity $n_j(z)$ is purely imaginary, the first term describes an exponentially attenuated wave and the second a wave that increases in amplitude exponentially. Violation of the geometric-optical approximation (22.25) in the $n_j^2(z) \approx 0$ layer leads to a coupling of the reflection type between waves being propagated towards positive and

[†] In accordance with the precise solution of the problem of wave propagation in a linear layer the expressions in the right-hand sides of the inequalities (22.23), (22.24) define (with an accuracy of up to a coefficient of the order of unity) the maximum wavelength $\lambda_{\max}$ reached in the plasma.

negative z.† This coupling is manifested in the connection between the coefficients c_1 and c_2 in the regions $z > 0$ and $z < 0$: for example, when a wave with an amplitude c_1 is incident on a linear layer from the interaction region a reflected wave with an amplitude $c_2 = c_1 e^{i\pi/2}$ goes into the $z < 0$ region, whilst only an exponentially attenuated field with an amplitude $c_1 e^{i\pi/2}$ exists in the region $z > 0$.

With inclined propagation of waves in a plane-layered plasma (at an angle to grad ε) the solution of the equations (22.6), (22.29) can be written in the form (Ginzburg, 1960b, section 19)

$$E = F(z) e^{-ik_x x - ik_y y}.$$

Without limiting the generality this expression can be put in the following form:

$$E = F(z) e^{-ik_x x}, \qquad (22.26)$$

if we rotate the system of coordinates so that the new x-axis runs along the component of the vector k in the plane xy. In (22.26) $k_x = $ const (i.e. it does not depend on the coordinates), and the variation of the field E along the z-axis is fully characterized by the function $F(z)$.

The equations describe the high-frequency electromagnetic fields with a vector E lying in the plane xz (in the plane of propagation) and with a vector E orthogonal to this plane split. This means that waves with these orientations of the electrical vector with inclined propagation in a plane-layered plasma remain completely independent. For electromagnetic waves with a vector E orthogonal to xz the function $F(z)$ is in general similar to the function $E(z)$ discussed above in the case of propagation along grad ε; the difference consists only in that the reflection point (in whose vicinity the oscillatory nature of the field variation is replaced by an exponential one) does not correspond to a value $n_j(z) = 0$ as before, but to a value

$$n_j(z) = \frac{ck_x}{\omega}. \qquad (22.27)$$

For waves with a vector E in the plane of propagation the function $F(z)$ is distinguished by greater singularity. The electromagnetic wave, it is true, is reflected as before from the point $n_{1,2}(z) = ck_x/\omega$. At the same time the approximation of geometrical optics for this wave also becomes invalid at the point $n_3(z) = ck_x/\omega$ where the plasma wave is reflected. Therefore between the electromagnetic and plasma waves in the vicinity

---

† Simultaneous violation of the geometrical optics approximation for two waves in a certain region is a necessary but insufficient condition for the appearance of effective coupling between them. In each individual case the presence of coupling can be established only as the result of special treatment (see sections 24 and 25).

of $n_3(z) = ck_x/\omega$ there appears a coupling accompanied by one type of wave changing into another. This effect will be discussed in detail in section 25.

The ray treatment of wave propagation is very convenient in a smoothly non-uniform medium within the framework of the approximation of geometrical optics. Here a ray is understood as the trajectory of the motion of the "centre of gravity" of a wave packet (a signal limited in space and in time). Within the framework of geometrical optics the direction of the packet's motion is determined by the group velocity vector $V_{gr} = d\omega/dk_j$ and is the same as the time-averaged direction of the energy flux in the wave. It follows from the equations (22.16) that in an isotropic plasma the direction of V_{gr} is also the same as the direction of the wave vector k_j. In its turn the direction of the wave vector k_j in the medium is determined by the condition $k_x = k_j \sin \varphi = \text{const}$ (where φ is the angle between k_j and grad ε) or, what is the same thing, by the condition

$$n_j(z) \sin \varphi = \text{const} \tag{22.28}$$

(the Descartes–Snellius refraction law). The constant in (22.28) will be fixed if we state the angle φ_0 between k_j and grad ε at a certain level $z = z_0$:

$$n_j(z) \sin \varphi = n_j(z_0) \sin \varphi_0. \tag{22.28a}$$

Often the z_0 layer chosen is the origin of the plasma layer $\varepsilon = 1$ where the change from a plasma to a vacuum occurs. The angle φ_0 in this region is called the angle of incidence of the wave on the layer; the value of $n_j(z_0)$ here is equal to unity for electromagnetic waves and $1/\beta_{th}$ for plasma waves.†

The behaviour of a ray with inclined incidence of the wave on a plane-layered plasma can be judged from the schematic Fig. 111a. The apex of the ray corresponding to the point of reflection z_{refl} can be determined from (22.28a) with the condition $\varphi = \pi/2$:

$$n_j(z_{refl}) = \sin \varphi_0,$$

or, what is the same thing,

$$\varepsilon(z_{refl}) \equiv 1 - \frac{\omega_L^2(z_{refl})}{\omega^2} = \sin^2 \varphi_0. \tag{22.29}$$

The last relation, as can easily be checked, is the equation (22.27) rewritten in different notations.

† The angle of incidence for plasma waves can, of course, be introduced only conventionally since these waves cannot exist in a vacuum, and in regions of the plasma far enough away from $\varepsilon(z) \approx 0$ (i.e. where $\lambda/2\pi \sim D$ and $n_3 \sim 1/\beta_{th}$) they are already sharply damped at a distance of the order of $\lambda/2\pi$, which makes propagation of the waves impossible in practice.

§ 22] Waves in an Isotropic Plasma

If in the plasma there is a source sending out electromagnetic waves in all directions, then because of refraction the radio emission is directional in nature as it leaves the plasma. The range of angles $\varphi_0 \leq \varphi_{max}$ in this case is easy to find if we remember that a ray leaving the plasma at the maximum angle $\varphi_0 = \varphi_{max}$ has an apex at the point $z = z_s$ where the source is located (Fig. 111b). Then in accordance with (22.29)

$$\varphi_{max} = \arcsin \sqrt{\varepsilon(z_s)} = \operatorname{arc\,sec}\left(\frac{f}{f_{Ls}}\right). \tag{22.30}$$

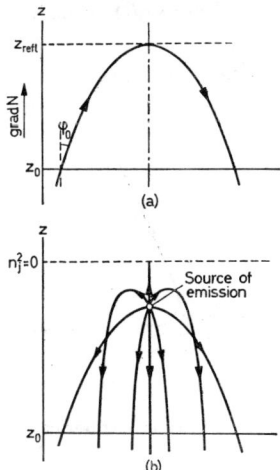

FIG. 111. Wave propagation in a plane-layered plasma with an increasing electron concentration N along the z-axis: (a) trajectory of a ray incident on the plasma at an angle φ_0; (b) emission of radio waves by a point source in a plasma

It follows from this relation that the directivity increases as the distance decreases between the source and the $\varepsilon(\omega) = 0$ level (either because of movement of an emission source of constant frequency into the plasma, or because of a decrease in the frequency generated by a motionless source).

The formula (22.30) makes it possible to judge the magnitude of ε in the generation region from the observed width $2\varphi_{max}$ of the angular distribution of the radio emission. For example, if $2\varphi_{max} \sim 40°$ (which is typical of the enhanced radio emission connected with sunspots), then in the generation region ε will be about 0·1 and $n_{1,2} = \sqrt{\varepsilon}$ will be about 0·3. The effect of the magnetic field on the value of the refractive index may, of course, considerably alter this estimate.

Because of its simplicity the plane-layered model of a non-uniform plasma with a steady variation in its parameters is very convenient for investigating the question of wave propagation in the solar corona as regards principles. In a number of cases, however—when studying the nature of the directional features of the emission, radio absorption along the ray, etc.—a more accurate treatment of the travel of rays in the corona allowing for its sphericity proves necessary.

The configuration of rays in a spherically symmetrical refractive medium is determined by the law of refraction, which is a generalization of the Descartes–Snellius law (22.23) for the plane-layered case (Ginzburg, 1960b, section 36):

$$n_j(R) R \sin \varphi(R) = \text{const.} \qquad (22.31)$$

Here $n_j(R)$ is the refractive index at a point at a distance R from the centre of symmetry (the centre of the Sun), φ is the angle between the direction

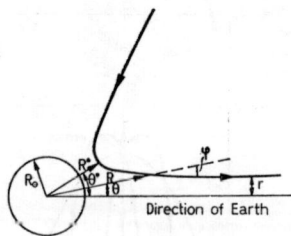

FIG. 112. Calculation of form of ray in a spherically symmetrical solar corona

of the ray and the radius vector $\mathbf{R}$. If we use R_∞ to denote a point on the ray trajectory outside the medium, then the refraction law can be stated as follows:

$$n_j(R) R \sin \varphi(R) = n_j(R_\infty) R_\infty \sin \varphi(R_\infty) = n_j(R_\infty) r, \qquad (22.31\text{a})$$

where r is the "aiming parameter" characterizing the distance from the ray entering the medium to the solar radius running parallel to it. For transverse electromagnetic waves the refractive index outside the corona is $n_j(R_\infty) = 1$; it is necessary to allow for the formal difference of $n_j(R_\infty)$ from unity only when dealing with the refraction of plasma waves. The typical travel of a ray in the corona with an electron concentration that decreases as R increases is shown in Fig. 112.

It is clear from (22.31) and this figure that an element of length along the ray is

$$dl = \frac{dR}{\cos \varphi} = \frac{dR}{\sqrt{1 - \dfrac{r^2}{n_j^2 R^2}}}. \qquad (22.32)$$

§ 22] Waves in an Isotropic Plasma

The ray trajectory in the polar coordinates R, θ is

$$\theta = \theta_\infty + \int_R^{R_\infty} \frac{r\,dR}{R\sqrt{n_j^2 R^2 - r^2}}. \tag{22.33}$$

The quantity θ_∞ is obviously the same as $\varphi(R_\infty)$. The reflection point (the "turning" point), where $\varphi = \pi/2$, $R = R^*$ and $\theta = \theta^*$, is defined by the equality

$$R^* n_j(R^*) = r. \tag{22.34}$$

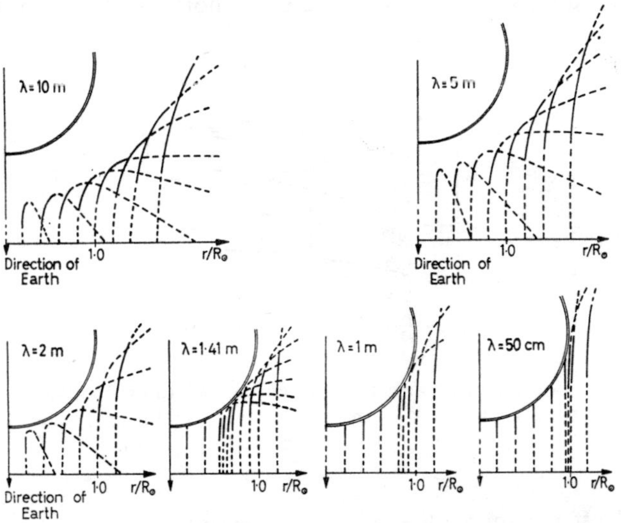

Fig. 113. Configuration of rays in the solar corona for wavelengths of 50 cm to 10 m (Reule, 1952).

The ray trajectory is symmetrical relative to the solar radius passing through the turn point.

Configurations of electromagnetic rays in the solar corona calculated by Reule (1952) from formula (22.33) on the assumption that the electron concentration distribution is subject to the Baumbach–Allen formula (1.1) are shown in Fig. 113. The travel of rays in the corona can also be judged if we know the coordinates R^*, θ^* of the turn points for different values of the parameter r; these coordinates are shown in the graphs of Fig. 114.

It is clear from Figs. 113 and 114 that at a given frequency the ray penetrates deeper into the corona at smaller values of the "aiming parameter" r. In other words, the turning point surface touching the level

$n_j(\omega) = 0$ (i.e. $\omega = \omega_L$) at the centre of the disk rises above this level as the distance r to the centre of the disk increases. Since this surface is the lower limit of the corona region from which radio emission at the given frequency can escape from it in a given direction, emission at the centre of the disk can be observed from sources that are deeper down than on the limb. As the frequency rises the turning-point surface moves into deeper layers of the corona; at the same time the region of the corona which can in principle be responsible for the creation of the observed radio emission expands.

Due to refraction the "position" of the source in the corona as seen in radio rays differs from the true one by a shorter distance to the centre

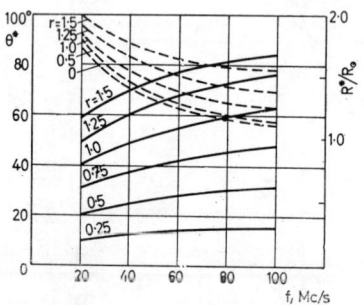

Fig. 114. Coordinates of turn point in corona R^* (dotted curves) and θ^* (solid curves) as a function of the frequency f for different values of the parameter r (Jaeger and Westfold, 1950).

of the disk. If the source lies on the turning-point surface, for a Baumbach–Allen corona the direction in which the radio emission is observed intersects the Sun's radius passing through the source near the $n_j(\omega) = 0$ level (Shain and Higgins, 1959).

If a source in the plasma generates radio emission of a pulsed nature (of the burst type) with a broad frequency spectrum, then because the group velocity is not the same at different frequencies the times of arrival of signals observed outside the plasma will be different. Since the group velocity of transverse waves is $V_{\text{gr}} = c\sqrt{\varepsilon}$, the group lag time (when compared with a signal propagated in a vacuum) is

$$\Delta t_{\text{gr}} = \frac{1}{c} \int \left(\frac{1}{\sqrt{\varepsilon}} - 1 \right) dl; \qquad (22.35)$$

the integral is taken along a path connecting the source and the point of reception. The maximum lag will obviously occur at a frequency ω for which

§ 22] Waves in an Isotropic Plasma

at the point where the source is located $n_{1,2} = \sqrt{\varepsilon} = 0$†. In the case of a linear function $\varepsilon(R)$ for a ray leaving the corona along a solar radius ($r = 0$) the maximum lag is

$$(\Delta t_{gr})_{max} = \frac{1}{c \text{ grad } \varepsilon}. \tag{22.36}$$

If $r \neq 0$, then it will be even less.

Under the conditions of the inner corona with values of grad $\varepsilon \sim 10^{-10}$ cm the group lag is not more than a second or fractions of a second. Such a small value of Δt_{gr} will not allow us to connect the observed frequency drift of type III bursts (not to mention the slower drift of type II bursts) with this effect (see sections 13 and 14).

We shall now make a few remarks about refraction in the planetary ionospheres.

The travel of rays in a spherically symmetrical model of a planetary ionosphere, generally speaking, is also defined by the relation (22.31). Unlike the solar corona, however, the extent of the ionosphere in height is much less than the radius of its curvature; therefore in (22.31) the change in the factor R when waves are propagated in the ionosphere can in many cases be neglected and a plane-layered model with a refraction law in the form of (22.28) can be used instead of a spherical ionosphere. When studying the propagation of radio waves in the ionosphere we must also allow for the fact that the dependence of the electron concentration on the altitude $N(z)$ is not monotonic: this quantity becomes greatest at a certain altitude z_{max} (at the maximum of the ionospheric layer) and falls as z moves away from z_{max}.

If a wave is incident on the ionospheric layer at an angle φ_0 it is reflected from the $z = z_{refl}$ layer where the plasma frequency is

$$\omega_L(z_{refl}) = \omega \cos \varphi_0 \tag{22.37}$$

(see (22.29)). As the frequency $\omega = 2\pi f$ increases the reflection point moves into the heart of the layer with greater values of $\omega_L = 2\pi f_L$. At a certain value of f equal to the ionospheric critical frequency for inclined incidence

$$f_{cr} = f_{Lmax} \sec \varphi_0 \tag{22.38}$$

---

† That V_{gr} becomes zero at this point does not prevent the escape of radio emission from the $n_{1,2} = 0$ layer since in this layer the approximation of geometrical optics is violated and formula (22.35) becomes invalid. Therefore in the vicinity of $n_{1,2} = 0$ we should, generally speaking, use a more accurate expression for the group lag time than (22.35). However, the contribution to Δt_{gr} from a thin layer with a thickness of the order of a few wavelengths is small when compared with the contribution from the extensive region in which geometrical optics are quite legitimate. This gives us some foundation for stating that the use of the formula (22.36) does not introduce any great error into the estimate of the quantity $(\Delta t_{gr})_{max}$.

it reaches the layer maximum, where the plasma frequency $\omega_{L\max} = 2\pi f_{L\max}$ corresponds to the electron concentration $N_{\max}$. In the region

$$f > f_{cr} = f_{L\max} \sec \varphi_0 \qquad (22.39)$$

reflection disappears and the wave passes freely through the ionospheric layer. At a fixed frequency f the inequality (22.39) defines the range of angles φ at which emission passing through the ionosphere leaves it; here the maximum angle

$$\varphi_{\max} = \text{arc sec}\left(\frac{f}{f_{L\max}}\right) \qquad (22.40)$$

decreases together with the ratio $f/f_{L\max}$, becoming zero when $f/f_{L\max} = 1$ (Ginzburg, 1960b, section 34).

The character of the directivity of an isotropic emitter placed in the ionosphere will differ according to its position in relation to the layer maximum. If the source of the emission is between the $z = z_{\max}$ level and the point of observation, then the maximum value of the angle $\varphi_{\max}$, just as in the case of the monotonic function $N(z)$, will be defined by the formula (22.30). It follows from this that the directivity at a frequency f depends upon the value of the plasma frequency at the point where the source is located, i.e. on the position of the source. If the source is separated from the observation point by a layer $z = z_{\max}$, then the directivity of emission coming from the layer will be defined by the formula (22.40). According to the latter the directivity at a frequency f does not depend in this case on the actual position of the source and is given by the value of the plasma frequency at the layer maximum.

Above we have been discussing the propagation of electromagnetic waves under the conditions of a uniform or a regularly non-uniform plasma, without making allowance for the fact that irregular variations in N are generally superimposed on the regular curve of the electron concentration. This leads to scattering of the electromagnetic waves as they are being propagated in the corona.

The nature of the scattering depends essentially on the scale of the inhomogeneities. Scattering on smooth large-scale inhomogeneities with a characteristic size of $l \gg \lambda$ and $\Delta n_j^2 \ll 1$ can be described in the approximation of geometrical optics everywhere with the exception of plasma layers with a refractive index of $n_j \approx 0$. This means that in this case scattering occurs with conservation of the normal wave type, without any noticeable change of the scattered emission into emission of another type. The whole effect actually comes down to broadening of the angular radio emission

§ 22] Waves in an Isotropic Plasma

spectrum caused by chaotic changes in direction of the radio beams in the corona.† However, as well as comparatively stationary large-scale inhomogeneities there are also non-stationary fluctuations with small values of l (in particular with $l \sim \lambda$) in the corona thanks to the presence of thermal motion.‡ Here the scattering can no longer be treated in the approximation of geometrical optics; as will be shown in section 25, scattering on thermal fluctuations is accompanied by the transformation of waves of one type into another.

The effect of scattering on large-scale inhomogeneities leads, in particular, to a rise in the angular size of discrete sources of radio emission during their eclipses by the solar corona. The scattering effect may in certain cases apparently also increase the observed angular size of local sources of the solar sporadic radio emission. This range of questions is discussed in the theoretical respect by Vitkevich (1956c, 1960b), Chernov (1958), Scheffler (1958), Ginzburg and Pisareva (1956), Pisareva (1958), Vitkevich and Lotova (1961). All we shall say here is that a definite indication of the effective part played by scattering of the radio emission of local sources in the metric band is the comparative consistency and constancy of the sizes of sources of the different components of the sporadic radio emission (generally 5–8′; see Table 4 in section 17). At the same time the recording of R-centres with angular sizes that vary within broad limits (from 3 to 9′; section 12) shows that it is very far from always that scattering occurs. We note that the scattering effect may in principle not only affect the size of local sources of sporadic radio emission; it also affects the radio brightness distribution over the "quiet" Sun's disk observed in the metric band. This question will be discussed in section 28.

Scattering of radio emission in the solar corona, generally speaking, also increases the duration of bursts of solar emission t_0 (Vitkevich, 1960b).

It is true that the effect of increasing the duration of a burst because of scattering in the supercorona is completely insignificant (Pisareva, 1958); however, if we bear in mind the scattering in the corona (at distances of $R < 4.5 R_\odot$), it turns out that under definite conditions (at the solar cycle maximum, at the long-wave end of the metric band or at decametric wavelengths) the contribution of the scattering to the value of t_0 for bursts of short duration may be noticeable.

---

† In essence to use the ray treatment of the scattering the condition $1 \gg \sqrt{b\lambda}$ (b is the distance from the point of observation to the effectively emitting region) must be added to the inequalities given above. This condition means that the size of the first Fresnel zone should be much less than the characteristic size of the inhomogeneities.

‡ The existence under the conditions of the corona of stationary inhomogeneities with sizes of the order of a wavelength is not possible. Suffice it to say that the mean free path of electrons in the corona is over 10^7 cm and in the supercorona over 10^{11} cm.

Broadening of Jupiter's bursts of radio emission because of scattering in the solar corona (at a period of conjunction) will proceed more effectively chiefly because of the great distance of the source of the emission from the scattering screen (Zheleznyakov, 1958a). It is quite probable that this effect may be found by careful comparison of the mean duration of Jovian bursts before and during Jupiter's eclipse by the solar corona.†

In conclusion we note that electron inhomogeneities in the corona may also cause certain fluctuations in the intensity of the solar radio emission at metric wavelengths (Ginzburg and Pisareva, 1956; Pisareva, 1958). In actual fact, even in the case of a stationary corona when the coronal inhomogeneities are fixed in a coordinate system linked with the Sun, the diffraction picture produced on Earth of solar radio emission scattered in the corona will, because of the Sun's rotation around its axis, move relative to an observer at a velocity $V \approx 400$ km/sec. Since the diffraction field of the scattered radio emission is non-uniform this displacement causes fluctuations of intensity at the point of observation. In principle they can be separated from fluctuations of different origin by comparing recordings of the radio emission intensity at two points at different longitudes: the intensity variations in question will be correlated with the time shift $t = L/V$, where L is the distance between the receiving stations. If, for example, $L \sim 4 \times 10^3$ km, then the lag is ~ 10 sec. However, the extreme changeability of the solar radio emission leaves little hope of finding this effect.

23. Propagation of Electromagnetic Waves in a Magnetoactive Coronal Plasma (Approximation of Geometrical Optics)

ELECTROMAGNETIC WAVES IN A HOMOGENEOUS PLASMA IN THE PRESENCE OF A CONSTANT MAGNETIC FIELD

In the presence of a magnetic field H_0 the properties of a plasma alter significantly: the plasma becomes a magnetoactive (anisotropic and gyrotropic) medium. The anisotropy shows itself in the dependence of the normal wave characteristics (refractive indices, polarization vectors) on the direction of propagation, and the gyrotropy in the elliptical polarization of these waves.

The values of the refractive indices $n_j = k_j(c/\omega)$ of normal waves in a magnetoactive plasma are defined by the dispersion equation (Ginzburg,

† Some preliminary data on eclipse observations of Jupiter are given by Shain (1956).

§ 23] Waves in a Magnetoactive Coronal Plasma

1960b, section 12)

$$\beta_{th}^2 (1-u\cos^2\alpha)n^6 - [1-u-v+uv\cos^2\alpha \\ + 2\beta_{th}^2(1-v-u\cos^2\alpha)]n^4 + [2(1-v)^2 \\ - u(2-v-v\cos^2\alpha) + \beta_{th}^2(1-2v+v^2-u\cos^2\alpha)]n^2 \\ + (1-v)[u-(1-v)^2] = 0, \quad (23.1)$$

where α is the angle between the wave vector k and the field H_0,

$$\beta_{th} = \sqrt{\frac{\varkappa T}{mc^2}}, \quad v = \frac{\omega_L^2}{\omega^2}, \quad u = \frac{\omega_H^2}{\omega^2} \quad (23.1a)$$

(ω_L is the plasma frequency, ω_H is the gyro-frequency).

Without taking thermal motion into consideration when $\beta_{th} = 0$, the equation (23.1) defines the refractive indices of the two normal waves that can be propagated in a uniform magnetoactive plasma (an in a non-uniform one in the approximation of geometrical optics):

$$n_{1,2}^2 = 1 - \frac{2v(1-v)}{2(1-v) - u\sin^2\alpha \mp \sqrt{u^2\sin^4\alpha + 4u(1-v)^2\cos^2\alpha}}. \quad (23.2)$$

Here the "plus" sign and the suffix "2" correspond to an ordinary wave, the "minus" sign and the suffix "1" to an extraordinary wave. Graphs of the function $n_{1,2}^2(v)$ in the case of longitudinal ($\alpha = 0$) and transverse ($\alpha = \pi/2$) wave propagation relative to the magnetic field are shown in Figs. 115 and 116. The function $n_{1,2}^2(v)$ with an arbitrary (but small) angle α is shown in Fig. 117.

The refractive indices become zero with $v = v_0$, where†

$$v_0 = 1, \quad v_0 = 1 \pm \sqrt{u}, \quad (23.3)$$

and become infinitely great with values of u, v that satisfy the relation

$$1 - u - v + uv\cos^2\alpha \equiv (1-u)(1-v) - uv\sin^2\alpha = 0, \quad (23.4a)$$

i.e. with $v = v_\infty$, where

$$v_\infty = \frac{1-u}{1-u\cos^2\alpha}. \quad (23.4b)$$

It follows from the figures that the character of the dispersion curves changes significantly on the change from $u < 1$ to $u > 1$. In the last case we must also distinguish two variants: $u\cos^2\alpha < 1$ and $u\cos^2\alpha > 1$. Since the parameter v cannot be negative the functions $n_{1,2}^2(v)$ have no poles in the region $u > 1$, $u\cos^2\alpha < 1$ (see (23.4b)). With small angles α this

† These values of v_0 do not change even when $\beta_{th} \neq 0$ (see (23.1)).

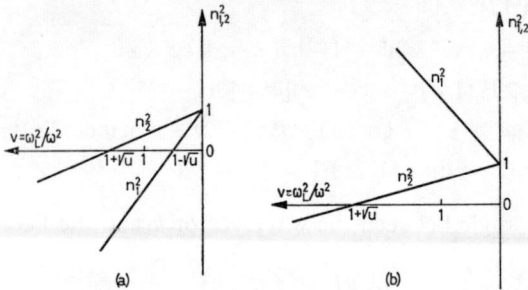

Fig. 115. Function $n_{1,2}^2(v)$ in the case of longitudinal propagation of waves relative to a magnetic field H_0 ($\alpha = 0$): (a) $u = \omega_H^2/\omega^2 < 1$; (b) $u = \omega_H^2/\omega^2 > 1$

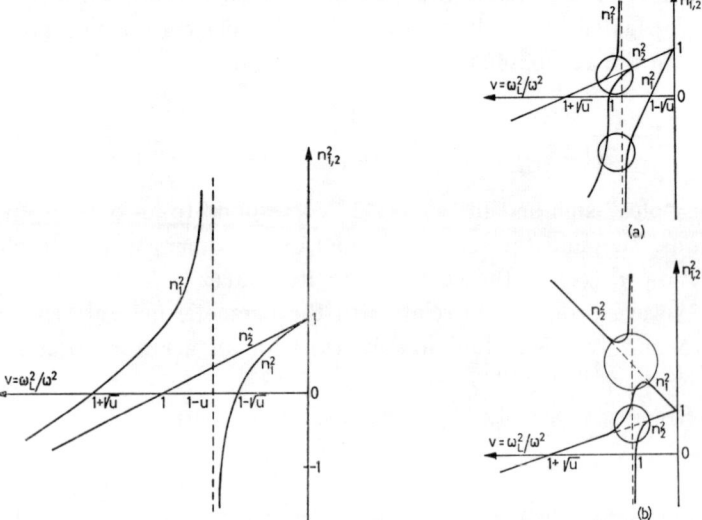

Fig. 116. Function $n_{1,2}^2(v)$ with transverse propagation of waves in a magnetic field H_0 ($\alpha = \pi/2$)

Fig. 117. Function $n_{1,2}^2(v)$ for an angle $\alpha \ll 1$: (a) $u < 1$; (b) $u > 1$ (to be more precise, $u \cos^2 \alpha > 1$). The circles show the regions where the waves interact

region is of little significance since it appears only in a very narrow range of values of the parameter u.

Figures 115–17 describe the variation in the refractive indices as a function of the plasma frequency ω_L for $\omega = $ const and $\omega_H = $ const. The variation of $n_{1,2}^2(v)$ with constant ω_L and ω_H is also of interest; the corresponding graph for one special case is shown in Fig. 118. With fixed values of ω_L and ω_H the square of the refractive index becomes infinity at

§ 23] Waves in a Magnetoactive Coronal Plasma

the two frequencies ω_∞ defined by the equation (23.4a):†

$$\omega_\infty^2 = \frac{\omega_H^2+\omega_L^2}{2} \pm \sqrt{\frac{(\omega_H^2+\omega_L^2)^2}{4} - \omega_H^2 \omega_L^2 \cos^2\alpha}. \qquad (23.4c)$$

The quantity ω_∞^2 is never negative, i.e. the two features of $n_{1,2}^2(\omega)$ remain for any values of the parameters ω_L, ω_H and α.

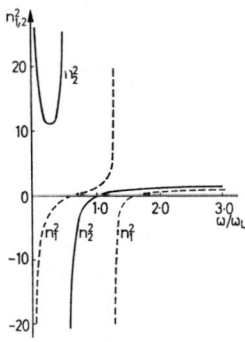

Fig. 118. Dispersion curves of $n_{1,2}^2(\omega)$ for constant ω_L and $\omega_H(\omega_H/\omega_L = 1, \alpha = 45°)$ (Ginzburg, 1960b, section 11)

Fig. 119. Orientation of vectors k and H_0 relative to the coordinate axes

The expression (23.2) for the refractive indices is rather complex. It can be simplified considerably, however, in two extreme cases—with the so-called "quasi-longitudinal" and "quasi-transverse" propagation (Ginzburg, 1960b, section 11) which are realized in a plasma with the conditions

$$\frac{u \sin^4\alpha}{4\cos^2\alpha} \ll (1-v)^2, \quad |1-\sqrt{u}\cos\alpha| \gg \frac{(1+v)u\sin^2\alpha}{2(1-v)^2} \qquad (23.5)$$

and

$$\frac{u\sin^4\alpha}{4\cos^2\alpha} \gg (1-v)^2, \quad \tan^2\alpha \gg (1+v) \qquad (23.6)$$

respectively. In the quasi-longitudinal approximation

$$n_{1,2}^2 = 1 - \frac{v}{1 \mp \sqrt{u}|\cos\alpha|}; \qquad (23.7)$$

in the quasi-transverse approximation

$$n_1^2 = 1 - \frac{v(1-v)}{1-v-u\sin^2\alpha}, \quad n_2^2 = 1-v. \qquad (23.8)$$

† In addition, as seen from (23.2), $n_{1,2}^2 \to \infty$ if $v \to \infty$ (i.e. $\omega \to 0$). In this case, however, we must allow for the motion of the ions, which was not taken into consideration when deriving (23.2).

Ordinary and extraordinary waves in a magnetoactive plasma, generally speaking, are elliptically polarized, the ratio of the electric field components being defined by the following relations:

$$\frac{E_{ay}}{E_{ax}} = iK_{1,2}, \quad \frac{E_{az}}{E_{ax}} = i\Gamma_{1,2}, \tag{23.9}$$

where the polarization coefficients are

$$K_{1,2} = -\frac{2\sqrt{u}\,(1-v)\cos\alpha}{u\sin^2\alpha \pm \sqrt{u^2\sin^4\alpha + 4u(1-v)^2\cos^2\alpha}}, \tag{23.10}$$

$$\Gamma_{1,2} = -\frac{v\sqrt{u}\sin\alpha + K_{1,2}uv\sin\alpha\cos\alpha}{1-u-v+uv\cos^2\alpha}. \tag{23.11}$$

Here E_{ax}, E_{ay}, E_{az} denote the components of the amplitudes $\mathbf{E}_a$ of the normal waves along the x, y, z axes of a cartesian system of coordinates orientated relative to $\mathbf{k}$ and $\mathbf{H}_0$ as shown in Fig. 119; the top sign relates to an extraordinary and the bottom one to an ordinary wave. The quantity $K_{1,2}$ characterizes the ratio of the axes of the ellipse described by the electrical vector in a plane orthogonal to the direction of the wave propagation, i.e. the degree of ellipticity of normal waves in the plasma: $p \equiv \tan\sigma = K_{1,2}$ (see section 6). In its turn the value of $\Gamma_{1,2}$ determines the relative magnitude of the longitudinal (with respect to $\mathbf{k}$) component of the electric field. Since $K_1 K_2 = -1$, the polarization ellipses of ordinary and extraordinary waves are similar, their long axes are at right angles to each other, and the signs of the curl of the vectors $\mathbf{E}$ are different. This means that ordinary and extraordinary waves have opposite polarizations.

With $\alpha = \pi/2$ the polarization is linear, the vector $\mathbf{E}$ in an ordinary wave being orientated along the constant magnetic field $\mathbf{H}_0$, whilst in an extraordinary one it lies in a plane orthogonal to $\mathbf{H}_0$. With $\alpha = 0$ the polarization becomes circular, with clockwise rotation in an extraordinary wave and anticlockwise in an ordinary one (if we look in the direction of propagation). This direction of rotation corresponds to propagation along $\mathbf{H}_0$; it is conserved until the wave vector $\mathbf{k}$ subtends an acute angle with $\mathbf{H}_0$. If the normal waves travel in the opposite direction to $\mathbf{H}_0$ (or the vector $\mathbf{k}$ subtends an obtuse angle with the direction $\mathbf{H}_0$) the signs of the rotation change to the opposite.[†]

With an arbitrary angle $\alpha \neq 0$ or $\pi/2$ the polarization becomes strictly linear only for the value $v = 1$, remaining elliptical with $v \neq 1$. However,

---

[†] In any case the direction of the rotation of the polarization vector in extraordinary waves is the same as, and in ordinary ones opposite to, the direction of rotation of the electron in a magnetic field.

§ 23] Waves in a Magnetoactive Coronal Plasma

the linear nature of the polarization is preserved over the whole range of angles α were there is quasi-transverse propagation, just as in the case of quasi-longitudinal propagation the polarization of ordinary waves is close to circular.

It is significant that ellipticity is also preserved in the change of v to zero, when $\Gamma_{1,2} = 0$ and

$$K_{1,2} = -\frac{2\sqrt{u}\cos\alpha}{u\sin^2\alpha \pm \sqrt{u^2\sin^4\alpha + 4u\cos^2\alpha}}. \quad (23.12)$$

The dependence in this case of the ratio of the polarization axes $K_{1,2}$ on the angle α for different values of the parameter u is shown in Fig. 120.

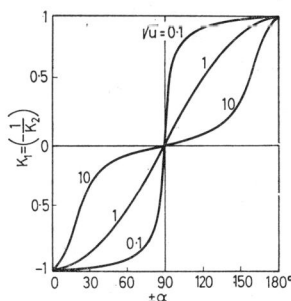

FIG. 120. Polarization coefficient $K_{1,2}$ as a function of the angle α

When allowing for thermal motion of the electrons ($\beta_{th} \neq 0$) in a magnetoactive plasma there are three normal waves corresponding to three dispersion curves $n_j^2(v)$ ($j =$ I, II, III). These curves are shown in Fig. 121 for the case of propagation at a small angle to the field H_0. Waves corresponding to individual branches of the dispersion curves separated by the circled regions are generally called extraordinary (n_1^2), ordinary (n_2^2) and plasma (n_3^2). Therefore the suffices 1, 2, 3 in the figure indicate the name of the normal wave and the numbers I, II, III indicate to which of the dispersion curves the normal wave relates.

In a uniform plasma the regions circled in Fig. 121 are not separated by anything; therefore the division of waves into extraordinary, ordinary and plasma waves is very much a convention. In a non-uniform plasma there is more justification for this, since the approximation of geometrical optics is violated in the above regions where dispersion curves with different numbers approach each other closely. This leads to a coupling of different types of waves belonging to different dispersion curves, and to a "transition" of waves from one curve to another. Here the individual branches of the dispersion curves are physically separated since they are connected with different geometrical optics solutions on different sides of the interaction regions in which these solutions lose their force (see sections 24 and 25).

It is clear from a comparison of the dispersion curves in Figs. 117 and 121 that thermal motion has a noticeable effect only on the nature of $n_j^2(v)$ in the region where $n_j^2(v) \gg 1$, leading to the appearance of plasma waves; with $\beta_{th} = 0$ the plasma branch of the dispersion curves corresponds to the high values of the refractive index near the poles (23.4) of the function $n_{1,2}(v)$. In the remaining regions any change in the nature of the waves' propagation (value of the refractive index, polarization coefficients)

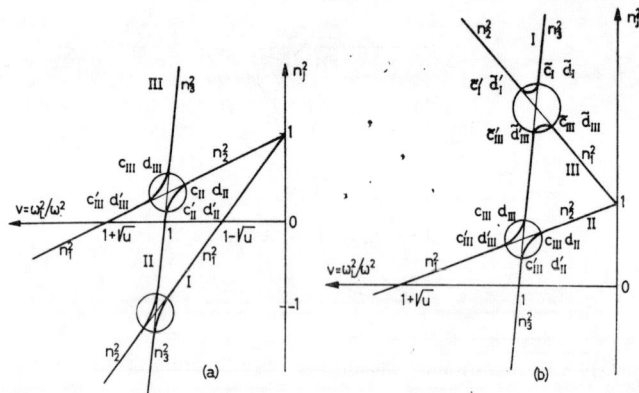

Fig. 121. Dispersion curves $n_j^2(v)$ ($j = $ I, II, III) with allowance made for thermal motion in the plasma ($\alpha \ll 1$): (a) $u < 1$; (b) $u > 1$. The circles show the regions of wave coupling and the thin lines show the function $n_j^2(v)$ in the case of purely longitudinal propagation. The notations c_I, d_I, etc. introduced here will be used in section 25

because of thermal motion is, as a rule, of little significance and in the majority of cases it can be neglected.

We notice that a magnetoactive plasma is analogous to an isotropic one in the respect that in both media three normal waves can be propagated in a given direction that differ in the nature of their polarization. The difference (and it is very significant) is as follows: with $H_0 \neq 0$ the refractive indices of all waves in the general case are different, whilst with $H_0 = 0$ two refractive indices (for transverse electromagnetic waves) are the same. That is why in an isotropic plasma the polarization of transverse normal waves may be any polarization (provided that it is opposite); in a magnetoactive plasma, on the contrary, the polarization is strictly fixed. When the magnetic field approaches zero the refractive indices of the ordinary and extraordinary waves come closer to each other and the latter

§ 23] Waves in a Magnetoactive Coronal Plasma

change into the transverse electromagnetic waves characteristic of an isotropic plasma; in just the same way the plasma waves of a magnetoactive medium change into the plasma waves of an isotropic medium.

Waves in a Non-uniform Magnetoactive Plasma

The dispersion curves in Figs. 117 and 121 provide (with an appropriate change in scale) the function $n_j^2(z)$ in a non-uniform plane-layered plasma if $v = \operatorname{grad} v \cdot z$ and $\operatorname{grad} v = \operatorname{const}$, $u = \operatorname{const}$ (linear layer). If $v(z)$ is a non-linear function but one that rises monotonically together with z, then the general nature of the dispersion curves (the relative position of the zeroes and the poles of n_j^2, the regions of interaction between the different types of waves) are preserved without any change.

Furthermore, in the case when $v(z)$ is a non-monotonic function that reaches its maximum value $v_{\max}$ at a point $z = z_{\max}$ and drops smoothly to zero as z moves away from $z_{\max}$, the behaviour of $n_j^2(z)$ will become more complex. In general it will be similar to that shown in Figs. 117 and 121 when v changes from 0 to $v_{\max}$ (over the range $z < z_{\max}$); the further progress of $n_j^2(z)$ can be judged by making a mirror reflection of the curves $n_j^2(v)$ in the range $0 < v < v_{\max}$ relative to the line $v = v_{\max}$. This case is of interest for planetary ionospheres where the variation in the field H_0 in the thickness of the layer can be neglected.

As in an isotropic plasma, in a non-uniform medium with $H_0 \neq 0$ the reflection of waves being propagated along the concentration gradient (grad ε) occurs in the vicinity of the points where $n_j^2 = 0$ (23.3). With inclined incidence of the wave onto the layer the change in direction of the wave vector k is determined as before by the refraction law (22.28);[†] however, n_j should now be taken from the dispersion equation (23.1). In accordance with (22.28), with incidence of the waves at an angle the reflection points move into a region with lower values of v, where $n_j^2 > 0$.

When analysing the conditions for the propagation of radio waves in the solar corona outside the active regions the effect of the magnetic field need not be taken into consideration (as was done above when discussing the form of rays in a spherically symmetrical corona), since the overall magnetic field of the Sun is small enough ($H_0 \sim 1$ oe). With this kind of field strength $\sqrt{u} = \omega_H/\omega \ll 1$ for frequencies of $f \gg 3$ Mc/s ($\lambda \ll 100$ m) and the refractive indices $n_{1,2}^2(v)$ defined by the dispersion equation (23.1)

[†] In a magnetoactive plasma the direction of the vector k no longer determines the direction of the ray since the group velocity vector $V_{gr} = d\omega/dk$ is orientated here at an angle to the vector k. Therefore the calculations of wave trajectories in a magnetoactive plasma are considerably complicated; the configuration of the rays for certain special cases is given by Ginzburg (1960b, section 29).

(valid only for $\beta_{th} \ll 1$) change in practice into (22.17) and (22.19). However, in active regions of the corona above sunspots the part played by magnetic fields becomes very significant since here the cases $\omega_H/\omega \sim 1$ and even $\omega_H/\omega \gg 1$ may easily be realized. For example, in the metric waveband at frequencies of $f \sim 100$ Mc/s values of $H_0 \gtrsim 30$ oe are sufficient for this.

Any investigation of the propagation of electromagnetic waves in the corona above centres of activity is further complicated by the fact that the electron concentration N (i.e. v) and the magnetic field H_0 (i.e. u) vary from point to point. At the same time when drawing the graphs given in the preceding section it was assumed that $u = $ const; therefore they do not reflect the true behaviour of the refractive index in the regions indicated in

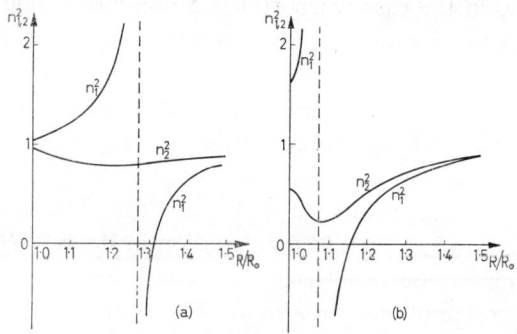

Fig. 122. Curves of $n_{1,2}^2(R)$ in the coronal plasma above a spot ($\alpha = 0$): (a) $H_{0b} = 2 \cdot 5 \times 10^3$ oe; (b) $H_{0b} = 2 \cdot 5 \times 10^2$ oe (Ginzburg and Zheleznyakov, 1959a)

the corona. In actual fact the distribution of the values of n_j^2 in the corona when the magnetic field is taken into consideration will be very complex. In order to get an idea of this distribution, let us examine, following Ginzburg and Zheleznyakov (1959a, 1958a and 1959b), the most simple case—the dependence of n_j^2 on the altitude $h = R - R_\odot$ above the photosphere on the axis of a unipolar sunspot, the model of whose magnetic field has been described in section 2 (see (2.1)). It is assumed here that the electron concentration N in the corona is given by the Baumbach–Allen formula (1.1).

Graphs of $n_{1,2}^2(R)$ in the corona with no allowance made for thermal motion are plotted in Figs. 122–4 for a frequency $\omega = 2\pi \times 10^8$ sec^{-1}, a magnetic pole area of $\pi b^2 = \pi \times 10^{19}$ cm^2 (which is equal to the area of a large spot in order of magnitude) and several fixed values for the magnetic field at the centre of the spot H_{0b} and the angle α between $\boldsymbol{H}_0$ and $\boldsymbol{k}$.

§ 23] Waves in a Magnetoactive Coronal Plasma

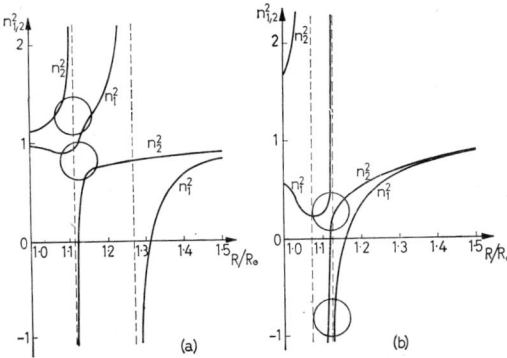

Fig. 123. The same as Fig. 122 for $\alpha = 15°$: (a) $H_{0b} = 2 \cdot 5 \times 10^3$ oe; (b) $H_{0b} = 2 \cdot 5 \times 10^2$ oe (Ginzburg and Zheleznyakov, 1959a)

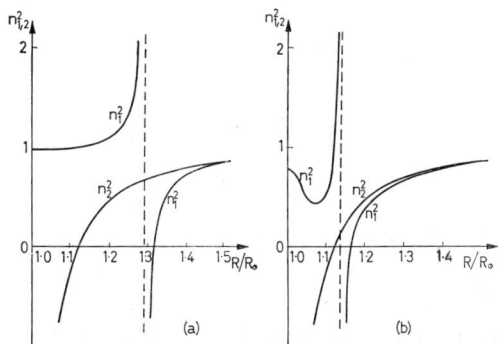

Fig. 124. The same as Fig. 122 for $\alpha = 90°$: (a) $H_{0b} = 2 \cdot 5 \times 10^3$ oe; (b) $H_{0b} = 2 \cdot 5 \times 10^2$ oe (Ginzburg and Zheleznyakov, 1959a)

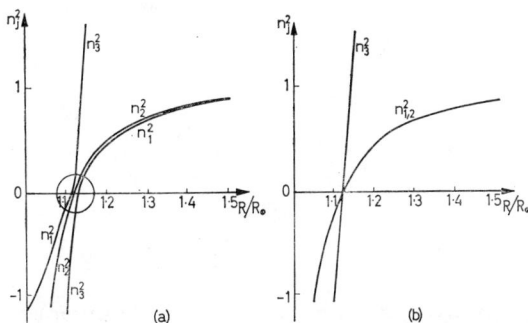

Fig. 125. Curves of $n^2(R)$ in a coronal plasma ($\alpha = 15°$): (a) in the presence of a weak magnetic field ($H_0 = 25$ oe); (b) without a magnetic field

The change to an isotropic coronal plasma is illustrated in Figs. 125 a, b. The first of them shows the function $n_j^2(R)$ in a weak magnetic field with $H_{0b} = 25$ oe (allowing for thermal motion in the quasi-hydrodynamic approximation). The second figure shows graphs of $n_j^2(R)$ for the same concentration $N(R)$ but in the absence of a magnetic field. In accordance with what we have said earlier, in a weak enough magnetic field the plasma wave branch n_3^2 corresponds in all its properties to the dispersion curve n_3^2 in an isotropic coronal plasma.†

The difference between the behaviour of the dispersion curves $n_j^2(R)$ in Figs. 123 a, b is connected with the fact that in the model of a spot with a strong magnetic field (Fig. 123a) the $\omega_H = \omega$ level is higher than the layer where $\omega_L = \omega$, whilst in the case of Fig. 123b, which was plotted for a model of a spot with a weaker magnetic field, the relative position of the levels is reversed. Thanks to this in the first model coupling between normal waves of different types occurs in the region where $\omega_H > \omega$ (i.e. $u > 1$, as in the case shown in Figs. 117b, 121b); in the second model the coupling region is located in a layer for which $\omega_H < \omega$ (i.e. $u < 1$; compare Figs. 117a, 121a). It should be noted that this difference in the nature of the dispersion curves above spots with strong and weak magnetic fields holds only for intermediate values of the angle α; with $\alpha = 0$ and $\alpha = \pi/2$ any change in the magnitude of the magnetic field does not lead to any qualitative adjustment of the dispersion curves (see Figs. 122 and 124).

The refractive index $n_j(R)$ with $\beta_{th} = 0$ becomes infinity in the layers of the corona for which the relation (23.4) is valid. From it follows the fact that at small angles α the quantity $n_j^2(R)$ has two poles at the $u \approx 1$ and $v \approx 1$ levels (see Fig. 123). It is easy to check by means of (23.4) that as the angle α increases the pole located closer to the photosphere and belonging to the ordinary wave moves into the inner layers of the corona and then disappears. On the other hand, the pole for the extraordinary wave that is further away from the solar surface exists for any angle α; as the latter increases it moves into the outer layers of the corona. For $\alpha = \pi/2$ this pole is located at the level $1-u-v = 0$, i.e. in the region where $u < 1, v < 1$.

For an extraordinary wave in the layer enclosed by the pole and located above (with $v = 1-\sqrt{u}$) the zero of the function $n_1^2(R)$ the values of the latter are negative (i.e. the refractive index n_1 is imaginary and the wave in this case is attenuated exponentially). This means that emission which is

---

† Figures 122–5 give a correct idea of the nature of the dispersion curves in the corona even with more general assumptions about the behaviour of N and H_0 than (1.1) and (2.1); all that is necessary is that N and H_0 in the corona should decrease monotonically as we move away from the photosphere.

§ 23] Waves in a Magnetoactive Coronal Plasma

generated in regions of the corona lying below the zero of n_1^2 (below the level $v = 1 - \sqrt{u}$) cannot in practice escape from the corona into the space around the Sun in the form of extraordinary waves.

The conclusion about the impossibility of escape of emission in the form of extraordinary waves from regions of the corona in which $v > 1 - \sqrt{u}$ does not depend within broad limits on the actual distribution of N and H_0 in the corona; all that is necessary is that the extent of the layer in which $n_1^2 < 0$ should not be too small, since otherwise the attenuation of the wave will become insufficient and extraordinary emission will escape from the region $v > 1 - \sqrt{u}$ because of strong "leakage" through the $n_1^2 < 0$ layer. As a detailed treatment of this question (which we shall not stop to do) shows, the "leakage" phenomenon is insignificant in the conditions of a stationary solar corona, although it may play an important part with the appearance of splits in the corona of the shock wave type.

Unlike the extraordinary emission which can escape beyond the solar corona only from regions located above the level $v = 1 - \sqrt{u}$ (where $n_1^2 = 0$), the region for direct escape of ordinary waves is more extensive: it is limited below by the layer in which $v = 1$ (and $n_2^2 = 0$). The electromagnetic waves corresponding to the other branches of the dispersion curves (in particular those located in the $v > 1$ layers) may escape from the corona in the form of ordinary emission only because of the coupling effect between different types of waves (ordinary and extraordinary)[†] in the region $v \approx 1$; as shown in section 25, this effect is generally noticeable only with wave propagation at small angles to the direction of the magnetic field H_0.[‡]

Faraday Effect in the Solar Corona

Electromagnetic emission in a plasma may be non-polarized or polarized (wholly or partly). Unpolarized (naturally polarized) emission is a superposition of non-coherent waves with opposite polarizations; in a magnetoactive plasma it is convenient to take normal waves (ordinary and extraordinary) as these waves, which corresponds to a physical statement of the problem. In its turn wholly polarized emission, i.e. emission with an entirely definite polarization (generally speaking, different from the polarization of normal waves in a plasma) can be broken down into a collection of coherent normal waves (ordinary and extraordinary). The polarization of

---

[†] Between ordinary, extraordinary and plasma waves when thermal motion is taken into consideration.

[‡] As can be seen from Fig. 123a the "leakage" phenomenon can in principle also occur for ordinary waves (in the layer between $n_2^2 = 0$ and $n_2^2 = -\infty$, where $n_2^2 < 0$ and the waves are exponentially attenuated). The part it plays in a stationary solar corona is just as small as for extraordinary waves.

this kind of emission may simply be the same as the polarization of one of the normal waves in a magnetoactive plasma. The partly polarized emission is composed of a purely polarized component with the properties indicated above and a completely independent (non-coherent with it) naturally polarized component (see Chandrasekar 1953, section 15).

We notice that whether the polarization coefficients of the wholly polarized emission (or polarized component of partly polarized emission) are different from or the same as the polarization coefficients (23.10), (23.11) of normal waves is obviously determined by the properties of the emitting object and the actual conditions in the radio wave propagation path (in particular by the interaction of different types of waves). Section 24 is devoted to a study of the part played by interaction in changing the polarization of radio waves (see also section 25); here we shall examine in the approximation of geometrical optics certain effects which are connected with the passage of polarized emission through a magnetoactive plasma.

Let the polarized emission be reduced to a single normal wave (ordinary or extraordinary); then its polarization coefficients in a non-uniform plasma are the same as (23.10), (23.11). If the emission is propagated in accordance with the laws of geometrical optics, then the polarization of ordinary and extraordinary waves escaping beyond the corona (into the region $v \ll 1$, $u \ll 1$) will in practice be circular for all angles α with the exception of $\alpha = \pi/2$ (see (23.12)). On the other hand, under the conditions of planetary ionospheres the polarization of normal waves on leaving the plasma may remain elliptical since $u \ll 1$ does not have to be the case here in the vicinity of the planet with $v \ll 1$. This is actually connected with the fact that the concentration of particles in the ionosphere falls more rapidly as we move away from the planetary surface than does the magnetic field; in the extensive solar corona the position is apparently otherwise.

However, according to Chapter IV, elliptical polarization is sometimes recorded when studying certain components of the Sun's sporadic radio emission (microwave bursts, types I and III bursts). It can obviously be assumed that in these cases it is not one circularly polarized wave (ordinary or extraordinary) that is escaping beyond the corona, but two coherent waves whose superposition does provide elliptically polarized emission. The coherent waves may occur in the emission source itself or in interaction regions where there is a partial change of waves of one type into another.

When a wave of one type is propagated in a non-uniform plasma the orientation of the polarization ellipse relative to the direction of propagation is determined only by the direction of the magnetic field H_0; in a smoothly non-uniform medium the polarization ellipse either keeps a fixed

§ 23] Waves in a Magnetoactive Coronal Plasma

position along the ray, or turns slowly in accordance with the orientation of the vector H_0. It is a different matter if ordinary and extraordinary waves with different refractive indices (different phase velocity) which are coherent with each other are being propagated in the plasma. Since in this case the orientation of the resultant polarization ellipse is determined not only by the form and orientation of the ordinary and extraordinary ellipses, but also by the phase shift between them, any change in the latter along the direction of propagation will lead (under certain conditions) to rotation of the resultant polarization ellipse in the process of propagation of the waves in the plasma—the Faraday effect. This effect as applied to the Earth's ionosphere is discussed briefly in section 7; here we shall discuss it in greater detail.

Let us assume that polarized emission with a polarization ellipse not corresponding to normal waves is coming from a certain region in a plasma. Then the emission is propagated further in the form of two coherent waves (extraordinary and ordinary) whose electrical vectors in the plane x, y orthogonal to the wave vector k can be written as follows:

$$E_x = a \sin(\omega t), \quad E_y = ap \cos(\omega t)$$

and

$$E_x = bp \sin(\omega t - \psi), \quad E_y = -b \cos(\omega t - \psi).$$

Here a and b characterize the amplitudes of the extraordinary and ordinary waves, ψ is the phase shift between them, $p = K_1$ is the ellipticity of the extraordinary wave. In accordance with (23.10), $-1 \leq K_1 \leq 0$; the expressions given allow for the fact that the ellipticity of an ordinary wave in a plasma is $K_2 = -1/K_1$.

Then the electric field components E_x and E_y in the emission in question vary in time as

$$\left. \begin{array}{l} E_x = a \sin(\omega t) + bp \sin(\omega t - \psi) = A \sin(\omega t - \Psi), \\ E_y = ap \cos(\omega t) - b \cos(\omega t - \psi) = B \sin(\omega t - \Phi), \end{array} \right\} \quad (23.13)$$

where

$$\left. \begin{array}{l} A^2 = a^2 + b^2 p^2 + 2abp \cos \psi, \\ B^2 = a^2 + b^2 p^2 - 2abp \cos \psi, \\ \tan \Psi = \dfrac{bp \sin \psi}{a + bp \cos \psi}, \quad \tan \Phi = \dfrac{ap - b \cos \psi}{b \sin \psi}. \end{array} \right\} \quad (23.14)$$

The ellipse defined by the relations (23.13), (23.14) is orientated at an angle

χ to the coordinate axis x and has a ratio of the half-axes $\tan \sigma$, where

$$\tan (2\chi) = \frac{2AB \cos (\Psi - \Phi)}{A^2 - B^2},$$

$$\tan^2 \sigma = \frac{A^2 + B^2 + \sqrt{(B^2 - A^2)^2 + 4A^2 B^2 \cos^2 (\Psi - \Phi)}}{A^2 + B^2 - \sqrt{(B^2 - A^2)^2 + 4A^2 B^2 \cos^2 (\Psi - \Phi)}}.$$

Substituting in this the expressions for A, B, Ψ and Φ (23.14) we obtain (Hatanaka, 1956):

$$\tan (2\chi) = \pm 2 \frac{1+p^2}{p} \cdot \frac{\sin \psi}{\left(\frac{1}{p} - p\right)\left(\frac{a}{b} - \frac{b}{a}\right) + 4 \cos \psi}, \qquad (23.15)$$

and

$$\sin (2\sigma) = \pm 2 \frac{b^2 - a^2 + ab(p^{-1} - p) \cos \psi}{(a^2 + b^2)(p^{-1} - p)}. \qquad (23.16)$$

The phase difference between the ordinary and extraordinary components of the emission that is acquired as the waves are propagated a distance L along $\mathbf{k}$ is

$$\Delta \psi = \frac{\omega}{c} \int_L (n_2 - n_1) \, dl. \qquad (23.17)$$

In the process of the radiation passing through a region where the sign of the difference $n_2 - n_1$ is constant the quantity

$$\psi = \psi_0 + \Delta \psi, \qquad (23.18)$$

where ψ_0 is the initial phase difference, rises monotonically; here the orientation and ratio of the axes of the resultant polarization ellipse vary in accordance with the expressions (23.15) and (23.16). If the position of the original ellipse is characterized by the angle χ_0, then the ellipse of a wave travelling along a path L in the plasma will be orientated at an angle

$$\chi = \chi_0 + \Delta \chi. \qquad (23.19)$$

In the simplest and often realized case of quasi-longitudinal propagation, when the polarization of the ordinary and extraordinary waves is close to circular (i.e. $p \approx -1$), the degree of ellipticity of the radiation mains constant, and the rotation of the resultant ellipse is described by the linear relation

$$\Delta \chi = \frac{\Delta \psi}{2}. \qquad (23.20)$$

§ 23] Waves in a Magnetoactive Coronal Plasma

This is a consequence of the equality $\tan(2\chi) = \tan\psi$ to which (23.15) is reduced in this case.

To find the explicit expression that connects the magnitude of the Faraday rotation $\Delta\chi$ with the parameters of the plasma and the frequency of the wave, we recall that, in accordance with (23.7), in the quasi-longitudinal approximation with $\omega_H \cos\alpha/\omega \ll 1$

$$n_2 - n_1 = \frac{n_2^2 - n_1^2}{n_2 + n_1} \approx \frac{2}{n_2 + n_1} \frac{\omega_L^2}{\omega^2} \frac{\omega_H}{\omega} |\cos\alpha|.$$

Then†

$$\Delta\chi \approx \frac{1}{\omega^2 c} \int \frac{\omega_L^2 \omega_H \cos\alpha}{n_2 + n_1} \, dl = \frac{4\cdot 7 \cdot 10^4}{f^2} \int \frac{NH_0 \cos\alpha}{n_2 + n_1} \, dl, \quad (23.21)$$

where f is in c/s, N in electrons/cm³, H_0 in oe and dl in cm. It follows from (23.21) that with $n_1 \approx n_2 \approx 1$ (i.e. in a fairly rarefied plasma) the angle $\Delta\chi$ is in inverse proportion to the square of the frequency f.

By using this formula it is not hard to see that the Faraday effect with quasi-longitudinal propagation in the solar corona is very great. In actual fact, even taking low estimates for the electron concentration, the magnetic fields and the extent of the "trajectory" of the radio waves in the corona ($N \sim 10^8$ electrons/cm³, $H_0 \sim 2$ oe, $L \sim 10^{10}$ cm) $\Delta\chi \sim 10^{23} f^{-2}$ radian, whilst under the conditions of the Earth's ionosphere $\Delta\chi$ is six orders less. It follows from this that in the corona $\Delta\chi \lesssim 1$ at wavelengths of $\lambda \lesssim 0\cdot 1$ cm and rises rapidly with the wavelength, reaching values of the order of 10^4 at $\lambda \sim 10$ cm and 10^6 at $\lambda \sim 1$ m. In actual fact the magnitude of the rotation in the corona (particularly above centres of activity with higher values of N and H_0) may be much higher.

We notice that in the case when the propagation is not quasi-longitudinal and the polarization of the ordinary and extraordinary waves becomes elliptical ($p \neq -1$), instead of the simple relation (23.20) we must use the more complex formula (23.15). An analysis of it shows that with $p \neq -1$ the Faraday rotation becomes uneven, although with a change in $\Delta\psi$ by an amount much greater than 2π the mean change in $\Delta\chi$ is as before the same as $\Delta\psi/2$. The unevenness of the rotation increases as the degree of ellipticity grows (as $|p|$ decreases) and finally, when

$$\left|\frac{1}{p} - p\right|\left|\frac{a}{b} - \frac{b}{a}\right| > 4, \quad (23.22)$$

† This formula is obtained here for the case when $\cos\alpha > 0$. It is not hard to see that the formula (23.21) also remains valid for $\cos\alpha < 0$, when the types of waves responsible for clockwise (right-handed) and anticlockwise (left-handed) polarization change places.

i.e. when the polarization ellipse of the normal waves differs enough from a circle and the ratio of the amplitudes of these waves differs enough from unity, rotation of the ellipse is replaced by its oscillation with an amplitude with respect to χ of not more than $\pi/4$.

It must be pointed out that no attention is generally paid to this fact although it follows directly from the formula (23.15); in the case of (23.22) the right-hand side of (23.15) remains finite for any ψ, whilst when the ellipse rotates, i.e. when χ passes through the values $-\pi/4$, $+\pi/4$, $+3\pi/4$, etc., both sides of the equation should become infinity. Then, provided that

$$\left|\frac{a}{b}-\frac{b}{a}\right| \gg 2\frac{1+p^2}{1-p^2}, \qquad (23.23)$$

as can easily be seen, the right-hand side of (23.15) becomes much less than unity; then χ takes up a fixed value:

$$\chi = -\frac{1+p^2}{1-p^2}\frac{\sin\psi}{\dfrac{a}{b}-\dfrac{b}{a}} \qquad (23.24)$$

or differs from (23.24) by an amount equal to or a multiple of $\pi/2$.[†]

Depolarizing Factors and the Question of Elliptical Polarization of Certain Bursts of Solar Radio Emission

It is clear from what has been said that the angle χ is a function of the path L travelled by the emission in an actively rotating medium. If the values of L are scattered in a narrow range ΔL near the mean value L_0 (for example, because of the finite size of the emission source), then the orientation of the polarization ellipse can be given in the form

$$\chi(L) \approx \chi(L_0) + \frac{\partial(\Delta\chi)}{\partial L}\bigg|_{L=L_0}(L-L_0). \qquad (23.25)$$

Then the orientation of the polarization ellipses for waves originating from different points of the generation region will be contained in the range

$$\delta\chi \approx \frac{\partial(\Delta\chi)}{\partial L}\bigg|_{L=L_0}\Delta L. \qquad (23.26)$$

[†] The fixed orientation of the polarization ellipse of the radiation in the case of (23.23) (with an accuracy of up to small oscillations relative to the position corresponding to the polarization ellipse of an ordinary or an extraordinary wave) is quite natural: the superimposing of the ellipse of a strong wave ellipse on that of a weak wave cannot make any significant alteration to the orientation of the resultant ellipse.

§ 23] Waves in a Magnetoactive Coronal Plasma

For a $\Delta\chi$ that can be described by the expression (23.21) this relation becomes the following (for $n_1 \approx n_2 \approx 1$):

$$\delta\chi \approx \frac{4\cdot7\cdot10^4}{f^2} N H_0 \cos\alpha\, \Delta L. \tag{23.27}$$

Here the product $NH_0 \cos\alpha$ relates to the region occupied by the source. It is clear that owing to the Faraday effect in this region elliptically polarized emission will escape beyond the latter only if $\delta\chi \lesssim 1$; in the opposite case the strong scatter in the ellipse orientations leads to a sharp reduction in the degree of linear polarization ϱ_l of the emission. Under generation conditions of this kind elliptically polarized emission becomes circularly polarized.†

The condition $\delta\chi \lesssim 1$ imposes rigid limitations on the size of sources in the solar corona:

$$\Delta L \lesssim \frac{f^2}{4\cdot7\cdot10^4 NH_0 \cos\alpha}. \tag{23.28}$$

In the centimetric band ($f \sim 10^{10}$ c/s) with very modest values of $N \sim 10^8$ electrons/cm³ and $H_0 \cos\alpha \sim 10$ oe the value is $\Delta L \lesssim 2\times10^6$ cm; at metric wavelengths it is two orders less.

Since the sizes of solar radio emission sources as a rule exceed these values of ΔL, depolarization because of the Faraday effect in the source will be practically complete, i.e. the sum of the elliptically polarized components is circularly polarized emission. This result is a full explanation of why circular polarization is generally observed during the reception of solar radio emission. In the comparatively rare cases when the ellipse of the recorded emission differs noticeably from a circle it can be assumed that the Faraday effect is in fact insignificant in the source. This is obviously caused by a difference in the efficiency of the generation or escape of extraordinary and ordinary waves beyond the source, so in practice there are only waves of one type making up the polarized part of the emission. In the first case rotation of the polarization ellipse does not generally occur in the source since the condition (23.22) will be satisfied because of the very small ratio of the intensities of the two types of waves; in the second case no kind of Faraday effect can appear in the region indicated since the propagation region of one of the normal waves is localized near the source.

As well as the depolarizing factor we have discussed, which is connected with the finite size of the source (or the different propagation paths of

---

† This depolarization process has been investigated (as applied to cosmic radio emission) by Getmantsev and Razin (1956); Razin (1956).

ordinary and extraordinary waves), there is a second factor which acts in the same direction and is caused by the finite bandwidth of the receiving equipment. In actual fact, it is clear from the above that the angle χ is a function of the frequency f even in the case that the original orientation of the polarization ellipse χ_0 does not depend on f. In a narrow frequency range, for example in the polarimeter band Δf_b, the angle χ can be represented in the form

$$\chi(f) \approx \chi(f_0) + \frac{\partial(\Delta\chi)}{\partial f}\bigg|_{f=f_0}(f-f_0), \qquad (23.29)$$

where f_0 is the working frequency of the equipment. Then the dispersion change in the orientation χ of the resultant ellipse in this band is

$$\delta\chi \approx \frac{\partial(\Delta\chi)}{\partial f}\bigg|_{f=f_0} \Delta f_b. \qquad (23.30)$$

In particular, with quasi-longitudinal propagation in the region $n_1 \approx n_2 \approx 1$, when $\Delta\chi \sim f^{-2}$ (see (23.21)), the coefficient $\partial(\Delta\chi)/\partial f|_{f=f_0} = -2f_0^{-1}\Delta\chi$ $(f=f_0)$, and the change in the magnitude of χ in the band Δf_p is

$$\delta\chi \approx 2f_0^{-1}\Delta\chi(f=f_0)\Delta f_b. \qquad (23.31)$$

Because of the difference in the positions of the ellipse at different frequencies the observed polarization of the emission will differ from the original by its lower degree of polarization ϱ and the greater ratio of the axes of the ellipse $|p|$. These effects as applied to the ionosphere have already been discussed in section 7; there, however, they were not discussed in detail since the Faraday rotation in the ionosphere at the frequencies of interest to us is not too great and the effect of dispersion of the orientations $\delta\chi$ on the values of ϱ and p can be reduced to nothing in the majority of cases with slight limitations on the polarimeter bandwidth.

These effects stand out more clearly in the solar corona because of the strong rotation of the plane of polarization. For example, if at $\lambda \sim 1$ m $\Delta\chi \sim 10^6$ radians, then the dispersion $\delta\chi$, in accordance with (23.31), will be much less than unity only in the very narrow band $\Delta f_b \ll 150$ c/s, which is not achieved in present-day polarization measurements. This leads to the necessity for allowing strictly for effects connected with the dispersion of the Faraday rotation in the solar corona. In discussing this question we shall follow Akabane and Cohen (1961) (see also Cohen (1958a)).

We first introduce the complex function

$$\mathcal{Q} = \frac{Q+iU}{(I^2-V^2)^{1/2}}, \qquad (23.32)$$

§ 23] Waves in a Magnetoactive Coronal Plasma

which can be reduced to the form

$$\mathcal{G} = \tilde{\mathcal{G}} e^{2i\chi}, \quad \tilde{\mathcal{G}} = \frac{\varrho \cos(2\sigma)}{[1 - \varrho^2 \sin^2(2\sigma)]^{1/2}}, \tag{23.33}$$

if we substitute the expressions (6.2) for the Stokes parameters I, Q, U, V in (23.32). Here ϱ characterizes the degree of polarization and $p = \tan \sigma$ the ellipticity of the radio emission. The modulus $\mathcal{G}$ is equal to zero for unpolarized emission and unity for wholly polarized emission.

On the other hand, by using the relations (6.12) $\mathcal{G}$ can be written as follows:

$$\mathcal{G} = \tilde{\mathcal{G}} e^{i\psi_{rl}}, \quad \tilde{\mathcal{G}} = \frac{\overline{E_{0r} E_{0l}}}{(I_r I_l)^{1/2}}. \tag{23.34}$$

According to (6.15) the real part of $\mathcal{G}$ is the same as the correlation coefficient $\mathcal{G}(t_0 = 0)$ between the radio emission components with clockwise and anticlockwise polarization, whilst the modulus $\tilde{\mathcal{G}}$, usually called the "degree of coherency" of the circular components, is equal to the amplitude of the correlation function $\mathcal{G}(t_0)$.

The Stokes parameters measured by a polarimeter are actually the result of integrating the corresponding quantities (6.2) related to a unit range of frequencies over the equipment bandwidth:

$$\left.\begin{aligned}
\bar{I} &= \int_{-\infty}^{+\infty} K(f) I \, df, \quad \bar{Q} = \int_{-\infty}^{+\infty} K(f) \varrho I \cos(2\sigma_0) \cos(2\chi) \, df, \\
\bar{U} &= \int_{-\infty}^{+\infty} K(f) \varrho I \cos(2\sigma_0) \sin(2\chi) \, df, \\
\bar{V} &= \int_{-\infty}^{+\infty} K(f) \varrho I \sin(2\sigma_0) \, df.
\end{aligned}\right\} \tag{23.35}$$

Here I is the spectral density of the received radio-emission intensity, σ_0 is the characteristic of the original ellipticity of the radio emission in the source, $K(f)$ is the frequency response of the polarimeter. The width of the function $K(f)$ determines the effective frequency band Δf_b of the equipment; we shall assume that $K(f)$ is standardized so that

$$\int_{-\infty}^{+\infty} K(f) \, df = 1.$$

Unless Δf_b is too large I and ϱ in (23.35) can be removed from the integrand with values of I_0, ϱ_0 at the working frequency f_0. Since σ_0 together with χ_0, by assumption, is not frequency-dependent within the limits of Δf_b we obtain:

$$\left.\begin{aligned}
\bar{I} &= I_0, \quad \bar{Q} = \varrho_0 I_0 \cos(2\sigma_0) \int_{-\infty}^{+\infty} K(f) \cos(2\chi) \, df, \\
\bar{U} &= \varrho_0 I_0 \cos(2\sigma_0) \int_{-\infty}^{+\infty} K(f) \sin(2\chi) \, df, \quad \bar{V} = \varrho_0 I_0 \sin(2\sigma_0).
\end{aligned}\right\} \tag{23.36}$$

Substituting the last expressions in (23.32) and remembering the relations (23.29) we find

$$\mathcal{G} = \frac{\varrho_0 \cos(2\sigma_0) e^{2i\chi(f_0)}}{[1-\varrho_0^2 \sin^2(2\sigma_0)]^{1/2}} \int_{-\infty}^{+\infty} K(f) e^{2i \frac{\partial(\Delta\chi)}{\partial f}\big|_{f=f_0}(f-f_0)} df. \quad (23.37)$$

The factor in front of the integral obviously corresponds to the value $\mathcal{G} = \mathcal{G}_0$, which corresponds to an infinitely narrow band Δf_b; at the same time it differs by a factor of only $e^{2i\Delta\chi(f_0)}$ from the value of the function $\mathcal{G}$ in the radio emission source. The relation (23.37) can be written in the simpler form

$$\mathcal{G} = \mathcal{G}_0 \int_{-\infty}^{+\infty} F(x) e^{ix\delta\chi} dx, \quad (23.38)$$

where

$$x = \frac{2(f-f_0)}{\Delta f_b}, \quad F(x) = \frac{K(f) \Delta f_p}{2} \quad (23.39)$$

and $\delta\chi$ is defined by the relation (23.30).

Therefore the function $\mathcal{G}$ for a finite frequency band in which the orientation of the polarization ellipse changes by $\delta\chi$ can be obtained by multiplying $\mathcal{G}_0$ by a Fourier-transformed non-dimensional frequency characteristic of the polarimeter $F(x)$. In particular, if the latter takes the form of a Gaussian curve

$$F(x) = \frac{1}{\sqrt{\pi}} e^{-x^2}, \quad (23.40)$$

then

$$\mathcal{G}^4 = \mathcal{G}_0^4 e^{-(\delta\chi)^2}, \quad \tilde{\mathcal{G}}^4 = \tilde{\mathcal{G}}_0^4 e^{-(\delta\chi)^2}. \quad (23.41)$$

Let us now see how the degree of polarization ϱ_0 and the ellipticity $p_0 = \tan \sigma_0$ alters when we change to a finite frequency band. To do this we use the relations (23.33) for the function $\mathcal{G}$ and (6.2), (23.36) for the parameter V. It follows from the first relation that

$$\tilde{\mathcal{G}} = \frac{\varrho \cos(2\sigma)}{[1-\varrho^2 \sin^2(2\sigma)]^{1/2}}, \quad \tilde{\mathcal{G}}_0 = \frac{\varrho_0 \cos(2\sigma_0)}{[1-\varrho_0^2 \sin^2(2\sigma_0)]^{1/2}},$$

whilst according to (6.2), (23.36)†

$$\varrho \sin(2\sigma) = \varrho_0 \sin(2\sigma_0).$$

† Since $\varrho \sin(2\sigma)$ is equal to ϱ_c—the degree of the circular polarization of the radio emission (see (6.16))—the relation $\varrho \sin(2\sigma) = \varrho_0 \sin(2\sigma_0)$ indicates that the dispersion of the Faraday rotation does not affect its magnitude in any way. All change in the degree of total polarization $\varrho = \sqrt{\varrho_c^2 + \varrho_l^2}$ occurs because of a decrease in the linear polarization ϱ_l

§ 23] Waves in a Magnetoactive Coronal Plasma

Combining these equations we obtain:

$$\varrho^2 = \varrho_0^2 \left[\frac{\tilde{\mathcal{G}}^2}{\tilde{\mathcal{G}}_0^2} + \sin^2(2\sigma_0) \left(1 - \frac{\tilde{\mathcal{G}}^2}{\tilde{\mathcal{G}}_0^2}\right) \right], \tag{23.42}$$

$$\tan^2(2\sigma) = \tan^2(2\sigma_0) \frac{\tilde{\mathcal{G}}^2}{\tilde{\mathcal{G}}_0^2}. \tag{23.43}$$

The nature of the radio emission depolarization defined by the formulae (23.42), (23.43) can be judged from Fig. 126 in which the contours of constant values of the ratio $|\tilde{\mathcal{G}}/\tilde{\mathcal{G}}_0|$ and the original ellipticity p_0 are plotted in the plane ϱ/ϱ_0 and p. Above the lines with the values of $|\tilde{\mathcal{G}}/\tilde{\mathcal{G}}_0|$ are also indicated the corresponding dispersions of the ellipse orientations $\delta\chi$ for the Gaussian form of the frequency response of the receiving apparatus (23.40). Unless the radio emission passes through a magnetoactive plasma or the frequency band Δf_b approaches zero, the character of the polarization is determined by a point on the line $\delta\chi = 0$ corresponding to the given value of p_0. Here the ratios $\tilde{\mathcal{G}}/\tilde{\mathcal{G}}_0$, ϱ/ϱ_0 and p/p_0 are obviously equal to unity. While the emission passes through the medium or as the frequency band increases the mapping point moves down along the line $p_0 = $ const, accompanied by a corresponding decrease in $\tilde{\mathcal{G}}/\tilde{\mathcal{G}}_0$, ϱ/ϱ_0 and p/p_0. The effect of dispersion of the rotation in the medium has less effect on radio emission with large p_0 values and disappears completely for emission that was originally circularly polarized ($p_0 = 1$).†

Polarization observations of solar radio emission with one value of the band Δf_b allow us to establish only the upper limits for the dispersion $\delta\chi$ of the angle of rotation and the value of the angle of rotation $\Delta\chi$ itself. Therefore measurements of the quantities $\tilde{\mathcal{G}}$ and ϱ determine only the lower limits and measurements of p the upper limits for the corresponding values of $\tilde{\mathcal{G}}_0$, ϱ_0 and p_0 in the emission source.

The position, however, changes in the case of simultaneous measurements of polarization in two bands $\Delta f_b'$ and $\Delta f_b''$ near the frequency f_0 ($\Delta f_b' < \Delta f_b''$). In actual fact we then have:

$$\left(\frac{\tilde{\mathcal{G}}'}{\tilde{\mathcal{G}}''}\right)^4 = e^{(\delta\chi'')^2 - (\delta\chi')^2},$$

† It is worth stressing that because of the complete analogy of the formulae (23.25), (23.26) on the one hand and (23.29), (23.30) on the other the relations (23.38), (23.41)–(23.43) obtained can be transferred *en bloc* to the case of depolarization caused by the finite dimensions of the source. Here x should be taken to mean the quantity $2(L-L_0)/\Delta L$, $F(x)$ to mean the function $K(L)\Delta L/2$ in which $K(L)$ characterizes the distribution of the emission intensity over the source. Further, $\delta\chi$ is then defined by the relation (23.27); ϱ_0 and p_0 have the meaning of the degree of polarization and the ellipticity of the emission generated by an individual element of the source.

where the primes and double primes relate to a narrow and a broad band respectively. Since $\delta\chi \propto \Delta f_b$,

$$\left(\frac{\tilde{Q}'}{\tilde{Q}''}\right)^4 = e^{(\delta\chi')^2 \left[\left(\frac{\Delta f_b''}{\Delta f_b'}\right)^2 - 1\right]}. \qquad (23.44)$$

In the relation (23.44) the bands $\Delta f_b'$ and $\Delta f_b''$ are known and $\tilde{Q}'$ and $\tilde{Q}''$ are measured. This allows us to find the quantity $\delta\chi'$, and then calculate by means of (23.41). It is not hard to find ϱ_0, p_0 from the ratio $\tilde{Q}/\tilde{Q}_0$ and the

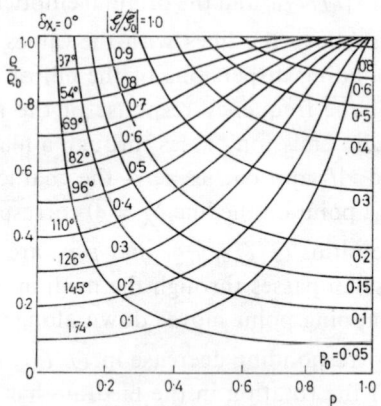

Fig. 126. Depolarization of emission because of dispersion of angle of Faraday rotation in a plasma (Akabane and Cohen, 1961)

measured values of ϱ, p (see (23.42), (23.43)). The rotation $\Delta\chi$ of the polarization ellipse in the corona at a frequency f_0 on the assumption of quasi-longitudinal propagation is then estimated (proceeding from the values of $\delta\chi'$, $\Delta f_b'$) from the formula (23.31).

Judging from the preliminary data given in section 14 on simultaneous measurements of the degree of coherency $\tilde{Q}$ in the bands $\Delta f_b' = 10$ kc/s and $\Delta f_b'' = 22$ kc/s near the frequency $f_0 = 200$ Mc/s, the ratio $\tilde{Q}''/\tilde{Q}'$ in the radio emission of type III bursts is 0·6–0·8 with a $\tilde{Q}'$ of not more than 0·3.[†] Insufficient observational accuracy made it impossible to obtain information on the smaller values of $\tilde{Q}''/\tilde{Q}'$. If we assume that the observed decrease in the degree of coherency is connected with the frequency dispersion of the Faraday rotation in the solar corona, then the above values

[†] The exponential decrease in the degree of coherency as $\delta\chi$ rises, i.e. as the frequency band Δf_b increases, explains the fact that the observations in the $\Delta f_b = 300$ kc/s band did not reveal any difference of $\tilde{Q}$ from zero within the limits of the measurement accuracy achieved.

§ 23] Waves in a Magnetoactive Coronal Plasma

of the ratio $\tilde{\mathcal{Q}}''/\tilde{\mathcal{Q}}'$ and the quantities $\tilde{\mathcal{Q}}'$ correspond, in accordance with what has been said above, to $\delta\chi' \approx 0\cdot7-0\cdot5$ radian, $\mathcal{Q} < 0\cdot34-0\cdot32$ and $\Delta\chi \approx (1\cdot5-1\cdot0)\times10^4$ radians.

The value obtained for the total rotation of the orientation of the polarization ellipse $\Delta\chi$ on its path from the radio emission source in the corona to the observation point on Earth is about two orders less than the expected value (on the assumption that the Faraday effect operates in the lower corona from the time the emission escapes beyond the generation region). The validity of what we have been saying is easy to check if we turn to the estimates of $\Delta\chi$ in the preceding sub-section. An even larger difference—of about four orders of magnitude—occurs for the elliptically polarized type III bursts which, to judge from Komesaroff's report, he observed at frequencies of 40–70 Mc/s with equipment having a 0·5 Mc/s band (see section 14). Here the Faraday effect must not exceed 40–70 radians for recording elliptical polarization. Recording of elliptical polarization of type I bursts at a frequency of 200 Mc/s with a polarimeter with a 100 kc/s band (section 12) was possible only provided that the angle of rotation in the corona was $\Delta\chi < 10^3$ radians.†

A similar position obtains not only in the metric band: for the elliptically polarized microwave bursts found by Akabane ($f_0 = 9500$ Mc/s, $\Delta f_b = 8$ Mc/s (Akabane, 1958b; Cohen, 1960)) the required quantity $\Delta\chi$ is less than 6×10^2 radians, and estimated by the formula (23.21) for $N \sim 2\times10^8$ electrons/cm³, $H_0 \sim 10$ oe and $L \sim 10^{10}$ cm is about 10^4 radians. This estimate is apparently still slightly too low since the strength of the magnetic field in a centre of activity above flocculi may be more than 10 oe.

What is the reason for this discrepancy? Since no basis can be seen for a re-examination of the estimates of the phase shift $\Delta\psi$ that are too low between waves of different types that they acquire as they are propagated from the source to the exit from the corona, it can be stated that the comparatively small values of $\Delta\chi$ obtained from an analysis of the results of observations of elliptically polarized bursts definitely indicate the absence of any Faraday effect over the majority of the wave trajectory in the corona. This kind of situation is created if the condition (23.22) is satisfied over the whole of the ray in the corona, the rotation of the polarization ellipse being replaced by its oscillation about a fixed angle χ.

If the intensities of ordinary waves leaving the generation region that are coherent between each other are comparable in value, then the criterion (23.22) is observed only in the case when the polarization of the normal

† These conditions for $\Delta\chi$ are obtained from the requirement that $\delta\chi$ (23.31) should be less than unity.

waves differs sufficiently from circular and the propagation from quasi-longitudinal. This usually occurs in a small range of angles α near $\pi/2$ (with $u \ll 1$, $1-v \sim 1$, which is known to obtain over the majority of the ray in the corona). Therefore in the case under discussion for the action of the Faraday effect to cease, the emission must be propagated in the corona almost across the magnetic field; that of course, is of little probability.†

Another, apparently more probable cause of the very limited action of the Faraday effect in the corona is that the intensities of the ordinary and extraordinary waves making up the polarized component of the radio emission leaving the generation region differ sharply (i.e. the polarized part of the emission is actually a wave of one type). In this case, however, circularly polarized emission should leave the corona provided that it does not pass on the way through a region of interaction and the partial transformation of one type of wave into another. Coherent ordinary and extraordinary waves leaving this region will provide the observed ellipticity in the radio emission of certain bursts. At the same time the Faraday rotation of the resultant polarization ellipse becomes (according to the experimental data) small enough if the interaction region in question is located in high enough layers of the corona. It will be shown in section 24 that the necessary transformation occurs as the result of waves passing through a transverse magnetic field region.

If in the first case it was necessary, in order to explain the observed degree of ellipticity, that the radio emission intersects the magnetic lines of force at angles of $\alpha \approx \pi/2$ over practically the whole path from the source to the exit from the corona, then the second possibility requires far less: all it needs is the existence on the ray path of a local layer where the case of transverse propagation obtains (with very special values of the parameters of this layer, it is true; for further detail see section 24).

24. Coupling of Electromagnetic Waves in a Plasma and Polarization of Solar Radio Emission

In the preceding section we have explained that during propagation in a non-uniform plasma the polarization of ordinary and extraordinary waves changes in accordance with the formulae (23.10) and (23.11) if we look upon the parameters $v = \omega_L^2/\omega^2$, $u = \omega_H^2/\omega^2$ and the angle α figuring

---

† In particular the criterion for the absence of the Faraday effect has to be satisfied here even in the actual source of the emission. In the opposite case the depolarization (change of elliptical into circular polarization) will be very strong even with very rigid limitations on the dimensions of the generation region (see p. 335).

§ 24] Coupling of Waves in a Plasma

in them as functions of the coordinates. This is valid until the approximation of geometrical optics loses its force, according to which a wave of each type is propagated in a plasma without being transformed into waves of another type.† In smoothly non-uniform media whose properties vary only a little over distances of the order of a wavelength this approximation breaks down only in certain limited regions. There the representation of an electromagnetic field in the form of the sum of terms of the (22.15) type, i.e. in the form of a superposition of ordinary, extraordinary and plasma waves, is invalid: the nature of the field becomes more complex, differing sharply from the geometrical optics solutions. If waves in the process of propagation pass through such a region coupling appears between them, consisting in the ratio of the amplitudes of the geometrical optics solutions taking up a fixed value that depends on the wave propagations conditions in the region where geometrical optics break down.‡ Without stopping to discuss the physical meaning of the coupling phenomenon, which we have already discussed in section 22, we shall content ourselves with saying that as a result of this phenomenon, when a wave of one type is incident on a coupling region, generally speaking, waves of two types leave it. In other words, in this region there occurs a partial change (transformation) of waves of one type into another, which is accompanied by a significant alteration in the polarization of the waves that cannot be described by the formulae (23.10) and (23.11).

An investigation of the coupling of waves of different types, to which this and the following section are devoted, shows that it occurs in the regions of a plasma where the dispersion curves $n_j^2(\omega)$ approach close to each other. This is quite natural since it is just in regions with close wave properties that they can "transfer" from one dispersion curve to another. From the example of Figs. 117 and 121 it is clear that this occurs in layers where $v \approx 1$, with propagation of waves at small angles α to the magnetic field H_0, and also in layers where $v \ll 1$.

The effective coupling of "normal" waves in the region $v \approx 1$, which, generally speaking, occurs in both a magnetoactive and an isotropic plasma, is of considerable importance for solving the problem of the escape of emission beyond the solar corona. This coupling will be discussed in detail in section 25. The transformation of ordinary waves into extraordinary ones and vice versa in a sufficiently rarefied plasma ($v \ll 1$) under

---

† The weak conversion of waves of one type into another appears here only in a higher approximation to the solution in the geometrical optics form (see section 25).

‡ The question of the coupling in a plasma of waves of different types (ordinary and extraordinary) was first studied by Ginzburg (1948) (see also Ginzburg 1960b, section 28) as applied to the so-called "tripling" effect of signals in the ionosphere.

certain conditions may have a very significant effect on the nature of the polarization of radio emission leaving the solar corona. We shall open the discussion of coupling effects in a plasma with this phenomenon.

LIMITING POLARIZATION OF EMISSION LEAVING THE CORONAL PLASMA

The statement of the question about the limiting polarization of waves leaving a magnetoactive plasma is connected with the following circumstance. It is physically clear that with small enough v and u, i.e. under weak anisotropy conditions, the effect of the magnetic field on the nature of the waves' polarization (the form of the polarization ellipse) should disappear. If $v \to 0$ the waves should be propagated as in a vacuum; when $u \to 0$—in the same way as in an isotropic plasma; therefore the variation in the polarization coefficients $K_{1,2}$ in a region of small v (or u) ceases and the emission escapes into a vacuum (or into an isotropic plasma) with approximately the same polarization ellipse as it had before entering the region of weak anisotropy. On the other hand, according to the approximation of geometrical optics the polarization ellipse should alter its form even here in accordance with the change in the parameters v and u. This means that with small v or u the geometrical optics solutions become inapplicable and the question of the nature of the limiting polarization when a wave moves from a region with high values of the parameters v, u into low values of v or u requires a closer examination on the basis of equations describing the propagation of electromagnetic waves in a non-uniform magnetoactive plasma.

We shall proceed from the system of equations (22.6) and (22.9). We shall neglect the thermal motion in the plasma, putting $T = 0$, and limit ourselves to discussing a plane-layered medium whose properties depend only on the single coordinate z. Then for plane waves being propagated along the z axis (i.e. for waves the fields in which depend only on this coordinate) this system will become (Ginzburg, 1960b, section 23)

$$\left. \begin{array}{l} \dfrac{d^2 E_x}{dz^2} + \dfrac{\omega^2}{c^2}(AE_x + iCE_y) = 0, \\[6pt] \dfrac{d^2 E_y}{dz^2} + \dfrac{\omega^2}{c^2}(-iCE_x + BE_y) = 0, \end{array} \right\} \quad (24.1)$$

where

$$\left. \begin{array}{c} A = \dfrac{u-(1-v)^2-uv\cos^2\alpha}{u-(1-v)-uv\cos^2\alpha}, \quad B = \dfrac{u(1-v)-(1-v)^2}{u-(1-v)-uv\cos^2\alpha}, \\[6pt] C = \dfrac{\sqrt{uv}(1-v)\cos\alpha}{u-(1-v)-uv\cos^2\alpha}. \end{array} \right\} \quad (24.2)$$

§ 24] Coupling of Waves in a Plasma

The equations (24.1) are written in the system of coordinates shown in Fig. 119.

We introduce instead of E_x, E_y the new variables U_1, U_2 by means of the relations:

$$E_x = \frac{\Pi_1}{\sqrt{1+K_1^2}} + \frac{\Pi_2}{\sqrt{1+K_2^2}}, \quad E_y = \frac{iK_1\Pi_1}{\sqrt{1+K_1^2}} + \frac{iK_2\Pi_2}{\sqrt{1+K_2^2}},$$
$$\Pi_1 = \frac{U_1(z)}{\sqrt{n_1}} e^{-i\frac{\omega}{2c}\int^z (n_2-3n_1)\,dz}, \quad \Pi_2 = \frac{U_2(z)}{\sqrt{n_2}} e^{i\frac{\omega}{2c}\int^z (3n_2-n_1)\,dz}, \quad (24.3)$$

where $n_{1,2}$ are defined by the formula (23.2) and the polarization coefficients $K_{1,2}$ by the formula (23.10). Then by starting from the system of two second-order equations (24.1) in a medium with parameters u and v, which vary slowly over distances of the order of a wavelength λ, we can arrive at a single second-order equation which will approximately describe the propagation of waves in the direction of negative z (Budden, 1952; see also Ginzburg 1960b, section 26):

$$\frac{d^2 U_{1,2}}{dz^2} + \left\{ \Psi^2 + \frac{1}{4}\frac{\omega^2}{c^2}(n_2-n_1)^2 - \frac{i\omega}{2c}\frac{d}{dz}(n_2-n_1) \right\} U_{1,2} = 0. \quad (24.4)$$

After a slightly different substitution from (24.1) we can also obtain a similar equation for describing waves travelling in the opposite direction.†
In (23.4)

$$\Psi(z) = \frac{1}{2}i\frac{d}{dz}\ln\frac{iK_2-1}{iK_2+1} = \frac{i}{4}\frac{d}{dz}\ln\frac{1-v+iq}{1-v-iq}, \quad (24.5)$$

where $q = \sqrt{u}\sin^2\alpha/2\cos\alpha$. Differentiating in Ψ with respect to z we find that

$$\Psi(z) = -\frac{1}{2}\frac{q\frac{dv}{dz}+(1-v)\frac{dq}{dz}}{(1-v)^2+q^2}. \quad (24.6)$$

The derivative dv/dz characterizes the non-uniform nature of the distribution of the electron concentration $N(z)$ in the plasma; the non-zero derivative dq/dz appears in the case of a non-uniform magnetic field $H_0(z)$.

† The fact that waves travelling in opposite directions can be described by independent equations means that the connection between them (coupling of the reflection type) can appear only where these equations become inapplicable. In a smoothly non-uniform plasma the latter occurs only at the points $n_1 = 0$ and $n_2 = 0$ where replacement of the variables of the (24.3) type cannot be made: Π_1 and Π_2 here become infinity with finite values of $U_{1,2}$. The coupling between different types of waves in layers where $n_1 \neq 0$, $n_2 \neq 0$ is not accompanied by reflection effects.

It is easy to see that the solutions of the equation (24.4) and the nature of the wave propagation are essentially different on different sides of a plasma layer in which

$$\Psi^2 \sim \left| \frac{\omega^2}{4c^2}(n_2-n_1)^2 - i\frac{\omega}{2c}\frac{d}{dz}(n_2-n_1) \right|. \tag{24.7}$$

In fact, in regions of a plasma where Ψ^2 is much less than the remaining terms in the brackets of the equation (24.4) the term Ψ^2 can be omitted. Then the solution of the equation (24.4), because of the slowness of the variation in the medium's properties (the difference n_2-n_1) will be the linearly independent functions

$$U_1 = C_1 e^{i\frac{\omega}{2c}\int^z (n_2-n_1)\,dz}, \quad U_2 = C_2 e^{-i\frac{\omega}{2c}\int^z (n_2-n_1)\,dz}$$

(C_1 and C_2 are constants). By substituting them in (24.3) we can see that the field E_x, E_y contains two independent components in the geometrical optics form; the first of them characterizes an extraordinary wave with a refractive index n_1 and a polarization coefficient K_1, and the second an ordinary wave with n_2 and K_2. In plasma layers where Ψ^2 is much greater than the rest of the terms in the brackets the character of the solution is different: in actual fact the solution is the same as in the case of an isotropic plasma, when $n_1 = n_2$, and the polarization of a wave given by the initial or boundary conditions of the problem remains constant in the process of propagation.

The transition from one form of propagation to another in a non-uniform plasma is achieved as follows. When a wave of one type, for example an extraordinary one, is propagated in the direction of escape from the magnetic field or from the plasma first the change in the polarization ellipse proceeds in complete accordance with the requirements of geometrical optics. Then, after passing through the characteristic layer (24.7), the nature of the polarization ceases to change and the wave is propagated further, preserving the polarization acquired in this layer. Therefore on escaping into a region $u = 0$ or $v = 0$ the limiting polarization of the emission will differ from that given by the formula (23.12), so that here the polarization ellipse can be looked upon as the sum of the ellipses characteristic of ordinary and extraordinary waves.[†] Accordingly when a wave of one type approaches a layer (24.7) coherent waves of two

---

[†] Changes in polarization do not occur in the case of strictly longitudinal and transverse propagation relative to a field H_0 with a fixed orientation, since in this case the polarization of the ordinary and extraordinary waves does not in general change, remaining circular ($\alpha = 0$) or linear ($\alpha = \pi/2$) with any u and v.

§ 24] Coupling of Waves in a Plasma

types escape into the isotropic plasma or vacuum; this is the process of wave transformation due to the coupling in the region where geometrical optics are not applicable.†

Let us now see where the transition layer (24.7) is localized in the actual conditions of the solar corona. It is clear from equations (24.6) and (24.7) that in a smoothly non-uniform plasma whose properties vary little over distances of the order of a wavelength the transition layer is located close to where n_1 is close to n_2. Therefore in equation (24.7) we can put (see also (23.2))

$$n_2 - n_1 \approx \frac{n_2^2 - n_1^2}{2n_1} = \frac{v\sqrt{u}\cos\alpha\sqrt{(1-v)^2 + q^2}}{n_1(1 - u - v + uv\cos^2\alpha)}.$$

Further, we take take into consideration the fact that with a slow enough variation in the plasma parameters in the transition layer the term with $d(n_2 - n_1)/dz$ is small when compared with the rest and it can be neglected. Then in the case

$$\left|\frac{1}{1-v}\frac{dv}{dz}\right| \gg \left|\frac{1}{q}\frac{dq}{dz}\right|, \qquad (24.8)$$

when the non-uniform nature of the magnetic field is insignificant, the relation (24.7) with small v becomes

$$\left(\frac{1}{v}\frac{dv}{dz}\right)^2 \sim 4\frac{\omega^2}{c^2}\cot^4\alpha\,\frac{\left(1 + \dfrac{u\sin^4\alpha}{4\cos^2\alpha}\right)^3}{(1-u)^2}. \qquad (24.9)$$

From this it is not hard to estimate the value of v (and the electron concentration N) within the transition layer. If the angle α approaches 0 or $\pi/2$, then $v \to 0$. The maximum v are obtained for intermediate angles $\alpha \sim 1$ and values of u for which $(1-u)^2 \sim 1$:

$$v \sim \frac{c}{\omega}\frac{dv}{dz} \sim \frac{\lambda}{2\pi L_N}, \qquad (24.10)$$

where L_N denotes $|\mathrm{grad}\,(v)|^{-1}$.

If

$$\left|\frac{1}{1-v}\frac{dv}{dz}\right| \ll \left|\frac{1}{q}\frac{dq}{dz}\right| \qquad (24.11)$$

then the relation (24.7) becomes ($v \ll 1$):

$$\left(\frac{1}{q}\frac{dq}{dz}\right)^2 \sim 4\frac{\omega^2}{c^2}v^2\cot^4\alpha\,\frac{(1+q^2)^3}{(1-u)^2}. \qquad (24.12)$$

† In this layer we must obviously include all the layer on the boundary with the isotropic plasma or vacuum, starting from the layer (24.7).

This last relation allows us to estimate the quantity $v\sqrt{\bar{u}}$ (and the product NH_0) in the transition layer. Since $dq/dz \to 0$ when $\alpha \to 0$, the quantity $v\sqrt{\bar{u}}$ in this case also approaches zero. The case of transverse propagation, in which q and dq/dz increase sharply, is exceptional; it will be discussed in detail in the next sub-section. With intermediate angles α and values of u for which $(1-u)^2 \sim 1$

$$v\sqrt{\bar{u}} \sim \frac{c}{\omega}\frac{dq}{dz} \sim \frac{1}{2\pi}\frac{\lambda}{L_H}, \qquad (24.13)$$

where $L_H = |\text{grad } q|^{-1}$.

In the solar corona, depending on the actual conditions, two cases apparently hold—(24.8) and (24.11). In accordance with (24.10) in the first case with $L_N \sim 10^{10}$ cm and $\lambda \sim 3$ cm the transition layer is localized in a region where the parameter $v \sim 5 \times 10^{-11}$; this corresponds to an electron concentration of $N \sim 10^2$ electrons/cm^3. As the wavelength increases the values of v and N decrease still further: for example, at $\lambda \sim 3$ m $v \sim 5 \times 10^{-9}$ and $N \sim 1$ electron/cm^3. In the second case, as follows from (24.13), with $L_H \sim 10^{10}$ cm and $\lambda \sim 3$ cm in the transition layer $v\sqrt{\bar{u}} \sim 5 \times 10^{-11}$ and $NH_0 \sim 2 \times 10^5$ oe electrons/cm^3. Since in the solar corona the magnetic field strength does not probably drop below about 1 oe, the transition layer should be located at a level where $N \lesssim 2 \times 10^5$ electrons/cm^3. At $\lambda \sim 3$m $v\sqrt{\bar{u}} \sim 5 \times 10^{-9}$, $NH_0 \sim 20$ oe electrons/cm^3 and $N \lesssim 20$ electrons/cm^3.

It is clear from these estimates that the region of limiting polarization for the band of radio frequencies being studied is either in the highest layers of the corona (at distances of not less than $3R_\odot$ from the centre of the Sun), or moves about in the circumsolar space. It is quite possible that in certain cases, especially at metric wavelengths, there is in general no such region on the propagation path from the Sun to the Earth and the polarization characteristics vary in complete accordance with the requirements of geometrical optics. However, even the passage of radio emission through a region of limiting polarization located so far from the solar surface is insignificant in practice: because of the low magnetic field strength in the outer corona $\sqrt{u} \ll 1$ and the polarization of ordinary and extraordinary waves is the same as circular (see (23.12)).

It follows from this that the elliptical polarization observed on Earth when receiving certain microwave bursts and types I and III bursts (see Chapter III) cannot be connected with the effect of coupling of the limiting polarization type in the solar corona. It can be explained only if the emission incident on the transition layer contains ordinary and extraordinary waves which are coherent with each other, one having

§ 24] Coupling of Waves in a Plasma

clockwise and the other anticlockwise polarization; only then will the resultant polarization be elliptical in nature.

The impossibility of explaining the ellipticity of the solar radiation by the effect of the limiting polarization can be very clearly seen from the example of microwave bursts. According to Akabane's observations (Akabane, 1958c; Cohen, 1960) at a frequency of 9500 Mc/s a whole series of microwave bursts were elliptically polarized, the ellipticity value averaged over many bursts being $\bar{p} \approx 0.8$ (i.e. the emission differs noticeably from circularly polarized emission). If we assume that in this case a wave of only one type approaches the limiting polarization region, then according to what we have been saying above the ellipticity of the radio emission that has left the corona will with sufficient accuracy for our purposes be defined by the formula (23.12) in which the values of the parameter u and the angle α relate to the transition layer (24.7):†

$$p = \frac{2\sqrt{u}\,|\cos\alpha\,|}{u\sin^2\alpha + \sqrt{u^2\sin^4\alpha + 4u\cos^2\alpha}}.$$

The angle α is unknown for each separate measurement since the orientation of the magnetic field $\boldsymbol{H_0}$ in the transition layer on the Sun is unknown. Since, however, the generation regions of the bursts studied belonged to different centres of activity scattered all over the Sun's disk it may be hoped that the assumption about the equal probability of all directions of the field $\boldsymbol{H_0}$ relative to the wave vector $\boldsymbol{k}$ will be quite permissible in this case. Then the observed values of p can be averaged over the directions (Gelfreich, 1962):

$$\bar{p} = \frac{1}{4\pi}\int_{4\pi} p\,d\Omega = \int_0^{\pi/2} p\sin\alpha\,d\alpha = \frac{1}{2}\left[1 + \sqrt{u}\ln\sqrt{u} + \frac{1-u}{\sqrt{u}}\ln(1+\sqrt{u})\right] \tag{24.14}$$

($d\Omega$ is an element of the solid angle in the direction $\boldsymbol{H_0}$).

According to (24.14) the observed value $\bar{p} \approx 0.8$ corresponds to $\sqrt{u} \approx 0.15$, i.e. $H_0 \approx 500$ oe. However, there is no basis for the existence of such strong fields in the corona at distances of $R > 3R_\odot$, where the region of extreme polarization may be located. Gelfreich (1962) avoided this difficulty by transferring the layer forming the extreme polarization of the emission to the inner corona (to the boundary of the emitting region) by the large gradients of v or u obtaining in non-stationary phenomena of the shock-wave type. Here, however, a fresh complication arises con-

---

† Here it is taken that the ellipticity p is the ratio of the short axis of the ellipse of polarization to the long one. With this definition the value of p is the same for ordinary and extraordinary waves and $0 \leq p \leq 1$.

nected with the great length of the path travelled by the emission from the transition layer until it leaves the corona: the strong dispersion of the angle of Faraday rotation would in this case make it impossible to record elliptical polarization on Akabane's polarimeter with a bandwidth of ~ 8 Mc/s (see the last subsection of section 23).

PRELIMINARY REMARKS ON THE EFFECT OF THE COUPLING OF WAVES IN THE REGION OF A QUASI-TRANSVERSE MAGNETIC FIELD

In the preceding section it was noted that when determining the position of the transition layer that gives the limiting polarization of waves when they leave the corona the case of quasi-transverse wave propagation is exceptional. This is connected with the sharp increase in the parameter q and the derivative dq/dz when $\alpha \approx \pi/2$ by comparison with their values in the region $\alpha \lesssim 1$; as will be shown below, the latter may lead to displacement of the transition layer into regions with higher values of v and $\sqrt{u}$ (i.e. the electron concentration N and the magnetic field H_0) and the appearance of interesting effects of great importance in interpreting the polarization features of solar radio emission.

The unusualness of the case of quasi-transverse propagation can easily be confirmed by starting with the relation (24.12) which is used to fix the position of the transition layer in a plasma with a non-uniform magnetic field. In actual fact, saying for the sake of simplicity that u and v are much less than unity, we obtain that in the transition layer

$$\left(\frac{dq}{dz}\right)^2 \sim 4\frac{\omega^2}{c^2}v^2q^8\cot^4\alpha\left(\frac{1}{q^2}+1\right)^3. \qquad (24.15)$$

The parameter $q \equiv \sqrt{u}\sin^2\alpha/2\cos\alpha$ changes because of the change in the magnetic field H_0 in magnitude and in direction (relative to the vector k):

$$\frac{dq}{dz} \approx \frac{1}{2\cos\alpha}\frac{d\sqrt{u}}{dz}+\frac{\sqrt{u}}{2\cos^2\alpha}\frac{d\alpha}{dz}.$$

The last equation takes into consideration the fact that for angles α close to $\pi/2$, $\sin\alpha \approx 1$ and $|\cos\alpha| \ll 1$. If in the transition layer

$$\frac{dq}{dz} \approx \frac{\sqrt{u}}{2\cos^2\alpha}\frac{d\alpha}{dz}, \qquad (24.16)$$

which with quasi-transverse propagation holds for a magnetic field in which†

$$\frac{d\alpha}{dz} \gg \left|\frac{1}{H_0}\frac{dH_0}{dz}\right|\cos\alpha, \qquad (24.17)$$

† This condition is obviously always realized when a wave passes through a layer where $H_0 \perp k$, being valid in a certain region around this layer. On the other hand, the

§ 24] Coupling of Waves in a Plasma

then the relation (24.15) will become the following:

$$\left(\frac{d\alpha}{dz}\right)^2 \sim \frac{1}{16}\frac{\omega^2}{c^2} v^2(4\cos^2\alpha + u)^3. \tag{24.18}$$

It is clear from this that the transition layer for waves of a frequency ω may be located at angles α not too close to $\pi/2$ (in particular with $4\cos^2\alpha \gg u$, when $q^2 \ll 1$) only in a rarefied plasma where the value of the parameter v is sufficiently small:

$$v \sim \frac{c}{\omega}\frac{d\alpha}{dz}\frac{1}{\cos^3\alpha} \sim \frac{\lambda}{2\pi L_H \cos^3\alpha}, \tag{24.19}$$

although it is greater than in the case of $\alpha \sim 1$ (see the formula (24.13)). If, however, the parameter v in the layer where $H_0 \perp k$ is greater than (24.19), then the transition layer moves into a region with lower values of $\cos^2\alpha$ (first into the region $q^2 \sim 1$, then $q^2 \gg 1$).† This holds until the value of v reaches its maximum defined by the relation

$$\left(\frac{d\alpha}{dz}\right)^2 \sim \frac{1}{16}\frac{\omega^2}{c^2} v^2 u^3. \tag{24.20}$$

As v rises further the transition layer disappears.

Introducing into the discussion (in accordance with (24.20)) the critical parameter

$$G = \frac{\omega v u^{3/2}}{4c\left|\frac{d\alpha}{dz}\right|} = \frac{e^5 N H_0^3}{16\pi^3 c^4 m^4 f^4 \left|\frac{d\alpha}{dz}\right|}, \tag{24.21}$$

where $f = \omega/2\pi$ is the frequency and the values of N, H_0 and $d\alpha/dz$ relate to the transverse magnetic field region, we shall have, in accordance with what has been said, the following picture of wave propagation through a quasi-transverse field region (Cohen, 1960).‡

Since the parameter G is a function of the frequency, the nature of the wave propagation through a transverse magnetic field region also depends

---

reverse inequality with $|\cos\alpha| \ll 1$ obtains in practice only in rare cases when the change in direction of the field H_0 relative to k is negligibly small when compared with the change of H_0 in magnitude. We shall not be discussing these cases.

† We recall that the inequalities $q^2 \ll 1$ and $q^2 \gg 1$ respectively characterize the regions of quasi-longitudinal and quasi-transverse propagation (see (23.5) and (23.6)).

‡ Cohen arrived at the expression (24.21) as the result of a clumsier treatment, since he based himself on more general equations than (24.1) and (24.2), which make it possible if necessary to investigate the propagation and interaction of waves while making allowance for "twisting" of the magnetic field H_0 (a change in the orientation of the component H_0 in the plane xy). Our equations are unsuitable for this since they are written for the case when the vector H_0 is located in a single plane yz in all points in space (see Fig. 119).

on the frequency. At low enough frequencies (such that $G \gg 1$) there is no transition layer and the geometrical optics approximation still holds. Here, as we know, the ordinary and extraordinary waves are propagated independently of each other; as it passes through a transverse magnetic field region each wave changes the sign of its rotation to the opposite because of the change in direction of the longitudinal component of the magnetic field.

On a rise in frequency (with $G \sim 1$) on both sides of the $\boldsymbol{H_0} \perp \boldsymbol{k}$ layer there appear transition layers that separate the regions where the geometrical optics are satisfied or not. In this case a wave in the range between these layers is propagated, as in an isotropic medium, preserving the degree of ellipticity and the sign of the polarization. This occurs even in a region of quasi-transverse propagation where, according to the laws of geometrical optics, the wave should become linearly polarized and change its sign of rotation on leaving this region.

At high frequencies, when $G \ll 1$, the transition layers are localized a comparatively long way from the $\boldsymbol{H_0} \perp \boldsymbol{k}$ level in regions where quasi-longitudinal propagation with circular polarization of the ordinary and extraordinary waves is valid. If in the process of propagation a wave of one type (let us say an ordinary one) enters a region contained between two transition layers, then it leaves it preserving circular polarization and the former sign of rotation. However, this wave will now correspond to another type (extraordinary), since the direction of the longitudinal component of the magnetic field $\boldsymbol{H_0}$ will become the opposite.

Therefore the case $G \ll 1$ corresponds to strong coupling of "normal" waves, a wave of one type changing completely into a wave of another type, and the case $G \gg 1$ to the absence of interaction, when the type of wave remains constant. At frequencies of

$$f_t \sim \left[\frac{e^5}{16\pi^3 m^4 c^4} \frac{NH_0^3}{\left|\frac{d\alpha}{dz}\right|} \right]^{1/4}, \qquad (24.22)$$

for which $G \sim 1$, the intermediate case occurs: if a wave of one type, let us say an ordinary one, approaches the interaction region, then two waves —an ordinary one and an extraordinary one—leave it. The ratio of the intensities of the two waves depends on the frequency and somewhere in the region $f \sim f_t$ it becomes equal to unity.

Since the transformed waves are coherent and have different signs of rotation the resultant polarization of the emission at the frequency $f \sim f_t$ will be purely linear. As we move away from this frequency the degree of linear polarization ϱ_l will decrease from 1 to 0, and the degree of circular polarization ϱ_c increase from 0 to 1. The values $\varrho_l \approx 0$, $\varrho_c \approx 1$ hold when $G \gg 1$ and when $G \ll 1$. The change in the polarization of waves during propagation through a quasi-transverse magnetic field region depending on the value of the characteristic parameter G is illustrated in Fig. 127.

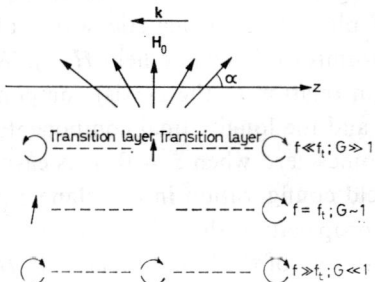

Fig. 127. Propagation of waves through a quasi-transverse magnetic field region

CALCULATIONS OF COUPLING BY THE PHASE INTEGRAL METHOD

The above discussion of the effect of wave interaction in a transverse magnetic field region is, of course, far from being exhaustive. It is based on estimates of the various terms in the wave propagation equations (24.4); this is not an entirely correct operation and may, generally speaking, lead to unreliable results. In addition, calculations of this kind make it possible only to estimate the characteristic ranges of frequencies at which strong and weak interaction occurs without allowing us to plot the frequency dependence of the interaction efficiency and the dependence of the degree of linear and circular polarization connected with it for emission leaving an interaction region. This problem has been solved by Zheleznyakov and Zlotnik (1963) where the interaction of "normal" waves in a plasma placed in a non-uniform magnetic field is investigated by the phase integral method. This method is particularly valuable since it allows us to use the geometrical optics approximation to find the connection between the amplitudes of interacting waves of different types a long way from the interaction region without knowing the precise solution of the equations (24.1) and using only certain of its general properties (in particular the analyticity in the complex plane z).

The phase integral method (Zwaan's method) was developed by Stückelberg (1932) for the solution of certain problems in quantum mechanics and then extended by Denisov (1955, 1957) to the theory of radio wave propagation in a non-uniform plasma. Here we shall describe the features of this method as briefly as possible with the example of the interaction of electromagnetic waves in a quasi-transverse magnetic field region, having in mind also the further applications of this method to the solution of the problem of the escape of electromagnetic waves from a magneto-active coronal plasma because of their interaction in a layer where $v \approx 1$ (see section 25).†

Therefore, following Zheleznyakov and Zlotnik (1963), let us examine the propagation of plane electromagnetic waves along the z axis in a plasma set in a non-uniform magnetic field $H_0(z)$. We select the latter so that its transverse (in relation to the z-axis) component does not depend on the coordinates, and the longitudinal component is a monotonic function of z which becomes zero when $z = 0$. It is easy to see that with this kind of magnetic field configuration in the plane $z = 0$ we have the case of interest to us of propagation through a transverse field region accompanied by a change in sign of the longitudinal field H_{0z}.‡

In the solution of the problem stated we can proceed from the system of equations (24.1). Here, however, it is not the equations for the field components E_x and E_y that are more convenient but the equations for their combinations $F_1 = E_x + iE_y$, $F_2 = E_x - iE_y$ (see, e.g., Zheleznyakov, 1958b):

$$\left.\begin{aligned}\frac{d^2 F_1}{dz^2} + k_0^2(C+B)F_1 &= -ik_0^2 A F_2, \\ \frac{d^2 F_2}{dz^2} + k_0^2(C-B)F_2 &= ik_0^2 A F_1,\end{aligned}\right\} \quad (24.23)$$

† Further detail on the phase integral method can be obtained from the papers by Stückelberg (1932), Denisov (1955, 1957), Zheleznyakov (1958b, 1959b); see also Ginzburg (1960b, section 28).

‡ It is appropriate to stress that this kind of field, strictly speaking, cannot exist in a plasma since for it div $H_0 \neq 0$ unlike the equation (22.1). The condition div $H_0 = 0$, it is true, is easily satisfied (preserving at the same time the configuration we need), unless H_0 depends on two coordinates instead of one. However, investigation of wave propagation in such a system becomes far more complicated. Therefore, as in the papers of Cohen (1960), Zheleznyakov and Zlotnik (1963), we shall model the magnetic field as indicated above, considering formally that the plasma is combined with a magnetic field which is noticeably magnetized by a stationary field H_0 (such that div $H_0 = -4\pi$ div M, where M is the magnetization vector), whilst at high frequencies there is a magnetic permeability of unity (div $H = 0$). The latter permits the existence in the plasma of any static magnetic field (with an appropriate choice of $M(z)$) and at the same time has no direct effect on the nature of the radio wave propagation.

It is clear from what we have said above that considerable interaction in a quasi-transverse magnetic field region can occur only in a sufficiently rarefied plasma ($v \ll 1$) where the refractive indices of the ordinary and extraordinary waves are close to unity. It may be taken, therefore, that results obtained with a one-dimensional model reflect with sufficient accuracy the nature of the interaction under the actual conditions of the solar corona, since the presence of a weak dependence of the magnetic field on the coordinates x, y with $n_{1,2} \approx 1$ does not essentially alter the structure of the waves being propagated.

§ 24] Coupling of Waves in a Plasma

where

$$A = -\frac{(1-\varepsilon)\omega_y^2}{2\omega_y^2-(1-\omega_z^2)\varepsilon}, \quad B = \frac{(1-\varepsilon)\omega_z\varepsilon}{2\omega_y^2-(1-\omega_z^2)\varepsilon},$$

$$C = 1 - \frac{(1-\varepsilon)(\omega_y^2-\varepsilon)}{2\omega_y^2-(1-\omega_z^2)\varepsilon}, \quad (24.24)$$

$$k_0 = \frac{\omega}{c}, \quad \varepsilon = 1-v, \quad \omega_x = \omega_y = \frac{eH_{0x}}{mc\omega}, \quad \omega_z = \frac{eH_{0z}}{mc\omega}.$$

The expressions given for A, B, C differ from (24.2) because the equations (24.23) are written in another system of coordinates rotated 45° around the z-axis relative to that shown in Fig. 119. As can be seen in Fig. 128, in the new system the field $\mathbf{H_0}$ is symmetrical relative to the x- and y-axes (i.e. $H_{0x} = H_{0y}$).

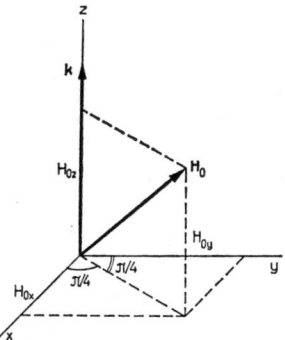

FIG. 128. Orientation of vectors k and $\mathbf{H_0}$ relative to coordinate axes

The coefficients A, B and C depend on z via the parameter ω_z, which in the case of a linear function $H_{0z}(z)$ can be represented in the form

$$\omega_z = \Lambda z, \quad (24.25)$$

where Λ is a constant equal to $(e/mc\omega)(dH_{0z}/dz)$. In the plane $z = 0$, therefore the magnetic field is at right angles to the directions of the wave's propagation.

Introducing the new independent variable $\zeta = \Lambda z$ and denoting the ratio k_0/Λ by ϱ_H,† we can write the system (24.23) in the following form:

$$\left.\begin{array}{l}\dfrac{d^2F_1}{d\zeta^2}+\varrho_H^2(C+B)F_1 = -i\varrho_H^2 AF_2, \\[6pt] \dfrac{d^2F_2}{d\zeta^2}+\varrho_H^2(C-B)F_2 = i\varrho_H^2 AF_1.\end{array}\right\} \quad (24.26)$$

† The parameter $\varrho_H \gg 1$ if there is little alteration in the magnetic field over a distance of the order of a wavelength.

By eliminating F_2 from these equations we obtain for F_1 a fourth-order equation with variable coefficients whose solution we shall find in the approximation of geometrical optics:

$$F_1 = e^{\varrho_H \left[S_0(\zeta) + \frac{1}{\varrho_H} S_1(\zeta) + \ldots \right]}. \tag{24.27}$$

Then for the function $(dS_0/d\zeta)^2$ we obtain two values n_1^2 and n_2^2:

$$\left(\frac{dS_0}{d\zeta}\right)^2_{1,2} \equiv n_{1,2}^2 = \frac{(\omega_y^2 - \varepsilon) + \varepsilon(\omega_z^2 + \omega_y^2) \pm (1-\varepsilon)\sqrt{\omega_y^4 + \varepsilon^2 \omega_z^2}}{2\omega_y^2 - \varepsilon(1-\omega_z^2)}, \tag{24.28}$$

which define the possibility of the existence in the plasma of two types of wave—ordinary and extraordinary. The function $S_1(\zeta)$ in its turn satisfies the equation (see Zheleznyakov and Zlotnik, 1963)

$$\left(\frac{dS_1}{d\zeta}\right)_{1,2} = -\frac{1}{2} \frac{\left(\frac{dS_0}{d\zeta}\right)_{1,2}}{S_{01,2}} \mp \frac{1}{2} \frac{\frac{d\sigma}{d\zeta}}{\sqrt{1+\sigma^2}} - \frac{1}{2} \frac{\sigma \frac{d\sigma}{d\zeta}}{1+\sigma^2}, \tag{24.29}$$

where

$$\sigma = -\frac{B}{A} = \frac{\varepsilon \omega_z}{\omega_y^2} = \frac{\zeta}{\zeta_0}, \quad \zeta_0 = \omega_y^2 (1-v)^{-1}. \tag{24.30}$$

Substituting (24.28), (24.29) in (24.27) we obtain the general solution for F_1 in the form

$$F_1(\zeta) = e^{-\frac{1}{2}\int \frac{\sigma\, d\sigma}{(1+\sigma^2)}} \left[(c_1 e^{i\varrho_H \int n_1\, d\zeta} + d_1 e^{-i\varrho_H \int n_1\, d\zeta}) e^{-\frac{1}{2}\int \frac{dn_1}{n_1} - \frac{1}{2}\int \frac{d\sigma}{\sqrt{1+\sigma^2}}} \right.$$
$$\left. + (c_2 e^{i\varrho_H \int n_2\, d\zeta} + d_2 e^{-i\varrho_H \int n_2\, d\zeta}) e^{-\frac{1}{2}\int \frac{dn_2}{n_2} + \frac{1}{2}\int \frac{d\sigma}{\sqrt{1+\sigma^2}}} \right]. \tag{24.31}$$

Since, as has already been pointed out in the preceding subsection, without allowing for reflection regions in which $n_{1,2}^2 \approx 0$, waves travelling in different directions are completely independent,[†] their simultaneous discussion complicates the solution without leading here to any fresh results. Therefore we shall limit ourselves to investigating waves travelling towards negative ζ and write the approximation of geometrical optics in the form[‡]

[†] That is, the constants c_1, c_2 and d_1, d_2 are not interconnected: interaction in a quasi-transverse magnetic field region occurs only between waves of different types being propagated in one and the same direction.

[‡] We recall that the electromagnetic field in these waves, by assumption, is time-dependent as $e^{i\omega t}$.

§ 24] Coupling of Waves in a Plasma

$$F_1(\zeta) = e^{i\varrho_H \int \frac{1}{2}(n_1+n_2)\,d\zeta - \frac{1}{2}\int \frac{\sigma\,d\sigma}{1+\sigma^2}}$$

$$\times \left(\frac{c_1}{\sqrt{n_1}} e^{-i\varrho_H \int \frac{1}{2}(n_2-n_1)\,d\zeta - \frac{1}{2}\int \frac{d\sigma}{\sqrt{1+\sigma^2}}} \right.$$

$$\left. + \frac{c_2}{\sqrt{n_2}} e^{i\varrho_H \int \frac{1}{2}(n_2-n_1)\,d\zeta + \frac{1}{2}\int \frac{d\sigma}{\sqrt{1+\sigma^2}}} \right). \tag{24.32}$$

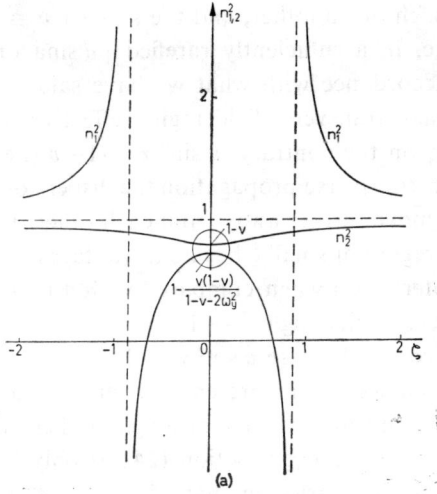

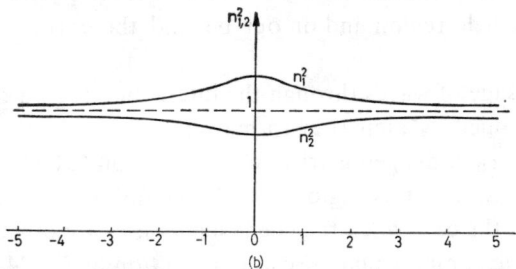

FIG. 129. $n_{1,2}^2$ as a function of ζ in a plasma with a non-uniform magnetic field of the (24.25) type: (a) $u \sin^2 \alpha < 1-v$, (b) $u \sin^2 \alpha > 1-v$ (Zheleznyakov and Zlotnik, 1963).

This solution for arbitrary c_1 and c_2 describes independent waves of two types (extraordinary and ordinary) and is an asymptotic representation of the solution with $\varrho_H \gg 1$ outside the regions of interaction and reflection. Since geometrical optics are violated in regions characterized by a sharp change in the amplitude of the waves in space, it is clear from (24.23) that this occurs (with large, but finite ϱ_H) in the vicinity of points where $n_1 = \infty$, $n_1 = 0$, $n_2 = 0$, and $n_1 = n_2$. The last equality corresponds to points $\sigma = \zeta/\zeta_0 = \pm i$ lying in the complex plane ζ.

Figure 129 shows the squares of the refractive indices $n_{1,2}^2$ as a function of the parameter $\omega_z = \zeta$. The shape of the dispersion curves is essentially different with $u \sin^2 \alpha = 2\omega_y \gtrless 1-v$ (see Fig. 129 a and b). If $u \sin^2 \alpha < 1-v$ (i.e. $\zeta_0 < \frac{1}{v}$), then the case of quasi transverse propagation which occurs in a region of small ζ is isolated in the respect that the dispersion curves here approach one another, and the smaller $v \equiv 1-\varepsilon$ is the closer they approach (i.e. in a sufficiently rarefied plasma or at high enough frequencies). In accordance with what we have said earlier, under these conditions in a quasi-transverse field region effective wave interaction is to be expected. If, on the contrary, $u \sin^2 \alpha > 1-v$ (i.e. $\zeta_0 > \frac{1}{2}$), then in the region of quasi-transverse propagation the dispersion curves $n_{1,2}^2(\zeta)$ do not approach but move further away from each other. It is clear from this that in this case a region of small ζ has no advantages in the sense of effective radio wave interaction when compared with a region of large ζ corresponding to values of the angle $\alpha \sim 1$.

Limiting ourselves to the case $u \sin^2 \alpha < 1-v$, we note that below we are making no allowance for the presence of zeroes and poles in the function $n_{1,2}^2(\zeta)$, considering that the actual magnetic field in the corona can be approximated by the linear function (24.25) only with small enough ω_z (and H_{0z}); then H_{0z} varies so that n_1^2 does not reach zero and an extraordinary wave, and likewise an ordinary one, passes through the quasi-transverse field region and on out beyond the corona without hindrance.

Since the passage of waves through the interaction region circled in Fig. 129a is accompanied by their transformation into waves of another type, the constants c_1, c_2 in the geometrical optics solution (24.32) will be different on the two sides of this region. The latter follows formally from the fact that only in the case when the coefficients c_1, c_2 are different in different regions of the complex plane ζ can the solution of F_1 (24.32), which is a combination of many-valued functions, approximate the precise solution of the system (24.26), which is an analytic (i.e. single-valued) function. It will be our task to find the connection between the values of the constants characterizing the amplitude of the ordinary and extraordinary waves on different sides of the interaction region.

It is known (Stückelberg, 1932) that in solutions like (24.32) the constants c_1, c_2 remain constant in regions bounded by the so-called Stokes lines (Stokes, 1904; Batson, 1949); when they cross these lines they change

§ 24] Coupling of Waves in a Plasma

with a jump. These lines, which can be determined from the condition

$$\arg\left[\pm i\varrho_H \int_{\pm i\zeta_0}^{\zeta} \tfrac{1}{2}(n_2-n_1)\,d\zeta\right] = \pi + 2\pi m \qquad (m=0,1,2\ldots), \quad (24.33)$$

are obviously characterized in that on them the functions

$$\exp\left[\pm i\varrho_H \int_{\pm i\zeta_0}^{\zeta} \tfrac{1}{2}(n_2-n_1)\,d\zeta\right],$$

contained in the approximation of geometrical optics (24.32) takes up (as ζ moves away from the points $\pm i\zeta_0$) real values that may be as large or as small as one likes, whilst the ratio of these functions becomes extremal.

As it passes through the Stokes line only one of the two constants changes with a jump, namely the constant in the solution with the exponentially decreasing function; in this case the magnitude of the jump $\Delta c_1 = c_1' - c_1$ is proportional to the coefficient c_2 with an increasing exponent:† $c_1' = c_1 + \alpha c_2$; $c_2' = c_2$. Similar relations also hold on the Stokes line where the exponent with the coefficient c_1 becomes an increasing one: $c_2' = c_2 + \beta c_1$; $c_1' = c_1$.

For plotting the Stokes lines we use the relation (24.28), according to which

$$n_2 - n_1 \equiv \frac{n_2^2 - n_1^2}{n_2 + n_1} = f(\zeta)\sqrt{\zeta^2 + \zeta_0^2}, \qquad (24.34)$$

where the function $f(\zeta)$ is analytical in the vicinity of the points $\zeta = \pm i\zeta_0$. In the case when $n_2 - n_1 = \sqrt{\zeta^2 + \zeta_0^2}$ the position of the Stokes lines on the plane ζ cut along the line connecting the points $i\zeta_0$ and $-i\zeta_0$ is shown in Fig. 130a. The appearance in the difference $n_2 - n_1$ of the analytical function $f(\zeta)$ causes a continuous deformation of the Stokes lines (such as that shown in Fig. 130b, let us say), but does not alter the general nature of these lines; a knowledge of their precise configuration is not necessary when discussing the coupling. We notice that when passing through the solid lines the coefficient c_2 jumps and when passing through the dotted ones c_1 jumps.

† This is quite natural since owing to the linearity of the original system of equations the coefficients on different sides of the Stokes line should be connected by a linear function of the $c_1' = \bar{\alpha}c_1 + \alpha c_2$ type. However, we must demand that $\bar{\alpha} = 1$, i.e. that there is no jump, to preserve the validity of the asymptotic approximation with $c_2 = 0$. At the same time the fact that α is non-zero (and therefore a jump appears in the term with the exponentially decreasing function with $c_2 \neq 0$) does not reduce the accuracy of the geometrical optics approximation to a rigorous solution, since on the Stokes lines the error in the exponentially increasing function becomes of the order of the absolute magnitude of the exponentially decreasing function.

Let us now assume that on the right of the interaction region the geometrical optics solution is of the form (24.32), the lower limit of the integrals in the exponents being taken at the point A. Then the initial values of the coefficients in front of the functions $\exp\left[\mp i\varrho_H \int \frac{1}{2}(n_2-n_1)\,d\zeta\right]$ will be c_1 and c_2 respectively. After going round the coupling points on the line

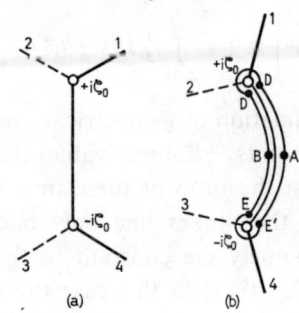

Fig. 130. Arrangement of the Stokes lines on the plane ζ: (a) for $n_2 - n_1 = \sqrt{\zeta^2 + \zeta_0^2}$; (b) for $n_2 - n_1 = f(\zeta)\sqrt{\zeta^2 + \zeta_0^2}$ (Zheleznyakov and Zlotnik, 1963)

shown in Fig. 130b the values of the coefficients in front of these functions at the point B will be

$$C_1 = M_2[M_1 c_1 + \beta(N_1 c_2 + \alpha M_1 c_1)]G_1, \\ C_2 = N_2[N_1 c_2 + \alpha M_1 c_1]G_1, \quad (24.35)$$

and likewise at the point A

$$c_1 = M_4[M_3 C_1 + \delta(N_3 C_2 + \gamma M_3 C_1)]G_2, \\ c_2 = N_4[N_3 C_2 + \gamma M_3 C_1]G_2. \quad (24.36)$$

By virtue of the analytical nature of the precise solution of the system (24.26) it (just like its asymptotic approximation (24.32)) cannot depend on the path of the circuit. In particular when passing along the contour around the points $\pm i\zeta_0$ the coefficients at the point A should be the same as their original values c_1 and c_2, which is allowed for in the relations (24.36). Substituting (24.35) in (24.36) we obtain

$$G_1 G_2[M_1 M_2 M_3 M_4(1+\alpha\beta)(1+\gamma\delta)c_1 + M_1 N_2 N_3 M_4 \alpha\delta c_1 \\ + N_1 M_2 M_3 M_4(1+\gamma\delta)c_2 + N_1 N_2 N_3 M_4 \delta c_2] = c_1, \\ G_1 G_2[M_1 M_2 M_3 N_4 \gamma(1+\alpha\beta)c_1 + M_1 N_2 N_3 N_4 \alpha c_1 \\ + N_1 M_2 M_3 N_4 \beta\gamma c_2 + N_1 N_2 N_3 N_4 c_2] = c_2. \quad (24.37)$$

§ 24] Coupling of Waves in a Plasma

These relations should be identically satisfied with any values of the constants c_1, c_2 and therefore

$$\left.\begin{aligned} M_1M_2M_3M_4(1+\alpha\beta)(1+\gamma\delta)+M_1N_2N_3M_4\alpha\delta &= (G_1G_2)^{-1}, \\ N_1M_2M_3M_4\beta(1+\gamma\delta)+N_1N_2N_3M_4\delta &= 0, \\ N_1M_2M_3N_4\beta\gamma+N_1N_2N_3N_4 &= (G_1G_2)^{-1}, \\ M_1M_2M_3N_4\gamma(1+\alpha\beta)+M_1N_2N_3N_4\alpha &= 0. \end{aligned}\right\} \quad (24.38)$$

These equations allow us to find the proportionality coefficients α, β, γ, δ between the size of the jump of one of the constants and the other constant in the solution (24.32) when crossing the corresponding Stokes lines 1, 2, 3, 4 (see Fig. 130b) and thus determine the values of the constants in the complex plane ζ with given values of c_1, c_2 at the point A.

In the relations (24.35)–(24.38)

$$\left.\begin{aligned} M_1 &= \exp\left[-i\varrho_H \int_A^D \frac{1}{2}(n_2-n_1)\,d\zeta - \frac{1}{2}\int_A^D \frac{d\sigma}{\sqrt{1+\sigma^2}}\right], \\ N_1 &= \exp\left[i\varrho_H \int_A^D \frac{1}{2}(n_2-n_1)\,d\zeta + \frac{1}{2}\int_A^D \frac{d\sigma}{\sqrt{1+\sigma^2}}\right]; \end{aligned}\right\} \quad (24.39)$$

the corresponding quantities for the ranges $D'B$, BE and $E'A$ are denoted by M_2, M_3, M_4 and N_2, N_3, N_4. At the same time

$$G_1 = \exp\left(-\frac{1}{2}\int_D^{D'} \frac{\sigma\,d\sigma}{1+\sigma^2}\right), \quad G_2 = \exp\left(-\frac{1}{2}\int_E^{E'} \frac{\sigma\,d\sigma}{1+\sigma^2}\right). \quad (24.40)$$

By stretching the integration contour to the points $\pm i\zeta_0$ and remembering that in this case it passes along the Stokes lines connecting $i\zeta_0$ and $-i\zeta$ it can be confirmed (see (24.33)) that the integral

$$-i\varrho_H \int (n_2-n_1)\,d\zeta \quad (24.41)$$

in (24.39) which is taken within the limits from A to D is a real and positive quantity which we shall denote by δ_0. Integration over the intervals $D'B$, BE, $E'A$ gives the same result, whilst the intervals DD' and EE' make no contribution to (24.41), so the characteristic parameter is

$$\delta_0 = -i\varrho_H \oint \frac{n_2-n_1}{4}\,d\zeta. \quad (24.42)$$

The integral

$$\int_A^D d\sigma/\sqrt{1+\sigma^2}$$

361

figuring in the expressions for M_1, N_1 is equal to $i\pi$. It keeps its value in the interval $D'B$ and changes its sign in the intervals BE and $E'A$. In G_1, G_2 the integrals $\int \sigma\, d\sigma/\sqrt{1+\sigma^2}$ along the contours round the points $i\zeta_0$ and $-i\zeta_0$ equal $i\pi$.

It follows from the above that the equations (24.38) are equivalent to the system

$$\left.\begin{array}{r}-(1+\alpha\beta)(1+\gamma\delta)\, e^{2\delta_0}+\alpha\delta=-1,\\ -\beta(1+\gamma\delta)\, e^{2\delta_0}+\delta=0,\\ \beta\gamma-e^{-2\delta_0}=-1,\\ \gamma(1+\alpha\beta)-\alpha e^{-2\delta_0}=0,\end{array}\right\} \quad (24.43)$$

whose solution

$$\alpha=\beta=\gamma=\delta=e^{i\pi/2}\sqrt{1-e^{-2\delta_0}} \quad (24.44)$$

allows us to find the values of the constants in the approximation of geometrical optics on the left of the interaction region (at the point B) from their values c_1 and c_2 at the point A (see (24.35)):†

$$\left.\begin{array}{l}c'_2 \equiv C_1 = e^{-\delta_0} c_1 + \sqrt{1-e^{-2\delta_0}}\, c_2,\\ c'_1 \equiv C_2 = \sqrt{1-e^{-2\delta_0}}\, c_1 - e^{-\delta_0} c_2.\end{array}\right\} \quad (24.45)$$

The squares of the constants in (24.45) obviously characterize the intensities of the corresponding waves. As was to be expected, the total intensity of the waves entering the interaction region is equal to the total intensity of the waves leaving this region:

$$|c_1|^2+|c_2|^2=|c'_1|^2+|c'_2|^2. \quad (24.46)$$

It is clear from the relations (24.25) that the connection between the ordinary and extraordinary waves as they pass through a quasi-transverse magnetic field region is determined by the characteristic parameter $2\delta_0$. At the limit with $2\delta_0 \ll 1$ the transformation effect is maximal—a wave of one type changes completely into a wave of another type ($c'_2 = c_1$; $c'_1 = c_2$), whilst with $2\delta_0 \gg 1$ there is no interaction ($c'_2 = c_2$; $c'_1 = c_1$).

The expression for δ_0 is obtained above in the form of the contour integral (24.42), whose calculation in the general case is difficult. The problem is simplified, however, with the condition $v \ll 1$, which is the only strong interaction expected. Here along the integration contour $n_1 \approx n_2 \approx 1$ and therefore

$$2\delta_0 \approx -i\varrho_H \oint \frac{n_2^2-n_1^2}{4}\, d\zeta.$$

---

† The replacement of the notations $C_2 \to c'_1$ and $C_1 \to c'_2$ is explained by the fact that when going round the branching point of the function n_2-n_1 (the point $+i\zeta_0$) the refractive index n_2 changes into n_1 and vice versa; thanks to this the constant C_2 characterizes to the left of the interaction region the amplitude of wave 1 (extraordinary) and C_1 the amplitude of wave 2 (ordinary).

§ 24] Coupling of Waves in a Plasma

Substituting the expressions (24.28) in this as $n_{1,2}^2$ we obtain (Zheleznyakov and Zlotnik, 1963):

$$2\delta_0 \approx \frac{i\varrho_H v}{2} \oint \frac{\sqrt{\zeta^2 + \zeta_0^2}}{\zeta^2 + 2\zeta_0^2 - 1} d\zeta = -\pi\varrho_H v \frac{\sqrt{1-2\zeta_0}+\zeta_0-1}{\sqrt{1-2\zeta_0}}. \quad (24.47)$$

Here $\zeta_0 = \omega_y^2/(1-v) \approx \omega_y^2$. In the case when $2\zeta_0 \approx 2\omega_y^2 \ll 1$ (i.e. with the condition that in the quasi-transverse field region $u = \omega_H^2/\omega^2 \ll 1$) the coupling parameter is

$$2\delta_0 \approx \frac{\pi\varrho_H v\omega_y^4}{2} = \frac{\pi}{2} \frac{\omega v \omega_y^4}{c \left|\frac{d\omega_z}{dz}\right|}. \quad (24.48)$$

It can also be written slightly differently if we remember that in the transverse field region $d\omega_z/dz = \sqrt{2}\omega_y(d\alpha/dz)$ and $2\omega_y^2 = u$:

$$2\delta_0 \approx \frac{\pi\omega v u^{3/2}}{8c\left|\frac{d\alpha}{dz}\right|} = \frac{e^5 NH_0^3}{32\pi^2 c^4 m^4 f^4 \left|\frac{d\alpha}{dz}\right|} \approx 10^{17} \frac{NH_0^3}{f^4 \left|\frac{d\alpha}{dz}\right|}. \quad (24.49)$$

The expression for $2\delta_0$ with an accuracy of up to a numerical factor of order unity is the same as the expression for the parameter G (24.21): $2\delta_0 = \pi G/2$.

Equations (24.45) and (24.49) can be used to plot the function of the degree of circular and linear polarization of the emission coming from the coupling region into the quasi-longitudinal propagation region if we know the characteristics of the emission incident on the first region.

It is known (see section 6) that the degree of circular polarization ϱ_c of the emission is defined by the equation $\varrho_c = (I_l - I_r)/(I_l + I_r)$, where I_l and I_r are the intensities of the anticlockwise and clockwise polarized components, equal in our case to $(c_2')^2$ and $(c_1')^2$ respectively. The degree of linear polarization ϱ_l in its turn is determined from the relation $\varrho_c^2 + \varrho_l^2 = \varrho^2$, where ϱ is the total extent of the emission's polarization.

If a wave of only one type (for example, an ordinary one) with an intensity c_2^2 of unity enters the quasi-transverse magnetic field region, then in accordance with (24.45) an ordinary wave leaves it with an intensity

$$1 - e^{-2\delta_0}$$

and an extraordinary wave with an intensity

$$|e^{-2\delta_0}.$$

The degree of circular polarization in this case is

$$\varrho_c = -1 + 2e^{-2\delta_0}, \quad (24.50)$$

and of linear polarization is

$$\varrho_l = 2e^{-\delta_0}\sqrt{1-e^{-2\delta_0}} \qquad (24.51)$$

(since $\varrho = 1$). The sign of ϱ_c and the direction of rotation in the resultant polarization ellipse should change when $\delta_0 = \tfrac{1}{2}\ln 2 \approx 0.35$. This value corresponds to the frequency

$$f_t = \left[\frac{e^5 N H_0^3}{32\ln 2 \cdot \pi^2 m^4 c^4 \left|\dfrac{d\alpha}{dz}\right|}\right]^{1/4} \approx \left[2\cdot 10^{17}\frac{NH_0^3}{\left|\dfrac{d\alpha}{dz}\right|}\right]^{1/4}, \qquad (24.52)$$

given previously with less accuracy (see (24.22)). The corresponding graphs of ϱ_c and ϱ_l as functions of f/f_t are given in Fig. 131 (curves I).

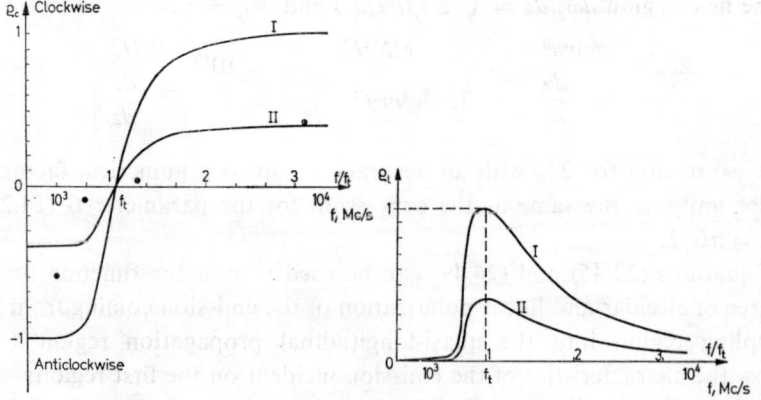

Fig. 131. Degrees of circular and linear polarization as functions of the ratio f/f_t

If the emission entering a transverse magnetic field region contains, as well as a polarized component (corresponding to a wave of one type), an unpolarized component, then the values of the degree of circular and linear polarization decrease by a factor of ϱ when compared with those given above:

$$\varrho'_c = \frac{I_l - I_r}{I_l + I_r + I_{\text{nat}}} = \varrho\varrho_c, \quad \varrho'_l = \sqrt{\varrho^2 - \varrho'^2_c} = \varrho\sqrt{1-\varrho_c^2} = \varrho\varrho_l. \qquad (24.53)$$

Here I_l, I_r and ϱ_c, ϱ_l relate only to the polarized part of the emission; the values of the latter as before are defined by the expressions (24.50) and (24.51). We note that, as can easily be checked by means of the relations (24.45), that when the unpolarized and polarized components pass through

§ 24] Coupling of Waves in a Plasma

an interaction region they retain their intensities I_{nat} and I_l+I_r. Therefore this interaction effect, although it alters ϱ'_c and ϱ'_l, in no way has any effect on the magnitude of the total degree of polarization ϱ.

CERTAIN FEATURES OF SOLAR RADIO EMISSION POLARIZATION AND THEIR INTERPRETATION ON THE BASIS OF WAVE COUPLING IN THE REGION OF A QUASI-TRANSVERSE MAGNETIC FIELD IN THE CORONA

The case of interaction investigated above has a significant part to play in explaining the observed polarization characteristics of solar radio emission and above all the change in the sense of rotation of the polarization

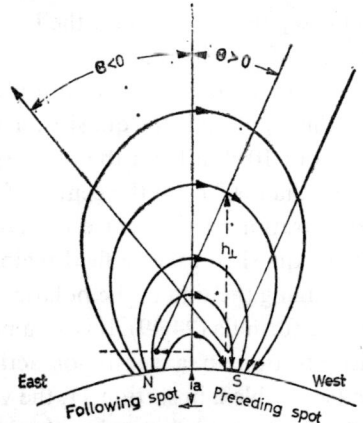

FIG. 132. Position of region of transverse wave propagation above a bipolar group of spots

according to the waveband, which was found by Tanaka and Kakinuma (1959), Kakinuma (1958) in two-thirds of the microwave bursts at frequencies of the order of 3000 Mc/s (see section 11). The explanation suggested by Cohen (1960) for this frequency dependence $\varrho_c(f)$ is quite reasonable on the basis of what has been said above, if under the conditions of the solar corona, namely in the layers where there is transverse propagation of radio emission from a source of bursts towards the Earth (Fig. 132), there occurs such a combination of the parameters N, H_0 and $d\alpha/dz$ that the frequency $f_t \sim 3000$ Mc/s. According to (24.52) the latter holds if $NH_0^3|d\alpha/dz|^{-1} \sim 4\times 10^{20}$. From this it is not hard to obtain an estimate for the magnetic field strength in the interaction region by putting $N \sim 3\times 10^8$ electrons/cm³ and $|d\alpha/dz| \sim 10^{-10}$ cm⁻¹; then $H_0 \sim 5$ oe. The value of H_0 is obviously rather weakly dependent on the value of the parameters N, $|d\alpha/dz|$ and far more strongly on the value of the characteristic frequency f_t. When the latter changes from 1000 to 9400 Mc/s (the scatter of f_t within these limits is actually observed in microwave bursts) the field changes approximately from 2 to 20 oe.

The position of the interaction region in the solar corona can be judged if we state a definite model of the magnetic field H_0 in the active region of the corona and the distribution of the electron concentration in the latter. A treatment of this kind has been carried out in the papers by Cohen and Dwarkin (1961), Cohen (1961a), where the magnetic field of the centre of activity is approximated by the field of a vertical magnetic dipole situated at a depth of $(0\cdot1-0\cdot2)R_\odot$ beneath the photosphere. A given configuration of the lines of force determines the position of the interaction region in the corona which, of course, also depends on the localization of the radio emission source in a centre of activity and the position of the latter on the Sun's disk. For example, for a source at an altitude of 5×10^4 km at a distance of 10^5 km from the dipole axis the quasi-transverse field regions are located at altitudes of about 10^5 km from the photosphere (for centres of activity at an angle of less than 60° from the centre of the disk). By putting an actual electron concentration at the centre of activity we can find the product $N|d\alpha/dz|^{-1}$ in the quasi-transverse field region. The field strength H_0 then necessary for changing the sign of the polarization at the frequency f_t can be obtained from the formula (24.49). It is assumed by Cohen (1961a) that the electron concentration in a centre of activity corresponds to Newkirk's distribution (see Table 2 in section 1); the values of H_0 obtained as a result, which in the final count determine the value of the magnetic dipole moment, are close to the estimates given above.

We notice the following as regards the actual nature of the frequency dependence of the degree of circular polarization of microwave bursts. If completely polarized emission (an ordinary wave) enters a transverse magnetic field region, then the expected dependence $\varrho_c(f)$ is defined by the curve I in Fig. 131. Polarization measurements by Tanaka and Kakinuma (1959) and Kakinuma (1958) show, however, that the degree of circular polarization is not more than 40%, even at frequencies which, to judge from this curve, correspond in practice to the absence of interaction ($f = 1000$ Mc/s) and a complete change of one wave into another ($f = 9400$ Mc/s). This can be explained if the emission leaving the generation region is only partly polarized; with $\varrho \approx 0\cdot4$ the frequency dependence of the degree of circular polarization $\varrho_c(f)$ (curve II in Fig. 131) corresponds better to the experimental results of Tanaka and Kakinuma (1959) marked by dots in the same figure (the values of ϱ'_c at the maximum of the burst of 15 July 1957 at frequencies of 1000, 2000, 3750 and 9400 Mc/s). It follows from Kakinuma's more detailed data (Kakinuma, 1958) that the degree of

§ 24] Coupling of Waves in a Plasma

circular polarization ϱ_c averaged over fifty-seven bursts at these frequencies is 36, 27, 10 and 20% respectively, the direction of rotation changing at $f_t \sim 3000$ Mc/s in the majority of the bursts. Therefore at the frequencies of 1000 and 9400 Mc/s that are far enough from f_t the value of ϱ_c is not the same, unlike the frequency dependence $\varrho_c(f)$ shown in Fig. 131. This circumstance can be explained by the fact that the degree of polarization ϱ of the emission entering the transverse field region changes according to the band.†

The expected frequency dependence of the linear polarization ϱ_l for $\varrho = 0.4$ is also given in Fig. 131 (curve II). It is clear from the figure that at frequencies of $f \approx f_t$ the values are $\varrho_l \approx 0.4$; such values can in principle be measured. However, in the process of subsequent propagation in the corona the plane of linear polarization will undergo strong rotation —by an angle $\Delta\chi \sim 4.7 \times 10^4 f^{-2} N H_0 L$ (23.21), where L is the length of a ray in the effectively rotating part of the corona. In the case of interest to us, when $f \sim 3 \times 10^9$ sec^{-1}, $N \sim 3 \times 10^8$ electrons/cm^3, $H_0 \sim 6$ oe and $L \sim 10^{10}$ cm, the angle $\Delta\chi$ is 3×10^4 radians. According to section 23 because of dispersion of the angle of rotation linear polarization can be found only with observations with a polarimeter having a narrow enough frequency band: $\Delta f_b < f/2\Delta\chi \sim 5 \times 10^4$ c/s. This fully explains the negative results of Kakinuma (1958) in measuring the degree of linear polarization of microwave bursts at a frequency of $f = 3750$ Mc/s where it should be close to the maximum (since $f \approx f_t$; Fig. 131): the polarimeter band here was too large (about 10 Mc/s).

---

† Other explanations have also been suggested for the change in the sense of the polarization that link the latter with features of the wave generation and propagation in the source itself. According to Kakinuma (1958) the change in the sense of the polarization at the frequency $f \sim f_t$ is caused by the fact that the reflection level of an extraordinary wave $v = 1 - \sqrt{u}$ (generated, by assumption, more efficiently than the ordinary component) is situated with $f \sim f_t$ in the middle of the source. Then with $f \gg f_t$ this level moves closer to the Sun's surface and waves of both types escape beyond the source without hindrance. The resultant polarization here will obviously correspond to an extraordinary wave. On the other hand, with $f \ll f_t$ the $v = 1 - \sqrt{u}$ level will rise higher than the source, so only ordinary emission with the opposite polarization will escape from it. Takakura (1959) developed a point of view close to this (with the difference only that instead of the reflection level of the extraordinary wave the level $u \approx 1$ of its absorption by coronal electrons was treated; see section 26). Explanations of this kind do not seem very convincing to us since they require very considerable magnetic fields in the source above the flocculi: since here $v \ll 1$, we should have $u \approx 1$ in both cases, i.e. $f_t \approx \omega_H/2\pi$. For example, at times inversion of the sign of the polarization is observed at frequencies $f_t > 3750$ Mc/s and therefore from the above point of view H_0 should be greater than 1350 oe. It also remains unclear how to match this kind of explanation with the fact that is sometimes noted of double inversion of the polarization sign according to the band.

At the same time it is well known that the polarization of certain components of the solar radio emission sometimes differs noticeably from circular. Since ordinary and extraordinary waves escaping beyond the corona are circularly polarized ellipticity can be the consequence only of coherence between these waves; in this case the resultant polarization ellipse will have considerable eccentricity only when the amplitudes of both types of wave are comparable. On the other hand the coherent nature of the ordinary and the extraordinary waves leads to effective rotation of the polarization ellipse in the corona and dispersion of the ellipse's orientation $\Delta\chi$ in the frequency band of the polarimeter. The latter masks elliptical polarization, so, according to section 23, it can be observed only if the quantity $\delta\chi$ (i.e. the value of the angle of rotation $\Delta\chi$) is far less than that expected when the Faraday effect is operative over the whole path of the radio emission source before leaving the corona.

The circumstances of the necessary limitation of the Faraday effect's sphere of action can be pictured as follows. Let the polarized part of the emission contain only waves of one type by virtue of some causes or other connected with the conditions of the generation or escape of the radio emission from the source.[†] Then the Faraday effect will be absent until, because of partial transformation of waves in a certain coupling region, coherent ordinary and extraordinary waves appear in the composition of the polarized component. If the transformation process takes place effectively enough it will lead to noticeable ellipticity of the emission leaving the corona; if at the same time this process takes place in high enough layers of the corona, then because of the low values of N and H_0 above these layers the small value of the Faraday rotation $\Delta\chi$ and therefore the dispersion $\delta\chi$ in the polarimeter band do not prevent the recording of elliptical polarization.

It was explained above that the efficiency of wave transformation in a coupling of the limiting polarization type is negligible under the conditions of the solar corona. Interaction in a quasi-transverse magnetic field region plays a more important part; as we shall now see, under certain conditions it can explain the appearance of coherent extraordinary and ordinary metric band waves in high enough layers of the corona. In the centimetric band, however, the interpretation of the elliptical polarization meets with certain difficulties.

In fact, in accordance with section 23, the value of $\Delta\chi$ should be less than 6×10^2 radians for recording of the elliptical polarization of the microwave bursts in Akabane's observations (frequency $f = 9500$ Mc/s,

---

† In the opposite case it is known that no ellipticity will appear (section 23).

§ 24] Coupling of Waves in a Plasma

band $\Delta f_b \approx 8$ Mc/s). It follows from (23.21) that this occurs in the case that in the region where the Faraday effect is operative

$$NH_0 < 10^{18}(L \cos \alpha)^{-1}. \tag{24.54}$$

On the other hand, to create the observed ellipticity $p \approx 0.8$ it is necessary for the ratio of the intensities of the two types of wave I_r/I_l to be about 0·01. The content of extraordinary and ordinary waves leaving the quasi-transverse field region is defined by the expressions $I_l = e^{-2\delta_0}$, $I_r = 1 - e^{-2\delta_0}$; their ratio will obviously be close to 0·01 if the characteristic interaction parameter $2\delta_0$ is equal to this value. Then in accordance with (24.49) in the coupling region

$$NH_0^3 \approx 8 \cdot 10^{20} \left| \frac{d\alpha}{dz} \right|. \tag{24.55}$$

The values of N and H_0 in (24.54) and (24.55) are naturally thought of as quantities of the same order and we take it that $\cos \alpha \sim 1$, $|d\alpha/dz|^{-1} \sim L \sim 5 \times 10^{10}$ cm. Then it follows from these relations that in the layers starting at the interaction region and above $H_0 > 28$ oe, $N < 7 \times 10^5$ electrons/cm³. There is little probability of such a combination of values of N and H_0 in the corona (see section 1). A more realistic combination ($H_0 > 2.8$ oe, $N < 7 \times 10^7$ electrons/cm³) can be obtained by putting $|d\alpha/dz|^{-1} \sim 5 \times 10^{11}$ cm, $L \sim 5 \times 10^9$ cm. These values of the characteristic dimensions cannot be recognized as ordinary, however. Therefore further careful measurements become attractive to check the actual fact of the existence of the elliptical polarization of the microwave bursts.

In the metric band the position is more favourable. According to sections 12 and 14 the polarization ellipses of types I and III bursts sometimes differ essentially from a circle. For an ellipticity of $p \approx 0.5$ the ratio of the intensities of two types of waves I_r/I_l and at the same time the quantity $2\delta_0$ take up values around 0·1. Then for type I bursts observed at a frequency $f = 200$ Mc/s in the band $\Delta f_b = 0.5$ Mc/s

$$NH_0^3 \sim 10^{15} \left| \frac{d\alpha}{dz} \right|, \tag{24.56}$$

and $\Delta \chi < 10^3$ radians, i.e. in the region of action of the Faraday effect

$$NH_0 < 10^{15}(L \cos \alpha)^{-1}. \tag{24.57}$$

With identical N and H_0 in (24.56), (24.57) and also with $L \cos \alpha \sim |d\alpha/dz|^{-1} \sim 5 \times 10^{10}$ cm it follows from these relations that $H_0 > 1$ oe, $N < 2 \times 10^4$ electrons/cm³. For the type III bursts studied by Akabane and Cohen at the same frequency the product NH_0^3 is of the same order as for the type I bursts. For this case it is known that $\Delta\chi \sim 10^4$ radians (see section 23), i.e.

$$NH_0 \sim 10^{16}(L\cos\alpha)^{-1}. \tag{24.58}$$

Then $H_0 \sim 0.3$ oe, $N \sim 7\times 10^4$ electrons/cm^3.

The requirements imposed on the values of N and H_0 when interpreting the elliptical polarization of types I and III bursts can apparently be satisfied in the outermost layers of the corona (in the supercorona). We note that if this explanation is correct, then elliptically polarized bursts should show inversion of the sign of the polarization according to the band. Unfortunately the polarization of bursts at metric wavelengths has not yet been studied from this standpoint.

The phenomenon of inverting the sense of the polarization found in observations of microwave bursts may in principle also occur for the slowly varying component of the solar radio emission whose polarized component is generated in the lower layers of the corona above spots. In order to examine the polarization characteristics of emission escaping beyond the corona in this case, let us examine, following Takakura (1961a), a model of a bipolar group of spots, the field H_0 above it being given by a horizontal magnetic dipole situated at a depth $a = 0.1R_\odot$ or $0.05R_\odot$ from the photosphere (see Fig. 132), whilst the distribution of the electron concentration is described by the Baumbach–Allen formula (1.1). Let emission, initially left-handed polarized, proceed from a point source situated at an altitude $h = 0.05R_\odot$ above the south magnetic pole, which for the sake of definition is considered to be the preceding spot, whilst the right-handed polarized emission proceeds from a point source situated at the same altitude above the north magnetic pole (succeeding spot).[†] The distance between the spots is taken as $0.2R_\odot$. Then it is not difficult to calculate the altitude of the quasi-transverse field $h_\perp$ and the frequency f_t as a function of the heliographic longitude Θ of the bipolar group (considering that the latter is located close to the solar equator and taking various values for the magnetic field H_{0b} at the base of the spots). Results of the calculations for $a = 0.1R_\odot$ are given in Fig. 133; in the case of $a = 0.05R_\odot$ the values of $h_\perp$ and f_t differ little from those given.

It is clear from Fig. 132 that in the model of the magnetic field used the emission from the preceding spot passes through the quasi-transverse field region only if the angle Θ is negative (i.e. the group is located in the western hemisphere); when $\Theta > 0$ the case of quasi-transverse propaga-

[†] This choice of polarization corresponds (in accordance with the observations; see section 10) to predominance of emission of the extraordinary type in the composition of the slowly varying component.

§ 24] Coupling of Waves in a Plasma

tion does not occur. For emission from the succeeding spot the situation is the opposite: it passes through a quasi-transverse field region when the angle Θ is positive, i.e. the group is localized in the eastern part of the solar disk. As $|\Theta|$ decreases, i.e. as the spots approach the central meridian, the altitude $h_\perp$ increases and the frequency f_t decreases. We recall that emission at a given frequency f changes its sign with propagation through a quasi-transverse field region until $f < f_t$; this inequality occurs only at large enough distances of the group from the central meridian, which are easy to determine from Fig. 133.

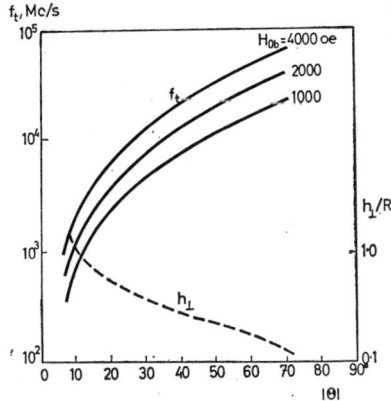 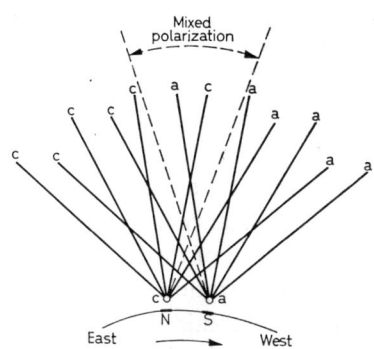

Fig. 133. Dependence of f_t and $h_\perp$ on the heliographic longitude above a bipolar spot (Takakura, 1961a)

Fig. 134. Sign of rotation of polarization of slowly varying component as a function of the position of the bipolar group on the solar disk; a—anticlockwise (left-handed); c—clockwise (right-handed) polarization

The distribution of the signs of rotation of both sources of polarized emission in the bipolar group according to the values of Θ obtaining as a result is shown in Fig. 134. It follows from it that when the group with the preceding S-spot appears on the eastern edge of the disk the emission from both spots has left-handed polarization; in the period when the group approaches the central meridian the emissions from the different spots have different senses of polarization; on moving further towards the western edge of the disk the polarization of both sources once more becomes the same (right-handed).

Therefore the direction of rotation of the polarization vector in the total emission produced by the bipolar group of spots as a whole should change when this group crosses the central meridian because of the interaction effect in the quasi-transverse magnetic field region. This kind of phenomenon is sometimes actually observed (section 10), although it is recorded far from often as could be expected on the basis of the model of the centre of activity taken above.

On the other hand, polarization observations (Tanaka and Kakinuma, 1958) of one of the bipolar groups of spots made during the eclipse of 19 April 1958 at frequencies of 2000, 3750 and 9400 Mc/s showed no change in the sense of the rotation according to the band in the emission connected with each spot separately: the sense of the rotation at all the frequencies corresponded, as usual, to preferential generation of extraordinary waves in the sources above the spots. It is significant that this occurred under conditions when, according to the dipole model of the magnetic field, the emission should have passed through a quasi-transverse field region with a value of the characteristic frequency f_t within the band studied. Since there is no doubt of the existence of the effects discussed above that accompany the passage of radio waves through a quasi-transverse magnetic field region, the only conclusion that can be drawn from this is that the dipole approximation of the magnetic field of bipolar groups is inaccurate at a sufficient distance from the spots: in actual fact in the majority of cases the configuration of the lines of force in the corona is such that the region of mixed polarization is far wider than that shown in Fig. 134. Another indication of this is apparently the usual constancy of the sense of the polarization of noise storms when their sources cross the central meridian.

In conclusion we note that the effect of coupling in a quasi-transverse field region also occurs under certain conditions in the exospheres of planets and the interplanetary and interstellar medium. The possible part played by this effect in the formation of the observed characteristics of Jupiter's sporadic radio emission will be discussed in section 32. As for the interstellar medium, strong interaction occurs there when radio emission passes through a system of clouds with different orientations of the magnetic fields over practically the whole range of frequencies that are of interest in radio astronomy. The characteristic frequency f_t above which propagation through a transverse field region does not alter the sense of rotation of the wave is of the order of 0·1 Mc/s here. This is easily checked by substituting parameter values typical of the galactic conditions in the formula (24.52): $H_0 \sim 3 \times 10^{-6}$ oe, $N \sim 0{\cdot}1$ electron/cm^3, $|d\alpha/dz|^{-1} \sim 3 \times 10^{20}$ cm.

§ 25] The Escape of Radio Emission from the Corona

25. Coupling of Electromagnetic Waves and the Problem of the Escape of Radio Emission from the Corona

PRELIMINARY REMARKS

The discussion of the problem of the escape of radio emission beyond the limits of the corona actually started as early as section 23, where the conditions for direct escape were explained in the framework of the approximation of geometrical optics. It turned out that the only waves to escape from the corona are those which correspond in Figs. 123–5 to the sections of the dispersion curves $n_j^2(R)$ contained between the values $n_j^2 = 0$ in the corona and $n_j^2 = 1$ outside it.† The condition $n_j^2(R) = 0$ determines the lower boundary of the region from which radio emission escapes from the corona without hindrance. For extraordinary waves at this boundary $v = 1 - \sqrt{u}$, for ordinary waves $v = 1$, i.e. in the latter case the boundary is located in the deeper-lying layers of the corona. The waves corresponding to the other branches of the dispersion curves are not able to escape beyond the limits of the corona. In this question the problem quite naturally arises of finding the conditions under which the "transition" of emission from these branches to sections of the dispersion curves corresponding to waves leaving the corona occurs and of the estimation of the efficiency of this kind of transition.

One of the possible ways of solving the problem of escape is the coupling action and mutual transformation of "normal" waves in the $v \approx 1$ region, which under certain conditions appears both in a magnetoactive and in an isotropic regularly non-uniform plasma. Another possibility is connected with the statistical non-uniformity of the plasma, which exists because of fluctuations in the electron concentration caused by the thermal motion of the plasma particles. In actual fact the non-uniformities, which have a characteristic spatial dimension of the order of a wavelength, cause noticeable scattering of the latter. This process obviously cannot be described in terms of geometrical optics, unlike the scattering of radio waves on large-scale coronal non-uniformities mentioned in section 22. Therefore the scattered emission contains, generally speaking, all the types of "normal" waves propagated in the corona, i.e. scattering on fluctuations in the electron concentration is accompanied by the transformation of waves of one type into another.

---

† However, even when escaping from these regions the emission may be considerably attenuated because of strong absorption in the plasma. The effect of absorption on the propagation and escape of emission from the corona is discussed in section 26.

The problem of the change of plasma waves into electromagnetic ones arises in connection with the circumstance that in a strictly uniform plasma both types of waves in the linear approximation are in no way connected and are propagated completely independently. This follows formally from the fact that plasma and electromagnetic waves in a uniform plasma are normal waves, i.e. linearly independent solutions of the Maxwell equations and the equations of the plasma motion.

This problem is particularly important for the "plasma hypothesis" of the origin of the solar sporadic emission (see section 31). According to this hypothesis the source of the observed radio emission is plasma oscillations (waves) excited by a certain agent in the solar corona at frequencies of $\omega \approx \omega_L$. An argued extension of this point of view to the origin of the sporadic radio emission depends on whether plasma waves under the actual conditions of the solar atmosphere can be transformed with sufficient efficiency or not into electromagnetic emission leaving the corona. The point is that the plasma waves themselves, being excited near the $\omega_L \approx \omega$ level, are completely absorbed in the more rarefied layers of the plasma because of the so-called Landau damping (see section 26).

Generally when looking for ways in which plasma waves change into electromagnetic ones and illustrating the difficulties connected with this, stress is laid on the fact that plasma waves in a uniform isotropic plasma are purely longitudinal and therefore have no magnetic field (curl $E = 0$, $H = 0$). Since in electromagnetic waves, on the contrary, $H \neq 0$ it becomes clear that with excitation in a certain region of a system of plasma waves no electromagnetic emission will escape from this region. From this follows the incorrectness of the "aerial theory" of plasma emission put forward by Feinstein (1952) and Sen (1955). Representing a plasma wave in the form of a certain secondary polarization $P^{sec} = E/4\pi$ (or secondary current $j^{sec} = \partial P^{sec}/\partial t$), they find the transverse electromagnetic waves "emitted" by a plasma wave in a uniform medium. It is known, however, that a distribution of secondary currents j^{sec} that is subject to the condition curl $j^{sec} = 0$ does not produce electromagnetic radiation, although it is this condition that should be satisfied both in a single plasma wave or in any combination of plasma waves (since in them curl $E = 0$). The appearance in Feinstein (1952), Sen (1955) of a non-zero Poynting vector characterizing the energy flux of the transverse electromagnetic waves is connected with the impermissible replacement of a plane plasma wave by a linear current (Zheleznyakov, 1958a).

The absence of a magnetic field in plasma waves being propagated in a uniform, isotropic and motionless plasma is a very obvious and sufficient but by no means necessary argument against their change into waves of

§ 25] The Escape of Radio Emission from the Corona

another type. An example is a uniform magnetoactive plasma where the plasma waves have a weak magnetic field but, while remaining normal waves, are propagated independently of the ordinary and extraordinary waves and do not change into the latter.

As another example we can point to the case of an isotropic plasma moving as a whole relative to a fixed coordinate system at a velocity $V_0 =$ const. Larenz (1955), it is true, has stated that under these conditions the change of plasma waves into electromagnetic emission does occur. This point of view, however, contradicts the relativity principle: if a plasma wave is propagated in a plasma at rest and there are no electromagnetic waves connected with it passing beyond the plasma (into a vacuum), then no transition into a coordinate system moving relative to the plasma can lead to the appearance of waves of this kind. The presence of a magnetic field in the plasma waves in a coordinate system where $V_0 \neq 0$ can in no way help in this respect (Zheleznyakov, 1958a). The ratio of the transverse component of the electric field to the longitudinal found by Larenz (1955) in the light of what has been said in no way characterizes the interaction between waves of different types which here are strictly independent; this quantity defines only the relative content of the longitudinal and transverse components in the plasma wave, which in a moving medium, generally speaking, is not longitudinal.†

A similar error is made by Hruška (1962) where the "interaction" of waves in a uniform plasma with an anisotropic velocity distribution of electrons is discussed. This anisotropy leads to anisotropy in the electron pressure. Thanks to this the normal waves cease to be purely longitudinal and transverse. In this case as well, however, they of course remain completely independent; when the pressure anisotropy disappears they change into the ordinary plasma and electromagnetic waves of an isotropic plasma. As a whole the situation is entirely similar to that occurring in a uniform magnetoactive plasma (right up to the nature of the change to the isotropic case; in this connection see section 23) and in a uniform moving plasma. It is therefore clear that finding the transverse components of an electromagnetic field from a given longitudinal component E under anisotropic pressure conditions will give an idea only of the connection between the field components (on the nature of the polarization) in each individual normal wave, but not of the interaction between them.

The position changes sharply under the conditions of a non-uniform plasma where the interaction and transformation of waves of different types, which in principle occur throughout the plasma, become particularly

† As far as can be judged Larenz (1962) has lately reviewed his opinion on the conditions under which wave interaction occurs.

effective in regions where there is simultaneous violation of the geometrical optics approximation for these types of waves. In this case the non-uniform nature of the plasma may be due to a change in space in the magnetic field H_0, the electron concentration N, the temperature T or the velocity V_0 of the plasma.[†]

CONVERSION OF PLASMA WAVES INTO ELECTROMAGNETIC WAVES IN A SMOOTHLY NON-UNIFORM ISOTROPIC PLASMA

The problem of the transformation of plasma waves into electromagnetic ones under the conditions of a smoothly non-uniform isotropic plasma (whose properties change because of the dependence of the concentration N_0 on the radius vector R) can be stated as follows. There is a plane-layered plasma where

$$V_0 = 0, \quad H_0 = 0, \quad T = \text{const}, \quad N_0 = N_0(z). \quad (25.1)$$

In it a plasma wave of frequency ω is propagated with a given intensity and a wave vector k lying in the plane yz. We have to find the intensity of an electromagnetic wave excited by a plasma wave when the former leaves the medium (Zheleznyakov, 1956; Zheleznyakov and Zlotnik, 1962 and 1963).[‡]

The system of initial equations for solving this problem by the quasi-hydrodynamic method consists of the Maxwell equations (22.6) and the equations of motion of the electrons (22.9) in which we must put $H_0 = 0$.

Starting from these equations we can confirm that the fields with the components E_x, H_y, H_z and H_x, E_y, E_z do not depend on each other. The presence in (22.9) of terms that allow for the electron pressure and make possible the propagation of plasma waves is in no way reflected in the field with the components E_x, H_y, H_z. The latter is therefore of no interest and we shall turn to investigate the fields containing H_x, E_y, E_z. By finding the solutions for them in the form

$$\left. \begin{array}{l} H_x = W(z) \exp\,(ik_0 n_3 \sin \varphi \cdot y + i\omega t), \\ E_y = V(z) \exp\,(ik_0 n_3 \sin \varphi \cdot y + i\omega t), \\ E_z = U(z) \exp\,(ik_0 n_3 \sin \varphi \cdot y + i\omega t), \end{array} \right\} \quad (25.2)$$

(where $k_0 = \omega/c$, φ is the angle between the z-axis and the wave vector of a plasma wave with the refractive index n_3) it can be shown that $n_3 \sin \varphi =$

[†] The question of the interaction of waves of different types in conditions when the non-uniform nature of the plasma is determined by the dependence of its velocity V_0 on the radius vector R has not yet been studied in detail when correctly stated.

[‡] The opposite problem of the conversion of an electromagnetic wave into a plasma wave has been discussed by Denisov (1954).

§ 25] The Escape of Radio Emission from the Corona

const (the refraction law), whilst W and U obey the equations (Zheleznyakov and Zlotnik, 1963)

$$\frac{d^2W}{dz^2} - \frac{\frac{d\varepsilon}{dz}}{\varepsilon' - \sin^2\varphi_0}\left(1 + \frac{\sin^2\varphi_0}{1-\varepsilon'}\right)\frac{dW}{dz} + k_0^2\left(\varepsilon' - \frac{\sin^2\varphi_0}{\beta_{\text{th}}^{'2}}\right)W$$

$$= \frac{\sin\varphi_0 \beta_{\text{th}}'}{\varepsilon' - \sin^2\varphi_0} \frac{\frac{d\varepsilon'}{dz}}{1-\varepsilon'} \frac{dU}{dz}, \qquad (25.3)$$

$$\beta_{\text{th}}^{'2}\frac{d^2U}{dz^2} + \frac{\varepsilon'\beta_{\text{th}}^{'2}}{1-\varepsilon'} \frac{\frac{d\varepsilon'}{dz}}{\varepsilon' - \sin^2\varphi_0} \frac{dU}{dz} + k_0^2(\varepsilon' - \sin^2\varphi_0)U$$

$$= -\frac{\sin\varphi_0 \beta_{\text{th}}'}{\varepsilon' - \sin^2\varphi_0} \frac{\frac{d\varepsilon'}{dz}}{1-\varepsilon'} \frac{dW}{dz} - k_0^2 \frac{\sin\varphi_0}{\beta_{\text{th}}'} W. \qquad (25.4)$$

In (25.3) and (25.4)

$$\varepsilon' = 1 - \frac{v}{1-iv}, \quad \beta_{\text{th}}' = \frac{V_{\text{th}}}{c\sqrt{1-iv}},$$

where

$$v = \frac{\omega_L^2}{\omega^2}, \quad V_{\text{th}} = \sqrt{\frac{\varkappa T}{m}}, \quad v = \frac{v_{\text{eff}}}{\omega}.$$

The quantity v in the conditions of the solar corona and the planetary ionospheres at the frequencies of interest to us is very small. For example if we put $\omega \sim 2\pi \times 10^8$ sec^{-1}, $v_{\text{eff}} \sim 6$ sec (which is typical for the corona), then $v \sim 10^{-8}$. Therefore ε' and β_{th}' are close to the dielectric permeability ε of an isotropic plasma in the absence of collisions and the ratio β_{th} of the thermal velocity to the velocity of light introduced in section 22 (unless we consider small imaginary additions of the order of v). Unlike the previous case, however, when the smallness of v made it possible to neglect the collisions immediately, this cannot be done here. The point is that with $v = 0$ the quantity ε' becomes real and the coefficients in the equations (25.3), (25.4) that contain the factor $1/(\varepsilon' - \sin^2\varphi_0)$ become infinity at the point where $\varepsilon = \sin^2\varphi_0$. Therefore in the solution of the system (25.3), (25.4) it is considered that $v \neq 0$ and only in the final expressions for the fields of an electromagnetic wave leaving the plasma does it make the transition to the limit $v \to 0$ (in β_{th}' we can immediately put $v = 0$, which is what is done below).

The quantity φ_0 in the equations under discussion is the formally introduced (in accordance with the refraction law) "angle of incidence" of a plasma wave corresponding to a layer where the concentration is $N_0 = 0$ and the refractive index of this wave is $n_3 = c/V_{\text{th}}$ (22.19a). In actual fact, because of Landau damping the plasma wave can be propagated only in a region where ω_L^2 is close to ω^2, i.e. $|\varepsilon'| \ll 1$ (see section 26).

The relations (25.3), (25.4) are a system of two connected second-order equations. With normal propagation ($\varphi_0 = 0$) the equations split, the first of them describing a transverse electromagnetic wave and the second a longitudinal plasma wave. If the angle φ_0 is non-zero, then the homogeneous equations corresponding to (25.3), (25.4) characterize an electromagnetic and a plasma wave whose electrical vectors are located in the plane of incidence (in the plane yz). The configuration of the rays for the electromagnetic and plasma waves is shown in Fig. 135. This allows

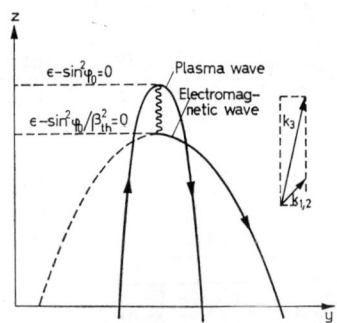

Fig. 135. Trajectory of rays in a plane-layered isotropic plasma for interacting plasma and electromagnetic waves.

for the fact that in accordance with the law of refraction $n_3 \sin \varphi = \text{const}$, where the constant is the same for both waves, the projections of the wave vectors $\mathbf{k}_{1,2}$ and $\mathbf{k}_3$ onto the y-axis coincide.

The approximation of geometrical optics for an electromagnetic wave is violated in the vicinity of the points $\varepsilon' - \sin^2 \varphi_0/\beta_{\text{th}}^2 = 0$ and $\varepsilon' - \sin^2 \varphi_0 = 0$; at the first the wave is reflected and at the second the solution has a singularity which shows itself in the sharp rise in the field $\mathbf{E}$ near this point. Since the geometrical optics approximation for a plasma wave is also violated in the layer $\varepsilon' - \sin^2 \varphi_0 \approx 0$ (it is reflected from this layer), it is clear that strong interaction and a mutual change of one type of wave into the other should be expected at the $\varepsilon' - \sin^2 \varphi_0 \approx 0$ level. This interaction is allowed for by the terms in the right-hand side of the equations (25.3), (25.4).

The rigorous solution of the system (25.3), (25.4) is difficult. We shall therefore try to find only its approximate solution that is valid for small enough $\sin \varphi_0$. From this standpoint we shall use the perturbation method,

§ 25] The Escape of Radio Emission from the Corona

taking as the zero approximation the solution of the equation (25.4) without the right-hand side: $U = U^{(0)}(z)$; $W = W^{(0)}(z) = 0$. This means that in the zero approximation only a plasma wave exists; since $W^{(0)} = 0$ the magnetic fields that appear in the process of the plasma wave propagation in the non-uniform medium are entirely neglected. The first approximation for these fields $W = W^{(1)}(z)$ can be found by substituting $U^{(0)}(z)$ in the right-hand side of the equation (25.3).

To find the functions $U^{(0)}(z)$ and $W^{(1)}(z)$ we make a simplifying assumption about the nature of the function $\varepsilon'(z)$. By noticing that in the conditions of a smoothly non-uniform medium the actual form of $\varepsilon'(z)$ is significant only in the regions where geometrical optics are inapplicable, $\varepsilon' - \sin^2 \varphi_0 \approx 0$, $\varepsilon' - \sin^2 \varphi_0/\beta_{th}'^2 \approx 0$, which with small enough φ_0 are localized near $\varepsilon' = 0$, we can approximate the actual function $\varepsilon'(z)$ by the linear function

$$\varepsilon'(z) = -z \,\text{grad}\, \varepsilon - iv. \tag{25.5}$$

The values of grad ε and ν in (25.5) are taken at the point $z = 0$ (i.e. at the point $\varepsilon \approx 0$). The condition for a slow enough change in the properties of the medium at distances of the order of $\lambda/2\pi = c/\omega$ as applied to the function (25.5) becomes the following:

$$\varrho_N \equiv \frac{\omega}{c \,\text{grad}\, \varepsilon} \gg 1.$$

It is satisfied well in the corona (and the ionospheres of the planets) because of the smallness of grad ε.†

For a linear layer the solution of the equation (25.4) without the right-hand side can be expressed in terms of the Eyrie function $v(\zeta)$ and $u(\zeta)$:‡

$$U^{(0)} = c_1 v(\zeta) + c_2 u(\zeta), \tag{25.6}$$

where

$$\zeta = \left(\frac{\varrho_N}{\beta_{th}}\right)^{2/3} \xi, \quad \xi = z \,\text{grad}\, \varepsilon + iv + \sin^2 \varphi_0.$$

The function $v(\zeta)$ decreases exponentially and $u(\zeta)$ rises without limit with $\zeta \to +\infty$ (Fok, 1946a). Since the plasma wave should disappear in

---

† For example, in a stationary coronal plasma with values of grad $\varepsilon \approx 10^{-10}$ cm^{-1} at frequencies of $\omega \approx 2\pi \times 10^8$ sec^{-1} the parameter $\varrho_N \sim 10^8$. The presence of inhomogeneities may reduce its value. In actual fact, in an isotropic plasma grad ε cannot exceed $1/l_{tr}$ in the time $t \gtrsim 1/\nu_{eff}$ ($l_{tr} = V_{th}/\nu_{eff}$ is the mean free path of the electrons). In the corona $V_{th} \sim 4 \times 10^8$ cm/sec, $\nu_{eff} \sim 10$ sec^{-1}; accordingly $l_{tr} \sim 4 \times 10^7$ cm and $\varrho_N \gtrsim (\omega/c)l_{tr} \sim 10^6$.

‡ This is true if in (25.4) the second term can be neglected. As shown by Zheleznyakov and Zlotnik (1963), this neglect is quite permissible with values of ν that are not too small, which are easily found in the conditions of the solar corona.

the depths of the plasma (when $z \to +\infty$, i.e. when Re $\varepsilon' \to -\infty$), in (25.6) the constant c_2 should be put equal to zero, so

$$U^{(0)} = c_1 v(\zeta). \tag{25.7}$$

Allowing for (25.7) the equation for the magnetic field in the first approximation is of the following form:

$$\left.\begin{aligned}\frac{d^2W}{d\xi^2} - \frac{1}{\xi}\left(1 + \frac{\sin^2\varphi_0}{1-\sin^2\varphi_0+\xi}\right)\frac{dW}{d\xi} - \varrho_N^2\left(\xi + \frac{\sin^2\varphi_0}{\beta_{th}^2}\right)W = G(\xi),\\ G(\xi) = \varrho_N^{2/3}\beta_{th}^{1/3}\sin\varphi_0\frac{v'(\zeta)}{\xi(1-\sin^2\varphi_0+\xi)}.\end{aligned}\right\} \tag{25.8}$$

The general solution of the equation (25.8) can be written by using the usual method of variation of a constant:

$$W^{(1)} = \gamma_1 W_1 + \gamma_2 W_2 - W_1 \int \frac{W_2 G}{\Delta} d\xi + W_2 \int \frac{W_1 G}{\Delta} d\xi. \tag{25.9}$$

Here W_1 and W_2 are the fundamental system of the functions of the equation (25.8); Δ is a Wronskian equal to $W_1(dW_2/d\xi) - W_2(dW_1/d\xi)$; γ_1 and γ_2 are constants determined from the boundary conditions. The latter are the requirements that in the depth of the plasma (with $z, \xi \to \infty$) the function $W^{(1)} \to 0$, and at the beginning of the layer the electromagnetic wave should be propagated towards the exit from the plasma (see Fig. 135). Without dwelling on the details of calculating $W^{(1)}$ (the choice of the fundamental functions, the calculation of the integrals, etc.) we shall give the final result at once.[†]

The solution of (25.9) can be represented in the form of the two terms W_{el} and W_{pl} so that with large $|\xi|$ the first of them describes the field of the electromagnetic wave and the second is the magnetic field of the plasma wave in the non-uniform medium. With simultaneous satisfaction of the inequalities

$$S_0^{2/3} \gg \beta_{th}^{2/3}, \quad S_0^{1/3} \ll \beta_{th}^{-2/3}, \tag{25.10}$$

where

$$S_0 = \frac{2}{3}\varrho_N\frac{\sin^3\varphi_0}{\beta_{th}^2}, \tag{25.11}$$

the amplitude of the magnetic field of an electromagnetic wave being propagated from an interaction region to the exit from the plasma

$$|W_{el}| \approx \frac{c_1^2}{\sqrt{\pi}} \frac{\varrho_N^{-1/6}\beta_{th}^{2/3}e^{-2S_0}|\xi|^{1/4}}{S_0^{1/2}H_{-2/3}^{(1)}(iS_0)} \tag{25.12}$$

[†] For details see Zheleznyakov and Zlotnik (1962 and 1963).

§ 25] The Escape of Radio Emission from the Corona

($H^{(1)}_{-2/3}$ is a Hankel function). Since the electric field in a transverse electromagnetic wave is connected with the magnetic field by the relation $H = \sqrt{\varepsilon} E$, the mean energy flux in it, in accordance with (22.18), (25.12),[†] is

$$S_{el} \approx \frac{c}{8\pi\sqrt{\varepsilon}} |W_{el}|^2 \approx \frac{c_1^2 c}{8\pi^2} \frac{\varrho_N^{-1/3}\beta_{th}^{4/3} e^{-4S_0}}{S_0[H^{(1)}_{-2/3}(iS_0)]^2}. \qquad (25.13)$$

The energy flux in a plasma wave incident on the interaction region $\varepsilon \approx \sin^2\varphi_0 \approx 0$ can easily be written with the help of the relations (22.21), (25.2) and (25.7). In the region $-\zeta \gg 1$, where the approximation of geometrical optics holds for a plasma wave $E_z = c_1 v(\zeta) e^{ik_0 n_3 \sin\varphi_0 \cdot y + i\omega t}$, the Eyrie function $v(\zeta)$ can be shown in the form (Fok, 1946a)

$$v(\zeta) \approx \frac{1}{|\zeta|^{1/4}} \sin\left(\frac{2}{3}|\zeta|^{3/2} - \frac{\pi}{2}\right)$$

$$= \frac{i}{2|\zeta|^{1/4}} e^{-i\frac{2}{3}|\zeta|^{3/2} - i\frac{\pi}{4}} - \frac{i}{2|\zeta|^{1/4}} e^{i\frac{2}{3}|\zeta|^{3/2} + i\frac{\pi}{4}}.$$

It is clear from this that the amplitude of the z-component of the electric field of a plasma wave incident on the layer where interaction occurs is

$$\frac{c_1}{2|\zeta|^{1/4}} = \frac{c_1}{2\varepsilon^{1/4}} \varrho_N^{-1/6}\beta_{th}^{1/6}. \qquad (25.14)$$

Since for small φ_0 the electric field of the plasma wave is practically the same as its z-component the energy flux is defined by the relation

$$S_p \approx \frac{c\sqrt{\varepsilon}}{8\pi} \beta_{th} E_a^2 \approx \frac{c_1^2 c}{32\pi} \varrho_N^{-1/3}\beta_{th}^{4/3}. \qquad (25.15)$$

In accordance with (25.13), (25.15) in the case when the inequalities (25.10) are satisfied the efficiency of the transformation of a plasma wave into electromagnetic radiation, defined as the ratio of the energy flux in an electromagnetic wave leaving the interaction region to the energy flux in a plasma wave incident on this region, will be

$$Q \approx \frac{4}{\pi} \frac{e^{-4S_0}}{S_0[H^{(1)}_{-2/3}(iS_0)]^2}. \qquad (25.16)$$

In special cases the efficiency of the transformation is defined by the

---

† When changing from (25.12) to (25.13) allowance is made for the fact that with small angles φ_0 the amplitude of the magnetic field is close to the amplitude of its x-component.

following relations:

$$Q \approx S_0^{1/3} e^{-4S_0} \quad (S_0 \ll 1),$$
$$Q \approx 1/4 \quad (S_0 \approx 1),$$
$$Q \approx 2e^{-2S_0} \quad (S_0 \gg 1).$$
(25.17)

It follows from the formulae given that the maximum interaction occurs when $S_0 \sim 1$. It is not hard to confirm that this condition is satisfied if the distance $\Delta z \approx |z| \approx \sin^2 \varphi_0/\mathrm{grad}\ \varepsilon \cdot \beta_{th}^2$ between the reflection point of the plasma wave $\varepsilon(z) \approx \sin^2 \varphi_0$ (in whose vicinity the interaction occurs) and the reflection point of the electromagnetic wave $\varepsilon \approx \sin^2 \varphi_0/\beta_{th}^2$ is of the order of the length of the electromagnetic wave in the region of its reflection $\lambda_{max} \approx (c/2\omega\sqrt{\mathrm{grad}\ \varepsilon})^{2/3}$.† If, however, this distance is much greater than λ_{max}, then $S_0 \gg 1$ and the transformation coefficient decreases, since the electromagnetic wave on leaving the interaction region undergoes strong attenuation until it reaches the level $\varepsilon \approx \sin^2 \varphi_0/\beta_{th}^2$. In the case of $S_0 \ll 1$ this attenuation is insignificant and the efficiency of the transformation is low because in the extreme case of extreme incidence ($\varphi_0 = 0$) there is in general no interaction.‡

The perturbation method used above remains correct only in the case that the zero approximation in the form of a standing plasma wave (25.7) with sufficient accuracy describes the electric field in a plasma. The latter is known not to hold in the case of $S_0 \sim 1$, when the flux of the transformed electromagnetic wave is comparable with the flux of the incident plasma wave.

For a stricter definition of the limits of applicability of the results obtained the next approximation of $U^{(1)}$ should be calculated for the function U and it should be compared with the zero approximation. Then the inequality $|U^{(1)}| \ll U^{(0)}$ will give us the necessary conditions for the validity of the perturbation method. The corresponding treatment, which is carried out by Zheleznyakov and Zlotnik (1962) shows that the expressions obtained for Q are correct in the region $S_0 \ll 1$. For angles φ_0 at which $S_0 \sim 1$ and $Q \sim 1$ the value of (25.17) is of importance only in estimating the order of magnitude. As for the formula (25.17) in the region $S_0 \gg 1$, the factor e^{-2S_0} is obviously retained in the precise expression for the transformation coefficient, since it defines the exponential decay of the electromagnetic wave as it leaves the interaction region; the inaccu-

---

† See the footnote on p. 308.
‡ With $S_0 \to 0$ the first of the expressions (25.17) becomes inapplicable because of violation of one of the conditions (25.10). In this case Q approaches zero as $3 \cdot 44 S_0^{2/3}$.

§ 25] The Escape of Radio Emission from the Corona

racy of the expression (25.17) with $S_0 \gg 1$ is therefore only in the factor in front of the exponent.

Therefore with optimum values of the "angle of incidence" of the plasma wave onto the layer when $S_0 \sim 1$ and accordingly

$$\varphi_0 \approx \sin \varphi_0 \sim \beta_{\text{th}} \left(\frac{3}{2} \frac{c \text{ grad } \varepsilon}{\omega} \right)^{1/3}, \qquad (25.18)$$

approximately one quarter energy flux of this wave transfers into the electromagnetic emission. The angles φ_0 (25.18) are very small since in a non-relativistic plasma $\beta_{\text{th}} \ll 1$; in addition, $(c/\omega) \text{grad } \varepsilon \ll 1$ because of the slowness of the variation in the medium's properties (see (25.6)). Putting in the corona grad $\varepsilon \sim 10^{-10}$ cm^{-1} and $\beta_{\text{th}} \sim 10^{-2}$, we obtain for $\omega \sim 2\pi \times 10^8$ sec^{-1} that the optimum angle is $\varphi_0 \sim 2 \times 10^{-5}$ radian. The values of φ_0 given relate to the beginning of the layer. In a plasma where the phase velocity of the plasma wave is $V_{\text{ph}} = V_{\text{th}}/\sqrt{\varepsilon}$ (for example, at the level where this wave is generated) in accordance with the law of refraction (22.28a) the angle φ_0 corresponds to the angle φ, where $\sin \varphi = \sin \varphi_0/\sqrt{\varepsilon} = \sin \varphi_0(V_{\text{ph}}/V_{\text{th}})$. Remembering (25.18), we obtain for the optimum angles φ that ensure maximum transformation efficiency:

$$\sin \varphi_{\text{opt}} \sim \frac{V_{\text{ph}}}{c} \left(\frac{3}{2} \frac{c \text{ grad } \varepsilon}{\omega} \right)^{1/3}. \qquad (25.19)$$

For the above values of ω and grad ε, $\sin \varphi_{\text{opt}} \sim 2 \times 10^{-3}(V_{\text{ph}}/c)$; in particular, if $V_{\text{ph}}/c \sim 1$, then $\varphi_{\text{opt}} \approx \sin \varphi_{\text{opt}} \sim 2 \times 10^{-3}$ radian.

If the "emitted" plasma waves occupy a broad angular spectrum (with wave vectors concentrated in the solid angle Ω), then the only waves that are effectively transformed into electromagnetic emission are those for which the angle φ between $\mathbf{k}$ and grad ε is less than or of the order of the optimum angle. This circumstance allows us to conclude that the mean transformation efficiency for plasma waves with a broad angular spectrum because of regular interaction in an isotropic plasma is

$$\bar{Q} \sim \frac{\pi}{\Omega} (Q \sin^2 \varphi)_{S_0 \sim 1} \sim \frac{\pi}{4\Omega} \frac{V_{\text{ph}}^2}{c^2} \left(\frac{3}{2} \frac{c \text{ grad } \varepsilon}{\omega} \right)^{2/3}. \qquad (25.20)$$

In a stationary corona for grad $\varepsilon \sim 10^{-10}$ cm, $\omega \sim 2\pi \times 10^8$ sec^{-1}, $\Omega \sim 2\pi$ and $V_{\text{ph}} \sim c$ the efficiency is $\bar{Q} \sim 5 \times 10^{-7}$ (Ginzburg and Zheleznyakov, 1958b).

Above, when investigating the transformation of plasma waves into electromagnetic ones, the only effects to be taken into consideration were those which were connected with the region $\varepsilon \approx \sin^2 \varphi_0$. Wave transformation in layers where in the first approximation geometrical optics are applicable was not taken into consideration because of the relative smallness of this effect, although the question of the magnitude of the latter may be of interest in the case that the level $\varepsilon \approx \sin^2 \varphi_0$ is not reached by the plasma waves. The corresponding transformation efficiency in an isotropic plasma when there is an electron concentration and temperature gradient has been found by Tidman (1960) (see also Tidman and Weiss, 1961). It follows from the expressions he obtained that the maximum possible transformation coefficient in the case $\lambda_{pl} \ll L$, where L is the characteristic dimension over which the plasma's density or its temperature changes, does not exceed

$$Q \sim \frac{\lambda}{L} \beta_{th} e^{-2\pi}. \tag{25.21}$$

Here λ is the length of the electromagnetic wave in a vacuum, connected with the length of the plasma wave in the medium by the relation $\lambda_{pl} = \lambda \beta_{th}/\sqrt{\varepsilon}$. As could be expected, $Q \ll 1$ (since $\lambda \beta_{th} \ll L$ with $\lambda_{pl} \ll L$); in particular Q is far less than that given (with optimum conditions) by interaction in the layer $\varepsilon \approx \sin^2 \varphi_0$.

It is also reasonable to compare the values of the mean transformation $\bar{Q}$ provided that the system of plasma waves has a broad angular spectrum in both cases. In both transformation regions a change occurs in practice only for those plasma waves which are propagated in a narrow cone with its axis running along the gradient. For transformation in the coupling region the angle of separation of this cone is defined by the relation (25.19) and $\bar{Q}$ by the relation (25.20). For transformation in the region of geometrical optics the corresponding angle is $\sin \varphi \sim V_{ph}/c$† and

---

† It follows from the above treatment (see in particular the formula (25.2)) that in a plane-layered medium the waves that have the same periodicity in a plane orthogonal to the gradient of the variation in the medium's properties (in the plane xy) are interconnected. In other words, for coupling waves the projections of the wave vectors onto this plane coincide:

$$\frac{\omega}{c} n_3 \sin \varphi = \frac{\omega}{c} n_{1,2} \sin \varphi'.$$

At the beginning of the layer this equality, in which φ' is the angle at which the electromagnetic wave is propagated, becomes

$$\sin \varphi_0 = \beta_{th} \sin \varphi_0'.$$

However, $\sin \varphi_0'$ cannot become more than unity; therefore $\sin \varphi_0 \leqslant \beta_{th}$ and $\sin \varphi = \sin \varphi_0(V_{ph}/V_{th}) \leqslant V_{ph}/c$.

§ 25] The Escape of Radio Emission from the Corona

accordingly

$$\bar{Q} \sim \frac{\pi}{\Omega} Q \sin^2 \varphi \lesssim \frac{\pi}{\Omega} \frac{\lambda}{L} \beta_{th}^2 e^{-2\pi \frac{V_{ph}^2}{c^2}}. \qquad (25.22)$$

The ratio of the mean efficiency coefficients $\bar{Q}$ in order of magnitude is known not to exceed $10e^{-2\pi}\beta_{th}^2(\lambda^{1/3}/L^{1/3})$, which is also much less than unity.

WAVE COUPLING IN A SMOOTHLY NON-UNIFORM MAGNETOACTIVE PLASMA

A full investigation of the interaction of waves in a plasma in the presence of a constant magnetic field H_0 is a very complex problem; this is quite natural since allowing for H_0 leads to a considerable complication of the quasi-hydrodynamic equations (22.9). Therefore among the problems of the interaction of electromagnetic waves in the region $v = \omega_L^2/\omega^2 \approx 1$ only the simplest variants are discussed at present, in which the magnetic field is assumed to be uniform, the medium to be plane-layered and the wave propagation to be along the electron concentration gradient.†

Interaction in the region $v \approx 1$ with normal propagation (along grad N_0) in a plasma with $u = \omega_H^2/\omega^2 < 1$ was first investigated by Ginzburg (1948) and in greater detail by Denisov (1954, 1957) as applied to the so-called effect of "tripling" of signals in the ionosphere (see also Ginzburg, 1960b, section 28). The possible part played by this kind of interaction in the solution of the problem of the escape of radio waves from the corona in the presence of a magnetic field has been indicated by Gershman and Zheleznyakov (1956). This phenomenon was then studied by Zheleznyakov (1958b and 1959b) (see also Zheleznyakov and Zlotnik, 1963) with $u < 1$ and with $u > 1$ (allowing for thermal motion). We notice that in the papers by Denisov (1957), Zheleznyakov (1958b, 1959b), and Denisov (1954) the whole of the treatment was by the phase integral method which was discussed in section 24. Below, therefore, we shall not give the detailed calculations (referring the reader to the original papers for them) and shall limit ourselves to giving the results obtained by Denisov (1957) and Zheleznyakov (1958b and 1959b).

As can be seen from Figs. 117 and 121, with $u < 1$ in a regularly non-uniform plasma there exists in practice only one interaction region; the other region in which $n_j^2 < 0$ is generally not reached by the waves since

---

† The last limitation is particularly undesirable since an analysis of the dispersion curves for the case of inclined propagation (Ginzburg, 1960b, section 29) shows that in a layer where the waves travel in a direction close to the direction of H_0 these curves approach for extraordinary and ordinary waves. This makes it permissible to expect here effective transformation of waves of one type into another.

the latter are exponentially damped as they approach it.† This circumstance noticeably simplifies the whole picture of the interaction; it is also significant that in the case when $u < 1$ thermal motion in the plasma can be completely ignored when studying the interaction since the shape of the dispersion curves in the region $v \approx 1$ with $\beta_{\text{th}} \to 0$ changes slightly. This allows us when solving the problem to proceed from the equations (24.1); however, since here, unlike section 24, the magnetic field is assumed to be uniform and the electron concentration to be varying along the z-axis, the coefficients A, B, C in (24.2) depend on z via the parameter $\varepsilon \equiv 1-v$, whilst $\omega_z = $ const, $\omega_y = $ const. For simplicity the function $\varepsilon(z)$ in the interaction region is considered to be linear: $\varepsilon(z) = z\,\text{grad}\,\varepsilon$. The part of the parameter ϱ_H in this case is played by $\varrho_N = k_0/\text{grad}\,\varepsilon$; in the weakly non-uniform plasma to which our discussion is limited $\varrho_N \gg 1$.

The solution of the interaction problem in the region $v \approx 1$ with $u < 1$ is exactly like that given in section 24 for the case of interaction in a quasi-transverse magnetic field region. The geometrical optics approximation, the position of the branching points and the nature of the Stokes lines in the complex plane ε—everything recalls the problem discussed in section 24. It is not surprising, therefore, that the connection between the constants in the solution that describes the propagation of extraordinary and ordinary waves is in general the same as (24.45) but with the interchanging of the suffices 1 and 2 in the constants in (24.45). This is because with quasi-transverse propagation in the vicinity of the interaction region $n_1^2 < n_2^2$, whilst in the case under discussion $n_1^2 > n_2^2$. Now, however, the characteristic interaction parameter $2\delta_0$ (we shall denote it by $2\delta_{01}$) is slightly different in form from (24.42):

$$2\delta_{01} = -i\varrho_N \oint \frac{n_2 - n_1}{2}\,d\varepsilon; \qquad (25.23)$$

the integration contour embraces the branching points $\varepsilon = \pm i\omega_y^2/\omega_z$ at

† Interaction in the region $n_j^2 < 0$ can be significant only if the magnetic field in the layer $v \approx 1$ is very small:

$$\sqrt{u} \lesssim \left(\frac{c}{\omega}\,\text{grad}\,\varepsilon\right)^{2/3}$$

(in the case of a weakly non-uniform plasma $(c/\omega)\,\text{grad}\,\varepsilon \ll 1$). When this criterion is satisfied an extraordinary wave leaving the interaction region is weakly damped in the overlying layers $1 > v > 1-\sqrt{u}$, where $n_1^2 < 0$: the point is that in this case the distance between the points $v = 1$ and $v = 1-\sqrt{u}$ will be less than or of the order of the length of an extraordinary wave in the reflection region. In the solar corona for $\omega \sim 2\pi \times 10^8\,\text{sec}^{-1}$ and grad $\varepsilon \sim 10^{-10}\,\text{cm}^{-1}$ this situation occurs only in the fields $H_0 \lesssim 10^{-4}$ oe. In actual fact $H_0 \gtrsim 1$ oe there, which requires an increase in grad ε by a minimum factor of 10^6; however, such high values of grad $\varepsilon \sim 10^{-4}\,\text{cm}^{-1}$ in the corona are very improbable, if we exclude the case of charp gradients in shock wave fronts.

§ 25] The Escape of Radio Emission from the Corona

which $n_1 = n_2$. The calculation of $2\delta_{01}$ with small enough angles α between H_0 and the direction of propagation of the waves gives (Zheleznyakov, 1959b):

$$2\delta_{01} \approx \frac{\pi \varrho_N \omega_y^2}{\omega_z^{1/2}(\omega_z+1)^{3/2}} \approx \frac{\pi}{2} \frac{\omega}{c \, \mathrm{grad}\, \varepsilon} \frac{\alpha^2}{\left(1+\dfrac{\omega}{\omega_H}\right)^{3/2}}. \qquad (25.24)$$

When changing to the last expression it is taken into consideration that $\varrho_N = \omega/c \, \mathrm{grad}\, \varepsilon$, $2(\omega_y^2/\omega_z^2) = \tan \alpha$ and that with small α we can put $\tan \alpha \approx \alpha$, $\cos \alpha \approx 1$ and $\omega_z \approx \omega_H/\omega$.

The allowance for thermal motion (i.e. the existence of plasma waves) does not alter the connection between the constants in the geometric-optical solution, so for waves being propagated towards negative ε (into the depth of the plasma) the following relations are as before valid (Zheleznyakov, 1959b):

$$\left. \begin{array}{l} c'_{\mathrm{II}} = \sqrt{1-e^{-2\delta_{01}}}\, c_{\mathrm{II}} - e^{-\delta_{01}}\, c_{\mathrm{III}}, \\ c'_{\mathrm{III}} = e^{-\delta_{01}}\, c_{\mathrm{II}} + \sqrt{1-e^{-2\delta_{01}}}\, c_{\mathrm{III}}. \end{array} \right\} \qquad (25.25)$$

For waves travelling in the opposite direction

$$\left. \begin{array}{l} d'_{\mathrm{II}} = \sqrt{1-e^{-2\delta_{01}}}\, d_{\mathrm{II}} - e^{-\delta_{01}}\, d_{\mathrm{III}}, \\ d'_{\mathrm{III}} = e^{-\delta_{01}}\, d_{\mathrm{II}} + \sqrt{1-e^{-2\delta_{01}}}\, d_{\mathrm{III}}. \end{array} \right\} \qquad (25.26)$$

The meaning of the notations c_{II}, c_{III}, d_{II}, d_{III} and c'_{II}, c'_{III}, d'_{II}, d'_{III} characterizing the amplitudes of the "normal" waves on both sides of the interaction region is clear from Fig. 121. The expression for the parameter $2\delta_{01}$, of course, changes although with the condition $\beta_{\mathrm{th}} = V_{\mathrm{th}}/c \ll 1$ that is usually satisfied the corrections to (25.24) will apparently be slight.

In the case when $u > 1$† both interaction regions correspond to the values $n_j^2 > 0$ (see Figs. 117 and 125); therefore when calculating the interaction in a strong magnetic field or at low enough frequencies both regions have to be taken into consideration. The treatment of the connection between different types of waves is complicated here by the fact that with no allowance made for thermal motion ($\beta_{\mathrm{th}}^2 = 0$) both regions are located in the vicinity of the branching points $\varepsilon = \pm i\omega_y^2/\omega_z$ and actually coincide with each other. Allowing for thermal motion, being connected with more cumbersome calculations, in essence simplifies the solution of the problem: with $\beta_{\mathrm{th}}^2 \neq 0$ spatial separation of the interaction regions occurs, which allows us to examine the interaction phenomenon in each region separately. A detailed investigation (Zheleznyakov, 1958b and 1959b) shows that the process of wave interaction and transformation

† To be more precise, when $u \cos^2 \alpha > 1$ (see the remark on pp. 319, 320).

in general outline proceeds here just as in the case of $u < 1$. The connection between the constants on the two sides of the lower interaction region (where the dispersion curves II and III approach close to each other, see Fig. 121) is defined as before by the formulae (25.25) and (25.26) with the characteristic parameter $2\delta_{01}$ (25.24). Similar relations between the constants in the geometric-optical solution hold on both sides of the upper interaction region (where the dispersion curves I and III are close):†

$$\left.\begin{array}{l}\tilde{c}'_I = \sqrt{1-e^{-2\delta_{02}}}\,\tilde{c}_I + e^{-\delta_{02}}\,\tilde{c}_{III}, \\ \tilde{c}'_{III} = -e^{-\delta_{02}}\,\tilde{c}_I + \sqrt{1-e^{-2\delta_{02}}}\,\tilde{c}_{III},\end{array}\right\} \quad (25.27)$$

$$\left.\begin{array}{l}\tilde{d}'_I = \sqrt{1-e^{-\delta_{02}}}\,\tilde{d}_I + e^{-\delta_{02}}\tilde{d}_{III}, \\ \tilde{d}'_{III} = -e^{-\delta_{02}}\,\tilde{d}_I + \sqrt{1-e^{-2\delta_{02}}}\,\tilde{d}_{III}.\end{array}\right\} \quad (25.28)$$

In these formulae with small α the characteristic interaction parameter is

$$2\delta_{02} \approx \frac{\pi \varrho_N \omega_y^2}{\omega_z^{1/2}(\omega_z-1)^{3/2}} \approx \frac{\pi}{2}\,\frac{\omega}{c\,\mathrm{grad}\,\varepsilon}\,\frac{\alpha^2}{\left(1-\dfrac{\omega}{\omega_H}\right)^{3/2}}. \quad (25.29)$$

It differs from the parameter $2\delta_{01}$ only by the sign in the denominator.

As an example of the use of the relations (25.25)–(25.28) let us examine one of the interaction variants with $u < 1$, which defines the efficiency of the change of plasma waves into electromagnetic emission leaving the corona and thus (as well as the similar case with $u > 1$) holds the greatest interest for the theory of the Sun's sporadic radio emission. Let only a plasma wave with an amplitude $c_{III} = 1$ (i.e. $c_{II} = 0$) approach an interaction region from the side of positive $\varepsilon \equiv 1 - v$ (i.e. on the right of it in Fig. 121). We shall be finding the amplitude of the ordinary wave d_I leaving this region.

In accordance with the conditions $c_{III} = 1$, $c_{II} = 0$, for waves leaving the interaction region and being propagated to the left of it

$$c'_{II} = -e^{-\delta_{01}}, \quad c'_{III} = \sqrt{1-e^{-2\delta_{01}}}.$$

Unless we take into consideration the reflection of extraordinary waves from the layer $v = 1+\sqrt{u}$ (let us say, because of strong absorption on the way to this layer and back)‡ the amplitude d'_{III} will be zero. At the same time the amplitudes c'_{II} and d'_{II} because of reflection of the ordinary wave from the point $v = 1$ are connected with each other by the relation

$$d'_{II} = c'_{II}\,e^{i\psi}, \quad \psi = -2\varrho_N \int n_{II}\,d\varepsilon + \frac{\pi}{2}.$$

† The sign $\sim$ above c and d is introduced to distinguish these coefficients from the corresponding amplitudes near the interaction region corresponding to smaller values of n_j^2.

‡ In the cases shown in Figs. 122–4 this layer does not generally exist within the solar corona.

§ 25] The Escape of Radio Emission from the Corona

Here $\varrho_N \int n_{II}$ is the change in phase on the way from the interaction region to the point $v = 1$, whilst the quantity $\pi/2$ allows for additional phase shift due to reflection of the wave. From this and from the relations (25.26) it follows that

$$\sqrt{1-e^{-2\delta_{01}}}\, d_{II} - e^{-\delta_{01}}\, d_{III} = c'_{II} e^{i\psi} = e^{-\delta_{01}+i\psi},$$

$$e^{-\delta_{01}}\, d_{II} + \sqrt{1-e^{-2\delta_{01}}}\, d_{III} = 0.$$

Solving these equations for d_{II} and d_{III}, we find:

$$d_{II} = \sqrt{1-e^{-2\delta_{01}}}\, e^{-\delta_{01}+i\psi}, \quad d_{III} = -e^{-2\delta_{01}+i\psi}. \tag{25.30}$$

Therefore if $u < 1$, then with incidence of a plasma wave on a coupling region located in layers $v \approx 1$ an ordinary wave propagated towards small v escapes into a region where $v < 1$. The relative intensity of this wave (the transformation coefficient Q) is defined in accordance with (25.30) by the formula

$$Q = e^{-2\delta_{01}}(1-e^{-2\delta_{01}}); \tag{25.31}$$

the relative intensity of a reflected plasma wave will be $e^{-4\delta_{01}}$.

By proceeding similarly it is not difficult to find the amplitudes and intensities of the waves for other forms of interaction. The corresponding expressions are obtained by Zheleznyakov (1959b) and given in the monograph by Ginzburg (1960b, section 28). It follows from them that if $u > 1$, then with incidence of a plasma wave on an interaction region $v \approx 1$ both ordinary and extraordinary waves come from the latter; the relative intensity of these waves, which define the transformation efficiency, are respectively

$$Q_1 = e^{-2\delta_{01}-2\delta_{02}}(1-e^{-2\delta_{01}}), \quad Q_2 = e^{-4\delta_{01}-2\delta_{02}}(1-e^{-2\delta_{02}}). \tag{25.32}$$

It is clear from what has been said that the efficiency of the interaction is of the order of unity if $2\delta_{01,\,02} \approx 1$; for example, for $u < 1$ the maximum value of $Q_{\max} = \tfrac{1}{4}$ is reached with the value $2\delta_{01} = \ln 2$. The efficiency drops as $2\delta_{01,\,02}$ decreases (i.e. as the angle α between H_0 and the direction of the waves' propagation decreases), becoming zero when $2\delta_{01,\,02} = 0$; it also falls very rapidly (exponentially) when $2\delta_{01,\,02}$ (i.e. the angle α) increases, starting at $2\delta_{01,\,02} \approx 1$. It follows from this that significant interaction occurs provided that the angle α is contained in the solid angle Ω around the direction of propagation, with†

$$\Omega \sim \pi(\alpha^2)_{2\delta_{01,\,02} \approx 1}. \tag{25.33}$$

† It was noted above that effective interaction exists not only in the case of propagation along grad ε, but also with inclined incidence (at an angle to grad ε). This possible form of interaction has not yet been examined quantitatively; there is no reason to assume, however, that the solid angle containing the directions of wave propagation corresponding to significant interaction with $Q \sim 1$ differs in order of magnitude from the value (25.33).

According to (25.24), 25.33) in the conditions of the solar corona for $\omega \sim 2\pi \times 10^8$ sec^{-1}, grad $\varepsilon \sim 10^{-10}$ cm^{-1} and field values of $H_0 \sim 1$ oe the solid angle is $\Omega \sim 2 \times 10^{-6}$. When H_0 rises the quantity Ω decreases, and with $H_0 \sim 36$ oe (when ω_H, while remaining less than ω, is close to it) $\Omega \sim 3 \times 10^{-8}$. With a further rise in H_0 (in the region $u > 1$) the angle Ω, as is clear from (25.29), becomes still smaller.†

If emission of one type (let us say plasma) approaching an interaction region has a broad angular spectrum corresponding to a solid angle of the order of unity, or in this field there are found different orientations of the magnetic field relative to the direction of propagation concentrated in the same solid angle, then the mean efficiency of transformation into waves of another type is‡

$$\bar{Q} \sim Q_{max} \pi(\alpha^2) 2\delta_{01,\,02} \approx 1. \tag{25.34}$$

For example, with $u < 1$ for the change of a plasma wave into an ordinary one $Q_{max} = \frac{1}{4}$ and therefore $\bar{Q}$ is about a quarter of the value estimated above for the effective angle Ω (25.33).

We note that as a result of wave interaction in a layer $v \approx 1$ only an ordinary wave escapes beyond the corona; the extraordinary component cannot pass through a layer bounded at the top by the level $v = 1 - \sqrt{u}$, since there $n_1^2 < 0$ and the wave is exponentially damped.

The results obtained, strictly speaking, are valid only in a plasma without collisions ($\nu_{\text{eff}} = 0$). The presence of dissipation in the interaction region, however, has no significant effect on wave transformation in a non-uniform magnetoactive plasma if

$$\nu_{\text{eff}} \ll \omega_H \frac{\sin^2 \alpha}{2 \cos \alpha} \tag{25.35}$$

(i.e. if $\nu \equiv \nu_{\text{eff}}/\omega \ll q \equiv \sqrt{u} \sin^2 \alpha / 2 \cos \alpha$). On the other hand, in the case of the opposite inequality

$$\nu_{\text{eff}} \gg \omega_H \frac{\sin^2 \alpha}{2 \cos \alpha}, \tag{25.36}$$

the presence of collisions determines the whole picture of the phenomenon.

---

† On the other hand as the field H_0 decreases in the region $\sqrt{u} \ll 1$ the effective solid angle increases and with $\sqrt{u} \sim (2c \text{ grad } \varepsilon/\omega)^{2/3}$ becomes of the order of unity. However, in fields of this kind we must allow for wave interaction in the region $n_j^2 < 0$ and the escape of an extraordinary wave from the latter. Here it is necessary, therefore, to investigate the interaction effect anew; a detailed examination has not yet been made. We note, moreover, that this relation with $\omega \sim 2\pi \times 10^8$ sec^{-1} and grad $\varepsilon \sim 10^{-10}$ cm^{-1} corresponds to fields $H_0 \sim 10^{-4}$ oe, whose existence is doubtful under the usual conditions of the solar corona.

‡ To be more precise the value of $\bar{Q}$ can be obtained by integrating the transformation coefficient Q (25.31) or (25.32) over the whole solid angle in which the values of α are concentrated.

§ 25] The Escape of Radio Emission from the Corona

This is connected with the fact that with $v \gg q$ the behaviour of the dispersion curves differs noticeably from that indicated in Fig. 117 for the angle $\alpha \neq 0$, and corresponds to the case of quasi-longitudinal propagation with which, obviously, the transformation of a plasma wave into other types of waves is strongly attenuated, and the change of extraordinary waves into ordinary ones and vice versa becomes practically complete.

Since effective interaction occurs in the solid angle $\Omega \approx \pi(\alpha^2)_{2\delta_{01,02} \approx 1}$ it is clear from (25.35) that allowing for collisions does not alter the estimates of the transformation efficiency Q if

$$v_{\text{eff}} \lesssim \frac{\omega_H \Omega}{2\pi}. \tag{25.37}$$

For fields $H_0 \sim 1$ oe and the values of ω and grad ε used previously in estimates of the angle Ω this means that in the layer $v \approx 1$ we should have $v_{\text{eff}} \lesssim 6 \text{ sec}^{-1}$. If $H_0 \sim 36$ oe, then the condition (25.37) is reduced to the following: $v_{\text{eff}} \lesssim 3 \text{ sec}^{-1}$. In the corona these conditions are in general satisfied, although there combinations of parameters are undoubtedly realized in which v_{eff} will noticeably exceed $\omega_H \Omega/2\pi$. In the latter case it must be expected that the efficiency of the transformation of plasma waves into electromagnetic ones will be lower than the values given previously.

In conclusion we note that whilst above we have discussed the interaction of waves in the extreme case of a smoothly non-uniform medium ($\varrho_N \gg 1$) in the papers by Field (1956), Kritz and Mintzer (1960), Tidman and Boyd (1962) there has been an investigation of the change of plasma waves into electromagnetic radiation at a sharp boundary in a plasma. This kind of approach is possible only in conditions when the characteristic dimension over which there is a significant alteration in the properties of the plasma is less than the length of a plasma wave, i.e. $\varrho_N \ll 1$. In an isotropic solar corona, where the mean free path is $l_{\text{fr}} \sim 10^8$ cm, such conditions rarely obtain in practice. In the presence of a magnetic field the gradients in the corona may be steeper; even in this case, however, it remains uncertain whether discontinuities with a front thickness of less than $\lambda_{\text{pl}}/2\pi$ (in particular shock waves) can exist in the plasma.

CONVERSION OF PLASMA WAVES INTO ELECTROMAGNETIC WAVES BECAUSE OF SCATTERING ON ELECTRON DENSITY FLUCTUATIONS (Zheleznyakov, 1959a; Ginzburg and Zheleznyakov, 1958b).

At the beginning of this section it was noted that when emission escapes beyond the corona, as well as interaction in a regularly non-uniform medium, a significant part is played by the effect of wave transformation because of their scattering on random (irregular) inhomogeneities of the plasma. Regular interaction, which is all that was discussed above, generally disappears in a uniform (on the average) medium; only the second effect remains—the change of plasma waves into electromagnetic ones and vice versa on fluctuations of the dielectric permeability $\delta\varepsilon$ connected with the variation in the electron concentration δN. The latter may be caused either by the thermal motion of the plasma particles or by the action of other causes (for example, by the presence of perturbations of the plasma wave type from a secondary source).

Starting with the case of an isotropic plasma ($H_0 = 0$) we shall show these variations in the electron concentration in the form $\delta N = \delta N' + \delta N''$. The term $\delta N' = \delta N_i$ is connected with the quasi-neutral fluctuations of the plasma density $\delta\varrho_i \approx m_i \delta N_i$; the second term allows for the change in the electron concentration accompanying fluctuations in the electric charge $\delta\varrho = -e\delta N''$ (here N_i and m_i are the concentration and mass of the ions in the solar corona, largely protons). By virtue of the inequality $m_i \gg m$ the fluctuations in the density ϱ_i and the fluctuations in the charge ϱ are independent during thermal motion, so the scattering of waves on fluctuations of these two types can be treated separately.

The nature of the variation of $\delta N'$ and $\delta N''$ in time determines the frequency spectrum of the scattered emission. Since changes in the plasma density proceed slowly when compared with the period of the oscillations in the plasma wave, the scattering of the latter on $\delta N'$ is not accompanied by any significant change in frequency (Rayleigh scattering); here $\Delta\omega \sim \omega_{Li}$, where $\omega_{Li} = (4\pi e^2 N_i/m_i)^{1/2}$. On the other hand, scattering on charge fluctuations, the long-wave part of whose spectrum ($\lambda/2\pi \gg D$, where D is the Debye radius; see section 22) is a combination of plasma waves, is accompanied by a considerable change in frequency $\Delta\omega \sim \omega_L = (4\pi e^2 N/m)^{1/2}$ (combination scattering).

Let us first examine the Rayleigh scattering of a plasma wave in which the electrical field is of the form $E_0 = E_{00} \cos(\omega t - kR)$. When this wave passes through a plasma in which there are fluctuations of the density ϱ_i and changes in the electron concentration $\delta N'$ connected with the latter, the field E causes additional polarization $\delta P^{sec} = (4\pi)^{-1}\delta\varepsilon E_0$, where $\delta\varepsilon$ is the change in the dielectric permeability connected with the fluctuations $\delta N'$. The polarization δP^{sec} in its turn is a source of scattered plasma and electromagnetic waves. The latter are of particular interest to us, since the scattering process in this case is one of the forms of transformation of waves not escaping directly from a plasma into waves that leave it without hindrance.

It is known (see, e.g., Landau and Lifshitz, 1960, section 67) that the

§ 25] The Escape of Radio Emission from the Corona

time-averaged total energy flux of the electromagnetic radiation created by a dipole moment $\gamma = \delta P^{\text{sec}} dV$ in a medium with a refractive index $n_{1,2}$ is

$$dS = \frac{2n_{1,2}}{3c^3} \overline{\left[\frac{d^2 \gamma}{dt^2}\right]^2} = \frac{n_{1,2}}{24\pi^2 c^3} \overline{\left[\frac{d^2}{dt^2}(\delta \varepsilon E_0 \, dV)\right]^2}. \qquad (25.38)$$

Here dV is an element of volume much less than $(\lambda_{\text{el}}/2\pi)^3$ and $(\lambda_{\text{pl}}/2\pi)^3$ (λ_{el} and λ_{pl} are the lengths of the electromagnetic and plasma waves); the bar over the brackets indicates time-averaging. Accordingly the flux of the energy scattered in an arbitrary volume V is of the form

$$S = \frac{n_{1,2}}{24\pi^2 c^3} \overline{\left[\frac{d^2}{dt^2} \int_V \delta \varepsilon E_{00} \cos(\omega t) \, dV\right]^2}. \qquad (25.39)$$

A simple expression of this kind for S (without phase factors to allow for the phase of the field E_0 of the plasma wave and the lag of the electromagnetic waves scattered by different elementary volumes) holds in the case that the fluctuations $\delta N'$ (and at the same time $\delta \varepsilon$) are independent at distances of $l \gtrsim \lambda_{\text{el}}/2\pi$, $\lambda_{\text{pl}}/2\pi$. Since in an isotropic plasma the correlation radius for $\delta N'$ is comparable with the Debye radius D, it is clear that the formula (25.39) is definitely valid with the condition $\lambda_{\text{pl}}/2\pi \gg D$.†

Since, as has already been pointed out, the density fluctuations take place slowly when compared with the oscillations of the field E, we can limit ourselves in (25.39) to differentiation of the factor $\cos(\omega t)$. As a result‡

$$S = \frac{n_{1,2}(\omega)\omega^4 E_{00}^2}{48\pi^2 c^3} \overline{\left[\int_V \delta \varepsilon \, dV\right]^2} = \frac{n_{1,2}(\omega)\omega^4 E_{00}^2 V^2}{48\pi^2 c^3} \overline{(\delta \varepsilon)_V^2}, \qquad (25.40)$$

where

$$(\delta \varepsilon)_V = V^{-1} \int_V d\varepsilon \, dV$$

are the fluctuations of the dielectric permeability averaged over the volume. To find the flux scattered on the fluctuations $\delta \varrho_i$ we must put into (25.40)

$$\overline{(\delta \varepsilon)_V^2} = \left(\frac{\partial \varepsilon}{\partial \varrho_i}\right)^2 \overline{(\delta \varrho_i)_V^2} = \left(\frac{\partial \varepsilon}{\partial \varrho_i}\right)^2 \frac{\varkappa T \varrho_i}{V} \left(\frac{\partial \varrho_i}{\partial p}\right)_T = \frac{\varkappa T(\varepsilon-1)^2}{\varrho_i V} \left(\frac{\partial \varrho_i}{\partial p}\right)_T. \qquad (25.41)$$

† Here automatically $\lambda_{\text{el}}/2\pi \gg D$ since $\lambda_{\text{el}} = \lambda_{\text{pl}} \beta_{\text{th}}^{-1}$, where $\beta_{\text{th}} \ll 1$. We notice that the condition $\lambda_{\text{pl}}/2\pi \gg D$ is also necessary for weak damping of plasma waves (in the region $\lambda_{\text{pl}}/2\pi \lesssim D$ they are not propagated in practice, being absorbed at a distance of the order of λ_{pl}; see section 26).

‡ A more correct derivation of the formula (25.40) can be found in Landau and Lifshitz (1957, sections 93, 95). There, it is true, it is assumed that scattered electromagnetic radiation appears when electromagnetic waves are propagated in a medium. It is not hard to confirm, however, that the whole treatment can be completely transferred to the case when electromagnetic radiation appears because of scattering of a longitudinal wave in a plasma.

When changing to the last expression we allowed for the fact that $\varrho_i(\partial\varepsilon/\partial\varrho_i)$ $= \varepsilon - 1$; the suffix T indicates that the derivative $\partial\varrho_i/\partial p$ (where p is the total pressure in the plasma caused by the thermal motion of the electrons and the ions) is taken with a constant temperature (see Landau and Lifshitz, 1957, section 96). Bearing in mind the relations given and considering at the same time that the plasma is subject to the equation for the state of an ideal gas with a number of particles per cubic centimetre $2N$: $p = 2N\varkappa T$, i.e. $(\partial\varrho_i/\partial p)_T \varrho_i^{-1} = 1/2N\varkappa T$, we obtain the final expression for the total flux of the electromagnetic energy with Rayleigh scattering of a plasma wave in the volume V (Ginzburg and Zheleznyakov, 1958b):[†]

$$S(\omega) = \frac{n_{1,2}(\omega)e^4 NV}{6m^2 c^3} E_{00}^2. \qquad (25.42)$$

In certain cases it is also convenient here to introduce the parameter Q characterizing the efficiency of the change of plasma waves into electromagnetic ones and defined by the expression

$$Q = \frac{S}{S_{pl}L^2}, \qquad (25.43)$$

in which L is the linear dimension of the scattering region, S is the total flux of the scattered emission, $S_{pl} \approx V_{th}\sqrt{\varepsilon} E_0^2/8\pi$ is the energy flux density in the plasma wave (22.21). For Rayleigh scattering at a frequency ω

$$Q \approx \frac{4\pi e^4 NL}{3m^2 c^3 V_{th}}. \qquad (25.44)$$

It follows from this that the efficiency of transformation of plasma waves into electromagnetic radiation because of scattering in the solar corona on fluctuations $\delta\varepsilon$ accompanying density fluctuations $\delta\varrho_i$ is of the order of 5×10^{-6} for $N \sim 10^8$ electrons/cm^3, $L \sim 4\times 10^9$ cm, $V_{th} \sim 4\times 10^8$ cm/sec ($T \sim 10^6$ °K).

Therefore in the usual conditions of an isotropic coronal plasma the Rayleigh scattering mechanism is in any case not less and is apparently

---

[†] If we assume that the scattering of the plasma waves occurs simply on free electrons (Thomson scattering), then for the emission flux we obtain a value of $S(\omega)$ that is twice (25.42). In the case under discussion, however, the electrons cannot be considered free, which is allowed for when splitting the fluctuations δN into two types. Thomson scattering in a plasma occurs only for waves shorter than the Debye radius D; since in an equilibrium plasma plasma waves with $\lambda_{pl}/2\pi \lesssim D$ are not propagated, this form of scattering may be of interest only for high-frequency electromagnetic waves. The transformation of plasma waves into electromagnetic radiation when they are scattered on free electrons in the corona has in fact been discussed (not explicitly, it is true) by Shklovskii (1946) when dealing with the plasma hypothesis of the origin of the Sun's sporadic radio emission.

§ 25] The Escape of Radio Emission from the Corona

more significant than the regular interaction mechanism, whose efficiency for plasma waves with a broad angular spectrum is generally less than or of the order of 5×10^{-7} and definitely does not exceed 10^{-5} with the most extreme assumptions about the value of grad ε. It must also be stressed that regular interaction can have a noticeable effect only for the plasma waves which are propagated towards grad ε and reach the level $\varepsilon \approx 0$. This kind of situation is far from always realized in a corona with an electron concentration that decreases as one moves away from the surface of the Sun. For example, with Cherenkov radiation of plasma waves by charged particles moving from a centre of activity into the upper, more rarefied layers of the corona the directions of these waves are concentrated in a cone whose generatrices subtend an acute angle with the direction of the velocity V of the particles (see section 26). In this case regular interaction and transformation of waves obviously play a very small part.

In a corona for transformation of plasma waves into electromagnetic ones combination scattering of these waves on fluctuations of the electric charge $\delta\varrho$ may also be significant.[†] If these fluctuations are caused by the thermal motion of particles of a plasma in a state of kinetic equilibrium the long-wave part of the spectrum of these fluctuations is a combination of plasma waves that satisfy the dispersion equation $\omega_{pl}^2 = \omega_L^2 + V_{th}^2 k_{pl}^2$ (ω_{pl} and k_{pl} are respectively the frequency and wave number of these waves; compare (22.16)). The level of plasma-type fluctuations in the region $k_{pl}D \ll 1$ (i.e. $\lambda_{pl}/2\pi \gg D$) is defined by the relation (Pines and Bohm, 1952)

$$\overline{(\delta N''_{k_{pl}})^2} = \frac{\varkappa T k_{pl}^2 NV}{m\omega_L^2}, \qquad (25.45)$$

in which

$$\delta N''_{k_{pl}} = \int_V \delta N''(R) e^{-ik_{pl}R}\, dV$$

(V is the volume of the region in which the changes $\delta N''$ occur).

The problem of scattering of electromagnetic waves on plasma fluctuations has been discussed by Akhiyezer, Prokhoda and Sitenko (1957); the more general question of the scattering of electromagnetic waves in the whole spectrum of electron concentration fluctuations $\delta N = \delta N' + \delta N''$ has been investigated by Feier (1960 and 1961). We are chiefly interested in the inverse problem of the scattering of plasma waves into electromagnetic ones; the efficiency of this kind of process has been found by Ginzburg and Zheleznyakov (1958b) (allowing for the results of the paper by Akhiyezer, Prokhoda and Sitenko, 1957 and then in Cohen, 1962). In Ginzburg and Zheleznyakov (1958b), however, all that was taken into

[†] This circumstance was noted by Panovkin (1957).

consideration was the scattering of a wave with an amplitude E_{00} on fluctuations of the electron concentration $\delta N''_{k_{pl}}$ produced by a plasma wave k_{pl}, ω_{pl}. In any fuller treatment attention should also be paid to a further effect—scattering of waves of fluctuation origin with k_{pl}, ω_{pl} on a passing plasma wave which E_0 also produces changes in the electron concentration. The expressions for the transformation coefficient that allow for the contribution from both these effects to the scattered electromagnetic radiation have been obtained by Terashima and Yajima (1964). Below we shall obtain these expressions by a simpler method. We shall dwell in detail on the intermediate calculations, having in mind the numerous applications of the method of calculating scattered emission used here (the so-called Hamilton method)[†] for finding the energy emitted by the individual particles in a plasma (see section 26).

Therefore, let a plasma wave, the electrical field in which varies as $E_0(R, t) = E_{00} e^{i\omega t - ikR}(E_0 \| k)$, be propagated in an isotropic and on the average uniform plasma. It is required to find the intensity of the electromagnetic radiation appearing because of scattering of this wave on fluctuations $\delta N''(R, t)$ connected with electric charge fluctuations $\delta\varrho = -e\delta N''$ and because of scattering of plasma waves of fluctuation origin connected with the charge $\delta\varrho$ on variations in the electron concentration under the effect of the plasma wave $E_0(R, t)$ being propagated.

The field of the scattered electromagnetic radiation is described by the Maxwell equations:

$$\operatorname{curl} H = \frac{1}{c}\frac{\partial D}{\partial t} + \frac{4\pi}{c}\frac{\partial(\delta P^{\text{sec}})}{\partial t}, \quad \operatorname{curl} E = -\frac{1}{c}\frac{\partial H}{\partial t},$$
$$\operatorname{div} H = 0, \quad \operatorname{div} D = -4\pi \operatorname{div}(\delta P^{\text{sec}}). \quad (25.46)$$

In (25.46) the vector D is connected with E by the relation $D = \varepsilon E$, where $\hat{\varepsilon}$ is the plasma's dielectric permeability operator, which in action on the monochromatic field $E_\omega \propto e^{i\omega t}$ is reduced to multiplying E_ω by $\varepsilon(\omega)$, defined by the formula (22.17):

$$D = \int \varepsilon(\omega) E_\omega \, d\omega, \quad E_\omega = \int E e^{-i\omega t} \, dt.$$

Further, P^{sec} is the polarization fluctuations occurring during the passage of a plasma wave; the derivative

$$\frac{\partial}{\partial t}(\delta P^{\text{sec}}) \equiv \delta j^{\text{sec}} = -eV \delta N''(R, t) - en_0 \delta V''(R, t)$$

[†] This method, which has long been known in quantum electrodynamics (see, e.g., Heitler, 1956), has been widely used by Ginzburg (1940) in macroscopic electrodynamics. We shall give only the basic stages when treating the Hamilton method. More detailed calculations are given by Ginzburg (1940), Kolomenskii (1953), Eidman (1958, 1959); the corresponding treatment with allowance made for absorption is given by Ryzhov (1959).

§ 25] The Escape of Radio Emission from the Corona

where δj^{sec} is the current density fluctuations. The plasma electron velocity V_0 acquired because of the action of the plasma wave passing through can be found from the equation of motion $\partial V_0/\partial t = -eE_0/m$:

$$V_0 = i\frac{e}{m\omega}E_0.$$

At the same time the variation in the electron concentration n_0 because of the action of the plasma wave E_0 can be found from the continuity equation $\partial n_0/\partial t + \text{div}(NV_0) = 0$:

$$n_0 = \frac{N}{\omega}kV_0 = i\frac{eN}{m\omega^2}kE_0.$$

From this it follows that

$$\frac{\partial}{\partial t}(\partial P^{\text{sec}}) \equiv \delta j^{\text{sec}}$$

$$= -i\frac{e^2}{m\omega}\left[\delta N'' E_{00} + \delta V'' \frac{N}{\omega}(kE_{00})\right]e^{i\omega t - ikR} \quad (25.47)$$

When finding the intensities of the scattered electromagnetic waves it is convenient to change from the fields E, H to the vector and scalar potentials A, φ:

$$H = \text{curl}\, A, \quad E = -\text{grad}\,\varphi - \frac{1}{c}\frac{\partial A}{\partial t}.$$

In the case of Coulomb calibration of $\text{div}\, A = 0$ the equations for the potentials become:

$$\left.\begin{array}{l} \Delta A - \dfrac{\varepsilon}{c^2}\dfrac{\partial^2 A}{\partial t^2} - \dfrac{\varepsilon}{c}\dfrac{\partial}{\partial t}\text{grad}\,\varphi = -\dfrac{4\pi}{c}\delta j^{\text{sec}}, \\ \hat{\varepsilon}\Delta\varphi = 4\pi\,\text{div}\,(\delta P^{\text{sec}}). \end{array}\right\} \quad (25.48)$$

We shall find the solution for A in the form of the expansion

$$A = \sum_\lambda q_\lambda(t) A_\lambda(R), \quad (25.49)$$

where

$$A_\lambda(R) = e_\lambda \left[\frac{4\pi c^2}{\varepsilon(\omega_\lambda)}\right]^{1/2} e^{-ik_\lambda R}. \quad (25.50)$$

In the last expression e_λ is a vector characterizing the polarization of the normal electromagnetic waves; for an isotropic plasma as e_λ we can take a pair of unit vectors orthogonal to each other and the vector k_λ. The value of the vector k_λ is connected with the frequency ω_λ by the dispersion relation $k_\lambda^2 = \varepsilon(\omega_\lambda)\omega_\lambda^2/c^2$; the set of values k_λ is fixed by the condition that over the volume chosen (which we shall consider to be a unit volume in order not to introduce it in the explicit form and thus simplify the writing of the formulae) there should be a whole number of wavelengths $\lambda = 2\pi/k_\lambda$. It follows from (25.50) that

$$\int A_\lambda A_{\lambda'}^* \, dV = \frac{4\pi c^2}{\varepsilon(\omega_\lambda)} \delta_{\lambda\lambda'}, \qquad (25.51)$$

where integration is carried out over the volume indicated; $A_{\lambda'}^*$ is a quantity complex-conjugate with $A_{\lambda'}$; $\delta_{\lambda\lambda'}$ is the Kronecker symbol ($\delta_{\lambda\lambda'} = 1$ with $\lambda = \lambda'$ and $\delta_{\lambda\lambda'} = 0$ if $\lambda \neq \lambda'$). Remembering (25.49)–(25.51),† we can find from (25.48) the oscillator equations for $q_\lambda(t)$:

$$\frac{\partial^2}{\partial t^2}(\hat{\varepsilon} q_\lambda) + \omega_\lambda^2 \varepsilon(\omega_\lambda) q_\lambda = \frac{\varepsilon(\omega_\lambda)}{c} \int A_\lambda^* \, \delta j^{\text{sec}} \, dV. \qquad (25.52)$$

Putting

$$\frac{1}{c} \int A_\lambda^* \, \delta j^{\text{sec}} \, dV = \sum_{\tilde\omega} b_{\tilde\omega} e^{i\tilde\omega t}, \qquad (25.53)$$

we obtain for the equation (25.52) with a right-hand side of the form $\varepsilon(\omega_\lambda) b_{\tilde\omega} e^{i\tilde\omega t}$ one of the solutions (namely the solution for which $q_\lambda = \dot{q}_\lambda = 0$ with $t = 0$):

$$q_\lambda(t) = \frac{b_{\tilde\omega}}{2}\left[\omega_\lambda^2 - \frac{\tilde\omega^2 \varepsilon(\tilde\omega)}{\varepsilon(\omega_\lambda)}\right]^{-1}\left\{2e^{i\tilde\omega t} - \left(1 - \frac{\tilde\omega}{\omega_\lambda}\right)e^{-i\omega_\lambda t} - \left(1 + \frac{\tilde\omega}{\omega_\lambda}\right)e^{i\omega_\lambda t}\right\}. \qquad (25.54)$$

The intensity of the scattered electromagnetic radiation of interest to us can be expressed in terms of $q_\lambda(t)$. In actual fact, the change in the energy of the field in a unit volume, as we know, is (for real values of E, D, H)

$$\frac{dW}{dt} = \frac{1}{4\pi} \int \left(E \frac{\partial D}{\partial t} + H \frac{\partial H}{\partial t}\right) dV.$$

Allowing for (25.49)–(25.51) this expression can be reduced to the following:

$$\frac{dW}{dt} = \sum_\lambda \frac{1}{4\varepsilon(\omega_\lambda)} \left[\dot q_\lambda \frac{\partial}{\partial t}(\hat\varepsilon q_\lambda^*) + \dot q_\lambda^* \frac{\partial}{\partial t}(\hat\varepsilon q_\lambda)\right] + \sum_\lambda \frac{\omega_\lambda^2}{4}[q_\lambda \dot q_\lambda^* + q_\lambda^* \dot q_\lambda].$$

Since $q_\lambda(t)$ is of the form (25.54), for dW/dt the valid formula will be

$$\frac{dW}{dt} = \sum_\lambda \frac{|b_{\tilde\omega}|^2}{2} \frac{\sin[(\tilde\omega - \omega_\lambda)t]}{\tilde\omega - \omega_\lambda}\left[2 + \frac{\omega_\lambda}{\varepsilon(\omega_\lambda)} \frac{d\varepsilon(\omega_\lambda)}{d\omega_\lambda}\right]^{-1},$$

† And also the relation $\int A_{\lambda'}^* \, \text{grad } \varepsilon \, dV = 0$, whose validity can be confirmed if we remember that the solution for the scalar potential φ will also be of a form similar to (25.49): $\varphi = \sum p_\lambda(t) e^{-ik_\lambda R}$.

§ 25] The Escape of Radio Emission from the Corona

which for large t changes into

$$\frac{dW}{dt} = \sum_\lambda \frac{|b_{\tilde\omega}|^2}{2} \pi \delta(\tilde\omega - \omega_\lambda) \left[2 + \frac{\omega_\lambda}{\varepsilon(\omega_\lambda)} \frac{d\varepsilon(\omega_\lambda)}{d\omega_\lambda}\right]^{-1}. \tag{25.55}$$

($\delta(\tilde\omega - \omega_\lambda)$ is a delta-function).

When summing with respect to the oscillators in (25.55) it must be borne in mind that for a unit volume the number of field oscillators for one polarization e_λ is $dZ = dk_\lambda/(2\pi)^3$, where $dk_\lambda = k_\lambda^2 \, dk_\lambda \, d\Omega$ ($d\Omega$ is an element of a solid angle along k_λ). However, $k_\lambda = \omega_\lambda \sqrt{\varepsilon(\omega_\lambda)}/c$, and therefore

$$dZ = \frac{1}{2(2\pi)^3} \frac{\omega_\lambda^2 \varepsilon^{3/2}(\omega_\lambda)}{c^3} \left[2 + \frac{\omega_\lambda}{\varepsilon(\omega_\lambda)} \frac{d\varepsilon(\omega_\lambda)}{d\omega_\lambda}\right] d\omega_\lambda \, d\Omega. \tag{25.56}$$

From (25.55), (25.56) we obtain that the intensity of the emission at a frequency $\tilde\omega$ from a system of given secondary currents δj^{sec} is (for a single polarization e_λ)

$$I(\tilde\omega) = \frac{1}{32\pi^2 c^3} \int \omega_\lambda^2 \, \varepsilon^{3/2}(\omega_\lambda) |b_{\tilde\omega}|^2 \delta(\tilde\omega - \omega_\lambda) \, d\omega_\lambda$$

$$= \frac{1}{32\pi^2 c^3} \tilde\omega^2 \, \varepsilon^{3/2}(\tilde\omega) |b_{\tilde\omega}|^2, \tag{25.57}$$

and the total emission intensity (for all frequencies $\tilde\omega$ and both polarizations $e_\lambda = e_{\lambda_1}, e_{\lambda_2}$)

$$I = \frac{1}{32\pi^2 c^3} \sum_{\tilde\omega} \sum_{e_\lambda} \tilde\omega^2 \varepsilon^{3/2}(\tilde\omega) |b_{\tilde\omega}|^2. \tag{25.58}$$

In our case $b_{\tilde\omega}$ can be found by substituting the expression for δj^{sec} (25.47) in (25.53)

$$b_{\tilde\omega} = -i \frac{e^2 \sqrt{\pi}}{m \sqrt{\varepsilon(\omega_\lambda)}} \left\{ \frac{(E_{00} \, e_\lambda) \, \delta N''_{k_{\mathrm{pl}},\, 0}}{\omega} + \frac{(\delta E_{k_{\mathrm{pl}},\, 0} \, e_\lambda) \, n_{00}}{\omega_{\mathrm{pl}}} \right\}. \tag{25.59}$$

In changing to this formula we have allowed for the fact that

$$\delta N''_{k_{\mathrm{pl}}}(t) = \int_V \delta N''(R, t) e^{ik_{\mathrm{pl}} R} \, dV \quad \text{and} \quad \delta E_{k_{\mathrm{pl}}}(t) = \int_V \delta E(R, t) e^{ik_{\mathrm{pl}} R} \, dV$$

(where $k_{\mathrm{pl}} = k_\lambda - k$) vary harmonically in time at a frequency ω_{pl} that satisfies the dispersion equation for plasma waves. Therefore the Fourier component of the electron concentration fluctuations is

$$\delta N''_{k_{\mathrm{pl}}}(t) = \delta N''_{k_{\mathrm{pl}},\, 0} \cos(\omega_{\mathrm{pl}} t) = \delta N''_{k_{\mathrm{pl}},\, 0} \frac{e^{i\omega_{\mathrm{pl}} t} + e^{-i\omega_{\mathrm{pl}} t}}{2};$$

a similar expression obtains for the Fourier component of the fluctuating electric field $E_{k_{\mathrm{pl}}}(t)$. There are therefore two combination frequencies in the spectrum of the "imposed force" $(\varepsilon(\omega_\lambda)/c) \int A_\lambda^* \, \delta j^{\mathrm{sec}} \, dV$ contained in the right-hand side of the oscillator equations (25.52):

$$\tilde{\omega} = \omega \pm \omega_{pl}. \tag{25.60}$$

The expression for $b_{\tilde{\omega}}$ (25.59) can be written slightly differently if we take into account the connection between the amplitudes of the variation in the electron concentration n_{00}, $\delta N''_{k_{pl},0}$ and the amplitudes of the electric field E_{00}, $\delta E_{k_{pl},0}$ respectively in a passing plasma wave and in fluctuating plasma waves. Since, as we indicated above, $n_0 = i(eN/m\omega^2)kE_0$ it follows from this that

$$n_{00} = i\frac{eN}{m\omega^2}kE_{00} = i\frac{eN}{m\omega^2}kE_{00}$$

($E_{00} \| k$). Likewise for plasma fluctuations

$$\delta N''_{k_{pl},0} = i\frac{eN}{m\omega_{pl}^2}k\delta E_{k_{pl},0},$$

i.e.

$$\delta E_{k_{pl},0} = -i\frac{m\omega_{pl}^2}{eN}\delta N''_{k_{pl},0}\frac{k_{pl}}{k_{pl}^2}.$$

Bearing these relations in mind we obtain

$$b_{\tilde{\omega}} = -i\frac{\sqrt{\pi}e^2 E_{00}\,\delta N''_{k_{pl},0}}{m\omega\sqrt{\varepsilon(\omega_\lambda)}}\left\{\frac{ke_\lambda}{k} + \frac{\omega_{pl}k}{\omega k_{pl}} \times \frac{k_{pl}e_\lambda}{k_{pl}}\right\}. \tag{25.61}$$

It follows from (25.58) and (25.61) that†

$$I = \frac{2}{(8\pi c)^3}\sum_{\tilde{\omega}=\omega\pm\omega_{pl}}\sum_{e_\lambda=e_{\lambda_1},e_{\lambda_2}}\frac{\omega_L^4\tilde{\omega}^2\sqrt{\varepsilon(\tilde{\omega})}}{N^2\omega^2}\overline{|\delta N''_{k_{pl}}|^2}$$

$$\times E_{00}^2\left\{\frac{ke_\lambda}{k} + \frac{\omega_{pl}k}{\omega k_{pl}}\frac{k_{pl}e_\lambda}{k_{pl}}\right\}^2 \tag{25.62}$$

Here ω_L is the plasma frequency, the term

$$\overline{|\delta N''_{k_{pl}}|^2} = \tfrac{1}{2}|\delta N''_{k_{pl},0}|^2$$

and is connected with the plasma parameters by the relation (25.45). The orientation in an isotropic plasma can be given arbitrarily to a certain extent; all that is necessary is that the three vectors e_{λ_1}, e_{λ_2} and $\tilde{k}$ should remain orthogonal. Let e_{λ_1} lie in the plane of the vectors k, $\tilde{k}$ and e_{λ_2} be orthogonal to it. Then, since the vector k_{pl} lies in the same plane, the scalar product $k_{pl}e_{\lambda_2} = 0$ in just the same way as $ke_{\lambda_2} = 0$. At the same time $ke_{\lambda_1} = k\sin\theta$, where θ is the angle between k and $\tilde{k}$, and $k_{pl}e_{\lambda_1} = -k_{pl}\sin\theta'$ and $k\sin\theta = k_{pl}\sin\theta'$, where θ' is the angle between

---

† When the expression for $b_{\tilde{\omega}}$ is substituted in (25.58) we must put $\omega_\lambda = \tilde{\omega}$, $k_\lambda = \tilde{k}_\lambda$ in the former since in (25.58) $b_{\tilde{\omega}}$ is removed from the integrand containing the delta-function $\delta(\tilde{\omega}-\omega_\lambda)$.

§ 25] The Escape of Radio Emission from the Corona

k_{pl} and $\tilde{k}$. As a result the intensity of the electromagnetic waves scattered in the direction k can be written in the form

$$I = \frac{2}{(8\pi c)^3} \sum_{\tilde{\omega} = \omega \pm \omega_{pl}} \frac{\omega_L^4 \tilde{\omega}^2 \sqrt{\varepsilon(\tilde{\omega})}}{N^2 \omega^2} |\delta N''_{k_{pl}}|^2 E_{00}^2 \left(1 - \frac{\omega_{pl}}{\omega} \frac{k^2}{k_{pl}^2}\right)^2 \sin^2 \theta$$

$$= \frac{2}{(8\pi c)^3} \sum_{\tilde{\omega} = \omega \pm \omega_{pl}} \frac{\omega_L^2 \tilde{\omega}^2 \sqrt{\varepsilon(\tilde{\omega})} \varkappa TV}{Nm\omega^2} E_{00}^2 k_{pl}^2 \left(1 - \frac{k^2}{k_{pl}^2}\right)^2 \sin^2 \theta. \quad (25.63)$$

In changing to the last expression we have taken into consideration the formula (25.45) and neglected the difference between ω_{pl} and ω which is less essential than the difference between k^2 and k_{pl}^2.

Therefore the combination frequencies of the scattered electromagnetic radiation differ significantly from the frequency of the plasma wave passing through ω and are $\tilde{\omega} = \omega \pm \omega_{pl}$. However, in an equilibrium plasma $\omega \approx \omega_L$, $\omega_{pl} \approx \omega_L$, so $\tilde{\omega} \approx 2\omega_L$ and $\tilde{\omega} \approx 0$. The last frequency is obviously not scattered in a plasma since for it the square of the refractive index is $n_{1,2}(\tilde{\omega}) < 0$. Therefore a combination scattering spectrum contains only the frequency $\omega \approx 2\omega_L$.

The wave vector k_{pl} of a fluctuating plasma wave on which scattering occurs is connected with the vectors of the plasma and scattered electromagnetic waves $k, \tilde{k}$ by the relation

$$k_{pl} = \tilde{k} - k$$

i.e.
$$k_{pl}^2 = \tilde{k}^2 + k^2 - 2\tilde{k}k \cos \theta, \quad (25.64)$$

where the quantity $\tilde{k} = (\tilde{\omega}/c) n_{1,2}(\tilde{\omega})$ is close to $\sqrt{3}\omega_L/c$ when $\tilde{\omega} \approx 2\omega_L$. Remembering (25.64) it is easy to obtain from (25.63) the polar diagram of the combination scattering:

$$I \sim \frac{(\tilde{k}^2 - 2\tilde{k}k \cos \theta)^2}{\tilde{k}^2 + k^2 - 2\tilde{k}k \cos \theta} \sin^2 \theta. \quad (25.65)$$

We note that if the phase velocity of the passing plasma wave is $V_{ph} \ll c/\sqrt{3}$, then $k = \omega/V_{ph} \approx \omega_L/V_{ph} \gg \tilde{k}$ and the vector $\tilde{k}$ in (25.64) is insignificant:

$$k_{pl} \approx -k. \quad (25.66)$$

Under these conditions the combination scattering of a plasma wave occurs on the component $\delta N''(R, t)$, $\delta V''(R, t)$ of the fluctuations which has the same wavelength and is propagated in the opposite direction. The polar diagram of the scattering becomes symmetrical (the intensity

I does not change when θ is replaced by $\theta + \pi$):

$$I \propto \tilde{4k}^2 \cos^2\theta \sin^2\theta. \qquad (25.66a)$$

In order to find the emission energy scattered in unit time in all directions from a volume V of plasma (25.63) must be integrated over a solid angle. It is difficult to do this in the general case since the values of k_{pl} and ω_{pl} depend on the angle θ. However, when $V_{ph} \ll c/\sqrt{3}$ the function $I(\theta)$ is simplified (see (25.66)); then

$$S(\tilde{\omega}) = 2\pi \int_0^\pi I(\theta) \sin\theta \, d\theta$$

$$\approx \frac{\omega_L^4}{120\pi^2 c^3} \frac{\tilde{\omega}^2}{\omega^2} \frac{n_{1,2}(\tilde{\omega})}{N^2} \overline{|\delta N'_{k_{pl}}|^2} E_{00}^2 \frac{\tilde{k}^2}{k_{pl}^2}. \qquad (25.67)$$

Putting $\tilde{\omega} \approx 2\omega_L$, $\omega \approx \omega_L$, $n_{1,2}(2\omega_L) = \sqrt{3}/2$, $\tilde{k} \approx (2\omega_L/c) n_{1,2}(2\omega_L) = \sqrt{3}\omega_L/c$ here and remembering the expression (25.45) for $\overline{|\delta N''_{k_{pl}}|^2}$ we obtain[†]

$$S(\tilde{\omega} \approx 2\omega_L) \approx \frac{4\sqrt{3}}{5} \frac{e^4 N V}{m^2 c^3} E_{00}^2 \frac{\varkappa T}{mc^2}. \qquad (25.68)$$

The ratio of the combination and Rayleigh scattering energies when a plasma wave passes through an equilibrium plasma is (see (25.42) and (25.68))

$$\frac{S''(\tilde{\omega})}{S'(\omega)} \approx \frac{24\sqrt{3}}{5} \frac{1}{n_{1,2}(\omega)} \frac{\varkappa T}{mc^2}. \qquad (25.69)$$

Considering here that $n_{1,2}(\omega) = \sqrt{3}(V_{th}/c) n_3(\omega)$ and remembering that the refractive index of a plasma wave is $n_3(\omega) = c/V_{ph}$, where V_{ph} is the phase velocity of this wave, we obtain

$$\frac{S''(\tilde{\omega} \approx 2\omega_i)}{S'(\omega \approx \omega_L)} \approx \frac{24}{5} \frac{V_{ph} V_{th}}{c^2}. \qquad (25.70)$$

Therefore the combination scattering is a noticeable fraction of the Rayleigh scattering although the former is generally less than the latter. In the conditions of the corona for plasma waves with $V_{ph} \sim 5 \times 10^9$ cm sec^{-1} ($V_{th} \sim 4 \times 10^8$ cm sec^{-1}) the reduced ratio is about 10^{-2}.

We notice that the process of scattering non-linear plasma waves having a second harmonic (frequency 2ω, phase velocity V_{ph}—the same as with

[†] In order of magnitude this formula can be used until $\tilde{k} \lesssim k$, i.e. for calculating the scatter of plasma waves with $V_{ph} \lesssim c/\sqrt{3}$.

§ 25] The Escape of Radio Emission from the Corona

the first harmonic) is accompanied by the appearance of new frequencies in the electromagnetic radiation spectrum. In this case the Rayleigh component of the radiation is

$$S'(2\omega \approx 2\omega_L) = \frac{e^4 N V}{4\sqrt{3} m^2 c^3} E_{\text{II}}^2, \qquad (25.71)$$

where E_{II} is the amplitude of the plasma wave's second harmonic (compare with (25.42)). The level of emission at the combination frequency $\bar{\omega}_1 \approx 3\omega$ is

$$S''(3\omega_- \approx 3\omega_L) \approx \frac{8\sqrt{2}}{45} \frac{e^4 N V}{m^2 c^3} \frac{\varkappa T}{mc^2}. \qquad (25.72)$$

It was assumed above in estimates of the transformation of plasma waves into electromagnetic ones because of scattering by fluctuations in the electron concentration that these fluctuations occur because of thermal motion under the conditions of an equilibrium plasma. If the velocity distribution of the plasma particles is not an equilibrium one the level of the fluctuations determining the value of $\overline{(\delta\varepsilon)_V^2}$ and $\overline{(\delta N''_{k_{\text{pl}}})^2}$ will differ from that accepted above, which, generally speaking, will lead to a change in the efficiency of the scattering process. In a whole number of cases, however, this difference is insignificant. For example a comparatively rarefied stream of particles moving in an equilibrium plasma at a velocity $V_s > V_{\text{th}}$ has practically no effect on the intensity of the Rayleigh scattering determined by the level of the fluctuations $\delta N'$ of the sonic and ion plasma wave type, since their phase velocity is $V_{\text{ph}} \ll V_{\text{th}}$. The combination scattering will also remain at the previous level if the velocity distribution function of the electrons corresponds to an equilibrium one for the velocity $V = V'_{\text{ph}} = (\omega/k_{\text{pl}}^2)k_{\text{pl}}$, where V'_{ph} is the phase velocity of the plasma-type fluctuations which ensure effective scattering at the frequency $\bar{\omega} \approx \omega + \omega_{\text{pl}}$. This kind of condition will be satisfied, let us say, when the phase velocity $V_{\text{ph}} \approx V_s$ of plasma waves excited by the stream and then scattered lies in the range $V_{\text{th}} \ll V_{\text{ph}} \approx V_s \ll c/\sqrt{3}$. Then, in accordance with what has been said earlier, $k_{\text{pl}} \approx -k$, $V'_{\text{ph}} \approx -V_{\text{ph}} \approx -V_s$ and the stream of particles with the velocity V_s and the dispersion $\Delta V \ll V_s$ has no effect on the form of the electron velocity distribution function in the vicinity of $V \approx V'_{\text{ph}} \approx -V_s$.

In a magnetoactive plasma the picture of the scattering becomes more complex (Ginzburg and Zheleznyakov, 1959a). The scattered emission, generally speaking, contains all three types of wave: ordinary, extraordinary and plasma; the absence of any of the waves can be connected only with the impossibility of its propagation ($n_j^2(\bar{\omega}) < 0$). Scattering on quasi-neutral fluctuations of the plasma density occurs without any significant change in frequency: $\tilde{\omega} \approx \omega$ (Rayleigh scattering). For preliminary estimates of this kind of scattering into ordinary and extraordinary waves it is natural to use the expression (25.42) obtained in an isotropic plasma with the appropriate replacement of the refractive index $n_{1,2}^2(\omega)$ figuring there. The accuracy of this estimate is improved when $\sqrt{u} \equiv \omega_H/\omega \ll 1$.

As for scattering on fluctuations of the electric charge, we should bear in mind here that three types of wave make their contribution to these fluctuations (unlike an isotropic medium where $\delta\varrho$ is connected only with fluctuations of the plasma wave type). This is explained by the fact that in a magnetoactive plasma in all the normal waves (including the ordinary and extraordinary waves) the density ϱ of the electric charge, generally speaking, is non-zero; this is connected with the presence in these waves of the longitudinal component of the electric field $\boldsymbol{E}_l \| \boldsymbol{k}$ ($\varrho = (4\pi)^{-1} \operatorname{div} \boldsymbol{E} = -ikE_l/4\pi \neq 0$). The frequencies ω_{ij} of the emission scattered in this direction can be found here from the law of the conservation of energy and momentum for the photons taking part in each elementary act of wave scattering with a frequency ω and a wave vector $\boldsymbol{k}$:

$$\hbar\tilde{\omega}_{ij} = \hbar\omega \pm \hbar\omega_j, \quad \hbar\tilde{\boldsymbol{k}}_{ij} = \hbar\boldsymbol{k} \pm \hbar\boldsymbol{k}_j. \tag{25.73}$$

($\hbar$ is Planck's constant). The values of $j = 1, 2, 3$ correspond to the three normal waves making up the fluctuations $\delta\varrho$ on which the scattering occurs, and the values $i = 1, 2, 3$ to the three types of scattered waves. Therefore the scattering is accompanied by a significant change in frequency (combination scattering); the connection between the frequencies and the wave vectors of the incident, fluctuation and plasma waves is similar to the case of an isotropic plasma (see above), although when there is a magnetic field present the scattering, as has been pointed out above, will occur not only on fluctuations of the plasma wave type but also on fluctuations corresponding to waves of the ordinary and extraordinary type.[†] However, at the limit (when $H_0 \to 0$ and the plasma becomes and isotropic) combination scattering occurs only on longitudinal (plasma) waves. With $\sqrt{u} \ll 1$ the estimates of the combination scattering intensity may be made by using the formulae derived previously for the isotropic case; if, however,

---

[†] This circumstance, however, was not taken into consideration in (Akhiyezer, Prokhoda and Sitenko, 1957), where the combination scattering of ordinary and extraordinary waves was discussed on the assumption that it occurs (just as in an isotropic plasma) only on plasma oscillations with a frequency $\omega_{\mathrm{pl}} \approx \omega_L \equiv (4\pi e^2 N/m)^{1/2}$. In actual fact scattering occurs on all three normal waves; at the same time when there is a field H_0 the frequency of the plasma waves that scatter most effectively is close to the frequency determined from the condition $n_{1,2}^2(\omega) = \infty$ (see (23.4)) and in the general case is not the same as ω_L.

§ 25] The Escape of Radio Emission from the Corona

$\sqrt{u} \gtrsim 1$, then the error may become considerable. A more detailed treatment is necessary here.

When discussing combination scattering above it was a question of the effects connected with the presence of longitudinal waves† of thermal (fluctuation) origin. In the presence of plasma waves of a different nature (for example, reflected from some inhomogeneity of waves with $k_{\text{pl}} \approx -k$) additional scattering of the wave passing through will also occur on them. If the spatial Fourier component of the electric field in these waves characterized by the wave vector k_{pl} is $E_{k_{\text{pl}}}$, then

$$\overline{|\delta N''_{k_{\text{pl}}}|^2} = \frac{V}{16\pi^2 e^2} \overline{|ik_{\text{pl}} E_{k_{\text{pl}}}|^2} = \frac{V}{16\pi^2 e^2} k_{\text{pl}}^2 \overline{|E_{k_{\text{pl}}}|^2} . \quad (25.74)$$

Here it is taken into consideration that in plasma waves $k_{\text{pl}} \parallel E_{k_{\text{pl}}}$. To find the intensity of the combination scattering the latter quantity must be substituted instead of (25.45) in the formula (25.67). The scattering energy can then be written in the form

$$S(\omega \approx 2\omega_L) \approx \frac{\sqrt{3}}{5\pi} \frac{e^4 NV}{m^3 c^5} \overline{|E_{k_{\text{pl}}}|^2} E_{00}^2. \quad (25.75)$$

It becomes clear from what has been said that transformation of plasma waves into electromagnetic radiation occurs if in a primary field excited somehow in the plasma there are plasma waves with corresponding values of the wave vectors k, k_{pl} which can satisfy a relation of the (25.64) type as well as the condition $\tilde{\omega} = \omega + \omega_{\text{pl}}$ with $n^2_{1,2}(\tilde{\omega}) > 0$. Only in this case, obviously, does the process of combination scattering of one plasma wave by another occur. We can approach this kind of transformation process slightly differently, however, without having recourse to expansion of the original field in the plasma into longitudinal waves with subsequent study of the propagation of one wave in a medium perturbed by another wave (as was done above), but by stating at once in the first (linear) approximation the distribution of the electric field $E(R, t)$ in the plasma, for which curl $E = 0$, finding in the second (non-linear) approximation the electromagnetic waves "radiated" by this kind of field distribution in the plasma. The condition curl $E = 0$ means that in the first approximation there is no magnetic field in the plasma, i.e. there are no electromagnetic waves. The latter appear only when the equations of motion of the electrons allow for the non-linear terms which reflect the perturbation of the medium by the primary field necessary for the transformation under discussion.

This approach to the problem of the transformation of plasma waves into electromagnetic ones (transformation because of non-linearity) has

† We have in mind an isotropic plasma.

been used by Tidman and Weiss (1961) (see also Burkhard, Fahl and Larenz, 1961). They proceeded from the equations (22.1), (22.2), assuming in the latter $\nabla p = 0$ (i.e. instead of the plasma waves appearing when the electron pressure is taken into consideration, simply plasma oscillations with a zero group velocity were discussed). Taking all the amplitudes of all the processes to be sufficiently small Tidman and Weiss (1961) find the solution of a non-linear system of equations in the form of expansions in powers of a certain small parameter μ:[†]

$$N = \sum_{l=0}^{\infty} N_l \mu^l, \quad E = \sum_{l=1}^{\infty} E_l \mu^l, \\ H = \sum_{l=1}^{\infty} H_l \mu^l, \quad V = \sum_{l=1}^{\infty} V_l \mu^l. \quad (25.76)$$

Substituting (25.76) in the original equations and equating the terms with the same powers of μ we obtain for $l = 1$

$$c^2 \operatorname{curl} \operatorname{curl} E_1 + \frac{\partial^2 E_1}{\partial t^2} + \omega_L^2 E_1 = 0. \quad (25.77)$$

We shall assume that in the first approximation the electric field in the plasma has the nature of a potential:

$$E_1 = \operatorname{grad} \varphi e^{-i\omega_L t} \quad (25.78)$$

(i.e. $\operatorname{curl} E_1 = 0$ and there are no electromagnetic waves). Then in the next approximation we obtain an inhomogeneous equation for curl E_2 which describes the electromagnetic waves "radiated" by the plasma oscillations (25.78):

$$\left(\frac{\partial^2}{\partial t^2} + \omega_L^2 - c^2 \Delta\right) \operatorname{curl} E_2 = \mathcal{F}(R) e^{-2i\omega_L t} \quad (25.79)$$

where

$$\mathcal{F}(R) = -\frac{2e}{m} \operatorname{curl} (\Delta \varphi \operatorname{grad} \varphi). \quad (25.80)$$

Since the right-hand side of the equation is proportional to $e^{-2i\omega_L t}$ the transformed electromagnetic waves will have a "combination" frequency $\tilde{\omega} = 2\omega_L$.

It must be pointed out, however, that transformation does not occur with just any configuration of the field E_1. For example, $\mathcal{F}(R) = 0$ if the potential $\varphi(R)$ is spherically symmetrical ($\varphi(R) = \varphi(R)$) or characterizes a plane wave ($\varphi(R) = e^{ikR}$). This is not unexpected, since in this case there is no second wave on which "scattering" could occur.

[†] In a medium where $H_0 = 0$ and $N_0 = \text{const.}$

§ 25] The Escape of Radio Emission from the Corona

As an example characterizing the efficiency of the change of plasma oscillations into electromagnetic radiation because of non-linearity we give the result of the solution of the equation (25.79) for

$$\varphi = \varphi_0 e^{-R/L}(1+aR), \qquad (25.81)$$

where the constant vector a is assumed to be so small that terms of the order of a^2 can be neglected. Then the part of the energy of the plasma oscillations (25.81) emitted in unit time in the form of electromagnetic waves is

$$S = \frac{2}{\sqrt{3}} \left(\frac{e\varphi_0}{mc^2}\right)^2 \left(\frac{c}{\omega_L L}\right)^3 \left(\frac{ac}{\omega_L}\right)^2 \omega_L |J|^2, \qquad (25.82)$$

where

$$J = 1 - \frac{1}{i\tilde{k}L} \ln\left(\frac{2+i\tilde{k}L}{2-i\tilde{k}L}\right).$$

The value of J depends on the ratio of the length of the electromagnetic wave $\lambda = 2\pi/\tilde{k}$ to the size of the region L occupied by the plasma oscillations. If $\lambda/2\pi \gg L$ ($\tilde{k}L \ll 1$), then $J \approx \frac{3}{4}(\omega_L L/c)^2$ and therefore

$$S \approx \frac{9}{8\sqrt{3}} \left(\frac{e\varphi_0}{mc^2}\right)^2 \left(\frac{\omega_L L}{c}\right) \left(\frac{ac}{\omega_L}\right)^2 \omega_L. \qquad (25.83)$$

In the case of the opposite inequality, when $\lambda/2\pi \ll L$ ($\tilde{k}L \gg 1$) the value is $J \approx 1$ and

$$S \approx \frac{2}{\sqrt{3}} \left(\frac{e\varphi_0}{mc^2}\right)^2 \left(\frac{c}{\omega_L L}\right)^3 \left(\frac{ac}{\omega_L}\right)^2 \omega_L. \qquad (25.84)$$

The maximum transformation efficiency is realized when $\lambda/2\pi \approx L$:

$$S_{\max} \approx \left(\frac{e\varphi_0}{mc^2}\right)^2 \left(\frac{ac}{\omega_L}\right)^2 \omega_L. \qquad (25.85)$$

In summarizing the results of this section we can conclude that in the main the problem of the transformation of waves of one type into another, which is closely connected with the problem of the escape of emission beyond the corona, is quite clear at present. A fairly large number of questions connected with the interaction and transformation of waves has also been solved in the quantitative respect, although there are still several problems (interaction in a moving medium, in a magnetoactive plasma when waves are propagated at an angle to H_0, transformation in the presence of streams of charged particles) which still await their solution. The major obstacle in the way is the considerable difficulty of calculation.

CHAPTER VII

Generation and Absorption of Electromagnetic Waves in the Solar Corona

IN THE preceding chapter we have discussed the propagation of electromagnetic waves, completely ignoring the questions connected with the change in their intensity because of processes of emission and absorption. These processes, as will become clear from what follows, depend essentially on the velocity distribution of the plasma particles. The nature of the absorption and emission in the conditions of an equilibrium plasma will be studied in section 26; we shall put off discussing the non-equilibrium case until section 27.

26. Emission and Absorption of Electromagnetic Waves in an Equilibrium Plasma

EMISSION TRANSFER EQUATION

In a uniform medium the variation in intensity of the emission on the jth normal wave I_j in the direction of propagation $V_{\text{gr}} = d\omega/dk_j$ is defined by the so-called transfer equation (see, e.g., Chandrasekar, 1950, section 6):

$$\frac{dI_j}{dl} = a_j - \mu_j I_j, \tag{26.1}$$

in which dl is an element of length along V_{gr}, μ_j is the absorption coefficient, a_j is the emissive power of the plasma. The quantity μ_j characterizes here the relative decrease in intensity dI_j/I_j in a unit length of the ray because of dissipation of electromagnetic energy, the quantity a_j is the energy emitted from a unit volume of plasma in unit time, frequency range and solid angle. We notice that in (26.1) it does not matter to which unit solid angle—in the direction of the wave vector k_j or of the group velocity

§ 26] Emission and Absorption

V_{gr}†—a_j and I_j relate. We shall agree to relate them to a solid angle along k_j, since in this case the actual expressions for a_j and I_j in terms of the plasma's parameters are simpler in form.

When changing to the case of a non-uniform plasma in the transfer equation we must allow for the fact that the emission intensity I_j does not remain constant even with $a_j = 0$ and $\mu_j = 0$. This is connected with the change in the solid angle $d\Omega$ in which the emission is being propagated because of refraction (with constant energy flux $I_j d\Omega = $ const). The refraction law of I_j in a non-uniform medium is easy to obtain by allowing for the change in $d\Omega$ by means of the Descartes–Snellius law of refraction (22.28). It then turns out (see, e.g., Leontovich, 1951, section 25) that when they pass through the interface of two media with different refractive indices the divergence (or convergence) of rays is such that

$$\frac{I_j}{n_j^2} = \text{const.} \tag{26.2}$$

This invariant holds also in the case of a smoothly non-uniform medium, as can easily be checked by breaking the latter down into thin uniform layers (transition between which proceeds in accordance with (26.2)) and then making the number of layers approach infinity. In a magnetoactive plasma, where the angle ϑ between the wave vector k_j and the group velocity $d\omega/dk_j$ generally speaking is non-zero, the relation (26.2) must be replaced by the following:

$$\frac{I_j |\cos \vartheta|}{n_j^2} = \text{const.} \tag{26.3}$$

Bearing in mind this circumstance we can write the equation for the transfer of emission in a non-uniform medium in the following form:

$$\frac{n_j^2}{|\cos \vartheta|} \frac{d}{dl} \left(\frac{I_j |\cos \vartheta|}{n_j^2} \right) = a_j - \mu_j I_j, \tag{26.4}$$

where n_j, a_j, μ_j are functions of the coordinates. In the case of a non-uniform isotropic medium $\cos \vartheta = 1$ and (26.4) changes into the well-known equation which is given by, for example, Woolley (1947) and Smerd (1950a).

It should be stressed that the equation (26.4) holds only within the framework of the approximation of geometrical optics, where the ray treatment of the process of emission propagation is possible (see section

† These directions are the same in an isotropic medium; however, in the presence of a magnetic field H_0 in the plasma they are different, generally speaking.

22). In particular, for (26.4) to be valid it is necessary that there should be no reflection or noticeable interaction of different types of waves in a smoothly non-uniform medium. The process should also be stationary since there are no time derivatives in the transfer equation. The transfer equation in the form (26.4) is applicable both to equilibrium and to non-equilibrium conditions (including the case when the plasma becomes unstable and the absorption coefficient μ_j becomes negative†).

Under conditions of complete thermodynamic equilibrium of a medium with emission the intensity of the emission in a magnetoactive plasma at the jth normal wave is (with $\hbar\omega \ll \varkappa T$)

$$I_j^{(0)} = \frac{n_j^2 \omega^2 \varkappa T}{8\pi^3 c^2 |\cos \vartheta|}, \qquad (26.5)$$

where $\hbar$ and $\varkappa$ are the Planck and Boltzmann constants, T is the temperature, ϑ is the angle between $\mathbf{k}_j$ and $d\omega/d\mathbf{k}_j$.‡ This expression can be easily obtained from the Rayleigh–Jeans equation for the energy density in the jth normal wave

$$\varrho_j^{(0)} \, d\omega \, d\Omega = \frac{n_j^2 \omega^2 \varkappa T}{(2\pi c)^3} \left| \frac{\partial(\omega n_j)}{\partial \omega} \right| d\omega \, d\Omega, \qquad (26.6)$$

if we remember that in a transparent (weakly absorbing) medium the emission intensity I_j is connected with the energy density ϱ_j by the relation

$$I_j = \varrho_j \left| \frac{d\omega}{dk_j} \right| = \varrho_j \frac{c}{\cos \vartheta \left| \frac{\partial(\omega n_j)}{\partial \omega} \right|}. \qquad (26.7)$$

The meaning of this relation becomes quite clear if we bear in mind that $d\omega/dk_j = V_{\mathrm{gr}}$ characterizes the transfer rate of the emission energy.

With thermodynamic equilibrium $I_j = I_j^{(0)}$, $T = \mathrm{const}$ and therefore the left-hand side of the equation (26.4) becomes zero; as a result the transfer equation is reduced to Kirchhoff's law

$$a_j = \mu_j I_j^{(0)}, \qquad (26.8)$$

where $I_j^{(0)}$ is given by the expression (26.5). This law defines the connection

---

† It is necessary, however, for the coefficient μ_j to retain its meaning of the relative change in intensity in unit length; according to section 27 this holds for convective instability of the plasma.

‡ In (26.3)–(26.5) $\cos \vartheta$ is governed by the sign of the modulus. In an isotropic plasma this circumstance is insignificant since there we always have $\vartheta = 0$. In a magnetoactive plasma (without allowing for thermal motion, and in certain cases even with the latter present) $\cos \vartheta$ is positive (Gershman and Ginzburg, 1962) and the sign of the modulus can also be omitted, as we shall be doing in what follows.

§ 26] Emission and Absorption

of the absorption coefficient with the emissive power of the equilibrium medium.

Kirchhoff's law provides a convenient way of finding the absorption coefficient in an equilibrium plasma if we know its emissive power a_j. The latter in its turn can be obtained by summation of the emission from the individual plasma particles making allowance for their velocity distribution. Data on the emission of the individual particles in a plasma are given in the next sub-section; the use of Kirchhoff's law to find the absorption coefficient μ_j in certain cases of most interest from the point of view of solar and planetary radio astronomy is given at the end of the present section.

Before proceeding to discuss these processes, however, let us examine the solutions of the transfer equation for non-equilibrium emission being propagated in a medium where there is kinetic velocity equilibrium of the particles. Then, generally speaking, $I_j \neq I_j^{(0)}$, but μ_j and a_j are connected by the Kirchhoff equation,† so the transfer equation (26.4) becomes as follows:

$$\frac{n_j^2}{\cos\vartheta}\frac{d}{dl}\left(\frac{I_j\cos\vartheta}{n_j^2}\right) = \mu_j(I_j^{(0)} - I_j), \qquad (26.9\text{a})$$

or

$$\frac{d}{d\tau_j}\left(\frac{I_j\cos\vartheta}{n_j^2}\right) = \frac{\omega^2\varkappa T}{8\pi^3 c^2} - \frac{I_j\cos\vartheta}{n_j^2}. \qquad (26.9\text{b})$$

In the last relation we have taken into consideration the expression (26.5) and introduced a new independent variable—the optical thickness of the medium on the section of the ray from l_0 to l:

$$\tau_j = \int_{l_0}^{l} \mu_j\, dl. \qquad (26.10)$$

The equation (26.9) has the solution

$$\frac{I_j\cos\vartheta}{n_j^2} = e^{-\tau_j}\int_0^{\tau_j}\frac{\omega^2\varkappa T}{8\pi^3 c^2}e^{\xi}\, d\xi + e^{-\tau_j}\left(\frac{I_j\cos\vartheta}{n_j^2}\right)_{l=l_0}, \qquad (26.11)$$

which defines the emission intensity at a point l for a given intensity at the point l_0 and optical thickness τ_j of the plasma in the range l, l_0. In the case when $n_j = $ const and $\cos\vartheta = $ const the emission intensity in the medium

---

† This can be explained by the weak action of the emission on the matter, the degree of connection between them being characterized by the small parameter $e^2/\varkappa c \approx 1/137$. Due to this the absorption μ_j and the emissive power a_j depend only on the velocity distribution of the plasma particles, but not on the emission intensity which may even be non-equilibrium. It is only necessary for I_j not to be too high, since with large I_j we must allow for non-linearity, which leads to μ_j depending on I_j.

is

$$I_j = e^{-\tau_j} \int_0^{\tau_j} I_j^{(0)} e^{\xi} d\xi + e^{-\tau_j}(I_j)_{l=l_0}. \tag{26.12}$$

If at the same time the plasma is thermally uniform ($T = $ const), then $I_j^{(0)} = $ const; here

$$I_j = I_j^{(0)}(1 - e^{-\tau_j}) + e^{-\tau_j}(I_j)_{l=l_0}, \tag{26.13}$$

where $I_j^{(0)}$ is the intensity of the jth component of the equilibrium emission in the medium.

In applications we are often interested in the intensity of emission coming from a thermally uniform layer into a vacuum:

$$I_j = I_j^{(0)}(1 - e^{-\tau_j}) + e^{-\tau_j} \left(\frac{I_j \cos \vartheta}{n_j^2} \right)_{l=l_0}. \tag{26.14}$$

Here $I_j^{(0)} = \omega^2 \varkappa T / 8\pi^3 c^2$ is the intensity of the jth component of equilibrium emission in a vacuum; the formula (26.14), unlike (26.13), is also valid if n_j and $\cos \vartheta$ vary from point to point in this layer.

The notation of the formulae given can be considerably simplified by using the concept of the effective temperature T_{eff} defined in the medium by the relation (compare (26.5))

$$I_j = \frac{n_j^2 \omega^2 \varkappa T_{\text{eff}}}{8\pi^3 c^2 \cos \vartheta}. \tag{26.15}$$

The latter is a generalization of the effective temperature introduced in section 4 for the intensity in a vacuum:

$$I_j = \frac{\omega^2 \varkappa T_{\text{eff}}}{8\pi^3 c^2}. \tag{26.16}$$

Then in accordance with (26.11) the effective temperature of the emission leaving a layer of a non-uniform plasma (no matter whether into a vacuum or into a medium with an index n_j) is

$$T_{\text{eff}}(l) = e^{-\tau_j} \int_0^{\tau_j} T e^{\xi} d\xi + e^{-\tau_j} T_{\text{eff}}(l_0), \tag{26.17a}$$

if the optical thickness of the layer is τ_j and emission with an effective temperature $T_{\text{eff}}(l_0)$ is incident on this layer. The first term here (and in (26.11)) relates to the eigen emission of the plasma layer; the second term describes the emission that has passed through the layer and been partly absorbed by it. The expression (26.17a) can be put in a slightly dif-

§ 26] Emission and Absorption

ferent form if instead of ξ (the current value of the optical thickness read from the point l_0) we introduce the new variable $\zeta = \tau_j - \xi$ (the optical thickness read from the point l):

$$T_{\text{eff}}(l) = \int_0^{\tau_j} Te^{-\zeta}\,d\zeta + e^{-\tau_j}T_{\text{eff}}(l_0). \qquad (26.17\text{b})$$

For a thermally uniform layer

$$T_{\text{eff}}(l) = T(1-e^{-\tau_j}) + e^{-\tau_j}T_{\text{eff}}(l_0). \qquad (26.18)$$

We notice that the region of applicability of the formulae (26.12), (26.13) is smaller than for the formulae (26.17), (26.18) that are similar in form: the former are valid only in a medium with $n_j = \text{const}$ and $\cos \vartheta = \text{const}$, whilst the latter together with (26.11) are free of this limitation.

According to (26.18) for an optically thick ($\tau_j \gg 1$), thermally uniform plasma T_{eff} is in practice the same as the kinetic temperature of this layer

$$T_{\text{eff}} \approx T_j; \qquad (26.19)$$

for an optically thin layer ($\tau_j \ll 1$)

$$T_{\text{eff}} \approx T\tau_j + T_{\text{eff}}(l_0). \qquad (26.20)$$

Therefore the eigen emission in the latter case has $T_{\text{eff}} \ll T$. Finally looking on T_{eff} in (26.17b) at the point l as a function of l_0 we can see that when the distribution of the temperature T is not too non-uniform it is created largely because of the eigen emission of the layer of the plasma l_0 in which the optical thickness is

$$\tau_j = \int_{l_0}^{l} \mu_j\,dl \sim 1.$$

Here $T_{\text{eff}} \sim T$, where T is the kinetic temperature of the layer in question.

ELECTROMAGNETIC WAVE EMISSION BY INDIVIDUAL PARTICLES

In a plasma the emission of individual particles, generally speaking is made up of the following elementary processes:

1. bremsstrahlung in close collisions of charged particles;
2. magneto-bremsstrahlung during the accelerated motion of particles in a magnetic field;†

---

† Magneto-bremsstrahlung caused by relativistic electrons is usually called synchrotron radiation; the radiation of non-relativistic electrons in a magnetic field is sometimes called gyro-frequency radiation (since it occurs at the gyro-frequency and its lower harmonics) or cyclotron radiation.

3. Cherenkov emission during motion of charged particles at a velocity greater than the phase velocity of the waves in the medium;
4. transition radiation during motion in a non-uniform medium and, finally;
5. emission during atomic and molecular processes (i.e. during transitions of free electrons to discrete energy levels, during transitions between discrete levels, etc.).

Without stopping to carry out the cumbersome derivation of the expressions for the emission intensity, all we shall give here is certain formulae which may be needed in theory of the radio emission of the Sun and the planets. For a more detailed knowledge the reader may turn to the appropriate sources; references to them are made below.

When a non-relativistic electron moves in an isotropic plasma with a refractive index $n_{1,2} = \sqrt{\varepsilon(\omega)}$ (22.17) the energy of its electromagnetic bremsstrahlung in unit time in the range of frequencies $d\omega$ is defined by the following expressions (Landau and Lifshitz, 1960, section 70):[†]

$$\frac{d\mathcal{E}_\omega}{dt} d\omega = \sqrt{\varepsilon} \frac{16e^6 Z^2 N_+}{3 V c^3 m^2} \ln \frac{2mV^3}{\gamma \omega Z e^2} d\omega \quad \left(\omega \ll \frac{mV^3}{Ze^2}\right), \quad (26.21)$$

$$\frac{d\mathcal{E}_\omega}{dt} d\omega = \sqrt{\varepsilon} \frac{16\pi e^6 Z^2 N_+}{3^{3/2} V c^3 m^2} d\omega \quad \left(\omega \gg \frac{mV^3}{Ze^2}\right). \quad (26.22)$$

Here V is the velocity of an electron, e and m are its charge and mass, $\gamma = e^C = 1\cdot 781$ (C is Euler's constant). The relations (26.21), (26.22) characterize the emission in electron collisions with positive ions (charge Ze, concentration N_+); they allow for the fact that the mass of an ion is $m_+ \gg Zm$. If there are negative ions with a mass $m_- \gg Zm$ in the plasma, then the formula (26.21) remains valid in this case also (with N_+ replaced by N_-) for bremsstrahlung in collisions with the latter; the formula (26.22) must be supplemented by introducing the factor exp $(-2\pi\omega Ze^2/mV^3)$ into it. In collisions of particles with identical charges and masses (for example, in electron–electron collisions) there is no bremsstrahlung in the dipole approximation.

The frequency spectrum of the bremsstrahlung is rather large: from $\omega = \omega_L$ (for which $n_{1,2}^2 = 0$) up to $\omega \sim \mathcal{E}_{\text{kin}} \hbar$, where $\mathcal{E}_{\text{kin}}$ is the kinetic energy of the radiating particle.

In the radio-frequency band ($\omega \lesssim 2 \times 10^{12}$ sec^{-1}) the formula (26.21) is valid for velocities of $V \gg 10^6$ cm/sec. This condition is satisfied well under cosmic conditions; suffice it to say that the value $V_{\text{th}} \sim 10^6$ cm/sec

---

[†] As well as the bremsstrahlung of electromagnetic (transverse) waves in collisions of charged particles, of course, plasma (longitudinal) waves also appear.

§ 26] Emission and Absorption

corresponds to a temperature of $T \sim 10°K$. In accordance with (26.21) under the conditions of the solar corona, when $\varepsilon \sim 1$, $\omega \sim 2\pi \times 10^8 \text{ sec}^{-1}$, $N_+ \sim 5 \times 10^7$ protons/cm³, $V \sim V_{th} \sim 4 \times 10^8$ cm/sec, the energy of an electron's bremsstrahlung is $\mathcal{E}_\omega \sim 4 \times 10^{-34}$. As the velocity increases it drops approximately as V^{-1}.

The Vavilov–Cherenkov effect for a particle moving uniformly and rectilinearly in the medium occurs at a frequency (see, e.g. Bolotovskii, 1957)

$$\omega = k_j V \qquad (26.23a)$$

or, what is the same thing, provided that

$$\beta n_j(\omega) \cos \theta = 1, \qquad (26.23b)$$

where V is the velocity of a charged particle, $\beta = V/c$, k_j is the wave vector of the jth normal wave emitted with a refractive index $n_j \equiv ck_j/\omega$, θ is the angle between k_j and V. According to (26.23b) Cherenkov emission appears only if $\beta n_j > 1$. It is clear from this that in an isotropic plasma there is no emission of electromagnetic (transverse) waves of this kind, since for them $n_{1,2}^2 = \varepsilon = 1 - \omega_L^2/\omega^2$ (22.17) and the condition $\beta n_j > 1$ cannot be satisfied with any $\beta < 1$. On the other hand, for plasma (longitudinal) waves at frequencies not too close to ω_L the square of the refractive index $n_3^2 = \varepsilon/\beta_{th}^2$ becomes greater than unity (see (22.19)), which in principle enables the Vavilov–Cherenkov effect to occur at these wavelengths.

It is clear from the condition $\beta n_j > 1$ that when an electron is moving in the plasma it generates only waves whose phase velocity is $V_{ph} = \omega/k \leqslant V$. On the other hand, as will be shown below, plasma waves with a wavelength of $\lambda/2\pi \lesssim D$ (D is the Debye radius) are not in practice propagated, being strongly attenuated by distances of the order of λ. Since for waves of this kind the ratio ω/k does not exceed V_{th}, it is clear that a particle with a velocity V generates plasma waves with a phase velocity in the range $V_{th} < V_{ph} \leqslant V$; if, however, the electron's velocity is $V < V_{th}$, then it hardly emits any plasma waves at all.

The direction θ in which a wave of frequency ω is generated is defined by the relation (26.23). For the range of phase velocities given above the values of $\cos \theta$ for plasma waves are between 1 and V_{th}/V.

The energy of the Cherenkov emission of an electron in unit time in the range $d\omega$ is (Andronov, 1961; Cohen, 1961b)

$$\frac{d\mathcal{E}_\omega \, d\omega}{dt} = \frac{e^2 \omega \, d\omega}{V\varepsilon(\omega)}. \qquad (26.24)$$

In order to find the total energy lost by a particle in unit time in the emission of plasma waves we must integrate $d\mathcal{E}_\omega/dt$ over all the frequencies being emitted

$$\left(\frac{d\mathcal{E}}{dt}\right)_{\text{tot}} = \int_{\omega_1}^{\omega_2} \frac{e^2\omega\, d\omega}{V\varepsilon(\omega)} = \frac{e^2\omega_L^2}{2V} \ln\left(\frac{2}{3} \frac{V^2}{V_{\text{th}}^2}\right). \qquad (26.25)$$

The lower limit in (26.25) is found from the equality $\beta n_3(\omega_1) = 1$, and the upper one from the condition $kD = kV_{\text{th}}/\omega_L \sim 1$, which is the criterion for strong plasma wave damping. Therefore we can put $\omega_2^2 = \omega_L^2 + 3V_{\text{th}}^2 k^2 \sim 4\omega_L^2$, which is allowed for in (26.25) when changing to the last expression.†

In the solar corona the efficiency of plasma wave emission is characterized by the quantity $d\mathcal{E}_\omega/dt \sim 5 \times 10^{-18}$ (with $\omega \sim 2\pi \times 10^8$ sec^{-1}, $V \sim V_{\text{ph}} = V_{\text{th}}/\sqrt{\varepsilon} \sim 5 \times 10^9$ cm/sec and $V_{\text{th}} \sim 4 \times 10^8$ cm/sec^{-1}; for these values of V_{ph} and V_{th} the value of ε is about 6×10^{-3}). This value of $d\mathcal{E}_\omega/dt$ exceeds the corresponding value for the bremsstrahlung of electromagnetic waves under the same conditions by ten orders or so of magnitude. However, unlike the bremsstrahlung of electromagnetic waves, Cherenkov plasma waves cannot directly escape beyond the non-uniform plasma and are significantly attenuated (by a factor of approximately 10^5) during their transformation into electromagnetic radiation (see section 25).

Still one more form emission, that disappears when grad ε approaches zero, appears in a non-uniform plasma. The existence of this kind of emission has been indicated by Ginzburg and Frank (1946); it is generally called "transition" radiation. Unfortunately we know of no papers in which there would be expressions for the energy emitted here in the case when the charged particle moves in a smoothly non-uniform medium with a monotonic function $\varepsilon(z)$. This case is obviously of most interest from the point of view of cosmic applications. Moreover, in the solar corona there may also be a certain interest in transition radiation of an electron in a medium with a periodically changing dielectric permeability $\varepsilon(z)$ created, for example by a plasma wave excited in a corpuscular stream. The effect of transition emission under cosmic conditions has not yet been examined; it may be thought, however, that all other things being equal the part it plays in the generation of solar radio emission is small when compared with the part played by bremsstrahlung, Cherenkov radiation and magneto-bremsstrahlung (for the latter see below).

The emission of a uniformly moving electron in a medium with random inhomogeneities has been discussed by Kapitza (1960) and Ter-Mikayelyan (1961) without taking spatial dispersion into consideration and by Tamoi-

† Oster (1959) gives a slightly different formula for the energy losses of an electron in plasma wave emission, differing from (26.25) by the logarithmic factor $\ln(1+2V^2/3V_{\text{th}}^2)$ instead of $\ln(2V^2/3V_{\text{th}}^2)$. Since the frequency ω_2, on whose value this factor depends, is determined only approximately, both formulae have the same accuracy.

§ 26] Emission and Absorption

kin (1963) with the latter taken into consideration. In this case the electromagnetic waves appear because of two effects: the scattering of Cherenkov plasma waves emitted by an electron on fluctuations $\delta\varepsilon$ and the transition radiation connected with the variation of the Coulomb field of a particle as it moves in a medium whose properties vary from point to point: $\delta\varepsilon = \delta\varepsilon(R)$. The second effect is significant at frequencies where there are no Cherenkov plasma waves; for these frequencies $\sqrt{\varepsilon(\omega)} < V_{th}/V$ (see (22.19a) and (26.23)). However in the region $\sqrt{\varepsilon(\omega)} > V_{th}/V$ the first effect will definitely prevail over the second if we have satisfaction of the inequality (Tamoikin, 1963)

$$\frac{\omega}{V} L \sqrt{\frac{\varepsilon V^2}{V_{th}^2} - 1} \gg 1 \qquad (26.26)$$

(L is the size of the region effectively scattering the plasma waves). Bearing in mind that in the case of the Vavilov–Cherenkov effect $V_{ph} = V\cos\theta$, where for the plasma waves $V_{ph} = V_{th}/\sqrt{\varepsilon}$, the preceding inequality can be put in the form $(\omega L/V)\tan\theta \gg 1$. Assuming that $\tan\theta \sim 1$ and putting for the sake of definition $\omega \sim 2\pi \times 10^8$ sec^{-1}, $V \sim 5 \times 10^9$ cm/sec, we obtain that transition emission on random inhomogeneities in the corona will be insignificant with $L \gg 10$ cm. This is undoubtedly satisfied for weakly damped waves with $V_{ph} \gg V_{th}$.

In the presence of a constant magnetic field H_0 the form of the expression for the emission intensity of a charged particle becomes far more complex. This is due to the change in the nature of the waves propagation and the trajectory of the particle motion under the action of the field H_0. The effect of anisotropy can usually be neglected in the case of a weak enough magnetic field when $\sqrt{\bar{u}} \equiv \omega_H/\omega \ll 1$. Distortion of the trajectory of the electron's motion need not be taken account when discussing bremsstrahlung, if the maximum aiming parameter in collisions, whose part in the plasma is played by the Debye radius, is much less than r_H where

$$r_H = \frac{V}{\omega_H \sqrt{1-\beta^2}}$$

is the gyro-radius of an electron in a magnetic field. Therefore the inequality $D \ll r_H$ is reduced to the following:

$$\omega_H \ll \frac{V}{V_{th}} \omega_L (1-\beta^2)^{-1/2}. \qquad (26.27)$$

In the case of Cherenkov emission the formulae given above for the intensity in an isotropic medium are obviously retained if the condition

$\omega_H \ll \omega$ is satisfied and at the same time the trajectory of the electron motion on a section with a length of the order of $\lambda = 2\pi c/n_j\omega$ can be considered rectilinear, i.e. (Ginzburg and Zheleznyakov, 1961)

$$\frac{r_H}{\lambda} = \frac{\beta n_j \omega}{\omega_H(1-\beta^2)^{-1/2}} \gg 1. \qquad (26.28)$$

When an electron moves in a magnetic field H_0 there is one more mechanism of emission—magneto-bremsstrahlung. If the trajectory of the particle is a helix with a radius r_H, then it emits at frequencies ω defined by the relation

$$\omega(1-\beta_\| n_j \cos \alpha) = s\omega^*, \qquad (26.29)$$

where $\omega^* = \omega_H\sqrt{1-\beta^2}$ is the frequency of rotation of an electron in the magnetic field, $\beta = V/c$, $\beta_\| = V_\|/c$ ($V_\|$ is the projection of the velocity V onto the direction H_0) α is the angle between the wave vector k_j and H_0, n_j is the refractive index of the jth normal wave; the index s takes up the values $0, \pm1, \pm2, \ldots$ The relation (26.29) follows from the classical treatment of the problem of the emission of an electron in a magnetic field (Eidman, 1959); at the same time it is not difficult to obtain it from the laws of the conservation of energy and momentum when photons are emitted by a charged particle in a field H_0 (see section 27).

Somewhat conventionally we can consider that the value $s = 0$ corresponds to Cherenkov emission of an electron in a magnetic field, although here the condition of the Vavilov–Cherenkov effect

$$\beta_\| n_j \cos \alpha = 1 \qquad (26.30)$$

differs from the corresponding condition (26.23) for the case of rectilinear motion of the electron. Further, emission connected with non-zero values of s is connected with the rotation of an electron in a magnetic field and is magneto-bremsstrahlung at frequencies ω defined by the Doppler equation

$$\omega = \frac{s\omega_H\sqrt{1-\beta^2}}{1-\beta_\| n_j \cos \alpha}. \qquad (26.31)$$

With $s > 0$ the so-called normal Doppler effect obtains, which is possible in the case

$$\beta_\| n_j \cos \alpha < 1. \qquad (26.32a)$$

The values $s < 0$, where

$$\beta_\| n_j \cos \alpha > 1, \qquad (26.32b)$$

correspond to the anomalous Doppler effect which appears only when a particle moves "faster than light" in a medium with $n_j > 1$ (Ginzburg

§ 26] Emission and Absorption

and Frank, 1947). We shall deal with the physical meaning of the difference between these forms of the Doppler effect later in section 27.

It follows from the relations (26.30)–(26.32) that magneto-bremsstrahlung is generated both in the region $\beta_{\|}n_j(\omega) < 1$ and in the region $\beta_{\|}n_j(\omega) > 1$, whilst the Vavilov–Cherenkov effect appears only when $\beta_{\|}n_j(\omega) > 1$.† It must be pointed out that in the presence of a magnetic field the Cherenkov emission corresponds not only to plasma waves, as was the case in an isotropic plasma, but also to ordinary and extraordinary waves the refractive index for which may be greater than unity.

The intensity of the emission of an electron in a magnetic field at the frequencies whose spectrum is given by the formula (26.29) can be found by the Hamiltonian method used in section 25 when discussing the combination scattering of plasma waves. In our case, however, as j^{sec} and ϱ^{sec} in the equations for the electromagnetic field of the (25.46) type we must put $-eV\delta(R-R_e)$ and $-e\delta(R-R_e)$ where V is the velocity of the emitting electron, R_e its radius vector and δ a delta-function. As a result for the Cherenkov emission intensity ($s = 0$) at the jth normal wave we obtain (Eidman, 1958 and 1959):

$$\frac{d\mathcal{E}_{j\omega}\,d\omega}{dt} = \frac{e^2\gamma_j^2}{2V_{\|}c^2} \frac{\{V_{\perp}J_0'(\xi)+\beta_j V_{\|} J_0(\xi)\}^2 \omega\,d\omega}{\left|1-\left(\frac{\partial n_j}{\partial \alpha}\right)\Big/n_j\sqrt{\beta_{\|}^2 n_j^2-1}\right|}, \quad (26.33)$$

where $\beta_{\|}n_j(\omega) > 1$, $V_{\perp} = \omega^* r_H$, $\xi = (\omega r_H/V_{\|})\sqrt{\beta_{\|}^2 n_j^2-1}$, $J_0(\xi)$ and $J_0'(\xi)$ are zero-order Bessel functions and its derivative with respect to the argument ξ. The frequency ω is given by the condition (26.30). If the electron moves uniformly along H_0, then $V_{\perp} = 0$ and the expression (26.33) become simpler:

$$\frac{d\mathcal{E}_{j\omega}\,d\omega}{dt} = \frac{e^2\gamma_j^2}{2c^2} \frac{\beta_j\omega\,d\omega}{\left|1-\left(\frac{\partial n_j}{\partial \alpha}\right)\Big/n_j\sqrt{\beta_{\|}^2 n_j^2-1}\right|}. \quad (26.34)$$

The general formula for the energy magneto-bremsstrahlung of an electron in unit time and in a unit solid angle $d\Omega$ at the harmonics $s \neq 0$

---

† The conventional nature of the division of the emission into magneto-bremsstrahlung and Cherenkov emission is particularly clear when the magnetic field strength decreases, i.e. with $\sqrt{u} \to 0$. In fact, when an electron moves in a circle (when $\beta_{\|} = 0$) only frequencies with $|s| \geq 1$ are emitted; we call this kind of emission magneto-bremsstrahlung. At the same time in a weak field, when (26.28) is satisfied and the trajectory of the particle over a distance of the order of λ is close to rectilinear, emission at an angle $\theta = \arccos(1/\beta n_j)$ in a medium with $n_j > 1$ will coincide in practice with the Cherenkov emission. The energy emitted will also be close to the Cherenkov emission energy in an isotropic medium (Tsytovich, 1951).

can be written as follows (Eidman, 1958, 1959):

$$\frac{d\mathcal{E}_{j\Omega}\,d\Omega}{dt} = \frac{n_j e^2 \omega^2 \gamma_j^2 \{\beta_\perp\, J'_s(\zeta) + [\alpha_j s \beta_\perp \zeta^{-1} + \beta_j \beta_\parallel]\, J_s(\zeta)\}^2\, d\Omega}{2\pi c \left| 1 - \beta_\parallel \cos\alpha \left(n_j + \omega\, \dfrac{dn_j}{d\omega} \right) \right|}. \quad (26.35)$$

Here $J_s(\zeta)$ and $J'_s(\zeta)$ are an s-order Bessel function and its derivative with respect to the argument $\zeta = k_j r_H \sin\alpha = s n_j \beta_\perp \sin\alpha/(1-\beta_\parallel n_j \cos\alpha)$; $\beta_\perp = V_\perp/c$, $k_j = \omega n_j/c$ is the wave number. The emission frequency is defined by the Doppler equation (26.31).

The parameters γ_j, α_j, and β_j are defined by the relations:

$$\gamma_j^2 = \frac{n_j^2(1-u)}{(1-u-v)(1+\alpha_j^2)+(1-v)(1-u)\beta_j^2 - 2v\sqrt{u}\,\alpha_j} \equiv \frac{1}{1+K_j^2}, \quad (26.36)$$

$$\alpha_j = K_j \cos\alpha + \Gamma_j \sin\alpha, \quad \beta_j = \Gamma_j \cos\alpha - K_j \sin\alpha, \quad (26.37)$$

$$K_j = \frac{2\sqrt{u}\,(1-v)\cos\alpha}{u\sin^2\alpha \pm \sqrt{u^2 \sin^4\alpha + 4u(1-v)^2 \cos^2\alpha}}, \quad (26.38)$$

$$\Gamma_j = \frac{v\sqrt{u}\sin\alpha + K_j uv \sin\alpha \cos\alpha}{1-u-v+uv\cos^2\alpha}, \quad (26.39)$$

where $v = \omega_L^2/\omega^2 = (4\pi e^2 N)/m\omega^2$, $u = \omega_H^2/\omega^2 = (e^2 H_0^2)/m^2 c^2 \omega^2$, and $n_j^2(\omega)$ is given by the formula (23.2). The top sign in K_j relates to an extraordinary wave ($j=1$) and the bottom one to an ordinary wave ($j=2$). The values of K_j and Γ_j characterize the polarization of the normal waves being emitted: $i\Gamma_j$ is the ratio of the longitudinal (along $\mathbf{k}_j$) component of the electric field $\mathbf{E}$ to the component of this field orthogonal to $\mathbf{H}_0$ and $\mathbf{k}_j$; the quantity iK_j in its turn is equal to the ratio of the transverse (orthogonal to $\mathbf{k}_j$) component of the field $\mathbf{E}$ lying in the plane $\mathbf{k}_j$, $\mathbf{H}_0$ to the component of the field at right angles to this plane.†

We would stress that the formulae (26.33)–(26.39) relate to the case of a magnetoactive plasma with no allowance for thermal motion in it. They therefore characterize only the emission of extraordinary and ordinary waves; these formulae are unsuitable for waves of the plasma type.

Equation (26.35) is rather complicated, We shall therefore consider a few simpler cases which are important for practical cases.

Here first of all we shall turn to the case when the effect of the plasma's anisotropy on the emission can be neglected, considering that an electron moving along a helix emits in an isotropic medium with a refractive index $n_{1,2}(\omega) = \sqrt{1-v}$. In the presence of a field $\mathbf{H}_0$ the plasma can be con-

† When comparing these expressions for K_j and Γ_j with the polarization coefficients (23.10), (23.11) it should be remembered that here the formulae are given for the case when in the normal waves the time dependence is taken in in the form $e^{-i\omega t}$ and not $e^{i\omega t}$ as in section 23.

§ 26] Emission and Absorption

sidered approximately isotropic if $\sqrt{u} = \omega_H/\omega \ll 1$ and v is not too close to unity: $1-v \sim 1$ (see section 23). In accordance with (26.31) the condition imposed on the quantity $\sqrt{u}$ will be observed at high enough harmonics

$$s \gg \frac{1-\beta_\| n_{1,2} \cos \alpha}{\sqrt{1-\beta^2}}. \tag{26.40}$$

In this case the magneto-bremsstrahlung intensity of an electron can be obtained from (26.35) by the extreme transition $\sqrt{u} \to 0$. Then, as is easily seen, $n_{1,2}^2 = 1-v$, $\gamma_j^2 = \frac{1}{2}$, $\alpha_j = \pm\cos \alpha$, $\beta_j = \mp\sin \alpha$ and the total intensity of both normal waves is

$$\frac{d\mathcal{E}_\Omega}{dt} = \sum_{j=1,2} \frac{d\mathcal{E}_{j\Omega}}{dt} =$$

$$\frac{n_{1,2}\, e^2 \omega^2 \{\beta_\perp^2\, [J_s'(\zeta)]^2 + (n_{1,2} \sin \alpha)^{-2}(\cos \alpha - n_{1,2}\beta_\|)^2\,[J_s(\zeta)]^2\}}{2\pi c \left|1 - \left(\dfrac{\beta_\|}{n_{1,2}}\right)\cos \alpha\right|}. \tag{26.41}$$

In a vacuum $n_{1,2} = 1$ and

$$\frac{d\mathcal{E}_\Omega}{dt} = \frac{e^2 \omega_H^2 (1-\beta^2)\, s^2 \{\beta_\perp^2\,[J_s'(\zeta)]^2 + \sin^{-2}\alpha(\cos \alpha - \beta_\|)^2\,[J_s(\zeta)]^2\}}{2\pi c\,|1-\beta_\| \cos \alpha|^3}. \tag{26.42}$$

When an electron moves in a circle ($\beta_\| = 0$) this expression changes into the well-known Schott formula (see, e.g., Landau and Lifshitz, 1960, section 74).

An analysis of (26.42) shows that the synchrotron emission of an electron with an energy $\mathcal{E} \gg mc^2$ is concentrated in a narrow conical layer $\Delta\alpha \sim mc^2/\mathcal{E}$ wide containing the cone described by the electron's velocity vector V when moving along the helix in the magnetic field. If the angle θ between V and H_0 is greater than $mc^2/\mathcal{E}$ the emission will be linearly polarized, with the electrical vector E at right angles to V and H_0. For example, when moving in a circle the plane of polarization will coincide with the plane in which the electron is moving.

As well as the relations given above for the intensity of emission in a given direction α at the sth harmonic there is considerable interest for practical applications in the expression for the energy $(d\mathcal{E}_\omega dt)\,d\omega$ emitted by a relativistic electron in unit time in a frequency ranged over all angles. Let us first examine the case of an electron moving in a circle ($\beta_\| = 0$; $\beta_\perp = \beta$). In accordance with the Doppler equation here the frequency of the emission at the sth harmonic $\omega = (s\omega_H mc^2)/\mathcal{E}$ and does not depend on the angle α. The emission spectrum is a system of discrete "lines" located at a short distance from each other: $\Delta\omega = (\omega_H mc^2)/\mathcal{E}$. Since the major contribution to the synchrotron emission is provided by harmonics with large s numbers, the frequency spectrum is quasi-continuous in nature ($\Delta\omega \ll \omega$); this allows us to change from a discrete spectrum to a continuous one, dealing with frequency ranges $d\omega$ containing many lines ($d\omega \gg \Delta\omega$). In order to obtain the energy $(d\mathcal{E}_\omega dt)/d\omega$ we must integrate (26.42) over the solid angles and then sum the result over all the harmonics in the range $d\omega$. We then arrive at the following formula, which defines the "smoothed" frequency spectrum of an electron's synchrotron emission (Vladimirskii, 1948):

$$\frac{d\mathcal{E}_\omega}{dt} d\omega = \frac{8}{\pi} \frac{e^2}{c} \omega_H \frac{mc^2}{\mathcal{E}} \left(\frac{\omega}{2\omega_H} \frac{\mathcal{E}}{mc^2}\right)^{1/3} Y(x) d\omega, \qquad (26.43)$$

where $\mathcal{E}$ is the electron energy, $\omega_H = eH_0/mc$ is the gyro-frequency, $x = (\omega/2\omega_m)^{2/3}$, $\omega_m = \omega_H(\mathcal{E}/mc^2)^2$. The function $Y(x)$ is tabulated in Vladimirskii (1948). With $x \ll 1$, $Y(x) \approx \frac{1}{4}$; with $x = 1$, $Y(x) \approx 5\cdot 5 \times 10^{-2}$; with $x \gg 1$, $Y(x) \approx [(2\pi)^{1/2}/16] x^{1/4} e^{-(4x/3)^{3/2}}$. The formula (26.43) can be written in a more convenient form by introducing the function

$$P\left(\frac{\omega}{\omega_m}\right) = \sqrt{x}\, Y(x). \qquad (26.44)$$

Then

$$\frac{d\mathcal{E}_\omega}{dt}(\omega, \mathcal{E}) = \frac{8}{\pi} \frac{e^2}{c} \omega_H P\left(\frac{\omega}{\omega_m}\right) \qquad (26.45)$$

and therefore the function $P(\omega/\omega_m)$ completely defines the form of the frequency spectrum of the synchrotron emission. With $\omega/\omega_m \ll 1$

$$P\left(\frac{\omega}{\omega_m}\right) \approx \frac{1}{4}\left(\frac{\omega}{\omega_m}\right)^{1/3}; \qquad (26.46)$$

with $\omega/\omega_m \gg 1$

$$P\left(\frac{\omega}{\omega_m}\right) \approx \frac{1}{16}\left(\frac{\pi\omega}{\omega_m}\right)^{1/2} e^{-2\omega/3\omega_m}. \qquad (26.47)$$

P obviously reaches its maximum value in the region $\omega \sim \omega_m$ (i.e. with $x \sim 1$); the width of the spectrum is of the order of ω_m. A graph of the function $P(\omega/\omega_m)$ is shown in Fig. 136; it allows us to determine more

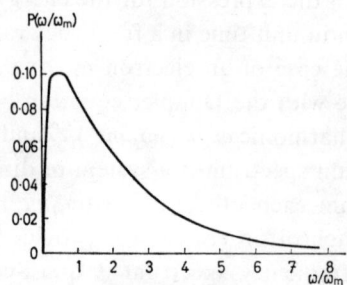

Fig. 136. Graph of the function $P(\omega/\omega_m)$

§ 26] Emission and Absorption

precisely the position and value of the maximum of the spectrum:

$$\omega_{max} \approx \frac{\omega_m}{2} = \left(\frac{\omega_H}{2}\right)\left(\frac{\mathcal{E}}{mc^2}\right)^2, \quad P_{max} \approx 0{,}1. \tag{26.48}$$

It is clear from what has been said that the high harmonics $s \gg 1$ play the major part in the emission of a relativistic electron. This circumstance has already been pointed out above. The total emission energy *in vacuo* (summed over all the frequencies) is (Landau and Lifshitz, 1960, section 74)

$$\left(\frac{d\mathcal{E}}{dt}\right)^0_{tot} = \int_0^\infty \frac{d\mathcal{E}_\omega}{dt} d\omega = \frac{2e^2\omega_H^2\beta^2}{3c(1-\beta^2)}. \tag{26.49}$$

It follows from the formulae given that the spectral intensity of the magneto-bremsstrahlung in the solar corona is generally several orders less than the corresponding value for the Cherenkov emission of plasma waves. In actual fact, in accordance with (26.45) and (26.48) at the spectral maximum

$$\frac{d\mathcal{E}_\omega}{dt} \approx \frac{1.6}{\pi} \frac{e^2}{c} \omega \left(\frac{mc^2}{\mathcal{E}}\right)^2. \tag{26.50}$$

so that $d\mathcal{E}_\omega/dt \lesssim 2 \times 10^{-21}$ at a frequency of $\omega \sim 2\pi \times 10^8$ sec^{-1}, whilst the intensity of the Cherenkov emission is characterized by the value $d\mathcal{E}_\omega/dt \sim 5 \times 10^{-18}$ (see p. 416). In practice, however, the magneto-bremsstrahlung of one particle is more efficient than the Cherenkov emission (since the latter is strongly attenuated while leaving the corona in the process of wave transformation, whilst the former leaves the corona freely); the magneto-bremsstrahlung is also many orders more intense than the ordinary bremsstrahlung.

The relations discussed above, which characterize the frequency spectrum of a relativistic electron, are valid only for plane motion of a particle when there is no velocity along the magnetic field ($\beta_{\parallel} = 0$). If $\beta_{\parallel} \neq 0$, then in the formulae (26.43)–(26.49) we must replace ω_H by $\omega_H \sin\theta$, where θ is the angle between the velocity V of the electron and the magnetic field H_0 (i.e. substitute instead of H_0 the magnetic field component $H_{0\perp}$ orthogonal to the velocity V). This operation, however, is legitimate only until $\theta \gg mc^2/\mathcal{E}$, i.e. much greater than the range of angles $\Delta\alpha \sim mc^2/\mathcal{E}$, in which the synchrotron emission is concentrated.

In the case of the opposite inequality $\theta \ll mc^2/\mathcal{E}$, i.e. when an electron moves along a helix strongly stretched along H_0 at a transverse velocity $\beta_\perp \ll (mc^2/\mathcal{E})\beta_{\parallel} \approx mc^2/\mathcal{E}$, the magneto-bremsstrahlung in all directions α is in the nature of a dipole emission. The energy emitted in a unit solid angle in a vacuum can then be obtained from (26.42) by expanding the Bessel function and its derivative with respect to the argument $\zeta = s\beta_\perp \sin\alpha(1-\beta_\parallel \cos\alpha)^{-1}$. When $\cos\alpha = \beta_\parallel$ the quantity ζ reaches its maximum value of $s\theta\mathcal{E}/mc^2$. It follows from this that the argument ζ is small when compared with unity for any α at harmonics that do not have too large a number $s \ll mc^2/\mathcal{E}\theta$. Neglecting the higher terms of the expansion in powers of ζ in (26.42) we arrive at the following formula for the dipole magneto-bremsstrahlung of an electron in a vacuum:

$$\frac{d\mathcal{E}_\Omega^{dip}}{dt} = \sum_{j=1,2} \frac{d\mathcal{E}_{j\Omega}^{dip}}{dt} = \frac{e^2\omega^2\{s^2\beta_\perp^2(1+\cos^2\alpha)+\zeta^2\beta_\parallel^2\sin^2\alpha\}}{2\pi c(1-\beta_\parallel\cos\alpha)} \frac{\zeta^{2s-2}}{2^{2s}(s!)^2}.$$
(26.42a)

Since $\zeta \ll 1$ the emission intensity decreases rapidly as the number s of the harmonic rises, i.e. in the present case, $\theta \ll mc^2/\mathcal{E}$, the part played by the higher harmonics becomes negligibly small when compared with the lower harmonics. The latter makes the limitation of $s \ll mc^2/\mathcal{E}\theta$ imposed above insignificant in practice.

How do the synchrotron emission formulae (26.43)–(26.50) alter when the influence of the medium is taken into consideration? In an isotropic $\sqrt{\bar{u}} \ll 1$) and sufficiently rarefied $(1-n_{1,2} \ll 1)$ plasma for which

$$\frac{\omega_H}{\omega} \ll 1; \quad \frac{\omega_L^2}{2\omega^2} \ll 1;$$

$$n_{1,2}^2 \approx 1 - \frac{\omega_L^2}{\omega^2},$$

the expression (26.43) for $d\mathcal{E}_\omega/dt$ retains its value if the argument x of the function $Y(x)$ is replaced by the quantity (Razin, 1960)

$$x' = \left(\frac{\omega}{2\omega_m}\right)^{2/3}\left[1+(1-n_{1,2}^2)\left(\frac{\mathcal{E}}{mc^2}\right)^2\right].$$

In the case of

$$(1-n_{1,2}^2)\left(\frac{\mathcal{E}}{mc^2}\right)^2 \ll 1$$

the effect of the isotropic plasma on the synchrotron emission can be neglected. This effect can become significant only in a region where

$$(1-n_{1,2}^2)\left(\frac{\mathcal{E}}{mc^2}\right)^2 \gtrsim 1,$$

i.e. at frequencies of

$$\omega \lesssim \omega_L \frac{\mathcal{E}}{mc^2}.$$

§ 26] Emission and Absorption

When this range of ω is found at frequencies much less than ω_{max} (26.48):

$$\frac{2\omega_L}{\omega_H}\frac{mc^2}{\mathcal{E}} \ll 1, \tag{26.51}$$

the presence of the medium is felt only in the lowest frequencies of the spectrum; therefore the maximum of the intensity and the total emission energy (summed over all the frequencies) will in fact be defined by the previous formulae (26.48) and (26.49).

In the opposite case of

$$\frac{2\omega_L}{\omega_H}\frac{mc^2}{\mathcal{E}} \gg 1 \tag{26.51a}$$

the part played by the medium is very important. Here the parameter $x' \gg 1$ at all frequencies. Since the function $Y(x')$ decreases exponentially as x' rises when $x' \gg 1$ the synchrotron emission maximum comes at frequencies of $\omega \approx \omega'_{max}$ corresponding to the minimum value of x' as a function of ω (Zheleznyakov and Trakhtengerts, 1965)

$$\omega'_{max} \approx \sqrt{2}\,\omega_L \frac{\mathcal{E}}{mc^2}. \tag{26.52}$$

In accordance with (26.48), (26.51a) and (26.52)

$$\frac{\omega'_{max}}{\omega_{max}} \approx 2\sqrt{2}\,\frac{\omega_L}{\omega_H}\frac{mc^2}{\mathcal{E}} \gg 1. \tag{26.52a}$$

Therefore the effect of the surrounding plasma is to "suppress" the low synchrotron emission frequencies, resulting in a shift of the spectrum maximum towards the high frequencies. The level of emission at the spectrum maximum decreases; it can be found by substituting in (26.43) the value $\omega = \omega'_{max}$ and remembering that at this frequency $x' \gg 1$:

$$\frac{d\mathcal{E}_\omega}{dt} \approx \frac{1}{2\sqrt{\pi}\,6^{1/24}}\frac{e^2}{c}\omega_H\,\delta^{7/12}e^{-\delta}.$$

Here δ is used to denote the parameter $\sqrt{3}(\omega_L/\omega_H)mc^2/\mathcal{E}$. It follows from (26.51a) that the last formula is valid if $\delta \gg 1$.

It is also shown by Zheleznyakov and Trakhtengerts (1965) that the total synchrotron emission power $(d\mathcal{E}/dt)_{tot}$ in a plasma when $\delta \gg 1$ decreases sharply when compared with a vacuum:

$$\left(\frac{d\mathcal{E}}{dt}\right)_{tot} \approx \left(\frac{d\mathcal{E}}{dt}\right)^0_{tot}\frac{3}{4}\delta e^{-\delta}. \tag{26.53}$$

This circumstance will be used in section 30 when analysing the mechanism of type IV radio emission.

Let us now examine in greater detail the question of the intensity of the magneto-bremsstrahlung of a weakly relativistic electron (Zheleznyakov, 1964a). In this case we shall consider $\beta = V/c$ is sufficiently small and, in particular, assume that the inequalities

$$\beta^2 \ll 1, \quad |n_j \beta_\parallel| \ll 1, \quad |sn_j \beta_\perp \sin \alpha| \ll 1, \quad \left|\beta_\parallel \omega \frac{\partial n_j}{\partial \omega}\right| \ll 1 \quad (26.54)$$

are valid. In the small velocity approximation the emission frequency (26.31) does not depend on $V_\perp$ and is

$$\omega \approx \frac{s\omega_H}{1 - n_j \beta_\parallel \cos \alpha}, \quad (26.55)$$

which in its turn is close to $s\omega_H$. The latter follows from the second inequality (26.54), which at the same time means that all that is being discussed is emission in a region of normal Doppler effect $s > 0$ (see (26.32)). Further, we bear in mind that for a weakly relativistic electron the modulus of the argument of the Bessel function in (26.35) $\zeta = sn_j \beta_\perp \sin \alpha$ is much less than unity. This allows us to limit ourselves in (26.35) to only the first terms of the expansion of J_s, J_s' in powers of ζ.

As a result we obtain that the intensity of the dipole† emission of an electron in a magnetoactive plasma is

$$\frac{d\mathcal{E}_{j\Omega}^{\text{dip}}}{dt} d\Omega \approx \frac{n_j e^2 \omega^2 \gamma_j^2 \{s\beta_\perp (1+\alpha_j) + \zeta \beta_j \beta_\parallel\}^2}{2\pi c \left|1 - \beta_\parallel \cos \alpha \left(n_j + \omega \frac{\partial n_j}{\partial \omega}\right)\right|} \frac{\zeta^{2s-2} d\Omega}{2^{2s}(s!)^2}. \quad (26.56)$$

Allowing for the second and last inequalities of (26.54) and also for the form of ζ this expression can be represented in the form

$$\frac{d\mathcal{E}_{j\Omega}^{\text{dip}}}{dt} \approx \frac{n_j e^2 \omega^2 \beta_\perp^2}{4\pi c} \gamma_j^2 (1 + \alpha_j + n_j \beta_j \beta_\parallel \sin \alpha)^2 2^{-2s+1} s^2 (s!)^{-2} (sn_j \beta_\perp \sin \alpha)^{2s-2}. \quad (26.57)$$

Since $\omega \approx s\omega_H$, in (26.57) we put $n_j(\omega) \approx n_j(s\omega_H)$, $\gamma_j(\omega) \approx \gamma_j(s\omega_H)$, etc., so

$$\frac{d\mathcal{E}_{j\Omega}^{\text{dip}}}{dt} \approx \frac{e^2 \omega_H^2 \beta_\perp^2}{4\pi c} \left[n_j \gamma_j^2 (1+\alpha_j)^2 (sn_j \beta_\perp \sin \alpha)^{2s-2}\right]_{\omega = s\omega_H} 2^{-2s+1} s^4 (s!)^{-2}. \quad (26.58)$$

† The contribution of multipoles to the emission of an electron can be taken into consideration by retaining the subsequent terms of the expansion of J_s, J_s' in powers of ζ (for further detail see Zheleznyakov, 1964a).

§ 26] Emission and Absorption

When changing to the last expression for $(d\mathcal{E}_{j\Omega}^{\text{dip}})dt$ it was also taken that $1+\alpha_j+n_j\beta_j\beta_\| \sin\alpha \approx 1+\alpha_j$.

However the formula (26.58) even as an approximation will not be true with $s=1$ (at frequencies $\omega \approx \omega_H$), since for the fundamental harmonic with any values of the parameter $v = \omega_L^2/\omega^2$ and the angle α (with the exception of $\alpha = 0$ and $v = 0$)† we have $(1+\alpha_j)_{\omega=\omega_H} = 0$. Therefore for $s=1$ the factor

$$1+\alpha_j+n_j\beta_j\beta_\| \sin\alpha = \left(1-\frac{\omega_H}{\omega}\right)\frac{1+\alpha_j}{1-\omega_H/\omega} + \left(1-s\frac{\omega_H}{\omega}\right)\beta_j \tan\alpha$$

(see (26.57)) should be taken as being equal to

$$(1-\sqrt{u})\left(\frac{1+\alpha_j}{1-\sqrt{u}}+\beta_j \tan\alpha\right)_{\omega=\omega_H},$$

where $(1+\alpha_j)(1-\sqrt{u})^{-1}$ remains finite when $1-\sqrt{u}$ approaches zero. Therefore the intensity of the emission at the fundamental harmonic is

$$\frac{d\mathcal{E}_{j\Omega}^{\text{dip}}}{dt} \approx \frac{e^2\omega_H^2\beta_1^2}{8\pi c}(1-\sqrt{u})^2\left[n_j\gamma_j^2\left(\frac{1+\alpha_j}{1-\sqrt{u}}+\beta_j \tan\alpha\right)^2\right]_{\omega=\omega_H}, \quad (26.59)$$

where, in accordance with (26.55), the factor $(1-\sqrt{u})^2$ can be written in the form

$$(1-\sqrt{u})^2 = n_j^2(\omega_H)\beta_\|^2 \cos^2\alpha.$$

The applicability of the formula (26.59) in the vicinity of $\alpha = \pi/2$ is limited by the fact that we used the Doppler equation in the form (26.55) instead of (26.31) when changing from (26.59). It is not hard to see that with small β^2 and $n_j\beta_\| \cos\alpha$ this cannot occur in the case $2|n_j\beta_\| \cos\alpha| \gg \beta^2$, which is violated when $\alpha \approx \pi/2$, and also in regions of the plasma where n_j is close to zero.

At the same time for $s \geqslant 2$ the factor $(1+\alpha_j)^2_{\omega=s\omega_H}$ does not become zero and the relation (26.58) will be valid provided that

$$|1+\alpha_j| \gg |n_j\beta_j\beta_\| \sin\alpha|. \quad (26.60)$$

The second inequality (26.54) helps the satisfaction of this condition; however, (26.60) is definitely violated in the vicinity of values of the parameters where β_j becomes infinity. As can easily be checked the latter holds for ordinary waves with $\alpha = \pi/2$.

---

† This is valid only when thermal motion is not taken into account. In actual fact near $\alpha = 0$ and $v = 0$ finite intervals exist outside which $(1+\alpha_j)_{\omega=\omega_H} = 0$. The value of these intervals can be found in the kinetic treatment (see below).

Above, when we were discussing the conditions of applicability of the expressions for the intensity of a weakly relativistic electron no allowance was made for the effect of thermal motion in the plasma on the value of n_j, γ_j, α_j or β_j, which in certain cases becomes very significant. For example, in the vicinity of values of u, v and α where n_j^2 written in the form (23.2) becomes infinity,[†] it becomes necessary to allow for thermal motion and the formula (26.35) together with its corollaries becomes invalid. In the case of a weakly relativistic electron of interest to us the expression for the emission intensity of an extraordinary wave in the form (26.59) becomes untrue with small angles α, since with $\alpha \to 0$ and $\omega \to \omega_H$ the refractive index $n_j(\omega)$ becomes infinitely great. A similar position also holds with small values of v and arbitrary α, when the frequency ω corresponding to n_1^2 becoming infinity approaches the gyro-frequency ω_H, and also with certain selected values of v and α for which $n_j = \infty$ at frequencies $\omega \approx s\omega_H (s \geqslant 2)$. The emission of an electron in these cases has been discussed in detail by Pakhomov, Aleksin and Stepanov (1963, 1961), Pakhomov and Stepanov (1963).

However, with the exception of these special cases, allowing for thermal motion in the propagation of extraordinary and ordinary waves provides only corrections to n_j^2, γ_j^2, α_j, β_j of the order of β_{th} or β_{th}^2, which for the greater part do not in any significant manner affect the magneto-bremsstrahlung of an electron in a non-relativistic plasma where $\beta_{th} \ll 1$. The single exception in this sense is the case of emission at the fundamental harmonic: the formula (26.59) becomes inapplicable in the region of frequencies where[‡]

$$\left(\frac{1-\sqrt{u}}{n_j \beta_{th} \cos \alpha}\right)^2 \lesssim 2 \qquad (26.61)$$

(for further detail see Zheleznyakov, 1964a). Moreover, even in this case (26.59) retains its value for estimates of the order of magnitude; this kind of estimate can be obtained by extrapolating (26.59) into the region $(1-\sqrt{u})^2 \sim 2n_j^2 \beta_{th}^2 \cos^2 \alpha$:

$$\frac{d\mathscr{E}_{j\Omega}^{dip}}{dt} \sim \frac{e^2 \omega_H^2 \beta_\perp^2 \beta_{th}^2 \cos^2 \alpha}{4\pi c} \left[n_j^3 \gamma_j^2 \left(\frac{1+\alpha_j}{1-\sqrt{u}} + \beta_j \tan \alpha \right)^2 \right]_{\omega=\omega_H}. \qquad (26.62)$$

Finally, let us compare the intensities of emission in an isotropic and magnetoactive plasma with helical motion of a weakly relativistic electron. If the medium is isotropic, then the intensity of emission into both normal

---

[†] These values are determined from the equation (23.4).

[‡] That is at frequencies where, as follows from the Doppler relation (26.55), the contribution to the magneto-bremsstrahlung from thermal electrons moving at a velocity of β_{th} becomes significant.

§ 26] Emission and Absorption

waves can be determined from (26.41). As a result we obtain that for the magneto-bremsstrahlung of a weakly relativistic electron in an isotropic medium with an index $n_{1,2} = \sqrt{1-v}$

$$\frac{d\mathcal{E}_{j\Omega}^{\text{dip}}}{dt} \approx \frac{n_{1,2}e^2\omega^2\beta_\perp^2}{4\pi c}(1+\cos^2\alpha)2^{-2s+1}s^4(s!)^{-2}(sn_{1,2}\beta_\perp \sin\alpha)^{2s-2}. \quad (26.63)$$

Integrating over all solid angles we obtain the expression for the total energy of the magneto-bremsstrahlung in unit time (Zheleznyakov, 1964a):

$$\frac{d\mathcal{E}_{\text{tot}}^{\text{dip}}}{dt} = \int_{4\pi} \frac{d\mathcal{E}_\Omega^{\text{dip}}}{dt} d\Omega \approx \frac{2n_{1,2}^{2s-1}e^2\omega_H^2\beta_\perp^{2s}}{c} \frac{s^{2s+1}(s+1)}{(2s+1)!}. \quad (26.64)$$

By comparing the formulae (26.59), (26.62) with (26.63) we can see that in a magnetoactive plasma the emission intensity of an electron at the gyrofrequency decreases sharply when compared with the case of an isotropic medium (roughly speaking, by a factor of $\beta_\|^{-2}$ with $\beta_\| > \beta_{\text{th}}$ and a factor of β_{th}^2 with $\beta_\| \lesssim \beta_{\text{th}}$, unless we allow for the factors containing $\gamma_j, \alpha_j, \beta_j$, etc.). At the same time the emission intensity at the higher harmonics $\omega \approx s\omega_H$ ($s \geq 2$) does not vary radically (see (26.58), (26.63)). Therefore, if in an isotropic medium the value of the harmonics in the emission spectrum of a weakly relativistic electron decreases as the number s increases (as $\beta_\perp^{2s}$, starting at $s = 1$), then in a magnetoactive plasma this decrease actually occurs only with $s \geq 2$. The point is that because of the appearance of the factor $(1-\sqrt{u})^2 \sim \beta_\|^2$ in the expression for the intensity at the fundamental harmonic the ratio (with respect to the velocities) $\{d\mathcal{E}_{j\Omega}^{\text{dip}}(s=2)\}/dt / \{d\mathcal{E}_{j\Omega}^{\text{dip}}(s=1)\}/dt \sim \beta_\perp^2/\beta_\|^2$ and with $\beta_\perp \sim \beta_\|$, generally speaking, becomes of the order of unity.† Attention was drawn to this circumstance in the paper by Ginzburg and Zheleznyakov (1958b).

We note that in the paper by Twiss and Roberts (1958) the calculation of the magneto-bremsstrahlung of an electron in a plasma is carried out incorrectly. This follows if only from the fact that in their paper the emission intensity at $\omega \approx \omega_H$ is proportional to the square of the velocity of a weakly relativistic electron for an extraordinary wave and to the sixth power for an ordinary one. This differs from the expressions given above, according to which $d\mathcal{E}_{j\Omega}^{\text{dip}}/dt$ is proportional to the fourth power of the velocity for both components emitted at an angle $\alpha \neq 0$ to the magnetic field H_0.

---

† It is clear from what has been said that unlike the magneto-bremsstrahlung of a relativistic electron in whose make-up high harmonics ($s \gg 1$) predominate, in a weakly relativistic case the major contribution to the emission is made by the lower harmonics ($s \sim 1 \cdot 2$).

Looking ahead a little, we would stress that the function $d\mathcal{E}_{j\Omega}^{\text{dip}}/dt \propto \beta^2$ also contradicts the well-known fact that there is no absorption at the frequency $\omega \approx \omega_H$ in a plasma without collisions or thermal motion (see (26.94) where with $\nu_{\text{eff}} \to 0$ the absorption index $\varkappa_j \to 0$). In actual fact, in the case of $(d\mathcal{E}_{j\Omega}^{\text{dip}})/dt \propto \beta^2$ the emissive power of an equilibrium plasma in the range of frequencies $\Delta\omega$ is $a_j \Delta\omega \propto \beta^2 \sim \beta_{\text{th}}^2$, where $\Delta\omega \propto \beta \sim \beta_{\text{th}}$ because of the Doppler effect. Then, in accordance with Kirchhoff's law (26.8), the coefficient of resonance absorption in the region of the first harmonic is $\mu_j^{\text{res}} \propto a_j/T$, which in its turn is proportional to $1/\beta_{\text{th}}$. Therefore $\mu_j^{\text{res}} \to \infty$ with $\beta_{\text{th}} \to 0$ in contradiction to the elementary theory of wave propagation in a plasma. A similar error is also made when calculating the emission intensity of an electron in Oster's paper (Oster, 1959). On the other hand, with correct treatment of the problem when, in accordance with what has been said above, $d\mathcal{E}_{j\Omega}^{\text{dip}}/dt \propto \beta_\perp^2 \beta_\parallel^2 \sim \beta_{\text{th}}^4$, the quantity $\mu_j^{\text{res}} \propto \beta_{\text{th}}$ and approaches zero when $\beta_{\text{th}} \to 0$.

It was noted above that a magnetoactive plasma radically alters the nature of the emission at the fundamental harmonic only if the electron concentration in the plasma N (and at the same time the parameter v) is not too small. The concrete condition imposed here on the quantity v can be obtained by means of the kinetic equation. According to Stepanov and Pakhomov (1960) the conclusion drawn above about the sharp decrease in the emission intensity of a weakly relativistic electron in a magnetoactive plasma at the fundamental harmonic is valid if

$$v \gg \beta_{\text{th}} n_j \cos\alpha \, |w(z_{j1})|. \tag{26.65}$$

The function $w(z_{j1})$ here is of the form

$$w(z_{j1}) = e^{-z_{j1}^2/2}\left(1 + \frac{2i}{\sqrt{\pi}} \int_0^{z_{j1}/\sqrt{2}} e^{\xi^2}\,d\xi\right), \quad z_{j1} \equiv \frac{1-\omega_H/\omega}{\beta_{\text{th}} n_j \cos\alpha}.$$

On the other hand, in a sufficiently rarefied plasma, when

$$v \ll \beta_{\text{th}} \cos\alpha \, |w(z_{j1})|, \tag{26.66}$$

the emission at the fundamental harmonic in practice differs in no way from the emission in a vacuum.

ABSORPTION OF ELECTROMAGNETIC WAVES IN AN ISOTROPIC PLASMA

The presence of absorption leads to the change of the electromagnetic energy of a wave into the energy of the plasma thermal motion. Due to this the emission intensity decreases along the ray (i.e. in the direction of

§ 26] Emission and Absorption

the group velocity V_{gr}) as

$$I_j \propto e^{-\tau_j} = e^{-\int \mu_j \, dl}. \tag{26.67}$$

The field of a wave in an absorbing medium can be given in the form (22.15) (or (22.25)) by changing the refractive index n_j to the quantity $n_j - i\varkappa_j$, i.e. by putting†

$$\boldsymbol{k}_j = \frac{\omega}{c}(n_j - i\varkappa_j)\frac{\boldsymbol{k}_j}{k_j}. \tag{26.68}$$

The quantity $\varkappa_j$ is called the absorption index of a wave of the jth type. Comparing (26.67) with (22.25), (26.68) and remembering here that the intensity is a characteristic of the emission quadratic in the field strength, we obtain that the absorption coefficient is

$$\mu_j = 2\frac{\omega}{c}\varkappa_j \cos \vartheta. \tag{26.69}$$

The factor $\cos \vartheta$ allows for the fact (significant for a magnetoactive plasma) that μ_j determines the absorption in the direction $V_{gr} = d\omega/d\boldsymbol{k}_j$, whilst $\varkappa_j$ determines it in the direction $\boldsymbol{k}_j$. For the case of an isotropic plasma these directions are the same ($\cos \vartheta = 1$).

It was assumed above that the frequency ω is real. This corresponds to problems in which the propagation of waves from a certain source is being investigated, the efficiency of generation in which does not vary (or weakly varies) with time. If the source of emission is definitely non-stationary the picture becomes more complex since ω and $\boldsymbol{k}$ must be considered complex quantities. In solar and planetary astronomy there is also interest in a different way of putting the problem of absorption, when the nature is investigated of time attenuation in a plasma of a certain initial perturbation of the electromagnetic type. In a uniform medium this perturbation can be expanded in normal waves (22.10) with real $\boldsymbol{k}_j$; then the imaginary part ω will characterize the attenuation in time of the field's Fourier components. The attenuation coefficient of the wave in time determined by the relation $\gamma_j = \Delta I_j / \varrho_j$ is

$$\gamma_j = \mu_j \left| \frac{d\omega}{dk_j} \right|, \tag{26.70}$$

since $I_j = \varrho_j |d\omega/dk_j|$ (see (26.7)) and by definition $\mu_j = \Delta I_j / I_j$ (ΔI_j is the energy absorbed in unit time in a unit volume of the plasma).

---

† Here and below we have in mind the so-called uniform waves, for which $\boldsymbol{k}' \parallel \boldsymbol{k}''$ ($\boldsymbol{k} \equiv \boldsymbol{k}' - i\boldsymbol{k}''$). We are not discussing the case of non-uniform waves.

After these remarks, which are general in nature, let us proceed to a concrete discussion of absorption in a non-relativistic or weakly relativistic plasma ($\beta_{th}^2 \ll 1$).

In the preceding section it was noted that the spontaneous emission of individual particles in a plasma is made up of several elementary processes amongst which the major role in solar radio astronomy is played by bremsstrahlung in the process of "collisions" with other charged particles, the Vavilov–Cherenkov effect and magneto-bremsstrahlung during helical motion in a magnetic field. Each elementary process makes a definite contribution to the emissive power a_j; then it follows from the Kirchhoff equation (26.8) that the corresponding absorption is connected with each component of the spontaneous emission in an equilibrium plasma.

In the absence of a constant magnetic field, when there is no magnetic bremsstrahlung, the attenuation of waves in the plasma consists largely of absorption due to electron–ion collisions and Cherenkov absorption. The latter obviously only occurs for plasma waves since there is no Vavilov–Cherenkov effect for transverse electromagnetic waves in an isotropic plasma.†

The absorption because of collisions can be obtained by means of Kirchhoff's law if we know the emissive power of the plasma caused by the bremsstrahlung mechanism. The latter can be found by using the expressions given in the preceding sub-section for the bremsstrahlung intensity of individual electrons. In the majority of cases, however, absorption due to collisions is studied within the framework of elementary electromagnetic wave propagation theory by introducing the effective number of collisions v_{eff} into the quasi-hydrodynamic equations of motion of the charged particles (22.7) (see Ginzburg, 1960b, section 3)—an operation which to a certain extent allows for the presence of thermal motion in the plasma.

Allowing for v_{eff} leads to an alteration of the form of the dispersion equations that connect ω and k_j. For example in an isotropic plasma instead of (22.16) the following relations will hold:

$$\omega^2 = \frac{\omega_L^2}{1 - \frac{iv_{\text{eff}}}{\omega}} + c^2 k^2, \quad \omega^2 = \frac{\omega_L^2 + V_{th}^2 k^2}{1 - \frac{iv_{\text{eff}}}{\omega}} \quad (26.71)$$

† The intensity of electromagnetic waves along the ray also decreases because of their scattering on fluctuations of the electron concentration and transformation into plasma waves in a regularly non-uniform medium. A similar effect occurs in the propagation of plasma waves (see section 25). It must be pointed out, however, that the decrease in I_j connected with these phenomena is generally insignificant when compared with the absorption of waves because of collisions.

§ 26] Emission and Absorption

(the first of them relates to an electromagnetic wave and the second to plasma waves).

In the high-frequency case ($\omega^2 \gg \nu_{\text{eff}}^2$), which is known to occur in cosmic conditions at the frequencies of interest to us, it follows from (26.71) that the refractive indices as before are defined with sufficient accuracy by the formulae (22.17), (22.19); at the same time the absorption coefficients are

$$\varkappa_{1,2} \approx \frac{1-n_{1,2}^2}{2\omega n_{1,2}} \nu_{\text{eff}} \qquad (26.72)$$

for an electromagnetic wave and

$$\varkappa_3 \approx \frac{1}{2\omega n_3 \beta_{\text{th}}^2} \nu_{\text{eff}} \qquad (26.73)$$

for a plasma wave. Accordingly the absorption coefficients can be written in the form

$$\mu_{1,2} \approx \frac{1-n_{1,2}^2}{cn_{1,2}} \nu_{\text{eff}} = \frac{v}{c\sqrt{1-v}} \nu_{\text{eff}} = \frac{4\pi e^2 N \nu_{\text{eff}}}{mc\omega^2 \sqrt{1-\frac{4\pi e^2 N}{m\omega^2}}}, \qquad (26.74)$$

$$\mu_3 \approx \frac{1}{cn_3 \beta_{\text{th}}^2} \nu_{\text{eff}} = \frac{1}{V_{\text{th}}\sqrt{1-v}} \nu_{\text{eff}} = \frac{V_{\text{ph}}}{V_{\text{th}}^2} \nu_{\text{eff}}. \qquad (26.75)$$

Here the parameter $v = \omega_L^2/\omega^2 = 4\pi e^2 N/m\omega^2$ and in the expression for μ_3 it is taken into consideration that the phase velocity of the plasma waves is $V_{\text{ph}} = V_{\text{th}}/\sqrt{\varepsilon} = V_{\text{th}}/\sqrt{1-v}$ (see section 22).

The coefficients of the attenuation in time γ_j when the expressions for the group velocities $d\omega/dk_j$ are taken into consideration (section 22) prove to be

$$\gamma_{1,2} = v\nu_{\text{eff}}, \qquad \gamma_3 = \nu_{\text{eff}}. \qquad (26.76)$$

In the region $v \approx 1$, $\gamma_3 \approx \gamma_{1,2}$. We notice that the same expressions for $\gamma_j = 2\text{Im}\,\omega$ can, of course, be obtained directly from the equations (26.71) by putting Im $k = 0$ in them.

The ν_{eff} figuring in the formulae given (the effective number of collisions undergone by an electron in unit time) is defined by the following relation (Ginzburg, 1960b, sections 6 and 36):

$$\nu_{\text{eff}} = \nu_{ei} + \nu_{em}, \qquad (26.77)$$

where the first term allows for collisions with ions and the second for collisions with neutral molecules.

In a plasma

$$\nu_{ei} = \pi \frac{e^4}{(\varkappa T)^2} \bar{V} N \ln\left(0{,}37 \frac{\varkappa T}{e^2 N^{1/3}}\right) \approx \frac{5{,}5N}{T^{3/2}} \ln\left(220 \frac{T}{N^{1/3}}\right) \quad (26.78)$$

or

$$\nu_{ei} = \pi \frac{e^4}{(\varkappa T)^2} \bar{V} N \ln\left[\delta \left(\frac{me^4}{\hbar^2 \varkappa T}\right)^{1/3} \frac{\varkappa T}{e^2 N^{1/3}}\right] \approx \frac{5{,}5N}{T^{3/2}} \ln\left(10^4 \frac{T^{2/3}}{N^{1/3}}\right), \quad (26.79)$$

depending on whether the value of the electron temperature T is less or more than 3×10^5 °K. In the region $T \sim 3 \times 10^5$ °K both formulae give approximately the same result. It is assumed in (26.78), (26.79) that the number of ions is $N_i = N_+ = N$;† the mean arithmetic velocity of an electron is $\bar{V} = \sqrt{8 \varkappa T/m\pi}$; the factor $\delta \sim 1$; the electron temperature T can, generally speaking, differ from the temperature of the ions. The effective number of electron collisions decreases as the electron kinetic temperature T rises, being approximately proportional to $T^{-3/2}$.

For collisions with neutral molecules

$$\nu_{em} = \frac{4\pi}{3} a^2 \bar{V} N_m = 8{\cdot}3 \times 10^5 \pi a^2 \sqrt{T} N_m. \quad (26.80)$$

Here N_m is the concentration of the molecules, πa^2 is the effective cross-section for collisions of this kind. In air $\pi a^2 \approx 4{\cdot}4 \times 10^{-16}$ cm²; for rough calculations in planetary atmospheres it can be taken that πa^2 has the same order of magnitude. The number of collisions of electrons with molecules rises as T increases in proportion to $T^{1/2}$.

According to (26.78)–(26.80) the ratio $\nu_{em}/\nu_{ei} \sim 10^{-10} T^2 N_m/N$. In the solar corona with $T \sim 10^6$ °K because of the low concentration of neutral molecules ($N_m \sim 10^{-7} N$; see section 1) this ratio is very small ($\sim 10^{-5}$), which allows the collisions with neutral molecules to be neglected. As for the chromosphere, the temperature there is lower ($T \sim 10^4$ °K) and $\nu_{em}/\nu_{ei} \ll 1$ if the neutral molecule concentration is $N_m \ll 10^2 N$. It is clear from Table 1 in section 1 that this inequality is definitely satisfied in the outer chromosphere (at altitudes $h \gtrsim 6000$ km) and apparently in the hot elements of the lower chromosphere ($h \lesssim 4000$ km); in the cold elements ν_{em} is comparable with ν_{ei}. A similar situation, as far as can be judged from the example of the Earth (Ginzburg, 1960b, section 6), also obtains in the planetary ionospheres: in the upper ionosphere (the F-layer) it can be taken that $\nu_{\text{eff}} \approx \nu_{ei}$, whilst in the lower E and D layers the number of neutral molecules, in all probability, becomes sufficient to provide the

† The ions are considered to be ionized once. In a hydrogen plasma such as the corona basically is this always holds. Because of the comparatively low temperature in planetary ionospheres the content of highly-ionized ions is likewise small.

§ 26] Emission and Absorption

major contribution to the number of collisions with a comparatively small effective cross-section: $\nu_{\text{eff}} \approx \nu_{em}$.

An idea is given of the actual values of ν_{eff} in the solar corona by the graph of the effective number of collisions as a function of the distance R to the centre of the Sun shown in Fig. 137. The values of ν_{eff} are cal-

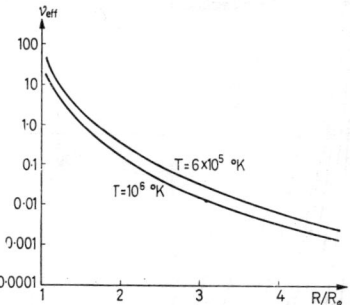

FIG. 137. Effective number of collisions ν_{eff} in the corona as a function of the distance R from the centre of the Sun.

culated here from the formula (26.79) for two values of the corona's temperature: $T = 6 \times 10^5\ °K$ and $T = 10^6\ °K$, on the assumption that the distribution of the electron concentration in the corona corresponds to the Baumbach–Allen formula (1.1). It follows from Fig. 137 that ν_{eff} in the corona rises rapidly as the Sun's surface is approached, reaching values around 5–20 sec^{-1} in the altitude range $h \sim (0 \cdot 1 – 0 \cdot 3) R_\odot$. In the chromosphere because of the low temperature and the high electron concentration the number of collisions increases sharply. For example, with $T \sim 2 \cdot 5 \times 10^4\ °K$ and $N \sim 10^{11}$ electrons/cm^3 (i.e. in the region of spicules at an altitude of $h = 6000$ km from the photosphere) the number of collisions ν_{eff}, according to (26.78), reaches a value of the order of 10^6 sec^{-1}.

Knowing ν_{eff} the formula (26.74) can be used to find the optical thickness of the corona

$$\tau_{1,2} = \int_{l_0}^{l} \mu_{1,2}\, dl$$

at different wavelengths. Figure 138 shows

$$\tau_{1,2} = \int_{R}^{\infty} \mu_{1,2}\, dR$$

as a function of R at waveengths of 50 cm to 20 m corresponding to the case of propagation of electromagnetic waves along a solar tadins. The graphs extend to values $R = R^*$ were the refractive index $n_{1,2}$ becomes zero.† The optical thickness of the chromosphere (and the corona) can also be judged from Fig. 139 (Smerd, 1950a). In the calculations here the following model of the solar atmosphere was used: Baumbach–Allen con-

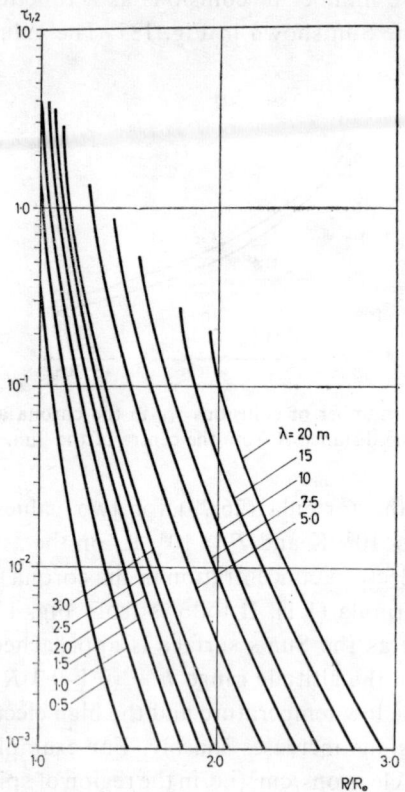

FIG. 138. Optical thickness of the corona in the case of radial propagation of radio waves (in the calculation the values of Fig. 137 with $T = 10^6\,°K$ were used for v_{eff}).

centration in the corona, temperature $10^6\,°K$; concentration in the chromosphere

$$N = 5\cdot7 \times 10^{11} \exp\,[-7\cdot7 \times 10^{-4}(h - 500)]\ \text{electrons/cm}^3 \quad (26.81)$$

(h in km), temperature constant at $3 \times 10^4\,°K$.

† Actually the formulae given above for the absorption coefficient μ_j become inapplicable in the vicinity of the wave reflection point (near $R \approx R^*$ with radial propagation), which is connected with the violation of the geometric-optical approximation in this region. This circumstance, however, which in principle affects the value of the coefficient of reflection of waves from a non-uniform plasma, is completely insignificant in the actual conditions of the solar atmosphere and the planetary ionospheres: the additional optical thickness introduced by the reflection region is generally small when compared with the total value τ_j (for further detail see Ginzburg, 1960b, section 36).

§ 26] Emission and Absorption

According to Figs. 138 and 139 the optical thickness of the corona read from the R^* level, where $n_{1,2} = 0$, rises as the wavelength decreases. This is connected with the corresponding displacement of the reflecting layer into the deeper layers of the corona and then the chromosphere with higher N^2 values and a lower temperature T (we note that in the layers with $n_{1,2} \approx 1$ the optical thickness caused by collisions of electrons with

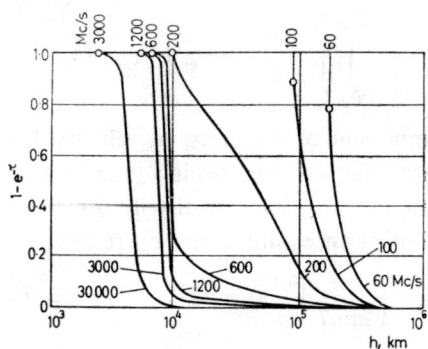

Fig. 139. Function $1 - e^{-\tau_{1,2}(h)}$ for radial propagation of radio waves in the chromosphere (Smerd, 1950a).

ions is proportional to $\lambda^2 \int N^2 T^{-3/2} \, dl$). Because of this the coefficient of reflection of electromagnetic waves from the solar atmosphere $e^{-2\tau_{1,2}(R^*)}$ falls as λ decreases.

Let us now examine the features of the absorption of plasma (longitudinal) waves in an isotropic plasma. It follows from the expression (26.75) that in the layers of the corona where $\nu_{\text{eff}} \sim 5 \text{ sec}^{-1}$ and $V_{\text{th}} \sim 4 \times 10^8$ cm/sec the value of the absorption coefficient connected with collisions for plasma waves with $V_{\text{ph}} \sim 4 \times 10^9$ cm/sec that is is 10^{-7} cm^{-1}. This means that the intensity of a plasma wave decreases by a factor of e over a distance of the order of 10^7 cm, i.e. over a distance that is small when compared, let us say, with the size of active regions on the Sun. The time that plasma oscillations take to be damped under the same conditions is fractions of a second ($\frac{1}{5}$ sec with $\nu_{\text{eff}} \approx 5 \text{ sec}^{-1}$).

As well as the attenuation because of collisions for plasma waves there exists specific damping not connected directly with the collisions. It has been noted above that the specific absorption is connected with the Cherenkov emission of plasma waves by the plasma electrons; therefore the corresponding expressions for the energy coefficients of Cherenkov attenuation γ_3 and absorption μ_3 can be obtained conveniently by using Kirchhoff's law (26.8) and the formula (26.24) for the intensity of Cherenkov emission of an electron in a plasma. The corresponding calculations (Andronov, 1961) are given below.

We first find the emissive power a_j by the appropriate averaging of the Cherenkov emission intensity over the electron velocity distribution function. Using the cylindrical system of coordinates $V_\parallel$, $V_\perp$, φ, we can write the energy of the plasma waves emitted by a unit volume of plasma in unit time in an element of solid angle $d\Omega$ as follows:

$$\frac{d\mathcal{E}_j}{dt}\,d\Omega = d\Omega \int_0^\infty \int_{V_\parallel}^\infty \int_0^{2\pi} \frac{d\mathcal{E}_{j\Omega}}{dt} f(V_\parallel, V_\perp, \varphi)\, V_\perp\, dV_\perp\, dV_\parallel\, d\varphi. \quad (26.82)$$

Here $V_\parallel$ is the component of the electron velocity V in the direction of emission k; $V_\perp, \varphi$ are polar coordinates in a plane orthogonal to the vector k; $f(V_\parallel, V_\perp, \varphi)$ is the electron velocity distribution function, $j = 3$. In the case under discussion of an equilibrium non-relativistic plasma

$$f(V) = N\left(\frac{m}{2\pi\varkappa T}\right)^{3/2} e^{-mV^2/2\varkappa T} \quad (V^2 = V_\perp^2 + V_\parallel^2) \quad (26.83)$$

(the Maxwell distribution).

The emission intensity of an electron in a unit solid angle $d\mathcal{E}_{j\Omega}/dt$ is easy to express in terms of the spectral density $d\mathcal{E}_{j\omega}/dt$ (26.24) by changing from the range $d\Omega$ to the range $d\omega$ by means of the well-known relation for the Vavilov–Cherenkov effect $kV = \omega$ (26.23a):

$$\frac{d\mathcal{E}_{j\Omega}}{dt} = \frac{d\mathcal{E}_{j\omega}}{dt} \frac{Vn_j^2}{2\pi c \dfrac{dn_j}{d\omega}}. \quad (26.84)$$

Then replacing the variable $V_\parallel = c/n_3(\omega)$ (see (26.23b)) in the integral (26.82) we obtain:

$$\frac{d\mathcal{E}_j\,d\Omega}{dt} = \frac{d\Omega}{2\pi} \int_0^\infty \int_\omega^\infty \int_0^{2\pi} V \frac{d\mathcal{E}_{j\omega}}{dt} f(\omega, V_\perp, \varphi)\, V_\perp\, dV_\perp\, d\omega\, d\varphi. \quad (26.85)$$

On the other hand it is obvious that

$$\frac{d\mathcal{E}_j}{dt}\,d\Omega = \int_\omega a_j\, d\omega\, d\Omega. \quad (26.86)$$

By comparing the last two expressions for $d\mathcal{E}_j/dt$ we can see that the emissive power of the plasma because of the Vavilov–Cherenkov effect is

$$a_j = \frac{1}{2\pi} \int_0^\infty \int_0^{2\pi} V \frac{d\mathcal{E}_{j\omega}}{dt} f V_\perp\, dV_\perp\, d\varphi. \quad (26.87)$$

§ 26] Emission and Absorption

Taking the actual form of $d\mathcal{E}_{j\omega}/dt$ and f into consideration this gives:

$$a_3 = \frac{e^2 N\omega}{(2\pi)^{3/2}\,\varepsilon(\omega)\,V_{\text{th}}} e^{-c^2/2 V_{\text{th}}^2 n_3^2(\omega)} \quad \left(V_{\text{th}}^2 = \frac{\varkappa T}{m}\right). \tag{26.88}$$

Substituting (26.88) in Kirchhoff's law we obtain that the coefficient of Cherenkov absorption of a plasma wave is

$$\mu_3 = \sqrt{\frac{\pi}{18}} \frac{\omega^3}{(kD)^4 \omega_L^2 V_{\text{th}}} e^{-\omega^2/2 V_{\text{th}}^2 k^2}. \tag{26.89}$$

In the change to (26.89) allowance is made for the fact that the square of the refractive index of the plasma waves is $n_3^2 = \varepsilon/3\beta_{\text{th}}^2$ and the Debye radius is $D = V_{\text{th}}/\omega_L$. In accordance with (26.70) the energy coefficient of the attenuation of a plasma wave in time will be of the form

$$\gamma_3 = \sqrt{\frac{\pi}{2}} \frac{\omega^2}{(kD)^3 \omega_L} e^{-\omega^2/2 V_{\text{th}}^2 k^2}, \tag{26.90}$$

if $\omega^2 = \omega_L^2 + 3 V_{\text{th}}^2 k^2$.

The last expression is the same as Landau's formula (Landau, 1946) which describes the attenuation of longitudinal waves in an equilibrium plasma without collisions. Landau (1946) obtained this on the basis of the kinetic equation with a self-consistent field which allows the presence of thermal motion to be allowed for better (when compared with the quasi-hydrodynamic equations in section 22).[†] However, the derivation given above is easier to follow and stresses the connection between Landau damping and the Vavilov–Cherenkov effect. This connection became clear after the appearance of the papers of Pines and Bohm (1952), Bohm and Gross (1949) and particularly of Shafranov (1958); it was indicated by, for example, Ginzburg and Zheleznyakov (1959b). Therefore Cherenkov damping and Landau damping are one and the same effect; only the methods of their investigation are different.[‡]

The formulae obtained above for μ_3 and γ_3 are valid only in the case of weak absorption, when $\mu_3 \ll k$, $\gamma_3 \ll \omega$. This follows even if only from the fact that the expression figuring in Kirchhoff's law for the equilibrium intensity $I_j^{(0)}$ holds only in transparent (weakly absorbing) media. The conditions for weak absorption in space and attenuation in time will

---

[†] Landau damping does not appear in the quasi-hydrodynamic treatment, which is a further indication of the insufficiency of the argument in this method.

[‡] The reservation should be made that at present Landau damping is the name usually given to all forms of wave attenuation in a plasma which exist even in the absence of collisions. Therefore in a magnetoactive plasma, for example, we shall start to give the name of Landau damping to not only the Cherenkov part but also the magneto-bremsstrahlung part of the dissipation.

obviously be satisfied with $k^2D^2 \ll 1$, i.e. for long enough plasma waves:

$$\left(\frac{\lambda}{2\pi}\right)^2 \equiv \frac{1}{k^2} \gg D^2, \tag{26.91a}$$

or, which is the same thing, for waves with a high enough phase velocity:

$$V_{\text{ph}}^2 \equiv \frac{\omega^2}{k^2} \gg V_{\text{th}}^2. \tag{26.91b}$$

The formula (26.90) was found in Landau's paper under the same conditions. We note that in the region in question the frequency of the plasma waves ω is close to the eigen frequency of the plasma ω_L.

If, however, $k^2D^2 \lesssim 1$, then the method used above for finding the value of the coefficient of specific absorption (attenuation) becomes unsuitable,[†] unlike the method of the kinetic equation with a self-consistent field (section 27) which retains its force even with $k^2D^2 \gtrsim 1$. According to Landau (1946) in this region of the values of k the attenuation of a plasma wave occurs in a time $t \lesssim 1/\omega$; accordingly in space a wave is absorbed over a distance $l \lesssim \lambda/2\pi$. The latter means that plasma waves with $\lambda/2\pi \lesssim D$ cannot in practice be propagated in a plasma.

The resultant absorption of plasma waves for which $k^2D^2 \ll 1$ and $\omega^2 \approx \omega_L^2$ consists of two components (bremsstrahlung and Cherenkov) which enter additively into the expressions for the total coefficients of absorption μ_3 and γ_3. For example, the coefficient

$$\gamma_3 = \nu_{\text{eff}} + \sqrt{\frac{\pi}{2}} \frac{\omega_L}{(kD)^3} e^{-1/2k^2D^2 - 3/2} \tag{26.92}$$

and the effect of Landau damping can be neglected (when compared with the effect caused by collisions) provided that

$$k^2D^2 < \left[2\ln\left(\sqrt{\frac{\pi}{2}}\frac{\omega_L}{\nu_{\text{eff}}(kD)^3}\right)\right]^{-1}. \tag{26.93}$$

In the inner corona, where $\omega_L \sim 2\pi \times 10^8$ sec^{-1} and $\nu_{\text{eff}} \sim 10$ sec^{-1}, this inequality is satisfied if $kD < 1/6$ (i.e. $V_{\text{ph}} > 6V_{\text{th}}$); in planetary ionospheres with $\omega_L \sim 2\pi \times 10^7$ sec^{-1} and $\nu_{\text{eff}} \sim 10^3$ sec^{-1}—if $kD < 2/9$ (i.e. $V_{\text{ph}} > 4\cdot5 V_{\text{th}}$).

ABSORPTION OF ELECTROMAGNETIC WAVES IN A MAGNETOACTIVE PLASMA

In the presence of a constant magnetic field H_0 the formulae for the bremsstrahlung and Cherenkov absorption become far more complicated; in addition, one more component—magneto-bremsstrahlung—appears.

[†] Extrapolation of the expressions given for μ_3 and γ_3 into the region $k^2D^2 \sim 1$ is permissible only for estimates of the order of magnitude of the absorption.

§ 26] Emission and Absorption

Absorption because of collisions in a non-relativistic magnetoactive plasma is characterized by the following relation (Ginzburg, 1960b, section 11):

$$(n-i\varkappa)^2_{1,2} = 1 - \frac{2v(1-v-iv)}{2(1-iv)(1-v-iv) - u\sin^2\alpha \mp \sqrt{u^2\sin^4\alpha + 4u(1-v-iv)^2\cos^2\alpha}}, \quad (26.94)$$

where

$$v = \frac{\omega_L^2}{\omega^2} = \frac{4\pi e^2 N}{m\omega^2}, \quad u = \frac{\omega_H^2}{\omega^2} = \left(\frac{eH_0}{mc\omega}\right)^2, \quad \nu = \frac{\nu_{\text{eff}}}{\omega},$$

α is the angle between the wave vector $\mathbf{k}$ and the field $\mathbf{H}_0$. The "plus" sign corresponds to an ordinary wave (n_2, $\varkappa_2$), the "minus" sign to an extraordinary wave (n_1, $\varkappa_1$). The expression (26.94) for the refractive index $n_{1,2}$ and the absorption index $\varkappa_{1,2}$ is simplified in the practically very important cases of quasi-longitudinal and quasi-transverse wave propagation† relative to $\mathbf{H}_0$. In fact, in the quasi-longitudinal approximation

$$(n-i\varkappa)^2_{1,2} = 1 - \frac{v}{1\mp\sqrt{u}|\cos\alpha| - iv}, \quad (26.95)$$

and in the quasi-transverse

$$(n-i\varkappa)^2_1 = 1 - \frac{v(1-v-iv)}{(1-iv)(1-v-iv) - u\sin^2\alpha}, \quad (n-i\varkappa)^2_2 = 1 - \frac{v}{1-iv}. \quad (26.96)$$

If at the same time $v \ll 1$, then in a wide range of frequencies $(1-\sqrt{u}\cos\alpha)^2 \gg v$ with quasi-longitudinal propagation the refractive index is defined by the formula (23.7) and the absorption coefficient by

$$\varkappa_{1,2} = \frac{v\nu_{\text{eff}}}{2\omega(1\mp\sqrt{u}|\cos\alpha|)^2 n_{1,2}}. \quad (26.97)$$

Accordingly in the range of frequencies for which $(1-\sqrt{u}\sin\alpha)^2 \gg v$, with quasi-transverse propagation $n_{1,2}$ will be given by the expressions (23.8) where

$$\varkappa_1 = \frac{v[(1-v)^2 + u\sin^2\alpha]\nu_{\text{eff}}}{2n_1\omega(1-v-u\sin^2\alpha)^2}, \quad (26.98)$$

and $\varkappa_2$ is the same as (26.72).

† The conditions in which quasi-longitudinal and quasi-transverse propagation occur are given in section 23 without allowance for collisions (formulae (23.5), (23.6)). The effect of the latter on the limits of the applicability of the formulae, which will be given below, have little effect with the inequality $\nu \ll 1$ that is usually satisfied in the cosmic plasma.

According to (26.94) the absorption index $\varkappa_{1,2}$ of a magnetoactive plasma rises sharply with values of ω for which the relation $1-u-v+uv\cos^2\alpha = 0$ (23.4), i.e. with the values of ω for which in the absence of collisions $n_{1,2}^2 \to \infty$. This frequency is the same as ω_H only with longitudinal propagation. If $\alpha \neq 0$, then $\omega \neq \omega_H$; in particular, with transverse propagation a sharp rise in absorption occurs as ω approaches $\sqrt{\omega_H^2+\omega_L^2}$.

At first glance this circumstance is very strange. In actual fact the "resonance" frequency of absorption should obviously be the same as the frequency at which each electron of the plasma is "pumped" most strongly by the field of the electromagnetic wave. In a non-relativistic plasma this kind of frequency would appear to be the gyro-frequency ω_H at which the electron rotates. This is not so, however; the "resonance" frequency of absorption in collisions is determined not only by the properties of a single electron (rotation at a frequency ω_H), but also by the parameters characteristic of the plasma as a whole (plasma oscillation frequency ω_L). In other words, the collective nature of the motion of the plasma electrons in a field of the wave leads to a change in the frequency of the most efficient absorption by an amount $\Delta\omega \lesssim \omega_L$.

The shift of the "resonance" absorption frequency in a magnetoactive plasma becomes particularly obvious when we examine the motion of one of the plasma electrons in the field of a normal wave (Ginzburg and Zheleznyakov, 1958c). The collective nature of the motion of the electrons in the plasma (the effect of the ambient medium on the absorbing electron) is reflected in the polarization of the normal waves in this kind of treatment. As an example let us examine the behaviour of one of the electrons of the plasma when acted upon by an extraordinary wave in the case $\alpha = \pi/2$. The linearized equation of motion of an electron in the alternating electric field of the wave (in the presence of a constant magnetic field $\boldsymbol{H}_0$) is of the following form:

$$m\dot{\boldsymbol{V}} + m\nu_{\text{eff}}\boldsymbol{V} = -e\boldsymbol{E} - \frac{e}{c}[\boldsymbol{V}\boldsymbol{H}_0].$$

The components of the velocity $\boldsymbol{V}$ that corresponds to the electron's forced oscillations are as follows:

$$V_x = \frac{ieE_x}{m\omega(1-iv)}, \quad V_y = -\frac{ie}{m\omega}\frac{(1-iv)E_y + i\sqrt{u}E_z}{u-(1-iv)^2},$$

$$V_z = -\frac{ie}{m\omega}\frac{(1-iv)E_z - i\sqrt{u}E_y}{u-(1-iv)^2}, \qquad (26.99)$$

(the field $\boldsymbol{H}_0$ runs along the x-axis). However, in the case of transverse propagation in an extraordinary wave $E_x = 0$ and the field components

§ 26] Emission and Absorption

E_y and E_z are connected by the relation $E_y v \sqrt{u} = iE_z[u+v(1-iv)-(1-iv)^2]$. Therefore

$$V_x = 0, \quad V_y = -\frac{ieE_y}{m\omega}\frac{v(1-iv)}{(1-iv)^2-v(1-iv)-u}, \quad V_z = \frac{ieE_z}{m\omega}\frac{1}{v},$$

so the mean energy transferred in unit time from the wave to an electron (and then changing into thermal motion energy) is ($v \ll 1$)

$$\text{Re}\left(-\frac{1}{2}eVE^*\right) = \text{Re}\left(-\frac{1}{2}eV_yE_y^*\right) \approx \frac{e^2E_yE_y^*}{2m\omega}\frac{u+(1-v)^2}{(1-v-u)^2}v.$$

It is easy to see that this expression has a maximum at the frequency $\omega = \sqrt{\omega_H^2+\omega_L^2}$ that satisfies the equation $1-v-u = 0$, and not at the gyrofrequency as could be expected.

The Cherenkov part of the Landau damping in a magnetoactive plasma has been investigated in detail by Gershman (1953) (see also Ginzburg, 1960b, section 12) on the basis of the kinetic equation. As shown by Ginzburg and Zheleznyakov (1959a), the conditions for strong attenuation $\gamma_j \gtrsim \omega$ can, however, be obtained elementarily without using the kinetic equation method.

The Cherenkov emission really will be very small at the frequencies which are emitted (and absorbed) by particles with a velocity $V \gg V_{\text{th}} = \sqrt{\varkappa T/m}$, since there are few such particles in an equilibrium plasma. On the other hand, absorption of the Cherenkov type becomes significant at frequencies corresponding to $V \lesssim V_{\text{th}}$. In the absence of a magnetic field it follows from the condition for the Vavilov–Cherenkov effect $\beta n_j \cos \theta = 1$ (θ is the angle between the particle velocity V and the wave vector k) and from the condition $V \lesssim V_{\text{th}}$ that strong attenuation is possible only in the case when $\beta_{\text{th}} n_j |\cos \theta| \gtrsim 1$, i.e. if

$$\beta_{\text{th}} n_j = \frac{kV_{\text{th}}}{\omega} \approx kD \gtrsim 1. \tag{26.100}$$

When changing to the last relation it was borne in mind that with a given direction of propagation of the wave the angle θ for the plasma particles takes up all possible values.

It is clear that the criterion (26.100) for strong attenuation in an isotropic plasma remains valid even when there is a weak enough magnetic field (in this connection we recall the remark made in the first subsection of this section that in weak fields H_0 the emission in a system of harmonics $s = 0, \pm 1, \pm 2, \ldots$ is reduced in practice to the Vavilov–Cherenkov effect for an isotropic medium!). At the same time for strong magnetic fields the condition of effective absorption will be different because the condition for the Vavilov–Cherenkov effect here is of a different form: $\beta_\parallel n_j \cos\alpha = 1$ (26.30). In this case the absorption obviously becomes significant only when

$$\beta_{\text{th}} n_j |\cos\alpha| \gtrsim 1 \qquad (26.101)$$

(we note that α, unlike θ, keeps a definite value with a given orientation of H_0 and k).

The criteria given above for strong Cherenkov attenuation can also be obtained from other considerations which allow us to determine more accurately the limits of applicability of the conditions (26.100), (26.101) and judge the magnitude of this attenuation. The perturbation that has appeared in the plasma corresponding to one of the normal waves is damped in a time $t \lesssim 1/\omega$ (i.e. in this wave $\gamma_j \gtrsim \omega$) if in this time the electrons, while taking part in the thermal motion, are displaced a distance $l \sim \lambda/2\pi$ along k. In fact, in this case the electrons in the period of the oscillations transfer an ordered velocity acquired under the action of the field of the wave into the region where the phase of the wave ψ differs by an amount $\Delta\psi \gtrsim 1$. It is clear that this circumstance is the cause of the sharp decrease in the wave amplitude in the time $t \lesssim 1/\omega$.

Therefore the criterion of strong attenuation is the relation $l/V_k \sim \lambda/2\pi V_k \sim t \lesssim 1/\omega$ or (what is the same thing) the condition $V_k/\omega \gtrsim \lambda/2\pi \equiv 1/k$, where V_k is the velocity at which the electrons move in the direction k. Since in an isotropic plasma $V_k \sim V_{\text{th}}$, the condition that has just been written becomes $\omega/k \equiv V_{\text{ph}} \lesssim V_{\text{th}}$ or $\beta_{\text{th}}^2 n_j^2 \gtrsim 1$. This is the same as the criterion (26.100) which obviously retains its value even in a magnetoactive plasma if the magnetic field does not prevent the displacement of the electrons by a distance $l \sim \lambda/2\pi$ along k at the thermal velocity. The latter holds if the radius of rotation of a thermal electron $r_H \sim V_{\text{th}}/\omega_H$ is greater than or of the order of $\lambda/2\pi \sin\alpha$, i.e.

$$\delta \equiv \frac{\beta_{\text{th}}^2 n_j^2 \sin^2\alpha}{u} \gtrsim 1. \qquad (26.102)$$

If, on the contrary, $\delta \ll 1$, then the electron can move $\lambda/2\pi$ along the normal to the wave only because of motion along the magnetic field H_0; in this case the velocity V_k of the advance along k is $V_{\text{th}} \cos\alpha$. It follows from this that with $\delta \ll 1$ the criterion of strong attenuation $\lambda/2\pi V_k \lesssim 1/\omega$ is of the form $\beta_{\text{th}}^2 n_j^2 \cos^2\alpha \gtrsim 1$ and therefore is the same as (26.101).

It is clear from (26.100), (26.101) that strong attenuation occurs with large $n_j^2 > \beta_{\text{th}}^{-2} \gg 1$. For example, in the Sun's atmosphere $\beta_{\text{th}} < 10^{-2}$ and therefore $n_j^2 > 10^4$. Similar values of n_j^2 correspond to the plasma branch of the dispersion curves, where the wave is close to longitudinal (i.e. the

§ 26] Emission and Absorption

component of the electric field along k is much greater than the transverse component).† The attenuation here becomes so strong that there is no sense in speaking of wave propagation.

Magneto-bremsstrahlung absorption at frequencies of $\omega \approx s\omega_H$ ($s = 1, 2, 3, \ldots$) has been treated kinetically in papers by Stepanov and Pakhomov (1960), Sitenko and Stepanov (1956), Stepanov (1958), Gershman (1960). However, to calculate this form of absorption as well as the kinetic equation method a simpler and clearer method based on Kirchhoff's law is also applicable in the majority of cases. The latter connects the absorption coefficient of the electromagnetic waves with the emissive power of the plasma, which in its turn can be found from the known emission intensity of an individual electron by summing the contributions from all the particles, making allowance for their velocity distribution. This approach to the solution of the problem of magneto-bremsstrahlung absorption in a weakly relativistic equilibrium plasma (or, as we shall call it, of gyro-resonance absorption) has been used by Zheleznyakov (1964a).‡

According to Zheleznyakov (1964a) the energy $d\mathcal{E}_{js}/dt$ of electron magneto-bremsstrahlung at the sth harmonic related to unit time, emitting volume and solid angle is defined by the relation (26.82) in which $d\mathcal{E}_{j\Omega}/dt$ should now be understood as the intensity of the magneto-bremsstrahlung into a unit solid angle. Since $d\mathcal{E}_{j\Omega}/dt$ (26.57) and f (26.83) do not depend on the polar angle φ in a plane orthogonal to $\boldsymbol{H}_0$ we can at once integrate in (26.82)

---

† The exception in this respect is an extraordinary wave when $\alpha = 0$ in the region where $u > 1$ (the branch n_1^2 in Fig. 115b) which remains transverse even with $n_1^2 \gg 1$. The criterion of strong attenuation for this wave is not the same as that indicated (see Stepanov and Pakhomov, 1960; Gershman, 1953), which is not surprising since in the case under discussion the absorption is not Cherenkov but magneto-bremsstrahlung. (This is connected with the absence of Cherenkov emission of transverse waves in the direction $\alpha = 0$ when an electron moves along a helix in a magnetic field.) It can be shown, however, that in this case when the relation (26.100) is satisfied the attenuation will be great if at the same time $v \lesssim \beta_{\text{th}}^{-2}$. Under cosmic conditions (the atmospheres of the Sun and the planets) for the range of frequencies of interest to us the latter condition is well satisfied.

‡ We have already spoken above of the resonance absorption of radio waves in a magnetoactive plasma (at frequencies for which $n_j(\omega) \to \infty$). The resonance nature of the absorption at these frequencies is connected with the collective effects during the motion of electrons in the field of a wave occurring even when the thermal motion of the particles in the plasma is neglected (with a fixed number of collisions ν_{eff}). On the other hand, gyro-resonance absorption is connected with the resonance of individual electrons at frequencies $\omega \approx s\omega_H$ that occurs only when thermal motion is taken fully enough into consideration. These types of absorption are caused by physically different phenomena: the first is connected with bremsstrahlung and is proportional to $T^{-3/2}$ (T is the plasma's electron temperature), whilst the second is connected with magneto-bremsstrahlung and, as follows from what is said below, is proportional to $T^{s-3/2}$ ($s \geqslant 2$) and $T^{1/2}$ ($s = 1$).

with respect to φ:

$$\frac{d\mathcal{E}_{js}}{dt} = 2\pi \int_0^\infty \int_{V_\parallel}^\infty \frac{d\mathcal{E}_{j\Omega}}{dt} f V_\perp \, dV_\perp \, dV_\parallel. \tag{26.103}$$

Noticing that the frequency of the magneto-bremsstrahlung (26.55) contained in $d\mathcal{E}_{j\Omega}/dt$ in the direction α is a function only of the longitudinal ($V_\parallel$) component of the electron velocity, we replace the variable $V_\parallel$ in (26.103) by ω:

$$\frac{d\mathcal{E}_{js}}{dt} = 2\pi \int_0^\infty \int_\omega^\infty \frac{d\mathcal{E}_{j\Omega}}{dt} f V_\perp \left|\frac{\partial V_\parallel}{\partial \omega}\right| dV_\perp \, d\omega. \tag{26.104}$$

It is clear from this that the emission power of a plasma a_{js} for the type of emission under discussion is of the form

$$a_{js} = 2\pi \int_0^\infty \frac{d\mathcal{E}_{j\Omega}}{dt} f \left|\frac{\partial V_\parallel}{\partial \omega}\right| V_\perp \, dV_\perp, \tag{26.105}$$

where, in accordance with (26.55),

$$\frac{\partial V_\parallel}{\partial \omega} \approx \frac{c}{\omega n_j \cos \alpha}. \tag{26.106}$$

The last equality is valid if at the same time as the conditions (26.54) the inequality

$$|n_j \cos \alpha| \gg |\beta_\parallel| \tag{26.107}$$

holds. With $\beta_\parallel \neq 0$ it can be satisfied only for angles α that are not too close to $\pi/2$. Substituting the expression for $d\mathcal{E}_{j\Omega}/dt$ (26.56),† f (26.83) and $dV_\parallel/d\omega$ (26.106) in (26.105) and integrating with respect to $V_\perp$ we obtain:

$$a_{js} \approx \frac{s^{2s}}{2^s (2\pi)^{3/2} s!} \frac{e^2 N \omega}{c^{2s}} \left(\frac{\varkappa T}{m}\right)^{s-1/2} \frac{\sin^{2s-2} \alpha}{|\cos \alpha|}$$

$$\times n_j^{2s-2} \gamma_j^2 (1 + \alpha_j + n_j \beta_j \beta_\parallel \sin \alpha)^2 \exp\left(-\frac{mV_\parallel^2}{2\varkappa T}\right). \tag{26.108}$$

---

† Therefore the discussion below relates only to the region of the normal Doppler effect ($s > 0$), where only this formula is valid for $d\mathcal{E}_{j\Omega}/dt$. In the region of the anomalous Doppler effect $n_j \beta_\parallel \cos \alpha > 1$ efficient absorption occurs at frequencies emitted by electrons with $\beta_\parallel \sim \beta_{\mathrm{th}}$, i.e. at frequencies where $n_j \beta_{\mathrm{th}} \cos \alpha > 1$. In the latter case, however, as is clear from (26.101), there is strong Cherenkov absorption as well as magneto-bremsstrahlung; allowing for the first effect obviously makes the conclusion of the impossibility of wave propagation with $n_j \beta_{\mathrm{th}} \cos \alpha \gtrsim 1$ more weighty.

§ 26] Emission and Absorption

From the known emissive power from Kirchhoff's law we find the energy coefficient of gyro-resonance absorption:

$$\mu_{js}(\omega) \approx B_{js} \frac{s^{2s}}{2^s s!} \frac{\omega}{c} \beta_{\text{th}}^{2s-3} \exp\left(-\frac{z_{js}^2}{2}\right) \cos\vartheta, \quad (26.109)$$

where

$$B_{js} = \sqrt{\frac{\pi}{2}} vn_j^{2s-4}\gamma_j^2[1+\alpha_j+\beta_j(1-s\sqrt{u})\tan\alpha]^2 \frac{\sin^{2s-2}\alpha}{|\cos\alpha|}, \quad (26.110)$$

$$z_{js} = \frac{\beta_\parallel}{\beta_{\text{th}}} = \frac{1-s\sqrt{u}}{n_j\beta_{\text{th}}\cos\alpha}, \quad (26.111)$$

$$\beta_{\text{th}} = \sqrt{\frac{\varkappa T}{mc^2}}, \quad v = \frac{\omega_L^2}{\omega^2}, \quad u = \frac{\omega_H^2}{\omega^2},$$

α is the angle between the wave vector $\mathbf{k}$ and the field $\mathbf{H}_0$, ϑ is the angle between $\mathbf{k}$ and the group velocity $d\omega/d\mathbf{k}$. The parameters α_j, β_j and γ_j are given by the formulae (26.36)–(26.39), the refractive index n_j by the formula (23.2). When changing from (26.108) to (26.109) the velocity component $V_\parallel = \beta_\parallel c$ is eliminated by using the Doppler equation in the form (26.55).

Because of the exponential factor in (26.109) the quantity $\mu_{js}(\omega)$ becomes very small at frequencies that are not too close to $s\omega_H$ (i.e. provided that $z_{js}^2 \gg 2$). Therefore the spectrum of effectively absorbed frequencies forms a number of discrete "lines" with centres at $\omega = s\omega_H$.

It is easy to see (see (26.54), (26.107)) that the expression obtained for the coefficient of gyro-resonance absorption is valid in the case when the following inequalities are satisfied:†

$$\left.\begin{array}{c} \beta_{\text{th}}^2 \ll 1, \quad \beta_\parallel^2 \ll 1, \quad |n_j\beta_\parallel| \ll 1, \\ |sn_j\beta_{\text{th}}\sin\alpha| \ll 1, \quad \left|\beta_\parallel\omega\frac{\partial n_j}{\partial\omega}\right| \ll 1, \\ |n_j\cos\alpha| \gg |\beta_\parallel|, \end{array}\right\} \quad (26.112)$$

where $\beta_\parallel = (1-s\sqrt{u})/n_j\cos\alpha$. In addition, at the fundamental harmonic $s = 1$ (i.e. at frequencies $\omega \approx \omega_H$) the expression for μ_{js} retains its force only at the edges of the absorption line (provided that $z_{j1}^2 \gg 2$), since in the region $z_{j1}^2 \lesssim 2$ the expression (26.59) for the intensity of an electron becomes inapplicable. Therefore with $z_{j1}^2 \sim 1$ the formulae (26.109)–(26.111) can be used only for estimates of the order of magnitude of the

† The first and fourth inequalities are obtained from the corresponding conditions (26.54) by replacing $\beta_\perp$ by β_{th}. This replacement allows for the fact that for a Maxwell distribution the contribution of electrons with $\beta_\perp \gg \beta_{\text{th}}$ can be neglected when compared with electrons for which $\beta_\perp \lesssim \beta_{\text{th}}$.

absorption coefficient; the rigorous expression for μ_{j1} in the region $z_{j1} \lesssim 1$ can be found only as the result of investigation on the basis of the kinetic equation. At the higher harmonics s, starting at the second, these limitations drop out and the formulae (26.109)–(26.111) for μ_{js} are applicable both at the edges of a line and inside it.

It is reasonable to simplify the expression for the coefficient of gyroresonance absorption in exactly the same way as was done for the dipole emission intensity of a weakly relativistic electron (see the second subsection of the present section). To do this for all harmonics $s \geqslant 2$ in the expression for B_{js} (26.110) we make the frequency ω equal to $s\omega_H$:

$$B_{js} \approx \sqrt{\frac{\pi}{2}} \{vn_j^{2s-4}\gamma_j^2(1+\alpha_j)^2\}_{\omega=s\omega_H} \frac{\sin^{2s-2}\alpha}{|\cos\alpha|}. \quad (26.113a)$$

At the first harmonic $s = 1$ this cannot be done† since $(1+\alpha_j)^2_{\omega=\omega_H} = 0$; since, however, $(1+\alpha_j)(1-\sqrt{u})^{-1}$ remains finite with $\sqrt{u} \to 1$ (i.e. with $\omega \to \omega_H$), the factor B_{js} can be shown (with $z_{j1}^2 \gg 2$) as follows:

$$B_{j1} \approx \sqrt{\frac{\pi}{2}} (1-\sqrt{u})^2 \left\{ vn_j^{-2}\gamma_j^2 \left(\frac{1+\alpha_j}{1-\sqrt{u}} + \beta_j \tan\alpha \right)^2 \right\}_{\omega=\omega_H} |\cos\alpha|^{-1}. \quad (26.114a)$$

An estimate of B_{j1} in the region $z_{j1}^2 \sim 2$ can be obtained by extrapolating (26.114):

$$B_{j1} \sim \sqrt{2\pi} \left\{ v\gamma_j^2 \left(\frac{1+\alpha_j}{1-\sqrt{u}} + \beta_j \tan\alpha \right)^2 \right\}_{\omega=\omega_H} \beta_{\text{th}}^2 |\cos\alpha|. \quad (26.115)$$

In the quasi-longitudinal approximation (23.5), which can obviously be satisfied in a broad enough range of angles α only with $s \geqslant 2$ (i.e. with $u \leqslant \frac{1}{4}$),

$$\left.\begin{aligned} B_{js} &\approx \sqrt{\frac{\pi}{8}} \left\{ vn_j^{2s-2} \left[1 - \frac{v(1\pm\sqrt{u}\cos\alpha)}{1-u} \right]^{-1} \right\}_{\omega=s\omega_H} \frac{(1\pm\cos\alpha)^2 \sin^{2s-2}\alpha}{|\cos\alpha|}, \\ n_j^2 &= 1 - \frac{v}{1\mp\sqrt{u}|\cos\alpha|} \end{aligned}\right\} \quad (26.116)$$

(see (23.7)). In the change from (26.113) to this formula allowance is made for the fact that in the quasi-longitudinal approximation

$$\gamma_j^2 \approx \frac{n_j^2}{2} \left[1 - \frac{v(1\pm\sqrt{u}\cos\alpha)}{1-u} \right]^{-1}, \quad \alpha_j \approx \pm\cos\alpha$$

† Elimination of the term $\beta_j(1-s\sqrt{u})\tan\alpha$ is impermissible even at the higher harmonics if it becomes greater than $1+\alpha_j$. Because of the smallness of $1-s\sqrt{u}$ in the vicinity of the absorption lines this circumstance is significant only with $\alpha \approx \pi/2$.

§ 26] Emission and Absorption

(the upper sign here and above relates to an extraordinary wave and the lower to an ordinary one). If $v \ll 1$, then it can be considered that $n_j \approx 1$, $\gamma_j^2 \approx \frac{1}{2}$ and $\cos \vartheta \approx 1$ (the case close to a vacuum); here the energy coefficient of gyro-resonance absorption becomes particularly simple:[†]

$$\mu_{js} \approx \sqrt{\frac{\pi}{8}} \frac{s^{2s}}{2^s s!} \frac{\omega}{c} v \beta_{\text{th}}^{2s-3} \frac{(1 \pm \cos \alpha)^2 \sin^{2s-2} \alpha}{|\cos \alpha|} \exp\left\{-\frac{z_{js}^2}{2}\right\}, \quad (26.117)$$

$$z_{js} = \frac{1 - s\sqrt{u}}{\beta_{\text{th}} \cos \alpha}.$$

At the fundamental harmonic $s = 1$ ($\omega \approx \omega_H$) at the edges of the absorption line for an ordinary wave with $v \ll 1$

$$n_2^2 \approx 1; \quad \cos \vartheta \approx 1;$$

$$\left\{\gamma_2 \left[\frac{1+\alpha_2}{1-\sqrt{u}} + \beta_2 \tan \alpha\right]\right\}_{\omega=\omega_H}^2 \approx \frac{\sin^4 \alpha}{\cos^2 \alpha} \frac{(1+2\cos^2 \alpha)^2}{(1+\cos^2 \alpha)^3}$$

and therefore

$$\left.\begin{array}{l}\mu_{21} \approx \sqrt{\dfrac{\pi}{8}} \dfrac{\omega}{c} \dfrac{(1-\sqrt{u})^2}{\beta_{\text{th}}} v \dfrac{\sin^4 \alpha (1+2\cos^2 \alpha)^2}{|\cos^3 \alpha| (1+\cos^2 \alpha)^3} \exp\left(-\dfrac{z_{21}^2}{2}\right), \\[6pt] z_{21} \approx \dfrac{1-\sqrt{u}}{\beta_{\text{th}} \cos \alpha}.\end{array}\right\} \quad (26.118)$$

Of course, the absorption coefficient in the form (26.118), just like the expression for the emission intensity of an electron at the fundamental harmonic (26.59), holds only for values of v (26.65) that are not too small. If v is small enough (see the criterion (26.66)), then the absorption at the fundamental harmonic is the same as in a vacuum. In the latter case for the calculation of μ_{j1} we can use, let us say, the formula (26.117) putting $s = 1$ in it.

The expressions for B_{js} can be given a different form which is more convenient when comparing the values obtained for the coefficients of gyro-resonance absorption with the corresponding results of the kinetic calculations. For this purpose we substitute in (26.113a), (26.114a) the explicit expressions for $\gamma_j, \alpha_j, \beta_j$ in terms of u, v, α and n_j; then after rather cumbersome transformations we find (Zheleznyakov, 1964a) that at all

---

[†] We notice that in the region of the values $1-v \sim 1$ for $s = 2$ the quasi-longitudinal approximation will be violated at angles $\alpha \gtrsim 75°$; for $s = 3$ this will occur if $\alpha \gtrsim 80°$, etc. This circumstance makes possible extensive use of of the formulae (26.116), (26.117) in actual calculations of the gyro-resonance absorption in the solar corona at harmonics with a number $s \geqslant 2$.

harmonics starting at the second

$$B_{js} = \sqrt{\frac{\pi}{2}} \frac{\sin^{2s-2}\alpha}{|\cos\alpha|} \left\langle vn_j^{2s-2}(1-u) \left\{ n_j^4 \sin^2\alpha - (1-v)(1+\cos^2\alpha)n_j^2 \right.\right.$$
$$+ 2\left(1 - \frac{v}{1+\sqrt{u}}\right)(1-v-n_j^2\sin^2\alpha) \right\} \left\{ 2(1-v)[(1-v)^2 - u] $$
$$\left.\left. + [(2-v)u - 2(1-v)^2 - uv\cos^2\alpha] \right\}^{-1} \right\rangle_{\omega = s\omega_H}, \quad (26.113\text{b})$$

and at the first harmonic at the edges of the absorption line

$$B_{j1} = \sqrt{\frac{\pi}{2}} \frac{(1-u)^2}{|\cos\alpha|} \left\langle \frac{1}{v} \left\{ (v\cos 2\alpha - 1)n_j^4 + \left(2 - 2v\cos^2\alpha - v^2\sin^2\alpha \right.\right.\right.$$
$$\left.\left. - \frac{v^2}{4}\tan^2\alpha\right)n_j^2 + (v-1)\left(1 - \frac{v^2}{4}\right) + \frac{v^2}{4}(2-v)\tan^2\alpha \right\} \left\{ 2(1-v)(2-v) \right.$$
$$\left.\left. + (2v-2-\sin^2\alpha)n_j^2 \right\}^{-1} \right\rangle_{\omega=\omega_H}. \quad (26.114\text{b})$$

At harmonics with a number $s \geq 2$ the absorption coefficient μ_{js} (26.113) is the same as that obtained by Stepanov (1958) on the basis of the kinetic equation differing by the factor of 2 introduced in Gershman (1960; see also Ginzburg, 1960b, section 12). However, μ_{js} (26.114) at the edges of the absorption line at the gyro-frequency differs from the corresponding expressions given by Sitenko and Stepanov (1956), Stepanov (1958), Gershman (1960); in particular, the result of Stepanov (1958) is the same as (26.114) only if we eliminate $\beta_j \tan\alpha$ in (26.114a) which, as we have seen, cannot be done at the first harmonic. At the same time the correctness of the formula (26.114) found by Zheleznyakov (1964a) is confirmed by the kinetic calculations made by Andronov, Zheleznyakov and Petelin (1964). We note that according to Gershman (1960; see also Ginzburg, 1960b, section 12) inside the absorption line at the fundamental harmonic ($s = 1$, $z_{j1}^2 \ll 2$)

$$\mu_{j1} \approx \sqrt{\frac{8}{\pi}} \frac{\omega}{c} \frac{\beta_{\text{th}}|\cos\alpha|}{v} (2v - 2 - \sin^2\alpha + 2n_j^2\sin^2\alpha)^{-1}$$
$$\times \left\langle \left\{ \left[1 - \left(1 - \frac{7}{4}\sin^2\alpha\right)v\right]n_j^4 - \left[2 + v\left(-\frac{5}{2} + \frac{7}{4}\sin^2\alpha\right)\right.\right.\right.$$
$$\left.\left. + \frac{v^2}{4}(2\cos 2\alpha - \tan^2\alpha)\right]n_j^2$$
$$\left.\left. + \left[1 - \frac{3}{2}v + \frac{v^2}{2}(1 - \tan^2\alpha) + \frac{v^3}{4}\tan^2\alpha\right]\right\}\cos\vartheta \right\rangle_{\omega=\omega_H}. \quad (26.119)$$

§ 26] Emission and Absorption

It is clear from the expressions given for B_{js} that in the general case this factor (and at the same time the coefficient of gyro-resonance absorption) depends in a complex manner on the parameters v, u, and the angle α. The nature of this dependence can be seen by turning to Fig. 140 in which are plotted the graphs of $B_{js}(v)/\beta_{th}^2 z_{js}^2$ for ordinary waves at the fundamental harmonic, and also to Figs. 141–4 which give graphs of the values of $B_{js}(v)$ for extraordinary and ordinary waves at the second and

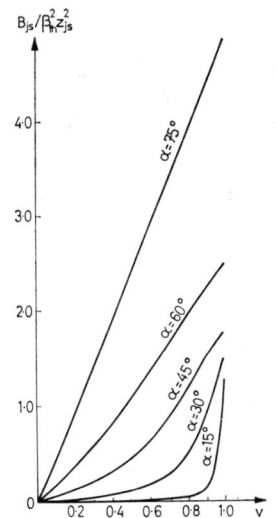

FIG. 140. Parameter $B_{js}/\beta_{th}^2 z_{js}^2$ as a function of $v = \omega_L^2/\omega^2$ for an ordinary wave at the first harmonic ($s = 1$) (Zheleznyakov, 1964a)

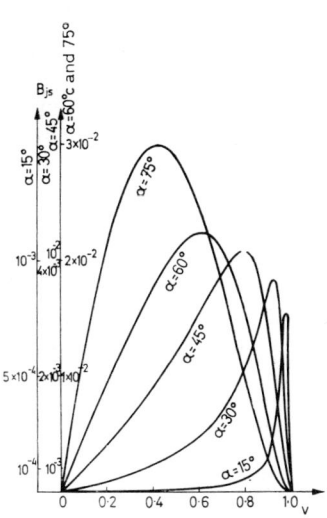

FIG. 141. Parameter B_{js} as a function of v for an ordinary wave at the second harmonic ($s = 2$) (Zheleznyakov, 1964a)

third harmonics. The ranges of values of v in the figures ($0 < v < 1-\sqrt{u}$ for an extraordinary wave and $0 < v < 1$ for an ordinary one) are selected with allowance made for the fact that in these ranges $n_j^2 > 0$.†

In conclusion it must be said that all the formulae given for gyro-resonance absorption are obtained without allowing for collisions in the plasma. The effect of ν_{eff} on the value of this type of absorption will be insignificant if $\nu_{\text{eff}}/\omega \ll 1$, $\nu_{\text{eff}}/\omega \ll \beta_{th} n_j \cos\alpha$ (although of course with the presence of collisions comes bremsstrahlung absorption which, com-

---

† With $s \geq 1$, i.e. with $u \leq 1$, for extraordinary waves there is one more region with positive values of n_1^2 (in layers $v > 1-\sqrt{u}$). However, for cases of practical interest of electromagnetic wave propagation in the solar corona and planetary ionospheres this region is of no great importance since immediate escape beyond the plasma from this region is impossible (see sections 23, 25). For the latter reason there is no graph here of $B_{js}(v)/\beta_{th}^2 z_{js}^2$ at the first harmonic for an extraordinary wave.

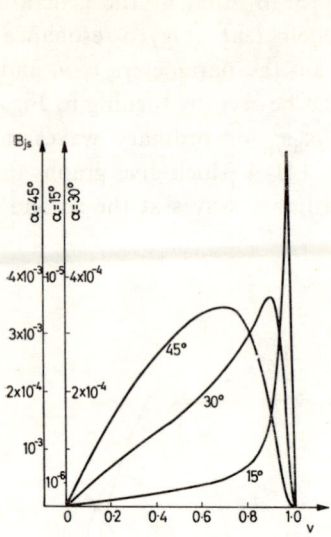
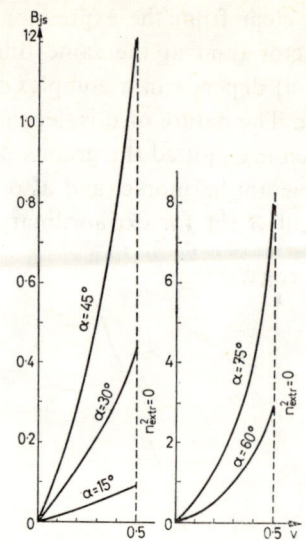

FIG. 142. The same as Fig. 141 for the third harmonic ($s = 3$) (Zheleznyakov, 1964a)

FIG. 143. Parameter B_{js} as a function of v for an extraordinary wave at the second harmonic ($s = 2$) (Zheleznyakov, 1964a)

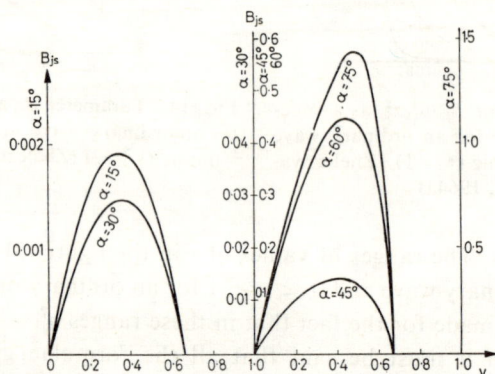

FIG. 144. The same as Fig. 143 for the third harmonic ($s = 3$) (Zheleznyakov, 1964a)

bining with magnetic-bremsstrahlung absorption, increases the resultant value of μ_j). As a rule these inequalities are found to hold well in cosmic conditions.

GYRO-RESONANCE ABSORPTION IN THE SOLAR CORONA

First of all let us compare the coefficients of bremsstrahlung and gyro-resonance absorption in the corona. To do this it is convenient to use the

§ 26] Emission and Absorption

formulae (26.117) and (26.118), which are valid with small values of the parameter v. It follows from them that, roughly speaking, the coefficient of magnetic-bremsstrahlung (gyro-resonance) absorption (Ginzburg and Zheleznyakov, 1959a) of ordinary ($s \geqslant 1$) and extraordinary ($s \geqslant 2$) waves with angles $\alpha \sim 1$ is

$$\mu_{j,\,s=1} \sim \frac{\omega}{c} v\beta_{\text{th}}, \quad \mu_{j,\,s\geqslant 2} \sim \frac{s^{2s}}{2^s s!} \frac{\omega}{c} v\beta_{\text{th}}^{2s-3}, \qquad (26.120)$$

so $\mu_{j,\,s=2} \sim \mu_{j,\,s=1}$. As has been explained earlier, this is caused by the weak emission of electrons at the gyro-frequency in a magnetoactive plasma. The estimates given for the value of μ_{js} relate to the region of frequencies within the absorption line $(\omega - s\omega_H)^2 \lesssim \omega^2 \beta_{\text{th}}^2 \cos^2 \alpha$; at the edges of a line, i.e. in the region $(\omega - s\omega_H)^2 \gg \omega^2 \beta_{\text{th}}^2 \cos^2 \alpha$, the coefficient μ_{js} decreases very rapidly (exponentially) as $(\omega - s\omega_H)^2$ rises. At the same time the coefficient of bremsstrahlung absorption connected with collisions with $v \ll 1$ and $(1 \mp \sqrt{u} \cos \alpha)^2 \sim 1$ is

$$\mu_j^{\text{coll}} \sim \frac{vv_{\text{eff}}}{c}$$

(see (26.69) and (26.97)).

These estimation formulae are sufficient for us to be sure that in the conditions of the solar corona the absorption coefficient at the lower harmonics in the resonance regions $\omega \approx s\omega_H$ is far greater than the coefficient of absorption due to collisions. In fact for the first and second harmonics $\mu_{js} \sim (\omega/c)v\beta_{\text{th}}$ and therefore

$$\frac{\mu_{j,\,s=1,\,2}}{\mu_j^{\text{coll}}} \sim \frac{\omega \beta_{\text{th}}}{v_{\text{eff}}}. \qquad (26.121)$$

In the solar corona (for $v_{\text{eff}} \sim 10 \text{ sec}^{-1}$, $\beta_{\text{th}} \sim 10^{-2}$) this ratio is very high—of the order of 6×10^5 at a frequency of $\omega \sim 2\pi \times 10^8 \text{ sec}^{-1}$ (the metric band); at higher frequencies it is still greater.

This circumstance, to which attention was drawn by Zheleznyakov (1959a), Ginzburg and Zheleznyakov (1959a), points to the very significant part played by resonance absorption in the total absorption of the coronal plasma in the radio band (when there is a strong enough magnetic field). The latter has a noticeable effect on the nature of the propagation in active regions of the corona and the amount of emission from these regions: as will be shown in section 29, the effective gyro-resonance emission and absorption play an important part in the solution of the problem of the origin of the slowly varying component of the solar radio emission.

At the same time when calculating resonance absorption in the solar corona it should be remembered that the magnetic field there is non-uniform (in particular varies with altitude, etc.; see section 2); therefore the nature of the frequency dependence of the absorption will differ from the case of a uniform field H_0 in which the absorption spectrum of the plasma consists of a series of discrete lines at frequencies $\omega \approx s\omega_H$. When the distribution of the field in the plasma is sufficiently non-uniform the absorption spectrum is smoothed out; however, now gyro-resonance absorption at a given frequency ω will not occur throughout the plasma (as in the uniform case), but in local layers where the magnetic field strength satisfies the relation $\omega \approx s\omega_H \equiv seH_0/mc$.

The optical thickness of the layer is obviously

$$\tau_{js} = \int \mu_{js} \, dl, \tag{26.122}$$

where dl is an element of length of the line along which the emission is propagated through the layer, the emission corresponding to a wave of the jth type (ordinary or extraordinary); the energy coefficient of gyro-resonance absorption μ_{js} is given by the formulae (26.109) and (26.110). Since the value of μ_{js} at the frequency ω in regions of the plasma where $z_{js}^2 \gg 2$ (26.111) is exponentially small, most of the contribution to the integral (26.122) is made by a thin layer located near the resonance level $\omega = s\omega_H$. The thickness of this layer L_{js} is easy to estimate if we remember that in this layer $z_{js}^2 \equiv (\omega - s\omega_H)^2/\omega^2 \beta_{th}^2 n_j^2 \cos^2 \alpha \lesssim 2$. If $L_{js} \ll L_H$, where L_H is the characteristic distance over which there is a significant change in the magnetic field's strength along the ray, then the function $\omega_H(l)$ can with sufficient accuracy be considered linear within the resonance layer.† Then obviously (Ginzburg and Zheleznyakov, 1959a)

$$L_{js} \sim 2\sqrt{2} \beta_{th} n_j \left| \frac{\omega_H \cos \alpha}{\frac{d\omega_H}{dl}} \right| = 2\sqrt{2} L_H \beta_{th} n_j |\cos \alpha|. \tag{26.123}$$

In changing to the last equation it is taken that $\omega_H |dl/d\omega_H| = L_H$.

From equations (26.120) and (26.123) we obtain the following estimates of the optical thickness of the gyro-resonance levels $\tau_{js} \sim \mu_{js} L_{js}$ (with $n_j \sim 1$, $\cos \alpha \sim 1$):

$$\tau_{j, s=1} \sim \tau_{j, s=2}, \quad \tau_{j, s \geq 2} \sim \frac{s^{2s}\sqrt{2}}{2^{s-1} s!} \frac{\omega}{c} v \beta_{th}^{2s-2} L_H. \tag{26.124}$$

In the solar corona with $\beta_{th} \sim 10^{-2}$, $L_H \sim 10^{10}$ cm at a frequency of $\omega \sim 2\pi \times 10^8$ sec^{-1}, $\tau_{j, s=1} \sim \tau_{j, s=2} \sim 10^5 v$; $\tau_{j, s=3} \sim 10^2 v$; $\tau_{j, s=4} \sim 10^{-1} v$, etc. Although these estimates are rough they leave no doubt that gyro-resonance absorption in the corona with v that are not too small

---

† In the conditions of the corona with $\beta_{th} \sim 10^{-2}$ for regions where $n_j \lesssim 1$ the linear approximation of $\omega_H(l)$ is fully justified: $L_{js} \lesssim 3 \times 10^{-2} L_H$.

§ 26] Emission and Absorption

causes very sharp attenuation of electromagnetic waves of a frequency ω passing through levels $\omega \approx \omega_H$, $\omega \approx 2\omega_H$ and $\omega \approx 3\omega_H$ in a direction $\alpha \sim 1$. This attenuation, of course, shows up only in active regions of the corona with a strong enough magnetic field: the harmonics $s = 1, 2, 3$ at frequencies $\omega \sim 2\pi \times 10^8$ sec^{-1} correspond to fields of $H_0 \sim 12$–36 oe; at higher frequencies the values of H_0 should be even greater. At the same time the overall field of the Sun is characterized by a strength of $H_0 \sim 1$ oe.

For more effective calculation of the optical thickness of the gyro-resonance levels in a plasma with a non-uniform magnetic field we must substitute the expression (26.109) for μ_{js} in (26.122) and integrate with respect to the resonance layer along the ray l. As a result we obtain:

$$\tau_{js} \approx \sqrt{2\pi} \frac{s^{2s}}{2^s s!} \frac{\omega}{c} \beta_{\text{th}}^{2s-2} n_j B_{js} L_H |\cos \alpha| \cos \vartheta, \qquad (26.125)$$

where at harmonics $s \geqslant 2$ the parameter B_{js} is given by the expressions (26.113). In the case of quasi-longitudinal propagation when B_{js} is comparatively simple in form (see (26.116))

$$\tau_{j,s} \approx \pi \frac{s^{2s}}{2^{s+1} s!} \left\{ \frac{\omega}{c} v n_j^{2s-1} \left[1 - \frac{v(1 \pm \sqrt{u} \cos \alpha)}{1-u} \right]^{-1} \right\}_{\omega = s\omega_H} \beta_{\text{th}}^{2s-2} L_H$$

$$\times (1 \pm \cos \alpha)^2 \sin^{2s-2} \alpha \cos \vartheta. \qquad (26.126)$$

If in addition $v \ll 1$, we can consider in (26.126) that the values of the quantities n_j, $[1 - v(1 \pm \sqrt{u} \cos\alpha)(1-u)^{-1}]^{-1}$ and $\cos \vartheta$ are equal to unity.

In a resonance layer corresponding to the first harmonic $s = 1$ the parameter B_{js} in the region $z_{js}^2 \sim 2$ is known only in order of magnitude. Here, therefore, we can give only an estimate of τ_{js} by substituting the expression for B_{js} in the form (26.115) in (26.125). With $v \ll 1$ we can use (26.118) by substituting $z_{js}^2 \sim 2$ there; as a result we obtain that

$$\tau_{21} \sim \pi \frac{\omega}{c} v \beta_{\text{th}}^2 \sin^4 \alpha \frac{(1 + 2\cos^2 \alpha)^2}{(1 + \cos^2 \alpha)^3} L_H. \qquad (26.127)$$

This estimate holds with v that are not too small when the criterion (26.65), in which $z_{21}^2 \sim 2$ is also put, is satisfied. Then this criterion becomes $v \gg \beta_{\text{th}} |\cos \alpha|$. In the case of $v \ll \beta_{\text{th}} |\cos \alpha|$ the optical thickness of the gyro-resonance layer $s = 1$ is the same as in a vacuum:

$$\tau_{j1} \approx \frac{\pi}{4} \frac{\omega}{c} v(1 \pm \cos \alpha)^2 L_H. \qquad (26.128)$$

It is curious that this quantity does not in general depend on the kinetic temperature.

By using (26.125) and the values of the parameter B_{js} in the graphs of Figs. 140–4 we can make a more accurate estimate of the optical thickness τ_{js} of gyro-resonance levels in the solar corona. Putting for the sake of argument $v \sim 0.3, \beta_{\text{th}} \sim 10^{-2}, L_H \cos \vartheta \sim 10^{10}$ cm and $\omega \sim 2\pi \times 10^8 \text{ sec}^{-1}$ (the metric band), we obtain the values of τ_{js} given in Table 5. Similar information on τ_{js} in the centimetric and decimetric wavebands are given in Table 6 (section 29). It is obtained for values of $v \ll 1$ using the formulae (26.126), (26.127).

TABLE 5

Harmonic	H, oe	τ_1 (extraordinary wave)		τ_2 (ordinary wave)	
		$\alpha = 30°$	$\alpha = 60°$	$\alpha = 30°$	$\alpha = 60°$
1	36			$\sim 1.8 \times 10^3$	$\sim 1.3 \times 10^4$
2	18	1.3×10^4	3.8×10^4	2.4×10	6.2×10^2
3	12	1.5	1.2×10	6.0×10^{-3}	5.3×10^{-1}

A comparison of the approximate values of τ_{js} given earlier with $\alpha \sim 1$ with the more accurate data of Table 5 for angles $\alpha = 30°$ and $\alpha = 60°$ shows that the expressions (26.124) correctly describe τ_{js} in order of magnitude when extraordinary waves pass through a gyro-resonance layer. For ordinary waves, on the contrary, the estimates from the formulae (26.124) are one or two orders too high.

It is clear from Table 5 that gyro-resonance absorption of extraordinary waves in the solar corona shows up chiefly when passing through layers where $\omega \approx 2\omega_H, 3\omega_H$[†]: in these layers $\tau_{js} \gtrsim 1$. Absorption of ordinary waves in regions $\omega \approx s\omega_H$ proceeds less effectively, which in the end is connected with the predominance of the extraordinary component in the magneto-bremsstrahlung of a weakly relativistic electron. Therefore under the conditions of the solar corona the absorption of an ordinary wave is significant only at the first and second harmonics (i.e. in layers where $\omega \approx \omega_H, \omega \approx 2\omega_H$); at the third harmonic, unlike the extraordinary component, it becomes relatively weak. A similar situation also holds in general in the centimetric and decimetric wavebands (see Table 6).

It should be noted, however, that gyro-resonance absorption plays this kind of significant part in layers corresponding to low harmonics only at angles α between $\boldsymbol{k}$ and $\boldsymbol{H}_0$ which are not too close to zero, since $\tau_{js} \to 0$ when $\alpha \to 0$, as $\sin^{2s-2} \alpha$ for an extraordinary wave and $(1-\cos \alpha)^2 \sin^{2s-2} \alpha$

[†] The $\omega \approx \omega_H$ is not being discussed since an extraordinary wave cannot escape from it beyond the corona without preliminary transformation into a wave of the ordinary type (see sections 23, 25).

§ 26] Emission and Absorption

for an ordinary one (see the formula (26.126), which is valid with the quasi-longitudinal propagation). According to the latter the range of angles α in which the resonance layer $s \geqslant 2$ becomes transparent can be determined from the condition $\alpha < \alpha_{cr}$, where the critical angle α_{cr} at which $\tau_{js} = 1$ is given by the following expressions: for extraordinary waves

$$\alpha_{cr} \approx \frac{\sqrt{2}}{s\beta_{th}} \left\langle \frac{1}{n_1} \left(\frac{s!\, cn_1}{\pi s^2 \omega L_H v} \right)^{\frac{1}{2s-2}} \right\rangle_{\omega = s\omega_H}, \qquad (26.129)$$

for ordinary waves

$$\alpha_{cr} \approx \frac{\sqrt{2}}{s\beta_{th}} \left\langle \frac{1}{n_2} \left(\frac{4s!\, c\beta_{th}^2 n_2^3}{\pi \omega L_H v} \right)^{\frac{1}{2s}} \right\rangle_{\omega = s\omega_H}. \qquad (26.130)$$

Here we have allowed for the fact that with small α, $\sin \alpha \approx \alpha$, $1 + \cos \alpha \approx 2$, $1 - \cos \alpha \approx \alpha^2/2$ and $\cos \vartheta \approx 1$.

In the corona with $\omega \sim 2\pi \times 10^8$ sec^{-1}, $v \sim 0.3$, $n_j \sim 1$ and the values taken above for L_H, β_{th} for the extraordinary component $\alpha_{cr} \sim 4 \times 10^{-3}$ ($s = 2$) and $\alpha_{cr} \sim 0.35$ ($s = 3$). For the ordinary component $\alpha_{cr} \sim 0.14$ ($s = 2$); estimates of the value of the transparency range at the first harmonic by means of the expressions (26.125) and (26.115) lead to the value $\alpha_{cr} \sim 6.4 \times 10^{-2}$.

If the emission incident on the layer $\omega \approx s\omega_H$ has a broad angular spectrum (let us say within the solid angle $\sim 2\pi$), then directional emission concentrated in the solid angle $\Omega \sim \pi \alpha_{cr}^2$ will leave the layer; the energy flux connected with this emission is obviously a fraction $\Omega/2\pi$ of its original value. For the estimates of α_{cr} given above the ratio $\Omega/2\pi \sim 10^{-5}$ ($s = 2$), $\Omega/2\pi \sim 6 \times 10^{-2}$ ($s = 3$) in the case of an extraordinary wave and $\Omega/2\pi \sim 2 \times 10^{-3}$ ($s = 1$), $\Omega/2\pi \sim 10^{-2}$ ($s = 2$) in the case of an ordinary wave. These values allow us to judge the extent of the effect of gyro-resonance layers on the escape of radio emission beyond the solar corona: the extraordinary component leaves without noticeable attenuation only the regions of the corona which are located above the level $\omega \approx 3\omega_H$; the attenuation becomes significant when radio emission is being propagated from the deeper-lying layers, whilst escape from regions lying below the layer $\omega \approx 2\omega_H$ proceeds with low efficiency. For the ordinary component noticeable attenuation shows from the $\omega \approx 2\omega_H$ level onwards; escape beyond the corona is actually limited by layers above the $\omega \approx \omega_H$ level.

We recall that all the estimates made above relate to the metric band; the corresponding data on the gyro-resonance absorption in the corona at centimetric and decimetric wavelengths are given in section 29, where they are also used in interpreting the slowly varying component of the solar radio emission. Absorption in the gyro-resonance layers of planetary ionospheres is discussed in section 32 taking Jupiter as an example. There is also a discussion of the part it plays in ensuring the directional nature of the sporadic Jovian radio emission.

In conclusion we should say that from a known optical thickness τ_{js} it is easy to find the intensity I_j of magnetic bremsstrahlung or the effective temperature T_{eff} connected with it of the gyro-resonance layers in the coronal plasma. The problem is simplified because of the small (as a rule) extent L_{js} of the gyro-resonance layers along the ray when compared with the characteristic size at which the electron concentration and kinetic temperature T change. (This circumstance has already been used earlier in finding the quantity T_{js}.) Therefore the values of I_j and T_{eff} can be determined by the simple formulae (26.13), (26.18); it follows from them that with optically thick ($\tau_{js} > 1$) gyro-resonance layers corresponding to low harmonics is connected emission which has an effective temperature $T_{\text{eff}} \approx T$. However, in the directions $\alpha < \alpha_{\text{cr}}$, where $\tau_{js} < 1$, the value of T_{eff} becomes less than T; in the range $\tau_{js} \ll 1$ it is equal to $T\tau_{js}$. A similar condition also occurs at higher harmonics where $\tau_{js} \ll 1$ for any directions α.

If the change in the magnetic field is so small that the thickness of the gyro-resonance layer is $L_{js} \gg L$ (L is the size of the plasma localization region), the plasma can be considered uniform. Then the frequency spectrum of the magneto-bremsstrahlung in a plasma is of the nature of "lines" which disappear only at high harmonics because of overlapping of the lines. The features of the gyro-frequency emission from a uniform plasma have been discussed by Stepanov and Pakhomov (1960), and by Trubnikov (1958), Trubnikov and Bazhanova (1958). (The results of the latter relate only to a strongly rarefied plasma in which with sufficient accuracy $n_j = 1$). Under solar conditions, however, it is highly unlikely that the inequality $L_{js} \gg L$ is satisfied. This can be seen particularly well if we imagine it slightly differently by means of the formula (26.123):

$$\frac{\Delta H_0}{H_0} \equiv \left(\frac{L}{\omega_H}\right) \left|\frac{d\omega_H}{dl}\right| \ll \beta_{\text{th}} n_j |\cos \alpha|,$$

where ΔH_0 is the change of the magnetic field in the region occupied by the plasma.

Magneto-bremsstrahlung (synchrotron emission) from an equilibrium relativistic plasma, as far as we know, has not yet been discussed by anyone; generally the case is discussed when the energy distribution of the electrons is exponential. The results of this can be found in the papers by Stepanov and Pakhomov (1960), Trubnikov (1958).

27. Emission, Absorption and Amplification of Electromagnetic Waves in a Non-equilibrium Plasma

THE KINETIC EQUATION METHOD AND THE EINSTEIN COEFFICIENTS METHOD. THE PROBLEM OF WAVE AMPLIFICATION AND INSTABILITY IN A PLASMA

If the particle distribution in a plasma is not an equilibrium one, then the change in the emission intensity along the ray in the majority of cases can be described as before by the transfer equation (26.4). Just as before, the intensity of emission from a uniform layer of thickness L will be

$$I_j = \frac{a_j}{\mu_j}(1 - e^{-\mu_j L}) + I_{j0} e^{-\mu_j L} \tag{27.1}$$

(I_{j0} is the intensity of the emission incident on the layer); in the case when $\mu_j > 0$ and the dimensions of the emitting system are large enough ($L \gg \mu_j^{-1}$)

$$I_j = \frac{a_j}{\mu_j}. \tag{27.2}$$

However, the expressions for the emissive power a_j and the coefficient of absorption μ_j will now be different from those in an equilibrium medium; in particular the connection between them is no longer defined by the Kirchhoff law (26.8).

As well as the necessity of finding the quantities a_j and μ_j the question arises of the limits of applicability of the transfer equation in the conditions of a non-equilibrium plasma; the method of the kinetic equation with a self-consistent field is of considerable help in the solution of this range of questions.

As applied to a plasma this method is based on the system of electrodynamic Maxwell equations (22.1) and the Boltzmann kinetic equation for the distribution function $f(\mathbf{R}, \mathbf{p}, t)$ (see, e.g., Silin and Rukhadze, 1961, section 10, and Sommerfeld, 1955, section 41):

$$\frac{\partial f}{\partial t} + \mathbf{V}\nabla_\mathbf{R} f - e\left(\mathbf{E} + \frac{1}{c}[\mathbf{VH}]\right)\nabla_p f + J = 0. \tag{27.3}$$

Here e is the value of the charge of particles with a velocity $\mathbf{V}$ and a momentum $\mathbf{p}$, $\mathbf{E}$ and $\mathbf{H}$ are the strengths of the electric and magnetic fields, J is the so-called collision integral which describes the change in the distribution function because of close collisions between particles. The distribution function characterizes the density of the particles at the point $\mathbf{R}, \mathbf{p}$ of the phase space at the time t, so the number of particles in an element

$dR\,dp$ is $dN = f(R, p, t)dR\,dp$ and the concentration of particles of the kind under discussion in the plasma is $N = \int f\,dp$.

The equation (27.3) is written for the case of negative particles (electrons) in a form that is also valid at relativistic velocities; there is also a similar equation for the positive ions. For reasons given in section 22, however, we shall consider the motion of the ions (their distribution function f_i) to be given;† therefore they require no kinetic equation. We notice finally that the current and charge densities figuring in the Maxwell equations can be expressed in terms of $f(R, p, t)$ by the relations

$$j = -e \int Vf\,dp + eN_iV_i, \quad \varrho = -e \int f\,dp + eN_i \qquad (27.4)$$

(compare with (22.3)).

The kinetic equation allows more closely for the thermal motion of the particles in the plasma than the quasi-hydrodynamic equations (22.2); therefore its applicability is far broader when studying processes in a plasma.

The equation (27.3) is non-linear which makes its solution very difficult. For weak electromagnetic perturbations, however, it can be linearized in just the same way as was done for the equations (22.2). We put the distribution function in the form $f = f_0 + f'$, where f_0 is the unperturbed distribution function that obeys the equation

$$\frac{\partial f_0}{\partial t} + V\nabla_R f_0 - e\left(E_0 + \frac{1}{c}[VH_0]\right)\nabla_p f_0 + J_0 = 0, \qquad (27.5)$$

and f' is a small correction to f_0 connected with the wave field E', H'. The linearized equation for f' is of the form

$$\frac{\partial f'}{\partial t} + V\nabla_R f' - e\left(E_0 + \frac{1}{c}[VH_0]\right)\nabla_p f' = e\left(E' + \frac{1}{c}[VH']\right)\nabla_p f_0 - J'. \quad (27.6)$$

Here E_0, H_0 are the stationary or quasi-stationary electric and magnetic fields in the plasma; below the field E_0 will not be taken into consideration. The part of the collision integral J' that acts on f' if f_0 is a Maxwell function can be written approximately as $I' = v(V)f'$ ($v(V)$ is the number of collisions of an electron having a velocity V with other particles); in the rest of the cases the form of J' is very complex. For simplification of notation the primes on the variables connected with the electromagnetic perturbations E', H' will be omitted in future.

The dielectric permeability tensor $\varepsilon_{il}(\omega, k)$ of a non-equilibrium uniform plasma without allowing for collisions ($J' = 0$) can be found conveniently

† The only exception being that discussed in the last subsection of this section.

§ 27] Emission, Absorption and Amplification

by using the equation (27.6) as follows (Shafranov, 1958; see also Sagdeyev and Shafranov, 1959). Instead of the independent variables R, V we introduce the new ones R_0, V_0, determining the connection between them from the equations of the unperturbed motion of individual particles:

$$\left.\begin{aligned}\frac{dR(t)}{dt} &= V(t), \quad R_0 = R(t=0), \\ \frac{dp(t)}{dt} &= -\frac{e}{c}[V(t)H_0], \quad V_0 = V(t=0).\end{aligned}\right\} \quad (27.7)$$

In the new variables (27.6) can be written in the form

$$\frac{\partial f}{\partial t} = e\left\{E(R(t), t) + \frac{1}{c}[V(t)H(R(t), t)]\right\}\nabla_{p(t)}f_0. \quad (27.8)$$

Provided that f approaches zero when $t \to -\infty$ its solution will be

$$f = e\int_{-\infty}^{t}\left\{E(R(\tilde{t}), \tilde{t}) + \frac{1}{c}[V(\tilde{t})H(R(\tilde{t}), \tilde{t})]\right\}\nabla_{p(\tilde{t})}f_0\, d\tilde{t}. \quad (27.9)$$

We can find the tensor $\varepsilon_{il}(\omega, k)$ if in the relation (22.12), which allowing for the equality $D = E - i(4\pi/\omega)j$ is written in the form

$$E_i - i\frac{4\pi}{\omega}j_i = \varepsilon_{il}(\omega, k)E_l, \quad (27.10)$$

we know the expression for the current j in terms of the components E_i, E_l of the electric field of a monochromatic wave $E = E_a e^{i\omega t - ikR}$. This expression is easy to determine from (27.9) if we remember that, as follows from the second Maxwell equation (22.6), the magnetic field of a wave is $H = (c/\omega)[kE]$:

$$j_i(R, t) = -e\int V_i f\, dp_0 = -e^2 E_{al} e^{i\omega t - ikR(t)}\int dp_0$$

$$\times \int_{-\infty}^{t} e^{i\omega(\tilde{t}-t)- ik[R(\tilde{t})-R(t)]} V_i(t)\left\{\left(1 - \frac{k_\alpha V_\alpha(\tilde{t})}{\omega}\right)\frac{\partial f_0}{\partial p_l(\tilde{t})} + \frac{k_\alpha V_l(\tilde{t})}{\omega}\frac{\partial f_0}{\partial p_\alpha(\tilde{t})}\right\} d\tilde{t}.$$

$$(27.11)$$

By making the replacement of the variable $\tilde{t} \to \xi = t - \tilde{t}$ we obtain in (27.11) an integral with respect to ξ in constant limits from 0 to ∞. This integral with any distribution function f should obviously not depend on t as on a parameter since the function of the change in the current j_i in time (which is the same as the corresponding dependence of the field E_l

on t) is already contained in the factor in front of the integral. It is clear that the integrand does not depend on t either, which now can be selected arbitrarily. Let $t = \xi$; then we obtain from (27.10), (27.11) that the tensor $\varepsilon_{il}(\omega, \mathbf{k})$ in a relativistic plasma is defined by the relation

$$\varepsilon_{il}(\omega, \mathbf{k}) = \delta_{il} + \sum i \frac{4\pi e^2}{\omega} \int d\mathbf{p}_0 \int_0^\infty V_i(t) e^{-i\omega t + i \int_0^t \mathbf{k} V(t') dt'}$$

$$\times \left[\left(1 - \frac{\mathbf{k} V_0}{\omega}\right) \frac{\partial f_0}{\partial p_{0l}} + \frac{k V_{0l}}{\omega} \frac{\partial f_0}{\partial \mathbf{p}_0} \right] dt. \qquad (27.12)$$

Here summation is carried out with respect to kinds of particles which have unperturbed distribution functions f_0; $\mathbf{p}_0$ and $\mathbf{V}_0$ are the momentum and velocity of these particles at the time $t = 0$, $\mathbf{V}(t)$ is the velocity of a particle's unperturbed motion as a function of time. In an isotropic plasma $\mathbf{V}(t)$ is the velocity of uniform and rectilinear motion and in a magnetoactive plasma $\mathbf{V}(t)$ is the velocity of helical motion with $|\mathbf{V}| = \text{const}$.

We recall the expression (27.12) is obtained provided that $f \to 0$ if $t \to -\infty$. This obviously means that the wave rises in time, i.e. $\text{Im}\,\omega < 0$. For damped waves ($\text{Im}\,\omega > 0$) the tensor $\varepsilon_{il}(\omega, \mathbf{k})$ can be treated as an analytical extension of (27.12) into the upper half-plane ω.

Substitution in (22.14) of the dielectric permeability tensor calculated by means of (27.12) for a plasma with a given velocity distribution f_0 of particles will give the dispersion equation for this medium; this defines (for a wave being propagated in a fixed direction $\mathbf{k}/k$) the connection between two complex quantities:

$$\omega = \text{Re}\,\omega + i\,\text{Im}\,\omega, \quad k = \text{Re}\,k + i\,\text{Im}\,k,$$

or, what is the same thing, between the four real variables $\text{Re}\,\omega$, $\text{Im}\,\omega$, $\text{Re}\,k$, $\text{Im}\,k$. This connection becomes definite if we can find the frequency

$$\omega = \varphi_1(k) \qquad (27.13)$$

from a given k or, contrariwise, the wave number

$$k = \varphi_2(\omega) \qquad (27.14)$$

from given ω. The original values of ω and k, generally speaking, are also complex; however, in the majority of cases depending on the statement of the problem we start either from real k and find complex ω by means of (27.13), or from real ω and find complex k from (27.14).

When we interest ourselves in the problem of the stability of electromagnetic perturbations (i.e. the problem of the change in the energy of

§ 27] Emission, Absorption and Amplification

these perturbations in time $\mathcal{E}(t)$), we must obviously operate with real values of k, considering that Im $k = 0$. In actual fact the question of the stability of a uniform plasma relative to high-frequency electromagnetic perturbations is put as follows. At the initial point in time let the plasma be perturbed from a state of equilibrium, i.e. a local perturbation appears in it (of fluctuation origin for example) with an electrical field that can be expanded into a Fourier integral in real values of $\boldsymbol{k}$:

$$\boldsymbol{E}(\boldsymbol{R}, t = 0) = \int \boldsymbol{g}_1(\boldsymbol{k}) e^{-i\boldsymbol{k}\boldsymbol{R}} \, d\boldsymbol{k}. \tag{27.15}$$

Then at the succeeding points in time each Fourier component will be a normal wave in which ω is connected with k by the dispersion relation (27.13).[†] Then

$$\boldsymbol{E}(\boldsymbol{R}, t) = \int \boldsymbol{g}_1(\boldsymbol{k}) e^{i\omega t - i\boldsymbol{k}\boldsymbol{R}} \, d\boldsymbol{k}, \quad \omega = \varphi_1(\boldsymbol{k}), \tag{27.16}$$

where, as follows from Fourier integral theory (Titchmarsh, 1937),

$$\int |\boldsymbol{E}|^2 \, dV = \int |\boldsymbol{g}_1|^2 e^{-2 \operatorname{Im} \omega t} \, d\boldsymbol{k}. \tag{27.17}$$

Here dV is an element of volume of the plasma.

Noting that the integral on the left is proportional to the perturbation energy $\mathcal{E}$, we come to the conclusion that the nature of the function $\mathcal{E}(t)$ is determined by the magnitude and sign of Im ω. Namely, $\mathcal{E} \to 0$ with $t \to \infty$ if with all $\boldsymbol{k}$ for which $\boldsymbol{g}_1 \neq 0$

$$\operatorname{Im} \omega > 0. \tag{27.18}$$

This means that the plasma is stable relative to perturbations containing normal waves with the above values of $\boldsymbol{k}$. On the other hand $\mathcal{E}$ will rise without limit with $t \to \infty$[‡] if among the values of $\boldsymbol{k}$ for which $\boldsymbol{g}_1 \neq 0$ we can find a region where

$$\operatorname{Im} \omega < 0. \tag{27.19}$$

The latter is a sign of the plasma instability relative to the perturbations which are made up by waves from the region indicated.

A more detailed study of the behaviour of the field (27.16) in space and in time allowed Sturrock (1958) to divide instability and stability into two clear-cut types—convective and absolute.[§] This classification is carried

---

[†] It is assumed for simplicity here that the function $\omega = \varphi_1(k)$ is unambiguous, i.e. in a plasma only normal waves of one type exist. In the opposite case in (27.16) we must carry out summation with respect to different types of waves.

[‡] In the linear approximation; actually this rise will be limited by non-linear effects which will be discussed below.

[§] In hydrodynamics this division was made earlier by Landau and Lifshitz (1953, section 29).

out as follows. Let the function $g_1(k)$ have a sharp peak with $k = k_0$ and be negligibly small a long way from k_0. Then the field (27.16) describes a packet in space in the sense that for all infinitely great R the field E becomes zero.† If the packet in space is at the same time a packet in time as well, i.e. if it can be put in the form

$$E(R, t) = \int g_2(\omega) e^{i\omega t - ikR} \, d\omega, \quad k = \varphi_2(\omega) \tag{27.20}$$

with real ω and limited values of $g_2(\omega)e^{-ikR}$, then by definition we are dealing with a convective type of instability or stability; in the opposite case the instability or stability will be absolute.

The difference between convective and absolute instability is well illustrated in Fig. 145 a, b where in the plane t, z (z is the coordinate read

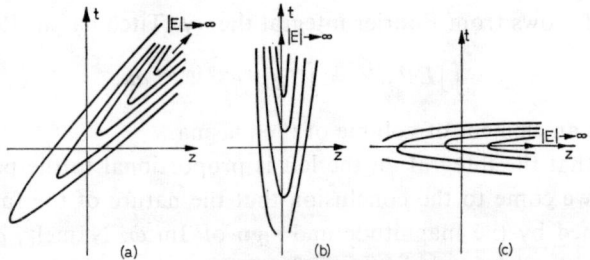

FIG. 145. Lines of equal field strength on the plane z, t: (a) packet in space is at the same time a packet in time (convective stability and amplification); (b) packet in space is not a packet in time (absolute instability); (c) packet in time is not a packet in space (disappearing waves).

along k_0) are plotted the lines of equal field strength $|E(z, t)| = \text{const}$. If in the system of finite size we have the case of convective instability, then the system can remain in a stationary state even with the presence of small chaotic perturbations since these perturbations are carried away from the region of their appearance as they rise. With absolute instability of the system the position is different since the amplitude of any perturbation will increase steadily in the region where it appeared; it is also possible that at the same time it will gradually embrace the whole system. The process of the system coming out of equilibrium will take place in an extremely non-stationary manner.

† The latter follows from Riemann's lemma (Titchmarsh, 1937), according to which the function $\Psi(z, t) \equiv \int_{-\infty}^{+\infty} \psi(k, t) e^{-ikz} \, dk \to 0$ with $z \to \pm \infty$ if $\psi(k, t)$ is limited for all real values of k.

§ 27] Emission, Absorption and Amplification

We note further that the contours $|E| = $ const for convective and absolute stability will be similar to those shown in Fig. 145a, b if we change the time axis to the opposite direction. Here the convective and absolute types differ in that the equilibrium state of the system is established after it has been upset by the action of some perturbation: in the first case a perturbation with a decreasing energy is carried out of the region where it appears, and in the second it is damped where it has appeared.

Unlike the problem discussed above of the conditions and types of stability and instability in which we had to take Im $k = 0$, for the problem of amplification and absorption (i.e. the problem of the variation in the emission intensity in space $I_j(R)$)† another way of stating the problem is reasonable—the assumption that Im $\omega = 0$. This condition must obviously be used when finding the absorption coefficient

$$\mu_j = -2 \operatorname{Im} k \cos \vartheta \qquad (27.21)$$

for emission with given spectral characteristics; a knowledge of this coefficient is necessary when studying the transfer equation (26.4).

In the case Im $\omega = 0$, Im $k \neq 0$ we can once again divide the waves into two types in a similar way to that done above for waves with Im $k = 0$, Im $\omega \neq 0$. Here we shall proceed from the expression for $E(R, t)$ in the form (27.20), assuming that $g_2(\omega)$ is a sharp function with a maximum when $\omega = \omega_0$ and is limited together with $\varphi_2(\omega)$ for all values of ω. Then in accordance with Riemann's lemma (27.20) will be a time packet. If this packet in time is also a packet in space (i.e. if it can be written in the form (27.16)), then a wave of frequency ω_0 is called amplified or absorbed depending upon whether it rises or falls in the direction of the group velocity vector $d\omega/dk$, i.e. depending on the sign of μ_j. If the time packet cannot be shown as a spatial packet the corresponding wave is called disappearing.

It is easy to see that for amplified waves the lines of equal field strength in the plane z, t will be the same as for convectively unstable waves (see Fig. 145a); likewise the contours for absorbed waves are the same as the contours of convectively stable waves. The close connection of amplification with convective instability and of absorption with convective stability is clear from this and the definitions given above. On the other hand, the lines of equal field strength for disappearing waves (Fig. 145c) differ sharply from those shown in Fig. 145a, b.

A special investigation is needed for classifying the waves in each individual case, since the possibility or impossibility of changing from the

---

† As applied to an equilibrium plasma this question has been discussed in section 26, largely on the basis of Kirchhoff's law, it is true, and not by the cumbersome kinetic equation method.

field representation (27.16) to the representation (27.20) depends on the actual form of the dispersion equation. The appropriate ways of solving this question can be studied in the papers of Sturrock (1958) and Fedorchenko (1962). An obvious, but far from simple method of classification is the study of the behaviour of the integrals (27.16) and (27.20) with infinitely great values of t and R (Landau and Lifshitz, 1953, section 29).[†] If $E(R, t)$ (27.16) approaches zero when $|t| \to \infty$, then the plasma stability or instability is convective; in the opposite case it will be absolute. In just the same way, if $E(R, t)$ (27.20) becomes zero with $|R| \to \infty$, then the waves are amplified or absorbed; if it does not become zero the waves are disappearing ones.

Unfortunately the question of the classification of waves in a plasma with different actual velocity distributions $f_0(V)$ of the particles is insufficiently studied at present, although there is reason to suppose that the waves being studied below belong to the amplified ($\mu_j < 0$) type and absorbed ($\mu_j > 0$) type or, what is the same thing, to the class of perturbations which have convective instability and stability.[‡] Outside our field will remain disappearing waves, an example of which are electromagnetic and plasma waves in the region $n_j^2 < 0$ and also waves having absolute instability. The part played by the former in cosmic conditions is clearly small; the latter may be more significant in the generation of certain components of the sporadic radio emission and their investigation would be highly desirable.

As is clear from the above the transfer equation method can be applied only to amplified and absorbed waves. The emissive power of the plasma a_j contained in this equation is calculated from the usual formulae for the emission intensity of individual particles by averaging this quantity over the distribution function $f_0(V)$. (Calculations of this kind for an equilibrium plasma have been carried out in section 26.) On the other hand the absorption coefficient in the transfer equation can be found if we know the dispersion equation for the waves in the plasma by means of the relation (27.21); here the dispersion equation is found by the kinetic equation method and in certain special cases by the quasi-hydrodynamic method. In section 26, however, we were able to be certain that weak absorption ($\lambda \operatorname{Im} k \ll 1$) in conditions of equilibrium velocity distribution is easier to

---

[†] Actually the behaviour of these integrals at infinity does not depend on the actual form of the functions $g_1(k)e^{-ikR}$, $g_2(\omega)e^{i\omega t}$ if they are analytical. Therefore in the investigation we can consider these functions to be unit ones and examine the integrals $\int e^{i\varphi_1(k)t} dk$ with $|t| \to \infty$ and $\int e^{-i(k/k)\varphi_2(\omega)R} d\omega$ with $|R| \to \infty$.

[‡] Data on the classification of waves in a plasma for certain forms of the function $f_0(V)$ are given by Sturrock (1958, 1960), Fedorchenko (1962), Fainberg, Kurilko and Shapiro (1961).

§ 27] Emission, Absorption and Amplification

obtain from a known emissive power a_j by Kirchhoff's law. In a non-equilibrium plasma the so-called Einstein coefficients method can be conveniently used for this purpose.

The starting point in this method is the concept of emission quanta in a medium with an energy $\hbar\omega$ and a momentum $\hbar k_j = (\hbar\omega/c)n_j(\omega, s)s$, where $\hbar$ is Planck's constant, ω is the frequency, $k_j = k_j s$ is the wave vector, c is the velocity of light in a vacuum, n_j is the refractive index of a type j wave being propagated in the medium. According to Einstein's quantum theory the interaction of an emission field and a system consisting of non-coherently emitting centres with two levels (m) and (n)[†] is characterized by three elementary processes: spontaneous emission, "true" absorption and stimulated emission. If the number of quanta emitted in 1 sec in a range $d\omega$ and a solid angle $d\Omega$ with transitions of particles from the state (m) to the state (n) is $N_m A_m^n \, d\omega \, d\Omega$ (spontaneous emission) and $N_m B_m^n \varrho_j \, d\omega \, d\Omega$ (stimulated emission) and the corresponding number of absorbed quanta is $N_n B_n^m \varrho_j \, d\omega \, d\Omega$, then it follows from the conditions of balance between these processes in a state of equilibrium that the Einstein coefficients A_m^n, B_m^n and B_n^m are connected by the following relations (Ginzburg and Zheleznyakov, 1958b):[‡]

$$\frac{B_n^m}{B_m^n} = 1, \quad \frac{A_m^n}{B_n^m} = \frac{n_j^2 \hbar \omega^3}{(2\pi c)^3} \frac{\partial(\omega n_j)}{\partial \omega} \qquad (27.22)$$

Here N_m, N_n are the numbers of particles in a unit volume in the states (m) and (n), ϱ_j is the emission density.

It must be stressed that the relations (27.22) that are characteristics of the actual emission centres are also valid in the absence of thermal equilibrium, although it is also introduced for finding the connection between A_m^n, B_m^n and B_n^m. This is not really necessary since the equalities (27.22) can also be obtained by another method (Heitler, 1956, section 17).

The absorption coefficient of the system under discussion is determined by the difference between the numbers of transitions accompanied by the processes of "true" absorption and stimulated emission (see Ginzburg and

---

[†] When we say the particle has only two states we have in mind only the energy levels between which transitions are connected with emission and absorption at a given frequency.

[‡] Finding the relations (27.22) in a medium with a refractive index n_j does not in principle differ in any way from the corresponding derivation in the case of a vacuum given, for example, by Blokhintsev (1961, section 5); we only need to remember that the expression for the equilibrium emission density $\varrho_j^{(0)}$ becomes different in the medium (see (26.6) and Landau and Lifshitz, 1957, section 91)).

Zheleznyakov, 1958b; Zheleznyakov, 1959c):

$$\mu_j \equiv \frac{\Delta I_j}{I_j} = \frac{\hbar\omega\,(N_n B_n^m \varrho_j - N_m B_m^n \varrho_j)}{\varrho_j \left|\dfrac{d\omega}{dk_j}\right|} = A_m^n N_m \frac{8\pi^3 c^2 \left(\dfrac{N_n}{N_m}-1\right)}{\omega^2 n_j^2}\,|\cos\vartheta|.$$
(27.23)

Here ΔI_j is the change due to absorption of the intensity value over a section of unit length, i.e. the emission energy absorbed in unit time in a unit volume: θ is the angle between $\mathbf{k}_j$ and the group velocity $d\omega/dk_j$. In the change to the last equation we have taken into consideration the relations (26.7), (27.22). At the same time the emissive power of the system in the direction of the wave vector $\mathbf{k}_j$ is

$$\alpha_j = A_m^n N_m \hbar\omega.\qquad(27.24)$$

According to (27.3) the absorption coefficient is proportional to the intensity of spontaneous emission (i.e. the emissive power a_j) and to the factor $(N_n/N_m - 1)$, which depends on the ratio of the number of particles in the upper (m) and lower (n) states for which the difference of the energies $\mathcal{E}_m$ and $\mathcal{E}_n$ is equal to $\hbar\omega$ (the energy of an emission quantum), and the difference of the particles' momenta in these states $\mathbf{p}_m$ and $\mathbf{p}_n$ is equal to the momentum of this quantum $\hbar\mathbf{k}_j$.

For a Maxwell velocity distribution of the plasma particles $(N_n/N_m) - 1 = e^{\hbar\omega/\varkappa T} - 1 \approx \hbar\omega/\varkappa T$ (in the region $\hbar\omega \ll \varkappa T$ of interest to us) and therefore the absorption coefficient is

$$\mu_j = A_m^n N_m \frac{8\pi^3 c^2 \hbar}{\omega n_j^2 \varkappa T}\,|\cos\vartheta|.\qquad(27.25)$$

As was to be expected, we find from (27.24), (27.25) that μ_j and a_j are connected by Kirchhoff's law, i.e. the emission intensity of an optically thick layer (27.2) is equal to its equilibrium value (26.5).

It has already been remarked above that (27.23), (27.24) are valid only for transitions between two states of emitting particles. In the case of a large number of similar states to obtain the absorption coefficient and the emissive power at the frequency ω we must sum these expressions over all transitions $(m) \rightleftarrows (n)$ corresponding to this frequency (Zheleznyakov, 1959c):

$$\mu_j = \frac{8\pi^3 c^2}{\omega^2 n_j^2}\,|\cos\vartheta|\sum_{(m)\rightleftarrows(n)} A_m^n N_m \left(\frac{N_n}{N_m}-1\right),\qquad(27.26)$$

$$\alpha_j = \hbar\omega \sum_{(m)\rightleftarrows(n)} A_m^n N_m.\qquad(27.27)$$

§ 27] Emission, Absorption and Amplification

The case

$$\sum_{(m)\neq(n)} A_m^n N_m \left(\frac{N_n}{N_m} - 1\right) > 0, \qquad (27.28)$$

when the absorption coefficient is positive, corresponds to reabsorption of emission in the system; the opposite case

$$\sum_{(m)\neq(n)} A_m^n N_m \left(\frac{N_n}{N_m} - 1\right) < 0, \qquad (27.29)$$

with which $\mu_j < 0$, corresponds to amplification of the emission in the system.

The physical meaning of the amplification is as follows. As is clear from (27.23), the coefficient μ_j is proportional to the difference between the intensities of true absorption and stimulated emission. Therefore in the case $\mu_j < 0$ stimulated emission processes predominate over true absorption so the particle system does not on the average take energy away from the emission but, on the contrary, adds it. This is due to the special nature of the distribution of the emitting particles with (27.29), when the major contribution to the sum over all the transitions corresponding to emission of a frequency ω is made by particles for which the population N_m of the state before the emission of a quantum is greater than the corresponding value for N_n in the state into which the particle makes a transition with the emission of a quantum $\hbar\omega$. Therefore with the condition (27.29) emission passing through the system or being emitted by the system will be amplified (a rise in the intensity I_j in space); if at the same time we remember that the absorption coefficient μ_j is connected with the attenuation coefficient in time γ_j by the relation (26.70) it becomes clear that in the case (27.29) the electromagnetic perturbations of a fluctuating nature existing in the system become convectively unstable (a rise in the energy of these perturbations $S(t)$ in time). In other words, the inequality (27.29) serves as a criterion of amplification and convective instability and the inequality (27.28) as a criterion of absorption and convective stability of the system of emitting particles.

The following can be said of the limits of applicability of the Einstein coefficients method (Zheleznyakov, 1959a and 1959c; Ginzburg and Zheleznyakov, 1965). It was pointed out above that it is valid only when studying amplification, absorption and the instability and stability of the convective type closely connected with these processes. At the same time it is clear that calculation of the coefficients μ_j, γ_j by this method (just as finding the emissive power a_j) is possible only if the original state of the system is non-coherent: this allows us to sum the number of emitted (absorbed) quanta when finding the emitted (absorbed) energy in the system of particles under discussion (see Fain, 1963).

From the classical point of view an increase or decrease in the amplitude of a wave in a plasma can be caused only by a change in the nature of the electron motion under the action of the field of this wave, so the original non-coherent nature of the electron emission process is violated. Since in the classical theory the relation between true absorption and stimulated emission is determined by the phase relations between the electrons and the passing wave, it can be concluded from this that the Einstein coefficients method (in the framework of its applicability) is similar to the classical treatment of the problem of instability in the linear approximation: it allows us to find the criteria for the amplification and instability of a system whose original state is non-coherent but, generally speaking, becomes unsuitable for treating processes of the rise and establishment of coherent emission in an unstable system. At the same time the quantum approach (just like the solution of the linearized problem) makes it possible to determine the steepness of the rise in the coherent emission—γ_j at the initial point in time, i.e. the derivative of the wave intensity at the time when the state of the system is still non-coherent. The same is, of course, valid also for the amplification coefficient μ_j at the original point in space which has the meaning of the derivative of the wave intensity along the ray.

Finally, the Einstein coefficients method can in practice be used only for finding comparatively weak amplification and absorption when $|\operatorname{Im} k| \ll \operatorname{Re} k = 2\pi/\lambda$, since it is assumed in this method that emission and absorption of particles in the first approximation proceeds in the same way as in a medium with $\operatorname{Im} k = 0$ characterized by a known refractive index n_j. The absorption coefficient μ_j (27.26), i.e. the quantity $\operatorname{Im} k$ (27.21), is then obtained as a small correction to the real part of k. It must be said that even in the kinetic equation method because of calculation difficulties $\operatorname{Im} k$ is also found in the majority of cases as a perturbation to $\operatorname{Re} k$.

The Einstein coefficients method was adduced by Ginzburg and Zheleznyakov (1958b), Zheleznyakov (1959c) and also by Twiss (1958)[†] to analyse questions of emission, absorption and, what is particularly important, to study the amplification and instability of waves in a non-equilibrium plasma. Later we shall see that quantum representations prove to be

---

[†] The same method has been used by Trubnikov (1958a and 1958b) for calculating the emission from a non-equilibrium system of charged particles in a magnetic field; this method has been used (Andronov, Zheleznyakov and Petelin, 1964) to study wave instability in a magnetoactive plasma with a complex electron velocity distribution (see the end of this chapter). A discussion of the part played by the Einstein coefficients method for analysing the problem of amplification and instability has also been given by Ginzburg, Zheleznyakov and Eidman (1962) and Ginzburg (1959).

§ 27] Emission, Absorption and Amplification

very fruitful for these purposes even in the cases when the problem is in essence classical and Planck's constant is not in the final formulae for μ_j and a_j.[†] The quantum treatment of the conditions of amplification and instability has undoubted advantages over the classical method of the kinetic equation in the sense of simplicity and clarity, whilst the latter far from always makes it possible to find without serious calculation difficulties the criteria for amplification and instability in an explicit form. This justifies the extensive application of the Einstein coefficients method in this section, although we shall not, of course, ignore the classical treatment where it is necessary.

REABSORPTION AND AMPLIFICATION OF PLASMA WAVES IN A NON-EQUILIBRIUM PLASMA WITH $H_0 = 0$ (QUANTUM TREATMENT)

We shall now apply the general expressions obtained for the coefficient of absorption and emissive power to studying reabsorption (self-absorption) of waves by the emitting particles and the emission intensity for a number of concrete variants of a non-equilibrium velocity distribution of the plasma particles. The choice of these variants is governed by the possibilities of their occurring in the conditions of the Sun and the planets.

Starting with the isotropic case ($H_0 = 0$) let us examine the emission of plasma waves by a stream of charged particles moving in a plasma (Ginzburg and Zheleznyakov, 1958b). Let a wave of frequency ω and with a wave vector $\mathbf{k}(k = (\omega/c)n_3)$ be emitted in a direction subtending an angle Θ with the vector of the mean ordered velocity $\mathbf{V}_s$. By virtue of the Cherenkov condition (26.23) this fixes the projection $V_k = c/n_3$ of the emitting particle's velocity V into the direction $\mathbf{k}$. Therefore all of the particles of the stream with a given projection V_k and with any values of the component $V_\perp$ at right angles to $\mathbf{k}$ make their contribution to the emission under discussion. It is clear from this that in determining μ_j a significant part is played by the form of the distribution function of the plasma particles with respect to the projections V_k, i.e. $f_s(V_k) = \int f_s(V) dV_\perp$, where $f_s(V)$ characterizes the distribution with respect to the velocities V.

Turning to the expression (27.26) for the absorption coefficient we see that in our case the ratio N_n/N_m is equal to $f_s(V_k^{(n)})/f_s(V_k^{(m)})$, where $V_k^{(m)}$ and $V_k^{(n)}$ are the values of V_k before and after the emission by a charged particle of a quantum of emission (plasmon). The difference of the momen-

---

[†] In the expression for a_j this is connected with the fact that in the region $\hbar\omega \ll \mathscr{E}$ ($\mathscr{E}$ is the energy of an individual particle) the motion of the particles is quasi-classical; in accordance with the principle of correspondence, this makes it possible to use the classical formulae for the particles' emission intensity when calculating the emissive power of the plasma (27.27).

ta of the particles in these states $mV_k^{(m)} - mV_k^{(n)}$ is equal to $\hbar k = (\hbar\omega/c)n_3$—the momentum of a plasmon; therefore in the case when the function $f(V_k)$ varies but little in the range $(\hbar\omega/mc)n_3$

$$\frac{N_n}{N_m} = \frac{f_s\left(V_k^{(m)} - \frac{\hbar\omega}{mc}n_3\right)}{f_s(V_k^{(m)})} \approx 1 - \frac{\hbar\omega n_3}{mc}\frac{\left(\frac{df_s}{dV_k}\right)_{V_k = V_k^{(m)}}}{f(V_k^{(m)})}, \quad (27.30)$$

where we can put $V_k^{(m)}$ equal to the phase velocity of the plasma waves $V_{\text{ph}} = c/n_3$. Substituting (27.30) in (27.26) and remembering (27.27) we find that

$$\mu_3 \approx \frac{8\pi^3 V_{\text{ph}} a_3}{\omega^2 m}\frac{df_s}{dV_k}\bigg|_{V_k = V_{\text{ph}}}. \quad (27.31)$$

The emissive power of the stream a_3 can easily be found from (26.87) by using the expression (26.24) for the intensity of plasma wave Cherenkov emission by an individual particle:

$$a_3 = \frac{1}{2\pi}\frac{e^2\omega}{\varepsilon}f_s(V_k = V_{\text{ph}}). \quad (27.32)$$

From this we finally obtain the absorption coefficient of the plasma waves in the form†

$$\mu_3 = -\frac{4\pi^2 e^2 V_{\text{ph}}^3}{3\omega m V_{\text{th}}^2}\frac{\partial f_s}{\partial V_k}\bigg|_{V_k = V_{\text{ph}}}. \quad (27.33)$$

In the expressions given the values of the square of the refractive index $n_3^2 = \varepsilon/3\beta_{\text{th}}^2$ and the thermal velocity $V_{\text{th}} = \beta_{\text{th}}c$ relate to a motionless equilibrium plasma (without allowing for a system of particles creating emission and absorption at the frequency ω).

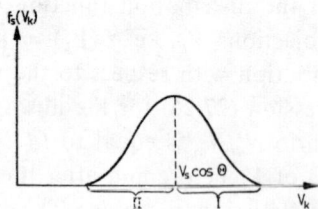

FIG. 146. Electron distribution in the stream with respect to the projections of the velocity V onto the direction k

† This formula (to be more precise the attenuation coefficient $\gamma_3 = \mu_3|d\omega/dk_3|$) is obtained kinetically by Bohm and Gross (1949b) and by Gertsenshtein (1952).

§ 27] Emission, Absorption and Amplification

Assuming for the sake of definition that in the stream of particles the distribution function is of the form (Fig. 146)

$$f_s(V) = N_s \left(\frac{m}{2\pi\varkappa T_s}\right)^{3/2} \exp\left[-\frac{m(V-V_s)^2}{2\varkappa T_s}\right], \qquad (27.34\text{a})$$

$$f_s(V_k) = N_s \left(\frac{m}{2\pi\varkappa T_s}\right)^{1/2} \exp\left[-\frac{m(V_k - V_s \cos \Theta)^2}{2\varkappa T_s}\right] \qquad (27.34\text{b})$$

(T_s, N_s and m are the temperature, concentration and mass of the particles in the stream), we arrive at the following expressions for a_3 and μ_3 (Andronov, 1961):

$$a_3 = \frac{\omega_{L_s}^2 \omega V_{\text{ph}}^2 m}{6(2\pi)^{5/2} V_{\text{th}}^2 V_{\text{th}_s}} e^{-(V_{\text{ph}} - V_s \cos \Theta)^2 / 2 V_{\text{th}_s}^2}, \qquad (27.35)$$

$$\mu_3 = \sqrt{\frac{\pi}{18}} \frac{\omega_{L_s}^2 V_{\text{ph}}^3}{\omega V_{\text{th}}^2 V_{\text{th}_s}^3} (V_{\text{ph}} - V_s \cos \Theta) e^{-(V_{\text{ph}} - V_s \cos \Theta)^2 / 2 V_{\text{th}_s}}, \qquad (27.36)$$

where $\omega_{L_s}^2 = (4\pi e^2 N_s)/m$, $V_{\text{th}_s}^2 = (\varkappa T_s)/m$.

According to (27.36) for plasma waves with a velocity from the range $V_{\text{ph}} > V_s \cos \Theta$, in which $df_s/dV_k < 0$ (region I in Fig. 146), absorption occurs ($\mu_3 > 0$). On the other hand, in the range $V_{\text{ph}} < V_s \cos \Theta$ (region II in the same figure) plasma waves will be amplified ($\mu_3 < 0$), which can be explained by the predominance of stimulated emission processes over true absorption. This is quite natural since here because of the positive sign of the derivative df_s/dV_k the number of electrons in a state before emission of a quantum $\hbar\omega$ is greater than the number of electrons in the state into which a particle passes after an act of emission.

The intensity of plasma waves emitted in region I from an optically thick layer is, according to (27.2), (27.35) and (27.36),

$$I_j = \frac{n_3^2 \omega^2 \varkappa T_s}{8\pi^3 c^2} \frac{V_{\text{ph}}}{V_{\text{ph}} - V_s \cos \Theta}. \qquad (27.37)$$

For the majority of the particles in region I, $V_{\text{ph}} - V_s \cos \Theta = V_k - V_s \cos \Theta \sim \sqrt{\varkappa T_s/m}$ and therefore in a thick layer (Ginzburg and Zheleznyakov, 1958b)

$$I_j \sim \frac{n_3^2 \omega^2 \varkappa}{8\pi^3 c^2} \sqrt{T_s T_0}, \qquad (27.38)$$

where $T_0 = m V_{\text{ph}}^2/\varkappa \sim m V_s^2/\varkappa$ with $V_{\text{ph}} \sim V_s$. The effective temperature of this kind of emission is $T_{\text{eff}} \sim \sqrt{T_s T_0}$. The expression (27.38) in actual fact defines only the lower limit of the intensity of non-coherent† plasma waves emitted by a stream of charged particles, since as V_{ph} approaches $V_s \cos \Theta$ the absorption coefficient decreases, whilst the intensity (27.37) increases correspondingly until the optical thickness $\tau_3 = \mu_3 L$ reaches values of the order of unity. At the same time in region *II* (i.e. at frequencies for which $V_{\text{ph}} - V_s \cos \Theta < 0$) the system becomes amplificatory, the rise in the intensity of the coherent emission being limited by non-linear effects (see below).‡

We notice that if we take into consideration the Cherenkov and bremsstrahlung absorption in a motionless equilibrium plasma (through which the stream of particles we have discussed is passing), then the resultant absorption coefficient will be equal to the sum of the expressions (26.75), (26.89) and (27.36). Moreover, with a large enough ratio V_s^2/V_{th}^2 the second term can be completely neglected by taking into consideration in the basic plasma only absorption because of collisions (see section 26 in this connection). This absorption obviously limits the range of frequencies in which amplification occurs; with a low enough concentration of electrons in the stream or a large number of collisions in a motionless plasma the plasma-stream system in general stops to amplify, i.e. the coefficient $\mu_3(\omega)$ will be positive with any values of ω.

Here high intensity of non-coherent emission obviously obtains if the system is on the brink of instability (amplification). In practice, however, this kind of state is possible only in a very narrow range of stream parameters (the values of T_s, N_s and V_s); therefore in the majority of cases in cosmic conditions two possibilities occur: either (1) the system of particles is far enough from a state of instability (amplification), or (2) it is unstable. In the second case as well as non-coherent emission a coherent component appears and the question arises of the relative content of the coherent and non-coherent components in the emission leaving its generation region. This question was put by Ginzburg and Zheleznyakov (1958b) and discussed in detail by the same authors (1961); the results obtained come down to the following.

Let us examine a layer of a non-equilibrium plasma of thickness L, the absorption coefficient in which $\mu_j(\omega)$ is characterized by the graph in Fig. 147. This kind of function $\mu_j(\omega)$ is typical not only when Cherenkov emission is present in a stream of charged particles penetrating a motionless

† The emission of a system at frequencies ω for which $\mu_j(\omega) > 0$ will be called non-coherent, unlike the coherent emission in the band where amplification occurs ($\mu_j(\omega) < 0$). For the meaning of this terminology see the beginning of Chapter VIII.

‡ Laboratory experiments on the generation of plasma waves by streams of charged particles can be studied, for example in the papers collected in *Ultra-High Frequency Oscillations in a Plasma*, 1961.

§ 27] Emission, Absorption and Amplification

plasma, but also in the case of magneto-bremsstrahlung emission in a non-equilibrium plasma (see below). The following discussion is therefore quite general in nature.†

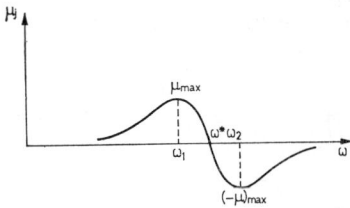

FIG. 147. Function $\mu_j(\omega)$ (schematic)

According to equation (27.1) the coherent emission flux from the layer will be

$$S(\mu_j < 0) = \int_{\omega^*}^{\infty} \left\{ \frac{a_j}{\mu_j}(1 - e^{-\mu_j L}) + I_{j0} e^{-\mu_j L} \right\} d\omega, \qquad (27.39)$$

and the non-coherent emission flux will be

$$S(\mu_j > 0) = \int_{0}^{\omega^*} \left\{ \frac{a_j}{\mu_j}(1 - e^{-\mu_j L}) + I_{j0} e^{-\mu_j L} \right\} d\omega. \qquad (27.40)$$

We shall show that the condition

$$(-\mu_j)_{\max} L \gg 1 \qquad (27.41)$$

is in practice sufficient for us to be able to neglect the quantity $S(\mu_j > 0)$ when compared with $S(\mu_j < 0)$.

In fact with the condition (27.41) an effective contribution to the integral (27.39) is made only by the region of frequencies $\Delta\omega$ near the point ω_2 at which $-\mu_j L$ differs from $(-\mu_j)_{\max} L$ by not more than unity: outside the range $\Delta\omega$ the value of the intensity $I_j(\omega)$ decreases rapidly (exponentially) as ω moves away from ω_2. Therefore in order of magnitude

$$S(\mu_j < 0) \sim \left\{ \frac{a_j}{(-\mu_j)_{\max}} + I_{j0} \right\} e^{(-\mu_j)_{\max} L} \Delta\omega. \qquad (27.42)$$

Turning to the expression (27.40) for $S(\mu_j > 0)$, we see that in the range (ω_1, ω^*) the optical thickness $\mu_j(\omega)L$ varies monotonically as ω changes, so

$$\int_{\omega_1}^{\omega^*} I_j \, d\omega = - \int_{0}^{\mu_j \max L} I_j \frac{d\omega}{d(\mu_j L)} d(\mu_j L).$$

† The function $\mu_j(\omega)$ may of course be also more complex (may have not one, but several maxima). This circumstance, however, has no effect at all on the results obtained below.

Then using the theorem of the mean we obtain:

$$\int_{\omega_1}^{\omega^*} I_j\, d\omega = -\tilde{a}_j L \overline{\frac{d\omega}{d(\mu_j L)}} \int_0^{\mu_{j\,\text{max}} L} \frac{1-e^{-\mu_j L}}{\mu_j L}\, d(\mu_j L)$$

$$-\tilde{I}_{j0} \overline{\frac{d\omega}{d(\mu_j L)}} \int_0^{\mu_{j\,\text{max}} L} e^{-\mu_j L}\, d(\mu_j L) = -\tilde{a}_j L \overline{\frac{d\omega}{d(\mu_j L)}} A(\mu_{j\,\text{max}} L)$$

$$-\tilde{I}_{j0} \overline{\frac{d\omega}{d(\mu_j L)}} (1 - e^{-\mu_{j\,\text{max}} L}), \qquad (27.43)$$

where

$$A(\mu_{j\,\text{max}} L) = -Ei(-\mu_{j\,\text{max}} L) + C + \ln(\mu_{j\,\text{max}} L) \qquad (27.44)$$

and the sign $\frown$ indicates that this quantity corresponds to a certain frequency from the range (ω_1, ω^*). In (27.44) $Ei(-\mu_{j\,\text{max}} L)$ is an integral exponential function of the argument $-\mu_{j\,\text{max}} L$, $C \approx 0.577$ is Euler's constant. In extreme cases

$$\left.\begin{array}{l} A(\mu_{j\,\text{max}} L) \approx \mu_{j\,\text{max}} L \quad (\mu_{j\,\text{max}} L \ll 1), \\ A(\mu_{j\,\text{max}} L) \approx C + \ln(\mu_{j\,\text{max}} L) \quad (\mu_{j\,\text{max}} L \gg 1). \end{array}\right\} \qquad (27.45)$$

Since a long way from ω^* a decrease in μ_j is accompanied, unlike the vicinity of ω^*, by a decrease in the emissive power a_j, it is clear that the contribution from the range $\omega < \omega_1$ to $S(\mu_j > 0)$ in order of magnitude will not exceed the contribution of the region (ω_1, ω^*). Therefore

$$S(\mu_j > 0) \sim a_j \frac{\Delta'\omega}{\mu_{j\,\text{max}}} A(\mu_{j\,\text{max}} L) + I_{j0} \frac{\Delta'\omega}{\mu_{j\,\text{max}} L} (1 - e^{-\mu_{j\,\text{max}} L}). \qquad (27.46)$$

Here we have allowed for the fact that $\tilde{d\omega}/d(\mu_j L) \sim -\Delta'\omega/\mu_{j\,\text{max}} L$, where $\Delta'\omega = \omega^* - \omega_1$, and $\tilde{a}_j$, $\tilde{I}_{j0}$ is of the order of a_j, I_{j0} figuring in the expression for $S(\mu_j < 0)$ (27.42).

By comparing (27.42) with (27.46) it is easy to see that the coherent emission will significantly exceed the non-coherent if at the same time we can satisfy the inequalities†

$$\frac{a_j \Delta\omega}{(-\mu_j)_{\text{max}}} e^{(-\mu_j)_{\text{max}} L} \gg \frac{a_j \Delta'\omega}{\mu_{j\,\text{max}}} [C + \ln(\mu_{j\,\text{max}} L)],$$

$$I_{j0} \Delta\omega e^{(-\mu_j)_{\text{max}}} \gg I_{j0} \frac{\Delta'\omega}{\mu_{j\,\text{max}} L}.$$

† When obtaining these inequalities it was taken into (27.46) that $\mu_{j\text{max}} L \gg 1$, which can only increase the non-coherent emission flux.

§ 27] Emission, Absorption and Amplification

They will definitely be satisfied if

$$\left.\begin{aligned}(-\mu_j)_{\max}L &\gg \ln\left\{\frac{(-\mu_j)_{\max}\Delta'\omega}{\mu_{j\max}\Delta\omega}[C+\ln(\mu_{j\,\max}L)],\right\}, \\ (-\mu_j)_{\max}L &\gg \ln\left\{\frac{\Delta'\omega}{\Delta\omega\mu_{j\max}L}\right\}.\end{aligned}\right\} \quad (27.47)$$

Since there are logarithms in the right-hand sides it is clear that in practice over a very wide range of values of the parameters $\mu_{j\,\max}L$, $\Delta'\omega/\Delta\omega$ and $(-\mu_j)_{\max}/\mu_{j\,\max}$ the condition (27.41) is a sufficient criterion for clear-cut predominance of coherent emission over non-coherent in a non-equilibrium plasma; according to this the amplification of waves at the dimensions of the system should be much more than unity.

The result obtained could, of course, have been foreseen right from the beginning since when there is amplification a major part in (27.39) is played by the factor $e^{|\mu_j|L}$, which is exponentially great with the condition (27.41). However by virtue of the dependence of the ratio $S(\mu_j < 0)/S(\mu_j > 0)$ on other factors it seemed to us good to make more detailed estimates.†

It should also be pointed out that in conditions when the coronal plasma is convectively unstable the condition $(-\mu_j)_{\max}L \gg 1$ is properly satisfied. As an example let us take the case of an isotropic motionless plasma with $V_{\text{th}} \sim 4\times 10^8$ cm/sec ($T \sim 10^6$ °K) penetrated by a corpuscular stream at a velocity $V_s \gg V_{\text{th}}$. It is this case which obtains when type III bursts are generated in the corona, when $V_s \gtrsim 5\times 10^9$ cm/sec (see sections 14 and 31). Here, as we have already said, the resultant coefficient of absorption of plasma waves with $V_{\text{ph}} \gg V_{\text{th}}$ is determined by the sum of (26.75) and (27.36), so that

$$(-\mu_j)_{\max} L \sim -\frac{V_{\text{ph}}}{V_{\text{th}}^2}\nu_{\text{eff}} L + \sqrt{\frac{\pi}{18}}\, e^{-1/2}\, \frac{\omega_{L_s}^2 V_{\text{ph}}^3}{\omega V_{\text{th}}^2 V_{\text{th}_s}^2}L. \quad (27.48)$$

When writing this formula allowance is made for the fact that at the frequency corresponding to $(-\mu_j)_{\max}$ the difference is $V_{\text{ph}} - V_s \cos\Theta \sim -V_{\text{th}_s}$. Putting $V_{\text{ph}} \sim V_s \sim 5\times 10^9$ cm/sec, $V_{\text{th}} \sim V_{\text{th}_s} \sim 4\times 10^8$ cm/sec, $\nu_{\text{eff}} \sim 10$ sec^{-1}, $L \sim 10^{10}$ cm, $\omega \sim 2\pi\times 10^8$ sec^{-1}, we obtain:

$$(-\mu_j)_{\max} L \sim -3\cdot 10^3 + 6\cdot 10^4 N_s. \quad (27.49)$$

† Here no allowance was made for the effect of non-linearity on the intensity of emission leaving a non-equilibrium plasma. This non-linearity, which is more significant with large intensities, reduces the ratio $S(\mu_j < 0)/S(\mu_j > 0)$ when compared with that given by the formulae (27.42), (27.46). However, it cannot change the criterion $(-\mu_j)_{\max} L \gg 1$, with which the non-coherent component of the emission definitely becomes less than its coherent component.

This quantity will definitely be much greater than unity if the electron concentration in the stream is $N_s \gg 5 \times 10^{-2}$ electron/cm^3, which in all probability is always satisfied in the generation region of type III bursts (see section 31). We note at the same time that the inequality $(-\mu_j)_{\max} L > 0$, i.e. $N_s > 5 \times 10^{-2}$ electron/cm^3, is the condition here where amplification in the plasma-stream system appears.

Amplification and Instability of Plasma Waves in a Non-equilibrium Plasma with $H_0 = 0$ (Classical Treatment)

Up to now we have been discussing the question of the conditions and magnitude of the amplification and convective instability of a non-equilibrium isotropic plasma by resorting only to the quantum method of Einstein's coefficients. Let us now see to what results the classical treatment of this problem leads.

From the classical point of view the instability of a plasma relative to perturbations of the plasma wave type emitted by the electron stream is connected with the presence in these waves of a longitudinal electrical field E. When a weak density grouping of electrons appears in the stream with a spatial period the same as the length of the plasma wave the initial changes in density under certain conditions will rise because of the action of this field. This kind of "bunching" causes gradual amplification of the coherent component in the system of Cherenkov plasma waves emitted by the stream electrons. In other words, this process leads to instability—increase of the plasma waves in the plasma-stream system.

Let us see how the behaviour changes of a plasma wave of length λ and frequency $\omega = kV$ (see (26.23)) emitted by an individual electron when a quasi-neutral stream passes through an isotropic plasma at a velocity V_s. If we ignore the dispersion of particle velocities because of thermal motion, then to describe the process of wave propagation in this system we can use the quasi-hydrodynamic method, considering that both the motionless plasma and the stream obey equations of the (22.7), (22.8) type, in which the electromagnetic fields are connected with each other by the Maxwell equations (22.6).

Limiting ourselves to investigation longitudinal waves ($E \| k$, $H = 0$) we leave only the third equation in the system (22.6). Then, for simplicity we assume that the temperature in the stream is $T_s = 0$ and make all the variables in the plasma wave proportional to $e^{i\omega t - ikR}$, arriving at a dispersion relation of the form

$$\frac{\omega_L^2}{\omega^2 - k^2 V_{\text{th}}^2} + \frac{\omega_{L_s}^2}{(\omega - kV_s)^2} = 1, \qquad (27.50)$$

§ 27] Emission, Absorption and Amplification

in which ω_L and ω_{L_s} are the natural frequencies of the oscillations of the plasma and the stream of particles. In order to find the explicit expression for the frequency $\omega(k)$ with real values of k we put $\omega = \omega_0 + \delta$, where $\omega_0 = kV_s$ is the frequency of the Cherenkov plasma wave without allowing for the stream ($\omega_{L_s}^2 \to 0$); the values of ω_0 and k are connected with each other by the dispersion relation $\omega_0^2 = \omega_L^2 + V_{th}^2 k^2$ (22.16). Then (27.50) is reduced to the equation

$$[(\omega_0+\delta)^2 - V_{th}^2 k^2]\delta^2 = \omega_L^2 \delta^2 + \omega_{L_s}^2[(\omega_0+\delta)^2 - V_{th}^2 k^2].$$

Taking $\omega_{L_s}^2$ (i.e. N_s) to be a small quantity we can find from it the frequency ω close to ω_0 (i.e. $\delta \ll \omega_0$). Bearing in mind that in this case when $\omega_{L_s}^2$ approaches zero the ratio $\omega_{L_s}^2/\delta^2 \equiv \omega_{L_s}^2/(\omega - kV_s)$ should also approach zero (see (27.50)) we obtain $\delta^3 = (\omega_L^2 \omega_{L_s}^2)/2kV_s$, and therefore the frequency of a plasma wave with the stream taken into consideration is[†]

$$\omega \approx kV_s\cos\Theta + \sqrt[3]{1}\left(\frac{\omega_L \omega_{L_s}^2}{2}\right)^{1/3}\left(1 - \frac{V_{th}^2}{V_s^2\cos^2\Theta}\right)^{1/6}. \quad (27.51)$$

In this expression $\sqrt[3]{1} = 1$ and $(-1\pm i\sqrt{3})/2$, so of the three solutions for plasma waves with a frequency $\omega \approx kV_s$ one corresponds to a wave increasing in time. The extent of the rise decreases as the angle Θ between k and V_s decreases. The corresponding expression for the wave number k obtained with real values of the frequency ω will be of the form

$$k \approx \frac{\omega}{V_s\cos\Theta} + \sqrt[3]{-1}\left(\frac{\omega_{L_s}^2 \omega_L^2}{\omega V_{th}^2 V_s \cos\Theta}\right)^{1/3}. \quad (27.52)$$

When trying to find the conditions of wave amplification and instability, generally speaking, we must not lose sight of the very important fact that the dispersion of the velocities of the electrons in the plasma and the stream due to thermal motion is not zero. To allow for this fact we must obviously introduce the kinetic equation method (see the first subsection of this section). It can then be shown (Landau, 1946; Gordeyev, 1954) that for plasma waves being propagated in a plasma consisting of several kinds of particles (for example electrons and protons) the connection between ω and k is defined by Landau's dispersion equation:[‡]

$$1 + \frac{4\pi e^2}{k}\sum_l \frac{1}{m_l}\int_C \frac{\partial f_{0l}}{\partial V_k}\frac{dV_k}{\omega - kV_k - i\nu_l} = 0. \quad (27.53)$$

[†] This formula is obtained by Ginzburg and Zheleznyakov (1958b); in the case of $\Theta = 0$ it is the same as the corresponding expression found earlier by Akhiyezer and Fainberg (1951).

[‡] It is assumed here, as when obtaining (27.50), that the plasma is infinite and uniform. We notice that the dispersion equation in the form (27.53) is valid even when we allow for the motion of ions in the wave's field.

The integration contour C for Im $k = 0$ passes along the real axis V_k from $-\infty$ to ∞ going round the pole $V_k = (\omega - i\nu_l)/k$ from above; in (27.53) f_{0l} is the distribution function standardized for a concentration of particles of a kind l with respect to the projections of the velocities V onto the direction $\mathbf{k}$; m_l and ν_l are the mass and effective number of collisions of particles of a kind l.

The dispersion equation (27.53) can be reduced to a form which is more convenient to use if we limit ourselves to the case when the unperturbed stationary velocity distribution function $f_{0l}(V_k)$ for each kind of particle is a Maxwell distribution in a system moving at a velocity V_{0l}. Then the relation $\omega(k)$ can be found from the equation (Gordeyev, 1954 and 1953; Fok, 1946b)

$$1 + \sum_l \frac{1 - A(\xi_l)}{k^2 D_l^2} = 0, \tag{27.54}$$

where

$$D_l = \left(\frac{\varkappa T_l}{4\pi e^2 N_l}\right)^{1/2}, \quad A(\xi_l) = \xi_l e^{-\frac{\xi_l^2}{2}} \int_{-i\infty}^{\xi_l} e^{\frac{z^2}{2}} dz, \tag{27.55}$$

$$\xi_l = \frac{\omega - kV_{0l}\cos\Theta - i\nu_l}{kV_{\text{th }l}}, \quad V_{\text{th }l}^2 = \frac{\varkappa T_l}{m_l}. \tag{27.56}$$

We note that for any values of ξ_l the inequality $0 < \mathrm{Re}\, A(\xi_l) \lesssim 1$ is satisfied, where $A(\xi_l)$ can be shown in the form

$$A(\xi_l) = B(\xi_l) + i\sqrt{\frac{\pi}{2}}\,\xi_l e^{-\frac{\xi_l^2}{2}}, \tag{27.57}$$

where $B(\xi_l)$ is a function which is real with real values of ξ_l and in the extreme cases is

$$B(\xi_l) \approx \xi_l^2 - \tfrac{1}{3}\xi_l^4 + \ldots \quad (|\xi_l^2| \ll 1), \tag{27.58}$$

$$B(\xi_l) \approx 1 + \frac{1}{\xi_l^2} + \frac{3}{\xi_l^4} + \ldots \quad (|\xi_l^2| \gg 1). \tag{27.59}$$

For simplicity we shall limit ourselves to the practically important case of one corpuscular stream passing through a motionless plasma; here, as usual, we neglect the motion of the ions. Then provided that the density of the electrons in the stream is $N_s \ll N$ (N is the concentration of the motionless plasma) and the parameter $\xi_l^2 = \omega^2/k^2 V_{\text{th}}^2$ in the motionless plasma with respect to the modulus is much greater than unity, we obtain (without

§ 27] Emission, Absorption and Amplification

allowing for collisions) the expression for the attenuation coefficient:

$$\gamma_3 \equiv 2\mathrm{Im}\,\omega = \sqrt{\frac{\pi}{2}} \frac{(V_\mathrm{ph} - V_s \cos\Theta)\omega_{Ls}^2 \omega}{k^2 V_{\mathrm{th}_s}^3} e^{-\frac{(V_\mathrm{ph}-V_s\cos\Theta)^2}{2V_{\mathrm{th}_s}^2}} \quad (27.60)$$

which is exactly the same (after recalculating for absorption in space $\mu_3 = \gamma_3(d\omega/dk)^{-1} = \gamma_3 V_\mathrm{ph}/3V_\mathrm{th}^2$) as the formula (27.36) found by the Einstein coefficients method.† In this case Re ω is close to the value of ω defined by the dispersion equation in a motionless plasma without a stream: $\omega^2 = \omega_L^2 + 3V_\mathrm{th}^2 k^2$.

We note further that for plasma waves with $|\xi_s^2| = |(\omega - kV_s \cos\Theta)^2 \times (kV_{\mathrm{th}_s})^{-2}| \gg 1$ the exponential term in (27.57) becomes small and the dispersion relation is actually reduced to an equation of the quasi-hydrodynamic type (27.50). Here the coefficients μ_j and γ_j are no longer obtained by the Einstein method: they will be proportional to $N_s^{1/3}$ (27.51), (27.52), whilst in the case (27.60) the coefficients $\mu_j, \gamma_j \propto N_s$.

This discrepancy is in the final analysis explained (Zheleznyakov, 1959a; Ginzburg and Zheleznyakov, 1965) by the fact that the applicability of the quasi-hydrodynamic formulae (27.51), (27.52) is limited below with respect to the stream concentration N_s, whilst the Einstein coefficients method treats absorption and growth in the stream as a small perturbation for waves being propagated in the absence of a stream of particles. It follows from (27.51) and the inequality $|\xi_s^2| \gg 1$ that plasma wave instability appears in the quasi-hydrodynamic approximation provided that‡

$$\left(\frac{\omega_{L_s}}{\sqrt{2}\,\omega_L}\right)^{4/3} = \left(\frac{N_s}{2N}\right)^{2/3} \gg \frac{V_{\mathrm{th}_s}^2}{V_s^2 \cos^2\Theta}. \quad (27.61)$$

For example when $V_{\mathrm{th}_s} \sim 5\times10^8$ cm/sec and $V_s \cos\Theta \sim 5\times10^9$ cm/sec (a case which is apparently typical in generation of type III bursts) N_s should be much greater than $2\times10^{-3}N \sim 2\times10^5$ electrons/cm³ (with $N \sim 10^8$ electrons/cm³).

In the theory of the Sun's sporadic radio emission as well as the case $V_s \gg V_\mathrm{th} \sim V_{\mathrm{th}_s}$ discussed above we are interested (possibly as applied to

---

† This once again indicates that the classical and quantum approaches to the problem of plasma instability and amplification are two different treatments of the same problem. Contrary to Twiss's statements (Twiss, 1958 and 1963) there are not two different types in principle of plasma wave instability and amplification in streams of charged particles: one obtained by the classical treatment using the kinetic equation, and another following from the quantum treatment by the Einstein coefficients method (for further detail see Ginzburg and Zheleznyakov, 1965; Ginzburg, Zheleznyakov and Eidman, 1962).

‡ Here we have already allowed for the fact that $kV_s \cos\Theta = \omega_0 \approx \omega_L$. We note that a detailed investigation of the dispersion equation with $|\zeta_s^2| \gg 1$, free of the limitation $N_s \ll N$, has been carried aut by Sumi (1958 and 1959).

the generation of type II bursts) in the case when the stream velocity V_s is less than the mean thermal velocity of the electrons in the corona V_{th}. Considering that the conditions $V_{\text{th}_s} \ll V_{\text{th}}$ and $V_{\text{th}_s} \ll V_s$ are satisfied as well as the inequality $V_s < V_{\text{th}}$, we can neglect the dispersion of the velocities in the stream by putting $V_{\text{th}_s} = 0$. In accordance with (27.54) here the longitudinal (plasma) waves are unstable if the phase velocity is close to the velocity of the stream ($\omega \approx kV_s$ with $\mathbf{k} \| \mathbf{V}_s$). If the concentration of the electrons N_s in the stream is small so that $\omega_{L_s} \ll kV_s$, then the growth coefficient is (Akhiyezer and Fainberg, 1951)

$$-\gamma_3 = -2\text{Im}\,\omega = \sqrt{\frac{\pi}{2}}\frac{V_s}{V_{\text{th}}}\omega_{L_s}\frac{(kD)^{-2}}{[1+(kD)^{-2}]^{3/2}} - \nu_{\text{eff}} \qquad (27.62)$$

(the values of V_{th}, D and ν_{eff} relate to a motionless plasma). The quantity $-\gamma_3$ reaches a maximum with $kD = 1/\sqrt{2}$ at the frequency

$$\omega_{\max} = \frac{\omega_L}{\sqrt{2}}\frac{V_s}{V_{\text{th}}}, \qquad (27.63)$$

where

$$(-\gamma_3)_{\max} = \frac{1}{3}\sqrt{\frac{2\pi}{3}}\omega_{L_s}\frac{V_s}{V_{\text{th}}} - \nu_{\text{eff}}. \qquad (27.64)$$

The effect of amplification and instability in the system under discussion obviously appears if the first term in (27.64) is greater than the second. With the usual velocity of sources of type II bursts $V_s \sim 10^8$ cm/sec and $V_{\text{th}} \sim 4 \times 10^8$ cm/sec, $\nu_{\text{eff}} \sim 10$ sec^{-1} this effect occurs for concentrations of $N_s \sim 6 \times 10^{-7}$ electron/cm^3.

MAXIMUM AMPLITUDE AND HARMONICS OF AMPLIFIED PLASMA WAVES

The problem of the generation of plasma waves in a non-equilibrium coronal plasma is not, of course, exhausted by the solution of the linear problem of amplification and instability. The harmonics in the radio emission of types II and III bursts (section 13 and 14) provide certain indications that the plasma waves amplified in the stream extend into the non-linear region. The effect of non-linearity ultimately leads to a limitation of the amplitude of the growing waves; it is natural, therefore, that if we are to judge the reasonableness of the plasma hypothesis of the origin of sporadic radio emission particular attention should be paid to a study of the establishment of plasma waves of maximum amplitude and a study of the nature of plasma waves that have settled down at the limit, in particular to finding their amplitude, their frequency and their harmonic content.

§ 27] Emission, Absorption and Amplification

The basic difficulty here is that the amplification conditions in a non-equilibrium plasma are generally satisfied not for a single wave with a phase velocity V_{ph} whose magnitude and direction are determined, but for a whole continuum of waves with different phase velocities (see, for example, the criteria given above for amplification when the dispersion of the velocities in the stream is taken into consideration). Therefore if the spontaneous emission of particles in a non-equilibrium plasma or the emission incident on it has a continuous frequency spectrum (which is the case), then a combination of waves with different V_{ph} will be amplified. In the linear approximation (with small amplitudes) this circumstance is completely insignificant since these waves are propagated independently of each other. However, upon further examination of the settling process, when the waves extend into the non-linear region we must allow for the non-linear interaction between the individual components in the system of growing plasma waves and between plasma waves and stream electrons as well; the steady state will also, generally speaking, be a combination of interacting plasma waves with different phase velocities.

If we assume, nevertheless, that the steady state solution is only one plasma wave of the form $E(R-V_{ph}t)$, then the finding of this solution is once more a far from simple problem. According to Sen (1955), Bohm and Gross (1949a), Bernstein, Greene and Kruscal (1957) it is reduced (without allowing for collisions) to a certain integro-differential equation for the electric potential $\varphi (E = -\text{grad } \varphi)$. The form of the function $\varphi(R-V_{ph}t)$ is determined in this case by the nature of the velocity distribution at the point where the wave potential becomes zero: $f(V)_{\varphi=0}$. The function $f(V)_{\varphi=0}$ is determined by the initial distribution of the electrons in the plasma before the start of the establishing process; it is clear from this that the importance of this equation is small, since finding the function $f(V)$ from its original value is a complex problem, including investigation of the settlement process.

At the same time the amplitude of a steady plasma wave in a system with a given unperturbed velocity distribution function can be estimated (allowing for collisions) from simple energy considerations (Bohm and Gross, 1949b). Let us assume that a motionless plasma is penetrated by an electron stream whose mean velocity is much greater than the mean thermal velocity of the electrons in the plasma and the stream, whilst the electron concentration N_s in the stream is small when compared with the concentration N in the basic plasma to such an extent that the inequality (27.61) is not satisfied. Under these conditions the linear amplification coefficient μ_3 is defined by the formula (27.33); dissipation of the plasma wave occurs chiefly by collisions in the motionless plasma, whilst the interaction of the electrons of the stream with the wave, whose phase velocity lies in the region where the derivative of the electron velocity distribution function $(df_s/dV_k)|_{V_k=V_{\text{ph}}} > 0$, leads to a transfer of energy from the stream of particles to the wave.

If an electron in the stream after colliding with other particles moves at a velocity V_k it will not be able to overcome the potential barrier created by the potential $\varphi = \text{grad } E$ in the steady plasma wave and will start to oscillate in a "potential hole", i.e. will be "captured" by the wave if the electron velocity is close enough to the phase velocity V_{ph}:

$$(V_k - V_{\text{ph}})^2 \leqslant \frac{2e\varphi_0}{m} \qquad (27.65)$$

(φ_0 is the amplitude of the potential in the plasma wave). Since the mean velocity of the captured electron is equal to V_{ph} the energy given up by the particle to the wave in the time between two successive collisions in the first approximation is $mV_{\text{ph}}(V_k - V_{\text{ph}})$. If $F(V_k)\,dV_k$ is the number of particles from a unit volume entering the range dV_k in unit time after the collisions, then the energy given up here to the wave will be

$$\left(\frac{d\mathcal{E}}{dt}\right)_1 \approx \int_{-(2e\varphi_0/m)^{1/2}}^{(2e\varphi_0/m)^{1/2}} mV_{\text{ph}}(V_k - V_{\text{ph}})\,F(V_k)\,d(V_k - V_{\text{ph}}). \qquad (27.66)$$

In the state of dynamic equilibrium that appears at the end of the settling process the number of particles entering a given range of velocities dV_k is equal to the number of particles leaving the latter as the result of collisions:

$$F(V_k)\,dV_k = v_s f_s(V_k)\,dV_k \qquad (27.67)$$

(v_s is the frequency of collisions for the stream of electrons).

If there is little change in the velocity distribution function $f_s(V_k)$ over the range of velocities (27.65) corresponding to "capture" of electrons by the wave (which occurs when $4e\varphi_0/m \ll \varkappa T_s/m$), then it can be expanded into a series in powers of $V_k - V_{\text{ph}}$. Limiting ourselves to terms of an order not greater than $V_k - V_{\text{ph}}$ and integrating (27.66), we obtain:

$$\left(\frac{d\mathcal{E}}{dt}\right)_1 \approx mV_{\text{ph}}\,v_s\,\frac{2}{3}\left(\frac{2e\varphi_0}{m}\right)^{3/2}\frac{df_s}{dV_k}\bigg|_{V_k=V_{\text{ph}}}. \qquad (27.68)$$

Unless the electrons transfer energy to the wave the latter is damped in a time $1/v_{\text{eff}}$, where v_{eff} is the number of collisions for electrons in a motionless plasma. It follows from this that the loss of energy in a plasma wave because of dissipation is

$$\left(\frac{d\mathcal{E}}{dt}\right)_2 \approx -v_{\text{eff}}\mathcal{E}, \qquad (27.69)$$

§ 27] Emission, Absorption and Amplification

where $\mathscr{E} \approx E_0^2/8\pi = k^2\varphi_0^2/8\pi$ is the energy density in this wave (see (22.20)). Since in the settled state the total change in the mean energy is

$$\left(\frac{d\mathscr{E}}{dt}\right)_1 + \left(\frac{d\mathscr{E}}{dt}\right)_2 = 0, \tag{27.70}$$

the amplitude of the electric field E_0^I in a plasma wave that has settled in a stream–plasma system with a fixed unperturbed distribution function $f_s(V_k)$ will be defined by the following expression:

$$E_0^I = k\varphi_0 \approx \frac{km}{2e}\left[\frac{64\pi}{3}\frac{e^2 V_{\text{ph}} v_s}{mk^2 v_{\text{eff}}}\frac{df_s}{dV_k}\bigg|_{V_k=V_{\text{ph}}}\right]^2. \tag{27.71}$$

The phase velocity here lies in the region V_k for which longitudinal wave amplification and stability are possible (i.e. in practice in the region $df_s/dV_k > 0$; see (27.33) and (27.49)). The maximum value of $(df_s/dV_k)_{V_k=V_{\text{ph}}}$ here will obviously be of the order of $N_s m/\varkappa T_s$; putting at the same time in (27.71) $V_{\text{ph}} \sim V_s$ and $k \sim \omega/V_s$ we obtain:

$$E_0^I \sim \frac{m\omega}{2eV_s}\left[\frac{16}{3}\frac{V_s^3\omega_{L_s}^2 v_s m}{\omega^2 v_{\text{eff}} \varkappa T_s}\right]^2. \tag{27.72}$$

Here V_s is the mean velocity of the stream, $\omega_{L_s} = 4\pi e^2 N_s/m$ is the characteristic frequency of plasma oscillations in the stream.

The relations (27.71) and (27.72) relate, of course, only to the fundamental (first) harmonic of the settled plasma wave. The non-harmonic nature of the motion of captured electrons even in a sine-wave electric field $E = E_0^I \sin(\omega t - kR)$ will lead to distortion of a longitudinal wave profile in space and in time. In other words, higher harmonics of the field $E(R, t)$ will appear propagated at the same velocity V_{ph}; their wave numbers are $k_s = s\omega/V_{\text{ph}}$, where s is the number of the harmonic and ω is the frequency of the fundamental harmonic.

When there is no stream of charged particles the non-linear nature of a plasma wave will be significant if the field E_0 is so great that the wave captures electrons from the motionless plasma. According to (27.65) this will be with the condition $V_{\text{ph}}^2 \lesssim 2eE_0/mk$, i.e.†

$$E_0 \gtrsim \frac{m\omega V_{\text{ph}}}{2e}. \tag{27.73}$$

It must be pointed out when there is a stream of particles the function $E(R - V_{\text{ph}}t)$ in a steady plasma wave becomes sharply anharmonic far

---

† It is not difficult to arrive at this relation by estimating the non-linear terms in the quasi-hydrodynamic equations (22.2) (Zheleznyakov, 1958a). Similar results are also obtained with a more detailed study of plasma wave propagation in a motionless equilibrium plasma (see Akhiyezer and Lyubarskii, 1951; Smerd, 1955, and also Amer, 1958).

earlier (with a high enough concentration N_s in the stream) since the particles of the stream with a velocity $V \sim V_s \sim V_{\text{ph}}$ are captured by the wave at lower values of the field E_0 than the particles of the motionless plasma. The amplitudes of the second harmonic in order of magnitude will then be

$$E_0^{\text{II}} \sim \frac{E_0^{\text{I}}}{4} \sqrt{\frac{eE_0^{\text{I}}}{2mk} \frac{\omega_{L_s}^2}{\omega^2}} V_{\text{ph}}^2 \frac{d^2 f}{dV_k^2}\bigg|_{V_k = V_{\text{ph}}}$$

$$\sim \frac{E_0^{\text{I}}}{4} \sqrt{\frac{eE_0^{\text{I}}}{2mk} \frac{\omega_{L_s}^2}{\omega^2}} V_{\text{ph}}^2 \left(\frac{m}{\varkappa T_s}\right)^{3/2}. \qquad (27.74)$$

We are not giving here the rather clumsy derivation of this formula which was obtained as a result of an approximate solution† of the kinetic equation together with the equation $\text{div } E = 4\pi\varrho$ provided the plasma wave is close to a sine wave (i.e. $E_0^{\text{II}} \ll E_0^{\text{I}}$). The formula (27.74) can also be used when $E_0^{\text{II}} \sim E_0^{\text{I}}$ for rough estimates. In this case the settled plasma waves will differ significantly from a sine wave; the presence of harmonics in a non-linear plasma wave will provide one of the possible ways of explaining the appearance of emission at the doubled frequency in types II and III bursts of solar emission (see section 31). The method of solution has been given briefly by Zheleznyakov (1959a).

The disadvantage of the estimate formulae (27.71), (27.72) and (27.74) is that they are obtained on the assumption that a single longitudinal wave with a given phase velocity V_{ph} is formed in the plasma.

To solve the question of the results of the establishment of amplified waves further investigation is absolutely necessary into this problem‡ which is of great importance for the theory of the Sun's sporadic radio emission and above all for the theory of types II and III bursts.

When using the formula (27.72), (27.74) for $E_0^{\text{I}}, E_0^{\text{II}}$ it is necessary to be certain that in the system under discussion the amplified plasma waves really do reach a steady state; the corresponding criterion can be found in the result of the correct calculation of the settling process. It may be thought, however, that the high amplification coefficient in the solar corona together with the great size of the corpuscular streams ensures that growing plasma waves change into a state with a steady amplitude in most of the source of emission. This allows us to use the formulae (27.72) and (27.74) for estimating the amplitude of plasma waves excited by streams of charged particles in the corona.

---

† This solution also allows us to find the expression for the amplitude of the first harmonic, which differs from (27.71) only in the numerical coefficient.

‡ In this connection see Vedenov (1963), Sturrock (1957) and Gross (1958).

§ 27] Emission, Absorption and Amplification

REABSORPTION AND AMPLIFICATION OF ELECTROMAGNETIC WAVES IN A NON-EQUILIBRIUM MAGNETOACTIVE PLASMA

Having in mind the application later of the Einstein coefficients method to study the effects of reabsorption and amplification of waves in a plasma placed in a magnetic field H_0, let us first examine the kinematics of the emission of a single electron, i.e. the conditions imposed upon the frequency, the direction of emission and the nature of the transition of the particle from one state to another by the laws of conservation of energy and momentum (Zheleznyakov, 1959c; see also Ginzburg, Zheleznyakov and Eidman, 1962).

When an electron moves in a constant magnetic field H_0 the emission of a quantum $\hbar\omega$ with a momentum $\hbar k = (\hbar\omega/c)n_j(k/k)$ is accompanied by a corresponding change in the electron state:

$$\mathcal{E}_m - \mathcal{E}_n \equiv \sqrt{m^2 c^4 + p_m^2 c^2} - \sqrt{m^2 c^4 + p_n^2 c^2} = \hbar\omega, \qquad (27.75)$$

$$p_{m\|} - p_{n\|} = \frac{\hbar\omega}{c} n_j \cos\alpha, \qquad (27.76)$$

where m is the particle mass at rest, $\mathcal{E}_m$, $\mathcal{E}_n$ and p_m, p_n are the energies and momenta of the electron before emitting a photon (state m) and afterwards (state n); the sign $\|$ indicates projection of the photon momentum, which subtends an angle α with the vector H_0, onto the direction of this vector. From the conservation laws (27.75), (27.76) we obtain (for further detail see Zheleznyakov, 1959c) the formula

$$\hbar\omega = \frac{\sqrt{1-\beta^2}\,(\varepsilon_m - \varepsilon_n)}{1 - n_j \beta_\| \cos\alpha} \qquad (27.77)$$

provided that $n_j \beta_\| \cos\alpha$ is not too close to unity:

$$\frac{n_j^2 \cos^2\alpha - 1}{(n_j \beta_\| \cos\alpha - 1)^2} \frac{2(1-\beta^2)}{mc^2} (\varepsilon_m - \varepsilon_n) \ll 1 \qquad (27.78)$$

and $\hbar\omega \ll mc^2$. Here ε_m, ε_n denote the quantities $p_{m\perp}^2/2m$, $p_{n\perp}^2/2m$, which have the meaning of the energy of transverse motion (for a non-relativistic electron); the sign $\perp$ indicates the component of the electron momentum at right angles to H_0; $\beta = V/c$ and $\beta_\| = V_\|/c$, where $V_\|$ is the component along H_0 of the total velocity V of the electron. Remembering that the square of the transverse momentum $p_\perp^2$ when the electron moves in a magnetic field takes up quantized values:

$$\varepsilon_m - \varepsilon_n \equiv \frac{p_{m\perp}^2}{2m} - \frac{p_{n\perp}^2}{2m} = \hbar\omega_H s, \qquad s \equiv m-n = \pm 1, \pm 2, \ldots, \qquad (27.79)$$

we obtain from (27.77) that

$$\hbar\omega = \frac{s\hbar\omega^*}{1 - n_j\beta_{||}\cos\alpha} \qquad (27.80)$$

($\omega^* = \omega_H\sqrt{1-\beta^2}$ is the frequency of rotation of the electron in the field H_0, $\omega_H = eH_0/mc$ is the gyro-frequency).

The formula (27.80) defines the magneto-bremsstrahlung frequency of an electron found previously (section 26) from the classical treatment. In addition, the conservation laws (27.75), (27.76) make it possible for an electron to emit a photon without changing its transverse momentum $p_\perp$, i.e. with $s = 0$, which corresponds to frequencies at which $n_j\beta_{||}\cos\alpha = 1$. This condition in its turn is the same as the classical equation for the Vavilov–Cherenkov effect in a magnetic field (26.30).

It is easy to see that in the case of the anomalous Doppler effect, i.e. if $n_j\beta_{||}\cos\alpha > 1$ (26.32b), the relation (27.80) is satisfied only with $s < 0$; therefore the difference $\varepsilon_m - \varepsilon_n$ can only be negative (since the frequency $\omega = \mathcal{E}/\hbar$, where $\mathcal{E}$ is the energy and is always positive). Therefore in this case the emission of a quantum is accompanied by the transition of the electron into a state with a large transverse momentum $p_\perp$; the energy necessary for this transition and the emission of a photon is drawn from the kinetic energy of the advancing motion.† For the normal Doppler effect, i.e. in the region $n_j\beta_{||}\cos\alpha < 1$ (26.32a), the parameter s is positive and the emission of a quantum $\hbar\omega$ is connected with a transition into a state with a lower value of $p_\perp$. The ranges of angles α with anomalous and normal Doppler effects are located respectively inside and outside the Cherenkov cone defined by the condition $n_j\beta_{||}\cos\alpha = 1$.

Let us now investigate the emission of a system of charged particles in a magnetoactive plasma using the Einstein coefficients method and the results obtained above for the emission of a single particle (Zheleznyakov, 1959c).

Let the momentum distribution of a system of electrons be axially symmetrical relative to the direction H_0:

$$dN_s = f_s(\boldsymbol{p})\,d\boldsymbol{p} = 2\pi f_s(p_\perp, p_{||})\,p_\perp\,dp_\perp\,dp_{||}. \qquad (27.81)$$

In the momentum space to one state (m) of a quasi-classical electron ($m \gg 1$) in a magnetic field belongs the volume

$$d\boldsymbol{p} = 2\pi p_\perp\,dp_\perp\,dp_{||} = 2\pi m\hbar\omega_H\,dp_{||}, \qquad (27.82)$$

since the range with respect to $\varepsilon = p_\perp^2/2m$ between two adjacent states is $\hbar\omega_H$ (see (27.79)). The number of electrons in a unit volume in a given

---

† Attention has been drawn to the unique nature of the emission in the region of the anomalous Doppler effect by Ginzburg and Frank (1947).

§ 27] Emission, Absorption and Amplification

quantum state (m) is
$$N_m = f_s(p_{m\perp}, p_{m\|}) \, 2\pi m \, \hbar\omega_H \, dp_\|, \tag{27.83}$$
and therefore the ratio of the populations in the formula (27.26) for the absorption coefficient μ_j is
$$\frac{N_n}{N_m} = \frac{f_s(p_{n\perp}, p_{n\|})}{f_s(p_{m\perp}, p_{m\|})} = 1 + \left(\frac{1}{f_s}\frac{\partial f_s}{\partial p_\perp}\right)_m \Delta p_\perp + \left(\frac{1}{f_s}\frac{\partial f_s}{\partial p_\|}\right)_m \Delta p_\|. \tag{27.84}$$
Here $\Delta p_\perp$ and $\Delta p_\|$ are the changes in $p_\perp$ and $p_\|$ during the transition of an electron from a state (m) into a state (n) with the emission of a quantum $\hbar\omega$; in accordance with equations (27.76) and (27.79), with magnetic bremsstrahlung and Cherenkov emission†
$$\Delta p_\perp = -\frac{sm\,\hbar\omega_H}{p_\perp}, \quad \Delta p_\| = -\frac{\hbar\omega n_j \cos\alpha}{c}. \tag{27.85}$$

If through a motionless plasma there passes a quasi-neutral non-relativistic stream of charged particles in which the velocity distribution of the electrons is of the form (27.34) with $V_s \| H$, then from (27.84) remembering (27.80), (27.85) we obtain:
$$\frac{N_n}{N_m} - 1 = \frac{\hbar\omega}{\varkappa T_s}(1 - n_j\beta_s \cos a), \tag{27.86}$$
where $\beta_s = V_s/c$; V_s corresponds to the velocity at the maximum of the distribution function $f_s(V)$. Since the ratio N_n/N_m is the same for all electrons emitting a frequency ω the absorption coefficient will be
$$\mu_j = \frac{8\pi^3 c^2}{\omega^2 n_j^2 \varkappa T_s} |\cos\vartheta|(1 - n_j\beta_s \cos\alpha)\,a_j \tag{27.87}$$
(see (27.26), (27.27)), and the intensity of the emission (magneto-bremsstrahlung or Cherenkov) of an optically thick plasma–stream system is
$$I_j = \frac{a_j}{\mu_j} = \frac{n_j^2 \omega^2 \varkappa T_s}{8\pi^3 c^2 (1 - n_j\beta_s \cos\alpha)\,|\cos\vartheta|}. \tag{27.88}$$
Using the concept of the effective temperature T_{eff} (26.15) for the emission characteristic we find that
$$T_{\text{eff}} = T_s(1 - n_j\beta_s \cos\alpha)^{-1}. \tag{27.89}$$
For an equilibrium velocity distribution ($\beta_s = 0$) we obtain a trivial result: $T_{\text{eff}} = T_s$. In the stream ($\beta_s \neq 0$) the quantity T_{eff} increases and with a velocity $\beta_s = (n_j \cos\alpha)^{-1}$ becomes infinite. As β_s increases further the effective temperature formally changes sign. The latter is because the coefficient μ_j here becomes negative.

† The first of these relations holds provided that $m - n \ll m, n$; it is only transitions of this kind that are being investigated here.

Therefore when a stream of charged particles is moving in a plasma the absorption coefficient becomes negative if

$$n_j \beta_s \cos \alpha > 1; \tag{27.90}$$

then the stream–plasma system with a magnetic field becomes amplifying and the electromagnetic waves being propagated in it become convectively unstable. In the case of $s = 0$ (Cherenkov amplification) the inequality (27.90) is actually satisfied at frequencies ω which are emitted by electrons in the stream with velocities of $V_\| < V_s$; this is clear from a comparison of (27.90) with the condition for the Vavilov–Cherenkov effect (26.30). For $s \neq 0$ (magneto-bremsstrahlung emission) this kind of limitation is not imposed on the velocity (see the Doppler formula (27.80)).

It is appropriate to stress that this amplification and instability criterion is not the same as the condition (26.32b) for the emission of particles in an anomalous Doppler effect region, although, generally speaking, it is close to it. This closeness can be explained by the fact that emission in a region of anomalous Doppler effect is accompanied by an increase in the transverse momentum $p_\perp$ (see p. 488); therefore for the distribution $f_s(p)$ used, in which the number of electrons in an element dp of momentum (or velocity) space decreases as $p_\perp$ rises, the number of electrons in the state before emission of a photon is greater than in the state into which the particle makes a transition after emission of a photon ($N_m > N_n$). This kind of population of the levels leads to amplification and instability. On the other hand, for electromagnetic perturbations in a region of normal Doppler effect the momentum distribution is stable, since here the emission is connected with a decrease in $p_\perp$ which leads to the opposite relation of the populations $N_m < N_n$.

The difference between the amplification and instability criterion (27.90) and the condition for the appearance of the anomalous Doppler effect is caused by the fact that when emitting a quantum $\hbar\omega$ an electron changes the value not only of $p_\perp$ but also of $p_\|$; owing to this the criterion for amplification $\mu_j < 0$ depends not only on the form of $f_s(p_\perp)$ but also on the nature of the change in $f_s(p_\|)$. In the case when the velocity dispersion in the stream is absent and all the electrons move at a velocity $\beta = \beta_s$, the inequality (27.90) is naturally the same as the condition for the appearance of anomalous Doppler effect.

The amplification and instability criterion (27.90) in a stream–plasma system with a magnetic field was obtained by Zheleznyakov (1959c) by the same method as above with an arbitrary angle α between $\boldsymbol{k}$ and $\boldsymbol{H}_0$, and also kinetically with $\alpha = 0$. Then the same problem (for any α) was studied by Kovner (1960a and 1960b) using the kinetic equation; the case

§ 27] Emission, Absorption and Amplification

when there is no dispersion of the velocities in the plasma has also been studied by Getmantsev and Rapoport (1960) and Rapoport (1960).

The amplification of electromagnetic waves in a magnetoactive plasma that we have just discussed occurs only with a special (although often found in solar conditions; see sections 13, 14 and 17) kind of non-equilibrium state when a stream of charged particles is moving in a motionless plasma at a velocity "greater than the velocity of light", i.e. at a velocity exceeding the phase velocity of these waves. At the same time there is no amplification and the plasma remains stable relative to perturbations whose velocity corresponds to the corpuscular stream moving at a velocity "up to the velocity of light". At the same time there is every reason to assume that the complex non-stationary processes in the Sun's atmosphere lead to the appearance of non-equilibrium states that differ from the very simple case of a stream in a plasma. For example it is quite possible that when a shock wave passes through in local regions of the corona the velocity dispersion of the electrons in the front of this kind of wave along the magnetic field and in the plane orthogonal to it will be different (for further details see the last subsection of this section).

The occurrence of electron velocity distributions characterized by temperature "anisotropy" in local regions of the coronal plasma is a very important circumstance because amplification of electromagnetic waves, and therefore also generation of coherent radio emission, is possible in these regions even in the cases when the mean ordered velocity of the electrons is zero. Amplification and instability under conditions of this kind can play an important part when interpreting certain components of the solar radio emission (type I bursts above all) and Jupiter's sporadic radio emission.

Among the first papers which discussed the instability of a magnetoactive plasma with an anisotropic electron temperature relative to high-frequency electromagnetic perturbations Sagdeyev and Shafranov (1960) is notable. This problem was then investigated in detail in a series of papers (Andronov, Zheleznyakov and Petelin, 1964; Zheleznyakov, 1960a, 1960b, 1961a and 1961b; Petelin, 1961) devoted to a more general distribution of the electrons with respect to the momenta p of the form

$$f_0(p)\,dp = A \exp\left[-\frac{(p_\parallel - p_\parallel^0)^2}{a_\parallel^2} - \frac{(p_\perp - p_\perp^0)^2}{a^2}\right] dp. \quad (27.91)$$

Here an element of the phase volume dp is equal to $p_\perp dp_\perp dp_\parallel d\psi$ ($p_\perp$ is the modulus of the component p orthogonal to the field H_0, $p_\parallel$ is the projection of p onto H_0, ψ is the angle in the cylindrical system of coordinates with the axis H_0); the factor A is determined from the standardization condition $\int f_0(p)\,dp = N$. The function (27.91) is sufficiently simple and at the same time reflects the most characteristic features of the actual electron distributions in cosmic conditions: the presence of dispersion ($a_\parallel \neq 0$, $a_\perp \neq 0$) and temperature "anisotropy" ($a_\parallel \neq a_\perp$), a finite mean particle velocity ($p_\parallel^0 \neq 0$), i.e. the presence of streams, etc.

According to Zheleznyakov (1960a) the propagation of ordinary and extraordinary waves in a uniform infinite plasma along a magnetic field H_0 is defined by the dispersion equation[†]

$$c^2k^2 - \omega^2 n_j^2(\omega) - \sum \pi \int_C \frac{\Omega_L^2}{N} \frac{(\omega m - kp_\parallel)\frac{\partial f_0}{\partial p_\perp} + kp_\perp \frac{\partial f_0}{\partial p_\parallel}}{\omega m - kp_\parallel \mp \Omega_H m}\,dp_\perp\,dp_\parallel = 0, \quad (27.92)$$

obtained by a combined solution of the relativistic kinetic equation and the Maxwell equations. In (27.92) $\Omega_L = 4\pi e^2 N/m$ is the Langmuir frequency, $\Omega_H = eH_0/mc$ is the electron gyro-frequency; the difference of these notations from the ω_L, ω_H used above indicates that m here depends on the particle energy. Furthermore, $n_j(\omega)$ is the refractive index of the medium in which the particle system under investigation is located; summation is carried out for the kinds of particles making up the plasma; the top sign relates to an extraordinary wave and the bottom to an ordinary one.

A study of plasma instability is reduced to finding the roots of the equation (27.92) located in the upper semi-plane ω. The corresponding calculations for establishing the instability criterion and finding the value of the growth coefficient are very complex; below we shall therefore chiefly give the results referring for details to Sagdeyev and Shafranov (1960), Zheleznyakov (1960a, 1960b, 1961a, 1961b).

If there is no dispersion of the electrons with respect to the momenta $p_\perp$ and $p_\parallel$, then the distribution function (27.91) is actually reduced to the form $f_0(p) = A\delta(p_\parallel - p_\parallel^0)\delta(p_\perp - p_\perp^0)$ and the dispersion equation (27.92) is reduced to the form (Zheleznyakov, 1960a)[‡]

$$c^2k^2 - \omega^2 n_j^2 + \Omega_L^2 \frac{\omega - kV_\parallel^0}{\omega - kV_\parallel^0 \mp \Omega_H} + \Omega_L^2 \frac{\beta_\perp^2}{2}\frac{c^2k^2 - \omega^2}{(\omega - kV_\parallel^0 \mp \Omega_H)^2} = 0, \quad (27.93)$$

[†] For the choice of the integration contour C in the dispersion equation see Zheleznyakov (1961a). We note that unlike the preceding one this equation is obtained for waves of the form $e^{ikR - i\omega t}$ and not $e^{i\omega t - ikR}$.

[‡] Here and in future the motion of the ions is not taken into consideration.

The dispersion relation (27.93), just like (27.92), is obtained on the basis of the relativistic kinetic equation (i.e. allowing for the velocity dependence of the electron mass). It is clear that it is necessary to allow for relativistic effects if the velocity of the plasma electrons is a noticeable fraction of the velocity of light. Even with small velocities, however, the dependence of m on p^2 must be taken into consideration since neglect of this circumstance will lead to an incorrect expression for the last term in (27.93).

§ 27] Emission, Absorption and Amplification

where $\beta_\perp = V_\perp^0/c$, $V_\perp^0$ and $V_\parallel^0$ are the components of the velocity v corresponding to the values of the momenta $p_\perp^0$ and $p_\parallel^0$. The same values also correspond to the quantities Ω_L, Ω_H. The equation (27.93) is of the fourth power in ω and the general expressions for the frequency $\omega(k)$ will be very cumbersome. We shall therefore limit ourselves to finding the instability conditions in the case when the eigen frequency Ω_L of the plasma's non-equilibrium component characterized by a distribution $\delta(p_\parallel - p_\parallel^0)\,\delta(p_\perp - p_\perp^0)$ is small enough.

Let $\omega = \omega_0 + \delta$, where δ is a correction to the frequency approaching zero together with Ω_L. Then with $\delta \ll \omega_0$ the dispersion equation under discussion is reduced to the following:

$$c^2k^2 - \omega_0^2 n_j^2(\omega_0) - \frac{d(\omega_0^2 n_j^2)}{d\omega_0}\delta + \Omega_L^2 \frac{\omega_0 - kV_\parallel^0 + \delta}{\omega_0 - kV_\parallel^0 \mp \Omega_H + \delta}$$
$$+ \Omega_L^2 \frac{\beta_\perp^2}{2} \frac{c^2k^2 - \omega_0^2 - 2\omega_0\delta - \delta^2}{(\omega_0 - kV_\parallel^0 \mp \Omega_H + \delta)^2} = 0. \tag{27.94}$$

Proceeding by the perturbation method we take as the zero approximation for the frequency ω the solution of the equation (27.94) when $\Omega_L^2 = 0$: $c^2k^2 - \omega_0^2 n_j^2(\omega_0) = 0$. Then in the region of the values of k for which

$$|\omega_0 - kV_\parallel^0 \mp \Omega_H| \ll |\delta|, \tag{27.95}$$

(27.94) can be written as follows:

$$-\frac{d(\omega_0^2 n_j^2)}{d\omega_0} + \Omega_L^2\left(1 - \frac{\beta_\perp^2}{2}\right) + \Omega_L^2 \frac{\omega_0 - kV_\parallel^0 - \omega_0\beta_\perp^2}{\delta} + \Omega_L^2 \frac{\beta_\perp^2}{2} \frac{c^2k^2 - \omega_0^2}{\delta^2} = 0. \tag{27.96}$$

It is assumed below that the second term here can be neglected when compared with the first. It is easy to see that in the case of (27.95) under certain conditions real k may correspond to complex ω with the sign of the imaginary part corresponding to waves which increase with time.

In actual fact let

$$|2\delta|\,|\omega_0 - kV_\parallel^0 - \omega_0\beta_\perp^2| \gg \beta_\perp^2|c^2k^2 - \omega_0^2|. \tag{27.97}$$

Then in (27.96) the last term can be neglected, so that

$$\delta^2 = \Omega_L^2 \omega_0 \frac{1 - \beta_\parallel n_j(\omega_0) - \beta_\perp^2}{\dfrac{d(\omega_0^2 n_j^2)}{d\omega_0}}, \quad \beta_\parallel = \frac{V_\parallel^0}{c}. \tag{27.98}$$

However in an equilibrium medium with a refractive index $n_j(\omega_0)$ the derivative $d(\omega_0^2 n_j^2)/d\omega_0 > 0$ (see Landau and Lifshitz, 1957, section 64) and

δ^2 becomes negative, and the system is unstable if

$$1-\beta_\| n_j(\omega_0) < \beta_\perp^2. \tag{27.99}$$

It follows from this that with $\beta_\perp = 0$ instability appears, in accordance with what has been said previously only with motion at a velocity "greater than that of light" (i.e. in a region of anomalous Doppler effect for the system's electrons: $n_j(\omega_0)\beta_\| > 1$). However, in the case when $\beta_\perp \neq 0$ the instability region increases and as well as the range corresponding to the anomalous Doppler effect includes a normal Doppler effect region: $0 < 1-\beta_\| n_j(\omega_0) < \beta_\perp^2$. We notice that with $n_j(\omega_0) = 1$ the instability criterion is reduced to the inequality $\beta_\| > 1-\beta_\perp^2$, which is possible only if $V_\|^0 \neq 0$, $V_\perp^0 \neq 0$.

In the opposite case to (27.97), when the penultimate term in (27.96) is insignificant

$$\delta^3 = \Omega_L^2 \frac{\beta_\perp^2}{2} \frac{\omega_0^2[n_j^2(\omega_0)-1]}{\dfrac{d(\omega_0^2 n_j^2)}{d\omega_0}} \tag{27.100}$$

and instability occurs both when the electrons are moving along the field H_0 at "superlight" velocities and at "sublight" velocities (in particular with a zero velocity $V_\|^0$). For this instability to occur we must have $\beta_\perp^2 \neq 0$†.

The second group of solutions of the equation (27.94) that approach zero together with Ω_L can be obtained if as the zero approximation for ω we take the solution of the equation $\omega_0 - kV_\|^0 \mp \Omega_H = 0$. Then when we satisfy the inequality

$$|c^2k^2 - \omega_0^2 n_j^2(\omega_0)| \gg \left|\delta \frac{d(\omega_0^2 n_j^2)}{d\omega_0}\right| \tag{27.101}$$

---

† The condition (27.97) and its opposite, when the formulae (27.98), (27.100) are applicable, can be shown in a slightly different form if we take into considerations the explicit expressions for the corrections to the frequency δ. The expressions (27.98) and (27.100) then prove to be valid if

$$\frac{4\Omega_L^2 |1-\beta_\| n_j - \beta_\perp^2|^3}{\omega_0 \dfrac{d(\omega_0^2 n_j^2)}{d\omega_0}} \gtrless \beta_\perp^4 (n_j^2-1)^2,$$

if, of course, the inequality (27.95) is satisfied here. We note, moreover, that in the opposite case to (27.95) the equation (27.94) becomes linear in δ; therefore the quantity δ will be real with real k and the system will be stable relative to perturbations of the type under discussion.

§ 27] Emission, Absorption and Amplification

the equation (27.94) becomes†

$$c^2k^2 - \omega_0^2 n_j^2(\omega_0) + \Omega_L^2\left(1 - \frac{\beta_\perp^2}{2}\right)$$
$$+ \Omega_L^2 \frac{\omega_0 - kV_\parallel^0 - \beta_\perp^2 \omega_0}{\delta} + \Omega_L^2 \beta_\perp^2 \frac{c^2k^2 - \omega_0^2}{2\delta^2} = 0. \quad (27.102)$$

Its solutions differ significantly from those studied previously. For example, with the condition (27.97) δ is real with real k, whilst in the case that we satisfy the opposite inequality to (27.97)

$$\delta^2 \approx -\frac{\Omega_L^2 \beta_\perp^2 (c^2k^2 - \omega_0^2)}{2\left[c^2k^2 - \omega_0^2 n_j^2(\omega_0) + \Omega_L^2\left(1 - \frac{\beta_\perp^2}{2}\right)\right]} \approx -\frac{\Omega_L^2 \beta_\perp^2 (c^2k^2 - \omega_0^2)}{2[c^2k^2 - \omega_0^2 n_j^2(\omega_0)]}.$$

(27.103)

In the change to the last expression we made the assumption that $|c^2k^2 - \omega_0^2 n_j^2(\omega_0)| \gg \Omega_L^2(1 - \beta_\perp^2/2)$. It follows from (27.103) that $\delta^2 < 0$ and the system is unstable if‡

$$\frac{\frac{c^2k^2}{\omega_0^2} - 1}{n_j^2(\omega_0) - 1} > 1, \quad (27.104)$$

The corresponding increase (attenuation) coefficients in a non-equilibrium plasma with a δ-distribution with respect to the momenta $p_\perp, p_\parallel$ for the case of electromagnetic wave propagation at an angle to the field H_0 have been obtained by Petelin (1961).

In the actual conditions of the Sun's atmosphere a study of the amplification and instability of a non-equilibrium plasma with allowance made for the electron velocity dispersion is, of course, of the greatest interest. As shown by Zheleznyakov (1960b), under certain conditions the dispersion equation (27.92) in a plasma where the electron distribution is characterized by the function (27.91) is actually reduced to the equation (27.93) studied above, the only difference being that instead of $\beta_\perp^2/2$ the parameter

$$Y = \frac{\overline{p_\perp^2} - a_i^2}{2c^2} \quad (27.105)$$

† In the opposite case to (27.101) the equation (27.94) is actually reduced to the form (27.96) discussed above.

‡ The criterion (27.97) for the applicability of the expression (27.103) can also be written slightly differently by substituting in it the explicit expression for δ:

$$\frac{2\Omega_L^2 |\omega_0 - kV_\parallel^0 - \omega_0 \beta_\perp^2|}{|c^2k^2 - \omega_0^2 n_j^2(\omega_0)|} \ll \beta_\perp^2 |c^2k^2 - \omega_0^2|.$$

figures in it. Here $\overline{p_\perp^2}$ is the mean square of the transverse momentum.†
It is clear from this that all the expressions following from (27.93) for δ, the limits of applicability of the latter, etc., remain valid when $\beta_\perp^2/2$ is replaced by Y.

The conditions under which this kind of replacement is permissible are basically the requirement that the change in the mass m should be small in the range $(p_\perp - p_\perp^0)^2 \lesssim a_\perp^2$, $(p_\| - p_\|^0)^2 \lesssim a_\|^0$, where the majority of the electrons are concentrated, and the inequality

$$|\xi_\gamma|^2 \equiv 2 \left| \frac{p_\| - p_\|^0}{a} \right| \gg 1, \qquad (27.106)$$

which should hold in the range $(p_\perp - p_\perp^0)^2 \lesssim a_\perp^2$. In (27.106) $p_\|$ is connected with ω and $p_\perp$ by the relation

$$\omega m - k p_\| \mp \frac{eH_0}{c} = 0, \qquad (27.107)$$

which is obviously the Doppler equation for the case under discussion. The function $p_\|(p_\perp)$ defined by (27.107) is generally speaking two-valued for each sign in front of eH_0/c, which also justifies the introduction of the suffix $\gamma = 1, 2$ in the quantity ξ_γ.

The meaning of the condition (27.106) is that we are discussing the instability of a plasma only at the frequencies which are emitted by individual electrons of the plasma in the "tails" of the distribution function $f_0(p)$ (27.91). The situation here is completely analogous to that holding in a stream–plasma system where there is no magnetic field (see (27.54)–(27.57) et seq.); only in the case of (27.106) do terms appear in the dispersion equation which have a singularity with $\omega \tilde{m} - k p_\|^0 \mp eH_0/c = 0$‡ and terms of the equation proportional to $e^{-\xi_\gamma^2/2}$ which explicitly contain an imaginary unit and which cause Landau damping in an isotropic plasma become exponentially small.

We notice, however, that even when $|\xi_\gamma^2| \gg 1$ it becomes essential to allow for terms with $e^{-\xi_\gamma^2/2}$ for investigating plasma instability if the solution $\omega(k)$ of the dispersion equation without these terms is real with real k. With $|\xi_\gamma^2| \lesssim 1$ the terms with $e^{-\xi_\gamma^2/2}$ play a definite part in the plasma instability. Plasma instability caused, as it were, by negative Landau damping has been investigated for a distribution $f_0(p)$ of the (27.91) type with longitudinal propagation $(k \parallel H_0)$ by Zheleznyakov (1961a

---

† If $p_\perp^0/a_\perp \ll 1$, then $\overline{p_\perp^2} = a_\perp^2$; when $p_\perp^0/a_\perp \gg 1$, $\overline{p_\perp^2} = (p_\perp^0)^2$.
‡ The sign $\sim$ denotes that the corresponding quantity is taken with the values $p_\| = p_\|^0, p_\perp = p_\perp^0$.

§ 27] Emission, Absorption and Amplification

and 1961b). The results of these papers, which were obtained by the kinetic equation method and the Einstein coefficients method, are in close agreement and are basically as follows.

The growth coefficient $\gamma_j = 2\,\mathrm{Im}\,\delta$ of the electromagnetic waves being propagated along the field H_0 is defined by the relation ($\omega = \omega_0 + \delta$; $|\delta| \ll \omega_0$)

$$\delta \frac{d(\omega_0 n_j^2)}{d\omega_0} = i\frac{\sqrt{\pi}}{Z_0} \int_0^\infty \zeta^2 e^{-(\zeta-\zeta_0)^2} \sum_\gamma \Upsilon \left[\mp \frac{\zeta-\zeta_0}{a_\parallel}\frac{eH_0}{c} - k\zeta\xi_\gamma \frac{a_\perp^2}{\sqrt{2}a_\parallel^2} \right]$$

$$\times \frac{\Omega_L^2}{k - \dfrac{\omega_0 p_\parallel}{mc^2}} e^{-\xi_\gamma^2/2}\, d\zeta, \qquad (27.108)$$

in which $\Upsilon = \mathrm{sgn}\,(k - [\omega_0 p_\parallel /mc^2])$, $n_j^2(\omega_0)$ is the refractive index in the plasma, $\zeta = p_\perp/a_\perp$, $\zeta_0 = p_\parallel^0/a_\perp$.† The parameter Z_0 is equal to

$$\int_0^\infty \zeta e^{-(\zeta-\zeta_0)^2}\, d\zeta;$$

the values of ξ_γ are given by the formulae (27.106), (27.107) with $\omega = \omega_0$. If $n_j^2(\omega_0)$ in the non-equilibrium system under discussion is close to the value for an equilibrium plasma, then $d(\omega_0^2 n_j^2)/d\omega_0 > 0$ and the sign of δ will be the same as the sign of the integral in (27.108).

It is appropriate to stress here that with longitudinal propagation in a plasma for ordinary waves $n_2 = ck/\omega_0 < 1$; therefore the relation (27.107) cannot be satisfied with $V_\parallel < c$ and for waves of this type $\mathrm{Im}\,\delta = 0$. This actually reflects the fact that ordinary waves with $k\|H_0$ can be emitted only by electrons moving along H_0 with a "superlight" velocity $V_\parallel > \omega_0/k$ (i.e., in a region of the abnormal Doppler effect), unlike extraordinary waves which are emitted only by "sublight" electrons with $V_\parallel < \omega_0/k$ (i.e. in a region of normal Doppler effect; see Zheleznyakov 1959c).

Having in mind only extraordinary waves we note that investigation of the expression (27.108) is difficult in the general case. The problem is simplified, however, in the two extreme cases (I and II): the first of them holds for a distribution $f_0(p)$ when waves of a frequency $\omega \approx \omega_0$ are emitted largely by electrons with close values of the longitudinal momentum $p_\parallel$ (i.e. ξ_γ) and different values of $p_\perp$; the second is when the frequency ω_0 is connected with particles in which $p_\perp \approx \mathrm{const}$ (i.e. $\zeta \approx \mathrm{const}$). These two possibilities are illustrated in Fig. 148, where the solid line shows the function $p_\parallel(p_\perp)$ with constant ω_0 and k defined by the Doppler equation and

---

† The corresponding absorption coefficient in space μ_j will be equal to $(\gamma_j/c) \times d(\omega_0 n_j^2)/d\omega_0$.

the dotted line shows the function $(p_\parallel - p_\parallel^0)^2/a_\parallel^2 + (p_\perp - p_\perp^0)^2/a_\perp^2 = 1$. At the points that satisfy the last relation the distribution function is $f_0(p) = $ constant, the majority of the plasma electrons being located in the region bounded by the dotted line.

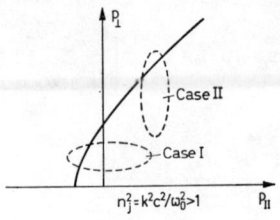

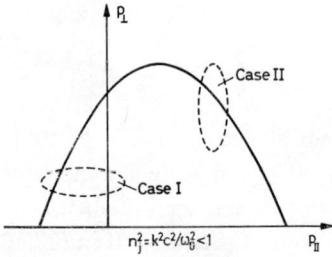

FIG. 148. Two extreme cases of the distribution function $f_0(p)$ when investigating the instability of a non-equilibrium plasma in a magnetic field.

For a non-relativistic plasma ($n_1 \gg p_\parallel/mc$) case I obtains provided that

$$\frac{a_\perp p_\perp}{m^2 c^2} \ll \min \left\{ \frac{n_1 a_\parallel}{mc} ; \frac{n_1^2}{|n_1^2 - 1|} \right\} \quad (27.109)$$

in the range $(p_\perp - p_\perp^0)^2 \lesssim a_\perp^2$. Then according to Zheleznyakov (1961b)

$$\text{Im } \delta = -\frac{\sqrt{\pi} \omega_L^2 m}{\frac{d(\omega_0^2 n_1^2)}{d\omega}} \frac{\overline{p_\perp^2}}{|k| a_\parallel^3} \left\{ \omega_0 - k V_\parallel^0 - \omega_H \left(1 - \frac{a_\parallel^2}{p_\perp^2} \right) \right\} e^{-\tilde{\xi}_y^2/2} . \quad (27.110)$$

Here $\tilde{\xi}_y$ is equal to ξ_y (27.106), (27.107) when $p_\perp = p_\perp^0$ and $\omega = \omega_0$. If $p_\perp^0 = 0$ and $a_\parallel = a_\perp$, then the expression (27.110) is the same as that investigated by Zheleznyakov (1959c); the case of temperature anisotropy, when $p_\perp^0 = 0$, $p_\parallel^0 = 0$, but $a_\perp \neq a_\parallel$, has been discussed by Sagdeyev and Shafranov (1960).

§ 27] Emission, Absorption and Amplification

It follows from (27.110) that electromagnetic waves propagated along H_0 are unstable when

$$\omega_0 - kV_\parallel^0 - \omega_H\left(1 - \frac{a_\parallel^2}{p_\perp^2}\right) < 0. \tag{27.111}$$

If the mean square of the transverse momentum is $\overline{p_\perp^2} = a_\parallel^2$, then this criterion is reduced to the instability condition already obtained earlier for a "superlight" stream (see (27.90) with $\alpha = 0$); this condition, however, ceases to be necessary for extra-ordinary waves with $a_\parallel^2 < \overline{p_\perp^2}$ if they are emitted by particles with a "sublight" velocity. From what has been said the conclusion can be drawn that in the direction H_0 the only waves that are unstable are extraordinary ones, chiefly in a "sublight" stream with $a_\parallel^2 < \overline{p_\perp^2}$.†

Case II occurs with a small enough dispersion of the longitudinal velocities $a_\parallel$; in a non-relativistic plasma, where $n_1 \gg p_\parallel/mc$, it occurs when

$$\frac{a_\perp p_\perp}{m^2c^2} \gg n_1 \frac{a_\parallel}{mc}. \tag{27.112}$$

This inequality should be satisfied in the range $(p_\parallel - p_\parallel^0)^2 \lesssim a_\parallel^2$ which makes the major contribution to Im δ when integrating. Then the quantity Im δ is defined by the relation (Zheleznyakov, 1961b)

$$\text{Im } \delta = \mp \frac{\pi\omega_L^2\omega_H}{\frac{d(\omega_0^2 n_1^2)}{d\omega_0}} \frac{m^2c^2}{Z_0} \tilde{p}_\perp(\tilde{p}_\perp - p_\perp^0)e^{-(\tilde{\zeta} - \zeta_0)^2}; \tag{27.113}$$

the values of $\tilde{p}_\perp$, $\tilde{\zeta}$ are found from the Doppler equation (27.107) for $p_\parallel = p_\parallel^0$ and $\omega = \omega_0$. It follows from this that in case II the instability condition of extraordinary waves is the inequality

$$\tilde{p}_\perp < p_\perp^0. \tag{27.114}$$

This result is quite natural from the quantum point of view, according to which instability and amplification are connected with the fact that in the state into which an electron passes after emission of a quantum $\hbar\omega$ the number of particles is smaller than in the state before the emission of a quantum. In case II the frequency ω is connected with electrons the majority of which have a transverse momentum $p_\perp \approx \tilde{p}_\perp$ and longitudinal

---

† If we exclude the special case when the stream's mean velocity is "superlight" and the waves emitted by the particles of the stream that move at a velocity "sublight" are unstable.

momenta in the range $(p_\parallel - p_\parallel^0)^2 \lesssim a_\parallel^2$. For an extraordinary wave with $\mathbf{k} \parallel \mathbf{H}_0$ emitted and absorbed by particles only in a region of normal Doppler effect the emission of a quantum $\hbar\omega$ is accompanied by the transition of an electron into a state with a lower $p_\perp \approx \tilde{p}_\perp$ (see p. 488). The density of the particles in this state will be lower than in the original state and the wave will become unstable if, in accordance with (27.114), $\tilde{p}_\perp < p_\perp^0$.

Instability and amplification of electromagnetic waves propagated at an arbitrary angle α to $\mathbf{H}_0$ in a plasma with an electron velocity distribution of the (27.91) type have been studied by Andronov, Zheleznyakov and Petelin (1964). The general treatment is rather cumbersome; therefore we shall give only certain results. The criteria of instability and amplification here can be obtained in a comprehensible form only in the two extreme cases I and II similar to those indicated above for longitudinal propagation. In case I, which is of most interest to us, provided that the maximum of the function $f_0(\mathbf{p})$ in a non-relativistic plasma is located on the axis $p_\parallel$ (i.e. $p_\perp^0 = 0$), the instability criterion for both ordinary and extraordinary waves in a normal Doppler effect region is written as follows:

$$a_\perp^2 k \cos\alpha (p_\parallel - p_\parallel^0) + a_\parallel^2 s \frac{eH_0}{c} < 0 \quad (s = 1, 2, 3, \ldots). \quad (27.115)$$

The corresponding amplification coefficient will be of the form ($\alpha \neq \pi/2$)

$$-\mu_j = \frac{\sqrt{\pi}\omega_L^2}{\frac{\partial Q}{\partial k}} \frac{m(u-1)}{\omega^2 \, a_\parallel k \cos\alpha} \left(\frac{a_\perp k \sin\alpha}{2m\omega_H}\right)^{2s-2} \left\{ s\omega_H + \frac{a_\perp^2}{a_\parallel^2}(\omega - kV_\parallel^0 \cos\alpha) \right.$$

$$\left. - s\omega_H \right\} \left\{ (n_j^2-1)(2-n_j^2 \sin^2\alpha) + v(2 - n_j^2 - n_j^2 \cos^2\alpha) + \frac{2v}{1+\sqrt{u}} \right.$$

$$\times (1-v-n_j^2 \sin^2\alpha) - 2(1-s\sqrt{u})n_j^2 \sin^2\alpha \left(n_j^2 - 1 + \frac{v}{1+\sqrt{u}}\right)$$

$$+ (1-s\sqrt{u})^2 \tan^2\alpha \left[(n_j^2-1)(1-n_j^2 \cos^2\alpha)\right.$$

$$\left.\left. + \frac{v}{1-u}(2-n_j^2-n_j^2 \cos^2\alpha - v)\right]\right\} e^{-m^2(\omega - kV_\parallel^0 \cos\alpha - s\omega_H)^2/a_\parallel^2 k^2 \cos^2\alpha},$$

$$(27.116)$$

where

$$Q = (1-u-v+uv\cos^2\alpha)n_j^4 - [2(1-v)^2 + uv(1+\cos^2\alpha) - 2u]n_j^2$$
$$- (1-v)[u-(1-v)^2],$$

$$n_j^2 = \frac{c^2 k^2}{\omega_0^2}.$$

It follows from (27.115) that in propapating at an angle to the field H_0, the instability occurs not only at the first but also at the higher harmonics s. Zheleznyakov and Suvorov (1968) have considered the character of a change of instability with the growth of the width of the distribution function of relativistic electrons $f_0(p)$ up to the value at which not one but many harmonics s give a contribution to the change of absorption at the frequency ω. In the last case there may occur the so-called synchrotron instability discovered by Zheleznyakov (1966) (see also NcGray (1966)). This instability is realized at the corresponding choice of the function $f_0(p)$ if the relativistic electrons exist in a sufficiently "cold" plasma. With this effect is associated the negative re-absorption of synchrotron radiation due to which there may occur, under definite conditions, the coherent synchrotron mechanism of radio emission.

For an estimate of the maximum amplitude reached in these cases by electromagnetic waves as a result of amplification a special treatment is required with allowance made for non-linearity.

THE APPEARANCE OF PLASMA WAVES IN SHOCK WAVE FRONTS

The fact established in this section of amplification and instability of plasma and electromagnetic waves in a non-equilibrium plasma makes us investigate particularly carefully the conditions under which in the solar corona particle velocity distributions occur which provide this kind of instability. One of these possibilities which is of great importance in the generation of type III bursts is the injection of streams of fast particles (with a velocity of $\sim 10^{10}$ cm/sec) into the coronal plasma from a chromospheric flare region. In an isotropic coronal plasma amplification of the plasma waves occurs in this case because of "beam" instability. The other possibility is the passage in the corona of shock waves ahead of plasma bunches ejected from the region of the bursts at a velocity of $\sim 10^8$ cm/sec. The latter possibility is of considerable interest above all for the theory of type II bursts.

We shall open the discussion of the instability of shock waves relative to plasma-type perturbations with an isotropic plasma when, according to Denisse and Rocard (1951) and Sen (1954), instability appears because of the twin-hump electron velocity distribution in the front of the shock wave.

According to Mott-Smith (1951), the particle velocity distribution in a shock wave front can be represented approximately in the form of the sum of two Maxwell terms whose temperatures and mean velocities are equal to the corresponding quantities on the two sides of the shock wave. This kind of representation has a clear physical meaning: a shock wave front with a thickness of the order of the mean free path l_{fr} consists of a mixture of particles coming in from the side of supersonic and subsonic streams. For a single-atom non-ionized gas with a thermal capacity ratio of $c_p/c_V = 5/3$ the velocity distribution function $f_0(V)$ becomes twin-humped if $M > 3$ (M is the Mach number defined as the ratio of the velocity of the shock wave' front to the velocity of sound).

Denisse and Rocard (1951) and Sen (1954) assumed that in a plasma the function $f_0(V)$ for the electrons also has two maxima when $M > 3$ and, therefore, the necessary conditions are created in a shock wave for the appearance of instability and amplification which ensure coherent emission of plasma waves.†

However, as shown by Zheleznyakov (1958a), this conclusion is based on a misunderstanding. In actual fact it is known that when shock waves are propagated in a plasma it is not the electrons, but ions with a mass $m_i \gg m$ that play the major part in determining the velocity of the subsonic and supersonic streams, density and temperature behind the front. It is the ions that determine the velocity of sound $V_{son} = \sqrt{(c_p/c_V)2\varkappa T/m_i}$ in an ionized gas with a pressure $p = 2N\varkappa T$ ($2N$ is the total number of particles in a unit volume). This velocity (which differs only by a factor of $\sqrt{2}$ from the corresponding value of a non-ionized gas) together with the M number gives the value of the above-mentioned parameters that characterize a shock wave in a plasma. Therefore in the case when $M > 3$ two maxima also appear in the shockfront, but only in the ion velocity distribution: as estimates show, the function $f_0(V)$ for the electrons will definitely have one maximum for Mach numbers of $M < 10^2$. This is quite understandable if we remember that the Maxwell distribution for the electrons is $\sqrt{m_i/m}$ times broader than the corresponding distribution for the ions (at the same temperature). This boundary increases still further if we allow for the enhanced electron velocity dispersion connected with the increase of their temperature in a wave front with high M.

† Without allowing for ion motion the plasma definitely remains stable in relation to longitudinal-type perturbations if the velocity distribution function $f(V^2)$ is characterized by a steady decrease as V^2 rises (see Akhiyezer and Lyubarskii, 1955). For instability several maxima must be present in the electron distribution function (Gertsenshtein, 1952) or there must be a shift of the maximum of the ion distribution function relative to the electron function, as the equations (27.54) will confirm. It is clear from what has been said why the variety of instability under discussion is sometimes called "twin-stream" or "beam" instability. It is appropriate to stress that the smallness of the ratio of the plasma wavelength λ to the width of the front of the shock wave $l_{sh} \sim l_{tr}$ provides a basis for assuming that in a non-uniform plasma such as the front of a shock wave the geometrical optics approximation is extensively applicable. This allows us when studying amplification to use the dispersion equation obtained for a uniform plasma, remembering at the same time that in this approximation the connection between k and ω at a given point in a non-uniform plasma is the same as the function $k(\omega)$ in a uniform medium with the same parameters (see section 22).

§ 27] Emission, Absorption and Amplification

On the other hand, it is easy to see that in an isotropic coronal plasma shock waves cannot exist with a Mach number of more than a few tens since for these waves the width of the front (of the order of l_{fr}) becomes comparable with the linear dimensions of the corona. It follows from what has been said that when there is no magnetic field H_0 shock waves in the solar corona cannot cause instability or amplification of plasma waves.

The dispersion equation (27.54) can be used to show that this result does not change when the motion of the ions is taken into consideration. However, under certain conditions the complex nature of electron and ion velocity distribution functions can apparently affect the stability of the shock wave front, but in relation to low-frequency perturbations of the ion plasma wave type.

The position changes considerably if a magnetic field H_0 is applied to the plasma (see, e.g., Sagdeyev's survey (1964) and also Vedenov, Velikhov and Sagdeyev (1961)). In this case the profile of a shock wave being

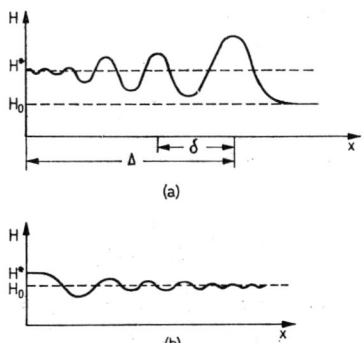

Fig. 149. Profile of a shock wave being propagated in a plasma: (a) across the magnetic field, (b) at an angle α that satisfies the condition $\sqrt{m/m_i} \ll \alpha \ll 1$.

propagated in a "cold" plasma ($N\varkappa T \ll H^2/8\pi$) across a magnetic field H at a velocity U is of the form shown in Fig. 149a; first H rises from a value H_0 ahead of the front until

$$H_{\max} = H_0(2M-1), \qquad (27.117)$$

then H oscillates around the value

$$H^* = \frac{H_0}{2}(-1+\sqrt{1+8M^2}), \qquad (27.118)$$

gradually damping until a field $H = H^*$ is established corresponding to the field strength behind the wave front. Damping of the oscillations occurs at a distance

$$l_{sh} \sim \frac{U}{\nu_{\text{eff}}},$$

where v_{eff} is the effective number of collisions of electrons with ions. In other words, the front is a system of so-called solitary waves ("solitons"), the distances between whose crests at the beginning of the front are

$$\delta \approx \frac{c}{\omega_L \sqrt{M-1}} \ln \frac{U \omega_L \sqrt{M-1}}{v_{\text{eff}} c}, \qquad (27.119)$$

and at its end are

$$\delta \approx \frac{c}{\omega_L \sqrt{M-1}}. \qquad (27.120)$$

Here

$$M = \sqrt{4\pi \varrho_0} \frac{U}{H_0} \qquad (27.121)$$

is the "magnetic Mach number" (the ratio of the velocity of the shock wave to the velocity of the magnetohydrodynamic wave); in shock waves we always have $M > 1$ (ϱ_0 is the plasma density before the shock wave front).

This kind of front structure exists only in a sufficiently rarefied plasma for which

$$l_{\text{sh}} \sim \frac{U}{v_{\text{eff}}} \gg \frac{c}{\omega_L}; \qquad (27.122)$$

in the opposite case the thickness of the front is characterized by the expression

$$l_{\text{sh}} \sim \frac{\eta_m}{U(M-1)}$$

well known in magnetohydrodynamics, where $\eta_m = c^2/4\pi\sigma$ is the "magnetic viscosity" (the conductivity is $\sigma = e^2 N/m v_{\text{eff}}$). The reservation must also be made that a front of the kind shown in Fig. 149a occurs only if the shock wave is propagated across the magnetic field. When propagated at an angle $\alpha \neq \pi/2$, i.e. in the case $\sqrt{m/m_i} \ll \alpha \ll 1$, the wave profile consists as before of solitary waves, but they will now be "rarefaction" waves (Fig. 149b). The field H inside the front reaches a minimum value; the amplitude of the oscillations gradually increases from the beginning of the front to its end, so the onset of oscillations is not sudden.

We note further that profiles of the type given in Fig. 149 exist only for "Mach numbers" of $M < 2$. In the case of a strong shock wave ($M > 2$) the position becomes more complicated: the ions of the oncoming stream are reflected from the potential barrier in the wave front, the motion becomes multi-stream and the velocity at a given point ceases to have a single value. The theory of strong shock waves has not yet been developed, although this case is also of considerable interest from the point of view of radio astronomy applications.

Under certain conditions in the front of a shock wave moving in a

§ 27] Emission, Absorption and Amplification

magnetoactive plasma instability appears relative to perturbations of the plasma wave type. Here notice should be taken above all of the possibility of instability caused by drift of the electrons relative to the ions in the plane of the front (Vedenov, Velikhov and Sagdeyev, 1961; Sagdeyev, 1964). The cause of the drift is the non-uniformity of the magnetic field in the x direction; in its turn the drift causes a current which ensures the self-consistent nature of the magnetic field—the variation in H along the x-axis. With large enough "Mach numbers":

$$M - 1 \gg \frac{3}{8} \left(\frac{8\pi N \varkappa T}{H^2} \right)^{1/3}, \qquad (27.123)$$

the velocity of the ions relative to the electrons (because of drift of the latter in a direction orthogonal to H and U) becomes greater than the thermal velocity of the electrons: $V_0 \gg V_{\text{th}} = \sqrt{\varkappa T/m}$. Then, as we shall now see, "beam" instability appears relative to perturbations of the plasma wave type.

When investigating the question of stability we can neglect the non-uniform nature of the plasma in the front of the shock wave if the length of the plasma waves $\lambda_{\text{pl}}/2\pi \approx V_{\text{ph}}/\omega_L$ is small when compared with the "width" of a solitary wave $\delta \sim c/\omega_L$ (i.e. $V_{\text{ph}} \ll c$), and the increase (or attenuation) time of the plasma waves $1/\text{Im}\,\omega$ is far less than the time taken by the plasma to pass through the region occupied by the "solitons":

$$\frac{1}{\text{Im}\,\omega} \ll \frac{\delta}{U} \sim \frac{c}{\omega_L} \frac{\sqrt{4\pi \varrho_0}}{M H_0}. \qquad (27.124)$$

If at the same time we neglect the effect of the magnetic field on the plasma waves, which is permissible at frequencies of $\omega \gg \omega_H$, then for investigating instability in the quasi-hydrodynamic approximation we can turn to the dispersion equation

$$\frac{\omega_L^2}{\omega^2 - k^2 V_{\text{th}}^2} + \frac{\omega_L^2 m}{m_i(\omega - k V_0 \cos\Theta)^2} = 1. \qquad (27.125)$$

This equation is similar to (27.50), but the stream of electrons is replaced by one of ions; it can be obtained by the quasi-hydrodynamic method (allowing for the motion of the electrons and the ions). It is also easy to see that the dispersion equation (27.54) found kinetically changes into the relation (27.125)† if the parameters ξ_l for the electrons and ions respectively are

$$\xi_e = \frac{\omega}{k V_{\text{th}}}, \quad \xi_i = \frac{\omega - k V_0 \cos\Theta}{k V_{\text{th}_i}},$$

† This is not entirely accurate; when changing from (27.54) to (27.125) the factor $3 V_{\text{th}}^2$ appears in the latter instead of V_{th}^2 (for reasons discussed in section 22).

and satisfy the inequalities

$$|\xi_e^2| \gg 1, \quad |\xi_i^2| \gg 1, \qquad (27.126)$$

whilst the temperatures of the electrons and the ions are the same. These inequalities also establish the limits of applicability of the equation (27.125).

Using the smallness of the ratio m/m_i we shall look upon the second term from the left in (27.125) as a perturbation. Proceeding in a similar way to the solution of the equation (27.30) we obtain:

$$\omega \approx kV_0 \cos\Theta + \sqrt[3]{1}\,\omega_L \left(\frac{m}{2m_i}\right)^{1/3} \left(1 - \frac{V_{\text{th}}^2}{V_0^2 \cos^2\Theta}\right)^{1/6}, \qquad (27.127)$$

where $[kV_0 \cos\Theta]^2 = \omega_L^2 + k^2 V_{\text{th}}^2$. The cube root has three values, one of which (with Im $\omega < 0$) corresponds to an increasing longitudinal wave. It follows from (27.127) that the phase velocity $V_{\text{ph}} \approx \omega/k \approx V \cos\Theta$; therefore the first of the inequalities (27.126) is satisfied provided that

$$V_0^2 \cos^2\Theta \gg V_{\text{th}}^2. \qquad (27.128)$$

This criterion is also sufficient for satisfaction of the second inequality, which allowing for (27.127), (27.128) can be represented (dropping factors of the order of unity) in the form

$$\omega_L^2 \left(\frac{m}{m_i}\right)^{2/3} \gg k^2 V_{\text{th}_i}^2 \quad \text{or} \quad V_0^2 \cos^2\Theta \gg V_{\text{th}}^2 \left(\frac{m}{m_i}\right)^{1/3}.$$

It is clear from what has been said that in the case of (27.123) in the front of a shock wave "beam" instability of plasma waves propagated in directions Θ for which $\cos^2\Theta \sim 1$ definitely occurs. Since the squares of the velocities figure in (27.128) the inequality (27.123) may be not particularly strong.

Another factor that ensures instability and amplification in the front of a shock wave in a plasma is the "anisotropy" of the electron temperature (Vedenov, Velikhov and Sagdeyev, 1961; Ginzburg and Zheleznyakov, 1960). The anisotropy appears because of the change in the strength of the magnetic field in the shock wave, when the characteristic dimension of the inhomogeneity H (i.e. the "width" of the solitary wave) is less than the distance U/ν_{eff} over which the electrons lose their ordered velocity. The electrons in the plasma change their transverse (in relation to H) velocity $V_\perp$, in accordance with the adiabatic invariant $V_\perp^2/H = \text{const}$, together with the change in the magnetic field strength in the shock wave front, whilst the dispersion of the longitudinal velocities remains constant. Therefore the "transverse" temperature rises as the field increases in the first solitary wave and then oscillates near a value corresponding to a field $H^* > H_0$.

§ 27] Emission, Absorption and Amplification

This, of course, holds over a range much less than U/ν_{eff} (from the beginning of the front); at a distance of the order of U/ν_{eff} the temperatures even out because of collisions (the "anisotropy" disappears).

The maximum degree of temperature "anisotropy" in this process (the ratio of the "transverse" temperature $T_\perp$ to the "longitudinal" $T_\parallel$) can be estimated if we remember, in accordance with the adiabatic invariant, that $T_\perp \propto H$:

$$\left(\frac{T_\perp}{T_\parallel}\right)_{\max} \approx \frac{H_{\max}}{H_0} = 2M - 1 \tag{27.129}$$

(see (27.117)). The degree of anisotropy in the field H^* is

$$\left(\frac{T_\perp}{T_\parallel}\right)^* \approx \frac{H^*}{H_0} = \frac{-1 + \sqrt{1 + 8M^2}}{2} \tag{27.130}$$

(see (27.118)). For example for $M \approx 2$ the values of $(T_\perp/T_\parallel)_{\max}$ and $(T_\perp/T_\parallel)^*$ will be close to 3 and 2·37 respectively.

The question of wave instability and amplification in a plasma under conditions when temperature "anisotropy" exists has been discussed in the preceding section. All that we shall say here (considering that in a shock wave $p_\perp^0 = 0$, $a_\perp^2/a_\parallel^2 = T_\perp/T_\parallel$) is the following: for values of the refractive index n_1 that are not too small, namely for

$$n_1 \gg \frac{a_\perp^2}{a_\parallel mc} \approx \frac{T_\perp}{\sqrt{T_\parallel mc^2/\varkappa}},$$

when electromagnetic waves are propagated along the magnetic field (parallel to the front of the shock wave) case I (27.109) obtains.

Instability and amplification of extraordinary waves then occur if

$$\sqrt{u} \equiv \frac{\omega_H}{\omega_0} > \frac{1}{1 - T_\parallel/T_\perp} \tag{27.131}$$

(see (27.111) with $V_\parallel^0 = 0$ and $\overline{p_\perp^2} = a_\perp^2$); the amplification coefficient will be

$$\left.\begin{array}{l} -\mu_1 = \dfrac{2\sqrt{\pi}\omega_L^2 m}{\omega_0 n_1^0} \dfrac{a_\perp^2}{a_\parallel^3} \left\{\dfrac{\omega_H}{\omega_0}\left(1 - \dfrac{a_\parallel^2}{a_\perp^2}\right) - 1\right\} e^{-\tilde{\xi}_\gamma^2/2}, \\[2mm] \dfrac{\tilde{\xi}_\gamma^2}{2} = m^2 \dfrac{(\omega - \omega_H)^2}{k^2 a_\parallel^2}. \end{array}\right\} \tag{27.132}$$

For electromagnetic waves propagated in a shock wave front at an angle to the field H (i.e. when $\alpha \neq 0$) the corresponding instability criterion and amplification coefficient can be obtained from the formulae (27.115), (27.116).

The results given will be used in sections 31 and 32 when discussing the problem of the origin of types I and II solar bursts and of Jupiter's sporadic radio emission.

CHAPTER VIII

Theory of the Sun's Thermal Radio Emission

THIS chapter and the ones following discuss the present state of the very important and complex problem of the origin of solar radio emission.

When discussing this problem it is best to divide the emission "mechanisms" in a high-temperature plasma (such as the solar corona and chromosphere are) into thermal and non-thermal and also into coherent and non-coherent—depending on the nature of the velocity (energy) distribution of the emitting particles (Ginzburg and Zheleznyakov, 1960 and 1961).

The emission mechanism is considered to be thermal if it is generated in a plasma layer with an equilibrium particle distribution; in the opposite case it is called non-thermal. The intensity of the thermal emission I_j, generally speaking, is not the same as the intensity of the equilibrium emission $I_j^{(0)}$: according to (26.19) $I_j \approx I_j^{(0)}$ only for an optically thick layer ($\tau_j \gg 1$), whilst for an optically thin layer ($\tau_j \ll 1$) $I_j \approx I_j^{(0)} \tau_j \ll I_j^{(0)}$.

In its turn the emission mechanisms can be divided into coherent and non-coherent as follows. If the resultant intensity of the emission leaving a given volume of plasma is

$$I_j \leq \sum I_j',$$

where I_j' is the intensity created by a single elementary emitter (electron, atom etc.), then we shall call this emission mechanism non-coherent; in the case when

$$I_j > \sum I_j'$$

we shall call the mechanism coherent. The inequality sign in the first relation between I_j and I_j' allows for self-absorption (reabsorption) in the plasma. As we explained in section 27, under certain conditions, determined in particular by the form of the distribution function $f(V)$, the reabsorption may become negative and the emission mechanism coherent. Therefore this definition of coherency is closely connected with the sign of the absorption coefficient μ_j in the emitting system: $\mu_j > 0$ corresponds to the non-coherent form and $\mu_j < 0$ the coherent form of emission mechanism.

Theory of the Sun's Thermal Radio Emission

This definition differs slightly from that generally accepted since only emission whose intensity is equal to the sum of the emission intensities of the individual particles is generally called non-coherent. From this point of view the emission mechanism cannot strictly speaking be considered non-coherent when allowing for reabsorption in the actual system (which is necessary when its optical thickness is $\tau_j \gtrsim 1$). However we shall give it this name even when there is noticeable reabsorption, leaving the term "coherent emission mechanism" only for the case of negative absorption (amplification) of waves in a plasma. This kind of demarcation is fully justified since qualitatively new phenomena appear in the case $\mu_j < 0$ (wave amplification is closely connected with instability and for an extensive enough system is limited to non-linear effects; see section 27).

The sign and magnitude of μ_j are determined, generally speaking, not only by the nature of the velocity (energy) distribution of the emitting particles but also by the elementary emission processes (see section 27). Therefore the subdivision of emission mechanisms into coherent and non-coherent also depends on these two factors, whilst the subdivision into thermal and non-thermal is not connected with the elementary emission processes. Thermal emission is always non-coherent no matter what the nature of the emission processes since in an equilibrium plasma $\mu_j > 0$.

It was noted in section 26 that emission leaving a plasma is made up in principle of the following elementary processes:

1. bremsstrahlung in close collisions of electrons with ions, i.e. in the process of free–free transitions of electrons in the Coulomb field of the ions;
2. magneto-bremsstrahlung during the helical motion of charged particles in a magnetic field;
3. Cherenkov emission when charged particles are moving "faster than light" (at a velocity $V > c/n_j$);
4. transition radiation during the motion in a non-uniform plasma;
5. emission of electrons in recombinations (because of free–free transitions) and monochromatic emission (in the process of transitions between discrete levels in the atoms and molecules).

To this, perhaps, we can add the emission which appears when there is a change in the plasma properties having a periodic nature or the nature of a "shock" which takes the plasma out of a state of equilibrium. In the first case we have in mind particularly the parametric excitation of electromagnetic waves during the passage of plasma waves;[†] in the second

---

[†] This mechanism has been studied by Stepanov (1963), Stepanov and Ostrovskii (1963); it has not yet been investigated under cosmic conditions.

emission in the process of the eigen oscillations of the plasma in the front or behind the front of a shock wave.

The five types of elementary processes given (which, generally speaking, may produce in combination both coherent and non-coherent emission) together with the emission caused by a change in the plasma parameters certainly exhaust all the possible mechanisms of radio emission in a high-temperature plasma. This makes it impossible to "invent" any new elementary mechanisms of solar radio emission which are not included in this list. The basic problem in the origin of the Sun's radio emission therefore consists of "dividing" these components among the types of emission listed above, developing the quantitative theory of solar radio emission and using it to obtain fresh information on the physics of the Sun's atmosphere by proceeding from the observed characteristics of the individual components of the solar radio emission and the physical conditions in the corona and the chromosphere. As will become clear from the contents of Chapters VIII and IX considerable progress has been made in the solution of this problem.

The reservation should be made that the choice of emission mechanisms is actually far smaller. The point is that coherent mechanisms producing a significant rise in intensity appear only with definite particle velocity distributions. For Cherenkov emission and magneto-bremsstrahlung these distributions are quite simple in form and occur in the coronal plasma during the passage of a stream of charged particles, shock waves, etc. (see section 27). On the other hand, bremsstrahlung amplification is achieved apparently only with very special relations between the number of electron collisions and the velocity. There is little probability that exceptional conditions of this kind can occur in the corona.

The action of the transition emission mechanism has not yet been studied under solar conditions. It can be assumed that, all other things being equal, the part it plays will be insignificant when compared with the Cherenkov emission mechanism, above all because of the small gradients of the dielectric permeability in the corona and the chromosphere. It is not impossible, however, that further investigations of transition emission, particularly of the possibility of the creation of coherent transition emission by streams of charged particles, may also be of interest to the theory of solar radio emission.

The effects connected with transitions of free electrons to discrete levels of atoms and with transitions between these levels are not in practice reflected in the overall solar radio emission balance because of the low density of the Sun's atmosphere. It is possible, of course, that we shall nevertheless succeed in finding in the spectrum of the Sun's radio

§ 28] Theory of the "Quiet" Sun's Radio Emission

emission emission or absorption lines caused by transitions between discrete levels, although at present the question of the presence of noticeable lines remains open (in this connection see sections 8 and 11).

It becomes clear from what has been said why in future when interpreting the solar radio emission we shall pay most attention to noncoherent bremsstrahlung and also to the coherent and non-coherent forms of magneto-bremsstrahlung and Cherenkov emission mechanisms.

28. Theory of the "Quiet" Sun's Radio Emission

RADIO EMISSION MECHANISM

The theory of the radio emission of the unperturbed Sun (the B-component) has as its main aim obtaining on the basis of a certain model of the solar corona and chromosphere characteristics of this emission component (frequency spectrum, radio brightness distribution over the disk, etc.) which agree with the observations. The model of the solar corona, of course, should correspond to the physical conditions in the chromosphere and the corona that are known from all the optical observations. In the case of a successful interpretation of the B-component it becomes possible, by analysing known data on the radio emission of the "quiet" Sun, to obtain fuller and more complete information on the distribution of the temperature and density in the corona and chromosphere, i.e. solve the problem which is the opposite of that set above.

We must first of all answer the question of the radio emission mechanism of the "quiet" Sun to fulfil this programme. It follows from the actual definition of the B-component as the minimum level of radio emission occurring in the absence of solar activity that this component is generated in a stationary solar atmosphere. The particle distribution in the coronal and chromospheric plasma is then close to an equilibrium (Maxwell) distribution (see Pikel'ner, 1950); therefore the radio emission of the "quiet" Sun is thermal in nature, no matter what the elementary emission processes, and therefore belongs to the non-coherent type of emission.

It is clear from the introduction to this chapter that the mechanism of the B-component must be chosen from bremsstrahlung, Cherenkov and magneto-bremsstrahlung. However, the last two mechanisms must be rejected on the following considerations. The frequency of the magneto-bremsstrahlung $\omega \sim \omega_H(\mathscr{E}/mc^2)^2$ (26.48) created by the electrons moving in the overall magnetic field of the Sun with a strength of $H_0 \sim 1$ oe ($\omega_H \sim 2 \times 10^7 \text{ sec}^{-1}$) belongs to waves of $\lambda \lesssim 3$ m (where the radio emission of the "quiet" Sun is largely observed) only for relativistic particle energies of $\mathcal{E} \gtrsim 5\,mc^2$. However, the mean energy of the thermal motion of particles in the corona and the chromosphere is far less: $\mathcal{E}_{\text{kin}} \lesssim 10^{-4}\,mc^2$.

At the same time an increase in the particle energy $\mathcal{E}_{\text{kin}} \sim \varkappa T$ to energies greater than or of the order of $5mc^2$, i.e. one that is completely unjustified from the point of view of the data of optical observations of the kinetic temperature of the upper layers of the Sun's atmosphere to values of $T \gtrsim 2\cdot5\times10^{10}\,°\text{K}$, is inefficient. In actual fact, a plasma heated to such a high temperature and generating emission with $T_{\text{eff}} \lesssim 10^6\,°\text{K} \ll T$ should be optically thin ($\tau_j \ll 1$); in this case the magneto-bremsstrahlung becomes strongly polarized because of the differing efficiency of the generation of ordinary and extraordinary waves by each individual electron. The radio observations, however, provide no indication of the presence of a considerable polarized component in the B-component (see section 8).

As for the Cherenkov emission, its contribution to the intensity of the "quiet" Sun's radio emission is negligibly small when compared with the contribution of the bremsstrahlung. This is connected in the end with the low efficiency of the transformation of plasma waves into electromagnetic waves, that leads to a weakening of the emission flux by a factor of hundreds of thousands (see section 25). The point is that both the bremsstrahlung and Cherenkov emission mechanisms, while acting under the normal conditions of the solar corona, are able to raise the effective emission temperature T_{eff} to the value of their kinetic temperature $T \sim 10^6-10^4\,°\text{K}$. The bremsstrahlung creates this kind of electromagnetic emission in layers from which it escapes without hindrance beyond the solar shell; on the other hand, the Cherenkov mechanism acts only in regions with $n_j > 1$ (n_j is the refractive index); the plasma waves emitted here cannot leave the limits of the corona without preliminary transformation into electromagnetic waves (see sections 23, 25 and 26). This process sharply weakens the emission leading to T_{eff} when leaving the corona being much less than the plasma temperature in the generation region.†

The idea of the bremsstrahlung mechanism and the thermal nature of the "quiet" Sun's radio emission was put forward by Ginzburg (1946), Shklovskii (1946) and Martyn (1946). Two years later Martyn published a paper (Martyn, 1948) in which he predicted the presence of "brightening" towards the edge of the Sun's disk in the radio band. This effect, which can be checked experimentally, was an important criterion of the correctness of the hypothesis of the bremsstrahlung origin of the solar

---

† The contribution made by bremsstrahlung of plasma waves to the B-component is insignificant for the same reason.

§ 28] Theory of the "Quiet" Sun's Radio Emission

radio emission's basic component. The existence of the brightening effect was later confirmed in numerous investigations of the radio brightness distribution over the Sun's disk (see section 9).

It must be pointed out that the calculations by Martyn (1948) were only approximate. More detailed investigations of the radio brightness distribution over the solar disk and of the radio emission spectrum of the unperturbed Sun on the basis of definite models of the corona and the chromosphere were afterwards made in a whole series of papers (Hagen, 1951; Panovkin, 1957; Reule, 1952; Scheffler, 1958; Unsöld, 1947; Denisse, 1949; Waldmeier and Müller, 1948; Nicolet, 1949; Giovanelli, 1948; Thomas, 1949; Burkhardt and Schlüter, 1949; Smerd, 1950a; Pikel'ner and Shklovskii, 1950; Coates, 1958).

THEORY OF THE B-COMPONENT IN THE SIMPLEST MODEL OF THE CHROMOSPHERE AND CORONA

Following Smerd (1950a) we shall first proceed from the simplest model of the solar atmosphere, assuming that the latter is a spherically symmetrical formation with a uniform distribution of the kinetic temperature in the chromosphere ($T_{ch} = 10^4 - 3 \times 10^4$ °K) and in the corona ($T_c = 5 \times 10^5 - 3 \times 10^6$ °K). The clear-cut boundary between the chromosphere and the corona is at an altitude of $h = 10^4$ km from the photosphere. The variation of the electron concentration $N(h)$ in the chromosphere at altitudes from 5×10^2 to 10^4 km is of the form

$$N = 5.7 \times 10^{11} e^{-7.7 \times 10^{-4} (h-500)} \quad \text{electrons/cm}^3 \tag{28.1}$$

(h is in km). In the corona the function $N(h)$ is characterized by the Baumbach–Allen formula (1.1).†

We notice that when discussing the emission and propagation of electromagnetic waves in the corona and the chromosphere the effect of the Sun's total magnetic field $H_0 \sim 1$ oe in the part of the radio band of interest to us ($\lambda < 5$ m) need not be taken into consideration in the first approximation since the parameter $\sqrt{u} \equiv \omega_H/\omega$ characterizing this effect is small when compared with unity. However, in a more accurate investigation and above all in the solution of the question of the polarization of the "quiet" Sun's radio emission we cannot ignore the presence of this field (see Smerd, 1950b).

† The same model of the corona was used by Reule (1952) when calculating the distribution of T_{eff} over the disk and the frequency spectrum; a model of the solar atmosphere close to this one was used by Pikel'ner and Shklovskii (1950) (see also Shklovskii, 1962, section 27) for obtaining the radio brightness distribution over the disk of the quiet Sun.

The trajectory of rays in this kind of spherically symmetrical medium depends on the distribution of the electron concentration and is defined by the formula (22.33) (see also Fig. 113), and the effective temperature of the thermal radio emission varies along the ray in accordance with the transfer equation, being given in the general case by the expression (26.17). The latter, however, can be simplified if we take into consideration the isothermal nature of the chromosphere and the corona in the model used: in accordance with (26.18) at a point l on a section of the ray located in a region with a uniform temperature T

$$T_{\text{eff}}(l) = T(1-e^{-\tau_j})+e^{-\tau_j}T_{\text{eff}}(l_0),$$

where $T_{\text{eff}}(l_0)$ is the effective temperature at another point on this ray and τ_j ($j = 1, 2$) is the optical thickness of the layer contained between the points l and l_0:

$$\tau_j = \int_{l_0}^{l} \mu_j \, dl.$$

In the framework of the bremsstrahlung mechanism the absorption coefficient μ_j is caused only by collisions of electrons with ions; in an isotropic plasma it is determined by the value of the temperature and concentration of the electrons (see the expression (26.74)). An element of the ray's path length dl is given by the formula (22.32); when this is allowed for, the expression for the optical thickness becomes

$$\tau_j = \int_{R_0}^{R} \frac{\mu_j(R)\, dR}{\sqrt{1-\dfrac{r^2}{n_j^2 R^2}}}. \qquad (28.2)$$

Here R and R_0 are respectively the distance of the points l and l_0 from the centre of the Sun. Therefore the effective temperature of bremsstrahlung escaping beyond the corona depends on the configuration of the ray in the corona and the chromosphere and the values of the absorption coefficient connected with collisions along this ray; in the final count it is determined by the distribution of the kinetic temperature and the electron concentration in the Sun's atmosphere.

It is clear from the expression given above for T_{eff} in a thermally uniform medium that if the trajectory of the ray passes only through the corona (ray 1 in Fig. 150) the effective radio emission temperature at a point A located outside the corona is

$$T_{\text{eff}} = T_c(1-e^{-2\tau_c(R^*)}). \qquad (28.3)$$

§ 28] Theory of the "Quiet" Sun's Radio Emission

Here $2\tau_c(R^*)$ is the optical thickness of the corona along ray 1; $\tau_c(R^*)$ is the optical thickness of half this ray—from the level $R = R^*$ (from the turn point) to departure from the corona. In accordance with (28.2)

$$\tau_c(R^*) = \int_{R^*}^{\infty} \frac{\mu_j(R)\,dR}{\sqrt{1 - \dfrac{r^2}{n_j^2 R^2}}}. \tag{28.4}$$

If a wave passes through the chromosphere (ray 2 in Fig. 150), then by applying the relation (26.18) three times—on the sections DC in the corona, CB in the chromosphere and BA once again in the corona—it is easy

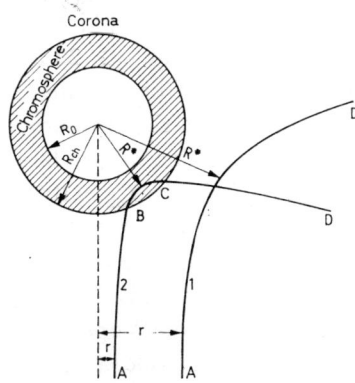

FIG. 150. Simplest model of the Sun's atmosphere: 1—ray passing only through the corona; 2—ray passing through the corona and the chromosphere. The figure is not to scale.

to obtain the expression for T_{eff} of emission leaving the corona (at the point A):

$$T_{\text{eff}} = T_c(1 - e^{-\tau_c(R_{\text{ch}})}) + T_{\text{ch}}\, e^{-\tau_c(R_{\text{ch}})}. \tag{28.5}$$

In (28.5) $\tau_c(R_{\text{ch}})$ is the optical thickness of the corona on the section BA (i.e. starting at the level R_{ch} where the emission leaves the chromosphere right up till it leaves the corona). In accordance with (28.2)

$$\tau_c(R_{\text{ch}}) = \int_{R_{\text{ch}}}^{\infty} \frac{\mu_j(R)\,dR}{\sqrt{1 - \dfrac{r^2}{n_j^2 R^2}}}. \tag{28.6}$$

In the expression given for T_{eff} we have already allowed for the fact that the chromosphere's optical thickness (section BC) is practically always such that $e^{-\tau_{ch}} \ll 1$.†

Calculations show (see section 26) that for radio waves whose trajectories pass through the chromosphere (i.e. for decimetric and shorter waves) the optical thickness of the corona $\tau_c(R_{ch})$ is small when compared with unity. Thanks to this circumstance the formula (28.5) becomes

$$T_{\text{eff}} \approx T_c \tau_c(R_{ch}) + T_{ch}, \qquad (28.7)$$

It is clear from this that the effective contribution of the corona to the emission received depends on the value of $\tau_c T_c/T_{ch}$ and may be significant even when $\tau_c \ll 1$ (since the ratio T_c/T_{ch} is large). For example, for $T_{ch} \sim 3\times 10^4$ °K and $T_c \sim 10^6$ °K the ratio $\tau_c T_c/T_{ch} \gtrsim 1$ when $\tau_c \gtrsim 3\times 10^{-2}$. In the centre of the solar disk values of this kind for the optical thickness of the corona obtain at wavelengths of $\lambda \gtrsim 20$ cm, and on the limb at wavelengths of $\lambda \gtrsim 10$ cm. It follows from this that in the millimetric band (and at the beginning of the centimetric) the corona is practically transparent and the observed radio emission comes largely from the chromosphere; its effective temperature is $T_{\text{eff}} \approx T_{ch} \sim 10^4$ °K. As the wavelength increases further the part played by the corona becomes more significant and is accompanied by a gradual increase in the effective radio emission temperature. Finally, at metric wavelengths the radio emission is generated chiefly in the corona: the corona becomes optically thick ($\tau_c \gg 1$) and the trajectories of the rays in this case, as a rule, avoid the chromosphere. Here T_{eff} reaches its maximum value of $T_c \sim 10^6$ °K.

This kind of function $T_{\text{eff}}(\lambda)$, obtained on the basis of the simplest isothermal model of the chromosphere and the corona, corresponds in its basic features to the observational data (see Chapter III). At the same time the obvious disadvantage of this model is that it cannot explain the complex behaviour of the frequency spectrum of the solar radio emission at millimetric wavelengths.

Continuing the analysis of the model used for the solar atmosphere let us see what radio brightness distribution over the disk characterizes this model. Above all the T_{eff} distribution will have circular symmetry because of the model's spherical symmetry. Furthermore, it is clear from the formulae (28.3), (28.5) that no brightening effect can occur towards the edge of the disk if $\tau_c (r = 0) \gg 1$, because in this case the central ray ($r = 0$) has the maximum possible temperature $T_{\text{eff}} = T_c$. This obstacle disappears if $\tau_c (r = 0) < 1$. Remembering that at metric wavelengths

† In the model of the chromosphere used this holds if the length of the section BC is greater than or of the order of 10^2 km. We recall in this connection that the extent of the chromosphere in altitude is 100 times greater.

§ 28] Theory of the "Quiet" Sun's Radio Emission

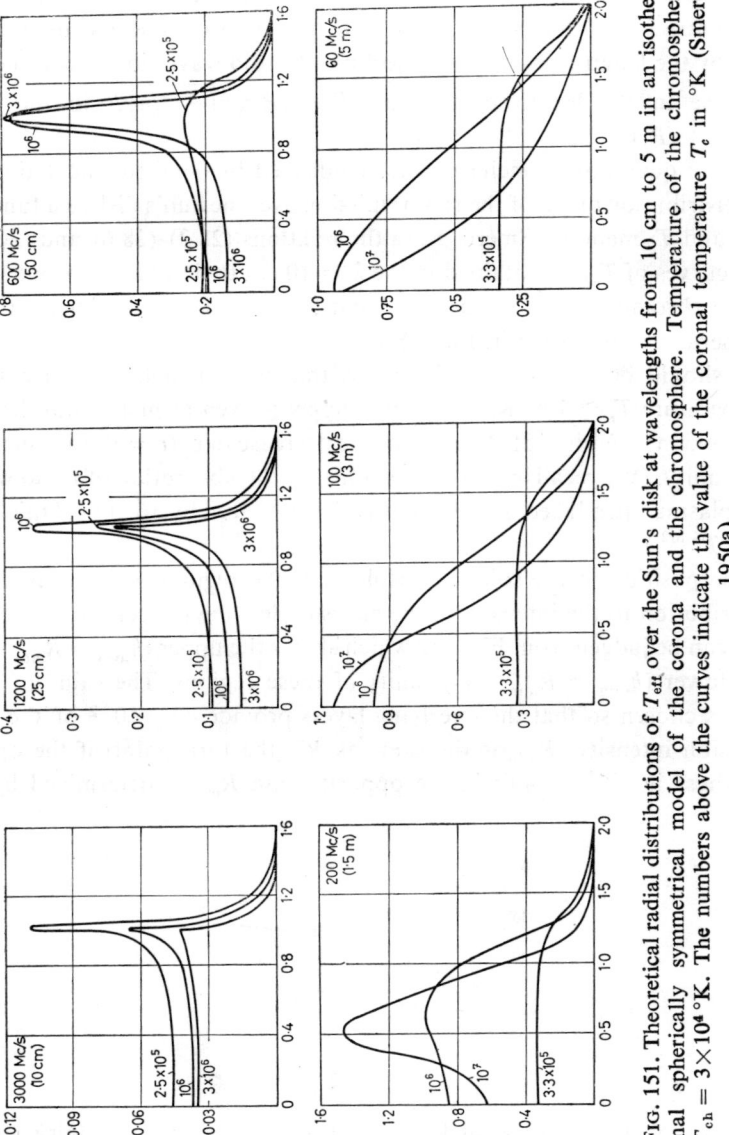

FIG. 151. Theoretical radial distributions of T_{eff} over the Sun's disk at wavelengths from 10 cm to 5 m in an isothermal spherically symmetrical model of the corona and the chromosphere. Temperature of the chromosphere $T_{ch} = 3 \times 10^4$ °K. The numbers above the curves indicate the value of the coronal temperature T_c in °K (Smerd, 1950a)

$\tau_c > 1$, whilst for shorter wavelengths $\tau_c < 1$, we conclude that the effect of an increase in brightness, generally speaking, can be observed only in the centimetric and decimetric wavebands.† Here it is connected with the rise in the optical thickness of the corona $\tau_c(r)$ (28.4) when moving from the centre of the Sun's disk ($r = 0$) to the limb ($r = R_\odot$) caused in its turn by the lengthening of the trajectory of radio waves in the corona (the increase in the length of the ray $dl = dR/\sqrt{1-r^2/n_j^2 R^2}$ in the layer $R-(R+dR)$).

These qualitative conclusions are confirmed by detailed calculations of the distribution of T_{eff} of the radio emission over the Sun's disk as a function of λ and T_c made by Smerd using the relations (28.3)–(28.6) and (26.74). The curves of $T_{\text{eff}}(r)$ obtained in the $\lambda = 10$ cm to 5 m band for one value of the chromosphere's temperature and three fixed values of the corona's temperature are shown in Fig. 151.

It should be pointed out, however, that with a high enough coronal temperature $T_c \gtrsim 10^6$ °K brightening appears even at metric wavelengths ($\lambda = 1.5$ m; see Fig. 151) because of the decrease in τ_c ($r = 0$) to values less than unity. (We recall in this connection that the absorption of radio waves in a plasma introduced by collisions is characterized by an optical thickness $\tau_j \propto T^{-3/2}$.)

The position in the solar atmosphere of the regions which make most contribution to the intensity of the emission leaving the central part of the disk can be judged from Fig. 152 which shows the upper ($h_{\text{max}} = R_{\text{max}} - R_\odot$) and lower ($h_{\text{min}} = R_{\text{min}} - R_\odot$) limits of these regions. The value of R_{max} here is chosen so that the overlying layers provide only 10^{-2} of the total emission intensity. R_{min} is the same as R^* (the turn point) if the optical thickness is $\tau(R^*) < 4.6$; in the opposite case R_{min} is determined by the

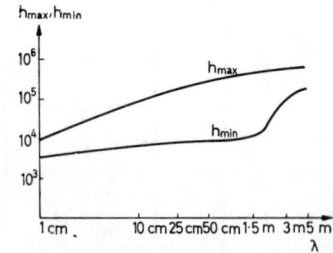

FIG. 152 Upper (h_{max}) and lower (h_{min}) limits of effectively emitting regions of the Sun's atmosphere in the centre of the disk at wavelengths of $\lambda = 1$ cm to 5 m ($T_{\text{ch}} = 3 \times 10^4$ °K; $T_c = 10^6$ °K) (Smerd, 1950a).

† See the end of this section for millimetric wavelengths.

§ 28] Theory of the "Quiet" Sun's Radio Emission

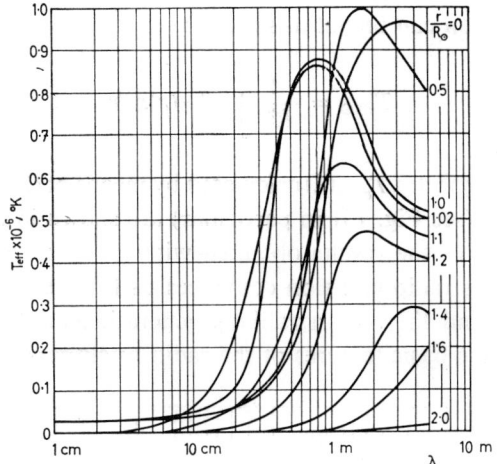

Fig. 153. Variation of T_{eff} with wavelength λ at different distances r from the centre of the Sun's disk ($T_{\text{ch}} = 3 \times 10^4\,°\text{K}$; $T_c = 10^6\,°\text{K}$) (Smerd, 1950a).

relation $\tau(R_{\min}) = 4\cdot 6$ in which $e^{-\tau} \approx 10^{-2}$. It is not hard to draw the conclusion from what has been said that the altitude of the effectively emitting layer increases as the wavelength rises and as we move away from the centre of the Sun's disk.

The frequency spectrum of the quiet Sun's radio emission $T_{\text{eff}}(\lambda)$ for a given local region of the solar disk can be found in principle from the combination of distributions $T_{\text{eff}}(r)$ for different wavelengths (of the type shown in Fig. 151). This kind of function $T_{\text{eff}}(\lambda)$, however, at different distances r from the centre of the disk can be followed more conveniently in the graphs of Fig. 153.

For the purpose of direct comparison with the results of measuring the spectral radio emission flux $S_\omega(\lambda)$ from all the unperturbed Sun as a whole (or, which is the same thing, the effective temperature $T_{\text{eff}\odot}(\lambda)$ reduced to the optical disk) it is also of interest to find the corresponding theoretical function $T_{\text{eff}\odot}(\lambda)$. Knowing the radio brightness distribution over the solar disk, it is not hard to find the value of $T_{\text{eff}\odot}$ from the relation (see (4.16))

$$T_{\text{eff}\odot} = \frac{1}{\pi R_\odot^2} \int T_{\text{eff}}\, d\sigma.$$

Here $\pi R_\odot^2$ is the area of the solar disk; the integral is taken over the whole area of the sky σ where the effective temperature of the B-component is non-zero. In the case of a radio brightness distribution with circular symmetry this equation can be put as follows:

$$T_{\text{eff}\odot} = \frac{2}{R_\odot^2} \int_0^\infty T_{\text{eff}}(r) r \, dr. \tag{28.8}$$

The frequency spectra $T_{\text{eff}\odot}(\lambda)$ calculated by means of (28.8) from known functions $T_{\text{eff}}(r)$ (see Fig. 151 in particular) are shown in Fig. 154.

A comparison of the results obtained with the observable features of the "quiet" Sun's radio emission convinces us that the hypothesis of its thermal nature and the bremsstrahlung generation mechanism gives the correct idea of the basic features of the radio brightness distribution over

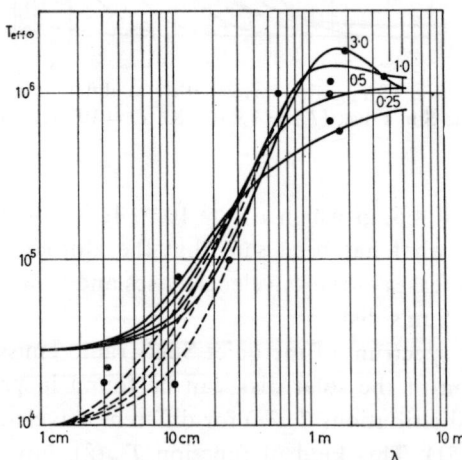

FIG. 154. $T_{\text{eff}\odot}$ as a function of the wavelength λ for different chromosphere and corona temperatures. Solid lines—$T_{\text{ch}} = 3 \times 10^4\,°\text{K}$, dotted lines—$T_{\text{ch}} = 10^4\,°\text{K}$. The values of $T_c \times 10^{-6}$ are shown by the figures above the curves. The dots are some experimental values of $T_{\text{eff}\odot}$ for the quiet Sun (Smerd, 1950a).

the disk and of the B-component's frequency spectrum even within the framework of the simplest model of the solar atmosphere. For example, it explains: (1) the rise in the effective temperature towards the edge of the disk in the centimetric–decimetric band (including $\lambda \sim 1\cdot 5$ m) and the monotonic curve of the radial distribution $T_{\text{eff}}(r)$ at longer wavelengths; (2) the great size of the "radio Sun" (when compared with the optical disk) in the metric band; (3) the increase of $T_{\text{eff}\odot}$ with a rise in wavelength from values around T_{ch} in the centimetric band to T_c at metric wavelengths.†

† We note in passing that the best agreement of the theoretical and experimental functions $T_{\text{eff}\odot}(\lambda)$ is achieved if we assume that the chromosphere temperature is $T_{\text{ch}} = 10^4\,°\text{K}$; when $T_{\text{ch}} = 3 \times 10^4\,°\text{K}$ the agreement is worse (see Fig. 154).

§ 28] Theory of the "Quiet" Sun's Radio Emission

At the same time on the basis of the theory developed above of the "quiet" Sun's radio emission we cannot interpret a whole series of very important details of the frequency spectrum and the distribution of the radio brightness over the disk, namely: (1) the complex curve of the function $T_{\text{eff}\odot}(\lambda)$ at millimetric wavelengths; (2) the increase in the brightness of the central part of the disk at $\lambda \sim 1$ cm; (3) the characteristic elongation of the radio-isophots in the equatorial direction (decimetric-metric band); (4) the absence of brightening towards the edge of the disk in the polar regions at $\lambda \sim 10\text{--}20$ cm; (5) the displacement of the maximum of T_{eff} from the limb towards the centre of the solar disk as λ rises at decimetric wavelengths; (6) the considerable decrease in T_{eff} in the centre of the disk as λ increases and the large effective radio diameter of the Sun at metric wavelengths when compared with the theoretical one.

INTERPRETATION OF CERTAIN FEATURES IN THE DISTRIBUTION OF THE RADIO BRIGHTNESS OVER THE SUN'S DISK ON THE BASIS OF MORE COMPLEX MODELS OF THE CORONA AND CHROMOSPHERE

The facts stated above, however, do not cast any doubt on the correctness of the hypothesis of the thermal nature and bremsstrahlung mechanism of the "quiet" Sun's radio emission. They are indubitably caused only by the imperfection of the model of the solar atmosphere used above[†] in which we ignore such very important features as the clear-cut non-isothermal nature and spicular structure of the chromosphere, the possible non-isothermal nature of the corona and the violation of the latter's spherical symmetry connected with the fall in the density of the coronal material at high latitudes and with the presence of coronal inhomogeneities.

Let us see what possibilities for the theory of the "quiet" Sun are opened up by allowing for these features.

Above all it is clear that unless we take into consideration the scattering of radio waves on coronal inhomogeneities, then the cause of the significant decrease in the metric band of the effective temperature of the central part of the disk as the wavelength rises may be only the fall in the corona's kinetic temperature with altitude (since for radial rays the optical thickness of the corona at metric wavelengths is greater than unity). Nevertheless,

---

† The high degree of brightening at decimetric wavelengths (Fig. 151) obtained in theory when compared with the observed values (Fig. 23) cannot be entirely ascribed to the model's drawbacks: it is quite possible (Smerd and Wild, 1957) that the data of the corresponding measurements are far too low because of the "smoothing" effect of the polar diagrams with insufficient angular resolution. This is also indicated by the systematically higher values of the degree of brightening found from eclipse observations (see sections 5 and 9).

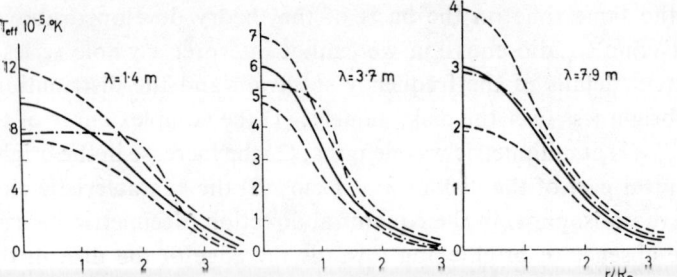

FIG. 155. Comparison of observed radial distributions of radio brightness at metric wavelengths with the theoretical ones: solid curves—O'Brien's experimental data (O'Brien, 1953a), dotted lines—limits of possible observational errors, dashes and dots—results of calculating $T_{eff}(r)$ in the O'Brien–Bell model of the corona.

according to O'Brien and Bell (1954), we cannot by merely changing T_c simultaneously explain the decrease found in $T_{\rm eff}$ in the centre of the disk and the great size of the radio Sun in the metric band. It follows from Fig. 155 that satisfactory agreement of the theory with the results of O'Brien's observations (1953a) can be achieved if at the same time as there is a decrease in temperature in the outer corona $(1\cdot 2R_\odot < R < 3R_\odot)$ as $T_c = 2\times 10^6 (R/R_\odot)^{-3}$ °K we increase the electron concentration in this region when compared with the Baumbach–Allen distribution by putting $N = 1\cdot 6\times 10^8 (R/R_\odot)^{-5}$ electron/cm³. The increase in the optical thickness of the corona's upper layers connected with this will lead to a slower fall in $T_{\rm eff}$ as we move away from the centre of the solar disk.

The suggestions about the distribution of N and T in the corona just mentioned are not necessary, however. As Scheffler has shown (Scheffler, 1958), the experimental results of O'Brien (1953a) can be explained well on the basis of a model of the corona with $T_c = $ const and a Baumbach–Allen distribution of the mean electron concentration $\bar{N}$, if we allow for the presence in the corona of inhomogeneities that cause effective scattering of metric waves. This, on the one hand, increases the radio size of the Sun;† on the other hand, the $T_{\rm eff}$ in the centre of the disk, since the total radio emission flux obviously remains constant when the radio diameter increases because of scattering, decreases. Without dwelling on the details of the complicated calculation of emission from an irregular corona we shall

† In the presence of non-uniformities the slower fall of $T_{\rm eff}$ as we move away from the centre of the disk may be connected not only with the scattering of the emission but also with the increase in the optical thickness $\tau_c \approx \int N^2\, dl$ when compared with the value of $\tau_c \approx \int (\bar{N})^2\, dl$ in the regular corona. This effect, however, was not taken into consideration by Scheffler who assumed that in the corona the condition $\bar{N^2} - (\bar{N})^2 \ll (\bar{N})^2$ is satisfied.

§ 28] Theory of the "Quiet" Sun's Radio Emission

proceed direct to Scheffler's results; they are shown in Fig. 156, which also gives the radial distributions of the effective temperature in the case of an isothermal model of the corona with a regular variation of the concentration $N(R)$ in accordance with (1.1). These distributions differ considerably

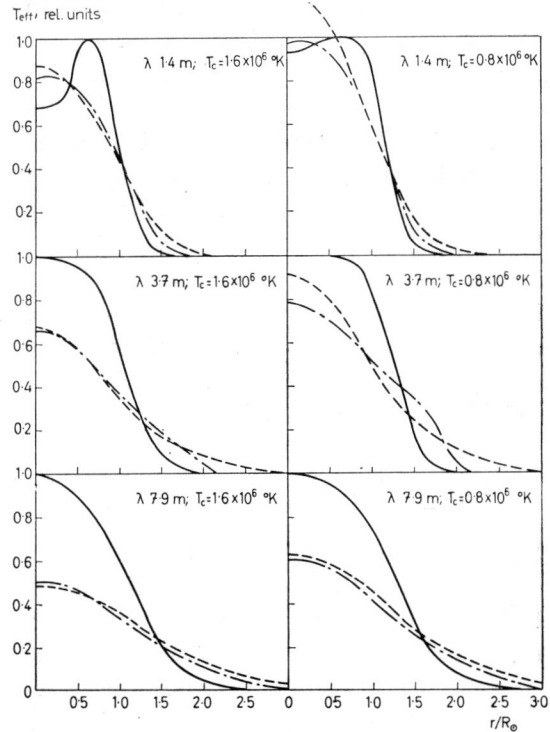

Fig. 156. Interpretation of the observed functions $T_{\text{eff}}(r)$ in the metric band on the basis of a corona model with an irregular distribution of the electron concentration. Solid curves—theoretical distributions of radio brightness in a Baumbach–Allen isothermal model with a regular $N(R)$ curve; dashed line—O'Brien's experimental results (1953a) normalized so that the radio emission flux from the whole Sun is the same as the value obtained from theory; dashes and dots—radio brightness distribution over the disk found with allowance made for the scattering of radio waves or coronal inhomogeneities (Scheffler, 1958).

from those obtained by O'Brien (1953b), but at the same time the results of O'Brien (1953b) agree closely with the functions $T_{\text{eff}}(r)$ found when scattering in the corona is taken into consideration.†

† It is clear from Fig. 156 that the agreement between experiment and theory becomes fuller with a corona temperature of $T_c = 1 \cdot 6 \times 10^6$ °K than with $T_c = 0 \cdot 8 \times 10^6$ °K.

Unfortunately the later (and more accurate) results of investigations into the radio brightness distribution at metric wavelengths given in section 9 have not yet been analysed from the point of view of their correspondence with one or other model of the corona. All that we shall say is that the distribution obtained by Firor in measurements at $\lambda = 1\cdot45$ m (Fig. 23) agrees closely with the distribution of Fig. 154 for the simplest model of the solar atmosphere with $T_c = 10^6$ °K (Firor, 1956). In a corona model with an irregular electron concentration distribution this means that the effect of scattering of radio waves on the nature of the radio brightness distribution over the disk of the "quiet" Sun at $\lambda \sim 1\cdot5$ m is insignificant. Since the basic scattering at $1\cdot5$ m occurs in the inner layers of the corona we conclude that at low altitudes in the corona the true value of the degree of scattering is smaller than that accepted by Scheffler.

The characteristic displacement of the maximum in the radio brightness distribution at these wavelengths (from the limb closer to the centre of the disk) can be explained in the framework of the simplest model of the solar atmosphere by the effect of refraction. A specific feature of the travel of rays at $\lambda \sim 1\cdot5$ m is that the turning point R^*, located in the chromosphere for central rays, reaches out into the corona for rays escaping beyond the solar shell at distances of $r \gtrsim 0\cdot6R_\odot$. Therefore as we move away from the centre of the disk T_{eff} first rises because of the increase in the optical thickness $\tau_c(R_{\text{ch}})$ read from the boundary of the chromosphere;† the latter in its turn is connected with the increase in the length of the ray in the corona (see (28.6)). However, after the turning point has moved into the corona (which occurs at a distance $r \sim 0\cdot6R_\odot$ from the centre of the disk; see Fig. 113) the effective temperature starts to decrease together with $\tau_c(R^*)$ because of the reduction of the ray's length. Formally this is caused by a decrease in the integration interval in the formula (28.4), which in the case of an isothermal Baumbach–Allen corona is only partially compensated by an increase in the integrand together with r.

A more complex problem is the interpretation of the considerable shift of the maximum of T_{eff} into the disk in the decimetric band (into the region $r \sim (0\cdot6-0\cdot8)R_\odot$ at $\lambda \sim 50$–60 cm; see Figs. 19 and 20). An explanation similar to that given above is unsuitable here if the electron concentration in the corona is given as before by the formula (1.1): refraction in the corona at 50–60 cm is insignificant and the turning point remains in the chromosphere right up to values of $r \approx R_\odot$ (see Fig. 113). This leads (in contradiction to the observations) to an increase in brightness at the limb and not inside the disk.

† Of course only if with $r = 0$ the optical thickness is $\tau_c(R_{\text{ch}}) < 1$, i.e. with a high enough corona temperature (Fig. 151, $\lambda = 1\cdot5$ m).

§ 28] Theory of the "Quiet" Sun's Radio Emission

The refraction effect will become sufficient for a shift of the radio brightness maximum into the region $r \sim 0.6 R_\odot$ at $\lambda \sim 50$ cm only in a corona whose density is 9 times higher when compared with the Baumbach–Allen distribution.[†] Such a high density is very improbable at a period of the corona's minimum phase. If the shift is in reality less (according to Fig. 20 T_{eff} at $\lambda = 60$ cm reaches a maximum with $r \approx 0.8 R_\odot$), then the density should be only quadrupled. However, it is difficult to be reconciled with this rise. We note that from this point of view the shift of the maximum at $\lambda \sim 50$–60 cm and brightening at $\lambda \sim 1.5$ m cannot be observed simultaneously; this statement can be checked experimentally.

Another possible way of explaining the shift of the maximum radio brightness into the disk at decimetric wavelengths has been discussed by Panovkin (1957) on the basis of a more complex (non-isothermal) model of the corona with a maximum kinetic temperature $T_{\max}$ in the layers $R \sim (1.1$–$1.3) R_\odot$. He proceeded from the following considerations. If at $\lambda \sim 50$–60 cm near the Sun's limb the corona's optical thickness τ_c reaches unity in layers $R \sim 1.3 R_\odot$ with a reduced temperature $T_c < T_{\max}$, then closer to the centre of the disk the level $\tau_c(R) \sim 1$ will obviously move into deeper layers of the corona—first into the region $R \sim (1.1$–$1.3) R_\odot$ with $T_c \sim T_{\max}$, and then still lower into the region with $T_c < T_{\max}$. Since the coronal levels at which $\tau_c(R)$ reaches unity compose an effectively emitting layer with a T_{eff} equal to the kinetic temperature T of this layer, it is clear from what has been said that a model of this kind can provide a complete explanation of the features of the radial radio brightness distribution with a maximum T_{eff} inside the disk. The high values of the optical thickness

$$\tau_c \sim \int N^2 \, dl$$

at $\lambda \sim 50$–60 cm necessary for interpreting this kind of $T_{\text{eff}}(r)$ distribution are obtained by Panovkin (1957) by assuming the corona to be highly non-uniform (i.e. the ratio $[\overline{N^2} - (\overline{N})^2]/(\overline{N})^2$ is large).

It remains unclear, however, to what extent this model of the corona can be made to agree with the quantitative characteristics of the radio brightness distribution in the decimetric band, where the increase in the effective temperature of the solar disk connected with the rise in τ_c will be very considerable. For example, in the model of Panovkin (1957) when

---

[†] The refraction is determined by the distribution of the value of the refractive index $n_{1,2}$ in the corona. Therefore the displacement of the T_{eff} maximum at $\lambda \sim 50$ cm into about the same region of the disk as at $\lambda \sim 1.5$ m will be achieved if the values of $n_{1,2}$ at decimetric wavelengths are the same as the values in a Baumbach–Allen corona at metric wavelengths. Since $1 - n_{1,2}^2 \propto N/\omega^2 \propto N\lambda^2$ (see (22.17)), we must increase N by a factor of ~ 9 for this case.

interpreting the shift of the radio brightness maximum at $\lambda \sim 50$ cm it is absolutely necessary that for the region $r \sim 0.6 R_\odot$ where T_{eff} reaches its maximum value the optical thickness of the corona $\tau_c(R_{\text{ch}})$ should be greater than or comparable with unity. Then at $\lambda \sim 20$ cm $\tau_c(R_{\text{ch}}) \gtrsim 0.16$† and therefore $T_{\text{eff}} > \tau_c T_c \gtrsim 1.6 \times 10^5$ °K (with $T_c \sim 10^6$ °K). The estimate obtained is not less than three times higher than the observed values of T_{eff}: according to Fig. 18a at $\lambda \sim 20$ cm $T_{\text{eff}} \approx 5 \times 10^4$ °K for sectors of the solar disk at a distance $r \lesssim 0.6 R_\odot$ from the centre. At metric wavelengths the enhanced values of τ_c will cause an increase in the radio diameter of the Sun when compared with Scheffler's model based on the assumption of the low degree of non-uniformity of the corona which agrees well with observational results. These circumstances, however, are not particularly significant; it is probable that with a more careful choice of model parameters here satisfactory agreement can be achieved between the data of theory and experiment.

As for the characteristic "elongated shape" of the Sun's radio isophots along the equator which has been established from observations in the metric and decimetric wavebands, it may be connected with the considerable decrease in the density in the polar regions of the solar corona when compared with the equatorial regions. In all probability this also causes the absence of an increase in brightness towards the edge of the disk in the directions of the poles at decimetric wavelengths which occurs when the rise in the corona's optical thickness

$$\tau_c \propto \int N^2 \, dl$$

The great length of a ray in the corona near the limb cannot compensate for the rapid decrease in τ_c because of the fall in the electron concentration at high latitudes (see Christiansen and Warburton, 1955; Swarup, 1961a and 1961b).‡ It is quite natural that the differences from circular symmetry of the radio isophots gradually disappear with the transition from the short waves of the centimetric band where the part played by the corona in creating the "quiet" Sun's radio emission becomes more and more meagre (see p. 516).

In a theoretical treatment of the B-component in the metric and decimetric bands where emission from the corona predominates we can limit ourselves to allowing very simply for the effect of the chromosphere in the framework of an isothermal and spherically symmetrical model of the latter (which has actually been done above). However when we move on

† The absorption with $n_{1,2} \approx 1$ connected with collisions is proportional to λ^2 (see section 26).

‡ The optical observations of the K-corona (section 1) indicate the decrease in the density of the coronal plasma in the polar regions.

§ 28] Theory of the "Quiet" Sun's Radio Emission

to the centimetric and particularly to the millimetric wavelengths, at which the chromosphere makes the major contribution to the "quiet" Sun's radio emission, this kind of model becomes inadequate. It is quite clear that it cannot be used to explain either the complex curve of the function $T_{\text{eff}\odot}(\lambda)$ in the millimetric waveband (with a minimum at $\lambda \sim 6$ mm and a maximum at $\lambda \sim 4$ mm; see Figs. 9 and 10) or the particular distribution $T_{\text{eff}}(r)$ established from observations at $\lambda = 8.6$ mm (with an increase in the brightness on the limb and at the centre of the disk; see Fig. 14). In fact the contribution from the corona here is negligibly small and the isothermal, optically thick chromosphere over the whole millimetric band and at all points on the disk generates radio emission with the same effective temperature $T_{\text{eff}} = T_{\text{ch}}$.

It is easy to explain brightening on the disk if we remember (in full accordance with the data of optical observations) that the temperature of the chromosphere rises as we move away from the solar surface. Then T_{eff} of the radio emission will increase from the centre of the disk towards the edge since as r rises the effectively emitting region moves into the higher and more strongly heated layers of the chromosphere.

The more complex curve of the radial distribution $T_{\text{eff}}(r)$ at $\lambda = 8.6$ mm in the framework of a spherically symmetrical model of the chromosphere may be caused only by the non-monotonic dependence of the kinetic temperature T on the altitude h in the layers which are responsible for creating the observed radio emission. The minimum of T_{eff} at a distance of $r \approx 0.9 R_\odot$ occurs provided that at the altitudes where the radio emission is generated that escapes from a ring of radius $r \approx 0.9 R_\odot$ there is a minimum of the kinetic temperature T. The enhanced values of T in the lower layers of the chromosphere will then cause an increase in brightness towards the centre of the disk and the high temperature of the upper layers brightening on the limb. However, the assumption about this law governing the variation in T in the layers of the chromosphere under discussion does not apparently correspond to the truth since in the opposite case the frequency spectrum of $T_{\text{eff}\odot}(\lambda) \approx T_{\text{eff}}(\lambda)$† should have a minimum in the region of $\lambda \sim 8.6$ mm. The known experimental data (Figs. 9 and 10) rather indicate a steady increase in the effective temperature with wavelength, although the absence of measurements in the millimetric band at $\lambda > 8.6$ mm does not allow us to state this absolutely categorically.

The assumption of the non-monotonic temperature distribution in the

---

† Since in the millimetric band the degree of non-uniformity in the distribution of T_{eff} over the disk is comparatively small and the size of the radio Sun is close to the size of the optical disk, making these temperatures identical will not produce a large error.

chromosphere becomes unnecessary when interpreting the observed function $T_{\text{eff}}(r)$ at the end of the millimetric band,[†] if we take into consideration the fine structure of the chromosphere—the presence of cold spicules disseminated in a hotter gas (section 1). It is easy to check that this kind of two-component model of the chromosphere introduced by Hagen (1956 and 1957) into the theory of the "quiet" Sun's radio emission agrees qualitatively with the observed function $T_{\text{eff}}(r)$ at the end of the millimetric band.

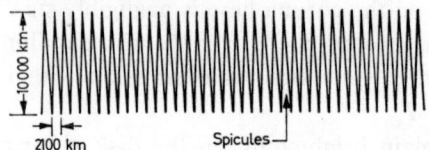

FIG. 157. Hagen's model of a two-component chromosphere (Hagen, 1956 and 1957).

In actual fact for the sake of simplicity let all the spicules have the form of cones with the same height and base diameter, the axes of the cones running vertically in relation to the solar surface (Fig. 157). Then in the centre of the disk the radio emission is created by the upper parts of the cold and dense spicules and by the deeper-lying layers of the hot and rarefied interspicular material. A contribution to the effective temperature averaged over the structural elements is made by kinetic temperatures of both regions mentioned, taken with an appropriate weighting which allows for the relative area of the emitting layers in the projection onto the solar disk. Nearer to the edge of the disk the spicules, being inclined to the line of sight, partly screen the emission from the hotter regions, increasing the relative contribution of the cold spicules to the value of T_{eff}. This leads to a decrease in the effective temperature as the distance from the centre of the disk increases. However, in the immediate vicinity of the limb a sharp increase in T_{eff} starts once again. Here it is connected with the rapid increase in the path travelled by the ray in the hot component above the spicules corresponding to the rise in the optical thickness and the intensity of emission from the upper layers of the chromosphere.

The radio brightness distribution in the millimetric waveband expected on the basis of a two-component model of the chromosphere has been

[†] We cannot do without this assumption, however, if we wish to explain the form of the frequency spectrum of the "quiet" Sun's radio emission at millimetric wavelengths. The complex nature of the observed variation of $T_{\text{eff}}(\lambda)$ undoubtedly indicates the non-monotonic nature of the dependence of the mean temperature over the structural elements on the altitide in the layers of the lower chromosphere responsible for creating radio emission at $\lambda \sim 4\text{–}6$ mm (see next section).

§ 28] Theory of the "Quiet" Sun's Radio Emission

calculated by Coates (1958d) and Athay (1959) (see also Thomas and Athay, 1961). The first of them approximated the spicules by cylinders $1{\cdot}5\times10^3$ km in diameter and from 3×10^3 to $1{\cdot}8\times10^4$ km high, assuming that the spicules are randomly distributed over the solar surface. The nature of the height distribution and the dispersion of their orientations

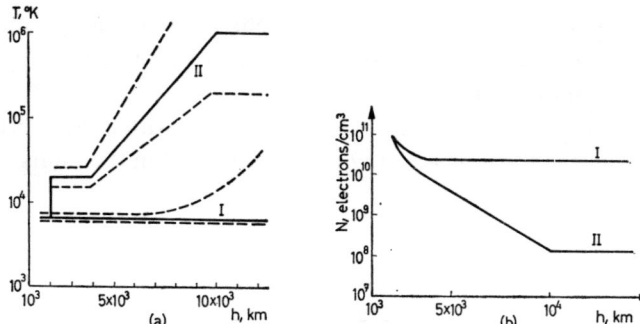

FIG. 158. Variation of the kinetic temperature (a) and electron concentration (b) in Coates's two-component model of the chromosphere (Coates, 1958d): I—in spicules, II—in the interspicular material.

relative to the radial directions were chosen by proceeding from the data of optical observations. The dependences of the temperature T and the electron concentration on the altitude above the photosphere h in the spicules and in the interspicular material used in the calculations are shown in Fig. 158. The temperature of the interspicular material at altitudes $h \sim (2\text{--}4)\times10^3$ km ($T = 1{\cdot}9\times10^4$ °K) and the temperature of the spicules ($T = 6{\cdot}4\times10^3$ °K) were selected so that the results of radio observations were best satisfied; at the same time the variation in the temperature with altitude in the regions between the spicules agrees in general outline with the optical data (section 1). The values of the electron concentration were found from the relation $p = 2N\varkappa T$ from known values of the temperature on the assumption that the pressure p in the chromosphere is the same in the spicules as well as in the interspicular space and does not vary with altitude.†

---

† For the values taken for T and N the absorption coefficient $\mu_{1,2}$ at $\lambda = 8{\cdot}6$ mm is $10^{-6}\text{--}6\times10^{-8}$ cm^{-1} in the spicules and $10^{-6}\text{--}2\times10^{-15}$ cm^{-1} in the interspicular material. (These values of $\mu_{1,2}$ are given for altitudes of $h \sim 2\times10^3\text{--}1{\cdot}8\times10^4$ km above the level of the photosphere.) It follows from this that the optical thickness $\tau_{1,2}$ in the spicules reaches unity over a path of only 10–200 km; if we remember that the diameter of the spicules is 1500–2000 km and $\tau_{1,2} \sim \lambda^2$, it becomes clear that the spicules remain optically thick right up to wavelengths of ~ 3 mm. The interspicular material, on the other hand, is transparent: the optical thickness reaches unity only at altitudes of about 2600 km (for $\lambda = 8{\cdot}6$ mm).

The results of calculating the radial distributions $T_{\text{eff}}(\lambda)$ in the millimetric band are shown in Fig. 159. A comparison of the theoretical function $T_{\text{eff}}(r)$ at $\lambda = 8\cdot 6$ mm with the experimental one (Fig. 14) shows that the spicular model of the chromosphere is in close agreement with the experimental data on the distribution of T_{eff} over the solar disk at $\lambda = 8\cdot 6$ mm. Here the theoretical distribution will correspond within the limits of the possible measurement errors to the observational data if the values of the temperature in the chromosphere remain within the limits marked by the dotted lines in Fig. 158a.

At $\lambda = 4\cdot 3$ mm the curve from scanning the Sun with a radio brightness distribution in the form of Fig. 159 is the same (within the limits of meas-

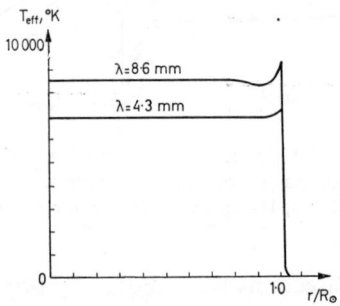

FIG. 159. Theoretical distribution of the radio brightness over the Sun's disk in Coates's model (1958d).

urement errors) as the similar experimental curve obtained by Coates (1958b) (see section 9). This, however, does not prove the presence of brightening on the limb following from the model of the chromosphere under discussion since the uniform distribution $T_{\text{eff}}(r)$ also agrees well with experiment. Below we shall show that, to judge from the data on the nature of the radio emission's frequency spectrum $T_{\text{eff}}(\lambda)$ at millimetric wavelengths, in the region of $\lambda \sim 5$ mm we can expect not an increase but a fall in the radio brightness on the Sun's limb.

We shall not discuss Athay's two-component model in detail (see Athay, 1959; Thomas and Athay, 1961). All that we shall say is that he proceeded from the model of Athay and Menzel (1956) which was compiled from the optical data and then modified by Thomas and Athay (1961). A number of parameters in this model, which were determined very roughly in experiment, were considered to be unknown; they were chosen so as best to satisfy the data of radio observations—a radio brightness distribution at $\lambda = 8\cdot 6$ mm and a steadily rising value of T_{eff} in the centre of the disk at $\lambda \sim 4$ mm–21 cm. The model obtained in this way differs significantly

§ 28] Theory of the "Quiet" Sun's Radio Emission

from Coates's model in the number of spicules and the temperature distribution; it also differs quite considerably from the original model (Athay and Menzel, 1956) which, as was pointed out by Shklovskii and Kononovich (1958), does not agree with the radio data.

CONSTRUCTION OF A MODEL OF THE SOLAR ATMOSPHERE FROM RADIO DATA

In the preceding subsections we have discussed the question of how far the different models of the Sun's atmosphere based chiefly on the results of observations in the optical part of the spectrum satisfy the known experimental data on the radio emission of the "quiet" Sun. We have been able to check that the idea of the thermal nature and the bremsstrahlung mechanism of this emission correctly conveys its basic features if we proceed from the information provided by optics on the structure and thermal conditions of the Sun's atmosphere. Therefore agreement with radio observational data may be a criterion of the correctness of one or another model of the Sun's atmosphere.

An obvious disadvantage of the above theory is the absence of a single model of the solar corona and chromosphere in whose framework successive argument of these concepts about the origin of the B-component would lead to a frequency spectrum and radio brightness over the Sun's disk corresponding to the observations. The agreement of the rather variegated models discussed (in each of which those characteristics of the solar atmosphere were stressed which were necessary for interpreting one or another feature of the "quiet" Sun's radio emission) and their combination into a unified model of the solar atmosphere is a far from simple problem. Considerable help in this respect may be provided (and is provided) by the solution of the converse problem of finding the parameters of the corona and chromosphere from the observed characteristics of the "quiet" Sun's radio emission.

In principle this problem can be stated as follows. At each point on the Sun's disk (and its immediate neighbourhood in the sky) let the function $T_{\text{eff}}(\lambda)$ be known.[†] We must find the distribution of the temperature T and the electron concentration N in the corona and the chromosphere—the two parameters which fully define the intensity of the "quiet" Sun's radio emission. The problem formulated in this way is ambiguous: from a known function of three variables $T_{\text{eff}}(\lambda, \theta, \xi)$ here we cannot find two functions of three variables $T(R)$ and $N(R)$ (θ, ξ are angular coordinates in the sky,

[†] We are ignoring here the circumstance that in reality the information available at present about the "quiet" Sun's radio emission is much more meagre: all that have been defined comparatively reliably are the radio brightness distributions for a small set of wavelengths and the frequency spectrum of the whole Sun $T_{\text{eff}\odot}(\lambda)$ (see Chapter III).

R is the radius vector of a point in the Sun's atmosphere).† It is clear from what has been said that by proceeding from the radio data we cannot at the same time also ignore the results of optical observations: only by combining them can we construct a model of the Sun's atmosphere that is adequate for the facts.

In the papers devoted to the question under discussion (see Hagen, 1951; Piddington, 1950; Shklovskii and Kononovich, 1958; De Jager, 1960; Piddington, 1954; Brooks and Oster, 1961; Zheleznyakov, 1964 d) the simpler problem is solved of finding the temperature distribution $T(h)$ from the function $T_{\text{eff}}(\lambda)$ at the centre of the disk known from radio observations (h is the altitude above the photosphere). According to (26.17b) in this case the effective radio emission temperature is

$$T_{\text{eff}} = \int_0^{\tau_L} T(h) e^{-\tau} \, d\tau, \qquad (28.9)$$

where τ_L is the optical thickness along the ray from the reflection point $n_{1,2}(\omega) = 0$ (i.e. from the level at which $\omega = \omega_L$) until escape from the corona, τ is the current value of the optical thickness from the altitude h until escape from the corona:

$$\tau(h) = \int_h^\infty \mu(h) \, dh. \qquad (28.10)$$

The bremsstrahlung absorption coefficient μ is of the form

$$\mu = \frac{0\cdot 58 a}{\omega^2 n_{1,2}(h)} \frac{N^2(h)}{T^{3/2}(h)} \qquad (28.11)$$

(see (26.74), (26.78) and (26.79)). Here a is the logarithmic factor that figures in the expressions for the effective number of collisions ν_{eff} and is weakly dependent on the quantities T and N.

The relation (28.9) together with (28.10) and (28.11) is a complex integral equation in $T(h)$, whose precise solution in analytical form (with a given electron concentration on the basis of optical data) is very difficult. The calculation becomes simpler, however, if instead of $N(h)$ we take the dependence of $N^2/T^{3/2}$ on altitude. In accordance with (28.10) and (28.11) this function determines the connection between τ and h if $a \approx \text{const}$ and $n_{1,2} \approx 1$; the latter is quite permissible in layers of the corona and chromo-

† In practice from experiment we can obtain only the function $T_{\text{eff}}(\lambda, \theta, \xi)$ averaged over the fluctuations of the effective temperature connected with the fine structure of the Sun's atmosphere (spicules and coronal inhomogeneities). As a result we lose the data on the scale of these non-uniformities and the arbitrariness in the choice of the distributions $T(R)$ and $N(R)$ that agree with experiment becomes still greater.

§ 28] Theory of the "Quiet" Sun's Radio Emission

sphere responsible for generation of radio emission at wavelengths of $\lambda \lesssim 1$ m.[†] This fact was used by Piddington (1950, 1954) (see also Shklovskii and Kononovich, 1958)) for calculating the kinetic temperature in the upper chromosphere.

Approximating the radio emission spectrum of the "quiet" Sun in the range from $\lambda \sim 4$ cm to $\lambda \sim 1.5$ m by the function[‡]

$$T_{\text{eff}}(\omega) = \frac{b}{\omega}, \qquad (28.12)$$

where $b = 9.4 \times 10^{14}$ degrees per cycle per second (see (8.2)), we obtain the integral equation for T in the form

$$b\omega = \int_h^\infty T e^{-\frac{\alpha}{\omega^2}} \, d\alpha, \qquad (28.13)$$

where

$$\alpha = 0.58 \int_h^\infty \frac{N^2}{T^{3/2}} \, dh \qquad (28.14)$$

((see (28.9)–(28.11))). In the change from (28.9) to (28.13) the limit τ_L is replaced by infinity since in the range of wavelengths of interest to us $\tau_L \gg 1$. The relation (28.13) can be looked upon as a Laplace transform for T; it follows from this that

$$T = \frac{b}{\sqrt{\pi\alpha}}, \qquad (28.15)$$

i.e.

$$T(h) = 1.2 \cdot 10^{15} \left(\pi a \int_h^\infty \frac{N^2}{T^{3/2}} \, dh \right)^{-1/2}. \qquad (28.16)$$

The range of temperatures T in which this expression is valid corresponds approximately to the range of effective temperatures $T_{\text{eff}} \sim 2 \times 10^4 - 7.5 \times 10^5$ °K in which an approximation like (28.12) is permissible.

When using (28.16) to find $T(h)$ Piddington based his work, on the optical observations of Wildt (1947), according to which in the chromosphere at

---

[†] At these wavelengths the effectively emitting layers with $\tau(h) \lesssim 1$ are located far higher in the Sun's atmosphere than the levels $n_{1,\,2}(\omega) = 0$.

[‡] This function differs from the approximation in the form $T_{\text{eff}} = a + b/\omega$ used by Piddington (1950, 1954). However, for this range of wavelengths $a \ll b/\omega$ and the first term may be neglected. The difference becomes noticeable in the millimetric band where, however, both approximations become unsuitable: as we showed in section 8, the frequency spectrum of the solar radio emission in this band is more complex in character.

altitudes h from $1 \cdot 5 \times 10^8$ to 10^9 cm

$$\frac{N^2}{T^{3/2}} \approx 1\cdot 4 \times 10^{16} e^{-1\cdot 2 \times 10^{-8} h}. \qquad (28.17\text{a})$$

Then
$$T(h) \approx 2\cdot 1 \times 10^2 e^{0\cdot 6 \times 10^{-8} h} \text{ degrees}. \qquad (28.18\text{a})$$

In the change to the last expression the logarithmic factor a is taken to be equal to 8·5 (see the formula (26.78)) with $T \sim 10^5$ °K, $N \sim 10^{11}$ electrons/cm^3). It follows from (28.17a) and (28.18a) that

$$N(h) \approx 6\cdot 6 \times 10^9 e^{-0\cdot 15 \times 10^{-8} h} \text{ electrons/cm}^3, \qquad (28.19\text{a})$$

which differs considerably, for example from the distribution (28.1). This is not surprising since the accuracy of the information about the quantity $N^2 T^{-3/2}$ leaves much to be desired. For example, if we proceed from the data of other radio observations (Athay et al. 1955) (in the continuous spectrum near $\lambda = 4700$ Å), according to which in the chromosphere at altitudes $h > 2\cdot 5 \times 10^8$ cm

$$\frac{N^2}{T^{3/2}} \approx 3\cdot 7 \times 10^{17} e^{-1\cdot 75 \times 10^{-8} h} \qquad (28.17\text{b})$$

(see Shklovskii and Kononovich, 1958), then

$$T(h) \approx 50 e^{0\cdot 875 \times 10^{-8} h} \text{ degrees}, \qquad (28.18\text{b})$$

$$N(h) \approx 1\cdot 1 \times 10^{10} e^{-0\cdot 22 \times 10^{-8} h} \text{ electrons/cm}^3. \qquad (28.19\text{b})$$

For constructing a model of the Sun's atmosphere we can use instead of the dependence of $N^2 T^{-3/2}$ on h the optical data on the distribution of the concentration N, although in this case the problem of finding $T(h)$ from the integral equations (28.9)–(28.11) becomes much more complicated. The following approximate and very simple method for its solution (Zheleznyakov, 1964d) can be suggested, however. We replace the relation (28.9) by the approximate equality

$$T_{\text{eff}}(\lambda) \approx T(h^*), \qquad (28.20)$$

in which h^* is the altitude where

$$\tau(h^*) = \int_{h^*}^{\infty} \mu(h)\, dh = 1; \qquad (28.21)$$

the quantity h^* obviously depends on the wavelength λ. Substituting the expression (28.11) in (28.21), we obtain:

$$\int_{h^*}^{\infty} \frac{0\cdot 58 a N^2}{T^{3/2} n_{1,2}}\, dh = \frac{4\pi^2 c^2}{\lambda^2}. \qquad (28.22)$$

§ 28] Theory of the "Quiet" Sun's Radio Emission

The factor $n_{1,2}$ here can be dropped for waves in which with sufficient accuracy $n_{1,2}(h \gtrsim h^*) = 1$. Then differentiating the equality (28.22) with respect to h^*, we obtain:

$$\left(\frac{0 \cdot 58 a N^2}{T^{3/2}}\right)_{h=h^*} = \frac{8\pi^2 c^2}{\lambda^3} \frac{d\lambda}{dh^*},$$

which with allowance made for (28.20) gives:

$$0 \cdot 58 (aN^2)_{h=h^*} \, dh^* = \frac{8\pi^2 c^2}{\lambda^3} T_{\text{eff}}^{3/2}(\lambda) \, d\lambda. \tag{28.23}$$

Integrating once more we arrive at the equality†

$$0 \cdot 58 \int_{\infty}^{h^*} a N^2 \, dh = 8\pi^2 c^2 \int_{\infty}^{\lambda} \frac{T_{\text{eff}}^{3/2}(\lambda)}{\lambda^3} \, d\lambda. \tag{28.24}$$

The parameter a depends on N and on T; however, this dependence is weak and it may be neglected entirely. The equality (28.24) defines the connection between h^* and λ, which can easily be found graphically if we know the functions $N(h)$ and $T_{\text{eff}}(\lambda)$. This connection together with the equality (28.20) allows us to obtain the height distribution of the temperature $T(h^*)$.‡

The limits of applicability of the above method for calculating $T(h)$ is determined by the conditions under which the replacement of the precise relation (28.9) by the approximate equality (28.20) becomes correct. In section 26 we have noted that this operation is permissible only if the observed radio emission is generated chiefly in layers with an optical thickness of $\tau(h) \sim 1$. For the "quiet" Sun's radio emission this kind of procedure will lead to considerable errors in the decimetric band, where we cannot neglect the imposing contribution made by the optically thin corona which is at a high temperature (see (28.7), and below).§ However, in the millimetric band the accuracy of this method is not very high, so that a better analysis is needed to obtain reliable quantitative data.

† Here and in the integrand we have replaced h^* by h.
‡ De Jager (1960) suggested another method for finding $T(h)$ from known $T_{\text{eff}}(\lambda)$ and $N(h)$ based on the application of the theorem of the mean to the integral (28.21); as a result the function $T(h)$ is obtained by the method of successive approximations. This method of calculation is clumsier than that given. In addition, it is definitely unsuitable in the case when $T_{\text{eff}}(\lambda)$ has a sharply non-monotonic character (for example, at millimetric wavelengths).
§ In the decimetric band we can assume (as was done above) that T_{eff} is inversely proportional to the frequency ω and use the relation (28.16) for calculating $T(h)$.

The distribution of the kinetic temperature in the chromosphere has been found by this method by Zheleznyakov (1964d). The function $T(h)$ in the lower chromosphere obtained there is shown in Fig. 160. In the calculation allowance was made for the latest experimental data on the complex behaviour of the function $T_{\text{eff}}(\lambda)$ at millimetric wavelengths, it being taken for the sake of definition that the frequency spectrum here is the same as the curve in Fig. 9. From the qualitative point of view this curve correctly characterizes the actual function $T_{\text{eff}}(\lambda) \approx T_{\text{eff}\odot}(\lambda)$ at millimetric wavelengths, although in the quantitative respect it undoubtedly

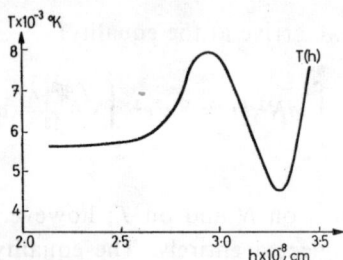

FIG. 160. Kinetic temperature T in the lower chromosphere as a function of the altitude h obtained by Zheleznyakov (1964d) on the basis of results of measurements of the frequency spectrum of the "quiet" Sun's radio emission $T_{\text{eff}\odot}(\lambda)$ provided that the electron concentration in the chromosphere is distributed as (28.25)

needs to be made more accurate; for this purpose we can adduce Kislyakov's results in the range $\lambda \sim 3.5$–7 mm obtained by the unified method (see Fig. 10). The function $T_{\text{eff}}(\lambda)$ in the centimetric–decimetric band was taken with allowance made for Fig. 11; the values of T_{eff} in these bands still make a noticeable contribution to the integral

$$\int_{\infty}^{\lambda} \lambda^{-3} T_{\text{eff}}^{3/2}(\lambda)\, d\lambda$$

when calculating $h^*(\lambda)$ at millimetric wavelengths.

The electron concentration distribution in the chromosphere for $h < 10^4$ km was taken in the form

$$N(h) = \exp(\alpha h^2 + \beta h + \gamma); \tag{28.25}$$

this function with

$$\alpha = 5.8 \times 10^{-18}\ \text{cm}^{-2}, \quad \beta = -1.28 \times 10^{-8}\ \text{cm}^{-1}, \quad \gamma = 27.24$$

gives a good approximation to the dependence of the electron concentration on altitude shown in Table 1 (section 1) for the cold elements of the

§ 28] Theory of the "Quiet" Sun's Radio Emission

lower chromosphere and the interspicular material of the upper chromosphere. The contribution of the higher layers of the Sun's atmosphere to is

$$\int_{\infty}^{h*} N^2\,dh$$

insignificant when calculating the function $h^*(\lambda)$ at millimetric wavelengths.

If we remember that Table 1 determines only a "working" model of the chromosphere, it becomes clear from everything that has been said that the graph of $T(h)$ in Fig. 160 is to a certain extent illustrative in nature; more and more extensive investigations into the solar emission in the radio and optical parts of the spectrum are definitely necessary to make it more accurate.

The values obtained for the kinetic temperature actually characterize a certain averaged† distribution $T(h)$ over the chromosphere's structural elements since the experimental values of $T_{\text{eff}}(\lambda)$ actually relate to the whole disk or to its parts that include a large number of chromospheric inhomogeneities. Therefore the shape of the curve $T(h)$ in Fig. 160 depends simultaneously on the altitude distribution and temperature and the relative area occupied on the Sun's disk by the hot and cold elements (to be fair, without allowing for differences between electron densities in these elements). The complex, non-monotonic curve of $T(h)$ with a minimum at an altitude $h \approx 3\cdot3\times10^3$ km‡ will really be connected with the complex height function of the kinetic temperature within each element, if the relative area of the hot elements is constant or rises together with the altitude. If this area decreases rapidly enough in the range $h \approx 3\cdot0-3\cdot3\times10^3$ km, the results of measurements of the frequency spectrum at millimetric wavelenghts, generally speaking, can also be explained even when the temperatures in the cold and hot elements rise monotonically with altitude. It is still unclear which possibility is the true one. We can apparently count on a certain choice only after further experimental and theoretical investigations of the solar chromosphere.

No matter what the actual causes of the non-monotonic curve of the averaged temperature $T(h)$ in the lower chromosphere are, this very fact leads to an interesting consequence touching on the radio brightness distribution over the Sun's disk. On the falling part of the spectrum (i.e. in the region $\lambda \approx 5$ mm), unlike the rest of the millimetric band, we must expect not brightening but a fall in the radio brightness on the Sun's limb. The point is that at these wavelengths the radio emission is generated in

---

† It is not difficult to see that this averaging is carried out over the level $\tau \approx 1$ which, however, is located at a different altitude in the cold and hot structural elements of the chromosphere.

‡ The existence of this minimum was indicated in the paper by Zheleznyakov (1958a).

layers where the mean temperature decreases with altitude; since on the limb the effectively emitting region where $\tau \approx 1$ is located in higher (and colder) layers than in the central part of the disk the effective radio emission temperature at the edge of the disk decreases. This effect still awaits experimental confirmation.

29. Origin of the Slowly Varying Component of the Sun's Radio Emission

THERMAL NATURE OF THE S-COMPONENT OF THE SPORADIC RADIO EMISSION

Unlike the "quiet" Sun's radio emission there is no basis for interpreting the sporadic radio emission at metric wavelengths as thermal emission of active regions of the corona, since to create emission with an effective temperature of up to 10^9–10^{10} °K and above, the plasma must be heated to the same temperatures which are three orders or more greater than the normal temperature of the corona $T \sim 10^6$ °K. At the same time we are quite justified in explaining the slowly varying component (the S-component) at wavelengths of $\lambda \sim 1$–30 cm with moderate values of $T_{\text{eff}} \lesssim 10^6$ °K as thermal radio emission of regions of the corona located above centres of activity (groups of sunspots and flocculi). These active regions are characterized by enhanced values of the electron concentration N, magnetic field strength H_0 and, probably, kinetic temperature T when compared with the surrounding regions of the corona (see section 2); the value of T lies apparently between the limits $(1-3) \times 10^6$ °K.

The observed upper limit $T_{\text{eff}} \approx 1.5 \times 10^6$ °K of local sources of the S-component (according to the data of Christiansen and Mathewson (1959) at a wavelength of $\lambda = 20$ cm; see section 10) can be explained naturally from this point of view by the fact that T_{eff} cannot be greater than the emitting layer's kinetic temperature. For example, the intensity and effective temperature of the eigen emission escaping from a thermally uniform ($T = $ const) layer of magnetoactive plasma into a vacuum are respectively

$$I_1 = I_1^{(0)}(1 - e^{-\tau_1}), \quad T_{\text{eff}\,1} = T(1 - e^{-\tau_1}) \tag{29.1a}$$

for an extraordinary wave and

$$I_2 = I_2^{(0)}(1 - e^{-\tau_2}), \quad T_{\text{eff}\,2} = T(1 - e^{-\tau_2}) \tag{29.1b}$$

for an ordinary one (see (26.14) and (26.18)). Here $I_{1,2}^{(0)} = \omega^2 \varkappa T / 8\pi^3 c^2$ is the intensity of equilibrium emission in a vacuum for a single polarization, $\tau_{1,2}$ is the optical thickness of the emitting layer. It is clear from this that $T_{\text{eff}} \ll T$ if the optical thickness of the layer for ordinary or extra-

§ 29] Origin of the Slowly Varying Component

ordinary waves is $\tau_{1,2} \ll 1$ and $T_{\text{eff}} \approx T$ with $\tau_{1,2} \gtrsim 1$. According to the expression given for $T_{\text{eff}1,2}$ the presence of a clear-cut boundary for the effective temperature of the S-component is an indication that the optical thickness $\tau_{1,2}$ in the emitting region reaches values greater than or of the order of unity.

The effective temperature of the isothermal layer depends only on the optical thickness $\tau_{1,2}$ (and, of course, on the kinetic temperature T). In the regions of the corona and the chromosphere from which the high-frequency electromagnetic waves escape without hindrance beyond the Sun's atmosphere the optical thickness $\tau_{1,2}$, as has been shown in section 26, is largely determined by the absorption in collisions of the electrons with the plasma ions and by the gyro-resonance absorption of electrons in the magnetic field:

$$\tau_{1,2} = \tau_{1,2}^{\text{coll}} + \tau_{1,2}^{\text{res}} = \int (\mu_{1,2}^{\text{coll}} + \mu_{1,2}^{\text{res}}) dl, \qquad (29.2)$$

where dl is an element of the ray trajectory, $\mu_{1,2}$ is the energy coefficient of absorption; integration is carried out along the ray in the plasma emitting layer. The first type of absorption is connected with bremsstrahlung of electrons in the process of accelerated motion of the latter in the Coulomb field of the ions. The second type of absorption occurs in layers where $\omega \approx s\omega_H$ ($s = 1, 2, 3, \ldots$); it is connected with the magneto-bremsstrahlung (gyro-frequency emission) of the electrons during their accelerated motion in the magnetic field.

It has been assumed in the majority of papers on the theory of the slowly varying component (Piddington and Minnett, 1951; Pawsey and Yabsley, 1949; Lehany and Yabsley, 1949; Waldmeier and Müller, 1950; Denisse, 1950) that the optical thickness $\tau_{1,2}$ of the local source of radio emission is connected only with absorption due to collisions and therefore the component under discussion is the bremsstrahlung of the corona's active regions. According to Waldmeier and Müller (1950) the intensity of the bremsstrahlung in these regions arises because of the high concentration of electrons in the so-called coronal condensations; Lehany and Yabsley (1949), Denisse (1950) explain the enhanced level of this emission by the effect of the local magnetic fields on the quantity $\tau_{1,2}^{\text{coll}}$. Piddington and Minett (1951) combine both these points of view. It should be pointed out that Waldmeier and Müller's ideas about the origin of the S-component have been accepted by others (see, e.g., the papers by Newkirk, 1959, 1961; Kawabata, 1960a).

On the other hand, it was noted by Zheleznyakov (1959a), Ginzburg and Zheleznyakov (1959a) that in the conditions of the solar corona the optical thickness of the gyro-resonance layers where $\omega \approx \omega_H, 2\omega_H, 3\omega_H$ is, generally speaking, comparable with unity or much greater (for more detail see the subsection of section 26 about gyro-resonance absorption in the corona and the results given below). This provided a basis for stating (Zheleznyakov, 1959a; Ginzburg and Zheleznyakov, 1961) that the magneto-bremsstrahlung of the coronal electrons at frequencies equal to or multiples of the gyrofrequency ω_H may prove to be very significant when interpreting the slowly varying radio emission.

It becomes clear from what has been said that the basic question in the theory of the S-component is what emission mechanism (bremsstrahlung or magneto-bremsstrahlung) makes the major contribution, and under what conditions, to the observed thermal emission of the local sources connected with the centres of activity. To answer this question we must obviously analyse the action of both mechanisms depending on the actual conditions in the lower layers of the solar corona above the spots and flocculi and see how far the expected characteristics of the bremsstrahlung and magneto-bremsstrahlung correspond to the observed features of the slowly varying component. This investigation has been carried out by Zheleznyakov (1962 and 1963);† the following exposition of the theory of the sporadic radio emission's S-component is also based on these papers.

Of particular importance in this respect is an analysis of the frequency spectrum of the S-component (including data from polarization observations): it allows us to answer the question of the prevailing mechanism of radio emission at the different wavelengths, making it depend on the actual conditions at the centre of activity (the strength of the magnetic field and the electron concentration). We can also use information about the S-component's directivity, etc., to investigate the relative parts played by its bremsstrahlung and magneto-bremsstrahlung generation mechanisms.

Data of this kind on the properties of the slowly varying component have been given in section 10. There we noted that for local sources connected with centres of activity containing spots the total emission flux as a function of the wavelength $S_\omega(\lambda)$ has a characteristic curve with a maximum in the region of $\lambda \sim$ 5–10 cm. For one source investigated that was connected with flocculi (among which there were no spots) the nature of $S_\omega(\lambda)$ differed sharply from that indicated: in the 3–21 cm range the flux was comparatively small and did not depend upon the wavelength. If we consider (in accordance with the data of section 10) that at decimetric

---

† The magneto-bremsstrahlung mechanism as applied to the generation of the S-component has also been discussed (based on Ginzburg and Zheleznyakov, 1959a) by Kakinuma and Swarup (1962a and 1962b). Their results agree basically with those obtained by Zheleznyakov (1962 and 1963).

§ 29] Origin of the Slowly Varying Component

wavelengths the area of the emitting regions is approximately the same as the area of the flocculi connected with them, and at centimetric wavelengths the same as the area of the spots,† and does not vary significantly with wavelength within these bands, then we can conclude from this that the curve of the intensity $I(\lambda)$ is generally similar to the function $S_\omega(\lambda)$. In accordance with this Fig. 161 shows the idealized frequency spectrum which will be compared below with the theoretical functions $I(\lambda)$.

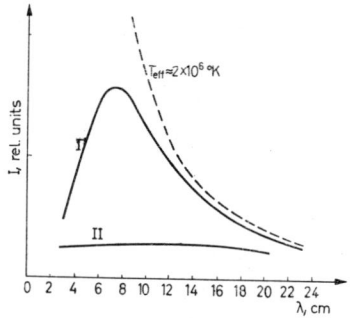

Fig. 161. Idealized frequency spectrum $I(\lambda)$ of a local source of the S-component: I—for sources connected with spots; II—for sources connected with flocculi without spots

The spectrum of Fig. 161 corresponds to the following law of the variation of the effective temperature with wavelength. In the decimetric band for large sources T_{eff} reaches values around 10^6 °K or slightly more ($\sim 2 \times 10^6$ °K). As the wavelength decreases right down to values of λ 5–10 cm I increases but T_{eff} varies weakly. On the other hand, at shorter wavelengths, where the intensity drops with the wavelength, T_{eff} decreases more rapidly than λ^2 (this is easy to understand, remembering that by definition $T_{eff} \propto I\lambda^2$). The nature of the spectrum at wavelengths of $\lambda < 3$ cm still remains unknown, although it is possible that the emission intensity in the millimetric band is comparable with the corresponding quantity at $\lambda = 3$ cm (see section 10).‡

---

† In the centimetric band the local sources sometimes have haloes—regions of weaker emission above the flocculi with $T_{eff} \sim 10^5$ °K surrounding a bright source with dimensions close to the spots' dimensions. According to different data, also given in section 10, the halo's T_{eff} does not exceed $(3-5) \times 10^4$ °K.

‡ In view of the particular importance of the question of the relative constancy of T_{eff} at $\lambda \sim 5$–10 cm and the fall in I with wavelength at $\lambda < 5$–10 cm for the theory of the S-component we must definitely find the extent of its reliability and determine more exactly the nature of the frequency spectrum (especially at wavelengths less than 5–10 cm, including the millimetric band). To do this we must carry out simultaneous investigations of local sources at different wavelengths with precise measurement of the radio emission fluxes and the sizes of the emitting regions.

BREMSSTRAHLUNG MECHANISM OF THE LOCAL S-COMPONENT SOURCES ABOVE SPOTS

Let us first examine the bremsstrahlung mechanism without making allowance for the effect of the gyro-resonance levels on the quantity $\tau_{1,2}$.

The absorption coefficient of electromagnetic waves because of collisions $\tau_{1,2}^{\text{coll}}$ is defined by the formula (26.74) when there is no constant magnetic field H_0 and by the formulae (26.69), (26.94) and (26.97) when there is H_0 in the plasma. It follows from the last expressions that the optical thickness of the emitting region in the quasi-longitudinal approximation is

$$\tau_{1,2}^{\text{coll}} \approx \int \frac{\omega_L^2 \nu_{\text{eff}}\, dl}{\omega^2 c n_{1,2} \left(1 \mp \frac{\omega_H |\cos \alpha|}{\omega}\right)^2} \qquad (29.3)$$

(the top sign relates to the extraordinary wave 1, the lower one to the ordinary wave 2). Here α is the angle between the magnetic field H_0 and the direction of wave propagation, ω_L is the plasma frequency, ω_H is the gyro-frequency, $n_{1,2}$ is the refractive index of the extraordinary and ordinary waves, ν_{eff} is the effective number of collisions. The quantities ω_L and ω_H are connected with the electron concentration and the magnetic field by the relations

$$\omega_L^2 \approx 3 \cdot 18 \cdot 10^9 N, \quad \omega_H \approx 1 \cdot 76 \cdot 10^7 H_0; \qquad (29.4)$$

the values of $n_{1,2}$ and ν_{eff} are given by the expressions (23.7) and (26.78), (26.79).

In the region of wavelengths of $\lambda \lesssim 10$ cm and ratios $\sqrt{u} \equiv \omega_H/\omega \lesssim \frac{1}{2}$, where the actual values of $\tau_{1,2}$ are significant for us, the refractive index $n_{1,2}$ is close to unity if the electron concentration is $N \ll 10^{11}$ electrons/cm^3. There is no basis for assuming that the latter inequality breaks down in the lower layers of the corona above centres of activity: according to Newkirk's data, at the maximum of the 11-year cycle the corresponding values are $N \sim 10^9$ electrons/cm^3; in the active regions of the corona when it has taken up its minimum form we should apparently expect even smaller values of the electron concentration. In the coronal condensations, which are generally looked upon as the source of the S-component,[†] the concentration is slightly higher: average—$4 \cdot 5 \times 10^9$ electrons/cm^3, maximum—about 8×10^9 electrons/cm^3 (section 2). However, even with these values of N the refractive index is as before close to unity.

Taking what has been said into consideration, we arrive at the following

† For critical remarks on this idea see below.

§ 29] Origin of the Slowly Varying Component

expression for the optical thickness of a uniform source with a linear dimension L along the line of sight:

$$\tau_{1,2}^{\text{coll}} \approx \frac{\omega_L^2 \nu_{\text{eff}} L}{\omega^2 c \left(1 \mp \frac{\omega_H |\cos \alpha|}{\omega}\right)^2} \approx \frac{7N^2 L}{\omega^2 T^{3/2} \left(1 \mp \frac{\omega_H |\cos \alpha|}{\omega}\right)^2}. \quad (29.5)$$

In the change to the last equation we allowed for the formulae (29.4) for ω_L and (26.79) for ν_{eff}. In the latter we assumed $\ln (10^4 T^{2/3}/N^{1/3}) \approx 12$, which corresponds to $T \approx 2 \times 10^6 \,°\text{K}$ and $N \approx 2 \times 10^9$ electrons/cm³. Thanks to the weak dependence of the logarithmic factor on the argument this quantity may also be used for values of N, T that differ considerably from those indicated.

In estimates that do not pretend to any great accuracy the formula (29.5) can also be used for a non-uniform source, in this case understanding by the parameters N, ω_H, L, etc., their characteristic (mean) values for the given source. If the temperature T in the source along the line of sight is constant and the ratio $\omega_H |\cos \alpha|/\omega$ either does not vary from point to point or remains sufficiently small when compared with unity, then the factor N^2 figuring in (29.5) takes on the meaning of the mean square of the electron concentration with respect to the source

$$\overline{N^2} = \int \frac{N^2 \, dl}{L} \quad (\text{see } (29.3)).†$$

We note finally that the quasi-longitudinal approximation in a magnetic field when $n_{1,2} \approx 1$ (i.e. $\omega_L^2/\omega^2 \ll 1$) will be satisfied if (see (23.5)) $\omega_H^2 \sin^4 \alpha \ll 4\omega^2 \cos^2 \alpha$; $|1 - (\omega_H |\cos \alpha|)/\omega| \gg (\omega_H^2 \sin^2 \alpha)/2\omega^2$. For small ω_H/ω these inequalities still hold over a wide range of angles α (with the exception of values of α close to 90°); however, even when $\omega_H/\omega \sim \frac{1}{2}$ the quasi-longitudinal approximation becomes inapplicable only in the case when $\alpha \gtrsim 75°$. For our purposes this is quite sufficient, although in accurate calculations of the bremsstrahlung from regions above groups of spots (in directions close to transverse in relation to the magnetic field) we should use the precise formula (26.94).

The frequency spectrum of a local source in which the bremsstrahlung generation mechanism is operative can be described (bearing in mind the remarks made) by the relations (29.1) and (29.5). If we neglect the effect of the magnetic field on the quantity $\tau_{1,2}^{\text{coll}}$ by making it equal to

$$\tau = \omega_L^2 \frac{\nu_{\text{eff}} L}{\omega^2 c} \equiv \frac{\lambda^2}{\lambda_{\text{cr}}^2}, \quad (29.6)$$

† The integral $\int N^2 \, dl$ is sometimes called the measure of emission.

where

$$\lambda_{cr}^2 = \frac{4\pi^2 c^3}{\omega_L^2 \nu_{eff} L} \approx \frac{5 \times 10^{21} T^{3/2}}{N^2 L} \quad (29.7)$$

then the emission intensity and the effective temperature for both polarizations are respectively

$$I = 2I^{(0)}(1-e^{-\tau}), \quad T_{eff} = T(1-e^{-\tau}), \quad (29.8)$$

The nature of the spectrum will be different on different sides of the critical wavelength λ_{cr} corresponding to the value $\tau = 1$ (see Fig. 162). In the

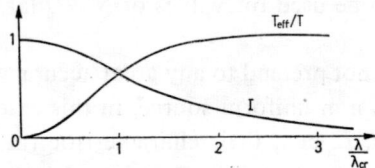

FIG. 162. Frequency spectrum of bremsstrahlung of an isotropic plasma

region $\lambda^2 \gg \lambda_{cr}^2$, where $\tau \gg 1$, the intensity is $I \approx 2I_j^{(0)}$ and varies as λ^{-2}, whilst the effective temperature is $T_{eff} \approx T$ and remains constant. In the region $\lambda^2 \ll \lambda_{cr}^2$, where $\tau \ll 1$, on the contrary, the intensity $I \approx 2I_j^{(0)}\tau$ is constant (since $I_j^{(0)} \propto \lambda^{-2}$, $\tau \propto \lambda^2$), and $T_{eff} \approx T\tau$ decreases with wavelength as λ^2.†

Since the frequency spectrum $I(\lambda)$ shown in Fig. 162 has no maximum a bremsstrahlung mechanism with no allowance made for the magnetic field cannot be responsible for the generation of the S-component over the whole range of wavelengths $\lambda \sim 3$–30 cm investigated. The theoretical spectrum at $\lambda > \lambda_{cr}$ corresponds to that observed in the $\lambda > 5$–10 cm range; however, at $\lambda < 5$–10 cm where I, to judge from the experimental data, falls as the wavelength decreases, agreement cannot be reached with any assumption about the magnitude of λ_{cr} since for $\lambda^2 > \lambda_{cr}^2$, $I \propto \lambda^{-2}$, and for $\lambda^2 < \lambda_{cr}^2$, $I \approx$ const.

Allowing for the effect of the magnetic field on the frequency spectrum of the bremsstrahlung does not save the situation (Molchanov, 1961). In actual fact let the ratio $\omega_H |\cos \alpha|/\omega$ be sufficiently small so that the expression for $\tau_{1,2}$ (29.5) can be expanded into a series in this quantity and

---

† We note that the inequalities $\lambda^2 \gtrless \lambda_{cr}^2$ when the expressions given for I and T_{eff} are valid are very weak. For example the relations $I \approx I_j^{(0)}\tau$, $T_{eff} \approx T\tau$ obtained by expanding $1-e^{-\tau}$ into a series with the first term retained have a relative accuracy of the order of $\tau/2$. If $\lambda \leqslant 0.5\lambda_{cr}$, then $\tau \leqslant 0.25$ and the accuracy will be about 10%, which is quite sufficient for us.

§ 29] Origin of the Slowly Varying Component

we can limit ourselves to only the first three terms:

$$\tau_{1,2}^{\text{coll}} \approx \frac{\lambda^2}{\lambda_{\text{cr}}^2}\left\{1\pm 2\frac{\omega_H}{\omega}|\cos\alpha|+3\left(\frac{\omega_H}{\omega}\cos\alpha\right)^2\right\}. \tag{29.9}$$

Then in the range $\lambda < \lambda_{\text{cr}}$ where the emitting region becomes optically thin the total emission intensity for both polarizations is[†]

$$I_1+I_2 \approx I_j^{(0)}\left(\tau_1+\tau_2-\frac{\tau_1^2}{2}-\frac{\tau_2^2}{2}\right) \approx 2I_j^0\frac{\lambda^2}{\lambda_{\text{cr}}^2}\left\{1-\frac{\lambda^2}{2\lambda_{\text{cr}}^2}+3\left(\frac{\omega_H}{\omega}\cos\alpha\right)^2\right\}. \tag{29.10}$$

In the change to the last equality we have left in the braces only terms with powers of $\lambda = 2\pi c/\omega$ not greater than squares.

As follows from (29.10), the sum I_1+I_2 rises as λ increases if

$$\frac{\lambda^2}{2\lambda_{\text{cr}}^2} < 3\left(\frac{\omega_H}{\omega}\cos\alpha\right)^2,$$

ie.

$$\lambda_{\text{cr}} > \frac{2\pi c}{\sqrt{6}\,\omega_H|\cos\alpha|} \tag{29.11}$$

(the factor $I_j^{(0)}\lambda^2/\lambda_{\text{cr}}^2$ does not depend on λ!). Then obviously I_1+I_2 with a certain value of $\lambda \sim \lambda_{\text{cr}}$ reaches a maximum, which is in qualitative agreement with the observations. The magnitude of this effect depends on the strength of the magnetic field: with high enough values of H_0 we can apparently obtain not only qualitative but also quantitative agreement with the observed frequency spectrum.

It should be remembered, however, that a random magnetic field strength cannot be chosen since it should correspond to the observed degree of polarization of the local sources of the S-component (less than or of the order of 30% at $\lambda \approx 3$ cm; see section 10). Since ordinary and extraordinary emission at the exit from the corona is circularly polarized (section 24) and generally only circular polarization is noted in the composition of the S-component, it can be assumed that the degree of polarization of the S-component is

$$\varrho_c = \frac{I_1-I_2}{I_1+I_2} \approx \frac{\tau_1-\tau_2}{\tau_1+\tau_2}. \tag{29.12}$$

[†] It follows from the expressions (29.9) that superimposing of a magnetic field on the plasma increases the intensity of its bremsstrahlung on an extraordinary wave but decreases it on an ordinary wave. The total emission intensity for both polarizations in an optically thin layer I_1+I_2 increases when compared with the corresponding value for $H_0 = 0$ (see (29.10)).

The latter relation is valid in the region $\lambda^2 \ll \lambda_{cr}^2$, where the source becomes optically thin. Remembering the expression for $\tau_{1,2}$ (29.9) we find that with $\omega_H |\cos \alpha|/\omega \ll 1$

$$\varrho_c \approx 2 \frac{\omega_H}{\omega} |\cos \alpha|. \qquad (29.13)$$

We note that in an optically thin region, where $\tau_{1,2} \ll 1$, the degree of circular polarization ϱ_c increases as the frequency ω decreases. Putting, in accordance with the observations at 3 cm, $\varrho_c \lesssim 0.3$ we find from (29.13): $\omega_H |\cos \alpha|/\omega \lesssim 0.15$, i.e. $\omega_H |\cos \alpha| \lesssim 10^{10}$ sec^{-1}, $H_0 |\cos \alpha| \lesssim 6 \times 10^2$ oe.

The frequency spectrum calculated when $\omega_H |\cos \alpha| = 10^{10}$ sec^{-1}—the dependence of $I_1 + I_2$ on λ for two values of λ_{cr} (6 and 10 cm)—is given in

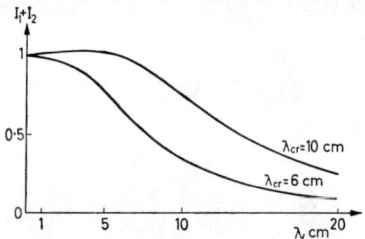

FIG. 163. Frequency spectrum of a bremsstrahlung from a plasma in the presence of a constant magnetic field H_0

Fig. 163. The maximum in the spectrum here occurs only for $\lambda_{cr} = 10$ cm since with $\lambda_{cr} = 6$ cm and $\omega_H |\cos \alpha| = 10^{10}$ sec^{-1} the above-mentioned condition for the appearance of a maximum is not satisfied. However, even with $\lambda_{cr} = 10$ cm, as a comparison of Figs. 161 and 163 shows, the increase in intensity expected in the framework of the bremsstrahlung mechanism does not correspond to the observed increase at all: it amounts to only about 2.5×10^{-2}, whilst according to Fig. 161 this effect is many times greater.†

It is clear from what has been said that only the bremsstrahlung mechanism is capable of explaining both the magnitude of the observed degree of polarization and the existence of a high maximum in the frequency spectrum of the slowly varying component. The dependence of the polar-

† We cannot make λ_{cr} much more than 10 cm because this will lead, in contradiction to the observations, to a significant displacement of the spectrum's maximum towards the long wavelengths. We note that the spectrum of Fig. 163 is calculated by the relations (29.1) and (29.9). The replacement of the latter by the more accurate expression (29.5) does not noticeably alter the dependence of $I_1 + I_2$ on wavelength, since at $\lambda \gtrsim \lambda_{cr}$ where the representation (29.9) becomes inapplicable $\tau_{1,2} \gtrsim 1$ and the ac ua form of $\tau_{1,2}$ is insignificant.

§ 29] Origin of the Slowly Varying Component

ization on the wavelength $\varrho_c(\lambda)$ obtained on the basis of the bremsstrahlung mechanism (the rise of ϱ_c together with λ in the region $\lambda < \lambda_{cr}$ with subsequent rapid decrease at wavelengths of $\lambda > \lambda_{cr}$) is not in agreement with experiment either, this generally indicating a systematic decrease in ϱ_c as λ increases.

We therefore come to the conclusion that the bremsstrahlung mechanism is not capable of explaining the features of the frequency spectrum of the S-component generated in regions of the corona above spots at wavelengths from 3 to 5–10 cm (at $\lambda < 3$ cm the situation remains unclear because of the absence of reliable observations). Since the intensity of the bremsstrahlung from these regions is in practice greatest at the short-wave boundary of the range under investigation, the value of the intensity at all wavelengths of $\lambda > 3$ cm can definitely not exceed the corresponding value at $\lambda \sim 3$ cm. This means that the upper boundary of the possible bremsstrahlung spectrum from these regions will correspond to curve I in Fig. 169 (see below). The lower boundary and the method of finding the precise level of bremsstrahlung will be indicated later; for now we shall give additional arguments against the basic role of the bremsstrahlung mechanism in the generation of radio emission above spots at centimetric wavelengths.

If the slowly varying component at $\lambda \sim 5$ cm is caused by the bremsstrahlung of electrons from a region with $T \sim 2 \times 10^6$ °K, then for the observed values of T_{eff} of the same order to occur at these wavelengths it is necessary that λ_{cr} should not exceed 5 cm. In accordance with (29.7) the latter will be satisfied if the emission measure $\overline{N^2}L$ is not less than 6×10^{29} electron^2/cm^5. With an emitting region thickness of $L \sim 5 \times 10^9$ cm and a uniform distribution of the concentration the value given for $\overline{N^2}L$ corresponds to a value of $N > 10^{10}$ electrons/cm^3. These values of N are far higher than the mean electron concentration in the coronal condensations ($4 \cdot 5 \times 10^9$ electrons/cm^3) or in the lower corona above a centre of activity ($\sim 10^9$ electrons/cm^3), although in the former case the break is insufficient for us to be able to draw a definite conclusion about the possibility of interpreting the S-component at $\lambda \sim 5$ cm as bremsstrahlung on the basis of data on the quantity N (particularly if we remember that because of non-uniform distribution of the concentration $\overline{N^2} > (\overline{N})^2$).†

At the same time lower concentrations are sufficient to explain the values of $T_{\text{eff}} \approx T$ at decimetric wavelengths. This is quite understandable: the optical thickness τ is proportional to $N^2\lambda^2$ and the value $\tau \sim 1$ is reached

---

† The magneto-bremsstrahlung mechanism discussed below does not need such high electron concentrations: values of $N \sim 5 \times 10^8$–10^9 electrons/cm^3 are quite sufficient for its effective action.

here with smaller N. For example, at $\lambda \sim 20$ cm we should have $N >$ 2×10^9 electrons/cm³, i.e. a quarter of that required at $\lambda \sim 5$ cm. This makes quite natural the satisfactory agreement established by Newkirk (1959) in simultaneous measurements of the electron concentration in active regions (from the scattering of the optical radiation in the corona) and of the effective temperature at a wavelength of $\lambda = 20$ cm of these regions on the assumption that all the observed radio emission is of bremsstrahlung origin.[†] It would obviously be of great interest to compare the values of N in active regions of the corona for wavelengths of $\lambda \sim 5\text{–}10$ cm corresponding to the maximum of the S-component's frequency spectrum since it is in this part of the spectrum that the bremsstrahlung mechanism meets with the greatest objections.

It should also be noted that the particular distribution of radio brightness over a source in the centimetric band—with a bright region above spots and a weak halo above the surrounding flocculi—in the case of bremsstrahlung origin of the radio emission[‡] occurs only if the measure of emission $\int N^2 \, dl$ at different points on the local source possess the same feature.[§] In accordance with this the intensity of the light scattering on the coronal electrons (the K-corona) should have a maximum of a width of the order of the spots' diameter above the spots. According to Newkirk (1959) and Christiansen et al. (1960) such a narrow maximum is not observed: the K-corona is stronger in the whole region above the flocculi and not above individual groups of spots. However, the measurements made relate to altitudes above the photosphere of about $0.125 R_\odot$, whilst the local regions of radio emission in the centimetric band are largely located at lower altitudes. Therefore photometry of the K-corona at shorter distances from the Sun's limb while finding spots on the limb would be very desirable.

Let us now see how the bremsstrahlung mechanism explains the charracteristic directivity of the S-component—the variation in the radio emission flux S_ω from a local source varies almost as $\cos \Theta$, where Θ is the helio-

[†] The successful interpretation by Newkirk (1959) of the slowly varying radio emission on the basis of a bremsstrahlung mechanism was also helped by the fact that the measurements related to local sources situated off the Sun's limb. Under these conditions the value of the radio emission flux is generally reduced (the effect of directivity), whilst the measure of emission may be increased when compared with sources in the central part of the Sun's disk.

[‡] Arguments in favour of the bremsstrahlung radio emission mechanism for the halo are given in the last subsection of this section.

[§] For example, at $\lambda \sim 3$ cm the ratio of T_eff of the halo to the T_eff of the bright region above the spots, which in a number of cases is as much as 10^6 °K, is less than or of the order of $1/10\text{–}1/20$; this means that the ratio of the values of the optical thickness and therefore also of the emission measure $\int N^2 \, dl$ in these regions does not exceed $1/10\text{–}1/20$ either.

§ 29] Origin of the Slowly Varying Component

graphic longitude of the emitting centre (see section 10). This directivity is obviously not connected with the effect of refraction in the corona, since there at the frequencies of interest to us the refractive index is $n_{1,2} \approx 1$.

Bearing in mind the relatively weak polarization of the slowly varying component we can neglect the effect of the magnetic field on the nature of the emission and absorption in the generation region. Then the emission flux from a source of area σ at a distance R_{S-E} away is

$$S_\omega \approx \sigma I R_{S-E}^{-2} = 2\sigma I_j^{(0)}(1-e^{-\tau}) R_{S-E}^{-2} = \frac{2\sigma a}{\mu R_{S-E}^2}(1-e^{-\mu L}) \quad (29.14)$$

(see (29.8)). Here we have allowed for the fact that in accordance with Kirchhoff's law (26.8) $I_j^{(0)} = a/\mu$, where a is the emissive power of a unit volume of plasma for a wave of one polarization (we are omitting the suffix j for simplicity of notation). If a local region of altitude h and extent d along the photosphere is located in the central part of the disk, then the emission flux is

$$S_\omega(\Theta \approx 0) \approx I d^2 R_{S-E}^{-2} = \frac{2 d^2 a}{\mu R_{S-E}^2}(1-e^{-\mu h}).$$

In the case, however, when this region is situated near the Sun's limb

$$S_\omega\left(\Theta \approx \frac{\pi}{2}\right) \approx I h d R_{S-E}^{-2} = \frac{2 h d a}{\mu R_{S-E}^2}(1-e^{-\mu d})$$

and therefore the ratio of the fluxes is

$$\frac{S_\omega(\Theta \approx 0)}{S_\omega\left(\Theta \approx \frac{\pi}{2}\right)} \approx \frac{d}{h} \frac{(1-e^{-\mu h})}{(1-e^{-\mu d})}. \quad (29.15)$$

The dependence of the ratio $S_\omega(\Theta \approx 0)/S_\omega(\Theta \approx \pi/2)$ on the value of the absorption coefficient μ (with $h < d$, $h = d$ and $h > d$) is shown in Fig. 164.

It is clear from the figure and directly from the formula (29.15) that for an optically thin region (when $\tau = \mu h \ll 1$ and $\tau = \mu d \ll 1$) the total flux remains constant as the heliographic longitude changes until the source starts to be hidden behind the Sun's disk. This result is not surprising: for an optically thin region reabsorption is insignificant and the emission flux in all directions is $2ahd^2$ (the product of the emissive power for two polarizations $2a$ and the volume of the emitting region hd^2).

It follows from the observations that the quantity $S_\omega(\Theta)$ decreases with the rise in $|\Theta|$, so $S_\omega(\Theta \approx 0)/S_\omega(\Theta \approx \pi/2)$ is always several times greater than unity (for further details see section 10). This kind of variation in

$S_\omega(\Theta)$ can occur (see Fig. 164) only when the altitude of the local region is less than its extent along the Sun's surface and the region becomes optically thick at least near the Sun's limb ($\tau = \mu d > 1$). Furthermore, the function $S_\omega \propto \cos \Theta$ will be approximately satisfied if the emitting region is optically thick not only on the limb but also near the central meridian ($\tau = \mu h > 1, \tau = \mu d > 1$) with a large enough ratio d/h. The value of d/h is easy to estimate since then it is equal to $S_\omega(\Theta \approx 0)/S_\omega(\Theta \approx \pi/2)$; the value of the latter ratio can be found in section 10.

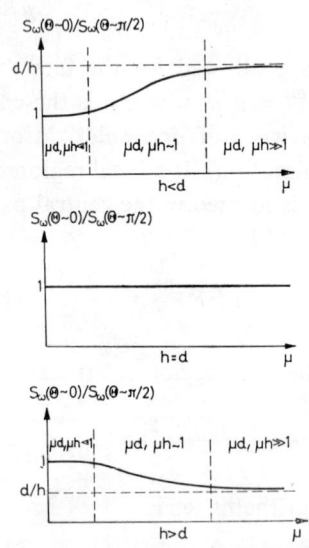

FIG. 164. Interpretation of the directivity of the S-component emission on the basis of the bremsstrahlung generation mechanism

Therefore the directivity depends on the optical thickness of the emitting region τ: it can hold only if $\tau \gtrsim 1$ even if near the Sun's limb; there is no directional feature for $\tau \ll 1$.

In the case of $\lambda > \lambda_{cr} \sim 5$–10 cm T_{eff}, as we know, is close to the temperature of the active parts of the corona $T \sim 2 \times 10^6$ °K. This means that $\tau \gtrsim 1$; the latter explains (in the case of a "plane" configuration of the emitting region) the observed nature of the directivity of the slowly varying component independently of its generation mechanism. In the other case, at wavelengths of $\lambda < \lambda_{cr} \sim 5$–10 cm for all local sources and at wavelengths of $\lambda > \lambda_{cr}$ for relatively weak sources $T_{eff} < T \sim 2 \times 10^6$ °K. Since in the framework of the bremsstrahlung mechanism the values of $T_{eff} < T$ are connected with a decrease in the optical thickness of the source to values of $\tau < 1$, this mechanism cannot provide the directivity

§ 29] Origin of the Slowly Varying Component

of the radio emission in cases when T_{eff} is noticeably less than T. As far as can be judged from the available experimental data there is always directivity, as a rule, in the range of wavelengths starting at 3 cm and above.

MAGNETO-BREMSSTRAHLUNG MECHANISM OF SLOWLY VARYING EMISSION

Let us turn now to the magneto-bremsstrahlung mechanism of the S-component. It has been shown in section 26 that in a coronal plasma with a Maxwell velocity distribution of the electrons and with a non-uniform magnetic field magneto-bremsstrahlung (gyro-frequency emission) at a frequency ω is generated in thin layers where H_0 satisfies the condition $\omega \approx s\omega_H$ ($s = 1, 2, 3, \ldots$ is the number of the harmonic, ω_H is the gyro-frequency of the electron). The optical thickness τ_{js} is given by the formula (26.125) together with (26.113). These expressions are rather complicated; for $s \geq 2$, however, they can be simplified in the case of quasi-longitudinal propagation, particularly when the refractive index is $n_{1,2} \approx 1$ (i.e. $v = \omega_L^2/\omega^2 \ll 1$):†

$$\tau_{js} \approx \pi \frac{s^{2s}}{2^{s+1} s!} \frac{\omega_L^2}{c\omega} \beta_{\text{th}}^{2s-2} L_H (1 \pm \cos \alpha)^2 \sin^{2s-2} \alpha \qquad (29.16)$$

(see (26.126)). In (29.16) ω_L is the plasma frequency, β_{th} is the ratio of the thermal velocity V_{th} to the velocity of light c, $L_H = \omega_H(dl/d\omega_H)$ is the characteristic dimension over which there is a change in the value of ω_H (the field H_0) along the line of sight, α is the angle between $\boldsymbol{H}_0$ and the normal to the wave. Thanks to the fact that $v \ll 1$ the expression for the optical thickness at the fundamental harmonic $s = 1$ also becomes simpler (see (26.127)).

Actual values of τ_{js} in active regions of the corona calculated from the formulae (26.127) and (29.16) are given in Table 6 (Zheleznyakov, 1963) ($T = 2 \times 10^6$ °K, $N = 2 \times 10^9$ electrons/cm^3; the scale of the variation in the magnetic field for all resonance layers is taken to be the same: $L = 4 \times 10^9$ cm). It is clear from the table that the optical thickness of the gyro-resonance layer for an extraordinary wave τ_{1s} will, all other things being equal, always be greater than the corresponding value of τ_{2s} for an ordinary wave.

For the values of the parameters α, N, T and λ given in the table the layers $\omega \approx \omega_H$, $\omega \approx 2\omega_H$ are opaque to extraordinary and ordinary waves. Since τ_{js} is large in these layers the levels $\omega \approx \omega_H$, $\omega \approx 2\omega_H$ remain optically thick while N and T vary within wide limits.

† For satisfaction of the equality $n_{1,2} \approx 1$ and realization of the quasi-longitudinal approximation under the conditions of interest to us here see the preceding subsection of this section.

TABLE 6

Harmonic No.	H, oe	$\lambda = 3$ cm			
		τ_1 (extraordinary wave)		τ_2 (ordinary wave)	
		$\alpha = 30°$	$\alpha = 60°$	$\alpha = 30°$	$\alpha = 60°$
1	3600			$2\cdot7\times10$	$9\cdot3\times10^3$
2	1800	$1\cdot2\times10^4$	$2\cdot4\times10^4$	$6\cdot5\times10$	$2\cdot7\times10^3$
3	1200	$8\cdot0$	$4\cdot7\times10$	$4\cdot1\times10^{-2}$	$5\cdot2$
4	900	$7\cdot6\times10^{-3}$	$1\cdot3\times10^{-1}$	$3\cdot9\times10^{-5}$	$1\cdot5\times10^{-2}$

		$\lambda = 20$ cm			
1	540			$1\cdot8\times10^2$	$6\cdot2\times10^4$
2	270	$8\cdot3\times10^4$	$1\cdot6\times10^5$	$4\cdot3\times10^2$	$1\cdot8\times10^4$
3	180	$5\cdot3\times10$	$3\cdot1\times10^2$	$2\cdot7\times10^{-1}$	$3\cdot5\times10$
4	135	$5\cdot1\times10^{-2}$	$8\cdot9\times10^{-1}$	$2\cdot6\times10^{-4}$	$9\cdot9\times10^{-1}$

Layers with $s > 4$ are transparent for both types of wave ($\tau_{js} \ll 1$). The same is also true, as a rule, for the level $\omega \approx 4\omega_H$ although in the decimetric band for extraordinary waves the absorption at $\omega \approx 4\omega_H$ becomes noticeable when there is an enhanced temperature and concentration of the coronal plasma.†

The degree of transparency of the layer $\omega \approx 3\omega_H$ (satisfaction of the inequalities $\tau_{js} \gtrsim 1$) in the conditions of the solar corona depends on the type of wave and actual values of N, T, α and λ. According to Table 6, for extraordinary waves in the centimetric and decimetric band this layer in the corona is optically thick; for ordinary waves it becomes opaque only with large angles between the directions of the field H and of the waves propagation.

Since τ_{js} at all harmonics s and for both types of wave decreases with α and when $\alpha = 0$ becomes zero,‡ regions of transparency corresponding to angles $\alpha < \alpha_{cr}$ nevertheless exist at optically thick gyro-resonance levels. By means of the formulae (26.127) and (29.16) it is easy to find that at $\lambda \sim 3$ cm in the corona above a centre of activity ($N \sim 2\times10^9$ elec-

† The increase in the temperature T has a particularly significant effect on the absorption in layers corresponding to large harmonics s, since $\tau_{js} \propto T^{s-1}$; in the fourth gyro-resonance layer $\tau_{js} \propto T^3$.

‡ With the exception of the optical thickness of the layer $\omega \approx \omega_H$ for an extraordinary wave. This exception is insignificant for us since from the level $\omega \approx \omega_H$ waves of the kind indicated do not escape beyond the corona.

§ 29] Origin of the Slowly Varying Component

trons/cm^3, $T \sim 2\times 10^6$ °K, $L_H \sim 4\times 10^9$ cm) the value of α_{cr} that separates the ranges of angles α with $\tau_{js} \gtrsim 1$ will be close to the following values: for an ordinary wave $\alpha_{cr} \approx 5°$ ($s = 1$), $\alpha_{cr} \approx 15°$ ($s = 2$), $\alpha_{cr} \approx 46°$ ($s = 3$); for an extraordinary wave $\alpha_{cr} \approx 14'$ ($s = 2$), $\alpha_{cr} \approx 17°$ ($s = 3$).

Since in all cases except the last for an ordinary wave the angles α are rather small in an approximate analysis of the generation conditions of magneto-bremsstrahlung in the coronal plasma we shall consider that the levels $\omega \approx 2\omega_H$, $\omega \approx 3\omega_H$ are always optically thick for extraordinary waves and the levels $\omega \approx \omega_H$, $\omega \approx 2\omega_H$ are optically thick for ordinary waves, and allow for the dependence of the degree of transparency on the angle α only for the ordinary component in the layer $\omega \approx 3\omega_H$. In a more accurate treatment, of course, we cannot but allow for transparency in the region of $\alpha < 15$–$17°$ of $\omega \approx 3\omega_H$ layers for extraordinary and $\omega \approx 2\omega_H$ for ordinary waves.

According to (29.1) the T_{eff} of radio emission from levels for which $\tau_{js} \gtrsim 1$ is close to the kinetic temperature T of the electrons in these layers; if, however, $\tau_{js} \ll 1$, then $T_{\text{eff}} \ll T$. It follows from this that the T_{eff} of the magneto-bremsstrahlung depends essentially on where the optically thick gyro-resonance levels are—in the "hot" corona or in the inner, "colder" layers (in the chromosphere or in the transition region from the chromosphere to the corona). To get an idea of the probable position of the gyro level above spots let us take a model of a unipolar spot with a field H_0 on the axis given by the formula (2.1). In this case the layers $\omega \approx s\omega_H$ ($s = 1, 2, 3, 4$) are localized as shown diagrammatically in Fig. 165.

It is clear from Fig. 165 and what has been said earlier about the optical thickness of gyro-resonance levels that the contribution from the layers $\omega \approx s\omega_H$ (where $s = 2$ and $s \geqslant 4$) to the extraordinary emission of the slowly varying component's local sources is insignificant overall: an extraordinary wave from the $\omega \approx 2\omega_H$ level does not escape beyond the corona, being absorbed in the optically thick layer $\omega \approx 3\omega_H$ which lies above (in a region with a weaker magnetic field), whilst only emission having $T_{\text{eff}} \ll T$ is connected with the optically thin layers with $\omega \approx 4\omega_H, 5\omega_H, \ldots$ It can therefore be taken that the part of the layer emitting effectively extraordinary waves is played by the layer $\omega \approx 3\omega_H$ with T_{eff} equal to T (the kinetic temperature of this layer).

Gyro-frequency emission from layers $\omega \approx \omega_H$ and $\omega \approx 4\omega_H, 5\omega_H, \ldots$ will be insignificant for ordinary waves (for the same reasons as for extraordinary waves). With $\alpha > \alpha_{cr}$, where α_{cr} is the critical angle in the layer $\omega \approx 3\omega_H$, this layer as before makes the major contribution to the ordinary emission as well; however, with $\alpha < \alpha_{cr}$ the part of the effectively emitting layer moves on to the $\omega \approx 2\omega_H$ level since the $\omega \approx 3\omega_H$ region becomes optically thin.

Furthermore, it is clear from Fig. 165 that above large spots with a strong magnetic field the effectively emitting levels $\omega \approx 2\omega_H$, $\omega \approx 3\omega_H$ in the decimetric band ($\lambda \sim 20$ cm) lie in the corona, in layers with more or less the same temperature $T \sim (1-3) \times 10^6$ °K. Therefore in this range the T_{eff} of the magneto-bremsstrahlung is close to the said value of the kinetic

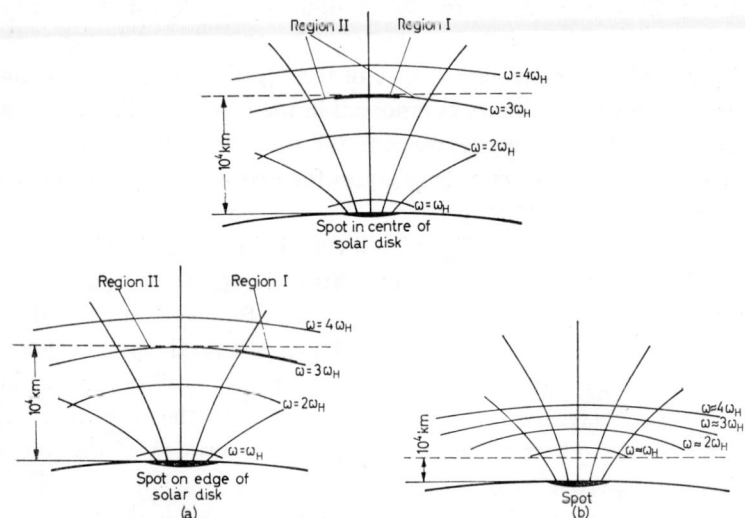

Fig. 165. Position of gyro-resonance layers in an active region above a spot (diagrammatic): (a) $\lambda = 3$ cm, (b) $\lambda = 20$ cm. Field at base of spot $H_{0b} = 2 \times 10^3$ oe, area of spot $\pi b^2 \sim \pi \times 10^{18}$ cm²

temperature of the corona's active regions. At the same time at shorter wavelengths ($\lambda \sim 4$ cm) the levels $\omega \approx 2\omega_H$, $\omega \approx 3\omega_H$ obviously lie in the transition region between the chromosphere and the corona, the higher-lying level $\omega \approx 3\omega_H$ corresponding to a higher kinetic temperature than the level $\omega \approx 2\omega_H$. Therefore we must expect here slightly lower values of T_{eff}, and then—as there is a further decrease in the wavelength λ and the displacement connected with it of the emitted levels into the "cold" chromosphere—a sharp drop in T_{eff}. The values of $T_{\text{eff}} \ll 10^6$ °K often observed in weak sources at wavelengths of $\lambda \gtrsim 3$ cm from the point of view of the magneto-bremsstrahlung mechanism are caused above all by the low strength of the spots' magnetic field (as a result of which the levels $\omega \approx 2\omega_H$, $3\omega_H$ move into the heart of the solar atmosphere).†

† A decrease in N to a value of 3×10^8 electrons/cm³ corresponding to the surrounding corona at its minimum phase, as is clear from Table 6, can have a significant effect on T_{eff} only at centimetric wavelengths.

§ 29] Origin of the Slowly Varying Component

In the centimetric band the $\omega \approx 2\omega_H$, $\omega \approx 3\omega_H$ levels are located high enough (in the "hotter" layers) only above sunspots with a strong magnetic field. A long way from the spots, where the magnetic field is weaker, the effectively emitting layers drop down and then generally disappear from the layers above the photosphere. Therefore the radio brightness distribution in the case of action of the magneto-bremsstrahlung mechanism should have a maximum above sunspots and drop sharply outside the spots (above the flocculi). This feature of the emission in the band under discussion is largely confirmed by observations (section 10). In the decimetric band the gyro-resonance levels $\omega \approx 2\omega_H$, $\omega \approx 3\omega_H$ correspond to considerably lower values of the magnetic field and are therefore located in the corona at greater distances from the spots. Since the distribution of the field strength H_0 over the Sun's disk in these layers is more uniform, as the wavelength rises the effect of brightening above the spots should show itself more weakly and this actually occurs.

We recall that the layers $\omega \approx s\omega_H$ ($s = 1, 2, 3$), which are optically thick when $\alpha \sim 1$, become transparent if $\alpha \approx 0$. This means that in the bright region above a spot with $T_{\text{eff}} \sim 10^6$ °K a minimum of T_{eff} may appear, since in the direction for which $\alpha \approx 0$ the slowly varying component is created by emission from deeper-lying layers with a lower kinetic temperature. We must count above all on wavelengths of 3–5 cm for which the $\omega \approx \omega_H$, $2\omega_H$ levels are far "colder" than the $\omega \approx 3\omega_H$ levels to find an effect of this kind. This is a difficult problem, however, because it requires observations to be made with a resolution of less than 1'.

Directivity of the slowly varying component of the $S_\omega \propto \cos \Theta$ type from the point of view of the magneto-bremsstrahlung mechanism is connected with the special configuration of the magnetic fields above the centre of activity, due to which the extent in height of the coronal region bounded at the top by an optically thick layer $\omega \approx 3\omega_H$ (and also $\omega \approx 2\omega_H$) is less than the extent of this region along the surface of the Sun (Fig. 165). Unlike the bremsstrahlung mechanism, the directivity of the magneto-bremsstrahlung is also preserved for weak sources since in the latter case the decrease in T_{eff} is caused chiefly by displacement of the effectively emitting layers (which remain optically thick) towards the photosphere.

Let us now see how the magneto-bremsstrahlung mechanism will explain the observed nature of the S-component's polarization and above all its dependence on the frequency and sign of rotation. According to Table 6, for the most probable values of N and T in the active regions of the corona the layer $\omega \approx 3\omega_H$ becomes optically thick both for ordinary and extraordinary emission with large enough angles $\alpha > \alpha_{\text{cr}}$ between the direction of the wave k and the field H_0. However, the values of α_{cr} are different for these types of emission: as was indicated above, with $N \sim 2 \times 10^9$ electrons/cm³, $T \sim 2 \times 10^6$ °K and $L_H \sim 4 \times 10^9$ cm at $\lambda \sim 3$ cm the value of α_{cr} for extraordinary waves in the third gyro-resonance layer is only 17°, unlike the ordinary waves for which α_{cr} is far greater (about 46°). It is clear from this that in the centimetric band the part of the $\omega \approx 3\omega_H$ layer in which $\alpha < \alpha_{cr}$ for ordinary waves emits polarized emission with the extraordinary component predominating (since the latter is emitted by the "hotter" $\omega \approx 3\omega_H$ layer, whilst the ordinary component is connected with the "colder" $\omega \approx 2\omega_H$ gyro-resonance layer; see Fig. 165). In the range $\alpha > \alpha_{cr}$ the position is different: here both components are generated in the optically thick $\omega \approx 3\omega_H$ layer, and because of the equality of their effective temperatures the total emission will not be polarized. As the wavelength rises the $\omega \approx 2\omega_H$ and $\omega \approx 3\omega_H$ levels balance out since they both move into the corona with an approximately uniform distribution of T. The consequence of this is a reduction in the degree of polarization of the local source emission as the wavelength increases. The dependence of the degree of polarization on the wavelength noted and the sign of the polarization agree with the observational data well.

In the case of effective action of the magneto-bremsstrahlung mechanism the size of a local source in polarized emission should, as a rule, be less than the size of the source in non-polarized emission (even without allowing for the halo). This effect is connected with the circumstance that the source of the polarized emission coincides with part of the third, gyro-resonance layer in which $\alpha < \alpha_{cr}$ for an ordinary wave, whilst the non-polarized emission comes from regions of the $\omega \approx 3\omega_H$ layer in which $\alpha > \alpha_{cr}$. It is clear from Fig. 165a that for a spot in the central part of the Sun's disk the local source polarized emission (region I) is located near the spot's axis and the source of non-polarized emission (region II) on the periphery of the spot (see also Fig. 166). In the process of the displacement of the centre of activity over the Sun's disk region I moves to the side of the spot's

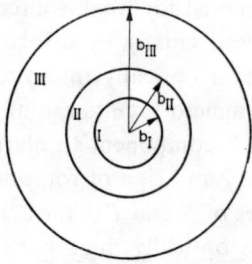

Fig. 166. The three characteristic regions of emission in a local source of the S-component (diagrammatic)

§ 29] Origin of the Slowly Varying Component

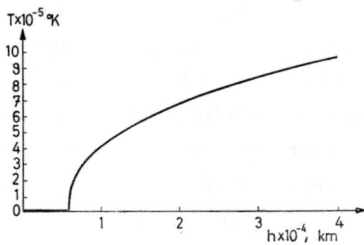

FIG. 167. Dependence of temperature T on altitude h in a model of an active region

axis where the magnetic field is weaker and the effectively emitting levels are located lower down, in colder layers. This leads to a decrease in the relative contribution of region I when compared with region II to the total emission of the local source. What has been said is sometimes the cause of the higher directivity sometimes observed of the polarized emission in the composition of the S-component when compared with the directivity of the emission of the local source as a whole.† We notice that this difference between the sizes of the polarized and non-polarized emission is apparently confirmed by the observations, although their accuracy is still insufficient for a definite conclusion about the existence of this effect.

When coming to discuss the frequency spectrum of the magneto-bremsstrahlung from the active regions of the inner corona and the upper chromosphere, we notice above all that the dependence on the wavelength of T_{eff} or $I_j \propto T_{\text{eff}} \lambda^{-2}$ for layers $\omega \approx 2\omega_H$ and $\omega \approx 3\omega_H$, which play a decisive part in the generation mechanism under discussion is determined by the nature of the distribution of H_0 and T above the spot. For a preliminary analysis of the question of the expected frequency spectrum we shall assume that the magnetic field varies as (2.1) both on the axis of the spot and on its periphery, whilst the kinetic temperature varies with altitude according to the graph in Fig. 167. The function $T(h)$ in the chromosphere, particularly in layers adjacent to the corona, is poorly known. For the sake of definition we have used here the law of variation of $T(h)$ given by Allen (1955).

It follows from the formula (2.1) that the level at which $\omega \approx s\omega_H$ is located at the altitude

$$h \approx b\left(1 - \frac{\omega}{s\omega_{H_{ob}}}\right)\left[\frac{\omega}{s\omega_{H_{ob}}}\left(2 - \frac{\omega}{s\omega_{H_{ob}}}\right)\right]^{-1/2} \qquad (29.17)$$

† A possible decrease in the temperature difference of the $\omega \approx 2\omega_H$, $\omega \approx 3\omega_H$ levels acts in the same direction (in the part which makes a contribution to the polarized emission, i.e. in region I) when the distance of the spot from the centre of the Sun's disk to the spot increases.

($\omega_{H_{0b}} = eH_{0b}/mc$ is the gyro-frequency corresponding to the magnetic field at the base of the spot H_{0b}); the kinetic temperature T at this altitude can be determined by the graph in Fig. 167. On the basis of these data it is not hard to calculate the approximate frequency spectrum of magneto-bremsstrahlung from layers $\omega \approx 2\omega_H$, $\omega \approx 3\omega_H$ in the case when they are optically thick: the effective emission temperatures are respectively

$$T^{(2)}_{\text{eff}} = T(h),$$

where h is found from the formula (29.17) with $s = 2$, and

$$T^{(3)}_{\text{eff}} = T(h),$$

where the values of h are obtained from (29.17) with $s = 3$.

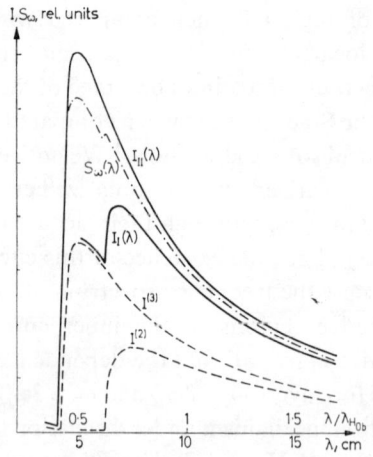

FIG. 168. Frequency spectra of magneto-bremsstrahlung from levels $\omega \approx 2\omega_H$ and $\omega \approx 3\omega_H$ in an active region of the corona (Zheleznyakov, 1963)

The frequency spectra found in this way for the intensities of the emission from these layers $I^{(2)} \propto T^{(2)}_{\text{eff}} \lambda^{-2}$, $I^{(3)} \propto T^{(3)}_{\text{eff}} \lambda^{-2}$ are shown in Fig. 168 (for a very large spot $\pi b^2 \approx \pi \times 10^{19}$ cm² in area). On the abscissa of the graph are plotted the values of $\lambda/\lambda_{H_{0b}} = \omega_{H_{0b}}/\omega$, where $\lambda_{H_{0b}}$ is the conventionally introduced wavelength corresponding to the gyro-frequency $\omega_{H_{0b}}$ at the base of the spot: $\lambda_{H_{0b}} = 2\pi c/\omega_{H_{0b}}$.

Since the function $I^{(3)}(\lambda)$ relates only to the points of a local source in which the layer $\omega \approx 3\omega_H$ is optically thick, it characterizes the frequency spectrum of the emission only in the region $\alpha > \alpha_{\text{cr}}$; in the region of the source where $\alpha < \alpha_{\text{cr}}$ this layer becomes transparent and makes no significant contribution to the emission. Since α_{cr} for extraordinary waves is small when compared with the α_{cr} of ordinary waves, we shall consider for

§ 29] Origin of the Slowly Varying Component

simplicity that $I^{(3)}(\lambda)$ is the frequency spectrum of the extraordinary emission over the whole area of the local source (with the exception of the halo) and the frequency spectrum of the ordinary emission on the periphery—in region II (in Fig. 166) where the magnetic field $\boldsymbol{H}_0$ subtends a certain angle with the line of sight $\alpha > \alpha_{cr}$.[†] In the central part of the source (in region I where $\alpha < \alpha_{cr}$) the spectrum of the ordinary emission is the same as $I^{(2)}(\lambda)$, i.e. it is determined by the deeper-lying second gyroresonance layer.

It follows from what has been said that the total frequency spectrum of the magneto-bremsstrahlung in region II of a local source

$$I_{II}(\lambda) = 2I^{(3)}(\lambda), \tag{29.18}$$

and in region I

$$I_I(\lambda) = I^{(2)}(\lambda) + I^{(3)}(\lambda). \tag{29.19}$$

These spectra are also shown in Fig. 168.

It can be seen that the frequency spectra $I^{(2)}(\lambda)$ and $I^{(3)}(\lambda)$ and at the same time the total spectra $I(\lambda)$ in the centre and on the periphery of the local source above the spot has a maximum when $\lambda_{max} \approx (0.5-0.7)\lambda_{H_{0b}}$. If the magnetic field at the base of the spot is $H_{0b} \sim 10^3$ oe,[‡] then $\lambda_{H_{0b}} \sim 10$ cm and $\lambda_{max} \sim 5-7$ cm, i.e. the position of the region of the maximum intensity values on the λ scale will correspond to that observed.

At wavelengths of $\lambda > \lambda_{max}$ the intensity of the magneto-bremsstrahlung falls as λ rises, whilst the effective temperature T_{eff} is close to the temperature of the active corona $T \sim 2\times10^6$ °K. At wavelengths less than λ_{max} the value of the intensity decreases rapidly because of the transition of the actively emitting levels $\omega \approx 2\omega_H$, $\omega \approx 3\omega_H$ from the corona to the chromosphere. The contribution of the magneto-bremsstrahlung definitely becomes insignificant in the millimetric band for which the levels $\omega \approx 2\omega_H$, $\omega \approx 3\omega_H$ are appropriate to magnetic fields with a strength $H_0 > 3600-4800$ oe; there is no reason to expect such strong fields in the layers above the photosphere. Actually the part played by the magneto-bremsstrahlung mechanism becomes negligibly small slightly earlier, in the short-wave part of the centimetric band, because of the sharp drop in temperature in the lower chromosphere. To judge from Fig. 168, in a model with $H_{0b} \sim 10^3$ oe and $T \sim 10^4$ °K with $h \lesssim 6\times10^3$ km this occurs at wavelengths of $\lambda < 4$ cm.

[†] We have in mind the α_{cr} of an ordinary wave.
[‡] This very moderate value of H_{0b} also corresponds to the scale of wavelengths λ (cm) in Fig. 168. We note that the quantity $\lambda_{H_{0b}}$ corresponding to the experimental values of λ_{max} depends essentially on the distribution used in the calculations of the magnetic field and the kinetic temperature in the active region above the spot. In particular, H_{0b} and $\lambda_{H_{0b}}^{-1}$ will increase if we reduce the spot area πb^2 in the model (2.1).

The strong attenuation of the centimetric radio emission in layers of the chromosphere lying at an altitude of $h \lesssim 10^4$ km also may have an effect on the position of the short-wave boundary of effective action of the magneto-bremsstrahlung mechanism because of absorption due to collisions (see Fig. 139).†

Direct comparison of the theoretical spectra of the type shown in Fig. 168 with the actual ones is possible only after measurements of the emission flux of the S-component at different wavelengths from areas of the solar disk which have a size much less than the diameter of the spots (i.e. much less than 1–2′). This is a complex problem requiring the use of aerials with exceptionally high directivity. For comparison with theory of the information at present available on the dependence on wavelength of the emission flux from local sources above spots, we must change from the intensity $I(\lambda)$ given by the formulae (29.18) and (29.19) to the expression for the flux $S_\omega(\lambda)$ of ordinary and extraordinary emission leaving the whole local source above the spot (i.e. from the regions I and II in Fig. 166). In a rough approximation when the intensities of the emission from the second and third gyro-resonance layers $I^{(2)}$, $I^{(3)}$ are considered to be constant over the part of the source above the spot

$$S_\omega(\lambda) \approx (I^{(2)}+I^{(3)})\pi b_I^2 R_{S-E}^{-2} + 2I^{(3)}\pi(b_{II}^2 - b_I^2) R_{S-E}^{-2} \qquad (29.20)$$

where b_I and b_{II} are the radii of regions I and II respectively, R_{S-E} is the distance of the Sun from the Earth. In a closely argued calculation of the characteristics of magneto-bremsstrahlung above spots the values of b_I and b_{II} will be determined by the actual model of the active region as a function of the wavelength λ; neglecting this dependence for the sake of simplicity we shall select b_I and b_{II} so that the degree of polarization corresponds to the experimental data. For this it suffices to put $b_{II} \approx b$ (b is the radius of the spot) and $b_I \approx b_{II}/2$. The function $S_\omega(\lambda)$ obtained as a result (Fig. 168) corresponds in its basic features to that observed at centimetric band wavelengths.

The choice made for the ratio b_I/b_{II} needs some explanation. At first sight it appears that in the framework of the magneto-bremsstrahlung mechanism

† The frequency spectra of Fig. 168 and the values of λ obtained from them corresponding to the intensity maximum and the short-wave boundary of the gyro-resonance emission can, of course, be used only as a guide. The correct calculation of the magnetic bremsstrahlung spectrum should be based on a better choice of active region model above groups of spots and contain a careful allowance made for the contribution of different gyro-resonance levels to the emission (and also allowance for attenuation of the latter because of magneto-bremsstrahlung and bremsstrahlung absorption in the process of propagation from the $\omega \approx s\omega_H$ level to escape from the corona). The papers of Ivanov–Kholodnyi and Nikol'skii (1961 and 1962), De Jager (1959) may be a considerable help in determining the function $T(h)$ in the upper chromosphere more accurately.

§ 29] Origin of the Slowly Varying Component

it is impossible to explain the degree of polarization $\varrho \approx \varrho_c \sim 30\%$ generally observed in the middle of the centimetric band since, as follows from Fig. 168, in region I of a local source the degree of polarization at $\lambda \sim 4\text{–}6$ cm reaches 100% because of the negligibly small intensity of the ordinary emission. However, when comparing the nature of the polarization it must be remembered that the data on the quantity ϱ_c relate not only to region I but are the result of some averaging over the local source. For example in the majority of polarized measurements of the centimetric band (interferometer or eclipse measurements) we obtain a one-dimensional distribution of the clockwise and anti-clockwise polarized emission over the local source (see section 5); therefore the value $\varrho_c \sim \frac{1}{3}$ found actually relates to the emission within a narrow band passing through the local source. The corresponding degree of polarization of the magneto-bremsstrahlung in a band passing through the centre of the source will then obviously be†

$$\varrho \approx \varrho_c \approx \frac{2(I^{(3)} - I^{(2)}) b_\mathrm{I}}{2(I^{(3)} + I^{(2)}) b_\mathrm{I} + 4 I^{(3)} (b_\mathrm{II} - b_\mathrm{I})} \qquad (29.21)$$

(the numerator contains the difference of the fluxes of the extraordinary and the ordinary emission, the denominator contains their sum within the band in question). At wavelengths of 4–6 cm, where ϱ_c is maximal, $I^{(2)} \ll I^{(3)}$. Allowing for this fact and putting $\varrho_c \approx \frac{1}{3}$ in (29.21), we find that $b_\mathrm{I} \approx b_\mathrm{II}/2$.

By combining the information given in the present and preceding subsections about the frequency spectra of bremsstrahlung and magneto-bremsstrahlung in local sources above spots we can see that in the millimetric band the observed radio emission with $T_\mathrm{eff} \ll 10^6$ °K should have a purely bremsstrahlung origin with a constant intensity over the band. The value of the intensity I at these wavelengths actually also determines the intensity of the bremsstrahlung in the centimetric and decimetric bands (Fig. 169), giving, moreover, values for the latter which are too high (see below). Therefore the real importance of this type of emission can be judged with certainty only from measurements of I in the millimetric band.‡ Unfortunately only single observations of this kind have been made to date (section 10).

† Without allowing for the halo.
‡ We note that it is measurements of the intensity I and the degree of polarization ϱ_c in the horizontal part of the spectrum at millimetric wavelengths (and possibly at the beginning of the centimetric band), where the emission is bremsstrahlung in origin and escapes from optically thin regions, which can give the information about the measure of emission $\int N^2 \, dl$ and the field strength H_0 in the lower corona above spots, which was obtained before in many papers (without sufficient basis for doing so) from results of observations at $\lambda > 3$ cm.

At centimetric wavelengths the part played by magneto-bremsstrahlung is very important: even if the intensity of the bremsstrahlung at $\lambda \lesssim 3$ cm reaches its upper limit (curve I in Fig. 169) the magneto-bremsstrahlung component of the emission is definitely determinative near λ_{max} (see in the figure the part of the spectrum relating to magneto-bremsstrahlung). At $\lambda \gtrsim \lambda_{max} \sim 5\text{--}10$ cm the contribution of the magneto-bremsstrahlung mechanism is actually greater than the shaded area in figure, since the intensity of the bremsstrahlung here will be less than the corresponding value in the millimetric band. This is caused by the reduction in the generation region of bremsstrahlung escaping beyond the corona when the optically thick levels $\omega \approx 2\omega_H$, $\omega \approx 3\omega_H$ rise from the chromosphere into the corona.

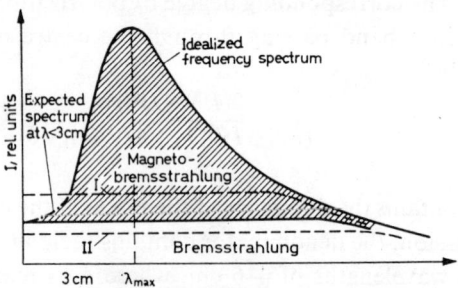

Fig. 169. Relative contribution of bremsstrahlung and magneto-bremsstrahlung mechanisms to the frequency spectrum of a local source above a spot (Zheleznyakov, 1963).

It is quite possible that even in the decimetric band the S-component above spots has a magneto-bremsstrahlung origin since the gyro-resonance levels $\omega \approx 3\omega_H$, $\omega \approx 2\omega_H$ (i.e. the fields $H_0 \sim 180\text{--}270$ oe for $\lambda \sim 20$ cm) are located at altitudes $h \sim 0.1 R_\odot$ (when $H_{0b} \sim 2.5 \times 10^3$ oe). As a result the bremsstrahlung generation region moves only in higher and more rarefied layers of the corona $h \sim 0.1 R_\odot$; its intensity decreases considerably because of a decrease in the emission measure $\int N^2 \, dl$. The observed values of the altitude of local sources at $\lambda \sim 20$ cm ($h \sim 0.03 R_\odot\text{--}0.15 R_\odot$, on the average about $0.06 R_\odot$) agree better with the assumption of the magneto-bremsstrahlung nature of the radio emission, although great significance cannot be given to this statement because of the uncertainty of the strength of the magnetic field in the corona above spots, particularly at considerable altitudes, on the one hand, and because of inaccuracy in the estimates of the effective altitudes of the local sources on the basis of experimental data on the other.

After the decay of the strong magnetic fields connected with the spots

§ 29] Origin of the Slowly Varying Component

the emission mechanism of local sources becomes bremsstrahlung no matter what it was before. The sharp decrease in the T_{eff} of local sources at centimetric wavelengths after the disappearance of spots once more indicates the predominance of magneto-bremsstrahlung from regions above spots in this band.

ORIGIN OF RADIO EMISSION OF HALOES AND LOCAL SOURCES ABOVE
FLOCCULI FREE OF SPOTS

If in regions above spots a very essential part is played by the magneto-bremsstrahlung mechanism, then the spectrum of a local source connected with flocculi free of spots ($I \approx$ const in the range $\lambda \sim$ 3–20 cm; see Fig. 161) can be fully explained on the basis of the bremsstrahlung mechanism alone, assuming that the critical wavelength λ_{cr} (29.7) in this case is far greater than 20 cm. The latter condition means that in the range studied a local source above flocculi (i.e. a region of the corona with enhanced density) is optically thin: $\tau \ll 1$. Then, in accordance with (29.8), the emission intensity is $I \approx 2I_j^{(0)}\tau =$ const and the effective temperature is $T_{\text{eff}} \approx T\tau \ll T$, where T is the temperature of the corona. This agrees well with the observations. We note, furthermore, that the criterion $\lambda_{\text{cr}} \gg$ 20 cm† gives an upper limit for the emission measure in the corona above flocculi (see (29.7)) of $\overline{N^2}L < 3 \times 10^{27}$ electron^2/cm^5 when $T \sim 10^6$ °K. For $L \sim 5 \times 10^9$ cm this corresponds to a concentration of $N < 8 \times 10^8$ electrons/cm^3. The actual value of N can be estimated from the observed value of T_{eff} at wavelengths of $\lambda^2 \ll \lambda_{\text{cr}}^2$.

When discussing the nature of the frequency spectrum of a local source above flocculi we have neglected the presence of magnetic fields in these regions. It is not hard to check that this is quite legitimate since the strength of the latter (judging from measurements at the level of the photosphere) does not exceed 100–200 oe (section 2). Then at $\lambda \sim 20$ cm the ratio $\omega_H^2/\omega^2 \lesssim 3 \times 10^{-2}$ and falls rapidly as the wavelength decreases; therefore the influence of the magnetic field does not have any significant effect on the form of the bremsstrahlung frequency spectrum (see (29.10)). Valuable information on the strength of the magnetic fields in local sources of the type under discussion could be provided by measurements of the polarization of the radio emission or even by data on its upper limit.

Such low values of the ratio ω_H/ω do not let us link the emission from regions above flocculi with the action of a magneto-bremsstrahlung mechanism at frequencies of $\omega \approx \omega_H, 2\omega_H, 3\omega_H$. It must be said that the latter also remains entirely valid for the flocculi surrounding groups of

---

† In practice it is sufficient for λ_{cr} to be only two or three times greater than 20 cm.

spots when explaining the radio emission at millimetric and centimetric wavelengths. This is quite understandable since the magneto-bremsstrahlung mechanism in these bands becomes effective only when there are strong magnetic fields in the lowest layers of the corona above the flocculi (for example at $\lambda \sim 3$ cm fields of $H_0 \sim 1200$–1800 oe). It follows from this that the weak halo around a bright source that can sometimes be observed in the centimetric band may be connected only with the bremsstrahlung of the coronal electrons, whose level increases because of the high density of the coronal plasma above centres of activity. Since the T_{eff} of the halo does not exceed 10^5 °K (whilst $T \sim 2 \times 10^6$ °K) it is clear that the region under discussion remains optically thin ($\tau \ll 1$) and the effective temperature T_{eff} remains equal to $T\tau = T\lambda^2/\lambda_{\text{cr}}^2$ at wavelengths of $\lambda^2 \ll \lambda_{\text{cr}}^2$; in the range where $\lambda^2 \gtrsim \lambda_{\text{cr}}^2$ the effective temperature is $T_{\text{eff}} \approx T$. If at $\lambda \sim 3$ cm the T_{eff} of the halo reaches 10^5 °K the value of λ_{cr} is obviously 13.5 cm. Then at $\lambda \gtrsim 13.5$ cm T_{eff} will be close to $T \sim 2 \times 10^6$ °K, i.e. is comparable with T_{eff} in the central part of the source in complete agreement with the observations. In the decimetric band the halo, as a region with a lower emission intensity, ceases to exist and the source with a more or less uniform distribution of radio brightness increases its size, which is comparable with the diameter of the spots or group of spots at centimetric wavelengths to the size of the accompanying group of flocculi (plages). In this case, as is clear from what has been said, the bremsstrahlung mechanism provides a complete explanation of the observed features of the emission outside spots above flocculi.

For lower values of the halo's T_{eff} the critical wavelength λ_{cr} moves further into the decimetric band. Then the sharply non-uniform distribution of the radio brightness over the source disappears only at longer wavelengths. With very small values of T_{eff} for the halo (less than $(3$–$5) \times 10^4$ °K at $\lambda \sim 3$–10 cm) λ_{cr} will lie in a range of wavelengths longer than about 50 cm. Here the bremsstrahlung mechanism of the radio emission of sources surrounding spots cannot explain the absence of a clear-cut difference between the values of T_{eff} above spots and flocculi in a centre of activity at decimetric wavelengths. In this case therefore we must admit the existence in the lower layers of the corona located above flocculi of magnetic fields $H_0 \sim 230$–350 oe, thanks to which the levels $\omega \approx 2\omega_H, 3\omega_H$ at wavelengths of $\lambda \gtrsim 15$ cm will be located in layers of the corona with $T \sim 2 \times 10^6$ °K: the magneto-bremsstrahlung of these levels with $T_{\text{eff}} \approx T$ will provide the necessary balance of radio brightness over the source in this band. On the other hand, at centimetric wavelengths this kind of magnetic field is insufficient for escape of the emitting levels $\omega \approx 2\omega_H, 3\omega_H$ into the corona and effective action of the magneto-bremsstrahlung mecha-

§ 29] Origin of the Slowly Varying Component

nism. Therefore at $\lambda \lesssim 10$ cm the halo radio emission will be bremsstrahlung as before.[†] The bremsstrahlung origin of the halo at centimetric wavelengths (and also in the millimetric band) makes it possible to find easily the value of the electron concentration N (to be more precise the measure of emission $\overline{N^2L}$) from known values of T_{eff} and T in the optically thin regions of the corona above flocculi.

It must be assumed that the density of these regions in any case is not greater than the density of the regions of the corona which are located directly above spots. Since the kinetic temperatures here have comparable values, it follows from what has been said that the intensity of bremsstrahlung above spots (without allowing for optically thick levels $\omega \approx \omega_H, 2\omega_H, 3\omega_H$ which prevent the escape of radio emission from the lowest layers of the corona) does not fall below the intensity of the halo. Because of this the latter quantity can be taken (with the reservation indicated) as the lower limit of the bremsstrahlung intensity in regions above spots (see curve II in Fig. 169).

In conclusion we note that in favour of the idea of the connection of the halo with a bremsstrahlung mechanism is the strong changeability of the value of the halo's T_{eff} and size from day to day, which is probably connected with variations in the value of $\overline{N^2L}$ in the optically thin source. The magnetic field in a centre of activity is more stable; therefore it is hard to explain the changeability of the halo in the framework of the magneto-bremsstrahlung mechanism. At the same time the action of the magneto-bremsstrahlung mechanism in regions above spots allows us to understand the characteristic stability of the S-component even at wavelengths of $\lambda < 5$–10 cm[‡] which exists despite the sharp variations in the electron concentration in the so-called lasting coronal condensations with $\overline{N} \sim 4 \times 10^9$ electrons/cm³.

Following Waldmeier and Müller (1950) condensations of this kind are generally taken as the sources of the slowly varying component. Since, however, the life of condensations is measured only in days and of the S-component in months, the high electron concentration characteristic of coronal condensations not being necessary when interpreting the emission under discussion, it is clear that this idea must be re-examined. It is more probable that the generation of the slowly varying component occurs

---

[†] At the same time noticeable polarization of the halo ($\varrho_e \approx 15\%$ at $\lambda \sim 3$ cm) is possible.

[‡] At $\lambda > 5$–10 cm, where T_{eff} is close to the temperature of active parts of the corona, emission leaves optically thick regions ($\tau_j > 1$) no matter what mechanism created it. Here, of course, variations in the field H_0 and concentration N are not reflected in the value of T_{eff} if the values of H_0 and N do not fall below a certain level determined by the condition $\tau_j \sim 1$.

simply in regions above centres of activity with a moderately enhanced concentration (up to 2×10^9 electrons/cm^3 at the solar cycle maximum) which exist for a time comparable with the spot's life. (It was regions of enhanced intensity in the corona of this kind that were investigated by Newkirk (1959).) At the same time the appearance of long-lived coronal condensations should essentially be reflected in the level of the radio emission of local sources in the millimetric wavelength region and in the halo emission in the centimetric band, i.e. in regions and conditions when the radio emission has a bremsstrahlung origin and is generated in optically thin layers. Unfortunately the correlation between these radio and optical phenomena has not yet been investigated.

$$* \ * \ *$$

The observed characteristics of the slowly varying component are therefore caused by the combined action of bremsstrahlung and magneto-bremsstrahlung mechanisms in regions with an enhanced plasma density and local magnetic fields above centres of activity. Bremsstrahlung alone is insufficient to explain the spectral and polarization features of the S-component, although at millimetric wavelengths above spots and in the millimetric–centimetric bands above the surrounding flocculi radio emission with $T_{\text{eff}} \ll 10^6$ °K has a bremsstrahlung origin with a constant intensity over the band. At centimetric wavelengths and particularly near the maximum of the spectrum ($\lambda \sim 5\text{--}10$ cm) magneto-bremsstrahlung of electrons in a magnetic field becomes very significant at frequencies that are multiples of the gyro-frequency: $\omega \approx 2\omega_H, 3\omega_H$; however, for sources with a relatively weak magnetic field the magneto-bremsstrahlung mechanism must yield pride of place to bremsstrahlung. It is possible that magneto-bremsstrahlung also plays an important part in the decimetric band. More definite conclusions on the relative contribution to the S-component from bremsstrahlung and magneto-bremsstrahlung for these wavelengths can be drawn only after accurate measurements of the intensity of the S-component in the millimetric band, since the latter to a considerable extent determines the level of bremsstrahlung at longer wavelengths as well.

The scheme developed above for the generation of the slowly varying radio emission, based on the combined action of magneto-bremsstrahlung and bremsstrahlung mechanisms, cannot, of course, be considered final, although it can be taken that basically it is true. At the same time further investigation of these phenomenon both experimentally and theoretically is absolutely necessary. Of particular interest here is the calculation of the characteristic curves of the S-component for actual models of a centre of

§ 29] Origin of the Slowly Varying Component

activity (distributions of T, N and $\boldsymbol{H}_0$) based in the final event on the data of optical observations; calculations of this kind allow us to judge with greater certainty the physical conditions in active regions of the corona and chromosphere.[†] As for experimental investigations, they should first be directed to determining more accurately the basic characteristics of the S-component (particularly the frequency spectrum, the polarization and the distribution of radio brightness over the source) and to studying the correlation of the spots' magnetic field strength with the altitude of the local sources' emitting region, and also of the connection between the altitude of the latter and the wavelength. Connections of this kind should exist if the ideas discussed above about the part played by magneto-bremsstrahlung in the generation of the S-component are correct, but they have not yet been investigated.[‡]

[†] This work has been done recently for one model of a centre of activity by Zlotnik (1968a, b).

[‡] See, however, section 10, where some data are given on the increase in altitude of a source of the S-component together with the wavelength.

CHAPTER IX

Theory of the Sun's Non-thermal Radio Emission

30. Generation of Continuum-type Sporadic Radio Emission

When analysing the solar radio emission the continuum-type components cover the components which have a comparatively "smooth" dynamic spectrum without any details: i.e. the basic and slowly varying components, microwave bursts, the enhanced radio emission connected with spots, types IV and V radio emission, and also events of the decimetric continuum and continuum storm type. The theory of the first two components has been given in sections 28 and 29; here we shall dwell on the problem of the origin of the remaining components, stressing that the relatively stationary nature of the radio emission under discussion and the broad spectrum of the frequencies occupied indicate a non-coherent generation mechanism.† The point is that coherent emission connected with amplification of the waves generally appears in limited frequency ranges and when there are special forms of particle velocity distribution which exist for only a limited time (see section 27).

As has already been pointed out in Chapter VIII, the non-coherent emission of charged particles in the coronal plasma is made up largely of bremsstrahlung, magneto-bremsstrahlung and Cherenkov emission. In the actual conditions of the solar corona the Cherenkov emission energy of an individual electron may in general be comparable with the magneto-bremsstrahlung energy; here both exceed the bremsstrahlung energy by many orders (section 26).

The low intensity of the bremsstrahlung may be used to explain the fact that the action of the bremsstrahlung is sufficient only for creating the "quiet" Sun's radio emission with an effective temperature $T_{\text{eff}} \approx 10^6 \, °\text{K}$, parts of the slowly varying component and possibly a certain fraction of the emission of microwave bursts (see below). Since as the velocity of the

---

† See the introduction to Chapter VIII for the division of the radio emission generation mechanisms into coherent and non-coherent.

§ 30] Continuum-type Sporadic Radio Emission

electrons increases the energy they emit in collisions decreases it becomes clear that the bremsstrahlung mechanism cannot provide generation in the corona of sporadic radio emission with $T_{\text{eff}} \gg 10^6$ °K. On the other hand, all other things being equal, non-coherent Cherenkov mechanism is less effective than the magneto-bremsstrahlung mechanism because of the strong attenuation of the Cherenkov emission due to transformation of waves as they leave the corona. It is therefore reasonable to assume that a non-coherent magneto-bremsstrahlung is most promising for interpreting the majority of the sporadic radio emission's components of the continuum type, although in certain cases (the S-component, microwave bursts) a certain contribution is also undoubtedly made by a bremsstrahlung mechanism.† A more definite conclusion about the part played by these mechanisms can be drawn only after comparing the observed characteristics of the continuum-type radio emission with the expected characteristics of the magneto-bremsstrahlung and bremsstrahlung in centres of activity. The present section is devoted to this problem.

ORIGIN OF MICROWAVE BURSTS AND CERTAIN PHENOMENA ACCOMPANYING THEM

As regards the different types of microwave burst (A, B and C) the most straightforward is the question of the nature and conditions of generation of type C radio emission ("gradual rise and fall") since, as has been remarked in section 11, this radio emission has much in common with the slowly varying component. According to section 29, this component is generated by coronal electrons by the combined action of bremsstrahlung and magneto-bremsstrahlung mechanisms in regions above a centre of activity having an enhanced electron density and a local magnetic field, and being heated to a temperature of $T \sim (1-2) \times 10^6$ °K. Type C bursts apparently appear in a similar manner and by the action of the same mechanisms, the only difference being that the S-component appears with a lengthy (tens of days) increase in the electron density relative to its unperturbed coronal value and the penetration of the spots' strong magnetic fields into the corona, whilst the bursts under discussion are connected with a further rise in the concentration above part of a centre of activity in short-lived (of the order of an hour) coronal condensations. The relative contribution of both mechanisms to the observed type C radio emission can be judged

---

† Of course, the magneto-bremsstrahlung mechanism will play such a big part only if strong magnetic fields exist in the generation region which are sufficient to transfer the magneto-bremsstrahlung frequency into the metric–centimetric band. Realization of this condition in the corona above centres of activity does not generally cause any particular difficulties.

only after careful investigations into the nature of the frequency spectrum, just as in the theory of the S-component (see section 29).

For type A events ("simple" bursts) the position is more complicated: the effective temperature of microwave outbursts reaches values of the order of 10^7–10^9 °K which cannot be obtained in a quasi-equilibrium plasma with $T \sim 10^6$ °K. Here, in all probability, the high T_{eff} owe their appearance in the flare (with which the microwave burst is connected) to energetic electrons whose magneto-bremsstrahlung in the local magnetic fields of the centre of activity produces the observed radio emission. The variation in the angular size of sources of type A bursts during their development and the attainment of a minimum value at the time of the minimum phase of the burst indicate that the acceleration of charged particles on the Sun is connected with some process or other of plasma compression in the region of the flare. The time of maximum compression then corresponds to maximum electron energy, maximum emission intensity and a minimum size of the generation region. The necessity for energetic electrons disappears for the generation of weak type A bursts (with $T_{eff} \sim 10^6$ °K): bursts of this kind, just as type C phenomena, may be caused by bremsstrahlung and magneto-bremsstrahlung of coronal electrons with a kinetic temperature $T \sim 10^6$ °K in the greater plasma densities above the flare.

The source of type B bursts ("post-bursts") is probably the region of enhanced density in the region of the flare which forms as the result of the plasma compression noted above; the emission mechanisms are similar to the mechanisms for type C events. Since the values of T_{eff} may be rather large at the time of the slow fall (up to 10^7 °K) it is also quite possible that a certain part in the generation of type B radio emission is also played by comparatively fast electrons diffusing into a region surrounding a type A source and losing a considerable part of their energy in magneto-bremsstrahlung during a "simple" burst. The decrease in the radio emission flux and the increase of the size of the source in the post-burst period becomes comprehensible from this point of view.†

The fact that bremsstrahlung plays a noticeable part in the process of generating microwave bursts follows from the analysis of the frequency spectra obtained by Hachenberg and Wallis (see section 11). They discovered that the spectrum of the majority of bursts studied (among which were type A, B and C bursts) can be described by the relation $S_\omega \propto \lambda^{-2}$ as far as wavelengths of $\lambda \sim 3$–15 cm; at shorter wavelengths the rise in the emission flux is replaced by an interval where S_ω is not frequency-

† See certain ideas on the generation mechanism of type B bursts by Kawabata, 1960b).

§ 30] Continuum-type Sporadic Radio Emission

dependent. According to Fig. 162 bremsstrahlung coming from a uniform volume of an isotropic plasma has this kind of spectrum; the region $S_\omega \approx$ const belongs to wavelengths of $\lambda < \lambda_{cr}$ where the layer of plasma is optically thin, and the region $S_\omega \propto \lambda^{-2}$ to wavelengths of $\lambda > \lambda_{cr}$ where the plasma becomes optically thick. The broadening of the range of frequencies in which $S_\omega \approx$ const observed when a burst is dying away can be explained from the standpoint of the bremsstrahlung mechanism by the displacement of λ_{cr} towards the long wavelengths because of a decrease in the electron concentration N in the source (see (29.7)).

The often observed maximum in the spectrum of microwave bursts may be caused by the effect of the strong magnetic field on the nature of the coronal electrons bremsstrahlung or be connected with the magneto-bremsstrahlung of the electrons at gyro-resonance levels just as in the case of the S-component. The formation of the spectral maximum may, of course, also be caused by the synchrotron radiation of relativistic electrons; here, as is easily seen from the formula (26.48), the requirements imposed on the strength of the magnetic field in the generation region will be far milder.[†]

In section 17 we were able to check that the region of a chromospheric flare and its vicinity are a source of electromagnetic radiation in a broad spectrum of frequencies, including microwave bursts, X-rays and gamma radiation. In addition, in certain vary rare cases the flare becomes visible in integrated light even against the background of the bright light of the photosphere. (The latter fact allows us to give the lower estimate of the flux of optical radiation from the flare of $S_{\text{light}} > 10^{-17}$ $\text{W m}^{-2}\text{c/s}^{-1}$ (Stein and Ney, 1963).) The theory of the electromagnetic radiation directly connected with chromospheric flares should explain from a unified point of view the origin of frequencies which are so widely separated on the scale yet are phenomena which are so closely connected with each other. At present there is not yet any theory developed in detail for the electromagnetic radiation from the vicinity of flares; below we shall discuss some possible hypotheses on the origin of the radiation which in future can be used as a basis for this theory as experimental data are accumulated and interpreted.

Starting with optical radiation in the continuous spectrum, we shall first of all indicate the possibility of explaining this radiation (as well as microwave emission) by the synchrotron mechanism of relativistic electrons.

---

† For the magneto-bremsstrahlung mechanism of microwave bursts see Takakura, (1959 and 1960), Zheleznyakov (1965a) and Gelfreich (1962) for the bremsstrahlung mechanism.

Gordon (1954) has drawn attention to this possibility; it has been discussed in detail by Stein and Ney (1963).

Let the maximum of the frequency spectrum of magneto-bremsstrahlung of electrons with an energy $\mathcal{E}$ in a field H_0 be in the optical frequency range $\omega \sim 4\times 10^{15}$ sec^{-1}; according to (26.48) this occurs if

$$\omega_m \equiv \omega_H \left(\frac{\mathcal{E}}{mc^2}\right)^2 \approx 2\omega \sim 8\times 10^{15} \text{ sec}^{-1}. \tag{30.1}$$

The assumption made is entirely reasonable since $S_{\text{light}} \gg S_f \sim 2\cdot 3\times 10^{-19}$ W m^{-2}c/s^{-1} (S_f is the spectral radio emission flux during the microwave burst).[†] Then the energy emitted by an electron in the optical band ($\omega \sim \omega_m/2$) in unit time and frequency is

$$\left(\frac{d\mathcal{E}_f}{dt}\right)_{\text{opt}} \approx 1\cdot 6 \frac{e^2}{c}\omega_H, \tag{30.2}$$

and in the radio band ($\omega \ll \omega_m/2$) is

$$\left(\frac{d\mathcal{E}_f}{dt}\right)_{\text{rad}} \approx 4\frac{e^2}{c}\omega^{1/3}\left(\omega_H \frac{mc^2}{\mathcal{E}}\right)^{2/3} \tag{30.3}$$

(see (26.45), (26.46) and (26.48)). Allowing for (30.1) the expression (30.3) will become (with $\omega \sim 2\pi \times 10^{10}$ sec^{-1}, i.e. at $\lambda \sim 3$ cm)

$$\left(\frac{d\mathcal{E}_f}{dt}\right)_{\text{rad}} \approx 8\times 10^{-2}\frac{e^2}{c}\omega_H. \tag{30.4}$$

If the system of emitting electrons remains optically thin in both the bands under discussion, then the ratio of the emission fluxes in the optical band and at radio frequencies is equal to the corresponding ratio $d\mathcal{E}_f/dt$ for a single electron, i.e. about 20 (see (30.2) and (30.4)). This does not contradict the experimental data, according to which the radiation flux in the optical range is one or two orders higher than the flux at centimetric wavelengths.

If follows from (30.1) that in fields of $H_0 \sim 5\times 10^2$ oe the maximum of the spectrum lies in the optical range with electron energies of $\mathcal{E} \sim 10^3 \times mc^2 \sim 5\times 10^8$ eV. This energy can be considered exceptionally high, but a flare in integrated light also is an exceptional phenomenon. Finally the total number of fast electrons $\int N_s dV$ required for creating radio emission with $S_f \sim 2\times 10^{-19}$ W m^{-2}c/s^{-1} and noticeable optical radiation can

---

[†] Unfortunately the information on the value of S_f during flares that can be seen in integrated light is very limited; here we have taken the radio emission flux at $\lambda = 3$ cm during the flare of 31 August 1956 when the spectral flux in the optical range was 2×10^{-17}–3×10^{-18} W m^{-2} c/s^{-1} (Hachenberg, 1960).

§ 30] Continuum-type Sporadic Radio Emission

easily be estimated from (30.4); it turns out here that

$$\int N_s \, dV \sim 10^{32} \text{ electrons.} \tag{30.5}$$

We have already noted above that hard radiation (X-rays and sometimes even gamma radiation) is recorded in flares as well as radio emission. An example of an event accompanied by a burst of gamma rays with an energy of 0·2–0·5 MeV per quantum is the flare of 20 March 1958 (see sections 11 and 17). Its radio emission is in all probability the synchrotron of relativistic electrons in local magnetic fields on the Sun. The close correspondence between the radio and the gamma bursts leads us to believe that the gamma radiation was caused by the same agent (relativistic electrons). Peterson and Winkler (1959) interpreted the gamma radiation as the bremsstrahlung of electrons with an energy of $\mathcal{E} \sim 1$ MeV. This point of view is also supported by Elwert (1961) and Takakura (1962) despite the difficulties pointed out by Peterson and Winkler. Then Gordon (1960) drew attention to two other possible mechanisms for the generation of the gamma-radiation: annihilation of proton–electron pairs and the "inverse" Compton effect. The latter mechanism has recently been discussed in detail in the paper of Zheleznyakov (1965a) in close connection with the synchrotron mechanism of radio emission; we shall be discussing later the bremsstrahlung and "annihilation" mechanisms of gamma-quantum generation.

The "inverse" Compton effect is the name sometimes given to the scattering of energetic electrons on "soft" photons accompanied by a transfer of energy from the electrons to the photons. When a relativistic electron interacts with an optical photon the latter acquires an energy (Ginzburg and Syrovatskii, 1964, section 8)

$$\Delta\mathcal{E} \approx \bar{\varepsilon} \left(\frac{\mathcal{E}}{mc^2} \right)^2 \tag{30.6}$$

($\mathcal{E}$ is the energy of the electron, $\bar{\varepsilon}$ is the mean energy of the light photon). The "Compton" energy losses of a relativistic electron in unit time can be described by the expression (Ginzburg and Syrovatskii, 1964, section 8)

$$\left(\frac{d\mathcal{E}}{dt} \right)_c \approx 2 \times 10^{-14} W_{\text{ph}} \left(\frac{\mathcal{E}}{mc^2} \right)^2 \text{ erg/sec.} \tag{30.7}$$

Here W_{ph} is the energy density of the light radiation expressed in ergs. The "inverse" Compton effect is of interest to us since with large ratios $\mathcal{E}/mc^2$ the photon energy rises to values characteristic of γ-rays; here the γ-radiation flux, because of the high density of the light emission near the

Sun's surface, is intense enough to explain the hard radiation connected with certain chromospheric flares.†

Considering the energy of the γ-quanta to be $\Delta\mathcal{E} \sim 0{\cdot}35$ MeV $\approx 5{\cdot}7\times 10^{-7}$ erg (flare of 20 March 1958) and making the mean energy of the photons in the photospheric optical radiation $\bar{\varepsilon} \approx 4\times 10^{-12}$ erg, we can at once obtain from the formula (30.6) the necessary energy of the relativistic electrons:

$$\frac{\mathcal{E}}{mc^2} \approx 4\times 10^2; \qquad \mathcal{E} \approx 2\times 10^8 eV. \qquad (30.8)$$

We can use (30.7) to find the total number of such electrons (denoted $\int N_s dV$) by proceeding from the value of the observed flux of γ-radiation on Earth $S_{\delta\gamma} \sim 2{\cdot}2\times 10^{-5}$ erg/cm^{-2}sec^{-1}. The latter value corresponds to a total γ-radiation power of $4\pi R_{S-E}^2 S_{\delta\gamma} \sim 6\times 10^{22}$ ergs/sec ($R_{S-E} \approx 1{\cdot}5\times 10^{13}$ cm is the distance of the Earth from the Sun), which leads to the estimate

$$\int N_s dV \sim 4\times 10^{30} \text{ electrons} \qquad (30.9)$$

(with an optical radiation energy density of $W_{ph} \approx 5$ ergs/cm^3). If we take the linear dimension of the quantum generation region to be equal to the diameter of the radio emission source at $\lambda \sim 3$ cm ($\sim 5'$, i.e. 2×10^{10} cm), then the relativistic electron concentration sufficient to produce the observed gamma-radiation flux is 0·5 electron/cm^3.

Let us now see whether we can interpret the microwave burst observed during the flare of 20 March 1958 in the framework of the synchrotron mechanism if we bring in for this purpose the electrons which are the source of the gamma radiation.

The spectral intensity of the synchrotron emission of relativistic electrons is defined by the formula (26.45). It follows from this formula that for a mono-energetic electron spectrum the radio emission intensity will (in accordance with the observations at $\lambda \gtrsim 3$ cm; see section 11) fall as the wavelength rises if $\omega/\omega_m \equiv (\omega/\omega_H)\times (mc^2/\mathcal{E})^2 \leqslant \frac{1}{2}$ at $\lambda \geqslant 3$ cm. The case $\omega/\omega_m = \frac{1}{2}$ does not in all probability occur in the corona since it requires too small a magnetic field: $H_0 \sim 5\times 10^{-2}$ oe when $\mathcal{E}/mc^2 \sim 4\times 10^2$; it may be expected that $H_0 \gtrsim 1$ oe and is most probably several hundreds of oersteds (in the region of a centre of activity). Therefore $\omega/\omega_m \ll \frac{1}{2}$; then the synchrotron energy of

---

† It will subsequently become clear that the effectiveness of the Compton effect on optical photons under solar conditions quite suffices for interpreting the observed gamma radiation. There is therefore no need to introduce infrared rays coming from the flare region instead of light quanta as was done by Gordon (1960). In the latter case for the same energy of generated gamma quanta the requirements for the energy of the relativistic electrons rise at once.

§ 30] Continuum-type Sporadic Radio Emission

one electron (in 1 sec and in a 1 c/s range) is

$$\left(\frac{d\mathcal{E}_f}{dt}\right)_m = 4\frac{e^2}{c}\omega^{1/3}\left(\omega_H\frac{mc^2}{\mathcal{E}}\right)^{2/3} \quad (30.10)$$

(see (26.45) and (26.46)).

We can judge the magnitude of H_0 more definitely if the life of a burst $t \sim 20$ sec is caused the energy losses of an electron in synchrotron. It follows from the formula (26.49) for synchrotron losses that a relativistic electron halves its energy in a time

$$t \approx \frac{3c}{2e^2\omega_H^2}\left(\frac{mc^2}{\mathcal{E}}\right)^2\mathcal{E}. \quad (30.11)$$

Putting, in accordance with (30.8), $\mathcal{E}/mc^2 \approx 4\times10^2$, $\mathcal{E} \approx 3\times10^{-4}$ erg and considering also that $t \approx 20$ sec we obtain $\omega_H \approx 4\cdot5\times10^9$ which corresponds to a value of $H_0 \approx 260$ oe.†

The total energy emitted by a system of $\int N_s dV \sim 4\times10^{30}$ electrons in a field $H_0 \approx 260$ oe at a wavelength $\lambda \sim 3$ cm can be found easily by means of (30.10); it is $2\cdot7\times10^{10}$ ergs sec^{-1} c/s^{-1}, which corresponds to a flux density on Earth of 10^{-21} W m^{-2} c/s^{-1}. The latter value is approximately a tenth of the value measured by Kundu and Haddock (1961) and Denisse (1959b). Great significance should not be attached to this difference, however, if we remember the low accuracy of the absolute measurements of the radio emission and gamma-ray fluxes on the one hand and the probable energy dispersion of the fast electrons on the other. (This difference can be reduced by an appropriate choice of electron distribution function with respect to the energies $\mathcal{E}$ since the efficiency of the radiation of an electron in the radio band and in gamma-rays depends differently on the energy of the particles.)

It should be noted that in the framework of the formula (30.10) the frequency spectrum of the synchrotron emission corresponds to the observed spectrum only qualitatively: the variation in intensity with frequency as $\omega^{1/3}$ proceeds too slowly. In experiment the spectrum dropped more steeply towards the long wavelengths, decreasing by a factor of 25 in the transition from 3 cm to 21 cm (see Denisse, 1959b). At the same time the "cut-off" of the spectrum at wavelengths of $\lambda \gtrsim 15$ cm becomes understandable if we remember the strong gyro-resonance absorption of these waves in the corona (in the presence of a field $H_0 \sim 260$ oe effective absorption at the third harmonic $\omega \approx 3\omega_H$ starts from $\lambda \approx 13\cdot5$ cm).‡

---

† It is not hard to calculate that the Compton losses in the case of interest to us are far less than the synchrotron losses.

‡ For gyro-resonance absorption in the corona see sections 26 and 29, Table 6 in particular.

Let us now discuss the bremsstrahlung mechanism of gamma-quantum generation (Peterson and Winkler, 1959).

The total losses to bremsstrahlung of an electron in an ionized plasma are (Ginzburg and Syrovatskii, 1964, section 8)

$$-\left(\frac{d\mathcal{E}}{dt}\right)_{brems} = 1.37 \times 10^{-16} \mathcal{E} N_+ \left(\ln \frac{\mathcal{E}}{mc^2} + 0.36\right). \quad (30.12)$$

We know that emission of gamma-quanta with a mean energy of $\mathcal{E} \sim$ 0·2–0·5 MeV will be ensured if the electrons have comparable energies: in the bremsstrahlung of very energetic electrons (for example of such as are necessary for the Compton mechanism) the mean energy per quantum, in contradiction to the observations, will be higher. With a fast electron energy of $\mathcal{E} \sim 2$ MeV the bremsstrahlung losses are

$$-\left(\frac{d\mathcal{E}}{dt}\right)_{brems} \sim 7.7 \times 10^{-22} \times N_+$$

calculated for one particle and

$$-\left(\frac{d\mathcal{E}}{dt}\right)_{brems} \int N_s dV \sim 7.7 \times 10^{-22} N_+ \int N_s \, dV \quad (30.13)$$

for the whole system emitting particles. On the other hand it follows from the observed gamma-ray flux on Earth (2.2×10^{-5} erg cm^{-2} sec^{-1}) that the value of (30.13) should be 6×10^{22} ergs/sec, which with the bremsstrahlung mechanism occurs only for

$$N_+ \int N_s dV \sim 7.8 \times 10^{43} \text{ cm}^{-3}. \quad (30.14)$$

In the corona with $N_+ \sim 4 \times 10^8$ cm^{-3} this is possible when $\int N_s \, dV \sim 2 \times 10^{35}$ electrons.

On the other hand, the value of $\int N_s \, dV$ can be estimated afresh from the requirement that the synchrotron flux of a system of fast particles at centimetric wavelengths should correspond to that observed. When $\mathcal{E} \sim 2$ MeV the maximum of the frequency spectrum is at $\lambda \sim 3$ cm in fields of $H_0 \sim 450$ oe (displacement of the maximum towards the shorter wavelengths or of the energy $\mathcal{E}$ towards the lower values is of low probability since this would lead to an unjustified rise in the required quantity H_0). Under these conditions the energy emitted by an electron in unit time and frequency is $(d\mathcal{E}_f/dt)_m \approx 1.6(e^2/c)\omega_H$. From this it is easy to find that the radio emission flux recorded on Earth from the flare of 20 March 1958 (about 10^{-20} W m^{-2} c/s^{-1} at $\lambda \sim 3$ cm) should have been generated in the corona by a system $\int N_s dV \sim 2.7 \times 10^{30}$ electrons. For a source with a

§ 30] Continuum-type Sporadic Radio Emission

volume of 8×10^{30} cm³ this corresponds to an electron concentration of $N_s \sim 0.3$ electron/cm³.

The difference of five orders between the number of electrons necessary for generating gamma-radiation (by the bremsstrahlung mechanism) and the number of electrons sufficient to create radio emission (on the basis of the synchrotron mechanism) proves the impossibility of a bremsstrahlung origin for the gamma-rays if the source of the latter is localized in the corona. The situation becomes more favourable, however, if we assume, following Takakura (1962), that the region where the gamma-quanta are born is in the lower layers of the Sun's atmosphere, since there the concentration of heavy particles (ions N_+ or neutral atoms N_a) is far higher than the coronal one. For example, in the case when the configuration of the magnetic fields in the centre of activity makes possible the penetration of fast electrons from a radio emission source in the lower corona to the level of the photosphere (thanks to which a concentration of relativistic electrons is maintained in it close to the concentration $N_s \sim 0.3$ electron/cm³ in the radio emission source), the total number of energetic electrons in a layer of the order of 10^8 cm thick near the photosphere will reach 10^{28} (when the area of the gamma and radio emission sources projected onto the photosphere is the same). Since in the photospheric layer $N_a \sim 4 \times 10^{15}$ cm⁻³ (see Table 1 in section 1), the product $N_a \int N_s dV$ will be of the order of 4×10^{43}, i.e. close to that required for the bremsstrahlung mechanism (30.14).†

Unfortunately on the basis of the data available at present we cannot choose between the Compton mechanism of gamma-quantum generation in the lower corona and the bremsstrahlung mechanism in the photosphere, although the first is the most natural in our opinion. For the flare of 20 March 1958 the choice depends upon the form of the radio emission's frequency spectrum at $\lambda < 3$ cm let us say at millimetric wavelengths: with emitting electron energies of $\mathscr{E} \sim 2 \times 10^8$ eV (Compton mechanism) the spectrum should rise with the frequency; with energies of $\mathscr{E} \sim 2$ MeV (bremsstrahlung mechanism), on the other hand, it should fall as the frequency increases. On 20 March 1958, however, no data were obtained on the amplitude of the microwave burst at millimetric wavelengths.

A third possible way gamma-quanta may be formed in the solar corona and chromosphere is the "annihilation" of electron–positron pairs (Gordon 1960). Most efficient here is the process of annihilation of slow electrons

---

† In the photosphere $N_+ \ll N_a$ and the contribution of N_+ to the bremsstrahlung is insignificant. When discussing the bremsstrahlung in collisions of electrons with neutral atoms we should, strictly speaking, use a formula that differs slightly from (30.12); this is not important, however, in our rough estimates.

and positrons ($\mathscr{E}_{kin} \ll mc^2$) with the appearance in each event of two gamma-quanta with energies $\hbar\omega = mc^2 \approx 0\cdot 5$ MeV. The "birth" of positrons in principle may occur either as the result of nuclear reactions or from collisions between energetic particles ($\mathscr{E}_{kin} > 1$ MeV). This possibility seems of low probability to us because of the small effective cross-sections of these processes, but it cannot be rejected without further study. If it turns out that the energy of the gamma-quanta is concentrated in a narrow range near mc^2 this will be a serious argument in favour of the "annihilation" hypothesis.

Above we have been discussing the generation mechanisms of gamma-rays and radiation in white light, i.e. phenomena which are to a certain extent exceptional. A more usual (or more easily detectable) event is the generation in the region of chromospheric flares of X-rays (with an energy of the order of ones and tens of keV per quantum, i.e. with an energy less than the energy of the gamma-rays but far greater than the energy of the light photons). The close correlation between X-rays and microwave bursts noted in section 17 is an argument in favour of a common source for both phenomena. As for the actual mechanism of the X-radiation, it is best sought among the synchrotron Compton and bremsstrahlung mechanisms that have been discussed above. Apparently the most probable will be the last mechanism which can always explain the observed level of emission by assuming high enough values for the concentration of ions N_+ or neutral atoms N_a in the generation region (see (30.12)), i.e. by assuming localization of this region in deep-down enough layers of the Sun's atmosphere.†

It is not excluded that a certain part is also played in the creation of X-radiation (and not only gamma radiation) by the Compton mechanism which from this point of view deserves careful study. Here, however, we cannot but allow for the sharp drop in the intensity of the Compton emission together with the decrease in the energy of the hard quanta $\Delta\mathscr{E}$ on the one hand and of the higher values of the X-ray flux when compared with the gamma-rays on the other hand (by about five orders during the flare of 31 August 1959, see Chubb, Friedman and Kreplin, 1960); both these factors will lead to a sharp increase in the concentration N_s of relativistic electrons. However, for smaller-scale phenomena the requirements imposed on the value of N_s may prove to be not so rigid. The synchrotron mechanism as applied to the X-radiation ($\lambda \sim 1$–10 Å) is effective only in very strong magnetic fields or with very high electron energies not reached under solar conditions.

† A preliminary discussion of the bremsstrahlung mechanism of the X-radiation of solar flares can be found in the papers by Kawabata (1960b) and Elwert (1961).

§ 30] Continuum-type Sporadic Radio Emission

ORIGIN OF THE ENHANCED RADIO EMISSION CONNECTED WITH SUNSPOTS (Ginzburg and Zheleznyakov, 1961).

Moving on to discuss the mechanisms of radio emission of the continuum group in the metric waveband we note first of all that the choice is actually limited to two mechanisms: bremsstrahlung and magneto-bremsstrahlung (see the beginning of this chapter). In actual fact, as applied to enhanced radio emission the following can be said about the bremsstrahlung mechanism. With reasonable assumptions about the nature of the magnetic field in the corona above spots the optical thickness $\tau_{1,2}^{coll}$ connected with bremsstrahlung absorption in collisions of electrons with ions does not exceed a few units at a temperature of $T \sim 10^6$ °K. Then $\tau_{1,2}^{coll} \propto T^{-3/2}$; as T increases the effective temperature $T_{\text{eff}} = T(1-e^{-\tau_{1,2}^{coll}})$ rises slightly (until $\tau_{1,2}^{coll} \gtrsim 1$) and then starts to decrease as $T^{-1/2}$ (with $\tau_{1,2}^{coll} \ll 1$; see section 26). It is easy to confirm that the maximum value of the effective temperature of the bremsstrahlung in an equilibrium coronal plasma definitely does not exceed 10^7 °K, which is clearly insufficient for interpreting the enhanced radio emission in the metric waveband (or the type IV radio emission). The criticism of Ryle's theory (Ryle, 1948) by Piddington (1953) was in essence based on this circumstance; Ryle interpreted the enhanced radio emission connected with sunspots as the thermal radio emission of local regions of the corona heated to 10^{10} °K.

If we remember, however, that a contribution is made to the quantity $\tau_{1,2}$ by magneto-bremsstrahlung (gyro-resonance) absorption in layers $\omega \approx s\omega_H$ ($s = 1, 2, 3, \ldots$) as well as by bremsstrahlung absorption, the dependence on T changes significantly. It has been established in section 26 that under normal conditions of the solar corona ($N \sim 10^8$ electrons/cm^3, $T \sim 10^6$ °K) the optical thickness is $\tau_{js} \gtrsim 1$ for the resonance layers of the corona $\omega \approx \omega_H, 2\omega_H$ (ordinary wave) and $\omega \approx 2\omega_H, 3\omega_H$ (extraordinary wave). As the temperature of the plasma increases the value of τ_{js} rises in proportion to T^{s-1} at the levels $\omega \approx s\omega_H$ ($s \geq 2$) and proportional to T in the gyro-frequency layer $\omega \approx \omega_H$ ($s = 1$). If in addition we remember that the levels $\omega \approx \omega_H, 2\omega_H$ in a strong magnetic field may be located higher than the layers of the corona where the refractive index is $n_j = 0$ for ordinary and extraordinary waves, it becomes clear that the high intensity of the sporadic radio emission at metric wavelengths with $T_{\text{eff}} \gtrsim 10^9$ °K can be explained in principle by the action of the magneto-bremsstrahlung mechanism which creates the thermal emission of regions of the corona with a local magnetic field of sufficient strength and temperature $T \sim T_{\text{eff}}$. There is no foundation for assuming, however, that conditions may arise in the corona under which heating of the coronal plasma is possible to temperatures that exceed the normal temperature of the corona by more than three orders. On the other hand, the impossibility of explaining the enhanced radio emission at metric wavelengths connected with sunspots by the magneto-bremsstrahlung of the coronal electrons (with energies corresponding to a kinetic temperature $T \sim 10^6$ °K) at the gyro-frequency is obvious (Kiepenheur, 1946). The corresponding objections against this idea have been raised by Ginzburg (1947).

The great optical thickness of the resonance levels $\omega \approx \omega_H$, $2\omega_H$ (ordinary wave) and $\omega \approx 2\omega_H$, $3\omega_H$ (extraordinary wave) in the normal conditions of the solar corona and the rise in gyro-resonance absorption (emission) as the plasma's kinetic temperature increases indicate that to explain the enhanced radio emission at metric wavelengths a sharp increase in the energy of all the particles in the emitting region is unnecessary. It is quite sufficient to raise to the value of T_{eff} the temperature T_s of the sub-system of electrons, the concentration N_s in which corresponds to values of the subsystem's optical thickness of the order of unity. When $T_{\text{eff}} \sim 10^{10}$ °K this leads to relativistic values of the electron energy of $E \sim 10^6$ eV.

The magneto-bremsstrahlung mechanism of enhanced radio emission in this form (with the introduction of relativistic electrons) was put forward by Getmantsev and Ginzburg (1952); it was then treated more fully in the papers of Gershman and Zheleznyakov (1956), Ginzburg and Zheleznyakov (1961), Zheleznyakov (1955). It must be pointed out that with lower values of T_{eff} the requirements imposed on the energy of the emitting electrons may be significantly reduced. For example with $T_s \sim 2 \cdot 5 \times 10^8$ °K (which is sufficient for interpreting enhanced radio emission with $T_{\text{eff}} \sim 2 \cdot 5 \times 10^8$ °K) the sub-system of energetic electrons under discussion can still be considered weakly relativistic ($\beta_{\text{th}_s}^2 = V_{\text{th}_s}^2/c^2 \sim 0 \cdot 1$; V_{th_s} is the mean thermal velocity of the fast electrons) and for estimating the optical thickness of the resonance level it is quite permissible to use the formula (26.124) with the replacement of N by N_s and β_{th} by β_{th_s}:†

$$\tau_{j,s} \sim \frac{s^{2s}}{2^{s-1}s!} \frac{\omega}{c} v_s (\beta_{\text{th}_s})^{2s-2} L_H \quad (s \geqslant 2), \qquad (30.15)$$

$$\tau_{j,s=1} \sim \tau_{j,s=2}. \qquad (30.16)$$

In the relation (30.15) $v_s = 4\pi e^2 N_s/m\omega^2$ and L_H is the characteristic dimension over which the value of the magnetic field changes significantly.

† The necessity for this replacement is easy to understand if we remember that in a non-uniform magnetic field with $T_s \gg T$ the coefficient of resonance absorption is $\mu_{js} \propto N_s \beta_{\text{th}_s}^{s-3}$ (for $s \geqslant 2$) and $\mu_{js} \sim N_s \beta_{\text{th}_s}$ (for $s = 1$) in a layer whose thickness is proportional to β_{th_s} (for further details see section 26). We notice that the formulae (30.15), (30.16) as applied to an ordinary wave are one or two orders too high in their values because numerical factors have not been allowed for.

§ 30] Continuum-type Sporadic Radio Emission

It follows from (30.15) that with $\beta^2_{\text{th}_s} \sim 0.1$, $L^1_H \sim 2\times 10^{10}$ cm and $\omega \sim 2\pi \times 10^8$ sec^{-1}, $\tau_j(\omega \approx 2\omega_H) \sim N_s$. Therefore the effective temperature of the emission of a sub-system heated to $T_s \sim 2.5 \times 10^8$ °K reaches the same value if $N_s \sim 1$ electron/cm³. Calculations by the more accurate formulae given in section 26 do not change this estimate in any essential way.

Above, when we were finding the optical thickness τ_j we allowed only for absorption by fast electrons with $\beta^2_{\text{th}_s} \sim 0.1$. The explanation of this is that the sharp absorption of radio waves in the layers $\omega \approx \omega_H$, $2\omega_H$ by the slower electrons of the corona ($\beta^2_{\text{th}} \sim 10^{-4}$) does not prevent the escape of the radio emission of the fast electrons located slightly above these layers. The point is that the thickness of the effectively emitting (absorbing) layer is proportional to the mean velocity of the electrons and for the subsystem of fast electrons is $\beta_{\text{th}_s}/\beta_{\text{th}}$ times greater than for the corona electrons (in this connection see the formula (26.123) for the thickness of the resonance layer L_{js}).

The nature of the polarization of the enhanced emission caused by magneto-bremsstrahlung depends on the height distribution of the fast electrons in the corona. For example, an ordinary component will predominate in radio emission with a frequency ω if the fast electrons with $\beta^2_{\text{th}_s} \sim 0.1$ are basically located below the level $\omega \approx 2\omega_H$, i.e. emit in a layer with a frequency $\omega \approx \omega_H$ from which only the ordinary component can escape beyond the corona. In the case when the energetic electrons are located largely above the level $\omega \approx \omega_H$ the extraordinary component can predominate in radio emission at a frequency ω, since emission from the $\omega \approx \omega_H$ is small and in the layers $\omega \approx 2\omega_H, 3\omega_H$, etc., the optical thickness for an extraordinary wave is greater than the corresponding value for an ordinary wave. Moreover, when the concentration of the emitting electrons is high enough, when the τ_j of the subsystem in the layer $\omega \approx 2\omega_H$ becomes greater than unity not only for an extraordinary wave but also for an ordinary one, the degree of polarization of the radio emission decreases significantly. Since the height distribution of the fast electrons depends on the actual conditions in the region of the spots (the nature and method of particle acceleration, the configuration of the magnetic fields holding the particles back, etc.) which, generally speaking, differ for different spots, the essential uncertainty in the sense of the polarization of the enhanced radio emission which is generally observed at metric wavelengths (section 12) becomes comprehensible.

The directivity of the enchanced radio emission is partly connected with the phenomenon of refraction in the corona; however, from the point of view of the magneto-bremsstrahlung mechanism directivity will also be noticeable if in the generation region and the overlying layers $n_j \sim 1$.

This is caused by the strong absorption in the corona in the $\omega \approx 2\omega_H$ layer of ordinary waves escaping from the $\omega \approx \omega_H$ level, and in the $\omega \approx 3\omega_H$ layer of extraordinary waves from the $\omega \approx 2\omega_H$ level when they are propagated at large enough angles to the field H_0.[†] The resultant degree of polarization depends essentially on the structure of the magnetic field (which, generally speaking, points away from the Sun above spots) and also on the height distribution of the energetic electrons.

The higher directivity for extraordinary waves connected with the more efficient absorption of the latter in the gyro-resonance layers may explain the predominance of the ordinary component in the composition of the enhanced radio emission which is nevertheless observed in observations of noise storms, despite the considerable uncertainty of the corresponding experimental data (see section 12).

The gyro-resonance layers at which there is sharp absorption of ordinary and extraordinary emission are located above large groups of spots with a strong magnetic field far higher than the layer where $\varepsilon = 1 - \omega_L^2/\omega^2 = 0$ or the layer for which the optical thickness caused by collisions is of the order of unity. From this we can understand why the region in which enhanced radio emission is generated is usually localized in high layers of the corona.

The mechanism of magneto-bremsstrahlung has been discussed above as applied to enhanced radio emission of comparatively low intensity ($T_{\text{eff}} \lesssim 2\cdot5\times10^8$ °K); emission of this kind can be provided by weakly relativistic electrons. In the rare cases when T_{eff} reaches values of the order of 10^9–10^{10} °K the energy of the emitting electrons should be raised to 10^5–10^6 eV and, apparently, their concentration should be slightly increased. Then in the emission detected the relative intensity of the high harmonics rises; however, both in the case of $\beta_{\text{th}_s}^2 \ll 1$ and with $\beta_{\text{th}_s}^2 \sim 1$ the harmonics cannot be resolved because of the considerable change in the strength of the magnetic field in a generation region with dimensions of $L \sim 10^{10}$ cm, and also because of the Doppler effect in the region $\beta_{\text{th}_s} n_j \sim 1$.

In the case when the enhanced radio emission is created by electrons on the edge of being relativistic the region of generation of emission at frequencies $\omega \approx s\omega_H$ will occupy a still higher position in the corona (because of the emission of harmonics with large values of s); there is a corresponding decrease in the degree of directivity when compared with the emission of weakly relativistic electrons. It follows in particular from this that with a large T_{eff} the distance from the enhanced radio emission generation region to the surface of the Sun, generally speaking, increases and the

[†] The dependence of τ_j on α is not reflected in the formulae (30.15), (30.16) which are suitable only for making estimates.

§ 30] Continuum-type Sporadic Radio Emission

degree of directivity decreases. Unfortunately, up to now, as far as we know, the nature of the dependence of the altitude of the generation region and the directivity on T_{eff} has not yet been investigated experimentally.

For more definite conclusions on the nature of the frequency spectrum, the polarization and the degree of directivity on the basis of the magneto-bremsstrahlung mechanism we must examine in detail the various possible distributions of the emitting electrons in the corona with respect to energies and coordinates. In this connection it should be noted that the model of relativistic electron emission used by Gershman and Zheleznyakov (1956) and Zheleznyakov (1955) requires considerable correction. This model assumes that above spots relativistic electrons exist in both the outer and the inner layers of the corona and even in the transition region from the corona to the chromosphere. In the latter region, however, electrons with an energy of 10^5–10^6 eV would make too large a contribution to the emission at $\lambda \sim 10$ cm (with $T_{eff} \sim 10^9$–10^{10} °K) which is not observed in reality.

As a whole what has been said above gives us reason to believe that the magneto-bremsstrahlung mechanism in solar conditions can explain the observed features of the enhanced radio emission connected with spots (intensity, strong polarization, directivity, close connection with the spots' magnetic fields, etc.) if we assume that in regions of the corona with a strong magnetic field there exist electrons with a concentration of $N_s \sim 1$ electron/cm³ (or slightly higher) and an energy corresponding to a kinetic temperature of $T \sim 10^8$–10^{10} °K. This assumption is realized if acceleration of charged particles (electrons) to energies of 10^5–10^6 eV occurs in the corona above sunspots. The corresponding arguments in favour of the existence of particles with energies of this order at the time of noise storms in regions of the corona above spots are given in section 17.

MECHANISM OF TYPE IV RADIO EMISSION

Let us now turn to type IV events whose basic characteristics are discussed in section 15. The discussion of the correlation of type IV bursts with a number of geophysical phenomena carried out in section 17 made it possible to establish that, in all probability, type IV radio emission is generated in a bundle of plasma ejected from the region of a flare and moving in the corona at a velocity of several thousand kilometres per second. This bundle with its "frozen-in" magnetic field also contains high-energy particles. Near the Earth these are largely protons, although in one case about which we spoke in the previous sub-section a high-energy electron component of solar corpuscular emission has been found. The polarized nature of type IV radio emission also indicates the existence of a magnetic field in the generation region. The absence in certain cases of polarization of the radio emission indicates either that the lines of magnetic force are chaotic or that there is quite a high concentration of emitting electrons, due to which the optical thickness of the system for both kinds of wave is greater than unity and the effective temperatures are comparable. It is clear that to retain fast electrons for about an hour (the life of type IV radio emission) in a region with a linear size of $L \sim 5 \times 10^{10}$ cm (L is the size of the type IV radio emission source) the magnetic field should be a good enough "trap".

What can we say about the generation mechanism of type IV emission? From the considerations given in the introduction to this chapter it is clear that the very high effective temperatures of the type IV radio emission ($T_{\text{eff}} \gg 10^6$ °K) make it impossible to interpret on the basis of the bremsstrahlung mechanism; at the same time the contribution of Cherenkov non-coherent emission, because of the low wave-transformation efficiency, is small when compared with the synchrotron emission. In favour of this mechanism is the absence of systematic frequency drift (similar to that observed in type II bursts) when the emitting region moves a distance of several solar radii in the corona—apparently because of the relative constancy of the magnetic field strength which determines the frequency spectrum of the synchrotron emission (see (26.45)).[†]

The hypothesis of the synchrotron origin of type IV radio emission, which has been given wide credence, was put forward by Boishot and Denisse (1957). The basic parameters of the system of emitting electrons for the case of type IV and without allowing for the effect of the ambient medium on the nature of the synchrotron emission are estimated in the paper of Ginzburg and Zheleznyakov (1961); for the conditions of type IV radio emission generation see also Takakura and Kai (1961), Akin'yan and Mogilevskii (1961) and Sakurai (1961).[‡]

The nature of the frequency spectrum of type IV radio emission (with a maximum at metric wavelengths and a smooth fall towards the shorter and longer wavelengths) allows us to assume that the energy distribution of the energetic particles $f(\mathcal{E})$ does not differ too sharply from an equilibrium one (in the sense of the value of the derivative $df/d\mathcal{E}$ which determines

---

[†] However, under conditions when the effect of the plasma on the nature of an electron's synchrotron emission becomes significant the frequency corresponding, for example, to the intensity maximum depends not only on ω_H but also on ω_L (i.e. on the concentration N). Then a certain displacement of the emitted frequencies becomes possible as the source moves into the upper layers of the corona (formula (30.18)).

[‡] Under similar conditions synchrotron emission had been discussed by Gordon (1954) as applied to outbursts accompanying chromospheric flares even before type IV radio emission had been recognized as a separate component.

§ 30] Continuum-type Sporadic Radio Emission

the intensity of emission from an optically thick system; see section 27). Then the observed values of T_{eff} allow us to establish the lower energy boundary of the electrons creating emission with this effective temperature $\mathscr{E} \sim \varkappa T_s \gtrsim \varkappa T_{\text{eff}}$. For the sake of definition let the effective temperature T_{eff} of the radio emission of type IV be 4×10^{10} °K (we have in mind very intense events, although in exceptional cases the scale of the phenomena is even greater: T_{eff} reaches 10^{12} °K; section 15). When $T_{\text{eff}} \sim 4 \times 10^{10}$ °K the energy is $\mathscr{E} \sim 4 \times 10^6$ eV. Particles with energies of this kind (not electrons, but protons) are recorded in the vicinity of the Earth after flares accompanied by type IV radio emission (see section 17). The usual absence of energetic electrons can be explained by the more effective retention of these particles in the solar magnetic fields and the greater energy losses to synchrotron radiation (when compared with the protons).

Furthermore, the strength of the magnetic field in a source can be estimated if we remember that relativistic electrons in a magnetic field emit a broad spectrum of frequencies $\Delta\omega \sim \omega_{\max}$ with a maximum at the frequency $\omega_{\max} \approx \frac{1}{2}\omega_H (E/mc^2)^2$ (see (26.48)). Putting $\omega_{\max} \sim 2\pi 10^8$ sec^{-1} ($\lambda \sim 3$ m) and $\mathscr{E} \sim 4 \times 10^6$ eV, we obtain that $H_0 \sim 1$ oe. Higher values of $\mathscr{E}$ will require lower values of the field H_0 in the source. The value of H_0, however, can definitely not fall below one oersted or fractions of an oersted, i.e. below the values of the field in the unperturbed corona. In estimates in future we shall therefore take $H_0 \sim 1$ oe.

Let us now estimate the concentration N_s of relativistic electrons required to create $T_{\text{eff}} \sim 4 \times 10^{10}$ °K in a region of an angular size of about 10' (a value typical for type IV events). At the maximum of the spectrum, where the formula (30.2) is valid, the energy emitted by one electron is 2×10^{-22} erg sec^{-1} c/s^{-1}. Unless we allow for reabsorption (which is permissible in a rough approximation for a system with an optical thickness $\tau_j \lesssim 1$) then the energy emitted by a system of $\int N_s \, dV$ electrons is $2 \times 10^{-22} \int N_s \, dV$. On the other hand it is easy to find by using the formulae (4.15) and (4.17) that a 10' source of radio emission with $T_{\text{eff}} \sim 4 \times 10^{10}$ °K creates a flux of 4×10^{-19} W m^{-2} c/s^{-1} on Earth. Considering the flux to be the same in all directions we find that the total energy is 10^{12} erg sec^{-1} c/s^{-1}. To find the total number of electrons that will ensure the appearance of events of this scale we must obviously compare this value with the one obtained above of $2 \times 10^{-22} \int N_s \, dV$. We then find that $\int N_s \, dV$ should be not less than 5×10^{33} electrons; for the size of source assumed this corresponds to a concentration of $N_s \sim 5 \times 10^2$ electrons/cm^3.

In the estimates above we ignored the effect of the medium on the nature of the synchrotron radiation. The part played by the coronal plasma in the generation of type IV radio emission has been discussed by Zheleznyakov and Trakhtengerts (1965). This part is insignificant if the condition (26.51) is satisfied in the generation region since in this case the intensity maximum and the total emission energy will be defined as before by the formulae valid in a vacuum. For the values of E and H_0 taken above in a type IV radio emission source the condition (26.51) will be observed if the plasma concentration is $N \ll 2\times 10^6$ electrons/cm³. The inequality obtained is quite rigorous: in an unperturbed Baumbach–Allen corona it obtains only at considerable altitudes starting at $h \sim R_\odot$ above the Sun's surface. Since the region where type IV radio emission is generated is generally at a distance of the order of several radii from the Sun the inequality $N \gg 2\times 10^6$ electrons/cm³ is perfectly possible. However, cases of higher concentrations in the source undoubtedly also occur, particularly at low altitudes.

When the emitting region moves into the upper layers of the corona not only does the concentration N change, but so do the magnetic field strength H_0, the energy of the relativistic particles $\mathcal{E}$ and the size of the source L. The question of the law governing the change of these quantities as the source moves (a plasma bundle with a "frozen in" magnetic field) in the corona remains essentially open although the answer to it affects the law governing the change of the parameter $\delta \equiv \sqrt{3}(\omega_L/\omega_H)mc^2/\mathcal{E}$, whose value determines whether or not it is necessary to allow for the effect of the medium on the generation of the radio emission.† If we assume that during motion of the source in the corona the energy is $\mathcal{E} \approx \text{const}$, and due to the "freezing in" $N \propto 1/L^3$, $H_0 \sim 1/L^2$, the parameter $\delta \propto L^{1/2}$.

The size of a type IV radio emission source as it moves in the corona rise quite weakly from 2′ to 10–12′, i.e. by a factor of about 6 (see section 15). The value of δ then increases altogether by a factor of only 2·5. The insignificant change in this characteristic parameter allows us to state that in all probability the medium has no effect on the generation of the radio emission throughout a type IV event unless this effect is felt in the upper layers of the corona. It is this case that we have discussed above.

At the same time another class of type IV radio emission sources is obviously possible for which $\delta \equiv \sqrt{3}(\omega_L/\omega_H)mc^2/\mathcal{E} \gg 1$ and the effect of the medium becomes decisive. It is clear from the relations (26.52)–(26.52a) that the effect of the coronal plasma leads to a shift of the maximum of the synchrotron emission towards the high frequencies and to a sharp decrease in the spectral power of the emission at the maximum, and also of the total power of the emission. All other things being equal the $\delta \gg 1$

---

† See in this connection the formulae (26.51)–(26.53), according to which allowing for the medium has little significance when $\delta \lesssim 1$ and becomes absolutely necessary if $\delta \gg 1$.

§ 30] Continuum-type Sporadic Radio Emission

variant obtains in sources with weaker magnetic fields. For example, when $\mathscr{E} \sim 4\times10^6$ eV the parameter δ will be greater than unity if in the process of rising into the upper layers of the corona the field in the source decreases to a value of $H_0 \sim 1$ oe earlier than the concentration N of the plasma bundle falls to 2×10^6 electrons/cm³. If, let us say, $H_0 \sim 1$ oe when $N \sim 10^8$ electrons/cm³ then $\delta \approx 7$. The frequency $\omega'_{\max}$ corresponding to the maximum of the frequency spectrum will then be an order greater, i.e. about $2\pi\times10^9$ sec⁻¹ ($\lambda \sim 30$ cm), the spectral power at the maximum decreases by a factor of $\sim 3\times10^2$ and the total power by a factor of $\sim 2\times10^2$ (with the same number $N_s L^3$ of relativistic electrons in the source as before).

The sharp drop in the emission efficiency in the case of $\delta \gg 1$ means that for the majority of type IV events recorded on Earth the conditions in the sources ensure that the relation $\delta \lesssim 1$ is satisfied. In other words, in observed type IV sources

$$\sqrt{3}\,\frac{\omega_L}{\omega_H}\,\frac{mc^2}{\mathscr{E}} \lesssim 1, \qquad (30.17)$$

there being generally no need to take the medium into consideration in the case of $\delta \ll 1$, just as in the case of $\delta \sim 1$ in the rough estimates. By combining the criterion (30.17) with the condition for maximum emission (26.48) we obtain that in the region where type IV radio emission is generated the following relations are satisfied:

$$\omega_L \lesssim \frac{2}{\sqrt{3}}\omega_{\max}\frac{mc^2}{\mathscr{E}}; \qquad (30.18)$$

$$\omega_L^2 \lesssim \frac{2}{3}\omega_{\max}\omega_H. \qquad (30.19)$$

In the case of $\delta \gg 1$, when the relations (30.17)–(30.19) are not satisfied, the emission drops sharply (approximately as $e^{-\delta}$) and the type IV source becomes very weak.

Above we have discussed the question of the origin of the three components making up the continuum group: microwave bursts, enhanced emission and type IV radio emission. The following can be said about the nature of the other, less studied events of this group: the decimetric continuum and the storm continuum. According to Zheleznyakov (1967) the noncoherent synchrotron mechanism discussed above cannot be used to explain the occurence of the outstanding type IV events with $T_{\text{eff}} \sim 10^{12}$ °K. Such bursts may be due to the coherent synchrotron mechanism of radio emission associated with a synchrotron instability effect in a system of relativistic electrons plus a "cold" plasma (Zheleznyakov, 1966).

The decimetric continuum, whose characteristics are close to the properties of microwave bursts (section 16), probably has a similar origin to them. A major part in the creation of intense decimetric radio emission is apparently played by the magneto-bremsstrahlung mechanism of energetic electrons which are injected from the flare region into the "trap" formed by the magnetic field of a bipolar group of spots. Cases of displacement of the emission maximum into the decimetric band are caused by the higher position of the generation source of the continuum in the corona, in a region with a weaker magnetic field than in the sources of microwave bursts. Still higher in the corona are localized the regions where the enhanced radio emission is generated (with a maximum at metric wavelengths) and the sources of the storm continuum—events which to a certain extent are similar to the enhanced emission at a period when it is not accompanied by type I bursts (section 16).

The ideas given in the present section on the generation conditions of continuum-type sporadic radio emission rest on the concept of the existence of charged particles on the Sun (chiefly electrons) with energies ranging from one to hundreds of MeV. It must be pointed out that the data available at present on solar corpuscles coming into the vicinity of the Earth undoubtedly indicate that in the region of chromospheric flares (and possibly also in the plasma bundles with a magnetic field ejected from it) a process is active in accelerating particles to comparable energies. However, mostly protons are found although it is highly probable that effective acceleration of electrons also occurs. A serious argument in favour of this is the finding of fast electrons with $\mathcal{E} \sim 10\text{--}35$ keV (Hoffman, Davis and Williamson, 1962) at the sudden onset of a magnetic storm (i.e. at the time a plasma bundle with a "frozen in" magnetic field ejected from a flare region arrives in the vicinity of the Earth). At the same time the proton energies proved to be far higher ($\sim 0\cdot1\text{--}5$ MeV). This should not disturb us, however, since it may be caused by the large energy losses of the electrons to magneto-bremsstrahlung in the corona.

In connection with what has been said particular importance in the theory of the sporadic radio emission is attached to the problem of the acceleration of charged particles in centres of activity and above all in the region of chromospheric flares and in the corona above spots. The appearance of enhanced radio emission in the development period of bipolar groups of sunspots indicates the action of an acceleration process in the last case. The way particles are accelerated in the region of flares is still not clear; certain possibilities in this direction connected with the action of the Fermi statistical acceleration mechanism† have been discussed by

---

† This mechanism is generally discussed in the theory of cosmic radio emission and the origin of cosmic rays (see, e.g., Ginzburg and Syrovatskii, 1964).

§ 31] Generation of Types I, II, and III Bursts

Ginzburg (1953). It is possible that a certain part is played in the acceleration of particles in flares and above spots by an induction mechanism when the particles acquire energy by the action of an electric eddy field appearing when the magnetic fields increase (see Zheleznyakov, 1958a; Riddiford and Butler, 1952). The energy rises here because the adiabatic invariant $p_\perp^2/H_0$ is constant ($p_\perp$ is the component of the particle momentum transverse to the field H). The possibility of acceleration by magnetohydrodynamic waves has also been discussed in published papers (Takakura, 1961b and 1962). As a whole, however, the problem of the acceleration of charged particles in the Sun's atmosphere and the questions connected with it of the retention of particles in local regions of the corona and escape from them, questions of the efficiency of acceleration of different kinds of particles (protons, electrons, heavy nuclei) in essence remain open. Here further investigation is definitely necessary, and results are of interest not only for radio astronomy but also for geophysics, physics of cosmic rays of solar origin, the theory of chromospheric flares, etc.

31. Generation of Types I, II and III Bursts

In this section we shall discuss the problem of the origin of types I, II and III bursts, i.e. of the components of the solar sporadic radio emission which differ from the emission of the continuum group in the complexity of their dynamic spectra, the clear-cut non-stationary nature and the narrow band of frequencies occupied. These features of the bursts indicate the coherent nature of the generation mechanisms, since it is coherent emission closely connected with wave amplification that occurs only in limited frequency ranges and with special forms of plasma particle velocity distributions which exist only for a limited time (section 27). We shall be able to confirm that the problem of the origin of types I, II and III bursts can really be solved on the basis of mechanisms of generation in a coherent form, whilst the non-coherent mechanisms which play a clear part in the interpretation of radio emission of the continuum group (section 30) are ineffective in this case.

The action of solar burst mechanisms studied in the present section (no matter what their actual form) is based on two phenomena: the generation in a non-equilibrium plasma of intense plasma waves and the partial conversion of the energy of these waves into electromagnetic radiation that escapes beyond the corona. The close connection of the radio bursts picked up on Earth with plasma waves in the corona and the nearness in certain cases of the radio emission frequency ω to the eigen frequency of the coronal plasma ω_L are the basic contents of the "plasma hypothesis" put forward by Shklovskii (1946) and Martyn (1947) in the very first years of the development of radio astronomy. Since then ideas about the methods and conditions of generation of bursts have, of course, changed considerably; however the idea of the connection of the sporadic radio emission with plasma waves in the solar corona has fully retained its importance as applied to types I, II and III bursts.

It is convenient to start a discussion of the nature and conditions of generation of narrow-band bursts with type III events, turning then to types II and I phenomena. When discussing the nature of "classical" rapidly drifting type III bursts and their different forms (U-bursts and type III fine structure in type II bursts) we shall also touch on the question of the generation features connected with the type III phenomena of reverse-drift pairs.

THEORY OF TYPE III BURSTS

The preliminary analysis of the experimental data on type III bursts carried out in section 14 has shown that these bursts owe their appearance to a certain agent coming from a flare region and moving in the corona at a velocity of $(0.2–0.8)c$ (c is the velocity of light). The streams of charged particles leaving the region of a flare at the time of its "explosive" phase (sudden expansion; see sections 2 and 14) are most probably this agent. Westfold's assumption (1957) about the connection between rapidly drifting bursts and magnetohydrodynamic shock waves being propagated in the corona from the flare is doubtful (Boishot, Lee and Warwick, 1960; Hughes and Harkness, 1963). In actual fact in accordance with (27.121) the wave front reaches a velocity of $u \gtrsim 0.2c$ only if $MH_0 \gtrsim 2.8 \times 10^2$ (M is the "magnetic Mach number", H_0 is the strength of the magnetic field). This condition, obtained for a particle concentration in the corona of $N \sim 10^8$ cm^{-3}, occurs (with moderate values of M) in fields $H_0 \sim 2.8 \times 10^2$ oe. At frequencies of $\omega \sim 2\pi \times 10^8$ sec^{-1} ($\lambda \sim 3$ m) these values of H_0 correspond to a ratio of $\omega_H/\omega \gtrsim 8$. However, as follows from the theory of the propagation of radio waves in the solar corona (see section 23), the emission generated in layers with high values of the parameter ω_H/ω escape beyond the corona only in the form of an ordinary wave. In other words, the radio emission will be strongly polarized in contradiction of the experimental data which point to comparatively weak polarization of type III bursts. In addition, the existence of such strong fields in the corona at a great altitude above the solar surface is also improbable. Neither is it clear how to explain from Westfold's point of view the inversion of frequency drift of U-bursts.

§ 31] Generation of Types I, II and III Bursts

Therefore we shall consider in future that type III radio emission is generated in streams of fast particles which penetrate the coronal plasma at a velocity which is a noticeable fraction of the velocity of light. The actual radio emission mechanism, which has been discussed in detail in the papers of Ginzburg and Zheleznyakov (1958b, 1958a, 1959b), is as follows. As was found in section 27, a stream–plasma system under certain conditions amplifies plasma waves incident on it or emitted by individual particles of the stream. When the stream is extensive enough the amplitude of these waves reaches a certain maximum (settled) value which no longer depends on the original intensity level of the waves being amplified. Furthermore, as a result of some transformation process (see section 25) part of the plasma wave energy changes into electromagnetic radiation, which is also observed on Earth in the form of type III bursts. We shall see that this generation scheme explains the features of type III bursts well; it is also fully applicable to the different forms of rapidly drifting bursts (U-bursts and type III fine structure in type II events).

We notice that in the framework of this kind of scheme the duration of a burst at a fixed frequency ω will obviously be made up of the time t_1 taken by the corpuscular stream to pass through the corona layer in which the radio emission of this frequency is generated and the time t_2 of the subsequent damping of the plasma waves. The ratio between these times depends on the stream parameters ($t_1 \sim L/V_s$, where L is the extent, V_s the velocity of the stream), on the one hand, and the parameters of the coronal plasma on the other ($t_2 \sim 1/\nu_{\text{eff}}$, where ν_{eff} is the effective number of collisions; see (26.76)). At frequencies of the order of 100 Mc/s the "life" t of a type III burst is determined basically by the stream, since here $t_1 \sim 5$–10 sec is significantly greater than $t_2 \sim 1/\nu_{\text{eff}} \sim 0.2$ sec. It is quite possible (Boishot, Lee and Warwick, 1960; Malville, 1962; Hughes and Harkness, 1963), however, that at low frequencies generated in more rarefied layers of the corona the duration of the bursts will depend chiefly on ν_{eff}, since the number of collisions decreases rapidly with the electron density. It must be stressed that the ideas developed below on the origin of type III radio emission remain valid for any relations between t_1 and t_2.

In the detailed discussion of the above generation mechanism of rapidly drifting bursts we shall not begin to allow for relativistic effects, limiting the discussion to streams whose velocity lies at the lower end of the $(0.2$–$0.8)c$ range characteristic of type III disturbances. Considering the parameter ω_H/ω to be sufficiently small† we shall also neglect the effect of the magnetic field H_0 on the radio emission generation process, re-

† In the Sun's overall magnetic field $H_0 \sim 1$ oe at frequencies $\omega \sim 2\pi \times 10^8$ sec^{-1} the parameter $\omega_H/\omega \sim 3 \times 10^{-2} \ll 1$.

membering, however, that it is the presence of a weak field H_0 in the source that leads to the appearance of the comparatively weak polarization of type III bursts.

If the bursts of radio emission are caused by the generation (amplification) of plasma waves in a stream–plasma system with subsequent transformation of the latter into electromagnetic radiation, then the effective temperature of such events can be estimated from the following considerations.

Denoting the energy of the electromagnetic waves "emitted" by the plasma waves in a frequency range $d\omega$ by $S_\omega\, d\omega$ we can find from simple considerations that the intensity of the radio emission escaping beyond the corona is

$$I \sim \frac{S_\omega e^{-\tau_{1,2}}}{\Delta\Omega L^2}. \tag{31.1}$$

The factor $e^{-\tau_{1,2}}$ here allows for the absorption of radio emission in the corona; $\Delta\Omega$ is the solid angle in which the emission leaving the corona is propagated; L is the linear size of the source of the burst. Comparing (31.1) with the intensity of the equilibrium emission in a vacuum $I^{(0)} = \omega^2 \varkappa T / 4\pi^3 c^2$ (for both polarizations see (4.3)) we obtain that the effective radio emission temperature related to the source area L^2 is

$$T_{\text{eff}} \sim \frac{4\pi^3 c^2}{\omega^2 \varkappa} \frac{S_\omega e^{-\tau_{1,2}}}{\Delta\Omega L^2} \sim \frac{4\pi^3 c^2}{\omega^2 \varkappa} \frac{S e^{-\tau_{1,2}}}{\Delta\omega \Delta\Omega L^2}. \tag{31.2}$$

The spectral density S_ω can be conveniently represented in the form $S/\Delta\omega$, where S is the total radio emission flux over the band and $\Delta\omega$ is the characteristic range of frequencies occupied by a burst; this has been done in changing to the last equation (31.2).

The value of S is connected with the amplitude of the plasma waves in the stream E_0 by a relation defining the efficiency of the process of transformation of plasma waves into electromagnetic ones. When there is no such transformation, as has been shown in detail in section 25, generation of electromagnetic waves by plasma oscillations is quite impossible.

The reservation must be made that in section 25 the problem of wave transformation and the escape of electromagnetic radiation beyond the corona was discussed only as applied to an equilibrium plasma. We obtained expressions for the coefficients of transformation of plasma waves into electromagnetic ones in the conditions of a regularly non-uniform and a statistically non-uniform plasma. (In the latter case the chaotic non-uniformities are thermal fluctuations of the electron concentration; transformation here occurs in the process of scattering of the plasma waves on these fluctuations.) The more complex case of a non-equilibrium medium,

§ 31] Generation of Types I, II and III Bursts

including the plasma–stream system, has hardly been studied, although from the point of view of applications in radio astronomy it undoubtedly deserves careful investigation. At the same time, as was remarked in section 25, the scattering of plasma waves into electromagnetic ones remains at the previous level even in the presence of a stream of charged particles if the velocity of the flux V_s satisfies the condition $V_{th} \ll V_s \ll c/\sqrt{3}$ and the phase velocity of the plasma waves V_{ph} is close to V_s. It is this situation that obtains when investigating the generation of type III bursts, in any case for $V_s \sim 0\cdot 2c \sim 6 \times 10^9$ cm/sec (V_{th} in the corona, as we know, is 4×10^8 cm/sec); here with sufficient accuracy for our purposes we can use the Rayleigh and combination scattering formulae (25.42), (25.67)–(25.72) obtained for an equilibrium plasma. The legitimacy of using the expressions for the transformation coefficient in a smoothly non-uniform equilibrium plasma becomes more problematical when there is a stream of particles. Below we shall base our work on the effect of wave transformation because of scattering but, since a change of plasma waves into electromagnetic ones by Rayleigh scattering with the frequency conserved determines only the lower limit to the efficiency of the transformation process, this efficiency may increase slightly because of transformation in a regularly non-uniform plasma.

Let us take it, therefore, that the electromagnetic flux S is connected with the amplitude of the plasma waves E_0 by the Rayleigh equation (25.42). Then after substituting the latter in (31.2) we obtain:

$$T_{\text{eff}} \sim n_{1,\,2}(\omega) \frac{\pi}{6} \frac{\omega_L^2}{\omega^2} \frac{e^2 E_0^2 L}{\varkappa mc} \frac{e^{-\tau_{1,\,2}}}{\varDelta\omega\varDelta\varOmega}. \tag{31.3}$$

We showed in section 27 that in a stream–plasma system the plasma waves whose phase velocity $V_{ph} = \omega/k$ is close to the velocity of the stream V_s are amplified. With a low enough particle concentration N_s in the corpuscular stream its effect on the connection of V_{ph} with frequency ω can be neglected by considering in the first approximation that $V_{ph} = V_{th}/\sqrt{\varepsilon} = V_{th}/n_{1,\,2}(\omega)$ (see (22.19a) and below). From the condition $V_{ph} \approx V_s \gg V_{th}$ follows the closeness of the frequency ω to the eigen frequency ω_L of the coronal plasma. As a result for the radio emission's T_{eff} we shall have:

$$T_{\text{eff}} \sim \frac{\pi^2}{6} \frac{V_{th}}{V_s} \frac{e^2 E_0^2 L}{\varkappa mc} \frac{e^{-\tau_{1,\,2}}}{\varDelta\omega\varDelta\varOmega} \sim 9 \times 10^{14} E_0^2. \tag{31.4}$$

In the change to the last expression allowance has been made for the fact that in our case $V_{th} \sim 4 \times 10^8$ cm/sec, $V_s \sim 6 \times 10^9$ cm/sec, $L \sim 2 \times 10^{10}$ cm, $\varDelta\omega \sim \omega/3 \sim 2 \times 10^8$ sec^{-1}. The optical thickness $\tau_{1,\,2}(\omega)$ characterizing the absorption of radio emission in the corona is taken to be 4 (although it is possible that it is slightly less; see Fig. 138). The solid angle $\varDelta\varOmega$ in which the radio emission is propagated on leaving the corona is (for small $\Delta\Omega$) $\pi n_{1,2}^2(\omega)$; this can easily checked from (22.30). For the generation mechanism being investigated $\Delta\Omega \approx \pi V_{\text{th}}^2/V_{\text{ph}}^2 \sim 1\cdot4\times10^{-2}$, i.e. the first harmonic of type III bursts leaves in a very narrow beam provided that the corpuscular stream is less dense than the coronal plasma and there are no significant irregularities in the distribution of the electron concentration in the corona. Violation of any of these conditions will lead to a rise of $\Delta\Omega$.

It was noted in section 14 that only in comparatively rare cases does the radio emission flux on Earth during type III events exceed 2×10^{-20} W m^{-2} c/s^{-1}. For a source with an area of $L^2 \sim 4\times10^{20}$ cm^2 this corresponds to an effective temperature of $T_{\text{eff}} \sim 10^{10}\,°$K (at $\lambda \sim 3$ m). According to the formula (31.4) this kind of radio emission level may be reached if the stream of charged particles in the coronal plasma will excite plasma waves with an electrical field amplitude of $E_0 \sim 3\cdot5\times10^{-3}$ c.g.s.u. ~ 1 V/cm. For weak bursts the requirements for the value of the field E_0 become milder.

The most important point in the theory of type III bursts is the estimate of the parameters of the corpuscular stream capable of bringing the level of the amplified plasma waves up to the value indicated. This problem, which is connected with allowing for the non-linear nature of the plasma waves, is very difficult; there is no complete clarity in the question of the nature and level of settled plasma waves, of the duration of the settling process, etc. Therefore we shall give here only a preliminary estimate of the concentration N_s in the stream on the basis of the formula (27.72), remembering that a more rigorous treatment of the range of problems under discussion may introduce corrections into the results obtained.

Therefore let the density of a quasi-neutral corpuscular stream be $N_s \ll N$ (N is the concentration in the coronal plasma) but nevertheless so great that the electrons of the stream basically collide with the stream particles and not those of the basic plasma. Furthermore, let us assume that the velocity dispersion in the stream is characterized by a temperature T_s which in order of magnitude is close to the coronal temperature T (an assumption which is quite natural in the generation of type III bursts). Then the ratio of the number of collisions for the electrons of the stream and the electrons of the corona $\nu_s/\nu_{\text{eff}} \sim N_s/N$. In this case, in accordance with (27.72), the amplitude of a settled plasma wave (to be more precise, its first harmonic) is

$$E_0^{\text{I}} \sim \frac{m\omega}{2eV_s}\left(\frac{16}{3}\frac{V_s^3 m}{\varkappa T_s}\frac{N_s^2}{N^2}\right)^2 \sim 6\cdot3\cdot10^6\frac{N_s^4}{N^4} \quad (31.5)$$

§ 31] Generation of Types I, II and III Bursts

for $\omega \sim \omega_L \sim 2\pi \times 10^8$ sec^{-1}, $V_s \sim 6 \times 10^9$ cm/sec and $T \sim 10^6$ °K. Putting $E_0^I \sim 3.5 \times 10^{-3}$ c.g.s.u., we find from this the necessary electron concentration in the corpuscular stream: $N_s \sim 5 \times 10^{-3} N \sim 5 \times 10^5$ electrons/cm^3.†

Therefore a stream of density $N_s \sim 5 \times 10^5$ electrons/cm^3 moving at a velocity of $V_s \sim 6 \times 10^9$ cm/sec can create radio emission at a frequency $\omega \approx \omega_L$ whose level corresponds to that observed during type III events. The nearness of the frequency emitted in this case to the value of the coronal plasma frequency and the fact that the agent exciting type III bursts moves into outer layers of the corona allow us to explain the rapid frequency drift of the bursts; we have already discussed this in detail in section 14.

In type III bursts as well as radio emission at the fundamental frequency ω emission is also observed at the double frequency 2ω (the second harmonic). In one of the first papers of Wild, Murray and Rowe (1954) devoted to type III bursts an entirely natural suggestion was put forward, according to which the second harmonic of the radio emission owes its appearance to the clear-cut non-linear nature of the plasma waves with a noticeable content of the second harmonic component. If we introduce the formula (27.74) for estimating the amplitude E_0^{II} of the field of the second harmonic in a plasma wave, it turns out that the ratio

$$\frac{E_0^{II}}{E_0^I} \sim \sqrt{\frac{eE_0^I}{32mk}} \frac{N_s}{N} V_{ph}^2 \left(\frac{m}{\varkappa T_s}\right)^{3/2} \tag{31.6}$$

(where $k = \omega/V_{ph}$) for the parameters given above is 0.1. Then $E_0^{II} \sim 3 \times 10^{-4}$ c.g.s.u. The change of the second harmonic 2ω of the plasma wave into electromagnetic radiation, similar to the transformation of the "fundamental" harmonic, may occur by Rayleigh scattering. The efficiency of the transformation in this case is defined by the formula (25.70); from it and the relation (25.42) if follows that

$$\frac{S'(2\omega)}{S'(\omega)} = \frac{\sqrt{3}}{2n_{1,2}(\omega)} \frac{E_{II}^2}{E_I^2} \approx \frac{\sqrt{3}}{2} \frac{V_{ph}}{V_{th}} \frac{E_{II}^2}{E_I^2}. \tag{31.7}$$

---

† In this connection we note that the stream's energy losses in generation of plasma waves and the electromagnetic radiation are relatively small. For example, the kinetic energy density in a proton–electron stream $N_s m_i \cdot V_s^2/2 \sim 15$ ergs/cm^3 is far more than the energy density in a plasma wave $(E_0^I)^2/8\pi \sim 5 \times 10^{-7}$ ergs/cm^3 and also the losses in maintaining a settled plasma wave for about $t \sim 20$ sec (the life of a burst): $(E_0^I)^2 \nu_{eff} t/8\pi \sim 10^{-4}$ ergs/cm^3. (In the expression for the loss density allowance is made for the fact that in the absence of a stream a plasma wave with $V_{ph} \gg V_{th}$ is damped in a time $1/\nu_{eff} \sim 10^{-1}$ sec; see section 26.) It is not hard either to check that the energy emitted during scattering in the form of electromagnetic waves is far less than the values given above.

In our case this ratio is approximately equal to 0·1; the value of $S'(2\omega)/S'(\omega)$ will, of course, vary depending on the stream's parameters, rising with the increase in the particle concentration N_s and the rise in the amplitude of the first harmonic E_1.[†] A very important conclusion follows from this: if the second harmonic in the radio emission is genetically connected with the second harmonic in plasma waves, the relative content of the second harmonic in intense type III events should rise on the average. In reality, however, no one, as far as we know, has observed this kind of effect. This circumstance is an argument against the above interpretation of the second harmonic, although the absence of any clear-cut connection between the content of the harmonics and the intensity of type III bursts apparently needs further checking.

In recent years currency has been given to another point of view put forward in the paper by Ginzburg and Zheleznyakov (1958b) on the origin of the second harmonic of the radio emission, according to which this harmonic is the result of combination scattering of plasma waves excited by the stream on plasma waves of thermal (fluctuation) origin. As was shown in section 25, the frequency of the radio emission is then close to $2\omega \approx 2\omega_L$ (where ω is the frequency of the plasma waves; in our case it is the same as the frequency of the burst's "fundamental harmonic"). If the "fundamental harmonic" in this case as before owes its appearance to the Rayleigh scattering of plasma waves, then the ratio of the scattered fluxes of electromagnetic radiation $S''(2\omega)/S'(\omega) \approx 10^{-2}$ (see the formula (25.70) and below). The corresponding ratio of the effective radio emission temperatures is (allowing for (31.2))

$$\frac{T_{\text{eff}}(2\omega)}{T_{\text{eff}}(\omega)} \sim \frac{1}{4} \left(\frac{S'' e^{-\tau_{1,2}}}{\Delta\omega \Delta\Omega} \right)_{2\omega} \left(\frac{S' e^{-\tau_{1,2}}}{\Delta\omega \Delta\Omega} \right)_{\omega}^{-1}. \tag{31.8}$$

Considering $S''(2\omega)/S'(\omega) \approx 10^{-2}$, $(\Delta\omega)_{2\omega}/(\Delta\omega)_\omega = 2$ and putting $\tau_{1,2} \approx 4$ at the basic frequency with $\Delta\Omega \sim 1\cdot 4 \times 10^{-2}$, and $\tau_{1,2} \approx 0\cdot 3$, $\Delta\Omega \sim 5\cdot 5$ at the second harmonic we obtain: $T_{\text{eff}}(2\omega)/T_{\text{eff}}(\omega) \sim 1\cdot 3 \times 10^{-4}$. The corresponding ratio of the intensities will be four times greater (since $I \sim T_{\text{eff}} \omega^2$), so $I(2\omega)/I(\omega) \sim 5 \times 10^{-4}$.

The value obtained obviously characterizes the minimum ratio $I(2\omega)/I(\omega)$ recorded provided that at the time of the generation of the burst the Earth is inside the polar diagram of the "fundamental harmonic". The relative content of the second harmonic will rise if the observer is on the edge of the polar diagram of the "fundamental harmonic"; only the second harmonic

[†] The ratio $S'(2\omega)/S'(\omega)$ still does not characterize the observed content of the second harmonic component in the bursts. It can be judged only after allowing for the difference in absorption of radio waves in the corona and the differing directivity of the emission. It is easy to make this kind of allowance on the basis of the formula (31.8).

§ 31] Generation of Types I, II and III Bursts

will be fixed outside this diagram. As a whole it follows from the picture given of the generation of type III bursts, in accordance with the observations, that the level of the second harmonic can be either more or less than the intensity of the "fundamental harmonic". Since the angular width of the polar diagram at the fundamental frequency is very small, it is clear that in the majority of cases only the second harmonic (without the first) is recorded; both harmonic components appear only in $(\Delta\Omega)_\omega/(\Delta\Omega)_{2\omega} \sim 0.25\%$ of the type III events provided that all the orientations of the radio emission diagram are of equal probability. The observations, however, indicate the more frequent appearance of two harmonic bands in several of the bursts (see section 14). This kind of discrepancy is not surprising, since above in estimates of the degree of directivity of the "fundamental harmonic" radio emission (the solid angle $\Delta\Omega$) we made no allowance for the effect of such factors as coronal inhomogeneities, etc. If we take the more realistic ratio $(\Delta\Omega)_\omega/(\Delta\Omega)_{2\omega} \approx 10\%$, then the minimum ratio $I(2\omega)/I(\omega)$ will become approximately 1/50. Wild, Smerd and Weiss (1963 and 1964) consider that in the solar corona the quantity $(\Delta\Omega)_\omega/(\Delta\Omega)_{2\omega}$ is still greater ($\sim 50\%$). In this case $I(2\omega)/I(\omega)$ rises to 1/10.

The ideas developed by Ginzburg and Zheleznyakov (1958b) on the connection between the second harmonic of rapidly drifting bursts and combination scattering of plasma waves in the corona explain the absence of higher harmonic components in type III radio emission (in particular the third harmonic). The combination origin of the second harmonic allows us to observe it at lower frequencies as well (tens of megacycles), where during most of the life of a burst the radio emission, in all probability, is due to damped plasma oscillations in the layers of the corona which have already been left by the corpuscular stream (see p. 591). On the other hand, the second harmonic here cannot be connected with the non-linear nature of the plasma oscillations, since in the absence of a stream of particles the harmonic content becomes noticeable only in fields $E_0 \gtrsim (m\omega V_{ph})/2e \sim 3.6$ c.g.s.u. (27.73) (when there is a stream—in fields of $E_0 \gtrsim 5\times 10^{-3}$ c.g.s.u.). The generation of such strong plasma waves in the corona would lead to the appearance of radio emission whose intensity exceeded the usually observed values by about six orders.

It is not impossible, however, that in certain cases the non-linear nature of the plasma waves amplified by the stream–plasma system nevertheless plays a significant part. As an example we can take the appearance of the third harmonic in certain strong U-bursts (section 14). In other cases the question of the formation mechanism of the second harmonic can be answered by investigating the mutual coherency of the radio emission of the first and second harmonics. This kind of coherency should be absent if the second harmonic of the radio emission is formed by combination scattering, since the phase of the second harmonic will be determined not only by the phase of the plasma wave amplified in the stream, but also by the random phase of the fluctuation plasma waves on which the combination scattering occurs. On the other hand, coherency will occur if the first and second harmonics of the radio emission are transformed (by Rayleigh scattering, let us say) into first and second harmonics of non-linear plasma waves. Unfortunately the corresponding experiments have not yet been made unless we consider Jennison's measurements (1959) which we discussed in section 13. Under the influence of the results of Jennison (1959), which indicated the presence of coherency, doubt was expressed by Kapitsa (1960) on the combination origin of the second harmonic of type III bursts. However, there is apparently no particular foundation for this doubt at present since the radio emission investigated by Jennison (1959) related to type II rather than to type III; in addition, according to Jennison's words, he is uncertain as to the results of his experiment.

It must be said that the hypothesis of the combination mechanism of the second harmonic's formation has proved to be very fruitful when explaining certain fine effects connected with type III bursts. Smerd, Wild and Sheridan (1962) (see also Wild, Smerd and Weiss, 1963 and 1964) used it to interpret the observed displacement of the source of the second harmonic relative to the "fundamental harmonic" in the plane of the solar disk. The idea of the second harmonic of the radio emission as the result of combination scattering of plasma waves made it possible to understand the reasons for the appearance of reverse-drift pairs (see Zheleznyakov, 1965b).

Let us examine the first effect to begin with. In section 14 we noted that in certain cases the position of the source of the second harmonic that can be "seen" in radio rays is displaced relative to source of the first harmonic towards the centre of the disk. Since the influence of refraction in the corona, which is particularly significant for the "fundamental harmonic", should lead to the opposite arrangement of the two sources, Smerd, Wild and Sheridan (1962) explained this displacement by the fact that it is not the region of generation of the radio emission that is observed at the second harmonic but its image in radio rays reflected from the deeper layers of the corona. For this it is obviously necessary that the radio emission at the second harmonic on leaving the generation region should be propagated towards the solar surface for preference. With the ordinary motion of a stream of particles away from the Sun this requirement means that plasma waves (with a wave vector k running along the velocity V_s)

§ 31] Generation of Types I, II and III Bursts

excited by the stream should in the process of being scattered on thermal fluctuations of a plasma type produce electromagnetic radiation whose intensity "backwards" (i.e. in directions subtending an obtuse angle with k) is greater than that "forwards" (in directions subtending an acute angle with k).

It is easy to confirm that it is this kind of situation which can obtain in reality. It follows from the theory of combination scattering of plasma waves developed in section 25 (see the formula (25.65)) that the distribution of intensity of the combination emission with respect to the angles θ read from the direction k is characterized by the relation

$$I(\theta) \propto \frac{(2\cos\theta - \sqrt{3}\,V_s/c)^2 \sin^2\theta}{1 - 2\sqrt{3}\,(V_s/c)\cos\theta + 3V_s^2/c^2}. \qquad (31.9)$$

In the change from (25.65) to (31.9) we have allowed for the fact that the wave number of a plasma wave is $k = \omega/V_{\text{ph}} \approx \omega_L/V_s$ and the wave number of an electromagnetic wave is $\tilde{k} \approx \sqrt{3}\omega_L/c$. The zeros in the polar diagram of the combination scattering $I(\theta)$ correspond to the angles

$$\theta = 0 \quad (\text{when} \quad V_s \neq c/\sqrt{3}); \quad \theta = \pi;$$
$$\theta = \pm \arccos\left(\frac{\sqrt{3}}{2}V_s/c\right). \qquad (31.10)$$

It is clear from this that the scattering diagram $I(\theta)$ has four lobes when $V_s \neq c/\sqrt{3}$ and three lobes if $V_s = c/\sqrt{3}$. It is significant that when $V_s > c/\sqrt{3}$ the lobes pointing "backwards" ($\cos\theta < 0$) are far larger than the lobes pointing "forwards" ($\cos\theta > 0$). In this case the second harmonic of type III bursts will be observed mostly in reflected beams and the "visible" position of the source of the second harmonic will be displaced relative to the first towards the centre of the solar disk. The same situation obtains when $V_s < c/\sqrt{3}$ if the Earth hits a null of the polar diagram. A different situation is possible, however, when the "backwards" emission is weaker than the "forwards"; the observed position of the second harmonic's source will coincide here with the position of the source of the "fundamental frequency" (here because of refraction the source of the "fundamental frequency" may be even closer to the centre of the disk than the second harmonic). This is often the situation with agent velocities of $V_s < c/\sqrt{3}$ when all the lobes of the diagram are of the same size.

If the interpretation suggested by Smerd, Wild and Sheridan (1962) for the effect of displacement of the second harmonic is correct, then it is quite possible that sometimes a second harmonic will be recorded simultaneously from two sources corresponding to direct and reflected rays. Furthermore, in favourable circumstances a sharp change in the position of the source of the second harmonic relative to the source of the "fundamental harmonic" is possible at the time when the sign of a U-burst's frequency drift changes (after which the stream moving away from the solar surface starts to approach it).

Let us now examine the reasons for the appearance of the reverse-drift pairs connected with type III radio emission. In section 16 which gives the characteristics of bursts of this kind it was noted that Roberts (1958) suggested that the pairs of bursts appear because of the reflection of radio emission from the lower layers of the corona. This kind of reflection will lead to noticeable "splitting" of the dynamic spectrum of the bursts if the frequency of the radio emission is $\omega \sim 2\omega_L$ (ω_L is the eigen frequency of the coronal plasma in the generation region). Roberts linked the reverse direction of the drift (towards the high frequencies) with the usual motion of the exciting agent away from the Sun at the time when this agent is passing through a local inhomogeneity in the corona, namely through the part of the inhomogeneity in which the electron concentration rises as one moves away from the Sun. The part of the stream which at this time is outside the local inhomogeneity generates type III radio emission that is weaker than the reverse-drift burst.

Two facts established from observations: (1) the absence in the majority of reverse-drift bursts of their continuation in the form of straight-drift radio emission, which would correspond to the period of the agent's motion after the electron density maximum in the local inhomogeneity, and (2) the higher intensity of reverse-drift bursts when compared with the accompanying type III radio emission (which also allows us to distinguish them against the background of type III events)—allow us to state that the generation conditions in one and the same frequency range depend essentially on the direction of the frequency drift; the latter is determined in its turn by the mutual orientation of the agent's velocity V_s and grad N. The level of the radio emission is far higher if the agent is moving towards an increase in the concentration N.

According to Zheleznyakov (1965b) this circumstance, which is highly essential for understanding the reasons for the appearance of reverse-drift bursts, can be explained by the fact that the plasma waves amplified by the stream are not scattered on weak thermal fluctuations in the corona but on the same plasma waves after their reflection. This kind of reflection becomes possible if the stream and the plasma waves amplified by it are propagated in the direction of grad N. The transformation of plasma waves into electromagnetic ones is then defined by the formula (25.75). It follows from it that the radio emission flux in combination scattering on reflected waves is $S''(2\omega_L) \propto E_0^4$ as opposed to the value $S''(2\omega_L) \propto$

§ 31] Generation of Types I, II and III Bursts

$E_{\text{pl}}^2 E_0^2$ (see (25.67)) which characterizes the process of scattering on thermal fluctuations of the electric charge. Here E_0 and E_{pl} are respectively the amplitudes of the amplified and fluctuation plasma waves; since $E_0^2 \gg E_{\text{pl}}^2$ the effect of scattering on reflected waves leads to a sharp increase in the level of radio emission.

The idea of reverse-drift pairs as the result of combination scattering on reflected waves of plasma waves amplified in the stream makes quite comprehensible the basic properties of phenomena of this kind, including the nature of the drift and the higher intensity when compared with type III radio emission. The considerable difference of the frequency from the eigen frequency of the coronal plasma ($\omega \approx 2\omega_L$) makes the appearance of a "radio echo" natural, and this leads to the splitting of the dynamic spectra of burst pairs. The close connection between bursts of the type under discussion and the second harmonic of type III helps us understand why Roberts did not once observe reverse-drift bursts at half the frequency at the same time as these phenomena.

Above, we have discussed the mechanism of type III radio emission and its different forms which connects the observed emission with the effect of plasma wave amplification in a corpuscular stream and coronal plasma system.

According to the terminology used in the introduction to Chapter VIII this mechanism can be called coherent Cherenkov. In the above system, however, as well as the coherent component of the Cherenkov emission at plasma wavelengths there is also a non-coherent component—in the range of phase velocities where the plasma wave absorption coefficient remains positive even in the presence of a stream of particles. This component (of which it can be said that it owes its appearance to a non-coherent Cherenkov mechanism) also makes a certain contribution to the radio emission. The question arises of the relative content of the coherent and non-coherent components in type III bursts; it has been discussed in detail in section 27, where we showed that in the actual conditions of the corona the non-coherent component is negligibly small when compared with the coherent.

It is interesting to note that the non-coherent Cherenkov mechanism was actually used by Shklovskii (1946) to interpret bursts displaying frequency drift. The transformation of plasma waves into radio emission, according to Shklovskii (1946) was due to the process of scattering of these waves on free electrons in the corona. We have spoken about the insufficient correctness of this approach to the effect of transformation of plasma waves with $V_{\text{ph}} \gg V_{\text{th}}$ (which gives, however, a correct estimate of the order of magnitude of the transformation coefficient) in section 25.

The mechanism of non-coherent Cherenkov emission was discussed later by Marshall (1956) in order to explain bursts of types II and III. She did not, however, allow for the considerable attenuation of the emission connected with wave transformation, and she also ignored the effect of plasma wave reabsorption on the intensity of the radio emission. The explanation of the harmonics in the bursts by Marshall (1956) also causes serious objections by Ginzburg and Zheleznyakov (1961). Marshall used the circumstance that in a magnetic field for particles with non-relativistic longitudinal velocities ($\beta_\| = V_\|/c \ll 1$) the Cherenkov condition holds only at frequencies where $n_j(\omega) \gg 1$. In a magnetoactive plasma, roughly speaking, $n_j(\omega) \gg 1$ for $\omega \sim \omega_L$ and $\omega \sim \omega_H$ (see section 23); therefore the Cherenkov emission will contain frequencies of ω and 2ω if the ratio between the eigen frequency and the gyro-frequency is close to two. It is improbable, however, that in the actual conditions—in a broad range of altitudes above the photosphere and for very different groups of spots—the ratio ω_H/ω_L remains constant and close to two. In addition, this kind of interpretation of the harmonics requires strong enough magnetic fields in the generation region ($\omega_H \sim \omega$); then the bursts would be strongly polarized, which is contradicted by the observational data.

Types III and II bursts cannot apparently be connected with non-coherent magneto-bremsstrahlung (as was done by Kruse, Marshall and Platt (1956)). The appropriate arguments can be found in the papers by Zheleznyakov (1958a), Ginzburg and Zheleznyakov (1961). The question of the origin of type III bursts has also been discussed by De Jager (1959b, section 90). In the papers by Weiss and Stewart (1965) and Zheleznyakov and Zaitsev (1968) the type V radio emission is shown to be connected with the capture of a part of the fast electrons responsible for type III emission in a magnetic trap above the bipolar sunspot group where they are multiply reflected from magnetic blockings. As a result, a system of two counterstreams which excite plasma waves with opposite directions of wave vectors in a coronal plasma, is formed in the trap. Such waves are scattered, effectively transforming each other into the radio emission at a frequency $\omega \approx 2\omega_L$. This fact makes it possible to understand the basic characteristics of the type V radio emission.

MECHANISM OF TYPE II BURSTS

Slowly drifting bursts have much in common with rapidly drifting type III events, differing from them chiefly in the time scales. Therefore the most probable mechanism of type II radio emission, just as type III bursts, is the coherent Cherenkov mechanism which ensures amplification of plasma waves with subsequent transformation into electromagnetic radia-

§ 31] Generation of Types I, II and III Bursts

tion. However, it must be assumed that the actual generation conditions will be different.

In actual fact, according to sections 13 and 17 type II bursts are excited by an agent moving in the corona at a velocity of $V_s \sim 10^8$ cm/sec (i.e. at a velocity far less than the velocity of the streams $V_s \sim 10^{10}$ cm/sec that generate type III radio emission and the mean thermal velocity of the coronal electrons $V_{th} \sim 4 \times 10^8$ cm/sec). A plasma bunch with a "frozen in" magnetic field ejected from the region of a flare is this kind of agent. During its motion this bunch should attract the surrounding ionized material; the transition from it to the motionless coronal plasma occurs via the front of the shock wave travelling ahead of the bunch. The magnetic field retains the relativistic particles which are the source of type IV radio emission (section 30); their escape from the magnetic "trap" in the form of streams of fast particles is the origin of the type III fine structure in type II bursts.

If we consider that the slowly drifting bands in the dynamic spectra identified with type II bursts consist wholly of rapidly drifting structural elements, then the problem of the origin of type II events is reduced simply to the question of the generation of type III bursts discussed in the previous subsection. Here only the problem of the escape of the fast particles from the "trap" and the question of whether the system of relativistic particles —the source of type IV radio emission—can maintain the fine structure in type II events at the observed level deserve particular attention. Actually the situation is far more complicated: as well as the rapidly drifting elements in type II events there is apparently a separate component which requires special discussion.

In a correct interpretation of type II radio emission an explanation is first sought based on the known possibility of plasma wave amplification and instability in a quasi-neutral plasma and quasi-neutral stream system even if $V_s \lesssim V_{th}$. Amplification appears here if the resultant electron velocity distribution function has two maxima; a variety of this kind of instability for a mono-kinetic stream has been discussed in section 27 (see the formulae (27.62)–(27.64)). This effect, however, is apparently insignificant since arguing the representation of a "two-stream" instability with $V_s < V_{th}$ as the cause of type II bursts meets with serious difficulties. Another disadvantage of the generation scheme under discussion for type II bursts is connected with the radical change in the nature of the scattering of plasma waves excited by the stream since their phase velocity $V_{ph} \sim V_s < V_{th}$ differs essentially from the $V_{ph} \sim V_s \gg V_{th}$ typical of type III bursts. Here the concept of Rayleigh and combination components of the scattered electromagnetic radiation concentrated in narrow frequency ranges near ω_L and $2\omega_L$—an idea which was so effective when interpreting type III bursts—becomes untrue. The point is that with $V_{\text{ph}} < V_{\text{th}}$ the scattering process occurs in free electrons; it is easy to check that the effect of a Doppler shift in frequencies will lead in this case to a clear-cut broadening of the radio emission spectrum which cannot be matched up with the data on the spectral observations of the slowly drifting bursts.

The latter difficulty disappears if the type II bursts, just like the type III phenomena, appear in streams of particles whose velocity is much greater than V_{th}. Before the appearance of the paper by Pikel'ner and Gintsburg (1963), however, it remained unclear how this condition could be observed when the agent was moving at a velocity less than V_{th}. Pikel'ner and Gintsburg (1963) refer to the circumstance known from plasma theory that in a shock wave front the electrons drift in the presence of a magnetic field relative to the ions at a velocity V_0 which with large enough "magnetic Mach numbers" becomes much greater than V_{th}. This effect may be of great importance for the generation process of type II bursts since in the case $V_0 \gg V_{\text{th}}$, $\omega_L \gg \omega_H$ a system of two streams—electrons drifting relative to ions—becomes unstable relative to perturbations of the plasma wave type. These waves have a frequency $\omega \simeq \omega_L$ in a coordinate system connected with the drifting electrons (see the last subsection of section 27).[†]

For drift at a rate $V_0 \gg V_{\text{th}}$ it is necessary for the values of M to satisfy the inequality (27.123). It will be satisfied in a corona with $N \sim 5 \times 10^7$ electrons/cm^3, $T \sim 10^6$ °K in fields of ~ 1.5 oe if $M - 1 \gg 0.18$. Actually under these conditions for a shock wave with a front velocity of $U \sim V_s \sim 10^6$ cm/sec the number is $M \sim 2$; when the field H_0 changes from 0.3 to 3 oe[‡] M changes from 1 to 10. It must be pointed out that the form of shock waves in a plasma has at present been studied only for values of M in the range from 1 to 2; no theory has yet been produced for $M > 2$ (see section 27). Therefore everything that has been said relates, strictly speaking, only to the case when a shock wave with $M < 2$ that is not too strong exists ahead of the bundle of matter ejected from the flare region.

The generation scheme for type II radio emission given above has not yet been studied in detail unfortunately; we shall therefore limit ourselves here to only a few remarks. Above all the effect of combination scattering

---

[†] General ideas on shock waves in the coronal magnetic field as the source of type II radio emission based on the agreement of the velocities of waves and agents of this kind which cause the slowly drifting bursts have been expressed by Westfold (1957) (see also Uchida, 1960).

[‡] Here we have chosen values of the field H_0 that are close in value to the total field strength of the Sun or slightly greater than the latter. With these values of H_0 the inequality $\omega_H/\omega \ll 1$ is satisfied at metric wavelengths; this is necessary to conserve the non-polarized (to be more precise, weakly polarized) nature of the type II bursts.

§ 31] Generation of Types I, II and III Bursts

of plasma waves having a phase velocity $V_{ph} \sim V_0 \gg V_{th}$ relative to the drifting electrons allows us to explain the appearance of the second harmonic and the usual absence of higher harmonic components. The strong side of the possible theory of type II radio emission under discussion is also the possibility of using it to interpret the characteristic "splitting" of each harmonic generally into two, sometimes into three bands; previously this effect had been linked (Wild, 1950a; Roberts, 1959b) without any foundation with a phenomenon like Zeeman splitting of the spectral lines.

The possibility of which we are speaking was pointed out in the paper of Zheleznyakov (1965b). It allows for the oscillating nature of the magnetic field H_0 (the presence of solitary waves) in the front of the shock wave and the presence of electron drift relative to the ions. According to section 27 the cause of the drift is the non-uniformity of the magnetic field. In this case the magnitude and direction of the drift rate are determined by the value and sign of the derivative dH_0/dx (x is a coordinate in a direction orthogonal to the plane of the shock wave front). Since instability of plasma waves disappears with low velocities V_0 it is clear from what has been said that the oscillations of the field H_0 will lead a system of alternating layers containing amplified plasma waves being established in the front. Because the drift rates in adjacent layers are opposite the radio emission that appears as the result of scattering on the electron fluctuations will have different frequencies. The frequency shift $\Delta\omega$ is determined here by the Doppler effect; in order of magnitude $\Delta\omega \sim (2\omega_L V_0)/c$ at the fundamental frequency and is twice as much at the second harmonic. At frequencies of about 30 Mc/s ($\omega_L \sim 6\pi \times 10^7$ sec^{-1}) the observed 5 Mc/s splitting occurs with a drift rate of $2 \cdot 5 \times 10^9$ cm/sec (i.e. $V_0/V_{th} \sim 6$). The small number of split bands from the point of view under discussion is connected with the small number of oscillations in the front since under the actual conditions the oscillations are "blurred" because of turbulence; the latter leaves only one or two of the strongest solitary waves well formed (see Galeyev and Karpman, 1963).

It was assumed above that the basic source of intense plasma waves in a shock wave front is the instability in the drift of the electrons relative to the ions. As well as this a significant part, generally speaking, may also be played by instability due to temperature "anisotropy", which also occurs with large values of the refractive index $n_j(\omega)$. The corresponding amplification and instability criteria and the amplification factors can be obtained by means of the formulae (27.129)–(27.132). Unfortunately a detailed treatment has not yet been made here; to judge from the criterion (27.131) this effect is not reflected in the generation process of type II bursts (in a source of type II radio emission $\omega_H < \omega$, whilst the inequality

(27.131) is satisfied only if $\omega_H > \omega$). A further investigation into the above idea of mechanism and generation conditions for the type II bursts has been performed by Zaitsev (1965). He has determined the shock wave parameters responsible for type II bursts and the value of the magnetic field in the corona (Zaitsev, 1968).

GENERATION OF TYPE I BURSTS

It follows from observations of type I radio emission that it originates from motionless sources localized in active regions of the corona above groups of spots with a strong magnetic field. With this circumstance are obviously linked the characteristic features of the narrow-band bursts that distinguish them from events of spectral types II and III—the high degree of polarization and the absence of drift covering the extensive frequency range. As for the actual generation mechanism of the emission under discussion, this question has not yet been studied in sufficient detail; we shall therefore limit ourselves to only certain preliminary ideas about the nature of type I bursts given in the paper of Zheleznyakov (1965b).

First of all it is not impossible that a considerable fraction of the type I bursts (basically weak surges which appear during noise storms against a background of intense enhanced emission) is generated in the same way as the enhanced emission, i.e. by a non-coherent magneto-bremsstrahlung mechanism (see section 30). The narrow band of frequencies characteristic of type I bursts can be obtained provided that the emitting electrons are localized in a small enough region of the corona with an almost constant magnetic field H_0, and their energy $\mathcal{E} \ll mc^2$; the rise in intensity then occurs mostly at the frequencies $\omega \sim \omega_H$, $2\omega_H$ and not in the wide range $\Delta\omega \sim \omega$ as for relativistic particles (section 26).

The generation process of the bursts can then be pictured as follows. We assume that in a limited region of the coronal plasma by virtue of certain causes (let us say under the action of shock waves in the magnetic field or the momenta of Alfvén waves) the fast electrons responsible for the enhanced radio emission either undergo additional acceleration or increase their concentration because of a decrease in the number of particles in the surrounding corona. The first effect leads to an immediate increase in the intensity of the emission from the region indicated; however, after a short time necessary for the escape of electrons of enhanced energy beyond this region the intensity returns in practice to its previous level. The same occurs with the second effect if the optical thickness of the system of fast electrons in the region in question is small enough ($\tau_s \ll 1$).

The generation scheme suggested for type I bursts deserves serious attention. It is attractive in that in accordance with the observations

§ 31] Generation of Types I, II and III Bursts

(section 12) the bursts are connected closely with the enhanced emission, and their polarization, position on the solar disk and level of generation in the corona are close to the corresponding characteristics of this emission. The weak point of this hypothesis, however, is the simultaneous rise in intensity at several multiple frequencies, in contradiction of the observations which show no simultaneous appearance of type I bursts at frequencies in the ratio of, say, 2:1. The non-coherent magneto-bremsstrahlung mechanism in the framework of this scheme cannot apparently ensure the creation of the intense type I bursts that appear against the background of weak enhanced emission or in the absence of the latter either (Ginzburg and Zheleznyakov, 1961). In actual fact during the generation of bursts with an effective temperature of $T_{\text{eff}} \sim 10^9$ °K the energy of the emitting particles (if their velocity distribution is not too far from an equilibrium one) is $\mathcal{E} \gtrsim \varkappa T_{\text{eff}} \sim 10^5$ eV. When there is no enhanced emission, i.e. in conditions when there are no energetic electrons in the corona above spots, the particles must be accelerated directly in the source in a time of the order of a second (the duration of a type I burst), starting from an original energy of a few hundred electron-volts. For a second the energy of part of the coronal electrons should rise by a factor of thousands right up to $\mathcal{E} \gtrsim 10^5$ eV. The possibility of acceleration of this kind is improbable. Therefore as well as the non-coherent magneto-bremsstrahlung mechanism we must also investigate other possible ways of generating type I bursts based on coherent emission mechanisms.

In the preceding subsections we have discussed coherent mechanisms of sporadic radio emission, assuming that the source of the emission (plasma waves) is excited and maintained by streams of charged particles (type III bursts) or bundles of plasma with a magnetic field "frozen in" (type II bursts). Apart from direct experimental data, the foundation for adducing special agents that maintain the plasma oscillations is the short attenuation time of free plasma waves in the corona with an equilibrium velocity distribution when compared with the total duration of types II and III events. The position is different for type I bursts since their life is comparable with the plasma wave attenuation time. The action of the agent causing the plasma oscillations may be of very short duration in this case, and be of the nature of a "push" after which the plasma oscilllations decrease their intensity exponentially with an attenuation constant γ_3 (26.76) if the length of the plasma wave is much greater than the Debye radius D. It follows from (26.76) that the characteristic life of type I bursts is then

$$t_0 \approx \frac{1}{\nu_{\text{eff}}} \tag{31.11}$$

and the emission frequency ω will be close to the Langmuir frequency ω_L (in an isotropic plasma). Since the optical thickness of the $\omega \approx \omega_L$ level rises as we move from the corona to the chromosphere the latter circumstance helps us to understand why type I bursts are observed largely at metric wavelengths: the reduction in number and the decrease in intensity of the bursts in the decimetric band is connected with the displacement of the $\omega_L \approx \omega$ level from the corona into the chromosphere. Furthermore, for bursts generated in the corona at an altitude of, let us say, $h \sim 0.4R_\odot$, where $\nu_{\text{eff}} \sim 1.4$ sec^{-1} with $T \sim 10^6$ °K (see Fig. 138), the time $t_0 \sim 0.35$–0.7 sec is in close agreement with the experimental data of section 12. We note that it is this agreement that convinces us that the time of the action of the agent exciting the plasma waves is here comparable in order of magnitude with the attenuation time of the plasma waves or less than it.

Allowing for the magnetic field does not introduce any significant change into the magnitude of the attenuation; however, it may noticeably alter the frequency of the plasma waves, particularly in the corona above large spots with a strong magnetic field. In actual fact in the presence of a magnetic field $\boldsymbol{H}_0$ the expression for γ_3 becomes more complex (Gershman, 1959):

$$\gamma_3 \approx \nu_{\text{eff}} \left[1 + \frac{2uv \sin^2 \alpha}{(1-u)^2}\right]\left[1 + \frac{uv \sin^2 \alpha}{(1-u)^2}\right]^{-1}. \tag{31.12}$$

Nevertheless the numerical values of γ_3 will remain close to the former ones since with any u and v the value of γ_3 lies in the range between ν_{eff} and $2\nu_{\text{eff}}$, and the values of t_0 correspondingly in the range from $1/\nu_{\text{eff}}$ to $\frac{1}{2}\nu_{\text{eff}}$. The frequency of the plasma waves in the magnetic field, generally speaking, is not the same as ω_L; for example, with transverse propagation it is close to $\sqrt{\omega_L^2 + \omega_H^2}$ (for further details see section 23). To judge from the values suggested for the magnetic field in the corona above spots, on the one hand, and from the observational data indicating the high degree of polarization of type I bursts, on the other, the condition $\omega_H/\omega \ll 1$ is not satisfied in the sources of radio emission. The latter may lead to a considerable difference between the frequency and the Langmuir frequency.

As for type II bursts a possible source of plasma waves in the generation of type I bursts is shock waves in the magnetic field. However, unlike the type II events, shock waves unsupported by the motion of plasma bunches are rapidly damped, which ensures the short duration and narrow-band nature of type I bursts. In addition, in the latter case we are dealing with shock waves in a strong magnetic field ($\omega_H/\omega > 1$), whilst in the sources of the practically non-polarized emission of type II the magnetic field is weak enough ($\omega_H/\omega \ll 1$). Therefore the results for type II bursts cannot

§ 31] Generation of Types I, II and III Bursts

be extended directly to type I; in particular the conditions for drift instability in the front of shock waves will not be the same as those found in section 27 (see (27.128)). The appropriate criteria have not yet been found; it is possible, however, that there is no particular need of them here, since another type of instability on plasma waves caused by temperature "anisotropy" in the shock wave (i.e. a coherent magneto-bremsstrahlung mechanism;† for further detail see section 27 and formula (27.131) in particular) apparently plays an important part in the generation of type I bursts. In the latter case we can understand the connection between the bursts and the enhanced emission since the presence of energetic electrons undoubtedly increases the amplification of the plasma waves caused by the temperature "anisotropy", without increasing at the same time the amplification at the expense of the beam instability. Unlike the non-coherent mechanism, however, we can hardly expect here close agreement of the polarization characteristics of the bursts and the enhanced emission because of the difference in the conditions for the escape of these components from the corona.

---

† The hypothesis of the coherent magneto-bremsstrahlung mechanism of type I bursts was put forward by Twiss (1958); the scheme of generation of bursts in the front of a shock wave because of temperature "anisotropy" was suggested by Zheleznyakov (1965b). For the generation of type I bursts, see also Takakura (1963a and 1956), Chertoprud (1963), and Trakhtengerts (1966).

CHAPTER X
Origin of Radio Emission of the Planets and the Moon

32. Hypotheses on the Mechanism of Jupiter's Sporadic Radio Emission

THE "THUNDERSTORM" HYPOTHESIS

Soon after the discovery of the radio emission of Jupiter a hypothesis was suggested for the origin of the bursts of radio emission based on an analogy between the atmospheric radio interference on Earth and Jupiter's sporadic emission (*Nature*, 1955). According to this hypothesis, the emission of this planet at frequencies of $f \sim 20$ Mc/s is caused by phenomena of a thunderstorm nature taking place in its atmosphere.

A decisive argument (Zheleznyakov, 1958c) against this suggestion is the small frequency band Δf occupied by each individual burst ($\Delta f/f \sim 1/20$), whilst the spectrum of an atmospheric discharge covers a broad band ($\Delta f/f \sim 1$). A comparison of Jupiter's radio emission with terrestrial atmospherics (Smith, F. G., 1955; Horner, 1957) leads to the following results. Typical radio emission at a frequency of 10 Mc/s from an isolated lightning discharge in the Earth's atmosphere consists of a burst up to 1 sec long with an amplitude of ~ 1 mV/m (at a distance of 5 km from the source of emission). The radio emission flux on Earth caused by a similar discharge on Jupiter is 5×10^{-28} W m^{-2} c/s^{-1}. At the same time the observations indicate that Jupiter's radio emission is far more intense (generally of the order of 10^{-20} W m^{-2} c/s^{-1}; see section 19). This discrepancy of seven orders could be partly explained by assuming that the emission of the planet is the result of superimposing numerous pulses that arise simultaneously. In this case, however, it is difficult to understand why Jupiter's radio emission consists of bursts that are separate but can be resolved well in time and frequency.

In a number of papers (see, e.g., Gallet, 1961) an attempt is made to connect the directional nature of the emission from lightning discharges

§ 32] Mechanism of Jupiter's Sporadic Radio Emission

with the effect of Jupiter's ionosphere, i.e. with the total internal reflection of radio waves from it (with large enough angles of incidence of the emission from a source located in the lower layers of this planet's atmosphere). There is little probability, however, that this is the correct explanation. In fact in the case when the source of the emission is below the N_{max} level of the planet's ionosphere the emission escapes beyond the ionosphere at angles of $\varphi_0 < \varphi_{0max}$ to the vertical. The quantity φ_{0max} is defined by the formula (22.40):

$$\varphi_{0\,max} = \text{arc sec}\, \frac{f}{f_{L\,max}}.$$

It follows from the formula that the directivity of the emission varies depending on the ratio $f/f_{L\,max}$ ($f_{L\,max}$ is the Langmuir frequency in the maximum of the layer where the electron concentration is N_{max}): the angle $\varphi_{0\,max}$ decreases together with f, becoming zero when $f = f_{L\,max}$.

The actual fact that $\varphi_{0\,max}$ decreases at lower frequencies does not directly contradict the experiment indicating a decrease as the frequency rises in the angle $\Delta\Theta$ which Jupiter turns in the time of recording intense radio emission (see Fig. 97). The point is that the quantity $\Delta\Theta$ is affected not only by the beam width of the radio emission $\Delta\Theta_{dir}$ but also by the extent of the local source $\Delta\Theta_{source}$ in longitude on Jupiter (see (19.5)). The situation is quite possible that $\Delta\Theta_{dir} \ll \Delta\Theta_{source}$ and the observed decrease in $\Delta\Theta$ at high frequencies is connected with a corresponding decrease in the size of the source. However, the fall of $\varphi_{0\,max}$ together with f as (22.40) proceeds so rapidly that it is not possible to match the observational data with the idea of the complete internal reflection in the Jovian ionosphere as the cause of the radio emission's directivity. For example, to judge from Fig. 97, at a frequency of $f = 27$ Mc/s $\Delta\Theta$ does not exceed 30°. Even with $\varphi_{0\,max} \approx \Delta\Theta/2 \approx 15°$ it follows that $f_{L\,max} \approx 26$ Mc/s and detection of the radio emission becomes impossible at frequencies lower than 26 Mc/s, which does not correspond to reality: at the same time as radio emission at 27 Mc/s bursts have also been observed at a far lower frequency (19·6 Mc/s). In the case of $\varphi_{0\,max} < \Delta\Theta/2$ the discrepancy with the experiment becomes even greater.

Since the original idea of the generation of bursts during electrical discharges between clouds in Jupiter's atmosphere cannot be made to agree with the observational data, attention is now being paid to other possible ways of explaining the sporadic Jovian radio emission. The problem of the origin of this emission has proved to be very complex; the difficulties met here have brought Strom and Strom (1962) even to the conclusion that the observed bursts are not genetically connected with Jupiter but are the emission of cosmic discrete sources focused on the Earth by Jupiter's ionosphere. One of the arguments against this hypothesis is that the bursts at the time of a noise storm are generated in one and the same region with a size of about the radius of Jupiter. During the refraction of remote sources in the planet's ionosphere one would expect that the observed positions of the local regions of radio emission will differ by an amount of the order of the diameter of Jupiter. As pointed out by Jelley (1963), for the Stroms' hypothesis to be valid we must also assume the presence of numerous discrete sources with anomalous properties; the existence of such sources is doubtful. Neither is any change observed in the frequency of the appearance of bursts as a function of the velocity of Jupiter's movement through the sky; the spectrum of the Jovian sporadic radio emission differs sharply from the spectrum of cosmic sources. The constancy of the sense of polarization of the bursts also appears incomprehensible from the point of view of the hypothesis under discussion, since the discrete sources are eclipsed just as often by the northern and southern hemispheres of Jupiter which have magnetic fields of different directions. See also the criticism of this hypothesis by Smith, Six, Carr and Brown (1963).

Landovitz and Marshall (1962) have recently put forward their own hypothesis, suggesting as the source of the sporadic radio emission stimulated spin transitions of electrons in the planet's magnetic field.

Mechanism of Plasma Oscillations

In our view the most probable source of the Jovian bursts of radio emission are plasma waves in Jupiter's ionosphere excited by some agent and then partially transformed into electromagnetic waves. There are various opinions about the actual nature of this agent which we shall discuss a little later. For now we shall examine the simplest variant of the plasma hypothesis developed by Zheleznyakov (1958a and 1958c), actually without allowing for Jupiter's magnetic field.

In this variant it is assumed that conditions can occur in Jupiter's ionosphere under which local violations appear of the quasi-neutral state of the plasma, the restoration of this state being accompanied by the appearance of eigen oscillations of the ionospheric plasma at frequencies $\omega \approx \omega_L = (4\pi e^2 N/m)^{1/2}$. The electromagnetic waves "emitted" by the plasma oscillations as the result of one transformation process or another (section 25) are detected on Earth in the form of Jupiter's sporadic bursts of radio emission.[†] Therefore this variant is similar to the mechanism of the eigen oscillations of type I solar bursts discussed in section 31.

[†] The possibility of the action of a plasma oscillation mechanism in the conditions of Jupiter has also been pointed out by Gardner and Shain (1958).

§ 32] Mechanism of Jupiter's Sporadic Radio Emission

Let us see what a systematic application of this hypothesis to the case of Jupiter's radio emission leads to.

According to section 22 the frequencies of plasma oscillations (plasma waves) damped in a time much greater than $1/\omega$ are close to $\omega_L = 2\pi f_L$. Since in the ionospheric layer the value of f_L does not exceed its value in the maximum layer $f_{L\,\text{max}}$ an upper limit should exist for the radio emission frequency. This is confirmed by observations, according to which Jupiter's activity generally appears only at frequencies of $f \lesssim 30$ Mc/s. It follows from this that the critical frequency $f_{L\,\text{max}}$ of Jupiter's ionosphere is, as a rule, around 30 Mc/s, i.e. $N_{\text{max}} \sim 10^7$ electrons/cm³.

The localization of the bursts' sources in longitude from the point of view of the hypothesis under discussion is connected with the violation of quasi-neutrality only in a few of Jupiter's most active regions. The narrow frequency spectrum of each individual burst can be explained only by assuming that the individual violations of the quasi-neutrality do not cover the whole thickness of the ionosphere, but are localized in altitude so that the relative variation in the electron concentration in the source is $\Delta N/N \ll 1$.

The observed radio emission appears either because of the transformation of plasma waves in a regularly non-uniform plasma (namely in the layer $\varepsilon \approx 0$ where the geometrical optics approximation is violated simultaneously for electromagnetic and for plasma waves), or by Rayleigh scattering of plasma waves on thermal fluctuations of the electron concentration (see section 25). Since these processes are achieved with conservation of frequency it is clear that the bursts appearing will be recorded on Earth only if they are generated above the maximum of the ionospheric layer. The latter eliminates from the plasma oscillation hypothesis the difficulties met by the idea of lightning discharges when explaining the directivity of the radio emission by refraction in the ionosphere. At the same time it is clear that radio emission whose source is plasma oscillations will be directional since it is generated in a region where $n_{1,\,2}^2(\omega \approx \omega_L) \ll 1$ (see section 22).

We notice by the way that combination scattering of plasma waves on plasma waves of fluctuation origin occurs as well as Rayleigh scattering. The result is non-directional (or weakly directional) radio emission at a frequency $\omega \approx 2\omega_L$, which explains the activity of Jupiter sometimes observed by Warwick in bands whose frequencies are in the ratio of approximately 2 : 1 (see section 19).

It was stressed in section 26, and also in section 31 when discussing the mechanism of the radio emission of type I bursts that the attenuation coefficient of plasma waves with small wave numbers is $\gamma = \nu_{\text{eff}}$ (26.76). In the ionosphere the effective number of collisions is $\nu_{\text{eff}} = \nu_{\text{ei}} + \nu_{\text{em}}$, where the number of collisions of electrons with ions ν_{ei} is defined by the formula (26.78) and with neutral molecules ν_{em} by the formula (26.80). It is clear from this that the time in which the intensity of the plasma oscillations decreases by a factor e and the duration of a burst of radio emission caused by damped oscillations of the ionospheric plasma brought out of a state of equilibrium are given by the expression

$$t_0 = \frac{1}{\nu_{\text{eff}}}.$$

It also remains valid when there is a magnetic field (with an accuracy up to a factor of the order of $\frac{1}{2}$; see (31.12)).

By basing our work on this relation we can find from the known duration of the Jovian bursts the value of ν_{eff} in Jupiter's ionosphere, provided, of course, that the action time of the process disturbing the ionospheric plasma from a state of equilibrium is small when compared with $1/\nu_{\text{eff}}$. Otherwise only a lower limit of ν_{eff} can be estimated by observing the duration t_0 of radio bursts. If there really are surges with $t_0 \sim 10^{-2}$–10^{-3} sec in Jupiter's sporadic radio emission (there is a certain doubt of this at present; see section 19), then $\nu_{\text{eff}} \gtrsim 10^2$–$10^3$ sec^{-1}. For definitely existing bursts with $t_0 \sim 0.7$ sec, $\nu_{\text{eff}} \gtrsim 1.4$ sec^{-1}.

If the number of collisions is determined by the collision of electrons with ions ($\nu_{\text{eff}} \approx \nu_{ei} > \nu_{em}$), then we obtain from the formula (26.78) that the kinetic temperature of Jupiter's ionosphere is $T \sim 2 \times 10^4$–4×10^3 °K for $N \sim 5 \times 10^6$ electrons/cm^3 and $\nu_{\text{eff}} \sim 10^2$–$10^3$ sec^{-1}. With the more probable limitation on the value of $\nu_{\text{eff}} \gtrsim 1.4$ sec^{-1} and with $N \sim 5 \times 10^6$ electrons/cm^3 we obtain the condition $T \sim 4 \times 10^5$ °K for the temperature of the ionosphere. Since it is extremely doubtful that temperatures of the order of 4×10^5 °K occur in the planetary ionospheres, we conclude from this that the duration $t_0 \sim 0.7$ sec of the bursts of radio emission cannot be explained on the basis of a plasma mechanism in an isotropic plasma with a finite oscillation damping time: it must be assumed that a major part is played here by the time of the process which disturbs the plasma from a state of equilibrium. This assumption is superfluous for the millisecond bursts since temperatures of a few thousand degrees in the ionosphere are quite possible.

Using the relations (Al'pert, 1960)

$$N_{\max} = \sqrt{\frac{J_{\max}}{\alpha_{\text{eff}}}}, \quad J_{\max} = \frac{S}{1 \cdot 36 \varepsilon_i L}, \quad L = \frac{2xT}{m_i g}, \tag{32.1}$$

which are valid for a simple ionospheric layer we can estimate the effective recombination coefficient α_{eff}, the ionizing capacity at the layer's maximum $J_{\max}$ and the layer's half-thickness L, knowing the ionization energy ε_i, the

§ 32] Mechanism of Jupiter's Sporadic Radio Emission

mass of the heavy particles m_i, the flux of the solar ionizing radiation S, the temperature in the layer T and the acceleration due to gravity on Jupiter g.

It was pointed out in section 3 that Jupiter's atmosphere contains the following gases: methane CH_4, ammonia NH_3, hydrogen H_2 and possibly helium He. The lightest components—hydrogen and helium—are apparently predominant at quite considerable altitudes (in the ionosphere). Assuming for the sake of definition that Jupiter's ionosphere is formed by the ionization of H_2 (molecular weight 2), we obtain that the half-thickness of the layer is $L \sim 6 \times 10^7$ cm if $g \sim 2 \cdot 6 \times 10^3$ cm/sec^2 and $T \sim 2 \times 10^3$ °K. Furthermore, the flux of ionizing solar radiation on Jupiter's orbit, whose radius is 5.2 a.u., is $(5 \cdot 2)^2$ times less than the corresponding values of S near the Earth. Following the data given by Al'pert (1960) (Chapter II) we shall take $S_\oplus \sim 4 \times 10^{-2} – 4 \times 10^{-1}$ erg/cm^2 sec; then $S_{2\!\!\!\downarrow}$ is approximately $1 \cdot 5 \times 10^{-3} – 1 \cdot 5 \times 10^{-2}$ erg/cm^{2}sec.† For an H_2 molecule the ionization energy is $\varepsilon_i = 15 \cdot 4$ eV $= 2 \cdot 5 \times 10^{-11}$ erg. Taking the above into consideration and putting $N_{\max} \sim 10^7$ electrons/cm^3, we find from (32.1) that the number of ions appearing in 1 sec in 1 cm^3 of the layer's maximum is $J_{\max} \sim 1–10$ ions/cm^3 sec, and the effective recombination coefficient is $\alpha_{\text{eff}} \sim 10^{-14} – 10^{-13}$ cm^3/sec. Allowing for possible dissociation of the hydrogen in Jupiter's upper atmosphere alters these estimates slightly.

By comparing the data obtained on Jupiter's ionosphere with the parameters of the F_2 layer on Earth ($N_{\max} \sim 10^6 – 2 \times 10^6$ electrons/cm^3, $\nu_{\text{eff}} \approx 3 \times 10^3$ sec^{-1}, $T \sim 2 \times 10^3$ °K, $\alpha_{\text{eff}} \sim 3 \times 10^{-10}$ cm^3/sec, $J_{\max} \sim 25–2000$ ions/cm^3 sec, $L \sim 10^7 – 3 \times 10^7$ cm), we can see that Jupiter's ionosphere from the point of view of the simplest plasma hypothesis (without allowing for the magnetic field) differs from the F_2 layer in its higher electron concentration with a far lower ionizing capacity of the solar radiation. This difference may be connected (in the respect under discussion) only with lower values of the recombination coefficient in Jupiter's ionosphere.

However values of $\alpha_{\text{eff}} \sim 10^{-14} – 10^{-13}$ cm^3/sec obviously cannot occur in the planets' atmospheres; actually the recombination coefficients are far higher. For example, the radiation recombination of electrons with atomic ions is characterized by a value $\alpha_{\text{eff}} \approx 10^{-12}$ cm^3/sec; for dissociative recombination of electrons with molecular ions α_{eff} is even higher—of the

---

† The value of S, of course, is not the same for the different gases and rises as the ionization potential decreases. The value taken for S corresponds to the ionization of the gases forming the F layer on Earth, i.e. N_2, N and O (there are few O_2 molecules in this layer since at great altitudes the oxygen is almost entirely dissociated). Since these molecules and atoms correspond to energies of $\varepsilon_i = 15 \cdot 8$ eV, $14 \cdot 5$ eV and $13 \cdot 5$ eV, close to the ionization energy of hydrogen, it is clear that the value $S_{2\!\!\!\downarrow} \sim 1 \cdot 5 \times 10^{-3} – 1 \cdot 5 \times 10^{-2}$ erg/cm^2 is fully capable of characterizing the radiation that ionizes the H_2 on Jupiter.

order of 10^{-7} cm^3/sec (see Von Engel, 1959, chapter 6). It is clear that the effective recombination coefficient, which is contributed to by different types of recombination process, is in any case not less than 10^{-12} cm^3/sec. Generally α_{eff} noticeably exceeds its minimum value; in the Earth's atmosphere, for example, this coefficient varies (depending on altitude) in the range 5×10^{-7}–3×10^{-10} cm^3/sec. The necessity for values of α_{eff} in Jupiter's ionosphere that are too low is an obvious drawback of the variant of the plasma hypothesis that has been discussed.

Another essential drawback to this variant is above all the requirement that the source of the bursts should be localized both as to its altitude in the ionosphere $\Delta N \ll N$)† and on the planet's disk ($\Delta \Theta_{\text{source}} \ll 180°$). If the latter condition can still be somehow justified on the basis of general ideas of the connection between sources and the most active regions on Jupiter's disk, then the limited extent in altitude in the ionosphere is not justified in any way. Furthermore, it is still not clear what process violates the quasi-neutrality of the isotropic plasma with subsequent restoration of the equilibrium via damped oscillations at a frequency $\omega \approx \omega_L$ (see the discussion of this question in section 27). Therefore we must move away from the representation of "shock" excitation of plasma oscillations (in accordance with the results of section 27) to the representation of generation of plasma waves as the result of instability of the ionospheric plasma, which is done below. And lastly, of course, generation in an isotropic plasma will not explain in any way the circular or elliptical nature of the polarization of Jupiter's sporadic radio emission.

PLASMA HYPOTHESIS OF THE ORIGIN OF JUPITER'S RADIO EMISSION WHEN THE PLANET'S MAGNETIC FIELD IS TAKEN INTO ACCOUNT

Allowing for the effect of Jupiter's magnetic field (whose existence is indicated by observations of Jupiter's decimetric and decametric radio emission) on the generation and propagation of the radio waves removes the above difficulties (Zheleznyakov, 1959a and 1965c).

It was pointed out in section 23 that the plasma waves in a magneto-active plasma correspond to frequencies of $\omega_{\text{pl}} \approx \omega_\infty$, where ω_∞ with fixed values of the gyro-frequency ω_H and the Langmuir frequency ω_L has two values defined by the expression (23.4c). It is clear from (23.4c) that when the angle α between the field H_0 and the wave vector k changes, the frequencies of the plasma waves (plasma oscillations) remain in the range

$$\max\{\omega_H, \omega_L\} \lesssim \omega_{\text{pl}} \lesssim \sqrt{\omega_L^2 + \omega_H^2}, \quad 0 \lesssim \omega_{\text{pl}} \lesssim \min\{\omega_H, \omega_L\} \quad (32.2)$$

† The condition $\Delta N \ll N$ means that the generation region should be situated inside the ionospheric layer; it cannot be localized in the beginning of the layer or in its uppermost part.

§ 32] Mechanism of Jupiter's Sporadic Radio Emission

and in any case ω_{pl} does not exceed $\sqrt{\omega_L^2 + \omega_H^2}$. In the range of frequencies that correspond to the higher values of the plasma wave frequencies

$$\omega_{\text{pl}} \approx \omega_L \quad \text{with} \quad \omega_L^2 \gg \omega_H^2, \tag{32.3a}$$

$$\omega_{\text{pl}} \approx \omega_H \quad \text{with} \quad \omega_L^2 \ll \omega_H^2 \tag{32.3b}$$

The case (32.3a) has been discussed above; we shall now turn to the more favourable case of (32.3b).

First of all we note that in the variant (32.3b) the frequency of the plasma waves, and therefore of the radio emission generated, is determined not by the Langmuir but by the gyro-frequency $\omega_H = eH_0/mc$. A burst at a frequency ω appears at the point on Jupiter's disk where the magnetic field strength (at the level of the ionosphere) is $H_0 \approx mc\omega/e$. Since the sporadic radio emission is observed between about 5 and 35 Mc/s we conclude that the magnetic field in the sources is 2–12 oe.

An estimate of the maximum value of the electron concentration can be obtained by requiring that the effective recombination coefficient α_{eff} should not drop below 10^{-12} cm^3/sec with an ionizing capacity of $J_{\text{max}} \sim$ 1–10 ions/cm^3 sec. According to (32.1) in this case $N_{\text{max}} \lesssim (1-3) \times 10^6$ electrons/cm^3. The reduced values of the electron concentration eliminate the difficulties with the value of the effective recombination coefficient in the Jovian ionosphere.

On the other hand, for generation of radio emission at frequencies around 5 Mc/s it is necessary for the Langmuir frequency not to exceed this value, i.e. that in the source we should have $N < 3 \times 10^5$ electrons/cm^3.†
If the generation process of each individual burst covers the whole thickness of the ionosphere over a given point on the surface, this leads to the condition $N_{\text{max}} < 3 \times 10^5$ electrons/cm^3.

It is not hard to see that with concentrations of $N_{\text{max}} < 3 \times 10^5$ electrons/cm^3 the band of frequencies taken up by an individual burst will remain narrow enough even if the size of the source in altitude is comparable with the thickness of the ionospheric layer (i.e. the variation ΔN in the source is of the order of N_{max}). It is clear from the first relation (32.2) that the band of frequencies generated with $\omega_L^2 \ll \omega_H^2$ because of the variation in the

---

† For the duration of a burst to be defined by the relation $t_0 \sim 1/\nu_{\text{eff}}$ (to be more precise, $t_0 \sim \frac{1}{2}\nu_{\text{eff}}$ in the magnetic field with $\omega \approx \omega_H$; see (31.12)) even lower values of N in the source are necessary (with a reasonable value of the kinetic temperature T). For example, for $T \sim 2 \times 10^3$ °K and $\nu_{\text{eff}} \approx 1 \cdot 4$ sec^{-1} the required concentration N will be of the order of 2×10^3 electrons/cm^3. Still lower values of N will lead (contradicting observations) to too long a burst duration; with higher concentrations the duration of a burst will no longer be determined by the effective number of collisions but by the duration of the action of the agent causing the plasma oscillations.

electron concentration in the generation region is

$$\Delta f \sim \frac{\Delta \omega_L^2}{4\pi \omega_H} \approx \frac{\Delta \omega_L^2}{8\pi^2 f}. \tag{32.4}$$

At a frequency of $f \sim 20$ Mc/s with $\Delta N \sim N_{max} < 3 \times 10^5$ electrons/cm³ the value is $\Delta f < 0.6$ Mc/s, whilst the observed value of Δf averages about 1 Mc/s.

The width of the frequency spectrum of an individual burst also depends on the variation of the magnetic field strength in the source; because of the latter

$$\frac{\Delta f}{f} \sim \frac{\Delta H_0}{H_0}. \tag{32.5}$$

From this we obtain that with $\Delta f \sim 1$ Mc/s at $f \sim 20$ Mc/s the permissible relative variation in the field is $\Delta H_0/H_0 \sim 5 \times 10^{-2}$; the corresponding absolute variation is $\Delta H_0 \sim 0.4$ oe. Therefore in an ionosphere with $N_{max} < 3 \times 10^5$ electrons/cm³ and a relative variation of the magnetic field in a radial direction of not more than 5×10^{-2}[†] the necessity for localization of the source in altitude to explain the narrow frequency spectrum of the individual bursts is eliminated.

Furthermore, the small extent of the local radio emission sources in longitude can be explained naturally (provided that $\omega \approx \omega_H$) by the fact that the magnitude of Jupiter's magnetic field varies with longitude, reaching a maximum with values of Θ corresponding to the centres of the local sources. The maximum value of $H_0 \sim 12$ oe ensures generation of radio emission at frequencies up to 35 Mc/s; the minimum value of $H_0 \sim 2$ oe is sufficient to create bursts at a frequency of 5 Mc/s. The estimate of the minimum magnetic field should be re-examined (by way of reducing it) if sporadic radio emission from Jupiter is found at frequencies below 5 Mc/s.[‡] Radio emission at the maximum frequency originates only from the centre of a source where the field H_0 is maximal; the emission at lower frequencies originates from regions surrounding the centre of the source. Here, obviously, the lower the frequency the greater the size of the source in

---

† This condition is not hard to satisfy. For example, if Jupiter's magnetic field is created by a dipole placed at the centre of the planet, then $H_0 \propto R^{-3}$, where R is the distance from the centre. The field variation in the range ΔR in the ionosphere is then $\Delta H_0/H_0 = 3\Delta R/R_{2\!\!\downarrow}$, where $R_{2\!\!\downarrow} \approx 7 \times 10^9$ cm is the radius of Jupiter. The ratio is $\Delta H_0/H_0 \lesssim 5 \times 10^{-2}$ if the thickness of the ionosphere is $\Delta R \lesssim 10^8$ cm (we recall that the half-thickness of the hydrogen ionized layer on Jupiter is $L \sim 6 \times 10^7$ cm when $T \sim 2 \times 10^{3} °K$).

‡ Therefore from the point of view of the plasma hypothesis in a magnetic field there should be both upper and lower limits to the frequency of the bursts of radio emission of Jupiter. Finding the lower limit would therefore be an additional argument in favour of the hypothesis under discussion.

§ 32] Mechanism of Jupiter's Sporadic Radio Emission

longitude. This circumstance apparently also explains the observed rise of $\Delta\Theta$ with wavelength (see Fig. 97). At low frequencies (5–10 Mc/s) radio emission may be generated in a broad range of longitudes, which helps us to understand the actual absence of sharply defined sources of radio emission at these frequencies.

Thanks to the directional nature of Jupiter's radio emission the observed bursts come from the part of the disk located near the planet's central meridian. As a result the range of frequencies in which Jupiter's radio emission is recorded will vary with the rotation of the planet in accordance with the variation of the magnetic field in longitude. As the local source approaches the central meridian Jupiter's activity will appear at the higher frequencies, whilst after the source has passed (to be more precise, its centre has passed) through the central meridian the activity will start to appear in a range that moves down the frequency scale. This feature of the sporadic radio emission is well confirmed by Warwick's observations (see Fig. 92). His data characterize the frequency drift for the main local source (longitude $\Theta \approx 180°$) and cover a range of longitudes from 85 to 285°. If Jupitre has a dipole magnetic field the required variation in the field strength H_0 over the surface (by a factor of 6–7) cannot be explained by setting the dipole in the centre of the planet: since the axis of the dipole is close to the rotational axis (and it is this orientation which is necessary when explaining the decimetric radio emission polarization observations; see section 20) the magnetic field H_0 is practically independent of the longitude.† On the other hand, with the formula (33.1) it is easy to check that the necessary variation $H_0(\Theta)$ (near the equator) will occur if the magnetic dipole is moved in the plane of the equator a distance of $\sim 0.3 R_{2\!\!\!\downarrow}$ from the centre (to a point with the longitude of the main local source $\Theta \approx 180°$). This kind of displacement is in complete agreement with the data obtained when investigating Jupiter's decimetric radio emission and given in section 20. We notice that the magnitude of the displacement alters slightly if the generation regions of the bursts observed on Earth are not in the equatorial belt but in the middle latitudes where the lines of magnetic force in the planet's ionosphere run along the line of sight (Warwick, 1961, 1961–2 and 1963). This siting of the sources is a natural consequence of using gyroresonance absorption to explain the directivity effect in the Jovian ionosphere (see below).

How are we to interpret the characteristic directivity of the emission in the framework of the plasma hypothesis while allowing for the magnetic field? Above we confirmed that if the hypothesis under discussion is valid

† And in general with a constant distance R from the centre of the dipole the value of H_0 varies by a factor of only 2 (see the formula (33.1)).

the electron concentration in the source should be lower than 3×10^5 electrons/cm³. Then the parameter $v = \omega_L^2/\omega^2 < 0.1$ at frequencies of $f \gtrsim 15$ Mc/s, where the directivity of the radio emission actually is observed. According to section 23 the refractive index of an ordinary wave is $n_2 \approx 1$; when $v \ll 1$; therefore the directivity cannot be connected with refraction in the ionosphere if the plasma waves are transformed for preference into ordinary waves and the condition $N < 3\times10^5$ electrons/cm³ (i.e. the equality $n_2 \approx 1$) is satisfied over the whole of the path from the source to escape from the ionosphere. The refraction becomes significant if there are enhanced values of N above the source; however, the interpretation of the directivity by the effect of total internal reflection meets here with the difficulties mentioned at the beginning of the chapter. On the other hand, if the plasma oscillations are transformed into extraordinary waves in the vicinity of the point $v = 1-\sqrt{u}$ (which with $v \ll 1$ is close to the level $u \approx 1$, i.e. $\omega \approx \omega_H$), then the radio emission may become directional since in the vicinity of $v = 1-\sqrt{u}$ the refractive index is $n_1 \approx 0$ (see section 23).

It is also possible that the directivity of the radio emission is caused by gyro-resonance absorption in the $\omega \approx 2\omega_H$ layer which increases as the angle between the direction $\mathbf{k}$ of the wave propagation and the field $\mathbf{H}_0$ increases (section 26). This layer is obviously located higher on Jupiter than the $\omega \approx \omega_H$ layer where the radio emission is generated (since the strength of the magnetic field falls with altitude); for example in the middle of the main local source at a distance of approximately $0.7R_{2\!\!\!\!\!+}$ from the magnetic dipole the magnetic field proportional to R^{-3} is halved at an altitude of $h \sim 0.2R_{2\!\!\!\!\!+}$.

To estimate the extent of gyro-resonance absorption in the $\omega \approx 2\omega_H$ layer we make use of the formula (26.126), putting $v \ll 1$ and $s = 2$ in it:

$$\tau_{j2} \approx \pi \frac{v\omega}{c} \beta_{\text{th}}^2 L_H (1 \pm \cos\alpha)^2 \sin^2\alpha. \tag{32.6}$$

The parameter $L_H \approx (dR/dH_0)H_0$ in the case of a dipole magnetic field where $H_0 \propto R^{-3}$ will be approximately $R/3$, which at an altitude of $h \sim 0.2R_{2\!\!\!\!\!+}$ is $0.4R_{2\!\!\!\!\!+}$. Then at a frequency $\omega \sim 4\pi \times 10^7$ sec⁻¹ ($f \sim 20$ Mc/s) the optical thickness of the $\omega \approx 2\omega_H$ layer is

$$\tau_{j2} \sim 1.3\times 10^{-9} NT(1\pm\cos\alpha)^2 \sin^2\alpha. \tag{32.7}$$

For creation of the observed directivity ($\Delta\Theta_{\text{dir}} \sim 45°$ at $f \sim 20$ Mc/s; see Fig. 97) it is necessary that τ_{j2} should be of the order of unity when $\alpha \approx \Delta\Theta_{\text{dir}}/2 \sim 22°$. We obtain from the formula (32.7) that this condition will be satisfied for an extraordinary wave if $NT \sim 1.5\times 10^9$, and for an ordinary wave if $NT \sim 7\times 10^{11}$. With a value of $T \sim 5\times 10^3$ °K, which is

§ 32] Mechanism of Jupiter's Sporadic Radio Emission

apparently a quite reasonable estimate of the kinetic temperature at altitudes of $h \sim 0.2R_{\mathrm{2\!\!\;l}}$, the electron concentration in the $\omega \approx 2\omega_H$ layer should reach 3×10^5 electrons/cm³ (extraordinary wave) and 1.4×10^5 electrons/cm³ (ordinary wave).

The over-high values of N for ordinary waves indicate the impossibility of using gyro-resonance absorption to explain the observed directivity if Jupiter's sporadic radio emission is waves of this type. On the other hand, the existence at an altitude $h \sim 0.2R_{\mathrm{2\!\!\;l}} \sim 1.4 \times 10^9$ cm of ionized layers with a concentration $N \sim 3 \times 10^5$ electrons/cm³ is quite possible. In fact, with $T \sim 2 \times 10^3$ °K, as we have seen, the half-thickness of the ionized layer formed by ionization of the hydrogen is 6×10^7 cm on Jupiter. The extent of the Chapman layer between the points where the concentration falls by a factor e from its maximum value is approximately $4L$, i.e. 2.4×10^8 cm. This is a fifth of the altitude of the $\omega \approx 2\omega_H$ level on Jupiter of interest to us. If, however, we remember that the decrease in the concentration N with altitude above the layer's maximum probably decreases far more slowly than that described by a simple Chapman layer (the example of the Earth's ionosphere convinces us of this) and the kinetic temperature can be over 2×10^3 °K, it becomes clear that an ionosphere $\sim 10^9$ cm in extent probably exists on Jupiter. In this case the radio emission should be generated only in the lowest layers of the ionosphere to provide the narrow frequency spectrum of the individual bursts; otherwise it will be "blurred" because of the variation in the magnetic field with altitude.

Therefore both the interpretations discussed above for the directivity of Jupiter's radio emission require that chiefly extraordinary waves should be generated in the ionosphere: in this case, in complete agreement with experiment, the radio emission acquires a sharply polarized nature. The observed clockwise (right-handed) rotation then points to the circumstance that in the generation region the magnetic field $\boldsymbol{H_0}$ is at an acute angle to the observer (if, of course, in the process of being propagated in Jupiter's exosphere the radio emission does not pass through a transverse magnetic field region, resulting in a possible change in the sense of rotation).†

† As shown in section 24, the sign of the rotation does not change when there is strong wave coupling in a transverse magnetic field region. We notice that the phenomenon of interaction of extraordinary and ordinary waves when passing through this region (namely the partial transformation of extraordinary emission into ordinary) can explain the elliptical nature of the polarization of Jupiter's decametric radio emission. For this it is necessary that the product NH_0^3 in the transverse field region should be of the order of 5×10^2 (see the formula (24.49) and below); with concentrations of $N \sim 10^3$ electrons/cm³ the latter condition gives reasonable field values of $H_0 \sim 1$ oe. The position here (and the difficulties with interpreting the ellipticity) is exactly the same as for the solar radio emission (see the last subsection of section 24).

The mechanism of transformation of plasma waves into electromagnetic waves should obviously ensure preferential radiation of extraordinary emission. Transformation by wave coupling with quasi-longitudinal propagation is ineffective here: it occurs with $v = \omega_L^2/\omega^2 \approx 1$, whilst in the model $v \ll 1$ (see section 25). Rayleigh scattering of plasma waves cannot be of any help in this respect either, since the effect of scattering into extraordinary emission has no advantages over effects of scattering into ordinary waves. On the other hand, the intensity of scattered ordinary emission here will probably be even higher since $n_2^2 > n_1^2$ (see, e.g., the formula (25.42)). The transfer of waves from the plasma branch onto the branch corresponding to an extraordinary wave is more promising; this change without allowing for thermal motion appears as the phenomenon of "leakage" of extraordinary waves from the branch corresponding to high values of the refractive index $n_1^2(v)$ when $v > 1-u$ into the branch $n_1^2(v)$ in the region $v < 1-\sqrt{u}$ (see, e.g., Fig. 116).

The coefficient of the emission "leakage" through the region $n_1^2 < 0$, which for us plays the part of the transformation coefficient, is obtained by Denisov (1958 and 1959; see also Ginzburg, 1960b, section 27) for the special case of transverse propagation ($\alpha = \pi/2$) in a constant magnetic field:

$$Q = e^{-\delta_0},$$

$$\delta_0 = 4u^{3/2} \frac{\omega}{c|\operatorname{grad} v|}(1-\sqrt{u})(1+\sqrt{u})^{1/2} \times$$

$$\int_0^1 \left[(1-t^2)\left(1+\frac{1-\sqrt{u}}{1+\sqrt{u}}t^2\right)\right]^{1/2} dt. \qquad (32.8)$$

Remembering that $v \ll 1$ and putting $u \approx 1$, $1-\sqrt{u} \sim v$, we obtain:

$$\delta_0 \approx 4\sqrt{2}\frac{\omega}{c} v |\operatorname{grad} v|^{-1} \int_0^1 \sqrt{1-t^2}\, dt = \sqrt{2}\pi \frac{\omega}{c} v |\operatorname{grad} v|^{-1}. \qquad (32.9)$$

In an unperturbed ionosphere the leakage is absolutely negligible: with $L \sim 6 \times 10^7$ cm and $|\operatorname{grad} v|^{-1} \sim L/v$ at frequencies of $f \sim 20$ Mc/s the parameter is $\delta_0 \sim 10^6$! However, in non-stationary conditions the position may alter significantly. For example, in the fronts of strong shock waves passing through the plasma $|\operatorname{grad} v|^{-1} \lesssim l_f/v$. In a magnetoactive plasma the width of the front l_f varies within wide limits depending on the strength of the magnetic field, reaching a value of U/v_{eff}, where U is the

§ 32] Mechanism of Jupiter's Sporadic Radio Emission

velocity of the shock wave (see (27.122)), in a fully ionized and sufficiently rarefied medium. In the latter case, however, the front consists of a series of solitary waves whose half-width ($\sim c/(2\omega_L\sqrt{M-1})$; M is the Mach number; see (27.119) and (27.120)) actually determines the characteristic distance over which the plasma parameters change. For shock waves with $M \sim 2$ the parameter is $|\text{grad } v|^{-1} \sim c/2\omega_L v$, so $\delta_0 \sim \pi\omega/\sqrt{2}\omega_L$. At $f = 20$ Mc/s with $N \sim 3\times 10^5$ electrons/cm³ we obtain: $\delta_0 \sim 8\cdot 6$; $Q \sim 1\cdot 5\times 10^{-4}$. The transformation coefficient given is apparently fully applicable; in any case it is higher than the values of the transformation coefficient in a regular solar corona (see section 25).

The reservation should be made that the estimates given for Q in a shock wave front are only by way of being a guide. This can be seen from the fact that the formulae (32.8) and (32.9) are valid only in a uniform magnetic field, whilst in a shock wave front there may be considerable variation of the field H. In addition, the estimates of the width of the shock wave front need to be made more accurate (allowing for incomplete ionization of the plasma, etc.).

We have thus come to the conclusion that the strong polarization and directivity of the Jovian bursts can be explained only provided that the transformation of the plasma waves is chiefly into extraordinary emission, which is possible when there are sharp drops in the density of the ionospheric plasma (and the magnetic field) of the shock wave type. The next very important problem is obviously the question of the conditions for the appearance of plasma waves at frequencies of $\omega \approx \omega_H$. When discussing in section 27 the problem of the appearance of plasma-type oscillations in the front or behind the front of a shock wave we remarked that there is no such effect in a plasma without a magnetic field.

In a magnetoactive plasma, however, instability and amplification of waves of the electromagnetic and plasma types can occur because of the non-equilibrium nature of the particle velocity distribution function in the shock wave front. In actual fact, as was made clear at the end of section 27, instability appears in the first place because of temperature "anisotropy" of the electrons when the magnetic field in the shock wave front changes[†] and, in the second place, because of a shift of the maximum of

---

[†] In this case the plasma wave excitation mechanism can with complete justification be called a coherent magneto-bremsstrahlung mechanism, since it leads to plasma wave amplification at the frequencies $\omega \sim \omega_H$ (where $n_3^2 \gg 1$) which at the same time satisfy the Doppler equation (26.31). We note that as applied to Jupiter's sporadic radio emission the magneto-bremsstrahlung mechanism of non-relativistic electrons (non-coherent) has been discussed by Ellis (1952 and 1963). The drawback of the non-coherent mechanism is the difficulties with explaining the high intensity and the burst nature of the decametric emission.

the ions' distribution function relative to the electron maximum; this kind of drift is also connected with the variation in the magnetic field in a shock wave. Warwick (1961, 1961–2, 1963) has indicated another possible way plasma waves can be generated in the Jovian ionosphere—because of instability of the ionospheric plasma penetrated by streams of particles from the radiation belts. In this case the shock waves can obviously play the part of a "starting mechanism" which will provide a sharp rise in the coefficient of the transformation of the plasma waves into electromagnetic radiation and as a consequence the effective escape of the radio emission recorded on Earth in the form of a burst from Jupiter's ionosphere.

It is still unclear which of the above possible ways of generating plasma waves plays the major part in the mechanism of Jupiter's sporadic radio emission. Warwick's point of view is attractive in that it allows us to connect (via the particle density in the planet's radiation belts) the observed variations in Jupiter's activity in the decametric band with the solar activity index. The wave amplification behind the shock wave front (in a temperature anisotropy region) also rises if in the ionospheric plasma there is an admixture of energetic electrons from the radiation belts, so Warwick's idea in this sense is not an exception.

A careful analysis of the possible ways of generating plasma waves in Jupiter's ionosphere, an estimate of the expected values of the amplitude of these waves, etc., now acquire first importance in the theory of Jupiter's sporadic radio emission.†

33. Origin of the Continuous Radio Emission of Jupiter and Saturn

RADIATION BELTS AS THE SOURCE OF JUPITER'S DECIMETRIC RADIO EMISSION

The closeness of the values of Jupiter's effective temperature $T_{\text{eff}\,2\!\!\!\downarrow}$ at wavelengths of $\lambda \sim 3$ cm to the temperature of the upper layer of clouds (determined by measurements in the infrared part of the spectrum) allow us to assume that the lower layers of the Jovian atmosphere play the major part in the generation of the radio emission in this band (and apparently also of the shorter wavelengths). The emission here is thermal in nature and is probably connected with molecular absorption in the components forming the planet's gaseous shell. It must be said, however, that the role of molecular absorption of Jupiter's atmosphere in the part located above

---

† We recall that the questions of the generation of plasma waves in the passage of shock waves in a plasma have also been discussed in the theory of types I and II bursts (section 31).

§ 33] Continuous Radio Emission of Jupiter and Saturn

the cloud layer has an effect only in the vicinity of $\lambda \approx 1\cdot 25$ cm where the lines of the ammonia inversion spectrum lie.†

The slightly higher values of $T_{\text{eff}\,2\!\!+}$ at $\lambda \sim 3$ cm when compared with the infrared temperature (see section 20) are possibly caused by the variation with altitude of the kinetic temperature in the planet's lower atmosphere. A certain part in the creation of this temperature difference may also be played by the emission of the radiation belts which is still noticeable in this range. An argument in favour of the last suggestion is the $\sim 30°$ change in Jupiter's effective temperature in 1957–8, and also the anomalously high values of $T_{\text{eff}\,2\!\!+} \approx 268°$K found on 30 April–1 May 1958 (Rose, Bologna and Sloanaker, 1963a and 1963b). In this connection investigations of the correlation between the intensity of Jupiter's radio emission at a wavelength of 3 cm and the intensity in the decimetric waveband, where the part played by the radiation belts is indisputable (see below), would be very valuable.

Observations of Jupiter in the decimetric band, which revealed a very high effective temperature (up to 10^4 °K and more), indicated that it is not the lower atmosphere that makes the major contribution to the radio emission investigated here but some other layers of the planet. Among the various possible ways of explaining such a high radio temperature (of the order of the temperature of the Sun's photosphere) discussed by Field (1959) and Roberts and Stanley (1959) there was also a hypothesis about Jupiter's radiation belts (similar to the Earth's Van Allen belts) being the sources of the observed decimetric radio emission. The decisive confirmation of this hypothesis was the results of radio measurements of the planet's radio diameter, which in the equatorial direction turned out to be three times greater than the optical diameter, and the discovery of linear polarization at decimetric wavelengths.

In the framework of the radiation belt hypothesis the decimetric radio emission is interpreted as the non-coherent magneto-bremsstrahlung of electrons retained by the planet's magnetic field. If the magnetic field at distances of one or two radii of Jupiter from its surface is by way of being a dipole, then with sufficient symmetry in the spatial distribution of the emitting electrons the circular polarization of the total emission disappears, whilst the linear polarization, generally speaking, remains.‡ It is clear from symmetry considerations that the plane of polarization in this case is either

---

† According to Naumov and Khizhnyakova (1965) the optical thickness of a column of atmosphere resting on the clouds is greater than or of the order of unity in the range $\lambda \approx 1\cdot 0$–$1\cdot 7$ cm.

‡ In this connection see the remarks about the polarization of the "quiet" Sun's radio emission made in section 8.

the same as the plane of the magnetic equator or will be orthogonal to the latter. What has been said is in complete agreement with the observations which did not display circular polarization and recorded only linear polarization in a plane close to the Jovian equator. The variations in the plane of polarization with the planet's rotational period and a maximum deviation of approximately ten degrees from the plane of the equator can be explained naturally by the deflection of the magnetic dipole's axis towards Jupiter's axis of rotation. The variations in the intensity of the radio emission with the same period are caused by the differing concentration of the emitting electrons in longitude. The last effect may be a consequence of changes in the injection conditions and losses of emitting electrons in the "trap" created by Jupiter's magnetic field at different longitudes; in its turn the most probable cause of this change is the displacement of the magnetic dipole towards the equator relative to the centre of the planet. We recall that the suggestion of the presence of this kind of displacement must also be introduced when interpreting Jupiter's decametric radio emission (see p. 619).

Furthermore, it is easy to see that the idea of the magneto-bremsstrahlung origin of Jupiter's decimetric radio emission does not contradict the observed weak dependence of the size of the source on the frequency (according to the data of Morris and Berge (1962) the planet's equatorial radio diameter rises in a ratio of 1·14 when the frequency decreases by a factor of 1·45; see section 20). For example, in the most unfavourable case when the radiation belt in the dipole magnetic field

$$H_0 = H_{00} \frac{R_{2\!\!\!\downarrow}^3}{R^3} (1+3 \sin^2 \varphi)^{1/2} \tag{33.1}$$

(φ is the magnetic latitude, H_{00} is the field in the equatorial plane at a distance R from the centre of the dipole equal to the planet's optical radius $R_{2\!\!\!\downarrow}$) has a great extent in R, the radio diameter of the planet $2R_{\text{eff}}$ at a frequency ω can obviously be determined from the following condition: R_{eff} is approximately equal to the distance at which the frequency (26.48)

$$\omega_{\max} = \frac{\omega_H}{2} \left(\frac{\mathscr{E}}{mc^2} \right)^2,$$

corresponding to the maximum intensity in the synchrotron spectrum, is the same as ω. It follows from this that[†]

$$R_{\text{eff}} \propto \sqrt[3]{\omega} \tag{33.2}$$

[†] The formula (33.2) obtained here for the magneto-bremsstrahlung of relativistic electrons is also valid, of course, for non-relativistic electrons for which, unlike (26.48), $\omega_{\max} = \omega_H$.

§ 33] Continuous Radio Emission of Jupiter and Saturn

(with constant φ and $\mathscr{E}$). Therefore when the frequency changes by a factor of 1·45, R_{eff} changes by a factor of only 1·12, which agrees closely with experiment.

Further realization of the non-coherent magneto-bremsstrahlung mechanism of Jupiter's radio emission is closely connected with the choice of the energy $\mathscr{E}$ of the emitting electrons and the strength of the planet's magnetic field. Drake and Hvatum (1959), Roberts and Stanley (1959) and Field (1959) without detailed discussion have suggested relativistic electrons as the source of the radio emission; as applied to Jupiter their emission has been discussed in detail by Chang and Davis (1962) and Korchak (1963a and 1963b). On the other hand, in a number of papers (especially in papers of Field, 1959, 1960 and 1961; see also Korchak and Lotova, 1963) a non-relativistic electron model has been strongly developed. The question of the choice between relativistic and non-relativistic electrons as the source of the radio emission has been discussed in the paper by Zheleznyakov (1965c), which gives the arguments against the non-relativistic electron model.

First of all the radio emission frequency ω in this model is equal to the gyro-frequency ω_H, which at a wavelength of 20 cm requires the existence in the radiation belt (i.e. in the equatorial plane at a distance of $R \sim 3R_{\jupiter}$ from the centre of the dipole) of a magnetic field with a strength $H_0 \sim 500$ oe. In accordance with (33.1) this value corresponds on the surface of Jupiter to a field of about $1·5 \times 10^4$ oe at the lower latitudes and 3×10^4 oe at the poles. It is completely unclear how such high magnetic field values, unacceptable in themselves, can be made to agree with the hypotheses of the origin of Jupiter's sporadic bursts of radio emission; as we have shown in the preceding section, the most probable ideas on the mechanism and conditions of generation of the bursts require the existence on the planet's surface of a far smaller field (about 5·5 at a distance $R \approx R_{\jupiter}$ from the centre of the magnetic dipole, which gives a value of $H_0 \sim 0·2$ oe in the region of the radiation belt $R \sim 3R_{\jupiter}$).

Further, it is appropriate to draw attention to the circumstance that the actual fact of the linear polarization of Jupiter's radio emission imposes definite limitations on the value of the planet's magnetic field, since with large enough magnetic fields depolarization of the observed radio emission will occur because of Faraday effect in the source.

As shown in section 23, the depolarization caused by rotation of the plane of polarization in the radio emission source will be insignificant only if

$$\frac{4·7 \cdot 10^4}{f^2} NH_0 \cos \alpha \, \Delta L < 1 \ .$$

(see the formula (23.28)). In the case of interest to us of Jupiter's radiation belts the dimension ΔL characterizing the difference in travel between waves originating from different points in the source can be taken as equal to $\sim 3R_{2\!\!\!\downarrow}$.[†] Then from the condition given we at once obtain that linear polarization at $\lambda \approx 30$ cm ($f \approx 10^9$ c/s) can be observed only if the magnetic field H_0 and the concentration of the plasma at the level of the belts (i.e. at a distance $R \sim 3R_{2\!\!\!\downarrow}$ from the centre of the planet) satisfy the inequality

$$NH_0 \cos \alpha < 10^3. \qquad (33.3)$$

The inequality (33.3) can be used to make a definite choice between the non-relativistic and relativistic electron models, giving preference to the latter. In fact, non-relativistic electrons require magnetic fields of $H_0 \sim 3.5 \times 10^2$ oe in the source for generation of emission at a frequency of $f \sim 10^9$ c/s. Then the condition (33.3) will be satisfied (with $\cos \alpha \sim 1$) only if the electron concentration of the plasma at a distance of up to two radii from the planet's surface is less than 3 electrons/cm^3. The permissible upper limit for N may rise at the expense of $\cos \alpha$; it is very improbable, however, that it exceeds 30 electrons/cm^3. Since the actual value of N is apparently higher, we come to the conclusion that the idea of a non-relativistic radiation belt contradicts the polarization measurements of Jupiter at $\lambda \sim 30$ cm. (It would be very interesting to carry out measurements of the polarization at longer wavelengths, let us say at $\lambda \sim 60$ cm; this would make it possible to impose a more rigid condition on N). At the same time for relativistic electrons and fields of $H_0 \sim 0.2$ oe the condition (33.3) is easily satisfied with quite reasonable limitations on the particle concentration: $N < 5 \times 10^3$ electrons/cm^3 for $\cos \alpha \sim 1$.

Taking as our starting-point a model of relativistic electrons emitting during motion in a Jovian magnetic field with a strength $H_0 \sim 0.2$ oe, we shall estimate the mean energy $\mathcal{E}$ and concentration N_s of these electrons which are necessary for interpreting the observed decimetric radio emission.

It is not hard to find the value of $\mathcal{E}$ from the formula (26.48), putting in it the gyro-frequency $\omega_H \sim 3.5 \times 10^6$ sec^{-1} ($H_0 \sim 0.2$ oe) and the frequency corresponding to the maximum intensity of the synchrotron emission $\omega_{\max} \sim 2\pi \times 10^9$ sec^{-1} (i.e. $\lambda \sim 30$ cm).[‡] Then

$$\mathcal{E} \sim 60 \, mc^2 \sim 30 \, MeV. \qquad (33.4)$$

[†] The quantity $3R_{2\!\!\!\downarrow}$ is the difference of travel for two rays—from the central part of Jupiter's disk and from the periphery of the source.

[‡] Apparently the maximum of the frequency spectrum of Jupiter's decimetric radio emission is at $\lambda \sim 20$–30 cm, although it is possible that it is displaced away from this value towards the longer wavelengths.

§ 33] Continuous Radio Emission of Jupiter and Saturn

Let us now discuss the concentration of the emitting particles. In accordance with (26.45) (see also Fig. 136) the energy of the electron's synchrotron emission at the maximum of the spectrum is

$$\frac{d\mathcal{E}_\omega}{dt} \approx \frac{0.8}{\pi}\frac{e^2}{c}\omega_H. \tag{33.5}$$

Then the emissive power of a unit volume in a 1 c/s range is $N_s(d\mathcal{E}_f/dt) = 2\pi N_s(d\mathcal{E}_\omega/dt)$, and the total energy of the emission of the whole radiation belt in the same range will be $N_s V_s(d\mathcal{E}_f/dt)$.[†] Considering for the sake of simplicity that this energy is evenly distributed in all directions, we find that the radio emission flux at a distance R from Jupiter is

$$S_f \approx \frac{N_s V_s}{4\pi R^2}\frac{d\mathcal{E}_f}{dt} \approx \frac{0.4\,N_s V_s}{\pi}\frac{e^2}{R^2\,c}\omega_H. \tag{33.6}$$

From this it is easy to find the value of N_s. Saying that the decimetric radio emission flux observed at a distance of $R = 5.2$ a.u. $\approx 7.8 \times 10^{13}$ cm is $S_f \approx 4.3 \times 10^{-26}$ W m^{-2} c/s^{-1} = 4.3×10^{-23} erg sec^{-1} cm^{-2} c/s^{-1}, and considering that the volume of the source is $V_s \sim 10 V_{2\!\!\!\;\text{l}} \sim 1.4 \times 10^{31}$ cm^3 ($V_{2\!\!\!\;\text{l}}$ is the volume of Jupiter), $\omega_H \sim 3.5 \times 10^6$ sec^{-1}, we obtain:

$$N_s V_s \sim 10^{29} \text{ electrons}, \quad N_s \sim 10^{-2} \text{ electron/cm}^3. \tag{33.7}$$

When investigating Jupiter's radiation belt of relativistic electrons there is also great interest in a more detailed comparison of the characteristics of their synchrotron emission (calculated on the basis of definite models of the belts) with the observed features of the decimetric radio emission. Without stopping to discuss this important question we refer the reader to Chang and Davis (1962 ; see also Korchak, 1963a and 1963b). All we shall say here is that the basic feature of these papers is finding the Stokes parameters for the total emission of a system of relativistic particles in a dipole magnetic field from the known Stokes parameters for the emission of an individual electron.[‡] The calculations are simplified here by discussing only a thin radiation belt, i.e. a belt whose extent along the planet's

---

[†] Here we are simply summing the intensity of the emission from all the particles making up the radiation belts without allowing for reabsorption. The energies of the emitting particles (~ 30 MeV) correspond to a temperature of $T \sim \mathcal{E}/\varkappa \sim 3 \times 10^{11}\,^\circ$K, whilst the effective radio emission temperature is seven orders lower. It is clear from a comparison of these temperatures that the radiation belt should be optically thin, which makes the reabsorption insignificant (see section 26). The difference of the particle distribution in the source from an equilibrium one does not alter this conclusion in practice because of the high value of the ratio $\mathcal{E}/\varkappa T_{\text{eff}}$.

[‡] For the Stokes parameters which fully characterize the radio emission intensity and polarization see section 6.

radius is small when compared with the distance from the belt to the magnetic dipole. If we consider in a good approximation that the electrons in the radiation belt move along the magnetic field's lines of force in conditions which ensure conservation of the adiabatic invariant $\sin^2 \psi / H_0 = $ const (ψ is the angle between the magnetic field $\boldsymbol{H}_0$ and the electron's velocity), then to obtain the distribution function of the particles $f(R, \psi, \mathcal{E})$ at each point it is sufficient to give it in Jupiter's equatorial plane. It is interesting to note that for the distribution

$$f_s(\mathcal{E}, \bar{\psi}) = \begin{cases} F(\mathcal{E}) \sin^p \bar{\psi} & \text{when} \quad |\bar{\psi}| > \bar{\psi}_{\min}, \\ 0 & \text{when} \quad |\bar{\psi}| < \bar{\psi}_{\min} \end{cases} \quad (33.8)$$

the particle density along a line of force remains constant only when $p = 1$ and $\bar{\psi}_{\min} = 0$, i.e. for an isotropic velocity distribution.[†] If, however, electrons predominate in the belt with velocities subtending large angles with the magnetic field (i.e. $p > 1$ or $\bar{\psi}_{\min} > 0$), then the density decreases towards the poles.[‡] This result is quite understandable since, in accordance with the adiabatic invariant, electrons moving in the plane of the equator at an angle $\bar{\psi}$ to $\bar{\boldsymbol{H}}_0$ cannot penetrate a strong field region, being "reflected" backwards along a line of force at the point $H_0 = \bar{H}_0 \sin^{-2} \bar{\psi}$.

According to Chang and Davis (1962), to achieve the observed degree of polarization ($\varrho_l \sim 20$–30%) it must be assumed that the majority of the electrons have almost circular orbits ($\bar{\psi} \sim \pi/2$). Then in accordance with what we have said the density of the emitting particles will decrease strongly towards the poles; this will lead (in accordance with experiment) to small dimensions of the source in the polar direction. The energy spectrum of the electrons, unlike the distribution with respect to the angles ψ, has a weak effect on the value of the resultant degree of polarization; however, the nature of the radio emission's frequency spectrum is dependent on it in the first place. If we assume that $f_s \propto \mathcal{E}^{-\gamma}$, then the observed emission flux is $S_f \propto \omega^{-\gamma'}$, where $\gamma' = (\gamma-1)/2$. To judge from the available experimental data the spectrum $S_f(\omega)$ is rather sloping and the value of γ' apparently lies in the range $-\frac{1}{2} < \gamma' < \frac{1}{2}$. It follows from this that the probable spectrum of the relativistic electrons (with an exponential approximation of the latter) corresponds to a value of $\gamma \approx 0$–2.

The question arises in connection with the above of the acceleration mechanism of the electrons comprising the radiation belts or of the external source supplying the relativistic electrons to Jupiter's magnetic field. This is still an open question at present; there is a preliminary discussion of Field (1959). It should be pointed out that it was the difficulties with the

---

[†] The bar on top shows that the quantity relates to the plane of the equator.
[‡] In the opposite case the density at high latitudes increases.

§ 33] Continuous Radio Emission of Jupiter and Saturn

solution of this problem that made Field turn to a non-relativistic model although, as we have shown above, it does not meet the facts. This is also recognized by Field (1961) (proceeding from different considerations).

Any further development of the representation of relativistic electrons in the Jovian radiation belts as the source of Jupiter's decimetric radio emission is obviously connected with the answer to the question of the reasons for the appearance of a large enough number of energetic electrons in the vicinity of Jupiter on the one hand, and with determining the characteristics of these electrons (above all their energy spectrum and distribution in space) more accurately on the other. Progress in the solution of the last problem is closely connected with obtaining fuller information about the frequency spectrum of Jupiter's continuous radio emission and about the radio brightness distribution over the source.

CONDITIONS OF GENERATION OF SATURN'S RADIO EMISSION

Jupiter's radio emission in the decimetric band is generated in radiation belts; it is linearly polarized in the plane of the equator with a 20–30% degree of polarization. The first polarization measurements of Rose, Bologna and Sloanaker (1963a and 1969b) also indicated the high degree of linear polarization of Saturn's radio emission. This fact (together with the enhanced level of radio emission at $\lambda \sim 10$ cm) assumed the observed radio emission as well as Jupiter's to arise from radiation belts of the planet. Later investigations (e.g. Davies, Beard and Cooper, 1964) did not confirm the the occurrence of linear polarization at $\lambda = 10$ cm; the source size at these wavelengths appeared to be not more than the optical diameter of the planet (Berge and Read, 1967). Therefore, at present Saturn's radio emission is believed to be associated with the hot subcloud layers of the planet.

It is necessary to empasize nevertheless that the situation can change at longer wavelengths ($\lambda \sim 20$ cm): it is not impossible that the radiation belts become noticeable in this band. However Drake (1962b) has expressed doubt that Saturn's radiation belts really exist at altitudes of $h < 2R_h$ since collisions of energetic charged particles with elements of the ring in the plane of the equator lead to rapid escape of these particles from the dipole "magnetic trap" and to breakdown of the radiation belts. A way out of these difficulties has been indicated by Zlotnik (1967). She has drawn attention to the fact that the effect of the magnetic field being drawn away by a plasma in the exosphere of Saturn (which was discussed previously by Zheleznyakov (1964b and 1964c)) may lead to a deformation of the planet's magnetic field which will make the existence of radiation belts possible. The point is that under certain conditions the planet's magnetic field, which is probably of the nature of a dipole near its surface (with the dipole pointing along the axis of rotation), is considerably deformed in the region of Saturn's rings. The lines of magnetic force slope considerably towards the plane of the rings (and do not remain at right angles to it as at low altitudes) because of the differential rotation of the plasma in the planet's exosphere. This kind of rotation may be connected with the plasma being drawn away by particles making up Saturn's rings. This deformation of the magnetic field leads to its strengthening in regions close to the equatorial plane: the latter obviously hinders charged particles from approaching the rings and from being absorbed in this region. It is clear that under these conditions Saturn's radiation belts may exist in "magnetic traps" between the polar and equatorial regions of the planet.

In order to approach the problem of the effect of Saturn's rings on the exosphere of the planet and in particular to find how realistic this suggestion is, let us first discuss the following idealized problem. In the plane $z = 0$ there is an external force $\mathcal{F} = \boldsymbol{F}\delta(z)$ acting on a conducting incompressible medium with a conductivity σ and a viscosity η (see Fig. 170). This force, which runs along x, characterizes the action of Saturn's rings on the plasma in the planet's exosphere. The plasma is in an external magnetic field $\boldsymbol{H}_0$ running along z. We are to find the resultant magnetic field established when the medium is moved by the force $\mathcal{F}$. We notice that an approximation of the function $\mathcal{F}(z)$ in the form of a δ-function is quite natural here since we are interested in changes of the field configuration on a scale of the order of the planetary radius, i.e. at intervals far greater than the thickness of Saturn's rings.

The initial system of magnetohydrodynamic equations is of the form (Landau and Lifshitz, 1957, section 51)

$$\operatorname{div} \boldsymbol{H} = 0, \quad \operatorname{div} \boldsymbol{V} = 0, \tag{33.9}$$

$$\frac{\partial \boldsymbol{H}}{\partial t} + (\boldsymbol{V}\nabla)\boldsymbol{H} = (\boldsymbol{H}\nabla)\boldsymbol{V} + \frac{c^2}{4\pi\sigma}\Delta\boldsymbol{H}, \tag{33.10}$$

$$\varrho\left\{\frac{\partial \boldsymbol{V}}{\partial t} + (\boldsymbol{V}\nabla)\boldsymbol{V}\right\} = -\nabla\left(p + \frac{H^2}{8\pi}\right) + \frac{1}{4\pi}(\boldsymbol{H}\nabla)\boldsymbol{H} + \eta\Delta\boldsymbol{V} + \mathcal{F}. \tag{33.11}$$

The equation $\operatorname{div} \boldsymbol{V} = 0$ is satisfied since the velocity $\boldsymbol{V}$ does not depend on x or y (like all the other variables) and has only an x-component. It follows from $\operatorname{div} \boldsymbol{H} = 0$ that

$$H_z = \text{const} = H_0. \tag{33.11a}$$

Further, the z-component of the equation (33.11) gives $p + H^2/8\pi = \text{const}$ and the x-components of the equations (33.10) and (33.11) can be written

§ 33] Continuous Radio Emission of Jupiter and Saturn

in the form

$$H_0 \frac{dV_x}{dz} + \frac{c^2}{4\pi\sigma} \frac{d^2H_x}{dz^2} = 0, \quad (33.12)$$

$$\eta \frac{d^2V_x}{dz^2} + \frac{H_0}{4\pi} \frac{dH_x}{dz} = -\mathcal{F}(z) = -F\delta(z). \quad (33.13)$$

Solving these equations for the condition that the lines of force of the magnetic field are symmetrical relative to the plane $z = 0$, and the velocity $V_x = 0$ when $z = \pm\infty$, we obtain (for further detail see Zheleznyakov, 1964b and 1964c):

when $z > 0$
$$H_x = -\frac{2\pi F}{H_0}(1 - e^{-\beta z}), \quad V_x = \frac{cF}{2H_0\sqrt{\eta\sigma}} e^{-\beta z}, \quad (33.14)$$

when $z < 0$
$$H_x = \frac{2\pi F}{H_0}(1 - e^{\beta z}), \quad V_x = \frac{cF}{2H_0\sqrt{\eta\sigma}} e^{\beta z}. \quad (33.15)$$

The parameter β figuring here is defined by the expression

$$\beta = \frac{H_0}{c}\sqrt{\frac{\sigma}{\eta}}. \quad (33.16)$$

It follows from (33.14), (33.15) that the values of the velocity in the plane $z = 0$ and of the magnetic field's x-component at infinity are connected with each other by the relation

$$H_x(z = \infty) = -\frac{4\pi}{c}\sqrt{\eta\sigma}\, V_x(z = 0). \quad (33.17)$$

Therefore outside the plane of the ring where $\mathcal{F}(z) = 0$ the velocity of the plasma V_x decreases exponentially as it moves away from $z = 0$, whilst the field H_x rises in absolute magnitude from zero when $z = 0$, approaching the constant value (33.17) exponentially also. The resultant configuration of the magnetic field's lines of force is shown in Fig. 170. It follows from it that the magnetic field changes its direction significantly only at large enough distances from the plane $z = 0$, i.e. when $|z| \gtrsim 1/\beta$.

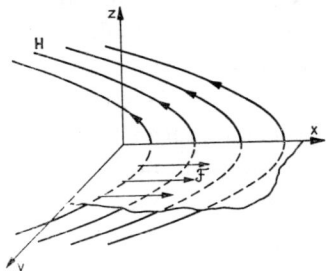

Fig. 170. Problem of magnetic field drawn away by plasma

We do not know the magnitude of the force F (see, however, the remarks at the end of the section); in the final event F is determined by the parameters of the ring (size of the structural elements, their form and number) and by the parameters of the gas that is being drawn away. In future we shall assume that due to interaction between the plasma and the ring the values of the gas velocities $V_x(z = 0)$ and of the ring relative to the magnetic field are close in order of magnitude (effective attraction). The velocity of the ring relative to the field H connected with the planet's solid body is about 7·3 cm/sec on the outer edge of ring A (i.e. at a distance of $1·4 \times 10^5$ km from the centre of Saturn) and is close to zero at a distance of $1·1 \times 10^5$ km (approximately in the region of Cassini's slit). Closer to the surface of Saturn the relative velocity changes sign, reaching values of about 4·75 km/sec on the inner edge of ring B (at a distance of 9×10^4 km). It follows from this that if the synchrotron component of Saturn's radio emission exists, it may be generated (from the point of view of the hypothesis under consideration) at distances of $R \sim 9 \times 10^4 - 1·4 \times 10^5$ km (in the equatorial direction), for preference in the middle of rings A and B, which is the only place where we can expect significant distortion of the planet's magnetic dipole field.

It is clear from what has been said that for significant strengthening of the magnetic field on both sides of the ring it is sufficient to satisfy two conditions simultaneously: (1) for the slope of the magnetic field to become maximal at distances from the plane of the ring less than the radius of Saturn $R_h \approx 6 \times 10^4$ km

$$\beta R_h \equiv \frac{H_0 R_h}{c} \sqrt{\frac{\sigma}{\eta}} \gg 1. \tag{33.18}$$

(2) for the maximum value of H_x to be large enough when compared with the unperturbed field H_0:

$$\frac{4\pi V_x(z=0)}{cH_0} \sqrt{\eta \sigma} > 1 \tag{33.19}$$

(see (33.14)–(33.17)). The first of these conditions becomes comprehensible if we bear in mind that any significant effect of the rings on the magnetic field will not actually spread a distance from the equatorial plane greater than or of the order of the width of the rings (i.e. $\gtrsim R_h$).

In a magnetoactive plasma (Saturn's exosphere) the conductivity σ and the viscosity η are anisotropic quantities. It is clear why the results (33.18)–(33.19), which are strictly speaking obtained for a conducting liquid, cannot be transferred directly to the plasma motion across a magnetic field of interest to us. Actually the anisotropy of the conductivity should be allowed for right from the start; however, for rough estimates for σ we

§ 33] Continuous Radio Emission of Jupiter and Saturn

use the formula (Pikel'ner, 1961, section 8)

$$\sigma = \frac{e^2 N}{m \nu_{\text{eff}}} \left[1 + \frac{A^2}{K_1 K_2 + K_1 K_3 + K_2 K_3} \right]^{-1}. \tag{33.20}$$

This expression characterizes the dissipation in a partially ionized gas provided that the current j is orthogonal to the field H. The choice of σ in the form (33.20) is apparently entirely reasonable for our problem: the "sliding" of the magnetic field's lines of force relative to the plasma is connected with dissipative processes, the resultant configuration of the field H in the case under discussion, as can be easily understood from the equation

$$\text{curl } H = \frac{4\pi}{c} j, \tag{33.21}$$

is created by currents j orthogonal to H.

The following notations are used in (33.20): A is a parameter characterizing the degree of ionization and is equal to the ratio of the concentration of neutral molecules N_m to the total concentration of heavy particles $N_m + N_i$ (N_i is the ion concentration), ν_{eff} is effective frequency of electron collisions, equal to $\nu_{ei} + \nu_{em}$,

$$K_1 = \frac{\nu_{ei}}{\omega_H}, \quad K_2 = \frac{\nu_{em}}{\omega_H}, \quad K_3 = \frac{\nu_{im}}{\omega_{H_i}}, \tag{33.22}$$

ν_{ei} and ν_{em} are the frequency of collisions of an electron with ions and molecules respectively, ν_{im} is the same for ions with molecules, $\omega_H = eH/mc$ and $\omega_{H_i} = eH/m_i c$ are the gyro-frequencies of the electrons and ions.

In stationary conditions the following expressions are valid for the collision frequencies (Ginzburg, 1960b, section 9):

$$\nu_{ei} = \frac{2 \cdot 8 N}{T^{3/2}} \ln \left(324 \frac{T}{N^{1/3}} \right), \tag{33.23}$$

$$\nu_{em} = \frac{3\pi}{8} \pi r^2 \bar{V} N_m, \tag{33.24}$$

$$\nu_{mm} = \nu_{im} = \frac{16\sqrt{2}}{3} \pi r^2 \bar{V}_i N_m, \tag{33.25}$$

where $\bar{V} = \sqrt{8 \varkappa T/\pi m}$, $\bar{V}_i = \sqrt{8 \varkappa T/\pi m_i}$, πr^2 is the cross-section of the molecular collision.

The formula (33.20) for the conductivity can be considerably simplified if we remember that $K_1 \gg K_2$, $K_3 \gg K_2$ (see (33.22)–(33.25)) and assume that $A \approx 1$:

$$\sigma \approx \frac{e^2 N}{m \nu_{\text{eff}}} \left[1 + \frac{1}{K_1 K_3} \right]^{-1} \approx \frac{e^2 N}{m \omega_H^2} \nu_{im} \frac{m_i}{m}. \tag{33.26}$$

In the change to the last expression we have borne in mind that for values of the magnetic fields that are not too small $\omega_H \gg \nu_{ei}$, $\omega_{H_i} \gg \nu_{im}$, and with values of the ratio N_m/N that are not too high the number of electron collisions is $\nu_{\text{eff}} \approx \nu_{ei}$.

As for the viscosity η, in a preliminary analysis it can be left the same as in an isotropic medium, since the velocity V varies in the direction of the magnetic field (for weak perturbations of the field $\boldsymbol{H}_0$):

$$\eta \approx \frac{N_m \varkappa T}{\nu_{mm}}. \tag{33.27}$$

Here we have already assumed that the concentration of molecules is $N_m > N$; in this case the contribution of the ions and electrons to the magnitude of the viscosity is insignificant and the molecules collide largely with molecules (number of collisions ν_{mm}).

Allowing for the remarks made above on the conductivity and viscosity of the plasma in Saturn's exosphere the necessary conditions for significant deformation of the magnetic field (33.18), (33.19) will become:

$$\sqrt{NN_m} \gg \frac{3}{64\sqrt{\pi r^2 R_\text{h}}}, \tag{33.28}$$

$$H_0^2 < 4\pi V_x(z=0)\sqrt{m_i \varkappa T N N_m}. \tag{33.29}$$

Putting in Saturn's exosphere, which apparently consists chiefly of hydrogen,† $\pi r^2 \approx 2{\cdot}2 \times 10^{-16}$ cm² (Von Engel, 1955), $m_i \approx 3{\cdot}3 \times 10^{-24}$ g, $T \sim 10^4\,°$K, we obtain (with $V_x(z=0) \sim 5 \times 10^5$ cm/sec and $R_\text{h} \approx 6 \times 10^9$ cm) that

$$\sqrt{NN_m} \gg 6{\cdot}3 \times 10^4 \text{ cm}^{-3}, \tag{33.30}$$

$$H_0^2 < 1{\cdot}3 \times 10^{-11} \sqrt{NN_m} \text{ oe}^2. \tag{33.31}$$

For example, in the case of $N/N_m \sim 0{\cdot}1$ the criterion (33.30) will be satisfied if $N \gg 2 \times 10^4$ electrons/cm³, $N_m \gg 2 \times 10^5$ particles/cm³. Requirements of this kind on the value of N and N_m at distances of about half to one radius of the planet from its surface are rather rigid. It is possible, however, that the gaseous component of Saturn's rings makes a significant contribution to the enhanced values of N and N_m.

---

† Here the estimates are made for molecular hydrogen. However, the position is not significantly altered if we consider that the hydrogen is not in a molecular (H_2) but in an atomic (H) state.

§ 33] Continuous Radio Emission of Jupiter and Saturn

We notice that the inequality (33.30) does not have to be very strong: for distortion of the magnetic field in most of the region $|z| \lesssim R_\hbar$ it is quite sufficient for the value of $\sqrt{NN_m}$ to be about four times greater than $6\cdot3\times10^4$ cm^{-3}, coming to around $2\cdot5\times10^5$ cm^{-3}. Then from (33.31) we obtain the following limitation on the value of the magnetic field in the equatorial plane of Saturn at altitudes of $\eta \sim R_\hbar$:

$$H_0 < 2\times10^{-3} \text{ oe.} \tag{33.32}$$

At the poles this corresponds to a field strength of less than approximately 3×10^{-2} oe. On the other hand, the field H_0 should not be too small either (let us say, fall to the interplanetary level or less);[†] in the opposite case it cannot act as a "trap" which will effectively retain the emitting relativistic electrons making up the radiation belts.

Furthermore, we must find the value of the force acting from the ring's structural elements on the surrounding gas. This allows us using a formula of the type

$$V_x(z=0) = \frac{cF}{2H_0\sqrt{\eta\sigma}}, \tag{33.33}$$

which follows from (33.14)–(33.15) to judge the extent to which the plasma is drawn out by the ring, which above was simply postulated. On the other hand, in the case of experimental confirmation of this configuration of the magnetic field the solution of the second problem will make it possible to judge such parameters of Saturn's ring as the number and size of the particles comprising it.

We shall now make a few remarks about the magnitude of the force F. Its precise calculation is complicated; as a guide, however, we can adduce the often used formula

$$f \approx \pi r_0^2 \varrho V_0^2, \tag{33.34}$$

in which f is the resistance force during the motion of a body with a cross-section πr_0^2 in a gas at a velocity V_0 ($\varrho \approx m_i N_m$ is the density of this gas). If we denote the number of particles in a ring in a unit volume by N_0, then obviously

$$F \approx fN_0 a \tag{33.35}$$

(a is the thickness of Saturn's rings).

[†] In the Earth's orbit the interplanetary magnetic field is of the order of 10^{-4} oe. In the vicinity of Saturn, which is at a distance 9·5 times further from the Sun, this field is undoubtedly far weaker.

At the same time it follows from (33.33) that for the medium to be drawn away by the ring at a velocity $V_x(z = 0)$ we need a force

$$F = \frac{2H_0\sqrt{\eta\sigma}}{c} V_x(z=0), \tag{33.36}$$

which satisfies the inequality

$$F > \frac{H_0^2}{2\pi} \tag{33.37}$$

(see (33.19)). In accordance with (33.32) $H_0 < 2\times 10^{-3}$ oe. For the sake of definition making the magnetic field in the region of the ring 10^{-3} oe, we obtain that the value of F should exceed $1\cdot 6\times 10^{-7}$ dyne/cm. It follows from (33.34), (33.35) that values of this kind for F will be reached if $N_0 r_0^2 > 3\times 10^{-7}$ (with $N_m \sim 10^6$ cm^{-3}, $m_i = 3\cdot 3\times 10^{-24}$ g, a ring thickness of $a \sim 5\times 10^6$ cm and a "sliding" velocity of the gas relative to the ring's particles of $V_0 \sim 10^5$ cm/sec).

Unfortunately there are no certain data at present on the concentration and size of the structural elements of Saturn's rings. We have available only approximate information obtained on the basis of the photometric theory of Saturn's rings (Bobrov, 1951, 1952, 1954, 1956); according to this information the mean values are $r_0 \sim 50$ cm and $N_0 \sim 2\times 10^{-9}$ cm^{-3}. The factor $N_0 r_0^2$ will then be of the order of 5×10^{-6}; this is quite sufficient for effective attraction of a gas with $N_m \sim 10^6$ cm^{-3} by the ring (compare with the estimate of $N_0 r_0^2 > 3\times 10^{-7}$ obtained above).

34. Sources of Venus's Radio Emission

At present the investigation of Venus's radio emission, as well as radar experiments, is rightly looked upon as one of the basic sources of information about this planet. This situation is rapidly changing, as these methods begin to yield pride of place to active study of Venus by systematic flights past the planet at close ranges, landing containers of scientific apparatus on the surface, etc.

The theory of the radio emission of Venus should connect its features with the physical conditions in the planet's outer layers (in the atmosphere and on its surface). This problem is very complex and still far from satisfactory solution, above all because of the limited extent of our knowledge of the nature of Venus obtained by optical and infrared observations.

The main point in the theory of Venus's radio emission is the question of the localization of the layers which are responsible for the creation of the observed radio emission.

§ 34] Sources of Venus's Radio Emission

When speaking of radio emission from Venus we note first of all that it cannot be connected with the planet's radiation belts. This is indicated both by the non-polarized nature of the emission and by the closeness of the planet's radio diameter to its optical diameter (see section 20).

If the radiation belts are not the source of the observed radio emission the latter can be generated only by the ionosphere, the lower atmosphere and the planet's surface. In this case the radio emission is undoubtedly thermal; the values of the effective temperature $T_{\text{eff}\,♀} \approx 660°K$ observed at $\lambda \sim 1\cdot5\text{–}40$ cm then indicate the presence on the planet of regions heated to a temperature of not less than $600°K$, i.e. far more strongly than follows from infrared observations of the cloud layer.

For interpreting the frequency spectrum of the Cytherean radio emission ($T_{\text{eff}\,♀} \approx 350°K$ at millimetric wavelengths and $T_{\text{eff}\,♀} \approx 600°K$ in the centimetric–decimetric band; see Fig. 102) basically two models are suggested, which are generally called the "ionospheric" model and the "hot" surface model. In the first of them it is assumed that the comparatively "cold" ($T \approx 350°K$) surface of the planet and the layer of the atmosphere adjacent to it create millimetric radio emission, whilst the "hotter" ionosphere ($T \approx 600°K$) provides the enhanced values of $T_{\text{eff}\,♀}$ at the longer wavelengths. In the second model the source of the centimetric–decimetric radio emission is the "hot" ($T \approx 600°K$) surface, whilst the "colder" layers ($T \approx 350°K$) of the dense atmosphere generate the radio emission at millimetric wavelengths. We shall speak later about certain other attempts to explain Venus's radio emission.†

THE "IONOSPHERIC" MODEL

A preliminary discussion of the "ionospheric" model by Zheleznyakov (1959a) showed that it can explain the observed frequency spectrum only with very high values of the electron concentration in Venus's atmosphere. This gave foundation for concluding that this kind of model is improbable. Subsequently the ionospheric model was investigated in a number of papers (Jones, 1961; Sagan, Siegel and Jones, 1961; Verozub, 1962; Danilov and Yatsenko, 1962 and 1963; Kuz'min, 1963) starting with Jones's paper (1961); the authors' basic efforts were directed at choosing the ionosphere's parameters so as to achieve agreement with the data of the radio observations and to finding the reasons for the upper atmosphere's high degree of ionization. The "hot" surface model was been investigated by Sagan (1960), Öpik (1961) and Barrett (1960 and 1961). The latter calculat-

---

† The radio emission of Venus is not, of course, connected with the visible layer of clouds, since $T_{\text{eff}\,♀}$ at radio wavelengths is higher than the temperature of this layer ($235°K$).

ed the expected ratio emission with a given atmospheric composition and density and also with a definite temperature distribution on the surface of Venus, whilst the two former chiefly studied the reasons for the high temperature of the planet's surface. A comparative analysis of the two models has been carried out by Kellogg and Sagan (1961).

In the framework of the "ionospheric" model the effective radio emission temperature is[†]

$$T_{\text{eff}} = T_s(1-R^2)e^{-\tau_i} + T_i(1-e^{-\tau_i}) \qquad (34.1)$$

where T_s and T_i are the temperatures of the surface and the ionosphere respectively, R^2 is the coefficient of reflection of radio waves from the planet's surface, τ_i is the optical thickness of the ionosphere. The latter is equal to $\int \tau_{1,2}\, dl$ where the absorption coefficient in the plasma because of collisions is

$$\mu_{1,2} \approx \frac{\omega_L^2 \nu_{\text{eff}}}{c\omega^2} \approx 2 \cdot 7 \times 10^{-3} \frac{N\nu_{\text{eff}}}{f^2} \qquad (34.2)$$

(when $n_{1,2} \approx 1$; see (26.74)).

It follows from (34.1) that with $\tau_i \gg 1$

$$T_{\text{eff}} \approx T_i, \qquad (34.3)$$

whilst with $\tau_i \ll 1$

$$T_{\text{eff}} \approx T_s(1-R^2). \qquad (34.4)$$

Since the optical thickness of the ionosphere, in accordance with (34.2), decreases as the frequency f rises, the case (34.3) occurs at the longer wavelengths and the case (34.4) at the shorter ones. It is clear from this that the function $T_{\text{eff}}(\lambda)$ described by the formula (34.1) will correspond to the observed frequency spectrum $T_{\text{eff}\,♀}(\lambda)$ if on Venus[‡]

$$T_s(1-R^2) \approx 350°K, \quad T_i \approx 600°K, \qquad (34.5)$$

and the characteristic wavelength λ^* for which the optical thickness of the ionosphere is $\tau_i = 1$ is about 1·3 cm.

From the last condition it is easy to find the measure $\int N^2\, dl$ of the emission of the Cytherean ionosphere and estimate the necessary electron concentration N. Considering that on Venus, as in the Earth's ionosphere, the

---

[†] On the subject of this formula see section 26, in particular the expression (26.18). The factor $1-R^2$ in (34.1) allows for the fact that part of the radio emission does not escape through the surface into the atmosphere but is reflected back.

[‡] Therefore the actual temperature of the surface in the "ionospheric" model differs slightly (on the high side) from the values of T_{eff} in the millimetric waveband because of the difference from zero of the reflection coefficient R^2.

§ 34] Sources of Venus's Radio Emission

effective number of collisions ν_{eff} is largely determined by the collisions of electrons with ions and not with neutral molecules, we have:

$$\nu_{\text{eff}} \approx \nu_{\text{ei}} = \frac{5 \cdot 5 N}{T_i^{3/2}} \ln\left(220 \frac{T_i}{N^{1/3}}\right) \tag{34.6}$$

(see (26.78)). Then, remembering (34.2), (34.6), we obtain:

$$\int N^2 \, dl \approx 6 \times 10^{22} \frac{T_i^{3/2}}{(\lambda^*)^2 \ln(220 T_i N^{-1/3})} \tag{34.7}$$

(the slowly varying logarithmic function is removed from the integrand here). Putting $T_i \approx 600°K$, $\lambda^* \approx 1 \cdot 3$ cm and taking in the logarithm $N \sim 10^9$ electrons/cm^3 we find:

$$\int N^2 \, dl \sim 10^{26} \text{ electron}^2/\text{cm}^5. \tag{34.8}$$

When the thickness of the ionospheric layer is $2 \cdot 5 \times 10^7$ cm this gives a value for the electron concentration of $N \sim 2 \times 10^9$ electrons/cm^3, which is three orders higher than the concentration at the F-layer maximum on Earth and one order greater than the value of N at the basis of the minimum form of the solar corona!

As we know, for a simple ionospheric layer the electron concentration at the layer's maximum is

$$N_{\max} = \sqrt{\frac{J}{\alpha_{\text{eff}}}},$$

where J is the number of electrons appearing in a unit volume in unit time because of the action of ionizing factors, α_{eff} is the effective recombination coefficient. Under terrestrial conditions the basic factor causing ionization of the upper atmosphere is the ultraviolet and X-ray part of the solar spectrum. There is no reason to assume that the ionizing ability of the Sun's rays on Venus is essentially different[†] from the value of J on Earth which from various estimates is 25–2000 (Al'pert, 1960; section 7). Therefore maintenance of the value of $N_{\max}$ in the ionosphere of Venus at a level of 2×10^9 electrons/cm^3 by solar radiation is possible only provided that $\alpha_{\text{eff}} \sim 10^{-18}$–$10^{-15}$ cm^3/sec. In section 32, however, we have already said that such values of α_{eff} are absolutely unrealistic: in actual fact α_{eff} never drops lower than 10^{-12} cm^3/sec (the value for radiation recombination), generally being noticeably greater than this value. For example, in the

---

[†] The fact that Venus is closer to the Sun leads to an increase in J by a factor of only 2.

Earth's ionosphere $\alpha_{\text{eff}} \sim 5\times10^{-7}\text{--}3\times10^{-10}$ cm³/sec depending on altitude (Al'pert, 1960, section 7).†

It is clear from what we have said that the presence on Venus of a large enough ionosphere for interpreting the data on the radio emission of this planet is possible only if there is an ionization source there many orders more efficient than the solar radiation.‡ With $\alpha_{\text{eff}} \sim 10^{-12}$ cm³/sec the value of J on Venus should be three to six orders greater than its terrestrial value, and for a more realistic recombination coefficient close to the α_{eff} in the Earth's ionosphere five to eleven orders greater!

Jones (1961) examined solar corpuscular streams as the possible ionizing agent of the upper atmosphere of Venus. On Earth these streams largely do not reach the ionosphere because of the screening action of the geomagnetic field (the high-latitude regions—the auroral zones—are an exception in this respect). It is possible that there is no such screening on Venus because of the low strength of its magnetic field.

If all the energy of the stream of particles (protons) is expended on ionizing the upper atmosphere of Venus, then the production J of electrons will obviously be close to $\mathcal{E} N_s V_s/\varepsilon_i L$, where $\mathcal{E}$ is the energy of one particle invading the ionosphere, V_s is the velocity of this particle, N_s is the concentration of the corpuscular stream, L is the thickness of the ionospheric layer being formed, ε_i is the energy expended in one act of ionization. Putting $N \sim 10^3$ electrons/cm³, $V_s \sim 10^8$ cm/sec, $\mathcal{E} \sim 10^3$ eV (see section 17), $\varepsilon_i \sim 15$ eV, $L \sim 2\times 10^7$ cm, we obtain $J \sim 3\times 10^5$ electrons/cm³ sec. It is clear from this that a corpuscular agent can sustain an ionosphere of the required concentration if $\alpha_{\text{eff}} \sim 10^{-13}$. There is still one order missing for the minimum value $\alpha_{\text{eff}} \sim 10^{-12}$ cm³/sec, but with a certain stretching of the imagination this source of ionization can be considered sufficient.

The position, however, becomes worse if, on the one hand, we take more probable values of α_{eff} and, on the other, allow for the fact that corpuscular

---

† In the opinion of Danilov and Yatsenko (1963) the three order difference between the concentrations N in the ionospheres of Venus and Earth may be connected with different values of α_{eff} on these planets with an ionizing capacity J more or less the same. It is clear from the relation $N_{\max} = \sqrt{J/\alpha_{\text{eff}}}$ that the recombination coefficients should differ by six orders for this. With $\alpha_{\text{eff}} \sim 10^{-12}$ cm³/sec in the Cytherean ionosphere this will happen if in the Earth's ionosphere $\alpha_{\text{eff}} \sim 10^{-6}$ cm³/sec, which is close to the lower limit of 5×10^{-7} cm³/sec given above and the value of α_{eff} for dissociative recombination ($\sim 10^{-7}$ cm³/sec). However, in the region of the F-layer on Earth, which plays a major part in the absorption (and therefore the emission) of radio waves, the recombination coefficient is far less: it does not exceed $3\times 10^{-10}-7\times 10^{-9}$ cm³/sec (Al'pert, 1962, section 7), so this explanation of the high values of $N_{\max}$ on Venus is unconvincing.

‡ This is the more necessary since the ionizing action of the solar radiation ceases on the night side of Venus to which the available data on the planet's radio emission largely relate.

§ 34] Sources of Venus's Radio Emission

streams with $N_s \sim 10^3$ electrons/cm^3 are to a certain extent an exceptional phenomenon: streams of this kind, as a rule, appear on the Sun only during large chromospheric flares and the types II–IV events in the radio band accompanying them (see section 17). Therefore the corpuscular streams can sustain an enhanced concentration in the ionosphere of Venus only sporadically and for a comparatively short time, whilst the Sun's permanent corpuscular emission (the so-called "solar wind") with $N_s \lesssim 10$ electrons/cm^3 can provide ionization only at a level definitely not greater than $N_{max} \lesssim 5 \times 10^7$ electrons/cm^3 (with $\alpha_{eff} \sim 10^{-12}$ cm^3/sec).[†]

Additional difficulties arise if we try to explain the results of radar experiments (Muhleman, 1961; Victor, 1961; Maling and Golomb, 1961; Kotel'nikov et al., 1962; Pettengill et al., 1962) at $\lambda = 12$–68 cm in the framework of the ionospheric model. We have shown above that to interpret the "bend" in the frequency spectrum of the radio emission of Venus at $\lambda \approx 1 \cdot 3$ cm we must assume that the optical thickness of the ionosphere τ_i at these wavelengths is equal to unity. By taking into consideration the function $\tau_i \propto \lambda^2$, we can see that at decimetric wavelengths in the radar sounding of Venus the coefficient of reflection decreases because of twofold attenuation in the ionosphere to values of less than $e^{-2\tau_i} \sim e^{-200}$. This sharply contradicts the radar data, according to which the effective coefficient of reflection from Venus is about 0·1.

In order to match up the results of observations of the radio emission of Venus and its radar observations on the basis of the ionospheric model we must assume (Kellogg and Sagan, 1961) that the ionosphere is not continuous: approximately 0·1 of the night-time surface of the planet from which at present reflected signals have been obtained[‡] is covered with an ionosphere which is transparent for decimetric waves because of the lower electron concentration (by a factor of 10–50 when compared with the value of 2×10^9 electrons/cm^3 given above). How can this gap be formed? It obviously appears if the action of the ionizing agent on the night side of Venus ceases or significantly weakens, whilst the concentration of 2×10^9 electrons/cm^3 on 0·9 of the unilluminated surface is created by the appearance of ionized particles from the daytime side of the planet.

[†] However, even in a period of intense corpuscular streams N_{max} in the ionosphere of Venus does not apparently rise to a level sufficient to have a noticeable effect on $T_{eff\, ♀}$. In the opposite case the radio emission received would be subject to considerable non-systematic variations which are not observed in reality (at $\lambda \lesssim 10$ cm; it is hard to say anything definite at the longer wavelengths because of the small number of observations). We notice that the maximum concentration in the Cytherean ionosphere can be judged from the minimum length of the wave at which variations of $T_{eff\, ♀}$ unconnected with the phase variation will be fixed.

[‡] This is connected with the fact that radar probing of Venus is generally carried out at a period near interior conjunction when the planet approaches the Earth.

The movement of the ionospheric particles may be connected with an ionospheric wind or diffusion. In the first case for movement of the particles a distance of the order of the radius of Venus $R_♀ \approx 6\times10^8$ cm a time of about 6×10^4 sec is required (with a wind velocity of $\sim 10^4$ cm/sec, i.e. the same as in the Earth's ionosphere see Al'pert, 1960, Chapter I)). For the electrons and ions really to fill most of the night side of Venus the characteristic recombination time $t_{\text{eff}} = 1/\alpha_{\text{eff}}N$ should not be less than 6×10^4 sec. When $N = 2\times10^9$ electrons/cm³ this, however, leads to an inadmissibly small recombination coefficient $\alpha_{\text{eff}} < 10^{-14}$ cm³/sec. Even less optimistic estimates are obtained for α_{eff} if the drift of the ionized particles proceeds by ambipolar diffusion. In this case the time to drift a distance $R_♀$ is $R_♀^2/6D_i$, where the diffusion coefficient is $D_i \sim V_{\text{th}_i}^2/\nu_{im}$ (V_{th_i} is the thermal velocity of the ions, ν_{im} is the number of collisions of ions with neutral molecules). Taking by analogy with the Earth $D_i \sim 10^9$ cm³/sec, we obtain that $\alpha_{\text{eff}} < 10^{-17}$ cm³/sec.

Kuz'min (1963) suggests that the ionized layer is located in the lower atmosphere of Venus where there are high values of the effective number of collisions ν_{eff} (because of collisions of electrons with neutral molecules). This made it possible to reduce the necessary values of N to $(4-8)\times10^6$ electrons/cm³. However, the possibility of the appearance of such values of N in the lower atmosphere of Venus is very doubtful. The ionizing part of the solar radiation does not penetrate to these layers; the sources of ionization suggested by Kuz'min (1963)—cosmic rays, radioactivity of the planet's atmosphere and surface—have little effect. In any case, on the Earth at altitudes of 12–15 km where the basic absorption of cosmic rays occurs the electron concentration does not exceed 3×10^{-3} electrons/cm³, which is nine orders less than that on Venus. This estimate is easy to obtain if we remember that the flux of primary cosmic rays in the Earth's orbit is about 1 particle/cm³ with a mean energy of 2×10^9 eV per nucleon. When the thickness of the region of effective ionization is $L \sim 10^6$ cm and the energy expended in one act of ionization is $\mathcal{E}_i = 15$ eV we obtain that the production of electrons by cosmic rays is $J \sim 10^2$ electrons/cm³ sec. Since at these altitudes the electrons live for only $\sim 3\times10^{-5}$ sec, after which they become attached to molecules forming negative ions, the electron concentration is altogether only $N \sim 3\times10^{-5}\times10^2 \sim 3\times10^{-3}$ electrons/cm³. The radioactivity of the Earth's atmosphere and surface does not produce any more electrons. Since there is no reason to assume that the stream of particles in the cosmic rays near Venus and the radioactivity of this planet are many orders greater than the corresponding values for the Earth, we arrive at the conclusion that the variant of the "ionospheric" model developed by Kuz'min (1963) is improbable.

§ 34] Sources of Venus's Radio Emission

THE "HOT" SURFACE MODEL

The serious difficulties with the "ionospheric" model force us to pay particular attention to the "hot" surface model, which is more probable in terms of radio astronomy, as we shall see. Results obtained by the space-probe "Venus-4" allow us to make a definite choice in favour of this model. In it we can use for the effective radio emission temperature (in a rough approximation) the formula

$$T_{\text{eff}} = T_s(1-R^2)e^{-\tau_a} + T_a(1-e^{-\tau_a}), \tag{34.9}$$

similar to (34.1), the only difference being that instead of T_i and τ_i there figure T_a and τ_a—the temperature and optical thickness of the lower atmosphere respectively. In (34.9) the atmosphere is assumed to be isothermal ($T_a = $ const); actually, however, $T_a = T_a(h)$, the value of the temperature decreasing with altitude h from a value of $T_a = T_s \gtrsim 600°$K when $h = 0$ to a value of $T_a = 235°$K in the upper layer of clouds. Therefore for a detailed calculation of the spectrum $T_{\text{eff}}(\lambda)$ we should use an expression of the type of (26.17) instead of the formula (34.9). Unlike the "ionospheric" model, in the case under discussion the basic absorption of the radio emission from the planet's surface proceeds in the lower atmosphere and not in the ionosphere. Since the molecular absorption or the absorption in aerosols can be significant only at wavelengths of $\lambda \lesssim 1$ cm, it is clear from the approximate formula (34.9) that the effective temperature of the radio emission of Venus in the centimetric–decimetric band is determined by the temperature of the surface ($\tau_a \ll 1$)[†]

$$T_{\text{eff}} \approx T_s(1-R^2). \tag{34.10}$$

With $T_{\text{eff}} = T_{\text{eff}♀} \approx 600°$K this requires temperatures of $T_s \approx 670°$K on the surface (this figure is given without allowing for the phase dependence of T_{eff} and on the assumption that the coefficient of reflection is $R^2 \approx 0.1$, i.e. close to the values of R^2 determined in the radar measurements). The decrease in $T_{\text{eff}♀}$ to $350°$K on the transition to millimetric wavelengths will be ensured provided that the optical thickness of the atmosphere here reaches values of $\tau_a \gtrsim 1$.

Before discussing the reasons for the existence of high temperatures on the surface and for large optical thicknesses of the planet's atmosphere we shall give the results of calculating the radio emission frequency spectrum

---

[†] Troitskii (1964) has drawn attention to the fact that the transparency of atmosphere of Venus at centimetric wavelengths allows us to use when analysing the observational data about the phase dependence of T_{eff} at $\lambda \approx 3$ cm the theory developed for the Moon's radio emission (see section 33). This theory can be applied to Mars and Mercury with even greater correctness.

of Venus obtained by Barrett (1960 and 1961) on the basis of the "hot" surface model (with concrete assumptions about the composition, density and thermal régime of the planet's atmosphere). He proceeded from the following chemical composition of the atmosphere: 75% CO_2, 22-25% N_2 and 0-3% H_2O. In order to obtain agreement with the observed value of $T_{\text{eff}\,♀}$ in the centimetric–decimetric band the surface temperature was considered to be 580°K. In the atmosphere the temperature, by assumption, decreases adiabatically with altitude (with a gradient $dT/dh = -9$ degrees/km)† right to the level $h = 33$ km at which $T = 285°$K. We notice that this value of T corresponds to the rotational temperature of carbon dioxide in the ranges $\lambda = 7820$ Å and $\lambda = 8689$ Å characterizing the inner parts of the cloud layer. Higher, the atmosphere was considered to be isothermal.

In the model used the atmospheric absorption is caused by water and carbon dioxide vapour; the effect of aerosols is not taken into account. We have spoken about absorption in water vapour in section 7; it is connected chiefly with rotational transitions of H_2O molecules having an electrical dipole moment at $\lambda = 1.34$ cm and $\lambda = 0.16$ cm. The CO_2 molecule is a symmetrical linear molecule without any constant dipole moment; because of collisions with other molecules, however, a dipole moment is induced in it which ensures non-resonance absorption in the radio band. The corresponding absorption coefficient is

$$\mu_{CO_2} \propto \frac{\omega^2 p_{CO_2}^2}{T^{3,7}} \left(1 + 0.378 \frac{p_{N_2}}{p_{CO_2}}\right) \text{cm}^{-1},$$

where the partial pressures p_{CO_2}, p_{N_2} are expressed in dyne/cm².

In accordance with the remarks made above the radio emission's frequency spectrum was calculated by formulae of the (26.17) type; the results of calculating the function $T_{\text{eff}\,♀}(\lambda)$ for different pressures at the surface of Venus are shown in Figs. 171 and 172.

It is clear from these figures that the experimental data are best satisfied by a model with a pressure at the base of the atmosphere of $p_0 \approx 20$–30 atm. We notice that with $p_0 \approx 30$ atm the pressure in the cloud layer is 1·1 atm, which agrees with the upper limit of 1 atm found by Herzberg (1952) by investigating broadening of the CO_2 lines. The value $p_0 \approx 30$ atm, however, contradicts the data on the CO_2 content above the cloud layer

---

† The adiabatic gradient $dT/dh = -g/c_p$ (g is the acceleration due to gravity, c_p is the thermal capacity of a unit mass of gas at constant pressure) gives a temperature equal to the temperature of the surrounding gas to a volume of gas that rises upwards and at the same time cools adiabatically. This is the maximum temperature gradient in a stable atmosphere; convection begins at high value of $|dT/dh|$ (Landau and Lifshitz, 1953, section 4).

§ 34] Sources of Venus's Radio Emission

(0·1–1 km of reduced atmosphere) since it requires an abundance of CO_2 for about 5 km of reduced atmosphere. From this point of view a pressure of $p_0 \approx 10$ atm is more acceptable since for the chosen atmospheric composition it leads to a carbon dioxide thickness of only 1·7 km.†

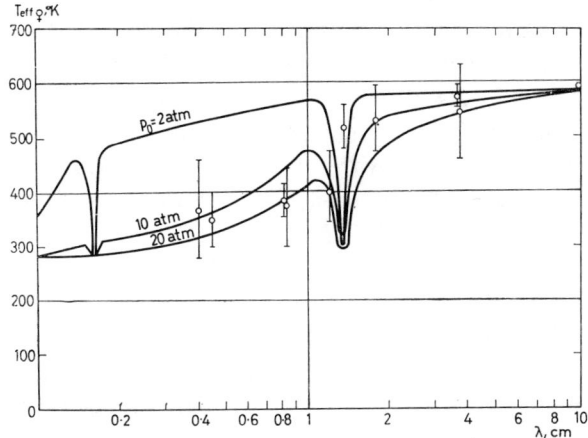

FIG. 171. Theoretical frequency spectrum for Venus's radio emission for the model of an atmosphere consisting of 75% CO_2, 24% N_2 and 1% H_2O (p_0 is the gas pressure at the planet's surface) (Barrett, 1960 and 1961). The figure also shows the results of the measurements of $T_{\text{eff}\,♀}$ transferred from Fig. 102a

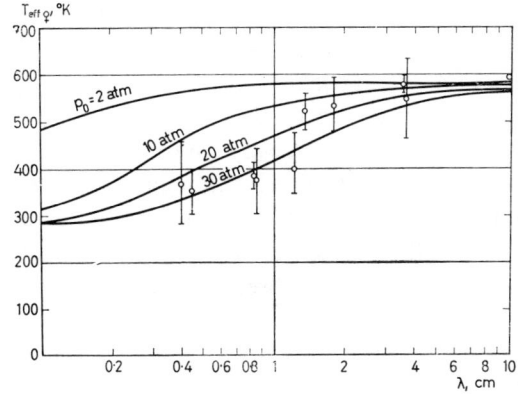

FIG. 172. The same as Fig. 171, but for an atmosphere without water: 75% CO_2 and 25% N_2 (Barrett, 1960 and 1761).

† Barrett's calculations become incorrect with pressures of tens of atmospheres: the formulae he used for the coefficient of molecular absorption are, strictly speaking, valid only for a lower density of gas.

647

Furthermore, in an atmosphere with a 1% water content clear-cut absorption lines should be observed at $\lambda = 1\cdot34$ cm and $\lambda = 0\cdot16$ cm; the values of $T_{\text{eff}\,\venus}$ in the centre of the line in this case drop to about 300°K. Since the recent observations of Gibson and Corbett (1963) at $\lambda = 1\cdot35$ cm have shown that $T_{\text{eff}\,\venus}$ is considerably higher here (520±40°K) it follows from this that the water content of the Cytherean atmosphere is far less than 1%. This agrees with Strong's data (see Gordon, 1954); he reported that the water vapour content above the clouds is 2·4 cm of reduced atmosphere (i.e. $2\cdot4 \times 10^{-2} - 2\cdot4 \times 10^{-3}$% in relation to the amount of CO_2).

Therefore the "hot" surface model can in all probability explain the observed frequency spectrum with the appropriate choice of composition, density and temperature distribution in the atmosphere of Venus that in general outline agrees with the data on the Cytherean atmosphere obtained by the space-probe "Venus-4". It is not impossible that the agreement will improve if we allow for the contribution of aerosols to the value of the atmosphere's optical thickness τ_a.

The next question that must be answered to put this model on a solid foundation relates to the reasons for the high temperature of Venus's surface.

In the opinion of Sagan (1960) the temperature $T_s \sim 600°$K is maintained by a "greenhouse effect"; the atmosphere of Venus is transparent to solar rays in the optical and near infrared parts of the spectrum which heat up the surface; at the same time the atmosphere holds back the long-wave infra-red emission of the surface, thus preventing its intense cooling by radiation.† According to the estimates of Sagan (1960) the degree of absorption of long-wave infrared rays necessary to heat the surface to 600°K is achieved (in an atmosphere with a high carbon dioxide content) if the H_2O content at the base of the atmosphere is 1–10 g/cm³. This amount of water vapour does not greatly contradict the known data on the H_2O content above the cloud layer, although the question as a whole is unclear (see in this connection Öpik's criticism of the "greenhouse theory" (Öpik, 1961), and also the data given in section 20 on the value of $T_{\text{eff}\,\venus}$ at $\lambda = 1\cdot35$ cm, which indicate the absence of an intense H_2O absorption line in the spectrum of Venus's radio emission).‡

Öpik has suggested another explanation for the high values of Venus's surface temperature based on the so-called "aeolospheric" model of the

---

† What is generally understood by the "greenhouse effect" has nothing in common with the actual rise in temperature in greenhouses, where it generally occurs because of the limited escape of hot air from the earth due to the glass cover.

‡ The question of the possibility of finding water in the atmosphere of Venus by radio methods has been discussed by Salomonovich (1964).

§ 34] Sources of Venus's Radio Emission

planet's atmosphere. Here it is assumed that the lower layers of the atmosphere are in a state of continuous mixing because of the action of winds blowing above the clouds. Because of the high dust content of the lower atmosphere the Sun's rays do not reach the planet's surface and are therefore not the direct source of the planet's heating; in the aeolospheric model this source is the winds which because of the viscosity of the gas generate a certain amount of heat near the surface which makes up for the radiation losses. These losses are small since the concentration of the dust component is assumed to be so high that the atmosphere becomes opaque at wavelengths of less than 1–3 cm.

The development of the aeolospheric model is not yet complete and at present it is hard to judge which of the models (aeolospheric or greenhouse) should be preferred. However, whatever the cause of the heating, the actual fact of the high surface temperature, which follows quite definitely both from theoretical considerations and from experimental data on the radio emission of Venus (see below), indicates that the planet's surface is an incandescent, badly illuminated stony desert; there is no water there in a liquid state and therefore no life in the biochemical forms which are known on Earth.

The conclusion drawn above in favour of the "hot" surface model and against the "ionospheric" model can be checked experimentally by radio-astronomy methods. According to the "ionospheric" model the optical thickness of the ionosphere is $\tau_i \sim 1$ at wavelengths of $\lambda \sim 1.5$ cm and decreases at shorter wavelengths. Therefore at wavelengths of $\lambda \sim 1-2$ cm we should expect an increase in brightness towards the edge of Venus's disk because of the rise in τ_i at the periphery, and therefore also an increase in the part played by the "hot" ionosphere (because of the great extent of the ionized layer along the line of sight on the edge of the disk; for further detail on effects of this kind under solar conditions see section 28). On the other hand, in a "hot" surface model with a negative temperature gradient in the atmosphere there should not be "brightening" but "darkening" towards the limb of Venus, since the rise in τ_i as we move away from the centre of the disk will lead to a decrease in the contribution from the hot surface to the observed radio emission (Kellogg and Sagan, 1961). At longer wavelengths the position becomes uncertain. In the ionospheric model when $\lambda > 2$ cm the optical thickness τ_i is noticeably greater than unity and all the radio emission both from the centre of the disk and from the periphery is created by the ionosphere. In this case brightening obviously occurs if the ionosphere's temperature T_i rises with altitude (as on Earth); otherwise there is darkening towards the edge of the disk. In the "hot" surface model at $\lambda > 2$ cm the radio emission comes from the planet's surface;

the distribution of T_{eff} over the disk will be characterized by a certain darkening towards the edge for the same reasons as on the Moon (a rise in the surface coefficient of reflection with inclined incidence of the radio waves; see section 35).

The experiments made to investigate the radio brightness distribution over the disk of Venus, of which we spoke in section 20, indicate a fall in T_{eff} towards the edge of the disk at $\lambda = 1 \cdot 9$ cm (measurements by the "Mariner II" space probe given by Barrett and Lilley (1963) and $\lambda = 3$ cm (terrestrial observations by Korol'kov et al. (1963)). To judge from the results at $\lambda = 1 \cdot 9$ cm preference should be given to the "hot" surface model; however, further experiments were necessary to choose definitely between the two models discussed. The final choice of model, in favour of the "hot" surface model, was made after the "Venus-4" flight.

Apart from the two basic models discussed above other possible ways have also been suggested latterly for explaining the high values of T_{eff} at centimetric and decimetric wavelengths. Scarf (1963) considers that the radio emission of Venus may be generated by the excitation of plasma waves in the ionosphere by a stream of particles (the solar wind). This hypothesis is quite unrealistic since the frequencies of these waves $\omega \sim \omega_L = (4\pi e^2 N/m)^{1/2}$ will correspond to the centimetric waveband only with concentrations of $N \sim 10^{12}$ electrons/cm³! However, densities of $N \sim 2 \times 10^9$ electrons/cm³ are sufficient for interpreting the radio emission of Venus as thermal emission of the ionosphere (see p. 641). Tolbert and Straiton (1962) have put forward a hypothesis according to which the generation of the centimetric and decimetric radio emission of Venus occurs in electrical microdischarges between particles suspended in the planet's atmosphere (for example, drops of water, etc). After recent developments these two hypotheses are only of historic interest.

35. Theory of the Moon's Radio Emission

BASIC RELATIONS

The theory of the Moon's radio emission connects the observed characteristics of this emission with the parameters of the surface layers of the Earth's satellite, namely: the thermal conditions of its surface, the thermal conductivity, the electrical conductivity and the dielectric permeability. Knowing these parameters allows us to make definite judgements on the composition and structure of the lunar rocks, and also on the heating conditions of the layers which are responsible for creating the radio emission being investigated.

§ 35] Theory of the Moon's Radio Emission

If in the direction ψ, ξ, where ψ, ξ are the selenographic longitude and latitude, the temperature distribution in depth is characterized by a function $T(y,)$ then in accordance with (26.17b) the effective temperature of the radio emission coming from inside to the lunar surface at an angle φ' to its normal will be

$$T_{\text{eff}} = \int_0^\infty T(y) e^{-\tau} d\tau,$$

Here the optical thickness is $\tau = \tau(y)$; if the absorption coefficient does not wary with the depth y, then

$$\tau = y\mu \sec \varphi'.$$

The quantity μ when

$$\varepsilon \gg \frac{4\pi\sigma}{\omega}$$

(ε and σ are the dielectric permeability and the electrical conductivity of the lunar substance at a frequency ω) is characterized by the relation

$$\mu = \frac{4\pi\sigma}{c\sqrt{\varepsilon}}. \tag{35.1}$$

Because of reflection from the Moon's surface of a part R^2 of the radio emission the effective temperature of an element of the lunar surface is

$$T_{\text{eff}} = (1-R^2) \int_0^\infty T(y) e^{-y\mu \sec \varphi'} \mu \sec \varphi' \, dy. \tag{35.2}$$

It follows from the law of refraction (22.28) that this will be the kind of temperature of radio emission leaving at an angle φ which is connected with φ' by the relation

$$\cos \varphi' = \frac{\sqrt{\varepsilon - \sin^2 \varphi}}{\sqrt{\varepsilon}}. \tag{35.3}$$

The factor $1 - R^2$ obviously depends on the orientation of the electrical vector of the radio waves emitted by a given sector of the lunar surface. For a linearly polarized component, in which the vector E makes an angle Φ with the wave's plane of incidence on the Moon's surface,

$$1 - R^2 = (1 - R_v^2) \cos^2 \Phi + (1 - R_h^2) \sin^2 \Phi. \tag{35.4}$$

Here R_v^2 and R_h^2 are the coefficients of reflection for vertical and horizontal polarization. According to Fresnel's reflection formulae

$$R_v = \frac{\varepsilon \cos \varphi - \sqrt{\varepsilon - \sin^2 \varphi}}{\varepsilon \cos \varphi + \sqrt{\varepsilon - \sin^2 \varphi}}, \quad R_h = \frac{\cos \varphi - \sqrt{\varepsilon - \sin^2 \varphi}}{\cos \varphi + \sqrt{\varepsilon - \sin^2 \varphi}}. \tag{35.5}$$

For emission coming to us from the central part of the lunar disk $\varphi = 0$ adn $R^2 = R_\perp^2$, where

$$R_\perp^2 = \frac{1-\sqrt{\varepsilon}}{1+\sqrt{\varepsilon}}. \tag{35.6}$$

For a radio emission component polarized in a plane parallel to the plane of the central lunar meridian and coming from areas located symmetrically to the lunar equator (their coordinates being ψ, ξ and ψ, $-\xi$) the factor $1-R^2$ will be respectively

$$(1-R_v^2)\cos^2\Phi + (1-R_h^2)\sin^2\Phi, \tag{35.7}$$

$$(1-R_v^2)\sin^2\Phi + (1-R_h^2)\cos^2\Phi. \tag{35.7a}$$

This is not hard to follow if we remember that the angles Φ for these elements of the lunar surface differ by $\pi/2$. The coefficients R_v^2 and R_h^2 for these areas will be the same if we neglect the differences in the value of ε for the "maria" and the "continents". To judge from the data of radio observations (see section 21) these differences are in actual fact insignificant. Let now an aerial be oriented towards the Moon so that it receives only emission with the above polarization and with equal efficiency from these areas. Then, provided that the radio emission approaching the lunar matter–vacuum interface has the same effective temperature, the total intensity of the emission from the elements with the coordinates ψ, ξ and ψ, $-\xi$ will be proportional to half the sum of (35.7), (35.7a), i.e. as the effective coefficient of reflection from the two areas we can take the quantity

$$R_{\text{eff}}^2 = \tfrac{1}{2}(R_v^2 + R_h^2). \tag{35.8}$$

Continuing the discussion of the formula (35.2) we notice that $T_{\text{eff}}(\psi, \xi, t)$, i.e. the radio brightness distribution over the lunar disk and the phase dependence of the effective temperature, can be found if we know the temperature distribution $T(y, \psi, \xi, t)$. The latter (in the case when there are no internal sources of heat on the Moon) can be determined unambiguously from a given temperature distribution of the lunar surface $T(0, \psi, \xi, t)$ which in principle can be found from observations in the infrared part of the spectrum.

It is natural to assume that the temperature of the illuminated part of the Moon's surface is a function of $\cos\varphi_S$, where φ_S is the zenith angle of the Sun at a point with coordinates ψ, ξ. In our case $\cos\varphi_S = \cos(\psi_S - \psi)\cos\xi$, where ψ_S is the hour angle of the Sun (the distance in longitude between points on the Moon's surface where the Sun and Earth are at zenith). According to Jaeger (1943), Barabashov (1952), Pettit and Nicholson (1930) for the illuminated part (i.e. for angles $|\psi_S - \psi| \leqslant \pi/2$)

§ 35] Theory of the Moon's Radio Emission

we can take it that

$$T(0, \psi, \xi, t) = \Theta \cos^{1/2}(\psi_S - \psi) \cos^{1/2} \xi + T_u, \tag{35.9}$$

where $\Theta = T_{max} - T_u$ (T_{max} is the temperature of the point beneath the Sun, T_u is the temperature of the unilluminated side of the Moon). For the shadow region (i.e. where $|\psi_S - \psi| > \pi/2$)

$$T(0, \psi, \xi, t) = T_u. \tag{35.9a}$$

In the general form

$$T(0, \psi, \xi, t) = \Theta \eta(\psi_S - \psi) \eta(\xi) + T_u. \tag{35.10}$$

Because of the motion of the Moon around the Earth the angle $\psi_S = \omega_\mathfrak{C} t$. Expanding (35.10) into a Fourier time series in the range $|\omega_\mathfrak{C} t - \psi| \leq \pi$, we obtain

$$T(0, \psi, \xi, t) = T_0(\xi) + \sum_{s=1}^{\infty} T_s(\xi) \cos(s\omega_\mathfrak{C} t - s\psi). \tag{35.11}$$

Here the constant component of the surface temperature is

$$T_0(\xi) = T_u + a_0 \Theta \eta(\xi), \tag{35.12}$$

and the amplitude of harmonics is

$$T_s(\xi) = a_s \Theta \eta(\xi), \tag{35.13}$$

where

$$a_s = \frac{1}{\pi} \int_{-\pi/2}^{+\pi/2} \eta(z) \cos(sz) \, dz \quad (z = \psi_S - \psi). \tag{35.13a}$$

If $\eta(z) = \cos^{1/2} z$, then $a_0 = 0 \cdot 382$, $a_1 = 0 \cdot 558$.

The temperature variation with depth is defined by the thermal conductivity equation

$$\frac{\partial T}{\partial t} = \chi \frac{\partial^2 T}{\partial y^2}, \tag{35.14}$$

where χ is the coefficient of temperature conductivity, equal to $K/\varrho c_V$ (K is the coefficient of thermal conductivity, ϱ is the density, c_V is the specific thermal capacity). The solution of (35.14) with the boundary condition (35.11) is of the form

$$T(y, \psi, \xi, t) = T_0(\xi) + \sum_{s=1}^{\infty} T_s(\xi) e^{-y\sqrt{s\omega_\mathfrak{C}/2\chi}} \cos\left(s\omega_\mathfrak{C} t - s\psi - y\sqrt{\frac{s\omega_\mathfrak{C}}{2\chi}}\right). \tag{35.15}$$

It follows from (35.15) that the constant component of the temperature on the assumption of the absence of internal sources of heat on the Moon does not vary with depth and is the same as the constant component of the surface temperature $T_0(\xi)$. On the other hand, the amplitude of the harmonics decreases exponentially as y rises; the depth of penetration of a thermal wave with a frequency $s\omega_\mathbb{C}$ is

$$l_{\text{th}}^{(s)} = \sqrt{\frac{2\chi}{s\omega_\mathbb{C}}}. \qquad (35.16)$$

The temperature oscillations also lag in phase behind the oscillations on the surface (by an angle $y\sqrt{s\omega_\mathbb{C}/2\chi}$).

In order to obtain the connection between the effective temperature of the Moon's radio emission and the temperature of the lunar surface we substitute (35.15) in (35.2). Then we find that

$$T_{\text{eff}}(\psi, \xi, t) = (1-R^2)T_0(\xi)$$
$$+(1-R^2)\sum_{s=1}^{\infty} \frac{T_s(\xi)\cos(s\omega_\mathbb{C} t - s\psi - \psi_s)}{\sqrt{1 + 2\delta_s \cos \varphi' + 2\delta_s^2 \cos^2 \varphi'}}, \qquad (35.17)$$

where

$$\tan \psi_s = \frac{\delta_s \cos \varphi'}{1 + \delta_s \cos \varphi'}, \qquad (35.18)$$

and the parameter

$$\delta_s = \frac{1}{\mu}\sqrt{\frac{s\omega_\mathbb{C}}{2\chi}} \qquad (35.19)$$

is the ratio of the depth of penetration of an electromagnetic wave (with normal incidence) $l_{\text{el}} = 1/\mu$ to the depth of penetration of a thermal wave $l_{\text{th}}^{(s)}$ (35.16); $T_0(\xi)$ and $T_s(\xi)$ are represented by the expressions (35.12), (35.13); $\cos \varphi'$ is connected with φ by the relation (35.3).

It is clear from this that the effective temperature of a fixed element of the lunar disk consists of constant and periodically varying components. The constant component of T_{eff} differs from the constant component of the Moon's surface temperature by a factor of only $1-R^2$. The distribution of this component is characterized by "darkening" towards the edge of the lunar disk because of the increase in the coefficient of reflection R^2 as φ increases. At the same time there is an additional fall in the effective temperature towards the poles because of the gentler heating of the Moon's polar regions by the Sun's rays.

The same kind of drop in radio brightness also occurs for the variable component. This component lags in phase behind the oscillations of the

§ 35] Theory of the Moon's Radio Emission

surface temperature by an angle ψ_s which obviously cannot exceed 45° for any δ_s. Since the higher harmonic content in the observed radio emission is generally low we can also neglect them in (35.17) by writing $T_{\text{eff}}(\varphi, \xi, t)$ in the form

$$T_{\text{eff}}(\psi, \xi, t) = (1-R^2)T_0(\xi) + \frac{(1-R^2)T_1(\xi)\cos(\omega_{\mathbb{C}} t - \psi - \psi_1)}{\sqrt{1+2\delta_1 \cos \varphi' + 2\delta_1^2 \cos^2 \varphi'}}. \quad (35.20)$$

This formula was obtained first by Piddington and Minnett (1949) (without allowing for the factor $1-R^2$). Troitskii (1954) has analysed the expression for $T_{\text{eff}}(\psi, \xi, t)$ and connected it with the observed integral characteristics of the Moon's radio emission.

As pointed out in section 21, in observations of the Moon's radio emission we make direct measurements of a certain radio temperature T_F averaged over the polar diagram which is connected with the distribution of the effective temperature $T_{\text{eff}}(\psi, \xi, t)$ over the disk by the relation (21.2). It follows from it that in observations with a narrow beam receiving emission in a solid angle $\Omega_A \ll \Omega_{\mathbb{C}}$, where $\Omega_{\mathbb{C}}$ is the solid angle subtended by the disk of the Moon, T_F is the same as the T_{eff} of the part of the disk at which the aerial's axis is pointing. If, however, $\Omega_A \gtrsim \Omega_{\mathbb{C}}$, then the change from the observed characteristics of the radio emission to the parameters of the lunar surface layer (for example, the quantity δ_s) requires special processing of the data obtained allowing for the width of the aerial beam. The appropriate methods are indicated by Troitskii (1954) and Krotikov (1963b).

All that we shall say here is that in the case of a broad beam we can find by experiment the mean effective temperature over the disk of the radio emission $T_{\text{eff }\mathbb{C}}$. It is interesting that the higher harmonics for this quantity are significantly less than the corresponding harmonics for T_{eff} in the centre of the disk. Thanks to this the phase dependence of $T_{\text{eff }\mathbb{C}}$ can be described well by the first harmonic even at millimetric wavelengths, where the asymmetry of the phase dependence for local values of T_{eff} becomes noticeable.

The formulae given above are valid for the so-called single-layer model based on the assumption of the uniformity in depth of the matter forming the covering of the Moon. Wesselink (1948) on the basis of the same idea has interpreted the results of Pettit's eclipse observations (Pettit, 1940) in optics, concluding that the experimental eclipse curve is close to the theoretical one if the thermal parameter is $\gamma = (K\varrho c_V)^{-1/2} \sim 1000$. However, according to Jaeger and Harper (1950), the agreement is better for a model in which the uniform material of the lunar surface (with $\gamma \sim 1000$) is covered with a thin layer (2–3 mm thick) of poorly conductive dust; certain data on the reflection of light from the lunar surface also make possible the assumption of the presence of a dust layer on the Moon. A two-layer model of the lunar surface has been used by Piddington and Minnett (1949) to explain the phase dependence of the Moon's radio emission at $\lambda = 1.25$ cm. This obliges us to examine a two-layer model as well as a one-layer one in the theory of the Moon's radio emission; in any case there is no foundation for rejecting one or the other model beforehand when interpreting the data of radio observation.

For a two-layer model the expression for $T_{\text{eff}}(\psi, \xi, t)$ is slightly more complicated. According to Piddington and Minnett (1949) (see also Troitskii, 1961 and 1962a) for the centre of the disk

$$\left.\begin{aligned} T_{\text{eff}} &= (1-R^2)T_0 + (1-R^2)\frac{T_1 \cos(\omega_\mathfrak{c} t - \psi - \psi_1 - \psi')}{\sqrt{1+2\delta_1+2\delta_1^2}\sqrt{1+2\delta'+2\delta'^2}}, \\ \tan \psi' &= \frac{\delta'}{1+\delta'}, \end{aligned}\right\} \quad (35.21)$$

where $\sqrt{1+2\delta'+2\delta'^2}$ and ψ' are the additional attenuation of a thermal wave and its phase shift in the upper (dust) layer. The parameter δ' here is determined from the formula

$$\delta' = l\frac{K}{K'}\sqrt{\frac{\omega_\mathfrak{c}}{2\chi'}},$$

in which K' and χ' are respectively the coefficients of thermal conductivity and temperature conductivity of an upper layer of thickness l. The formulae (35.21) are valid provided that the absorption of radio waves in the dust layer is insignificant and it is noticed only in the change in temperature of the "support"; this layer should obviously be sufficiently thin for this.

INTERPRETATION OF THE RESULTS OF OBSERVATIONS OF THE MOON'S RADIO EMISSION AND THE PHYSICAL CHARACTERISTICS OF ITS SURFACE

The obtaining of reliable information on the structure and parameters of the lunar surface by means of radio data depends above all on the answer to the question of the choice between the single-layer and two-layer models of the Moon.

It followed from the early observations of Piddington and Minnett (1949) at $\lambda = 1.25$ cm (see section 21) that the phase shift was $\psi_0 = \psi_1 + \psi' \approx 45°$. This can occur in the single-layer model if $\delta_1 \gg 1$ and therefore the amplitude of the oscillations of T_{eff} is very small. Since the ratio $T_{F\sim}/T_{F=}$ was rather large (about 0·17) in their experiments Piddington

§ 35] Theory of the Moon's Radio Emission

and Minnett decided in favour of the two-layer model. In Troitskii (1956), however, it was noted that the phase dependence of T_F observed at 1·25 cm can also be well approximated by a curve with a phase shift of 35°; in this case the experimental data also agree with the single-layer model of the lunar surface. Subsequent observations (Salomonovich, 1958; Gibson, 1958; Zelinskaya, Troitskii and Fedoseyev, 1959a and 1959b) at $\lambda =$ 0·8–1·6 cm did not produce a definite result either, since the data obtained on the phase dependence of the Moon's radio emission within the limits of observational errors satisfied both the single-layer and the two-layer models. The situation has become clearer, thanks chiefly to the measurements of radio astronomers from Gor'kii—Troitskii, Krotikov, Kislyakov *et al.*—which were made by a single method over a wide range of wavelengths. The corresponding analysis of the experimental data, which made it possible to judge the nature and physical state of the Moon's surface layers, has been carried out by Troitskii. The results he obtained are contained in the survey by Krotikov and Troitskii (1963a) (see also Troitskii, 1961 and 1962a); it is these papers that we shall basically follow in what follows.

When answering the question of the structure of the Moon's surface layer it is best to base ourselves first of all on those data which do not depend on the accuracy of absolute measurements (since the latter may contain considerable errors). In radio observations during lunations these data will obviously be the phase shift ψ_0 and the relative change in the effective temperature in the course of the lunation $T_{F\sim}/T_{F=}$; the values of $T_{F=}$, $T_{F\sim}$ and ψ_0 are indicated in Figs. 104–6.

According to (35.21) the theoretical values of the phase shift ψ_0 and the ratio M of the constant component of the radio temperature to its variable part are respectively

$$\psi_0 = \psi_1 + \psi' = \arctan\frac{\delta_1}{1+\delta_1} + \psi', \quad M = \vartheta\vartheta'\frac{T_0}{T_1}, \quad (35.22)$$

where $\vartheta = \sqrt{1+2\delta_1+2\delta_1^2}$, $\vartheta' = \sqrt{1+2\delta'+2\delta'^2}$. Eliminating δ_1 from (35.22) we obtain:

$$\psi_0 = \frac{-1+\sqrt{1+2[(MT_1/\vartheta'T_0)^2-1]}}{1+\sqrt{1+2[(MT_1/\vartheta'T_0)^2-1]}} + \psi'. \quad (35.23)$$

Putting $T_0/T_1 \approx 1\cdot5$ (on the basis of the optical data on the temperature of the lunar surface and also in accordance with the results following from Fig. 175; see below), we plot the graphs of the function $\psi_0(M)$ with different assumptions about the parameters of the dust layer δ' and ψ' (Fig. 173). Curve 1 corresponds to a single-layer model ($\vartheta' = 1$, $\psi' = 0$); curve

2 to a two-layer model with $\vartheta' = 1\cdot 1$ and $\psi' = 5°$. As the layer of dust lying on the substratum thickens the curves of $\psi_0(M)$ will shift to the right and upwards; for example, for a thicker layer with $\vartheta' = 1\cdot 4$ and $\psi' = 15°$ the function $\psi_0(M)$ corresponds to curve 3.

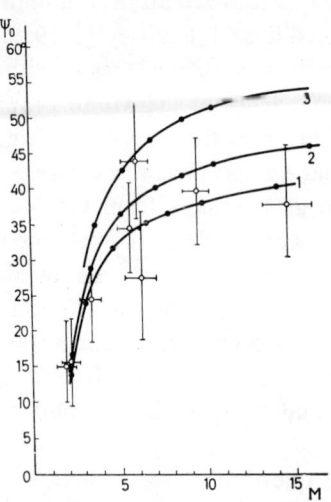

Fig. 173. Values of the phase lag ψ_0 of the first harmonic of the Moon's radio temperature as functions of the ratio M of the constant component of the radio temperature to the amplitude of the first harmonic (Krotikov and Troitskii, 1963a; Troitskii, 1961 and 1962a)

In order to be able to connect with each value of $\psi_0(M)$ a definite wavelength λ, we must state for the substratum (in the two-layer model) and for the uniform layer (in the one-layer model) the connection between δ_1 and λ. It is known, for example, that in dielectrics δ_1 usually varies linearly with wavelength. (A similar situation apparently exists for lunar rock.) We are convinced of this by Fig. 174, which shows the values of δ_1/λ averaged for each wavelength over all the available experimental data;[†] although the dispersion of the points is rather great the function $\delta_1(\lambda)$ can be approximated by the relation (Zelinskaya, Troitskii and Fedoseyev, 1959a and 1959b; Troitskii, 1956)

$$\delta_1 = 2\lambda, \quad l_{el} = 2\lambda_{\text{th}}^{(1)}. \tag{35.24}$$

It follows from this that the depth of penetration of an electromagnetic

---

[†] The values of δ_1 are calculated in the framework of the single-layer model from experimentally measured values of M and a known ratio T_0/T_1.

§ 35] Theory of the Moon's Radio Emission

wave into the Moon (and thus the thickness of the effectively emitting layer) rises in proportion to the wavelength λ.

An examination of the relations (35.21) with allowance made for the proportionality between δ_1 and λ makes understandable the observed dependences of the constant and variable components of the Moon's radio temperature and also the phase shift on the wavelength (Figs. 104–6). For example, the comparative constancy of $T_{F=}$ over a broad range of wave-

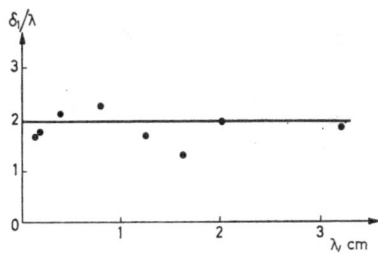

FIG. 174. Ratio δ_1/λ as a function of wavelength λ. The black spots are the values of δ_1/λ averaged over all the experimental data for the corresponding wavelengths (Troitskii, 1956)

lengths is because with small values of the coefficient of reflection R^2 the value of $T_{F=}$ is determined only by the value of the surface temperature T_0 (we shall speak later about the weak rise of $T_{F=}$ with wavelength). The decrease in amplitude of the oscillations of the radio temperature $T_{F\sim}$ as λ rises is connected with the corresponding decrease in the oscillations of a thermal wave as we get further away from the surface into the Moon; this connection becomes obvious if we remember that the effective thickness of the emitting layer rises with λ. This last circumstance also leads to an increase in the phase angle ψ_0 at longer wavelengths right up to its maximum value of $45°$ (see the formula (35.16) for a single-layer model).

Returning to Fig. 173 we note that the black points on curves 1 and 2 correspond to wavelengths of $\lambda = 0\cdot4, 0\cdot8, 1\cdot25, 1\cdot63$ and $3\cdot2$ cm provided that $\delta_1 = 2\lambda$; at the same time the points on curve 3 correspond to the same wavelengths, but with the proviso that $\delta_1 = 1\cdot25\lambda$. In addition, a number of experimental values of M and ψ_0 are marked on the figure in the form of light circles with an indication of the possible measurement errors.

It is clear from Fig. 173 that the available experimental data on the Moon's radio emission cannot as a whole be matched up with the two-layer model of the surface which has rather high values of $\vartheta'-1$ and ψ'.

The single-layer model is the most probable, although the assumption of the presence of a thin upper layer (for example with $\vartheta'-1 \leqslant 0\cdot1$ and

$\psi' \leqslant 5°$) leads to results which do not differ in the limits of error from the data of radio observations.

In many papers (see, e.g., Troitskii, 1954; Salomonovich, 1958, 1962a) the preferability of the single-layer model is judged from the agreement of the values of δ_1 found by (35.17) from the measured amplitude and phase of the effective temperature. This method was actually used in Fig. 173 with, however, the essential difference that the analysis is extended to a broad range of wavelengths.

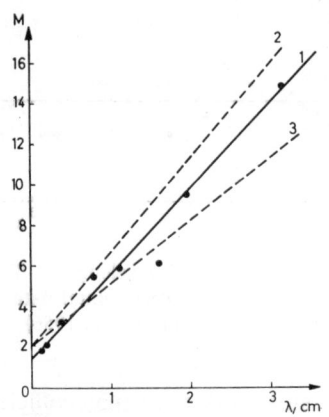

FIG. 175. Ratio M as a function of wavelength λ: 1—single-layer model, $\vartheta' = 1, \delta_1 = 2\lambda$; 2—two-layer model, $\vartheta' = 1\cdot5, \delta_1 = 1\cdot5\lambda$; 3—two-layer model, $\vartheta' = 1\cdot5, \delta_1 = \lambda$ (Krotikov and Troitskii, 1963a; Troitskii, 1961 and 1962a)

A conclusion in favour of uniform structure of the lunar surface can also be drawn from a comparison of the observed values of M at different wavelengths with the theoretical function $M(\lambda)$. It can be seen from Fig. 175 that the experimental values of M agree better with curve 1 for the one-layer model than with curves 2, 3 that characterize two-layer models and are plotted from formula (35.22) for different values of the ratio δ_1/λ. Since as λ decreases the radio emission comes from a thinner and thinner layer (whilst at the limit $\lambda \to 0$ even in the framework of the single-layer model—from the surface of the Moon) the value of M will approach the ratio T_0/T_1 characterizing the oscillations of the surface temperature. According to Fig. 175 this ratio is close to 1·5.

The structure of the Moon's upper layers can also be judged from the results of radio observations during lunar eclipses. Unfortunately, there are very few published data on eclipse measurements with a corresponding analysis of the results. Here we can only point to the paper (Gibson, 1961) in which Gibson was able to match up eclipse measurements at $\lambda = 8\cdot6$ mm,

§ 35] Theory of the Moon's Radio Emission

which showed no drop in the effective temperature by a value greater than 1°, with the data of radio observations during lunations only on the basis of a three-layer model of the surface. His analysis, however, is not free of objections (see Krotikov and Troitskii, 1963a); therefore at present it is very important to carry out careful eclipse radio observations with subsequent processing of the results on the basis of different models of the surface. This will permit us to answer the question of the extent of agreement in the theory of the Moon's radio emission between the eclipse data and the data on the phase dependence of T_{eff} and on the structure of the uppermost layer of the lunar surface.

By using the actual data that have been accumulated to date on the Moon's radio emission we can obtain a whole quantity of information on the thermal properties, the density and the dielectric constant of the lunar rock. In this case, in accordance with what we have said, we should base our work on a single-layer model which corresponds best to the results of observations of lunar radio emission during lunations.

Starting with a discussion of the thermal properties of the Moon's surface we notice first of all that an analysis of the optical data generally leads to values of the parameter $\gamma = (K\varrho c_V)^{-1/2}$ lying between 200 and 1000. Just as broad limits for γ have been found by Salomonovich (1962a) with the introduction of the data of radio observations (300–1000). The large dispersion of the values in the latter case is connected with the low accuracy of the radio measurements. More rigorous estimates of the thermal parameter γ have been obtained on the basis of the infrared measurements of Sinton (1962) and the observations of the Gor'kii radio astronomers which were made with higher accuracy.

Sinton found that the night-time temperature of the lunar surface is $123 \pm 3°$K. On the other hand, this temperature can be obtained by calculating the thermal conditions of the Moon from the known value of the solar constant and with different assumptions about the value of the parameter γ. A calculation of this kind has been made by Krotikov and Shchuko (1963); it follows from it that the value $123 \pm 3°$K corresponds to $\gamma \approx 350–430$.

The value of γ can also be estimated from the values of the constant component of the temperature of the lunar surface at the centre of the disk T_0 and the ratio T_0/T_1, bringing in for definition the latest methods of radio astronomy (Krotikov and Troitskii, 1963b). It follows from Fig. 104b that the constant component of the Moon's mean radio temperature over the disk at $\lambda = 3 \cdot 2$ cm is $211 \pm 2°$K. The corresponding calculation for the effective temperature of the centre of the Moon's disk gives a value of $T_{\text{eff}} = 227 \pm 5°$K. For a coefficient of reflection (with normal incidence) of $R_\perp^2 \approx 0 \cdot 01–0 \cdot 04$ this leads to a constant component of the surface temperature of $T_0 \approx 229\text{–}236°\text{K}$ (see the formula (35.20)). On the other hand, T_0 can be found by examining the thermal conditions of the lunar surface for different γ; the corresponding calculations (Krotikov and Shchuko, 1963) then lead to the conclusion that $T_0 \approx 229\text{–}236°\text{K}$ corresponds to values of $\gamma \approx 250\text{–}450$. Furthermore, the radio astronomy observations indicate that the ratio $T_0/T_1 \approx 1·5 \pm 0·1$.[†] Calculation of the thermal conditions shows that this ratio obtains when $\gamma \approx 270\text{–}550$.

The values given for γ differ from the value of $\gamma \approx 1000$ obtained in optical observations of lunar eclipses. It is not clear at present with what this discrepancy, which is apparently beyond the limits of measurement errors, is connected. It is not impossible that this circumstance is connected with the presence of dust on the surface, the layer of which is so thin that it has no significant effect on the results of radio observations during lunations but, however, noticeably alters the curve of an eclipse in infrared rays. It is also possible (Krotikov and Troitskii, 1963a) that the temperature dependence of the lunar rock's properties (K and c_V) plays a part here.

Let us in future take a parameter γ of 350 with an accuracy of 20–25%. Then for a lunar rock density of $\varrho \approx 0·5$ g/cm³[‡] and a thermal capacity of $c_V \approx 0·2$ we obtain a value for the coefficient of thermal conductivity of $K \approx 10^{-4}$ cal cm^{-1} sec^{-1} degree^{-1}. We can then estimate the depth of penetration of a thermal wave (35.16):

$$l_{\text{th}}^{(1)} = \sqrt{\frac{2\chi}{\omega_\zeta}} = \sqrt{\frac{2K}{\varrho c_V \omega_\zeta}} \approx 25 \text{ cm}; \quad (35.25)$$

in accordance with (35.24) the penetration depth of electromagnetic wave- and the thickness of the emitting layer at centimetric–decimetric wave lengths is

$$l_{\text{el}} \approx 50\lambda. \quad (35.26)$$

We note that since the adequacy of the single-layer for the results of radio observations during lunations was found from the phase dependence of the radio emission at wavelengths of $\lambda \lesssim 3$ cm it follows from (35.26) that the Moon has a uniform structure right to a depth of about 150 cm (with certain reservations about the topmost layer).

Let us now examine the question of the density ϱ and the dielectric permeability ε of the lunar material.

In many papers (Gibson, 1958; Baldwin, 1961; Gibson, 1961; Sinton, 1962) the values of these constants are taken simply by analogy with ter-

---

[†] We give the error indicated by Troitskii, (1961, 1962a). To judge from Fig. 175, however, it is possibly greater, which will also considerably increase the error in determining γ from the ratio T_0/T_1.

[‡] For the density of the lunar rock see below.

§ 35] Theory of the Moon's Radio Emission

restrial rocks ($\varrho \approx 2$, $\varepsilon \approx 4$–5). Better founded values of ϱ and ε can be obtained, however, if we bring in the data on the Moon's radio emission, in particular the value given above for the thermal parameter γ. The point is that the thermal conductivity of any material is determined by its mean density ϱ, which is determined in its turn by the material's porosity $p = 1 - \varrho/\varrho_0$, where ϱ_0 is the density in a non-porous state. Therefore the thermal conductivity of the material is $K = K(\varrho)$; knowing this function, we can then also determine ϱ from a known γ since $\gamma = [K(\varrho)\varrho c_V]^{-1/2} = \gamma(\varrho)$ (Troitskii, 1962b).

The form of the function $K(\varrho)$, generally speaking, depends on the chemical composition of the material; in addition, the functions $K(\varrho)$ may be different for materials with a different structure—hard foamy or granular flowing. If the lunar rocks consist of the usual silicates known on Earth, then when choosing the function $K(\varrho)$ we can base our work on experimental data on the thermal conductivity of silicates, remembering, however, that on the Moon the silicates are in a vacuum and not in air. In accordance with these data, for foamy and for flowing silicates the thermal conductivity in air is

$$K(\varrho) = \alpha\varrho = 6\cdot 10^{-4}\varrho \quad (0\cdot 4 \leqslant \varrho \leqslant 1\cdot 5). \tag{35.26a}$$

There is very little information on experiments in a vacuum. If we assume that the proportionality between K and ϱ is preserved in a vacuum—only the coefficient α changes—then it turns out that for foamy materials (with a porosity of $p > 30\%$)

$$K(\varrho) \approx 2\cdot 10^{-4}\varrho, \tag{35.27}$$

and for flowing ones (with grains that are not too small)

$$K(\varrho) \approx 5\cdot 10^{-5}\varrho. \tag{35.27a}$$

Then, taking $\gamma = 350$ and $c_V = 0\cdot 2$ we obtain

$$\varrho = \gamma^{-1}(\alpha c_V)^{-1/2} \approx 0\cdot 45 \text{ g/cm}^3 \tag{35.28}$$

for foamy and

$$\varrho = \gamma^{-1}(\alpha c_V)^{-1/2} \approx 0\cdot 9 \text{ g/cm}^3 \tag{35.28a}$$

for flowing silicates on the Moon's surface.[†]

The dielectric permeability of the lunar rocks can be determined in several ways, one using the Moon's reflecting power: knowing the coefficient of reflection R^2 at the centre of the disk and considering the surface to be smooth enough, we can find ε by the formula (35.6). In accordance with

---

[†] The accuracy of these quantities is no greater than 20–25%, i.e. no better than the accuracy of determining γ, and apparently is actually far worse because of the uncertainty in the vacuum values of the function $K(\varrho)$.

(35.21) R^2 is found in relation to the constant components of the effective temperature and the true temperature on the Moon's surface. The determination of these temperatures in order to determine ε requires the carrying out of very accurate absolute measurements. In the paper by Krotikov and Troitskii (1962) the values of the radio temperature found with enhanced accuracy by the "artificial Moon" method and of the surface temperature obtained from infrared measurements led to a value of $\varepsilon \approx 1 \cdot 5$.

Another method (Salomonovich and Losovskii, 1962; Koshchenko, Losovskii and Salomonovich, 1962) consists of measuring the distribution of the radio brightness's constant component over the Moon's disk, i.e. the function $1 - R^2(\psi, \xi, \varepsilon)$. The advantage of this method is that it is connected with relative measurements, although high-directivity aerials are required. Processing of the available radio brightness distributions leads to $\varepsilon = 1-2$.

We can also arrive at the value of ε by measuring the degree of polarization of the emission coming from a local area of the lunar surface (Troitskii and Tseitlin, 1960). Since the constant components of the effective temperature of the horizontally and vertically polarized emission are $T_h = T_0 (1-R_h^2)$ and $T_v = T_0(1-R_v^2)$, where R_h and R_v are connected with ε by the Fresnel formulae (35.5), it becomes clear that the degree of polarization

$$\varrho_l = \frac{T_v - T_h}{T_v + T_h}$$

will be a function of ε. This method is connected only with relative measurements, but also requires the use of highly directional aerials. This is because noticeable polarization of the Moon's radio emission occurs only near the edge of the disk where the angle of incidence is close to the Brewster angle. A determination of the dielectric permeability by this method was made by Soboleva (1962); it turned out that $\varepsilon \approx 1 \cdot 6$.

Since the quantity ε, as well as the thermal conductivity, depends on the material's porosity, the possibility arises of finding the density of the lunar rocks from a known dielectric constant. It is known (Odelevskii, 1951; Levin, 1954) that for a substance with a foamy structure

$$\varepsilon = \varepsilon_0 \left(1 - \frac{3p}{\frac{3\varepsilon_0 + 1}{\varepsilon_0 - 1} + p}\right), \tag{35.29}$$

where ε_0 is the permeability of a non-porous specimen, p is the degree of porosity. Since ε_0 for silicate rocks is approximately the same, to determine the density ϱ of the lunar rock we need to know only ε and ϱ_0—the

§ 35] Theory of the Moon's Radio Emission

density in a non-porous state. The empirical formula

$$\sqrt{\varepsilon-1} = A\varrho, \quad A \approx 0.5 \text{ cm}^3 \cdot \text{g}^{-1}, \tag{35.30}$$

which is valid for different dry rocks, is preferable, however, since here we need know only ε to find ϱ. Using the value $\varepsilon \approx 1.5$ we find from the formula (35.30) that $\varrho \approx 0.5$ g/cm^3. Comparing this value with the value determined from the known thermal parameter γ (see (35.28), (35.28a)) we can see that it corresponds to a foamy rather than a flowing state of the material. The reservation should be made, however, that this conclusion, which is very important for ideas on the structural state of the lunar rocks, needs further checking on the basis of more accurate data. When finding the density $\varrho \approx 0.45$ g/cm^3 (35.28) here, however, we used the function (35.27) based only on single experiments on the connection of the thermal conductivity and density of materials in a vacuum.

The study of the composition of lunar rocks is based on a comparison of various characteristics of the lunar and terrestrial rocks. Until recently only optical characteristics were used for this purpose. Unfortunately these characteristics provide information only on the uppermost, very thin layer on the surface of the Moon, this information (colour, scatter indicatrix, polarization) depending to a considerable extent on either a negligible quantity of impurities or on structural features of the surface—the extent to which it is broken up, the nature of the unevennesses, etc.

Radio-astronomy investigations have opened up fresh possibilities in this direction—we can compare the electrical characteristics of the lunar and terrestrial rocks (the dielectric permeability ε and the tangent of the loss angle tan $\varDelta$). The position, however, is complicated by the fact that ε and tan $\varDelta$ are strongly dependent on the density of the material, whilst it is clear from the above that the difference in the density of the rocks on Earth and on the Moon is rather great. We should therefore compare only the characteristics which are not dependent on ϱ: according to Troitskii (1961 and 1962a) this kind of parameter can be tan $\varDelta/\varrho$, which varies chtefly with the variation in the chemical composition of the materials but not with the density of the latter.

We note first of all that from the formula (35.1) that describes dielectrics with small losses, and from the relation (35.24) that is valid for lunar rocks in a layer not less than 150 cm thick, follows the proportionality between the conductivity σ and the frequency ω and thus the frequency dependence of tan $\varDelta \equiv 4\pi\sigma/\varepsilon\omega$. Aluminosilicate-based inorganic dielectrics have this kind of constancy of the tangent of the angles of losses.

Furthermore, it follows from (35.1), (35.19) and (35.24) that for the lunar rock

$$\sqrt{\frac{\omega_\iota}{2\chi}} = \frac{8\pi\sigma\lambda}{c\sqrt{\varepsilon}}.$$

Allowing for the expressions for $\tan \Delta$ and χ we find:

$$\frac{\sqrt{\varepsilon}\tan \Delta}{\varrho} = 8.8\times 10^{-5} c_V \gamma, \qquad (35.31)$$

where $\gamma = (K\varrho c_V)^{-1/2}$. Putting $c_V = 0.2$, $\varepsilon = 1.5\pm 0.3$, $\gamma = 350\pm 75$, we obtain that for the Moon's covering

$$\frac{\tan \Delta}{\varrho} = (0.5\pm 0.3)10^{-3} \text{ cm}^3 \cdot \text{g}^{-1}. \qquad (35.32)$$

In order to compare this value with the corresponding values for terrestrial rocks Krotikov (1962) (see also Fensler *et al.*, 1962; Olte and Siegel, 1961) made measurements of $\tan \Delta/\varrho$ for different minerals, meteorites and tektites. It turned out that terrestrial rocks with differing SiO_2 contents or, as they say, differing basicity, fall in the range of values of $\tan \Delta/\varrho$ corresponding to the material on the Moon: acid rocks (liparite, granite), medium ones (syenite, diorite) and basic ones (iolite, gabbro). We notice that almost all tektites have the same specific losses $\tan \Delta/\varrho$ as the lunar rocks. This obviously indicates the closeness of their chemical composition which is characterized by the presence of the following components: 60–65% quartz, 15–20% aluminium oxide, 20% of the oxides of potassium, sodium, calcium, iron and magnesium. We recall, however, that the structure of the lunar rock differs strongly from that of the terrestrial ones by its very high porosity which ensures lower values of the density ϱ.

The parameters of the Moon's surface layers given above are the result of averaging over large areas of the lunar disk and even over all the lunar disk as a whole (because of the considerable width of the aerials' polar diagrams). The local characteristics of the lunar rocks may, of course, differ from those given. This circumstance should be borne in mind when comparing the lunar rocks with the terrestrial ones, and in particular in conclusions on the chemical composition of the upper layers of the Moon.

Above, the data on the Moon's radio emission were interpreted on the assumption that there are no internal sources of heat on the Moon, so the constant component of the temperature T in the depths of the Moon is the same as the corresponding value at the surface. This assumption was fully justified in an analysis of observations made with an absolute accuracy of the order of 10%; therefore the measurements of enhanced accuracy made

§ 35] Theory of the Moon's Radio Emission

by the Gor'kii radio astronomers in the 0·4–50 cm range definitely indicate a systematic rise with wavelength of the constant component $T_{\mathbb{C}=}$ averaged over the Moon's disk. As can be seen from Fig. 104b, the quantity $dT_{\mathbb{C}=}/d\lambda$ decreases gradually and very slowly as λ rises, averaging 1 degree/cm.

The consequences following from this circumstance have been analysed by Troitskii (1962c) and Krotikov and Troitskii (1963c). The authors came to the conclusion that the observed rise in the radio temperature together with λ cannot be connected with different subsidiary factors whose role increases at longer wavelengths (the effect of the ionosphere and the cosmic radio emission); this rise probably cannot be the consequence of these causes (the temperature dependence of the thermal properties of the lunar rocks and the rise in the emissive power of the Moon's surface $1-R^2$ with wavelength). Therefore we can state that the curve in Fig. 104b reflects the actual frequency dependence of the lunar radio temperature caused by a rise in temperature (to be more precise, its constant component) in the depths of the Moon. This rise obviously indicates the presence of a considerable flow of heat from the Moon's core. The value of the flux is easy to find if we know the temperature gradient dT/dh. It is obviously defined by the following expression:

$$\frac{dT}{dh} \approx \frac{dT_{\text{eff}}}{d\lambda} \frac{d\lambda}{dh} \frac{1}{1-R_\perp^2}, \tag{35.33}$$

where T_{eff} is the effective radio emission temperature at the centre of the lunar disk. The quantity h in the derivative $d\lambda/dh$ has the meaning of the effective thickness of the layer from which radio emission at a wavelength λ (i.e. $h \approx l_e$) escapes. The connection between l_e and λ for the first one and a half metres from the Moon's surface and wavelengths of up to 3 cm is given by the linear relation (35.24) ($d\lambda/dl_e \approx$ const); when this is allowed for the formula (35.33) will become

$$\frac{dT}{dh} \approx \frac{dT_{\text{eff}}}{d\lambda} \frac{1}{2l_{\text{th}}^{(1)}(1-R_\perp^2)}. \tag{35.34}$$

Putting $dT_{\text{eff}}/d\lambda \approx 1$ degree/cm, $1-R_\perp^2 \approx 1$ and $l_{\text{th}}^{(1)} \approx 25$ cm, we obtain that the temperature gradient is about 2 degrees/cm. The corresponding density of the heat flux from the Moon's core

$$q = K\frac{dT}{dh} = \frac{dT_{\text{eff}}}{dh} \frac{\sqrt{\frac{\omega_{\mathbb{C}}}{8}}}{\gamma(1-R_\perp^2)} \tag{35.35}$$

when $\gamma \approx 350$ then reaches a value of 1.6×10^{-6} cal cm^{-2} sec^{-1}, which is close to the density of the flux from the Earth. The total heat flux from the Moon's core through its surface is $\sim 2\times 10^{19}$ cal/year. At the same time the estimates of Jaeger (1959), Levin and Mayeva (1960), McDonald (1959) of the possible heat flux from the heart of the Moon by radioactive decay in chondrite rocks lead to values 4–6 times less than the experimental ones. This discrepancy is an argument against the hypothesis of the Moon's formation from meteoritic material of the chondrite type.

A paper by Troitskii (1967) reports most results of investigations into the emission from the Moon and gives a description of the latest ideas about the lunar surface layer composition. In particular, he quotes more exact values of

$$\gamma \approx 600 \text{ cal}^{-6} \text{ cm}^2 \text{ sec}^{1/2} \text{ degree and}$$
$$q \approx (0\cdot 85 \pm 0\cdot 2)\times 10^{-6} \text{ cal cm}^{-2} \text{ sec}^{-1}.$$

References*

AARONS, J., BARROW, W. R. and CASTELLI, J. P. (1958) *Proc. IRE* **46**, 325.
AKABANE, K. (1955) *Proc. Japan Acad.* **31**, 161.
AKABANE, K. (1958a) *Proc. IRE* **46**, 194.
AKABANE, K. (1958b) *Publ. Astron. Soc. Japan* **10**, 99.
AKABANE, K. (1958c) *Annals Tokyo Astron. Obs.*, Ser II, **6**, 1.
AKABANE, K. and COHEN, M. H. (1960) *Astron. J.* **65**, 49.
AKABANE, K. and COHEN, M. H. (1961) *Ap. J.* **133**, 258.
AKHIYEZER, A. I. and FAINBERG, YA. B. (1951) *Zh. Eksp. i Teoret. Fiz.* **21**, 1262.
AKHIYEZER, A. I. and LYUBARSKII, G. YA. (1951) *Doklady Akad. Nauk SSSR* **80**, 193.
AKHIYEZER, A. I. and LYUBARSKII, G. YA. (1955) *Tr. Fiz.-mat. Fak-ta Khar'k. Gos. Un-ta* **6**, 13.
*AKHIYEZER, A. I., PROKHODA, I. G. and SITENKO, A. G. (1957) *Zh. Eksp. i Teoret. Fiz.* **33**, 750.
AKIN'YAN, S. T. and MOGILEVSKII, E. I. (1961) *Geomagnetizm i Aeronomiya* **1**, 156.
ALEKSEEV, U. J. and VITKEVICH, V. V. (1959) in *Paris Symposium on Radio Astronomy*, Ed. R. N. BRACEWELL, Stanford Univ. Press, p. 259.
ALFVÈN, H. (1950) *Cosmical Electrodynamics*, Oxford Univ. Press.
ALLEN, C. W. (1955) *Astrophysical Quantities*, Univ. London.
ALLEN, C. W. (1957) in *Radio Astronomy*, Ed. H. C. VAN DE HULST, Cambridge Univ. Press, p. 253.
ALON, I., ARSAC, J. and STEINBERG, J. L. (1953) *Acad. C. R. Sci. Paris* **237**, 300.
ALON, I., ARSAC, J. and STEINBERG, J. L. (1955) *Acad. C. R. Sci. Paris* **240**, 595.
AL'PERT, YA. L. (1960) *Propagation of Radio Waves in the Ionosphere* (in Russian), Akad. Nauk SSSR.
AL'PERT, YA. L., GINZBURG, V. L. and FEINBERG, YE. L. (1953) *Propagation of Radio Waves* (in Russian), Gostekhizdat.
ALSOP, I. E., GIORDMAINE, J. A., MAYER, C. H. and TOWNES, C. H. (1959a) in *Paris Symposium on Radio Astronomy*, Ed. R. N. BRACEWELL, Stanford Univ. Press, p. 69.
ALSOP, I. E., GIORDMAINE, J. A., MAYER, C. H. and TOWNES, C. H. (1959b) *Astron. J.* **64**, 332.
AMBARTSUMYAN, V. A., MUSTEL', E. R., SEVERNYI, A. B. and SOBOLEV, V. V. (1952) *Theoretical Astrophysics* (in Russian), Gostekhizdat (English Translation, Pergamon Press, 1958).
AMENITSKII, M. A., LI TS'IN FANG, SALOMONOVICH, A. YE., KHANGIL'DIN, U. V. and CH'ENG CHUN LIAN (1958) *Solnechnyye Dannyye*, No. 7, 169.
AMER, S. (1958) *J. Electr. and Control* **5**, 105.
ANDERSON, K. A. (1958) *Phys. Rev. Lett.* **1**, 335.
ANDERSON, K. A. and ENEMARK, D. C. (1960) *J. Geophys. Res.* **65**, 2657.
ANDRONOV, A. A. (1961) *Izv. Vuzov, Radiofiz.* **4**, 861.
ANDRONOV, A. A., ZHELEZNYAKOV, V. V. and PETELIN, M. I. (1964) *Izv. Vuzov, Radiofiz.* **7**, 251.

* It should be noted that there exist English translations of many of the Russian articles quoted here. Some of these are indicated by an asterisk.

References

APPLETON, E. V. and HEY, J. S. (1946) *Phil. Mag.* **37**, 73.
ARAKAWA, D. (1936) *Rep. Radio Res. Japan* **6**, 31.
ARTEM'YEVA, G. M., BENEDIKTOV, YE. A. and GETMANTSEV, G. G. (1961) *Izv. Vuzov, Radiofiz.* **4**, 831.
ATHAY, R. G. (1959) in *Paris Symposium on Radio Astronomy*, Ed. R. N. BRACEWELL, Stanford Univ. Press, p. 98.
ATHAY, R. G., MENZEL, D. H. (1956) *Ap. J.* **123**, 285.
ATHAY, R. G., MENZEL, D. H., PECKER, J. C. and THOMAS, R. N. (1955) *Ap. J. Suppl.* **1**, 505.
ATHAY, R. G. and MORETON, G. E. (1961) *Ap. J.* **133**, 935.
AVIGNON, Y., BOISHOT, A. and SIMON, P. (1959) in *Paris Symposium on Radio Astronomy*, Ed. R. N. BRACEWELL, Stanford Univ. Press, p. 240.

BABCOCK, H. D. (1959) *Ap. J.* **130**, 364.
BAILEY, D. K. (1959) *Proc. IRE* **47**, 255.
BALDWIN, J. E. (1961) *MN* **122**, 513.
BARABASHOV, N. P. (1952) *Investigation of Physical Conditions on the Moon and the Planets* (in Russian), Khar'kov.
BARRETT, A. H. (1960) *J. Geophys. Res.* **65**, 1835.
BARRETT, A. H. (1961) *Ap. J.* **133**, 281.
BARRETT, A. H. and LILLEY E. (1963) *Sky and Telescope* **25**, 192.
BARROW, C. H. (1960) *Nature* **188**, 924.
BARROW, C. H. and CARR, T. D. (1958) *J. Brit. Astron. Assoc.* **68**, 63.
BARROW, C. H., CARR, T. D. and SMITH, A. G. (1957) *Nature* **180**, 381.
BAUM, F. A., KAPLAN, S. A. and STANYUKOVICH, K. P. (1958) *Introduction to Space Gas Dynamics* (in Russian), Gostekhizdat, Chap. II, sect. 8.
BENEDIKTOV, YE. A. and GETMANTSEV, G. G. (1961) *Izv. Vuzov, Radiofiz.* **4**, 244.
BENEDIKTOV, YE. A., GETMANTSEV, G. G. and GINZBURG, V. L. (1961) *Isk. Sput. Zemli*, No. 7, 3.
BENEDIKTOV, YE. A. and MITYAKOV, N. A. (1961) *Izv. Vuzov, Radiofiz.* **4**, 44.
BEN'KOVA, N. P., TURBIN, R. I. and FLIGEL', M. D. (1961) *Geomagnetizm i Aeronomiya* **1**, 842.
BERGE, G. L. and READ R. B. (1967) The microwave emission of Saturn, Preprint 9 of the Owens Valley Observatory, Caltech.
BERNSTEIN, I. B., GREENE, J. M. and KRUSCAL, M. D. (1957) *Phys. Rev.* **108**, 546.
BESPROZVANNAYA, A. S. and DRIATSKII, V. M. (1960) *MGG, Issled. Ionosfery*, No. 5, 7.
*BIBINOVA, V. P., KUZ'MIN, A. D., SALOMONOVICH, A. YE. and SHAVLOVSKII, I. V. (1962) *Astron. Zhurn.* **39**, 1083.
BILLINGS, D. E., PECKER, C. and ROBERTS, W. O. (1954) *C. R. Acad. Sci. Paris* **238**, 1194.
BLACKWELL, D. E. (1956) *MN* **116**, 56.
*BLOKHINTSEV, D. I. (1961) *Fundamentals of Quantum Mechanics* (in Russian), Vysshaya Shkola.
BLUM, E. (1953) *Ann. Univ. Paris* **23**, 136.
BLUM, E. J., DENISSE, J. F. and STEINBERG, J. L. (1951) *C. R. Acad. Sci.* **232**, 387.
BLUM, E. J., DENISSE, J. F. and STEINBERG, J. L. (1952a) *Ann. d'Astrophys.* **15**, 184.
BLUM, E. J., DENISSE, J. F. and STEINBERG, J. L. (1952b) *C. R. Acad. Sci. Paris* **234**, 1957.
BLUM, E. J., DENISSE, J. F. and STEINBERG, J. L. (1958) *Proc. IRE* **46**, 39.
BOBROV, M. S. (1951) *Doklady Akad. Nauk SSSR* **77**, 581.
BOBROV, M. S. (1952) *Astron. Zhurn.* **29**, 334.
BOBROV, M. S. (1954) *Astron. Zhurn.* **31**, 41.
BOBROV, M. S. (1956) *Astron. Zhurn.* **33**, 161.
BOGORODSKII, A. F. and KHINKULOVA, N. A. (1950) *Byull. Komissii po Issled. Solntsa*, No. 5–6.

References

Bohm, D. and Gross, E. P. (1949a) *Phys. Rev.* **75**, 1851.
Bohm, D. and Gross, E. P. (1949b) *Phys. Rev.* **75**, 1864.
Boishot, A. (1957) *C. R. Acad. Sci. Paris* **244**, 1326.
Boishot, A. (1958) *Ann. d'Astrophys.* **21**, 273.
Boishot, A. (1959) in *Paris Symposium on Radio Astronomy*, Ed. R. N. Bracewell, Stanford Univ. Press, p. 186.
Boishot, A. and Denisse, J. F. (1957) *C. R. Acad. Sci. Paris* **244**, 2194.
Boishot, A., Fokker, A. D. and Simon, P. (1959) in *Paris Symposium on Radio Astronomy*, Ed. R. N. Bracewell, Stanford Univ. Press, p. 263.
Boishot, A., Haddock, F. T. and Maxwell, A. (1960) *Ann. d'Astrophys.* **23**, 478.
Boishot, A., Lee, R. H. and Warwick, J. W. (1960) *Ap. J.* **131**, 61.
Boishot, A. and Simon, P. (1959) in *Paris Symposium on Radio Astronomy*, Ed. R. N. Bracewell, Stanford Univ. Press, p. 140.
Boishot, A. and Simon, P. (1960) *Ann. d'Astrophys.* **23**, 1006.
Boishot, A. and Warwick, J. W. (1959) *J. Geophys. Res.* **64**, 683.
Bolotovskii, B. M. (1957) *Uspekhi Fiz. Nauk* **62**, 201.
Bondar', L. N., Zelinskaya, M. R., Porfir'yev, V. A. and Strezhneva, K. M. (1962) *Izv. Vuzov, Radiofizika* **5**, 802.
Booker, H. G. (1958) *Proc. IRE* **46**, 298.
Boorman, J. A., McLean, D. J., Sheridan, K. V. and Wild, J. P. (1961) *MN* **123**, 87.
Bossome, F., Blum, E. J., Denisse, J. F., Leroux, E. and Steinberg, J. L. (1951) *C. R. Acad. Sci. Paris* **233**, 917.
Bracewell, R. N. (1955) *Australian J. Phys.* **8**, 15.
Bracewell, R. N. (1956) *Australian J. Phys.* **9**, 3.
Brissenden, P. and Erickson, W. C. (1962) *Ap. J.* **136**, 1140.
Brooks, C. and Oster, L. (1961) *Ap. J.* **134**, 942.
Brown, R. H. (1953) *Observatory* **73**, 105.
Brown, R. R. and D'Arcy, R. G. (1959) *Phys. Rev. Lett.* **3**, 390.
Budden, K. G. (1952) *Proc. Roy. Soc.* **215**, 215.
Budejický, J. and Švestka, Z. (1958) *Bull. Astr. Inst. Czechoslov.* **9**, 48.
Burke, B. F. (1960) in *The Radio Noise Spectrum*, Ed. D. H. Menzel, Harvard Univ. Press, p. 129.
Burke, B. F. and Franklin, K. L. (1955) *J. Geophys. Res.* **60**, 213.
Burkhardt, G., Fahl, C. and Larenz, R. W. (1961) *Z. Physik* **161**, 380.
Burkhardt, G. and Schlüter, A. (1949) *Z. Astrophys.* **26**, 295.

Carr, T. D. (1959) *Astron. J.* **64**, 39.
Carr, T. D., Smith, A. G. and Bollhagen, H. (1960) *Phys. Rev. Lett.* **5**, 418.
Carr, T. D., Smith, A. G., Bollhagen, H., Six, N. F. and Chatterton, N. E. (1961) *Ap. J.* **134**, 105.
Carr, T. D., Smith, A. G., Chatterton, N., Six, F. and Bollhagen, H. (1960) *Astron. J.* **65**, 485.
Carr, T. D., Smith, A. G., Pepple, R. and Barrow, C. H. (1958) *Ap. J.* **127**, 274.
Castelli, J. P., Ferioli, C. P. and Aarons, J. (1960) *Astron. J.* **65**, 485.
Chandrasekar, S. (1950) *Radiative Transfer*, Oxford.
Chang, D. B. and Davis, L. (1962) *Ap. J.* **136**, 567.
*Charakhch'yan, A. N., Tulinov, V. F. and Charakhch'yan, T. N. (1961) *Zh. Eksp. i Teoret. Fiz.* **41**, 735.
*Chertoprud, V. Ye. (1963) *Astron. Zhurn.* **40**, 48.
Chikhachev, B. M. (1956) *Proceedings of the 5th Congress on Questions of Cosmogony* (in Russian), Akad. Nauk SSSR, p. 245.
Christiansen, W. N. and Hindman, J. V. (1951) *Nature* **167**, 635.
Christiansen, W. N. and Mathewson, D. S. (1958) *Proc. IRE* **46**, 127.

References

CHRISTIANSEN, W. N. and MATHEWSON, D. S. (1959) in *Paris Symposium on Radio Astronomy*, Ed. R. N. BRACEWELL, Stanford Univ. Press, p. 108.
CHRISTIANSEN, W. N., MATHEWSON, D. S. and PAWSEY, J. L. (1957) *Nature* **180**, 944.
CHRISTIANSEN, W. N., MATHEWSON, D. S., PAWSEY, J. L., SMERD, S. F., BOISHOT, A., DENISSE, J. F., SIMON, P., KAKINUMA, T., DODSON-PRINCE, H. and FIROR, J. (1960) *Ann. d'Astrophys.* **23**, 75.
CHRISTIANSEN, W. N. and WARBURTON, J. A. (1953) *Austral. J. Phys.* **6**, 190, 262.
CHRISTIANSEN, W. N. and WARBURTON, J. A. (1955) *Austral. J. Phys.* **8**, 474.
CHRISTIANSEN, W. N., WARBURTON, J. A. and DAVIES, R. D. (1957) *Austral. J. Phys.* **10**, 491.
CHRISTIANSEN, W. N., YABSLEY, D. E. and MILLS, B. Y. (1949) *Austral. J. Sci. Res.* **A2**, 506.
CHUBB, T. A., FRIEDMAN, H. and KREPLIN, R. W. (1960) *J. Geophys. Res.* **65**, 1831.
COATES, R. J. (1957) *Astron. J.* **62**, 90.
COATES, R. J. (1958a) *Proc. National Electronics Conference, Chicago, October 1957* **13**, 353.
COATES, R. J. (1958b) *Proc. IRE* **46**, 122.
COATES, R. J. (1958c) *Ap. J.* **128**, 33.
COATES, R. J. (1958d) *Ap. J.* **128**, 83.
COATES, R. J. (1960) *Report on the XIII General Assembly URSI*, London.
COATES, R. J. (1961) *Ap. J.* **133**, 723.
COATES, R. J., GIBSON, J. E. and HAGEN, J. P. (1958) *Ap. J.* **128**, 406.
COHEN, M. H. (1958a) *Proc. IRE* **46**, 172.
COHEN, M. H. (1958b) *Proc. IRE* **46**, 183.
COHEN, M. H. (1960) *Ap. J.* **131**, 664.
COHEN, M. H. (1961a) *Ap. J.* **133**, 978.
COHEN, M. H. (1961b) *Phys. Rev.* **123**, 711.
COHEN, M. H. (1962) *J. Geophys. Res.* **67**, 2729.
COHEN, M. H. and DWARKIN, M. (1961) *J. Geophys. Res.* **66**, 411.
COHEN, M. H. and FOKKER, A. D. (1959) in *Paris Symposium on Radio Astronomy*, Ed. R. N. BRACEWELL, Stanford Univ. Press, p. 253.
CONWAY, R. G. (1956) *Observatory* **76**, 106.
CONWAY, R. G. and O'BRIEN, P. A. (1956) *MN* **116**, 386.
CONWAY, R. G., CHUTER, W. L. H. and WILD, P. A. T. (1961) *Observatory* **81**, 106.
COOK, J. J., CROSS, L. G., BAIR, M. E. and ARNOLD, C. B. (1960) *Nature* **188**, 393.
COPELAND, J. and TYLER, W. C. (1964) *Ap. J.* **139**, 409.
COUTREZ, R. (1960) *Radioémission d'origine solaire, Rendiconti della Scuola Internazionale di Fisica "Enrico Fermi"*, Corso XII, Radioastronomia Solare, Bologna.
COUTREZ, R., HUNAERTS, J. and KOECKELENBERGH, A. (1958) *Proc. IRE* **46**, 274.
COVINGTON, A. E. (1948) *Proc. IRE* **36**, 454.
COVINGTON, A. E. (1949) *Proc. IRE* **37**, 407.
COVINGTON, A. E. (1951a) *J. Roy. Astron. Soc. Canada* **45**, 157.
COVINGTON, A. E. (1951b) *J. Roy. Astron. Soc. Canada* **45**, 15, 49.
COVINGTON, A. E. (1954) *J. Geophys. Res.* **59**, 163.
COVINGTON, A. E. (1958) *J. Roy. Astron. Soc. Canada* **52**, 161.
COVINGTON, A. E. (1959) in *Paris Symposium on Radio Astronomy*, Ed. R. N. BRACEWELL, Stanford Univ. Press, p. 159.
COVINGTON, A. E. and BROTEN, H. W. (1954) *Ap. J.* **119**, 569.
COVINGTON, A. E. and MEDD, W. J. (1954) *J. Roy. Astron. Soc. Canada* **48**, 136.
COVINGTON, A. E., MEDD, W. J., HARVEY, C. A. and BROTEN, N. W. (1955) *J. Roy. Astron. Soc. Canada* **49**, 235.

DAS, A. K., SETHUMADHAVAN, K. and DAVIES, R. D. (1953) *Nature* **172**, 446.

References

DANILOV, A. D. and YATSENKO, S. P. (1962) *Geomagnetizm i Aeronomiya* **2**, 363.
DANILOV, A. D. and YATSENKO, S. P. (1963) *Geomagnetizm i Aeronomiya* **3**, 585, 594.
DAVIES, R. D. (1954) *MN* **114**, 74.
DAVIES, R. D., BEARD, M. and COOPER, B. F. C. (1964) *Phys. Rev. Lett.* **13**, 325.
DAVIES, R. D. and JENNISON, R. C. (1960) *Observatory* **80**, 74.
DE GROOT, T. (1960) *Rendiconti della scuola internazionale di fisica "Enrico Fermi"*, Corso XII, Radioastronomia Solare, Bologna, p. 371.
DE JAGER, C. (1958) *Probl. Plasmas in Phys. and Astron.*, Berlin, Acad.-Verl., p. 38.
DE JAGER, C. (1959a) in *Paris Symposium on Radio Astronomy*, Ed. R. N. BRACEWELL, Stanford Univ. Press, p. 89.
DE JAGER, C. (1959b) in *Handbuch der Physik*, Vol. 52, Astrophysik III, Ed. S. FLÜGGE, Springer-Verlag.
DE JAGER, C. (1960) *Rendiconti della scuola internazionale di fisica "Enrico Fermi"*, Corso XII, Radioastronomia Solare, Bologna, p. 313.
DE JAGER, C. and VAN'T VEER, F. (1957) *Radio Astronomy*, Ed. H. C. VAN DE HULST, Cambridge Univ. Press, p. 366.
DENISOV, N. G. (1954) Thesis (in Russian), Gor'kii, Gor'k. Gos. Univ.
*DENISOV, N. G. (1955) *Zh. Eksp. i Teoret. Fiz.* **29**, 380.
DENISOV, N. G. (1957) *Uch. Zap. Gor'k. Un-ta, Ser. Fiz.* **35**, 3.
*DENISOV, N. G. (1958) *Zh. Eksp. i Teoret. Fiz.* **34**, 528.
DENISOV, N. G. (1959) *Radiotekhn. i Elektron.* **4**, 388.
DENISSE, J. F. (1949) *C. R. Acad. Sci. Paris* **228**, 751.
DENISSE, J. F. (1950) *Ann. d'Astrophys.* **13**, 181.
DENISSE, J. F. (1953) *Ciel et Terre* **59**, No. 3–4, 53.
DENISSE, J. F. (1959a) in *Paris Symposium on Radio Astronomy*, Ed. R. N. BRACEWELL, Stanford Univ. Press, p. 81.
DENISSE, J. F. (1959b) in *Paris Symposium on Radio Astronomy*, Ed. R. N. BRACEWELL, Stanford Univ. Press, p. 237.
DENISSE, J. F. and KUNDU, M. R. (1957) *C. R. Acad. Sci. Paris* **244**, 45.
DENISSE, J. F. and ROCARD, Y. (1951) *J. Physique et Radium* **12**, 893.
DICKE, R. H. and BERINGER, R. (1946) *Ap. J.* **103**, 375.
DMITRENKO, D. A., KAMENSKAYA, S. A. and RAKHLIN, V. L. (1964) *Izv. Vuzov, Radiofiz.* **7**, No. 3.
DOBROVOL'SKII, O. V. (1961) Non-stationary processes in comets and solar activity (in Russian), *Tr. In-ta Astrofiz. Akad. Nauk Tadzh. SSR*, **8**.
DODSON, H. W. (1954) *Ap. J.* **119**, 564.
DODSON, H. W. (1957) in *Radio Astronomy*, Ed. H. C. VAN DE HULST, Cambridge Univ. Press, p. 327.
DODSON, H. W. (1958) *Proc. IRE* **46**, 149.
DODSON, H. W. and HEDEMAN, E. R. (1957) *Ap. J.* **125**, 827.
DODSON, H. W., HEDEMAN, E. R. and COVINGTON, A. E. (1954) *Ap. J.* **119**, 541.
DODSON, H. W., HEDEMAN, E. R. and OWREN, L. (1953) *Ap. J.* **118**, 169.
DORMAN, L. I. (1957) *Cosmic Ray Variations* (in Russian), Gostekhizdat, pp. 10, 11.
DORMAN, L. I. and KOLOMEYETS, YE. V. (1961) *Geomagnetizm i Aeronomiya* **1**, 655.
DOUGLAS, J. N. (1960) *Astron. J.* **65**, 487.
DOUGLAS, J. N. and SMITH, H. J. (1963) *Astron. J.* **68**, 163.
DOUGLAS, J. N., and SMITH, H. J. (1967) *Ap. J.* **148**, 885
DRAKE, F. D. (1962a) *Publ. NRAO* **1**, No. 11, 165.
DRAKE, F. D. (1962b) *Nature* **195**, 893.
DRAKE, F. D. (1962c) *Nature* **195**, 894.
DRAKE, F. D. (1963) *Sky and Telescope* **25**, 73.
DRAKE, F. D. and EWEN, H. T. (1958) *Proc. IRE* **46**, 53.
DRAKE, F. D. and HVATUM, S. (1959) *Astron. J.* **64**, 329.
DRAVSKIKH, A. F. (1960a) Paper read at the Plenum of the Radioastronomy Commis-

References

sion of the U.S.S.R. Academy of Sciences Astro-Council.
DRAVSKIKH, A. F. (1960b) *Izv. Glav. astron. obs.* **21**, No. 164, 128.
DRAVSKIKH, Z. V. and DRAVSKIKH, A. F. (1961) *Solnechnyye Dannyye* No. 10.

EDELSON, S., COATES, R. J., SANTINI, N. and MCCULLOUGH, T. F. (1959) *Astron. J.* **64**, 330.
EDELSON, S. and GRANT, C. (1960) *Astron. J.* **65**, 488.
EDELSON, S., GRANT, C. G. and CORBETT, H. (1960) *Astron. J.* **65**, 51.
EDELSON, S., MCCULLOUGH, T., SANTINI, N. and COATES, R. J. (1960) *Report on XIII General Assembly URSI*, London.
*EIDMAN, V. YA. (1958) *Zh. Eksp. i Teoret. Fiz.* **34**, 131.
*EIDMAN, V. YA. (1959) *Zh. Eksp. i Teoret. Fiz.* **36**, 1335.
EIGENSON, M. S., GNEVYSHEV, M. N., OL', A. I. and RUBASHEV, B. M. (1948) *Solar Activity and its Manifestations on Earth* (in Russian), Gostekhizdat.
ELGARØY, Ø. (1957) *Nature* **180**, 862.
ELGARØY, Ø. (1959) in *Paris Symposium on Radio Astronomy*, Ed. R. N. BRACEWELL, Stanford Univ. Press, p. 248.
ELLIS, G. R. A. (1962a) *Nature* **194**, 667.
ELLIS, G. R. A. (1962b) *Austral. J. Phys.* **15**, 344.
ELLIS, G. R. A. (1963) *Austral. J. Phys.* **16**, 74.
ELLIS, G. R. A. and MCCULLOUGH, P. M. (1963) *Nature* **198**, 275.
ELLISON, M. A. (1955) *The Sun and its Influence*, Macmillan.
ELSÄSSER, H. (1957) *Mitt. Astr. Ges.* **2**, 61.
ELSMORE, B. (1959) in *Paris Symposium on Radio Astronomy*, Ed. R. N. BRACEWELL, Stanford Univ. Press, p. 47.
ELWERT, G. (1961) *J. Geophys. Res.* **66**, 391.
EPSTEIN, E. E. (1959) *Nature* **184**, 52.
ERICKSON, W. C. (1959) *Phys. Rev. Lett.* **3**, 365.
ERICKSON, W. C. (1961) *J. Geophys. Res.* **66**, 1773.
ERICKSON, W. C. and BRISSENDEN, P. (1962) *Ap. J.* **136**, 1138.
EVANS, J. V. (1956) *Proc. Phys. Soc.* B **69**, 953.

FAIN, V. M. (1963) *Izv. Vuzov., Radiofiz.* **6**, 3.
FAINBERG, YA. B., KURILKO, V. I. and SHAPIRO, V. D. (1961) *Zh. Tekh. Fiz.* **31**, 633.
FEDORCHENKO, A. M. (1962) *Radiotekhn. i Elektron.* **7**, 1455.
FEDOSYEV, L. N. (1963) *Izv. Vuzov, Radiofiz.* **6**, 655.
FEIER, J. A. (1960) *Canad. J. Phys.* **38**, 1114.
FEIER, J. A. (1961) *Canad. J. Phys.* **39**, 716.
FEINSTEIN, J. (1952) *Phys. Rev.* **85**, 245.
FENSLER, W. E., KNOTT, E. F., OLTE, A. and SIEGEL, K. M. (1962) *IAU Symposium 14*, Acad. Press.
FIELD, G. B. (1956) *Ap. J.* **124**, 555.
FIELD, G. B. (1959) *J. Geophys. Res.* **64**, 1169.
FIELD, G. B. (1960) *J. Geophys. Res.* **65**, 1661.
FIELD, G. B. (1961a) *J. Geophys. Res.* **66**, 1395.
FIELD, G. B. (1961b) *Astron. J.* **66**, 283.
FIROR, J. (1955) *Astron. J.* **60**, 161.
FIROR, J. (1956) *Ap. J.* **123**, 320.
FIROR, J. (1957) in *Radio Astronomy*, Ed. H. C. VAN DE HULST, Cambridge Univ. Press, p. 294.
FIROR, J. (1959a) in *Paris Symposium on Radio Astronomy*, Ed. R. N. BRACEWELL, Stanford Univ. Press, p. 107.
FIROR, J. (1959b) in *Paris Symposium on Radio Astronomy*, Ed. R. N. BRACEWELL, Stanford Univ. Press, p. 136.

References

FLEISCHER, R. and OSHIMA, M. (1961) *Astron. J.* **66**, 43.
FLÜGGE, S. (Ed.) (1959) *Handbuch der Physik*, Vol. 52, Astrophysik III, Springer-Verlag.
FOK, V. A. (1946a) *Tables of Airy Functions* (in Russian).
FOK, V. A. (1946b) *Diffraction of Radio Waves* (in Russian), Akad. Nauk SSSR.
FOKKER, A. D. (1957) in *Radio Astronomy*, Ed. H. C. VAN DE HULST, Cambridge Univ. Press, p. 371.
FOKKER, A. D. (1960a) *Rendiconti della scuola internazionale "Enrico Fermi"*, Corso XII, Radioastronomia Solare, Bologna, p. 374.
FOKKER, A. D. (1960b) *ibid.*, p. 385.
FRANKLIN, K. L. (1959) *Astron. J.* **64**, 37.
FRANKLIN, K. L. and BURKE, B. F. (1956) *Astron. J.* **61**, 177.
FRANKLIN, K. L. and BURKE, B. F. (1958) *J. Geophys. Res.* **63**, 807.
FREIER, P. S., NEY, E. P. and WINCKLER, J. R. (1959) *J. Geophys. Res.* **64**, 685.
FURRY, W. H. (1947) *Phys. Rev.* **71**, 360.

*GALEYEV, A. A. and KARPMAN, V. I. (1963) *Zh. Eksp. i Teoret. Fiz.* **44**, 592.
GALLET, R. M. (1958) *Report of URSI-USA Nation. Comm.*, *XII General Assembly*, Nat. Acad. Sci., Nation. Res. Council Pub. No. 581, 143.
GALLET, R. M. (1961) in *Planets and Satellites. Solar. System. III*, ed. G. P. KUIPER and B. M. MIDDLEHURST, Univ. Chicago Press.
GALLET, R. M. and BOWLES K. L. (1956) *Astron. J.* **61**, 194.
GARDNER, F. F. (1957) *Symposium on Radio Astronomy, Sydney, September 1956*, Melbourne, p. 38.
GARDNER, F. F. and SHAIN, C. A. (1958) *Austral. J. Phys.* **11**, 55.
GARSTANG, R. H. (1958) *J. Brit. Astron. Association* **68**, 155.
GARY, B. (1963) *Astron. J.* **68**, 568.
GEBBIE, H. A. (1957) *Phys. Rev.* **107**, 1194.
GELFREICH, G. B. (1962) *Solnechnyye Dannyye*, No. 5, 67.
GELFREICH, G. B., IKHSANOVA, V. N., KAIDANOVSKII, N. L., SOBOLEVA, N. S., TIMOFEYEVA, G. M. and UMETSKII, V. N. (1959) in *Paris Symposium on Radio Astronomy*, Ed. R. N. BRACEWELL, Stanford Univ. Press, p. 218.
GELFREICH, G. B., KOROL'KOV, D., PICHKOV, N. and SOBOLEVA, N. (1959) in *Paris Symposium on Radio Astronomy*, Ed. R. N. BRACEWELL, Stanford Univ. Press, p. 125.
GEOFFRION, A. R., KORNER, M. and SINTON, W. M. (1961) *Lovell Observatory Bull.* **5**, 106.
Geomagnetizm i Aeronomiya (1961) **1**, 844.
GERSHMAN, B. N. (1953) *Zh. Eksp. i Teoret. Fiz.* **24**, 659.
*GERSHMAN, B. N. (1959) *Zh. Eksp. i Teoret. Fiz.* **37**, 695.
*GERSHMAN, B. N. (1960) *Zh. Eksp. i Teoret. Fiz.* **38**, 912.
GERSHMAN, B. N. and GINZBURG, V. L. (1962) *Izv. Vuzov, Radiofiz.* **5**, 31.
GERSHMAN, B. N. and ZHELEZNYAKOV, V. V. (1956) *Proc. of the 5th Congress on Questions of Cosmogony* (in Russian), Akad. Nauk SSSR, p. 273.
GERTSENSHTEIN, M. YE. (1952) *Zh. Eksp. i Teoret. Fiz.* **23**, 669.
GETMANTSEV, G. G. and GINZBURG, V. L. (1952) *Doklady Akad. Nauk SSSR* **87**, 187.
GETMANTSEV, G. G., GINZBURG, V. L. and SHKLOVSKII, I. S. (1958) *Uspekhi Fiz. Nauk* **66**, 157.
GETMANTSEV, G. G. and RAPOPORT, V. O. (1960) *Zh. Eksp. i Teoret. Fiz.* **38**, 1205.
GETMANTSEV, G. G. and RAZIN, V. A. (1956) *Proc. of the 5th Congress on Questions of Cosmogony* (in Russian), Akad. Nauk SSSR, p. 496.
GIBSON, J. E. (1958) *Proc. IRE* **46**, 280.
GIBSON, J. E. (1961) *Ap. J.* **133**, 1072.
GIBSON, J. E. and CORBETT, H. H. (1963) *Astron. J.* **68**, 74.

References

GIBSON, J. E. and MCEWAN, R. J. (1959) in *Paris Symposium on Radio Astronomy*, Ed. R. N. BRACEWELL, Stanford Univ. Press, p. 50.
GINZBURG, V. L. (1940) *Zh. Eksp. i Teoret. Fiz.* **10**, 601.
GINZBURG, V. L. (1946) *Doklady Akad. Nauk SSSR* **52**, 491.
GINZBURG, V. L. (1947) *Uspekhi Fiz. Nauk* **32**, 26.
GINZBURG, V. L. (1948) *Zh. Eksp. i Teoret. Fiz.* **18**, 487.
GINZBURG, V. L. (1953) *Doklady Akad. Nauk SSSR* **92**, 527.
GINZBURG, V. L. (1959) *Uspekhi Fiz. Nauk* **64**, 537.
GINZBURG, V. L. (1960a) *Izv. Vuzov, Radiofiz.* **3**, 341.
*GINZBURG, V. L. (1960b) *Propagation of Electromagnetic Waves in a Plasma* (in Russian), Fizmatgiz.
GINZBURG, V. L. and FRANK, I. M. (1946) *Zh. Eksp. i Teoret. Fiz.* **16**, 25.
GINZBURG, V. L. and FRANK, I. M. (1947) *Doklady Akad. Nauk SSSR* **56**, 583.
GINZBURG, V. L. and GETMANTSEV, G. G. (1950) *Zh. Eksp. i Teoret. Fiz.* **20**, 347.
GINZBURG, V. L. and PISAREVA, V. V. (1956) *Proc. of the 5th Congress on Questions of Cosmogony* (in Russian), Akad. Nauk SSSR, p. 229.
GINZBURG, V. L. and SYROVATSKII, S. I. (1964) *The Origin of Cosmic Rays*, English transl. Pergamon Press.
GINZBURG, V. L. and ZHELEZNYAKOV, V. V. (1958a) *Izv. Vuzov, Radiofiz.* **1**, No. 5-6, 9.
GINZBURG, V. L. and ZHELEZNYAKOV, V. V. (1958b) *Astron. Zhurn.* **35**, 694.
GINZBURG, V. L. and ZHELEZNYAKOV, V. V. (1958c) *Izv. Vuzov, Radiofiz.* **1**, No. 2, 59.
GINZBURG, V. L. and ZHELEZNYAKOV, V. V. (1959a) *Astron. Zhurn.* **36**, 233.
GINZBURG, V. L. and ZHELEZNYAKOV, V. V. (1959b) in *Paris Symposium on Radio Astronomy*, Ed. R. N. BRACEWELL, Stanford Univ. Press, p. 574.
GINZBURG, V. L. and ZHELEZNYAKOV, V. V. (1960) Paper read at the Plenum of the Radio Astronomy Commission of the U.S.S.R. Acad. of Sci., Moscow.
*GINZBURG, V. L. and ZHELEZNYAKOV, V. V. (1961) *Astron. Zhurn.* **38**, 3.
GINZBURG, V. L. and ZHELEZNYAKOV, V. V. (1965) *Phil. Mag.* **11**, 197.
GINZBURG, V. L., ZHELEZNYAKOV, V. V. and EIDMAN, V. YA. (1962) *Phil. Mag.* **7**, 451.
GIORDMAINE, J. A., ALSOP, L. E., MAYER, C. H. and TOWNES, C. H. (1959) *Proc. IRE* **47**, 1062.
GIOVANELLI, A. (1948) *Austral. J. Sci. Res.* A **1**, 360.
GIOVANELLI, A. (1958) *Austral. J. Phys.* **2**, 350.
GIOVANELLI, A. (1959) in *Paris Symposium on Radio Astronomy*, Ed. R. N. BRACEWELL, Stanford Univ. Press, p. 214.
GIOVANELLI, R. G. and ROBERTS, J. A. (1958) *Austral. J. Phys.* **11**, 353.
GIOVANELLI, R. G. and ROBERTS, J. A. (1959) in *Paris Symposium on Radio Astronomy*, Ed. R. N. BRACEWELL, Stanford Univ. Press, p. 201.
*GNEVYSHEV, M. N. (1960) *Astron. Zhurn.* **37**, 227.
GOLDBERG, L. (1958) *Astron. J.* **63**, 366.
GOLDBERG, L. (1960) *Astron. J.* **65**, 539.
GOLDSTEIN, S. J. (1959) *Ap. J.* **130**, 393.
GOL'NEV, V. YA., IKHSANOVA, V. N., LESNIK, G. E. and SOBOLEVA, N. S. (1961) *Solnechnyye Dannyye* No. 10.
GOL'NEV, V. YA., PARIISKII, YU. N. and SOBOLEVA, N. S. (1963) *Scientific Conference on Radio Astronomy, Gor'kii, 26 February to 2 March 1963, Abstracts of Papers* (in Russian), Akad. Nauk SSSR.
GOODMAN, J. and LEBENBAUM, M. (1958) *Proc. IRE* **46**, 132.
GORDEYEV, G. V. (1953) *Zh. Eksp. i Teoret. Fiz.* **24**, 445.
GORDEYEV, G. V. (1954) *Zh. Eksp. i Teoret. Fiz.* **27**, 19, 24.
GORDON, I. M. (1954) *Doklady Akad. Nauk SSSR* **94**, 813.
*GORDON, I. M. (1960) *Astron. Zhurn.* **37**, 934.
GORDY, W., DITTO, S. J., WYMAN, J. H. and ANDERSON, R. S. (1955) *Phys. Rev.* **99**, 1095.

References

*GORELIK, A. G. and KOSTAREV, V. V. (1959) *Doklady Akad. Nauk SSSR* **125**, 59.
GORELIK, G. S. (1950) *Oscillations and Waves* (in Russian), Gostekhizdat.
GOROKHOV, N. A., DRYAGIN, YU. A. and FEDOSEYEV, L. I. (1962) *Izv. Vuzov, Radiofiz* **5**, 413.
GOSACHINSKII, I. V., YEGOROVA, T. M. and RYZHKOV, N. F. (1961) *Solnechnyye Dannyye*, No. 7, 70.
GRANDJEAN, J. and GOODY, R. M. (1955) *Ap. J.* **121**, 548.
GRANT, G. R. and CORBETT, H. H. (1962) *Astron. J.* **67**, 115.
GREBENKEMPER, C. J. (1958) NRL Report No. 5151.
GROOT, T. DE, (1959) in *Paris Symposium on Radio Astronomy*, Ed. R. N. BRACEWELL, Stanford Univ. Press, p. 245.
GROSS, E. P. (1958) *Proc. Symp. Electronic Waveguides*, N.Y., p. 43.
GUTMANN, M. and STEINBERG, J. L. (1959) in *Paris Symposium on Radio Astronomy*, Ed. R. N. BRACEWELL, Stanford Univ. Press, p. 123.
HACHENBERG, O. (1960) *Rendiconti della scuola Internazionale "Enrico Fermi"*, Corso XII, Radioastronomia Solare, Bologna, p. 217.
HACHENBERG, O., FÜRSTENBERG, F. and PRINZLER, H. (1956) *Z. Astrophys.* **39**, 232.
HACHENBERG, O. and KRÜGER, A. (1959) *J. Atm. Terr. Phys.* **17**, 20.
HACHENBERG, O. and VOLLAND, H. (1959) *Z. Astrophys.* **47**, 69.
HACHENBERG, O. and WALLIS, G. (1960) *Report on XIII General Assembly URSI*, London.
HADDOCK, F. T. (1957) in *Radio Astronomy*, Ed. H. C. VAN DE HULST, Cambridge Univ. Press, p. 273.
HADDOCK, F. T. (1958) *Proc. IRE* **46**, 3.
HADDOCK, F. T. (1959a) *Amer. Rocket Soc.* No. 794.
HADDOCK, F. T. (1959b) in *Paris Symposium on Radio Astronomy*, Ed. R. N. BRACEWELL, Stanford Univ. Press, p. 188.
HAGEN, J. P. (1949) NRL Report No. 3504.
HAGEN, J. P. (1951) *Ap. J.* **113**, 547.
HAGEN, J. P. (1956) *Solar Eclipses and Ionosphere* (Special suppl., vol. 6 to *J. Atmosph. Terr. Phys.*), Pergamon Press, p. 253.
HAGEN, J. P. (1957) in *Radio Astronomy*, Ed. H. C. VAN DE HULST, Cambridge Univ. Press, p. 263.
HAGEN, J. P., HADDOCK, F. T. and REBER, G. (1951) *Sky and Telescope* **10**, 111.
HAGEN, J. P. and HEPBURN, N. (1952) *Nature* **170**, 244.
HAKURA, Y. and GOH, T. (1959) *Radio Res. Lab. Japan* **6**, 635.
HAKURA, Y., TAKENOSHITA, Y. and OTSUKI, T. (1958) *Resp. Ionosph. Res. Japan* **12**, 459.
HATANAKA, T. (1956) *Publ. Astron. Soc. Japan* **8**, 73.
HATANAKA, T. (1957a) in *Radio Astronomy*, Ed. H. C. VAN DE HULST, Cambridge Univ. Press.
HATANAKA, T. (1957b) *Ibid.*, p. 358.
HATANAKA, T., AKABANE, K., MORIYAMA, F., TANAKA, H. and KAKINUMA T. (1956) *Solar Eclipses and Ionosphere*, Pergamon Press, p. 264.
HATANAKA, T. and MORIYAMA, F. (1952) *Rept. Ionosphere Res. Japan* **6**, 99.
HATANAKA, T., SUZUKI, S. and TSUCHIYA, A. (1955a) *Proc. Japan Acad.* **31**, 81.
HATANAKA, T., SUZUKI, S. and TSUCHIYA, A. (1955b) *Astron. J.* **60**, 162.
HEIGHTMAN, D. W. (1938) *Wireless World* **42**, 356.
HEITLER, W. (1954) *The Quantum Theory of Radiation* Oxford, The Clarendon Press
HERZBERG, (1952) in *The Atmospheres of the Earth and Planets*, Ed. G. P. KUIPER, Univ. Chicago Press.
HEWISH, A. (1957) in *Radio Astronomy*, Ed. H. C. VAN DE HULST, Cambridge Univ. Press, p. 298.
HEWISH, A. (1958) *MN* **118**, 534.

References

HEWISH, A. (1959) in *Paris Symposium on Radio Astronomy*, Ed. R. N. BRACEWELL, Stanford Univ. Press, p. 268.
HEY, J. S. (1946) *Nature* **157**, 47.
HEY, J. S. (1957) in *Radio Astronomy*, Ed. H. C. VAN DE HULST, Cambridge Univ. Press, p. 278.
HEY, J. S. and HUGHES, V. A. (1955) *MN* **115**, 605.
HEY, J. S. and HUGHES, V. A. (1958) *Proc. IRE* **46**, 119.
HEY, J. S., PARSONS, S. J. and PHILLIPS, J. W. (1948) *MN* **108**, 354.
HOFFMAN, R. A., DAVIS, L. R. and WILLIAMSON, J. M. (1962) *J. Geophys. Res.* **67**, 5001.
HÖGBOM, J. A. (1959) in *Paris Symposium on Radio Astronomy*, Ed. R. N. BRACEWELL, Stanford Univ. Press, p. 253.
HORNER, F. (1957) *Nature* **180**, 1253.
HOWARD, W. E., BARRET, A. H. and HADDOCK, F. T. (1961) *Astron. J.* **66**, 287.
HOWARD, W. E., BARRET, A. H. and HADDOCK, F. T. (1962) *Ap. J.* **136**, 995.
HRUŠKA, A. (1962) *Byull. Astron. Institutov Chekhoslovakii* **13**, 125.
HUGHES, M. P. and HARKNESS, R. L. (1963) *Ap. J.* **138**, 239.

IKHSANOVA, V. (1959) in *Paris Symposium on Radio Astronomy*, Ed. R. N. BRACEWELL, Stanford Univ. Press, p. 171.
IKHSANOVA, V. (1960a) *Izv. Gos. Astron. Obs.* **21**, No. 5, 62.
IKHSANOVA, V. (1960b) Paper read at the Plenum of the Radio Astronomy Commission, Moscow.
*IVANOV-KHOLODNYI, G. S. and NIKOL'SKII, G. M. (1961) *Astron. Zhurn.* **38**, 45.
*IVANOV-KHOLODNYI, G. S. and NIKOL'SKII, G. M. (1962) *Astron. Zhurn.* **39**, 777.

JAEGER, J. C. (1943) *Proc. Camb. Phil. Soc.* **49**, 355.
JAEGER, J. C. (1959) *Nature* **183**, 1316.
JAEGER, J. C. and HARPER, A. F. (1950) *Nature* **166**, 1026.
JAEGER, J. C., WESTFOLD, K. C. (1950) *Austral. J. Sci. Res.* **A3**, 376.
JELLEY, J. V. (1963) *Observatory* **83**, 61.
JELLEY, J. V. and PETFORD, A. D. (1961) *Observatory* **81**, 104.
JENNISON, R. C. (1959) *Observatory* **79**, 111.
JONES, D. E. (1961) *Planet. Space Sci.* **5**, 166.
JORAND, M. (1953) *Ann. Astrophys.* **16**, 151.

KAIDANOVSKII, N. L. (1960) *Radio Emission of the Moon*. In coll. *The Moon* (in Russian), Ed. A. V. MARKOVA, Fizmatgiz.
KAIDANOVSKII, N. L., ICHSANOVA, B. N., APUSHKINSKII, G. P., SHIVZIS, O. N. (1961) *Izv. Vuzov, Radiofiz.* **4**, 428.
KAIDANOVSKII, N. L., MIRZABEKYAN, E. G. and KHAIKIN, S. E. (1956) *Proc. of the 5th Congress on Questions of Cosmogony* (in Russian), Izd. Akad. Nauk SSSR, p. 113.
KAIDANOVSKII, N. L., MOLCHANOV, A. P. and PETEROVA, N. G. (1960) Paper read at the Plenum of the Radio Astronomy Commission, Moscow.
KAIDANOVSKII, N. L., TURUSBEKOV, M. T. and KHAIKIN S. E. (1956) *Proc. of the 5th Congress on Questions of Cosmogony* (in Russian) Izd. Akad. Nauk. USSR, p. 347.
KAKINUMA, T. (1955) *Proc. Res. Inst. Atm. Nagoya Univ.* **3**, 96.
KAKINUMA, T. (1956) *Proc. Res. Inst. Atm. Nagoya Univ.* **4**, 78.
KAKINUMA, T. (1958) *Proc. Res. Inst. Atm. Nagoya Univ.* **5**, 71.
KAKINUMA, T. and SWARUP, G. (1962a) *Stanford Radio Astron. Inst. Publ.* No. 14.
KAKINUMA, T. and SWARUP, G. (1962b) *Ap. J.* **136**, 975.
KAMENSKAYA, C. A., SEMENOV, B. I., TROITSKII, V. S., PLECHKOV, V. M. (1962) *Izv. Vuzov, Radiofiz.* **5**, 882.
KAMIYA, Y. (1961) *J. Geomagnetism and Geoelectricity* **13**, 33.
KAMIYA, Y. (1962) *J. Phys. Soc. Japan* **17**, Suppl. A-II, 391.

References

*KAPITSA, S. P. (1960) *Zh. Eksp. i Teoret. Fiz.* **39**, 1367.
KAWABATA, K. (1954) *Rep. Ionosphere Res. Japan* **8**, 143.
KAWABATA, K. (1960a) *Publ. Astron. Soc. Japan* **12**, 513.
KAWABATA, K. (1960b) *Rep. Ionosph. Res. Japan* **14**, 405.
KELLOGG, W. W. and SAGAN, C. (1961) *The Atmospheres of Mars and Venus*, Nat. Acad. Sci.
KERBLAI, T. S. and KOVALEVSKAYA, YE. M. (1960) *MGG, Issled. Ionosf.*, No. 3, 22.
KHAIKIN, S. E. (1960) Paper read at the Plenum of the Radio Astronomy Commission, Moscow.
KHAIKIN, S. E. and CHIKHACHEV, B. M. (1947) *Doklady Akad. Nauk SSSR* **58**, 1923.
*KHAIKIN, S. E. and KAIDANOVSKII, N. L. (1959) *Pribory i Tekhnika Eksperimenta* **2**, 19.
KHAIKIN, S. E., KAIDANOVSKII, N. L., YESEPKINA, N. A. and SHIVRIS, O. N. (1960) *Izv. GAO* **21**, No. 164, 3.
KHANIN, YA. I. and YUDIN, O. I. (1955) *Astron. Zhurn.* **32**, 439.
KIEPENHERZ, K. O. (1946) *Nature* **158**, 340.
KIESS, C. C., CORLISS, C. H. and KIESS, H. K. (1960) *Ap. J.* **132**, 221.
KISLYAKOV, A. G. (1961a) *Izv. Vuzov, Radiofiz.* **6**, 655.
*KISLYAKOV, A. G. (1961b) *Astron. Zhurn.* **38**, 561.
KISLYAKOV, A. G. (1961c) *Izv. Vuzov, Radiofiz.* **4**, 433.
KISLYAKOV, A. G. (1961d) *Izv. Vuzov, Radiofiz.*, **4**, 760.
KISLYAKOV, A. G., KUZ'MIN, A. D. and SALOMONOVICH, A. YE. (1961) *Izv. Vuzov, Radiofiz.* **4**, 573.
*KISLYAKOV, A. G., KUZ'MIN, A. D. and SALOMONOVICH, A. YE. (1962) *Astron. Zhurn.* **39**, 410.
KISLYAKOV, A. G., LOSOVSKII, B. YA. and SALOMONOVICH, A. YE. (1963) *Izv. Vuzov, Radiofiz.* **6**, 192.
KISLYAKOV, A. G. and PLECHKOV, V. M. (1963) *Izv. Vuzov, Radiofiz.* **6**, No. 6.
KISLYAKOV, A. G. and SALOMONOVICH, A. YE. (1963a) *Astron. Zhurn.* **40**, 229.
KISLYAKOV, A. G., SALOMONOVICH, A. YE. (1963b) *Izv. Vuzov, Radiofiz.*, **6**, 431.
KOLOMENSKII, A. L. (1953) *Zh. Eksp. i Teoret. Fiz.* **24**, 167.
KOLOMEYETS, YE. V. (1961) *Geomagnetizm i Aeronomiya* **1**, 41.
KOMESAROFF, M. (1958) *Austral. J. Phys.* **11**, 201.
KONTOROVICH, V. M. and KUTIK, I. N. (1963) *Izv. Vuzov, Radiofiz.* **6**, 1129.
KORCHAK, A. A. (1963a) *Geomagnetizm i Aeronomiya* **3**, 394.
*KORCHAK, A. A. (1963b) *Astron. Zhurn.* **40**, 994.
KORCHAK, A. A. and LOTOVA, N. A. (1963) *Geomagnetizm i Aeronomiya* **3**, 37.
KOROL'KOV, D. V. (1962) Thesis (in Russian), Gor'kii.
*KOROL'KOV, D. V., PARIISKII, YU. N., TIMOFEYEVA, G. M. and KHAIKIN, S. E. (1963) *Doklady Akad. Nauk SSSR* **149**, 65.
KOROL'KOV, D. V. and SOBOLEVA, N. S. (1957) *Solnechnyye Dannyye* No. 1, 149.
*KOROL'KOV, D. V. and SOBOLEVA, N. S. (1961) *Astron. Zhurn.* **38**, 647.
KOROL'KOV, D. V., SOBOLEVA, N. S. and GELFREICH, G. B. (1960) *Izv. GAO* **21**, No. 5, 81.
KOSCHENKO, V. M., LOSOVSKII, B. YA., SALOMONOVICH, A. YE. (1961) *Izv. Vuzov, Radiofiz.* **4**, 396.
KOSCHENKO, V. N., KUZ'MIN, A. D. and SALOMONOVICH, A. YE. (1961) *Izv. Vuzov, Radiofiz.*, **4**, 425.
KOTELNIKOV, V. A. (1961) *J. Brit. IRE* **22**, 293.
*KOTEL'NIKOV, V. A., DUBROVIN, V. M., KISLIK, M. D., KORENBERG, YE. B., MINASHIN, V. P., MOROZOV, N. A., NIKITSKII, N. I., PETROV, G. M., RZHIGA, O. N. and SHAKHOVSKII, A. I. (1962) *Doklady Akad. Nauk SSSR* **145**, No. 5.
KOVNER, M. S. (1960a) *Izv. Vuzov. Radiofiz.* **3**, 631.
KOVNER, M. S. (1960b) *ibid.* 746.
KOZYREV, N. A. (1954) *Izv. Krymsk. Astrofiz. Observ.* **12**, 169, 177.

References

KRAUS, J. D. (1946a) *Nature* **178,** 33, 103.
KRAUS, J. D. (1956b) *Nature* **178,** 159.
KRAUS, J. D. (1956c) *Astron. J.* **61,** 182.
KRAUS, J. D. (1957a) *Nature* **179,** 371.
KRAUS, J. D. (1957b) *Astron. J.* **62,** 21.
KRAUS, J. D. (1958a) *Astron. J.* **63,** 55.
KRAUS, J. D. (1958b) *Proc. IRE* **46,** 266.
KRAUS, J. D. (1960) *Nature* **186,** 462.
KRISHNAN, T. and LABRUM, N. R. (1961) *Austral. J. Phys.* **14,** 403.
KRISHNAN, T. and MULLALY, R. F. (1961) *Nature* **192,** 58.
KRISHNAN, T. and MULLALY, R. F. (1962) *Austral. J. Phys.* **15,** 86.
KRITZ, A. H. and MINTZER, D. (1960) *Phys. Rev.* **117,** 382.
KROTIKOV, V. D. (1962a) *Izv. Vuzov, Radiofiz.* **5,** 604.
KROTIKOV, V. D. (1962b) *Izv. Vuzov, Radiofiz.* **5,** 1057.
KROTIKOV, V. D. (1963a) *Izv. Vuzov, Radiofiz.* **6,** 1087.
KROTIKOV, V. D. (1963b) *Izv. Vuzov, Radiofiz.* **6,** 889.
KROTIKOV, V. D. and PORFIR'YEV, V. A. (1963 *Izv. Vuzov, Radiofiz.* **6,** 242.
KROTIKOV, V. D., PORFIR'YEV, V. A. and TROITSKII, V. S. (1961a) *Izv. Vuzov, Radiofiz.* **4,** 1004.
KROTIKOV, V. D., PORFIR'EV, V. A., TROITSKII, V. S. (1961b) *Izv. Vuzov, Radiofiz.* **4,** 759.
*KROTIKOV, V. D. and SHCHUKO, O. V. (1963) *Astron. Zhurn.* **40,** 297.
*KROTIKOV, V. D. and TROITSKII, V. S. (1962) *Astron. Zhurn.* **39,** 1089.
*KROTIKOV, V. D. and TROITSKII, V. S. (1963a) *Uspekhi Fiz. Nauk* **81,** 589.
*KROTIKOV, V. D. and TROITSKII, V. S. (1963b) *Astron. Zhurn.* **40,** 158.
*KROTIKOV, V. D. and TROITSKII, V. S. (1963c) *Astron. Zhurn.* **40,** 1076.
KRÜGER, A., KRÜGER, W. and WALLIS, G. (1964) *Z. Astrophys.* **59,** 37.
KRUSE, U. E., MARSHALL, L. and PLATT, J. R. (1956) *Ap. J.* **124,** 601.
KUIPER, G. P. (Ed.) (1947) *The Atmosphere of the Earth and Planets*, Univ. Chicago Press.
KUIPER, G. P. (Ed.) (1953) *The Sun*, Univ. Chicago Press.
KUIPER, G. P. and MIDDLEHURST, B. M. (Eds.) (1961) *Planets and Satellites*, Univ. Chicago Press.
KUNDU, M. R. (1958) *C. R. Acad. Sci. Paris* **246,** 2740.
KUNDU, M. R. (1959a) in *Paris Symposium on Radio Astronomy*, Ed. R. N. BRACEWELL, Stanford Univ. Press, p. 222.
KUNDU, M. R. (1959b) *Ann. d'Astrophys.* **22,** 1.
KUNDU, M. R. (1960) *J. Geophys. Res.* **64,** 4308.
KUNDU, M. R. (1961a) *Ap. J.* **134,** 96.
KUNDU, M. R. (1961b) *J. Geophys. Res.* **66,** 4308.
KUNDU, M. R. and FIROR, J. W. (1961) *Ap. J.* **134,** 389.
KUNDU, M. R. and HADDOCK, F. T. (1960) *Nature* **186,** 610.
KUNDU, M. R. and HADDOCK, F. T. (1961) *IRE Trans. on Antennas and Propagation* AP-9, No. 1, 82.
KUNDU, M. R., ROBERTS, J. A., SPENCER, C. L. and KUIPER, J. W. (1961) *Ap. J.* **133,** 255.
KURNOSOVA, L. V., RAZORENOV, L. A. and FRADKIN, M. I. (1960) *Isk. Sputniki Zemli*, No. 6.
KUZ'MIN, A. D. (1963) *Izv. Vuzov, Radiofiz.* **6,** 1090.
*KUZ'MIN, A. D. and SALOMONOVICH, A. YE. (1960) *Astron. Zhurn.* **37,** 297.
*KUZ'MIN, A. D. and SALOMONOVICH, A. YE. (1961) *Astron. Zhurn.* **38,** 1115.
*KUZ'MIN, A. D. and SALOMONOVICH, A. YE. (1963) *Astron. Zhurn.* **40,** 154.
KUZNETSOVA, G. V., PARIISKII, YU. N., SOBOLEVA, N. S. and KHANBERDIYEV, A. (1961) *Solnechnyye Dannyye* No. 4, 65.

References

LABRUM, N. R. (1960) *Austral. J. Phys.* **13**, 700.
LAFFINEUR, M. and HOUTGAST, J. (1949) *Ann. d'Astrophys.* **12**, 137.
LAFFINEUR, M., MICHARD, R., PECKER, J. C. and VANQUOIS, B. (1954) *Ann. d'Astrophys.* **17**, 358.
*LANDAU, L. D. (1946) *Zh. Eksp. i Teoret. Fiz.* **16**, 576.
*LANDAU, L. D. and LIFSHITZ, YE. M. (1953) *Mechanics of Continuous Media* (in Russian), Gostekhizdat.
LANDAU, L. D. and LIFSHITZ, YE. M. (1960a) *Electrodynamics of Continuous Media*, Pergamon, Oxford.
*LANDAU, L. D. and LIFSHITZ, Ye. M. (1960b) *Field Theory* (in Russian), Fizmatgiz.
LANDOVITZ, L. and MARSHALL, L. (1962) *Nature* **195**, 1186.
LARENZ, R. W. (1955) *Z. Naturforsch.* **10A**, 901.
LARENZ, R. W. (1962) *Symposium on Electromagnetic Theory and Antennas*, Copenhagen.
LEHANY, F. J. and YABSLEY, D. E. (1949) *Austral. J. Sci. Res.* A **2**, 48.
LEIGHTON, R. B. (1959) *Ap. J.* **130**, 366.
LEONTOVICH, M. A. (1951) *Introduction to Thermodynamics* (in (Russian), Gostekhizdat.
*LEVIN, B. YU. and MAYEVA, S. V. (1960) *Doklady Akad. Nauk SSSR* **133**, 44.
LEWIN (1951) *Advanced Theory of Waveguides*, Iliffe.
LILLEY, A. E. (1961) *Astron. J.* **66**, 290.
LITTLE, A. G. and PAYNE-SCOTT, R. (1951) *Austral. J. Sci. Res.* A **4**, 489.
LONG, R. J. and ELSMORE, B. (1960) *Observatory* **80**, 112.
LOUGHEED, R. E., ROBERTS, J. A. and MCCABE, M. K. (1957) *Austral. J. Phys.* **10**, 483.
LOVELL, A. C. B. (1959) *Proc. Roy. Soc.* A **253**, 494.
LYNN, V. L., MEEKS, M. L. and SOHIGIAN, M. D. (1963) *Astron. J.* **68**, 284.

McCRAY, R. (1966) *Science* **154**, 1320.
McCLAIN, E. F. (1960) *Astron. J.* **65**, 560.
McCLAIN, E. F. (1962) *Astron. J.* **67**, 675.
McCLAIN, E. F., NICHOLS, J. H. and WAAK, J. A. (1960) *Report at XIII General Assembly URSI*, London.
McCLAIN, E. F., NICHOLS, J. H. and WAAK, J. A. (1962) *Astron. J.* **67**, 724.
McCLAIN, E. F. and SLOANAKER, R. M. (1959) in *Paris Symposium on Radio Astronomy*, Ed. R. N. BRACEWELL, Stanford Univ. Press, p. 61.
McCREADY, L. L., PAWSEY, J. L. and PAYNE-SCOTT, R. (1947) *Proc. Roy. Soc.* A **190**, 357.
McDONALD, C. J. F. (1959) *J. Geophys. Res.* **64**, 1967.
MACHIN, K. E. (1951) *Nature* **167**, 889.
MACHIN, K. E. and O'BRIEN, P. A. (1954) *Phil. Mag.* **45**, 973.
MACHIN, K. E. and SMITH, G. (1951a) *Nature* **168**, 599.
MACHIN, K. E. and SMITH, G. (1951b) *Nature* **170**, 319.
McLEAN, D. J. (1959) *Austral. J. Phys.* **12**, 404.
MALINGE, A. M. (1960a) *Ann. d'Astrophys.* **23**, 574.
MALINGE, A. M. (1960b) *C.R. Acad. Sci. Paris* **250**, 1186.
MALLING, L. R. and GOLOMB, S. W. (1961) *J. Brit. IRE* **22**, 297.
MALVILLE, J. M. (1962) *Ap. J.* **136**, 266.
MALTBY, P. (1959) *Nature* **184**, 1391.
MARKEYEV, A. K. (1961) *Geomagnetizm i Aeronomiya* **1**, 999.
MARKOV, A. V. (Ed.) (1960) *The Moon* (in Russian), Fizmatgiz.
MARSHALL, L. (1956) *Ap. J.* **124**, 469.
MARTYN, D. F. (1946) *Nature* **158**, 632.
MARTYN, D. F. (1947) *Nature* **159**, 26.
MARTYN, D. F. (1948) *Proc. Roy. Soc.* A **193**, 44.
MARTYN, D. F. (1951) *Nature* **167**, 92.

References

MAYER, C. H. (1959) *Astron. J.* **64**, 43.
MAYER, C. H., MCCULLOGH, T. P. and SLOANAKER, R. M. (1958a) *Ap. J.* **127**, 1.
MAYER, C. H., MCCULLOGH, T. P. and SLOANAKER, R. M. (1958b) *Proc. IRE* **46**, 260.
MAYER, C. H., MCCULLOGH, T. P. and SLOANAKER, R. M. (1958c) *Ap. J.* **127**, 11.
MAYER, C. M., MCCULLOUGH, T. P. and SLOANAKER, R. M. (1960 *Report on XIII General Assembly URSI*, London.
MAYER, C. H., SLOANAKER, R. M. and HAGEN, J. P. (1957) in *Radio Astronomy*, Ed. H. C. VAN DE HULST, Cambridge Univ. Press, p. 269.
MAXWELL, A. (1951) *Observatory* **71**, 72.
MAXWELL, A., HOWARD, W. E. and GARMIRE, G. (1960) *J. Geophys. Res.* **65**, 3581.
MAXWELL, A., HUGHES, M. P. and THOMPSON, A. R. (1963) *J. Geophys. Res.* **68**, 1347.
MAXWELL, A. and SWARUP, G. (1958) *Nature* **181**, 36.
MAXWELL, A., SWARUP, G. and THOMPSON, A. R. (1958) *Proc. IRE* **46**, 142.
MAXWELL, A. and THOMPSON, A. R. (1962) *Ap. J.* **135**, 138.
MAXWELL, A., THOMPSON, A. R. and GARMIRE, G. (1959) *Planet. Space Sci.* **1**, 325.
MEDD, W. J. and BROTEN, N. W. (1961) *Planet. Space Sci.* **5**, 307.
MEZGER, P. G. (1958) *Z. Astrophys.* **46**, 234.
MEZGER, P. G. and STRASSL, H. (1960) *Veröffentlichungen der Universitäte Sternwarte zu Bonn* No. 59, 87.
MILLER, A. C. and GARY, B. L. (1962) *Astron. J.* **67**, 727.
MILLS, B. Y. and LITTLE, A. G. (1953) *Austral. J. Phys.* **6**, 272.
MILLS, B. Y., LITTLE, A. G., SHERIDAN, K. V. and SLEE, O. B. (1958) *Proc. IRE* **46**, 67.
MINNIS, C. M. and BAZZARD, G. H. (1958) *Nature* **181**, 1796.
MINNETT, H. C. and LABRUM, N. R. (1950) *Austral. J. Sci. Res.* A **3**, 60.
MITCHELL, F. M., WHITEHURST, R. N. (1958) *Univ. Alabama Phys. Dept. Rep. under OOR contract.*
MITRA, S. K. (1952) *The Upper Atmosphere*, Asiatic Soc., Calcutta.
MOGILEVSKII, E. I. and AKIN'YAN, S. T. (1961) *Geomagnetizm i Aeronomiya* **1**, 843, 922.
MOISEYEV, I. G. (1960) *Izv. Krymsk. Astrofiz. Observ.* **24**, 3
*MOISEYEV, I. G. (1961) *Astron. Zhurn.* **38**, 541.
MOLCHANOV, A. P. (1956) *Proc. of the 5th Congress on Problems of Cosmogony* (in Russian), Akad. Nauk SSSR.
MOLCHANOV, A. P. (1960) *Izv. GAO* **21**, No. 5, 114.
MOLCHANOV, A. P. (1961a) Paper read at the XIth Congress of the International Astronomical Union, Berkely, U.S.A.
*MOLCHANOV, A. P. (1961b) *Astron. Zhurn.* **38**, 849.
MOLCHANOV, A. P. (1962) *Solnechnyye Dannyye* No. 2, 53.
MOLCHANOV, A. P., CH'ENG TAN YUNG, WAN SHOU KUANG, KOROL'KOV, D. V., MIRZABEKYAN, E. G. and SALOMONOVICH, A. YE. (1959) in *Paris Symposium on Radio Astronomy*, Ed. R. N. BRACEWELL, Stanford Univ. Press, p. 174.
MOLCHANOV, A. P. and KOROL'KOV, D. V. (1961) *Solnechnyye Dannyye* No. 4, 62.
MOLCHANOV, A. P. and PETEROVA, N. G. (1961) *Solnechnyye Dannyye* No. 10.
MOORE, P. (1959) *The Planet Venus*, Norton.
MORETON, G. E. (1960) *Astron. J.* **65**, 494.
MORIMOTO, M. (1961) *Publ. Astron. Soc. Japan* **13**, 285.
MORIMOTO, M. and KAI, K. (1961) *Publ. Astron. Soc. Japan* **13**, 294.
MORRIS, D. and BERGE, G. L. (1962) *Ap. J.* **136**, 276.
MOTT-SMITH, H. M. (1951) *Phys. Rev.* **82**, 885.
MUHLEMAN, D. O. (1961) *Astron. J.* **66**, 292.
MULLALY, R. F. (1961) *Austral. J. Phys.* **14**, 540.
MULLALY, R. F. and KRISHNAN, T. (1963) *Austral. J. Phys.* **16**, 8.
MÜLLER, H. (1956) *Z. Astrophys.* **39**, 160.
MÜLLER, H. G., PRIESTER, W. and FISCHER, G. (1957) *Naturwiss.* **44**, 392.

References

Mustel', E. R. (1957) Physics of corpuscular streams and their effect on the earth's upper atmosphere (in Russian), *Proc. Conf. Comm. on Solar Research, Akad. Nauk SSSR*, p. 8.

Nakagami, M. and Miya, K. (1939) *Electrotech. J. Japan* **3**, 216.
Nature (1955) **175**, 1074.
Nature (1958) **181**, 542.
Naumov, A. N. (1963) *Izv. Vuzov, Radiofiz.* **6**, 847.
*Naumov, A. P. and Khizhnyakova, I. P. (1965) *Astron. Zhurn.* **42**, 629.
Newkirk, G. (1959) in *Paris Symposium on Radio Astronomy*, Ed. R. N. Bracewell, Stanford Univ. Press, p. 149.
Newkirk, G. (1961) *Ap. J.* **133**, 983.
Neylan, A. A. (1959) *Austral. J. Phys.* **12**, 399.
Nicolet, M. (1949) *Misc. Inst. Roy. Meteor. Uccle* **35**.
Nikol'skii, G. M. (1962) *Geomagnetizm i Aeronomiya* **2**, 3.

Obayashi, T. (1959) *J. Radio Res. Lab. Japan* **6**, 375.
O'Brien, P. A. (1953a) *MN* **113**, 597.
O'Brien, P. A. (1953b) *Observatory* **73**, 106.
O'Brien, P. A. and Bell, C. J. (1954) *Nature* **173**, 219.
O'Brien, P. A. and Tandberg-Hanssen, E. (1955) *Observatory* **75**, 11.
Odelevskii, V. I. (1951) *Zh. Eksp. i Teoret. Fiz.* **21**, 667.
Olte, A. and Siegel, K. M. (1961) *Ap. J.* **133**, 706.
Öpik, E. J. (1961) *J. Geophys. Res.* **66**, 2807.
Oster, L. (1959) *Phys. Rev.* **116**, 474.
Ovsyankin, M. A. and Panovkin, B. N. (1956) *Radiotekhn. i Elektron.* **1**, 886.
Owren, L. (1954) *Radio Astr. Report of Cornell Univ.* No. 15, 74.

*Pakhomov, V. I., Aleksin, V. F. and Stepanov, K. N. (1961) *Zh. Eksp. i Teoret. Fiz.* **31**, 1170.
*Pakhomov, V. I., Aleksin, V. F. and Stepanov, K. N. (1963) in coll. *Plasma Physics and Problems of Controlled Thermonuclear Synthesis* (in Russian), Akad. Nauk Ukr. SSR, Kiyev, p. 40.
*Pakhomov, V. I. and Stepanov, K. N. In coll. *Plasma Physics and Problems of Thermonuclear Synthesis* (in Russian), Akad. Nauk Ukr. SSR, Kiyev, p. 70.
Panovkin, B. N. (1957) *Astron. Zhurn.* **34**, 505.
Pawsey, J. L. (1946) *Nature* **158**, 633.
Pawsey, J. L. (1957) in *Radio Astronomy*, Ed. H. C. Van de Hulst, Cambridge Univ. Press, p. 284.
Pawsey, J. L. and Bracewell, R. N. (1955) *Radio Astronomy*, Oxford.
Pawsey, J. L. and Smerd, S. F. (1953) Radio emission of the sun. In *The Solar System. I. The Sun*, ed. G. P. Kuiper, Univ. Chicago Press.
Pawsey, J. L. and Yabsley, D. E. (1949) *Austral. J. Sci. Res.* A **2**, 198.
Payne-Scott, R. (1946) *Nature* **158**, 633.
Payne-Scott, R. and Little, A. G. (1951) *Austral. J. Sci. Res.* A **4**, 508.
Payne-Scott, R. and Little, A. G. (1952) *Austral. J. Sci. Res.* **5**, 32.
Payne-Scott, R., Yabsley, D. E.. and Bolton, J. G. (1947) *Nature* **160**, 256.
Peck, B. M. (1958) *The Planet Jupiter*, Faber & Faber, London.
Petelin, M. I. (1961) *Izv. Vuzov, Radiofiz.* **4**, 455.
Peters, B. (1954) in coll. *Cosmic Ray Physics*, 1 (Russian ed.).
Peterson, L. E. and Winkler, J. R. (1959) *J. Geophys. Res.* **64**, 697.
Pettengill, G. H., Briscoe, H. W., Evans, J. V., Gehrels, E., Hyde, G. M., Kraft, L. G., Price, R. and Smith, W. B. (1962) *Astron. J.* **67**, 181.
Pettit, E. (1940) *Astron. J.* **91**, 408.

References

Pettit, E. and Nicholson, S. R. (1930) *Ap. J.* **71**, 102.
Pick-Gutmann, M. (1960) *C. R. Acad. Sci. Paris* **250**, 2127.
Pick-Gutmann, M. (1961) *Ann. d'Astrophys.* **24**, 183.
Piddington, J. H. (1950) *Proc. Roy. Soc.* A **203**, 417.
Piddington, J. H. (1953) *Proc. Phys. Soc.* **66** B, 97.
Piddington, J. H. (1954) *Ap. J.* **119**, 531.
Piddington, J. H. and Davies, R. D. (1953) *MN* **113**, 582.
Piddington, J. H. and Hindman, J. V. (1949) *Austral. J. Sci. Res.* A **2**, 524.
Piddington, J. H. and Minnett, H. C. (1949a) *Austral. J. Sci. Res.* A **2**, 539.
Piddington, J. H. and Minnett, H. C. (1949b) *Austral. J. Sci. Res.* A **2**, 63.
Piddington, J. H. and Minnett, H. C. (1951a) *Austral. J. Sci. Res.* A **4**, 130.
Piddington, J. H., Minnett, H. C. (1951b) *Austral. J. Sci. Res.* A **4**, 459.
Pikel'ner, S. B. (1950) *Izv. Krymsk. Astrofiz. Observ.* **5**, 34.
Pikel'ner, S. B. (1961) *Fundamentals of Cosmic Electrodynamics* (in Russian), Fizmatgiz.
*Pikel'ner, S. B. and Gintzburg, M. A. (1963) *Astron. Zhurn.* **40**, 842.
Pikel'ner, S. B. and Shklovskii, I. S. (1950) *Izv. Krymsk. Astrofiz. Observ.* **6**, 29.
Pines, D. and Bohm, D. (1952) *Phys. Rev.* **85**, 338.
Pisareva, V. V. (1958) *Astron. Zhurn.* **35**, 112.
Poloskov, S. M. (1953) *Astron. Zhurn.* **30**, 68.
Priester, W. and Dröge, F. (1955) *Z. Astrophys.* **37**, 132.
Proc. of the Conf. of the Comm. on Solar Res. (in Russian) (1957) Physics of solar corpuscular fluxes and their effect on the Earth's upper atmosphere, Akad. Nauk SSSR.

Rabben, H. H. (1960) *Rendicontidella Scuola Internazionale di Fisica "Enrico Fermi"*, Corso XII, Radioastronomia solare, Bologna, p. 395.
Radhakrishnan, V. and Roberts, J. A. (1960a) *Phys. Rev. Lett.* **4**, 493.
Radhakrishnan, V. and Roberts, J. A. (1960b) *Astron. J.* **65**, 498.
Rapoport, B. O. (1960) *Izv. Vuzov, Radiofiz.* **3**, 737.
Razin, V. A. (1956) *Radiotekhn. i Elektron.* **1**, 846.
Razin, V. A. (1960) *Izv. Vuzov, Radiofiz.* **3**, 584.
Razin, V. A. and Fedorov, V. T. (1963) *Izv. Vuzov, Radiofiz.* **6**, 1052.
Reber, G. (1944) *Ap. J.* **100**, 279.
Reber, G. (1946) *Nature* **158**, 945.
Reber, G. (1955) *Nature* **175**, 132.
Reid, G. C. and Leinbach, H. (1959) *J. Geophys. Res.* **64**, 1801.
Report U.S. Commission V URSI, Univ. of Alabama (XIII General Assembly URSI, London, Sept. 1960).
Reule, A. (1952) *Z. Naturforsch.* A **7**, 234.
Riddiford, L. and Butler, S. T. (1952) *Phil. Mag.* **43**, 447.
Roberts, J. A. (1958) *Austral. J. Phys.* **11**, 215.
Roberts, J. A. (1959a) *Austral. J. Phys.* **12**, 327.
Roberts, J. A. (1959b) in *Paris Symposium on Radio Astronomy*, Ed. R. N. Bracewell, Stanford Univ. Press, p. 194.
Roberts, J. A. (1963) *Planet. and Space Sci.* **11**, 221.
*Roberts, J. A. (1964) *Uspekhi Fiz. Nauk* **83**, No. 3.
Roberts, W. O. (1955) *Scientific American* **192**, No. 2, 40.
Roberts, J. A. and Stanley, G. J. (1959) *Publ. Astron. Soc. Pacific* **71**, 485.
Roederer, J. G., Manzano, J. R., Santochi, O. R., Necurkar, N., Troncoso, O., Palmeira, R. A. R. and Schwachheim (1961) *J. Geophys. Res.* **66**, 1603.
Rose, W. K., Bologna, J. M. and Sloanaker, R. M. (1963a) *Astron. J.* **68**, 78.
Rose, W. K., Bologna, J. M. and Sloanaker, R. M. (1963b) *Phys. Rev. Lett.* **10**, 123.
Rothwell, P. and McIlwan, C. (1959) *Nature* **184**, 138.

References

RYDBECK, O. E. H. (1953) *Atti Convegni Accad. Naz. Lincei* No. 11, 290.
RYLE, M. (1948) *Proc. Roy. Soc.* A **195**, 82.
RYLE, M. (1952a) *Proc. Roy. Soc.* A **211**, 351.
*RYLE, M. (1952b) *Uspechi Fiz. Nauk*, **45**, 508.
RYLE, M. and VONBERG, D. D. (1946) *Nature* **158**, 339.
RYLE, M. and VONBERG, D. D. (1948) *Proc. Roy. Soc.* A **193**, 98.
RYZHOV, YU. A. (1959) *Izv. Vuzov, Radiofiz.* **2**, 869.

SAGAN, C. (1960) Calif. Institute Techn. Jet Propulsion Lab., Tech. Rep. No. 32-34.
SAGAN, C., SIEGEL, K. M. and JONES, D. E. (1961) *Astron. J.* **66**, 52.
*SAGDEYEV, R. Z. (1964) in coll. *Questions of Plasma Theory* (in Russian), 4, Gosatomizdat.
SAGDEYEV, R. Z. and SHAFRANOV, V. D. (1959) *Proc. II Internat. Conf. Peaceful Use of Atomic Energy* (Geneva, 1958) (Russian ed.), p. 202.
*SAGDEYEV, R. Z. and SHAFRANOV, V. D. (1960) *Zh. Eksp. i Teoret. Fiz.* **39**, 181.
SAKURAI, K. (1961a) *Geomagnetism and Geoelectricity* **12**, 59.
SAKURAI, K. (1961b) *Geomagnetism and Geoelectricity* **12**, 70.
SALOMONOVICH, A. YE. (1958) *Astron. Zhurn.* **35**, 129.
*SALOMONOVICH, A. YE. (1960) *Astron. Zhurn.* **37**, 969.
*SALOMONOVICH, A. YE. (1962a) *Astron. Zhurn.* **39**, 79.
*SALOMONOVICH, A. YE. (1962b) *Astron. Zhurn.* **39**, 260.
SALOMONOVICH, A. YE. (1964) *Izv. Vuzov, Radiofiz.* **7**, 51.
SALOMONOVICH, A. YE. and KOSCHENKO, V. N. (1961) *Izv. Vuzov, Radiofiz.* **4**, 591.
SALOMONOVICH, A. YE. and LOSOVSKII, B. YA. (1962) *Astron. Zhurn.* **39**, 1074.
SALOMONOVICH, A. YE., PARIISKII, YU. N. and KHANGIL'DIN, U. V. (1958) *Astron. Zhurn.* **35**, 659.
SANDER, K. F. (1947) *Nature* **159**, 506.
SCARF, F. L. (1963) *J. Geophys. Res.* **68**, 141.
SCHEFFLER, H. (1958) *Z. Astrophys.* **45**, 113.
SEEGER, C. L., STUMPERS, F. L. H. M. and VAN HURCK, N. (1959/60) *Philips Tech. Rev.* **21**, No. 11, 317.
SEEGER, C. L., WESTERHOUT, G. and CONWAY, R. G. (1957a) IAU Circular No. 1599.
SEEGER, C. L., WESTERHOUT, G. and CONWAY, R. G. (1957b) *Ap. J.* **126**, 585.
SEN, H. K. (1954) *Austral. J. Phys.* **7**, 30.
SEN, H. K. (1955) *Phys. Rev.* **97**, 849.
SEVERNYI, A. B. (1957) *Izv. Krymsk. Astrofiz. Observ.* **17**, 129.
SEVERNYI, A. B. (1958) *Astron. Zhurn.* **35**, 335.
SEVERNY, A. V. and BUMBA, V. (1958) *Observatory* **78**, 33.
SHAIN, C. A. (1955) *Nature* **176**, 836.
SHAIN, C. A. (1956) *Austral. J. Phys.* **9**, 61.
SHAIN, C. A. and HIGGINS, C. S. (1959) *Austral. J. Phys.* **12**, 357.
SHAIN, C. A. and SLEE, O. B. (1957) *Observatory* **77**, 204.
*SHAFRANOV, V. D. (1958a) *Zh. Eksp. i Teoret. Fiz.* **34**, 1475.
*SHAFRANOV, V. D. (1958b) in coll. *Plasma Physics and the Problem of Controlled Thermonuclear Reactions* (in Russian), 4, Akad. Nauk SSSR, p. 416.
SHARONOV, V. V. (1958) *The Nature of the Planets* (in Russian), Fizmatgiz.
SHERIDAN, K. V. and ATTWOOD, C. F. (1962) *Observatory* **82**, 155.
SHERIDAN, K. V. and TRENT, G. H. (1961) *Observatory* **81**, 71.
SHERIDAN, K. V., TRENT, G. H. and WILD, J. P. (1959) *Observatory* **79**, 51.
SHERRILL, W. M. and CASTLES, M. P. (1963) *Ap. J.* **138**, 587.
SHKLOVSKII, I. S. (1946) *Astron. Zhurn.* **23**, 333.
SHKLOVSKII, I. S. (1956) *Cosmic Radio Emission* (in Russian), Gostekhizdat (English translation, Harvard Univ. Press. 1960).
SHKLOVSKII, I. S. (1962) *Physics of the Solar Corona* (in Russian), Fizmatgiz (English Translation, Pergamon Press, 1965).

References

SHKLOVSKII, I. S. and KONONOVICH, E. V. (1958) *Astron. Zhurn.* **35,** 37.
SILIN, V. P., RUKHADZE, A. A. (1961) *Electromagnetic properties of plasma and plasma-like media* (in Russian) Atomizdat.
SIMON, P. (1955a) *C. R. Acad. Sci. Paris* **240,** 940.
SIMON, P. (1955b) *C. R. Acad. Sci. Paris* **240,** 1056.
SIMON, P. (1955c) *C. R. Acad. Sci. Paris* **240,** 1192.
SIMON, P. (1956) *Ann. Univ. Paris* **26,** 260.
SIMON, P. (1957) in *Radio Astronomy*, Ed. H. C. VAN DE HULST, Cambridge Univ. Press.
SIMON, P. (1960a) *Rendiconti della Scuola Internazionale di fisica "Enrico Fermi"*, Corso XII, Radioastronomia solare, Bologna, p. 403.
SIMON, P. (1960b) *Ann. d'Astrophys.* **23,** 102.
SINNO, K. (1959) *Radio Res. Lab. Japan* **6,** 17.
SINNO, K. and HAKURA, Y. (1958a) *Rep. Ionosph. Res. Japan* **12,** 296.
SINNO, K. and HAKURA, Y. (1958b) *Rep. Ionosph. Res. Japan* **12,** 285.
SINTON, W. M. (1952) *Phys. Rev.* **82,** 464.
SINTON, W. M. (1955) *J. Opt. Soc. America* **45,** 975.
SINTON, W. M. (1956) *Ap. J.* **123,** 325.
SINTON, W. M. (1962) *Physics and Astronomy of the Moon*, Acad. Press.
*SITENKO, A. G. and STEPANOV, K. N. (1956) *Zh. Eksp. i Teoret. Fiz.* **31,** 642.
Sky and Telescope (1963) **25,** 89.
SLEE, O. B. (1961) *MN* **123,** 223.
SLEE, O. B. and HIGGINS, C. S. (1963) *Nature* **197,** 781.
SLEE, O. B. and HIGGINS, C. S. (1966) *Austral. J. Phys.* **19,** 167.
SLOANAKER, R. M. (1959) *Astron. J.* **64,** 346.
SLOANAKER, R. M. and BOLAND, J. W. (1961) *Ap. J.* **133,** 649.
SMERD, S. F. (1950a) *Austral. J. Sci. Res.* A **3,** 34.
SMERD, S. F. (1950b) *Austral. J. Sci. Res.* A **3,** 265.
SMERD, S. F. (1955) *Nature* **175,** 297.
SMERD, S. F. and WILD, J. P. (1957) in *Radio Astronomy*, Ed. H. C. VAN DE HULST, Cambridge Univ. Press, p. 290.
SMERD, S. F. WILD, J. P. and SHERIDAN, K. V. (1962) *Austral. J. Phys.* **15,** 180.
SMITH, A. G. and CARR, T. D. (1959) *Ap. J.* **130,** 641.
SMITH, A. G., CARR, T. D., BOLLHAGEN, H., CHATTERTON, N. and SIX, F. (1960) *Nature* **187,** 568.
SMITH, A. G., CARR, T. D., SIX, N. F., MOCK, W. and BOLLHAGEN, H. (1963) *Astron. J.* **68,** 292.
SMITH, A. G., SIX, N. F., CARR, T. D. and BROWN, G. W. (1963) *Nature* **199,** 267.
SMITH, F. G. (1955) *Observatory* **75,** 252.
SMITH, H. J. (1959) *Astron. J.* **64,** 41.
SMITH, H. J. and DOUGLAS, J. N. (1957) *Astron. J.* **62,** 247.
SMITH, H. J. and DOUGLAS, J. N. (1959) in *Paris Symposium on Radio Astronomy*, Ed. R. N. BRACEWELL, Stanford Univ. Press, p. 53.
SMITH, H. J., LASKER, B. M. and DOUGLAS, J. N. (1960) *Astron. J.* **65,** 501.
SMITH, W. B. (1963) *Astron. J.* **68,** 15.
*SOBOLEVA, N. S. (1962) *Astron. Zhurn.* **39,** 1124.
Solnechnyye Dannyye (1958) No. 1–2, 4.
Solnechnyye Dannyye (1960) No. 1, 8.
SOMMERFELD, A. (1956) *Thermodynamics and Statistical Physics*, Academic Press.
SOUTHWORTH, G. C. (1945) *J. Franklin Inst.* **239,** 285.
SPITZER, L. (1960) *Astron. J.* **65,** 539.
STAELIN, D. H., BARRETT, A. H. and KUSSE, B. R. (1963) *Astron. J.* **68,** 294.
STANIER, H. M. (1950) *Nature* **165,** 354.
STEIN, W. A. and NEY, E. P. (1963) *J. Geophys. Res.* **68,** 65.
STEINBERG, J. L. (1953) *Onde Electrique* **33,** 274.

References

Steinberg, J. L. and Lequeux, J. (1960) *Radioastronomie*, Dunod.
Steljes, J. F., Carmichael, H. and McCracken, K. G. (1961) *J. Geophys. Res.* **66**, 1363.
*Stepanov, K. N. (1958) *Zh. Eksp. i Teoret. Fiz.* **35**, 283.
Stepanov, K. N. (1963) *Izv. Vuzov, Radiofiz.* **6**, 461.
*Stepanov, N. S. and Ostrovskii, L. A. (1963) *Zh. Eksp. i Teoret. Fiz.* **45**, 1473.
*Stepanov, K. N. and Pakhomov, V. I. (1960) *Zh. Eksp. i Teoret. Fiz.* **38**, 1564.
Stewart, R. T. (1962) *Time Delays and Frequency Ratios between Fundamental and Second Harmonics in Bursts of Spectral type III.* CSIRO, Radio Phys. Div. Report.
Stokes, G. G. (1904) *Mathematical and Phys. Papers* 4.
Straiton, A. W. and Tolbert, C. W. (1960) *Proc. IRE* **48**, 898.
Straiton, A. W., Tolbert, C. W. and Britt, C. O. (1958) *J. Appl. Phys.* **29**, 776.
Strezhneva, K. M., Troitskii, V. S. (1961) *Izv. Vuzov, Radiofiz.* **4**, 600.
Strom, S. and Strom, K. (1962) *Ap. J.* **136**, 307.
Stückelberg, E. C. (1932) *Helv. Phys. Acta* **5**, 369.
Sturrock, P. A. (1957) *Proc. Roy. Soc.* A **242**, 1230, 277.
Sturrock, P. A. (1958) *Phys. Rev.* **112**, 1488.
Sturrock, P. A. (1960) *Phys. Rev.* **117**, 1426.
Su Shih Weng, Hsiao Kuan Chia, Wu Huai Wei, Tun Wu, Wu Ching Tzu, Troitskii, V. S., Rakhlin, V. L., Strezhneva, K. M. and Zelinskaya, M. R. (1962) *Izv. Vuzov, Radiofiz.* **5**, 807.
Sumi, M. (1958) *J. Phys. Soc. Japan* **13**, 1476.
Sumi, M. (1959) *J. Phys. Soc. Japan* **14**, 653.
Suzuki, S. (1961) *Ann. Tokyo Astron. Obs.* **7**, 75.
Suzuki, S. and Tsuchiya, A. (1958) *Proc. IRE* **46**, 190.
Swarup, G. (1960) *Stanford Radio Astron. Inst.* Publ. No. 8.
Swarup, G. (1961a) Sci. Rep. No. 13, AFI8 (603)-53, Stanford Electron. Lab., Stanford Univ.
Swarup, G. (1961b) *Astron J.* **66**, 296.
Swarup, G. and Parthasarathy, R. (1955) *Austral. J. Phys.* **8**, 487.
Swarup, G. and Parthasarathy, R. (1958) *Austral. J. Phys.* **11**, 338.
Swarup, G., Stone, P. H. and Maxwell, A. (1960) *Ap. J.* **131**, 725.

Takakura, T. (1953) *Nature* **171**, 445.
Takakura, T. (1954) *Publs. Astron. Soc. Japan* **6**, 185.
Takakura, T. (1956) *Publ. Astron. Soc. Japan* **8**, 182.
Takakura, T. (1959) in *Paris Symposium on Radio Astronomy*, Ed. R. N. Bracewell, Stanford Univ. Press, p. 562.
Takakura, T. (1960) *Publ. Astron. Soc. Japan* **12**, 325, 352.
Takakura, T. (1961a) *Publ. Astron. Soc. Japan* **13**, 312.
Takakura, T. (1961b) *Publ. Astron. Soc. Japan* **13**, 166.
Takakura, T. (1962) *J. Phys. Soc. Japan* **17**, 243.
Takakura, T. (1963a) *Publ. Astron. Soc. Japan* **15**, 462.
Takakura, T. (1963b) *Publ. Astron. Soc. Japan* **15**, 327.
Takakura, T. and Kai, K. (1961) *Publ. Astron. Soc. Japan* **13**, 94.
Tamoikin, V. V. (1963) *Izv. Vuzov, Radiofiz.* **6**, 258.
Tanaka, H. (1964) *Proc. Res. Inst. Atm., Nagoya Univ.* **11**, 41.
Tanaka, H. and Kakinuma, T. (1955) *Proc. Res. Inst. Atm. Nagoya Univ.* **3**, 102.
Tanaka, H. and Kakinuma, T. (1958) *Rep. Ionos. Res. Japan* **12**, 273.
Tanaka, H. and Kakinuma, T. (1959) in *Paris Symposium on Radio Astronomy*, Ed. R. N. Bracewell, Stanford Univ. Press, p. 215.
Tandberg-Hanssen, E. (1955) *Ap. J.* **121**, 367.
Taubenheim, J. (1958) *Abhandl. Dtsch. Acad. Wiss. Berlin, Kl. Math. Phys. und Tech.*, No. 3, 112.
Terashima, Y. and Yajima, N. (1964) *Prog. Theoret. Phys.* **30**, 443.

References

TER-MIKAELYAN, M. L. (1961) *Izv. Akad Nauk ArmSSR* **14**, 103.
TERLETSKII, YA. P. (1949) *Zh. Eksp. i Teoret. Fiz.* **19**, 1059.
THOMAS, R. N. (1949) *Ap. J.* **109**, 480.
THOMAS, R. N. and ATHAY, R. G. (1957) in *Radio Astronomy*, Ed. H. C. VAN DE HULST, Cambridge Univ. Press, p. 279.
THOMAS, R. N. and ATHAY, R. G. (1961) *Physics of the Solar Chromosphere*, Interscience Publ.
THOMPSON, A. R. (1959) *Paris Symposium on Radio Astronomy*, Stanford Univ. Press, p. 210.
THOMPSON, A. R. (1961) *Ap. J.* **133**, 648.
THOMPSON, A. R. (1962) *J. Phys. Soc. Japan* **17**, Suppl. A-II, 198.
THOMPSON, A. R. and MAXWELL, A. (1960a) *Nature* **185**, 89.
THOMPSON, A. R. and MAXWELL, A. (1960b) *Planet. Space Sci.* **2**, 104.
THOMPSON, A. R. and MAXWELL, A. (1962) *Ap. J.* **136**, 546.
TIDMAN, D. A. (1960) *Phys. Rev.* **117**, 366.
TIDMAN, D. A. and BOYD, J. M. (1962) *Phys. Fluids* **5**, 213.
TIDMAN, D. A. and WEISS, G. H. (1961a) *Phys. Fluids* **4**, 703.
TIDMAN, D. A. and WEISS, G. H. (1961b) *Phys. Fluids* **4**, 866.
TITCHMARSH, E. C. (1937) *Introduction to the Theory of Fourier Integrals*, Oxford.
TOLBERT, C. W. and STRAITON, A. W. (1961) *Ap. J.* **134**, 91.
*TRAKHTENGERTS, V. YU. (1966) *Astron. Zhurn.* **43**, 356.
TOLBERT, C. W. and STRAITON, A. W. (1962) *J. Geophys. Res.* **67**, 1741.
TRANTER, K. J. (1956) *Integral Transformations in Mathematical Physics* (Russian ed.) Gostekhizdat, chap. 1.
TROITSKII, V. S. (1954) *Astron. Zhurn.* **31**, 511.
TROITSKII, V. S. (1956) *Proc. 5th Congress on Questions of Cosmogony* (in Russian), Akad. Nauk SSSR, p. 325.
TROITSKII, V. S. (1961) *Izv. Kommissii po Fiz. Planet* No. 3, 16.
*TROITSKII, V. S. (1962a) *Astron. Zhurn.* **39**, 73.
TROITSKII, V. S. (1962b) *Izv. Vuzov, Radiofiz.* **5**, 885.
TROITSKII, V. S. (1962c) *Izv. Vuzov, Radiofiz.* **5**, 602.
TROITSKII, V. S. (1964) *Izv. Vuzov, Radiofiz.* **7**, 208.
*TROITSKII, V. S. (1967) *Izv. Vuzov, Radiofiz.* **10**, 1266.
TROITSKII, V. S. and TSEITLIN, N. M. (1960) *Izv. Vuzov, Radiofiz.* **3**, 1127.
TROITSKII, V. S., ZELINSKAYA, M. R. (1955) *Astron. Zhurn.* **32**, 550.
TROITSKII, V. S., ZELINSKAYA, M. R., RAKHLIN, V. L. and BOBRIK, V. G. (1956) *Proc. 5th Congress on Questions of Cosmogony* (in Russian), Akad. Nauk SSSR, p. 182.
*TRUBNIKOV, B. A. (1958a) *Doklady Akad. Nauk SSSR* **118**, 913.
TRUBNIKOV, B. A. (1958b) in coll. *Plasma Physics and the Problem of Controlled Thermonuclear Reactions* (in Russian), 4, Akad. Nauk SSSR, p. 305.
*TRUBNIKOV, B. A. and BAZHANOVA, A. YE. (1958) in coll. *Plasma Physics and the Problem of Controlled Thermonuclear Reactions* (in Russian), 4, Akad. Nauk SSSR, p. 121.
TSUCHIYA, A. and MORIMOTO, M. (1961) *Publ. Astron. Soc. Japan* **13**, 303.
TSYTOVICH, V. I. (1951) *Vestnik Mosk. Gos. un-ta, Ser. Fiz.-matem. i Est. Nauk*, No. 11, 27.
TU LENG YAO, MALAKHOV, A. N., PLECHKOV, V. M., RAZIN, V. A., RAKHLIN, V. L., STANKEVICH, K. S., STREZHNEVA, K. M., T'AN SHOU P'EI, TROITSKII, V. S., KHRULEV, V. V. and TSEITLIN, N. M. (1959) *Izv. Vuzov, Radiofiz.* **2**, 151.
TWISS, R. Q. (1958) *Austral. J. Phys.* **2**, 564.
TWISS, R. Q. (1963) *Phil. Mag.* **8**, 1249.
TWISS, R. Q. and ROBERTS, J. A. (1958) *Austral. J. Phys.* **2**, 424.
TYAS, J. P. I., FRANKLIN, C. A. and MOLOZZI, A. R. (1959) *Nature* **184**, 785.
TYLER, W. C. and COPELAND, J. (1961) *Astron. J.* **66**, 56.

References

UCHIDA, Y. (1960) *Publ. Astron. Soc. Japan* **12**, 376.
Ultra-High Frequency Oscillations in a Plasma (1961) Izd. Inostr. Lit.
UNSÖLD, A. (1947) *Naturwiss.* **34**, 194.

VAN DE HULST, H. C. (1947) *Ap. J.* **105**, 471.
VAN DE HULST, H. C. (1953) in *The Sun*, Ed. G. P. KUIPER, Univ. Chicago Press.
VAN SABBEN, D. (1953) *J. Atm. Terr. Phys.* **3**, 194.
VAUCOULEURS, G. DE (1951) *Physique de la Planète Mars*, Paris.
VAUQUOIS, B. (1955) *Observatory* **75**, 259.
VAUQUOIS, B. (1959a) in *Paris Symposium on Radio Astronomy*, Ed. R. N. BRACEWELL, Stanford Univ. Press, p. 143.
VAUQUOIS, B., (1959b) *Ann. d'Astrophys.* **22**, 189.
VAUQUOIS, B., CUPIAC, P. and LAFFINEUR, M. (1953) *C. R. Acad. Sci. Paris* **237**, 1630.
*VEDENOV, A. A. (1963) in coll. *Questions of Plasma Theory* (in Russian) 3, Atomizdat.
VEDENOV, A. A., VELIKHOV, YE. P. and SAGDEYEV, R. Z. (1961) *Nuclear Synthesis* (in Russian) **1**, 82.
VEISIG, G. S. and BOROVIK, V. N. (1961) *Solnechnyye Dannyye* No. 6, 61.
VEROZUB, L. V. (1962) *Uch. Zap. Khar'kovsk. un-ta, Tr. Astron. Obs.* **14**, 86.
VETUKHNOVSKAYA, YU. N., KUZ'MIN, A. D., KUTUZA, B. G., LOSOVSKII, B. YA. and SALOMONOVICH, A. E. (1963) *Izv. Vuzov, Radiofiz.* **6**, 1054.
VICTOR, W. K. (1961) *R. Stivens Sci.* **134**, 46.
VITKEVICH, V. V. (1951) *Doklady Akad. Nauk SSSR* **77**, 585.
VITKEVICH, V. V. (1952) *Astron. Zhurn.* **29**, 450.
VITKEVICH, V. V. (1955) *Doklady Akad. Nauk SSSR* **101**, 229.
VITKEVICH, V. V. (1956a) *Proc. 5th Congress on Questions of Cosmogony* (in Russian), Akad. Nauk SSSR, p. 149.
VITKEVICH, V. V. (1956b) *ibid.* p. 312.
VITKEVICH, V. V. (1956c) *Astron. Zhurn.* **33**, 62.
VITKEVICH, V. V. (1957a) *Astron. Zhurn.* **34**, 217.
VITKEVICH, V. V. (1957b) in *Radio Astronomy*, Ed. H. C. VAN DE HULST, Cambridge Univ. Press. p. 363.
VITKEVICH, V. V. (1957c) Physics of Solar Corpuscular Streams and their Action on the Earth's Upper Atmosphere. *Proc. Conf. on Solar Res.* (in Russian), Akad. Nauk SSSR.
VITKEVICH, V. V. (1958) *Astron. Zhurn.* **35**, 52.
VITKEVICH, V. V. (1959) in *Paris Symposium on Radio Astronomy*, Ed. R. N. BRACEWELL, Stanford Univ. Press, p. 275.
*VITKEVICH, V. V. (1960a) *Astron. Zhurn.* **37**, 32.
VITKEVICH, V. V. (1960b) *Izv. Vuzov, Radiofiz.* **3**, 595.
VITKEVICH, V. V. (1961) *Radio Astronomy Investigations of the Sun's Supercorona* (in Russian), Fiz. Inst. Akad. Nauk SSSR.
VITKEVICH, V. V. (1962a) *Izv. Vuzov, Radiofiz.* **5**, 402.
VITKEVICH, V. V. (1962b) *Izv. Vuzov, Radiofiz.* **5**, 404.
*VITKEVICH, V. V. and GORELOVA, M. V. (1960) *Astron. Zhurn.* **37**, 622.
VITKEVICH, V. V., KAMENEVA, Z. I. and KOVALEVSKII, D. V. (1956) *Radiotekhn. i Elektron.* **1**, 864.
*VITKEVICH, V. V., KUZ'MIN, A. D., SALOMONOVICH, A. YE. and UDAL'TSOV, V. A. (1958) *Doklady Akad. Nauk SSSR* **118**, 1091.
VITKEVICH, V. V., KUZ'MIN, A. D., SALOMONOVICH, A. YE. and UDAL'TSOV, V. A. (1959) in *Paris Symposium on Radio Astronomy*, Ed. R. N. BRACEWELL, Stanford Univ. Press, p. 129.
VITKEVICH, V. V., LOTOVA, N. A. (1961) *Izv. Vuzov, Radiofiz.* **4**, 415.
VITKEVICH, V. V. and MATHEWSON, D. S. (1959) in *Paris Symposium on Radio Astronomy*, Ed. R. N. BRACEWELL, Stanford Univ. Press, p. 116.

References

VITKEVICH, V. V. and SIGAL, M. I. (1956) *Radiotekhn. i Elektron.* **1**, 861.
VITKEVICH, V. V. and SIGAL, M. I. (1957) *Astron. Zhurn.* **34**, 716.
VITKEVICH, V. V. and UDAL'TSOV, V. A. (1958) *Astron. Zhurn.* **35**, No. 5.
VLADIMIRSKII, V. V. (1948) *Zh. Eksp. i Teoret. Fiz.* **18**, 393.
VON ENGEL, A. (1955) *Ionised Gases*, Oxford.
WAAK, J. A. (1961) *Astron. J.* **66**, 298.
WALDMEIER, M. (1953a) *Atti Convegni Acad. Naz. Lincei* No. 4, 283.
WALDMEIER, M. (1953b) *Z. Astrophys.* **32**, 116.
WALDMEIER, M. (1956) *Z. Astrophys.* **40**, 221.
WALDMEIER, M. (1959) in *Paris Symposium on Radio Astronomy*, Ed. R. N. BRACEWELL, Stanford Univ. Press, p. 118.
WALDMEIER, M. and MÜLLER, H. (1948) *Astron. Mitteil. Eidg. Sternwarte* 155.
WALDMEIER, M. and MÜLLER, H. (1950) *Z. Astrophys.* **27**, 58.
WARWICK, J. W. (1957) *Ap. J.* **125**, 811.
WARWICK, J. W. (1960) *Science* **132**, 1250.
WARWICK, J. W. (1961) *Ann. New York. Acad. Sci.* **95**, 39.
WARWICK, J. W. (1961-2) Scientific Report No. 1, High Altitude Observatory, Boulder, Colorado.
WARWICK, J. W. (1963a) *Ap. J.* **137**, 41.
WARWICK, J. W. (1963b) *Science* **140**, 814.
WARWICK, J. W. (1967) *Space Science Review* **6**, 841.
WARWICK, C. and WARWICK, J. W. (1959) in *Paris Symposium on Radio Astronomy*, Ed. R. N. BRACEWELL, Stanford Univ. Press, p. 203.
WARWICK, C. and WOOD, M. (1959) *Ap. J.* **129**, 801.
WATSON, G. N. (1944) *Theory of Bessel Functions*, Cambridge.
WEAVER, R. R., MITCHELL, F. H. and WHITEHURST, R. N. (1958) *Bull. Amer. Phys. Soc.*, Ser. II, **3**, 301.
WEISS, A. A. and SHERIDAN, K. V. (1962) *J. Phys. Soc. Japan* **17**, Suppl. A-II, 223.
WEISS, A. A., STEWARD, R. T. (1965) *Austral. J. Phys.* **18**, 143.
WESSELINK, J. C. (1948) *Bull. Astron. Inst. Netherl.* **10**, 351.
WESTERHOUT, C. (1958) *Bull. Astron. Inst. Netherl.* **14**, 215.
WESTFOLD, K. C. (1957) *Phil. Mag.* **2**, 1287.
WHITEHURST, R. N., COPELAND, J. and MITCHELL, F. H. (1957) *J. Appl. Phys.* **28**, 295.
WHITEHURST, R. N. and MITCHELL, F. H. (1956a) *Astron. J.* **61**, 192.
WHITEHURST, R. N. and MITCHELL, F. H. (1956b) *Proc. IRE* **44**, 1879.
WHITFIELD, G. R. and HÖGBOM, J. (1957) *Nature* **180**, 602.
WHITFIELD, G. R. and HÖGBOM, J. (1959) in *Paris Symposium on Radio Astronomy*, Ed. R. N. BRACEWELL, Stanford Univ. Press, p. 56.
WILD, J. P. (1950a) *Austral. J. Sci. Res.* A **3**, 399.
WILD, J. P. (1950b) *ibid.* 541.
WILD, J. P. (1951) *Austral, J. Sci. Res.* A **4**, 36.
WILD, J. P. (1952) *Ap. J.* **115**, 206.
WILD, J. P. (1957) in *Radio Astronomy*, Ed. H. C. VAN DE HULST, Cambridge, p. 321.
WILD, J. P. (1960a) *Rendiconti della Scuola Internazionale di fisica "Enrico Fermi"*, Corso XII, Radioastronomia Solare, p. 281.
WILD, J. P. (1960b) *ibid.*, p. 296.
WILD, J. P. (1962) *J. Phys. Soc. Japan* **17**, Suppl. A-II, 249.
WILD, J. P. and MCCREADY, L. L. (1950) *Austral. J. Sci. Res.* A **3**, 387.
WILD, J. P., MURRAY, J. D. and ROWE, W. C. (1953) *Nature* **172**, 533.
WILD, J. P., MURRAY, J. D. and ROWE, W. C. (1954) *Austral. J. Phys.* **7**, 439.
WILD, J. P., ROBERTS, J. A. and MURRAY, J. D. (1954) *Nature* **173**, 532.
WILD, J. P. and SHERIDAN, K. V. (1958) *Proc. IRE* **46**, 160.
WILD, J. P., SHERIDAN, K. V. and NEYLAN, A. A. (1959) *Austral. J. Phys.* **12**, 369.

References

WILD, J. P., SHERIDAN, K. V. and TRENT, G. H. (1959) in *Paris Symposium on Radio Astronomy*, Ed. R. N. BRACEWELL, Stanford Univ. Press, p. 176.
WILD, J. P., SMERD, S. F. and WEISS, A. A. (1963) Solar bursts, *Ann. Rev. of Astronomy and Astrophysics* 1.
*WILD, J. P., SMERD, S. F. and WEISS, A. A. (1964) *Uspekhi Fiz. Nauk* **84**, No. 1.
WILD, J. P. and ZIRIN, H. (1956) *Austral. J. Phys.* **9**, 315.
WILDT, R. (1947) *Ap. J.* **105**, 36.
WINCKLER, J. R., PETERSON, L. E., HOFFMAN, R., ARNOLDY, R. and ANDERSON, K. A. (1959a) *J. Geophys. Res.* **64**, 1133.
WINCKLER, J. R., PETERSON, L. E., HOFFMAN, R., ARNOLDY, R. and ANDERSON, K. A. (1959b) *Bull. Amer. Phys. Soc.* **4**, No. 4, 238.
WOOD, M. B. (1961) *Austral. J. Phys.* **14**, 234.
WOOLLEY, R. (1947) *Austral. J. Sci. Res. Suppl.* 10(2), i.
WORT, D. J. H. (1962) *Nature* **195**, 1288.

YOUNG, C. W., SPENCER, C. L., MORETON, G. E. and ROBERTS, J. A. (1961) *Ap. J.* **133**, 243.
YUDOVICH, L. A. and FEL'DSHTEIN, YA. I. (1958) *Solnechnyye Dannyye* No. 3, 69.

ZAITSEV, V. V. (1965) *Astron. Zhurn.* **42**, 740.
ZAITSEV, V. V. (1968) *Astron. Zhurn.* **45**, 766.
ZAKHAROV, A. V., KROTIKOV, V. D., TROITSKII, V. S. and TSEITLIN, N. M. (1964) *Izv. Vuzov, Radiofiz.* **7**, No. 3.
ZELINSKAYA, M. R. and TROITSKII, V. S. (1956) *Proc. of the 5th Congress on Questions of Cosmogony* (in Russian), *Izd. Akad. Nauk. SSSR* p. 99.
ZELINSKAYA, M. R., TROITSKII, V. S., FEDOSEEV, L. N. (1959a) *Izv. Vuzov, Radiofiz.* **2**, 506.
ZELINSKAYA, M. R., TROITSKII, V. S., FEDOSEEV, L. N. (1959b) *Astron. Zhurn.* **36**, 643.
ZHELEZNYAKOV, V. V. (1955) *Astron. Zhurn.* **32**, 33.
ZHELEZNYAKOV, V. V. (1956) *Radiotekhn. i Elektron.* **1**, 840.
ZHELEZNYAKOV, V. V. (1958a) Radio emission of the Sun and planets, *Uspekhi Fiz. Nauk* **64**, 113.
ZHELEZNYAKOV, V. V. (1958b) *Izv. Vuzov, Radiofiz.* **1**, No. 4, 32.
ZHELEZNYAKOV, V. V. (1958c) *Astron. Zhurn.* **35**, 230.
ZHELEZNYAKOV, V. V. (1959a) Thesis (in Russian), Gor'kii Univ.
ZHELEZNYAKOV, V. V. (1959b) *Izv. Vuzov, Radiofiz.* **2**, 858.
ZHELEZNYAKOV, V. V. (1959c) *Izv. Vuzov, Radiofiz.* **2**, 14.
ZHELEZNYAKOV, V. V. (1960a) *Izv. Vuzov, Radiofiz.* **3**, 57.
ZHELEZNYAKOV, V. V. (1960b) *Izv. Vuzov, Radiofiz.* **3**, 180.
ZHELEZNYAKOV, V. V. (1961a) *Izv. Vuzov, Radiofiz.* **4**, 619.
ZHELEZNYAKOV, V. V. (1961b) *Izv. Vuzov, Radiofiz.* **4**, 849.
*ZHELEZNYAKOV, V. V. (1962) *Astron. Zhurn.* **39**, 5.
*ZHELEZNYAKOV, V. V. (1963) *Astron. Zhurn.* **40**, 829.
ZHELEZNYAKOV, V. V. (1964a) *Izv. Vuzov, Radiofiz.* **7**, 67.
*ZHELEZNYAKOV, V. V. (1964b) *Astron. Zhurn.* **41**, No. 5.
ZHELEZNYAKOV, V. V. (1964c) *Voprosy Kosmogonii* **10**, "Nauka".
*ZHELEZNYAKOV, V. V. (1964d) *Astron. Zhurn.* **41**, 1021.
*ZHELEZNYAKOV, V. V. (1965a) *Astron. Zhurn.* **42**, 96.
*ZHELEZNYAKOV, V. V. (1965b) *Astron. Zhurn.* **42**, 244.
*ZHELEZNYAKOV, V. V. (1965c) *Astron. Zhurn.* **42**, 798.
*ZHELEZNYAKOV, V. V. (1966) *Zh. Eksp. i Teoret. Fiz.* **51**, 570.
*ZHELEZNYAKOV, V. V. (1967) *Astron. Zhurn.* **44**, 42.
*ZHELEZNYAKOV, V. V. and TRAKHTENGERTS, V. YU. (1965) *Astron. Zhurn* **42**, 1005.
*ZHELEZNYAKOV, V. V., ZAITSEV, V. V. (1968) *Astron. Zhurn.* **45**, 19.
ZHELEZNYAKOV, V. V. and ZLOTNIK, YE. YA. (1962) *Izv. Vuzov, Radiofiz.* **5**, 644.

References

*Zheleznyakov, V. V. and Zlotnik, Ye. Ya. (1963a) *Astron. Zhurn.* **40**, 633.
Zheleznyakov, V. V., Zlotnik, Ye. Ya. (1963b) *Izv. Vuzov, Radiofiz.* **6**, 634.
Zhigalov, L. N. (1960) *Solnechnyye Dannyye*, No. 5, 69.
*Zirin, G. (1961) *Astron. Zhurn.* **38**, 861.
Zirin, H. and Severny, A. (1961) *Observatory* **81**, 155.
*Zlotnik, Ye. Ya. (1967) *Astron. Zhurn.* **44**, No. 2.
*Zlotnik, Ye. Ya. (1968a) *Astron. Zhurn.* **45**, 310.
*Zlotnik, Ye. Ya. (1968b) *Astron. Zhurn.* **45**, 585.

Index

Absorption
  by gyroresonance in solar corona 452–458
  in Earth's atmosphere 63–65
  of e.m. waves
    in isotropic plasma 430–440
    in magnetoactive plasma 440–452
Aerial
  Christiansen 40–41
  knife-edge 31
  Mills cross 42–44
  parabolic 30–31
  requirements 30–44
  temperature 23–26
Altitude of local sources on Sun 116–119
Amplification of e.m. waves in plasma 459–482
  amplification of harmonics 482–486
  classical treatment 478–482
  magnetoactive plasma 487–507
  quantum treatment 459–478
Anomalous uctu flations in signal 66–68
Atmosphere
  of Earth
    absorption in 63–65
    effect on observations 63–72
    refraction in 65–68
  of Sun 1–8
    temperature 3–7
    theoretical model 531–538
Aurora 224

Basic extraterrestrial radio emission 20–72
B-component of Sun's emission 73–98, 511–513
  complex theory 521–530
  simple theory 513–521
Blackouts due to Sun's emission 215, 230–235
Bremsstrahlung emission of e.m. waves 413–430, 579–583

Bremsstrahlung mechanism for S-component of Sun's emission 509–512, 541–551
Brightness distribution, determination 36–39
Brightness variation over Sun's disk 83–98

Cherenkov radiation 415–423, 509–512
Christiansen aerial 40–41
Chromosphere of Sun 2–3
Coherent emission 508–509
Comets, radio emission of 249–250
Compton effect 573–578
Continuum emission
  of Jupiter 270–277
  of Jupiter and Saturn, theory 624–631
  of Mars 277
  of Mercury 283
  of the Moon 283–296
  of planets 269–283
  of Saturn 269–270
  of Sun
    decimetric 202–204
    quiet Sun 73–98
    storms 208–210
  of Venus 277–282
Conversion of plasma waves to e.m. waves 373–385
  in isotropic plasma 376–385
Corona of Sun 3–8
  coronal condensation 14–15
  electron concentration in 170
  temperature 5–7

Decimetric continuum of Sun's emission 202–204
Decimetric outbursts of Sun's emission 204–208
Depolarisation of waves in solar corona 334–342
Detectors, frequency response of 21–22

Index

Directionality of disturbed Sun's S-emission 113–115
Disturbed Sun's S-emission 101–119
Dynamic spectrum of a source 28–29

Earth's atmosphere, refraction in 65–68
Eclipse observations 44–46, 84
Effective temperature
  of emission 23–26
  of Jupiter 259, 270–276
  of Mars 277
  of Mercury 283
  of Moon 284
  of Saturn 269
  of Sun 74–81, 83–88, 98, 110, 155, 199, 238–240
  of Venus 278–282
Einstein coefficients method for e.m. emission 467–471
Electron concentration in Sun's atmosphere 3–7, 170, 186
  fluctuations 170, 186
Emission
  effective temperature of 23–26
  transfer equation 408–413
E.m. wave propagation
  in isotropic plasma 298–318
  in magnetoactive plasma 318–342 487–501
  in solar corona 297–407
E.m. waves
  emission and absorption
    in equilibrium plasma 408–458
    in non-equilibrium plasma 458–507
  emission by individual particles 413–430
  in solar corona
    conversion from plasma waves 373–407
    depolarization 334–342
    emission 372–407
    Faraday effect 329–334
    polarisation 342–372
    interactions in plasma 342–407
Energy flux of sources 20
Event of 4 November 1957 210–211

Fade-outs due to Sun's emission 215, 231–237
Faraday effect in solar corona 329–334
Flares 12–14
Flocculi on Sun 8–9
  radio emission of 563–567

Frequency-drifting outbursts of Sun 154–194, 205–208
Frequency drift
  of type II bursts 167–176
  of type III bursts 185–191
Frequency response of detectors 21–22
Frequency spectrum
  of disturbed Sun's S-emission 110–113
  of Jupiter's emission 253–257
  of microwave bursts 125–128
  of Moon's emission 285–292
  of noise storms 137–139
  of quiet Sun emission 74–83
  of sources 21–22, 27–29
  of type V emission 196–197

γ-radiation accompanying solar bursts 222
Generation of e.m. waves
  behaviour in non-equilibrium plasma 459–507
  emission and absorption in equilibrium plasma 408–458
  in solar corona 408–609
  theory of Sun's non-thermal emission 568–609
  theory of Sun's thermal emission 508–567
Geomagnetic disturbances 224–230
Geophysical effects of Sun's emission 215–243
Gyroresonance absorption in solar corona 452–458

Haloes, radio emission of 563–567
Hot surface model for radio emission of Venus 645–651
Hydrogen lines 82–83

Intensity of emission 20–22
Interaction of waves in a plasma 342–407
  calculation by phase integral 353
  explanation for polarisation 365–372
  in presence of transverse field 350–353
Interference polarimeter 55–59
Interferometers 31–44
  knife-edge 31
  Mills Cross 42–44
  multi-element 39–44
  phase-switching 34
  two-element 31–39

Index

Ionospheric absorption 64
Ionospheric disturbances due to Sun 218–224
Ionospheric model for Venus's radio emission 639–644
Isotropic plasma, wave propagation in 305–318
  scattering due to irregularities 316

Jupiter
  physical condition of 17–18
  radio emission of 244–247, 250–268
Jupiter's emission
  connection with solar activity 266–268
  continuous spectrum of 270–271
  effective temperature 270–271
  frequency spectrum 253–257
  local sources 258–266
  polarization 257–258
  theory of continuous emission 624–638
  theory of sporadic emission 610–624
  time dependence of 250–253

Kinetic equation method for e.m. emission 459–466
Knife-edge aerial 30–31

Mars
  physical condition of 16–17
  radio emission of 277
Magnetic disturbances on Earth 224–230
Magnetic field of sun-spots 9–12, 553–555
Magnetic plasma mechanism for Jupiter's sporadic emission 616–624
Magnetoactive plasma
  non-uniform, propagation of waves in 325–329
  uniform, propagation of waves in 318–325
Magneto-bremsstrahlung emission of e.m waves 413–430, 509–512, 579–583
Magneto-bremsstrahlung mechanism for S-component 551–563
Mercury
  physical condition of 15
  radio emission of 283
Microwave bursts from Sun 120–135
  correlation with flares 130–134
  frequency spectrum 125–128

origin 569–578
  polarisation 128–130
  size of sources 123
  types 120–122
Mills Cross aerial 42–44
Multichannel interferometers 39–44
Multichannel receivers 27–29
Moon
  physical condition of 18–19
  radio emission of 244, 283–296
Moon's emission
  brightness distribution 292–296
  frequency spectrum 285–292
  interpretation of results 657–668
  theory 651–668

Noise storms of Sun 135–154
  correlation with optical features 137–143
  directionality 144
  frequency spectrum 137–139
  polarization 148–454
  position of sources 144–148
  time scale 136–137
Non-coherent emission 508–509

Origin of slowly-varying component 538–567

Parabolic aerials 30–31
Phase-switching interferometer 34
Photosphere of Sun 1–2
Physical conditions
  of Moon and planets 15–19
  of Sun 1–15
Plages on Sun 8–9
  connection with radio emission 106–107
Plasma
  interaction of waves in 342–407
  isotropic, propagation of waves in 298–318
  magnetoactive, propagation of waves in 318–342
Plasma oscillation mechanism for Jupiter's sporadic emission 612–624
Plasma waves
  conversion to e.m. 373–407
  in shock wave fronts 501–507
Polar blackouts 215, 230–235
Polarimeters 53–62
  for centimetric waves 59–62

695

Index

Polarimeters *(cont.)*
  interference 55–59
  two-aerial 53–55
Polarization
  of disturbed Sun's S-emission 115–116
  of emission 46–62
    leaving coronal plasma 344–350
  of Jupiter's emission 257–258
  of noise storms 148–154
  of quiet Sun emission 81–82
  of Sun's emission, theoretical explanation 365–372
  of Sun's microwave bursts 128–130
  of type I bursts (noise storms) 148–154
  of type II bursts 156
  of type III bursts 180–181
  of type IV emission 199
  of waves in coronal plasma 342–372
  measurement 51–62
  Stokes parameters 47
Prominences on Sun 9, 174–175
Pulkovo radio telescope 31

Quasihydrodynamic method for wave propagation in plasma 298–305
Quiet Sun emission 73–98
  complex model 521–531
  frequency spectrum 74–83
  hydrogen lines 83
  polarisation 81–82
  simple model 513–521
  theory 508–538
  variation with wavelength 77–80

Radiation, Cherenkov 415–423, 509–512
Radio brightness
  of noise storms 147
  of Sun, variation with wavelength 104–112
  over Sun's disk 83–98
Radio emission
  extraterrestrial, basic features 20–72
  of specific sources, see under source
Radio spectrographs 27–29
Receivers, multichannel 27–29
Refraction in Earth's atmosphere 65–68
Response of radio telescopes 21
Reverse-drift pairs in Sun's emission 212–215
  theory 600–602

Reversing layer of Sun 1

Satellites, observations from 72
Saturn
  physical conditions of 17–18
  radio emission of 248, 269
    conditions of generation 631
    theory 624–638
S-component of Sun's emission 101–119, 538–567
  bremsstrahlung mechanism 541–551
  magneto-bremsstrahlung mechanism 551–563
  thermal nature 538–541
Shock wave fronts, plasma waves in 501–507
Solar corona 3–8
  wave propagation in 297–407
Spectrographs 27–29
Spicules of Sun 2–3, 528–529
Sporadic emission of Sun 99–243
  effect on Earth 215–243
  effect on Jupiter 206–208
  microwave bursts 120–135
    origin 569–578
  noise storms 135–154
    origin 606–609
  other bursts 202–215
  S-component 101–119
    origin 538–567
  summary 237–243
  table 238–240
  theory 568–609
  types 100
  type I bursts (noise storms) 135–154
    theory 606–609
  type II bursts 154–176
    theory 602–605
  type III bursts 176–194
    theory 589–602
  type IV emission 194–201
    theory 583–589
  type V emission 201–202
  U-bursts 191–194
  wide-band bursts 212
Sun
  activity of 8–14
  atmosphere of 1–8, 531–538
  chromosphere of 2–3
  corona of 3–8
  effective temperature of 74–81
  electron concentration in 3–7
  emission when quiet 73–98

Index

Sun *(cont.)*
  photosphere of 1–2
  physical condition 1–15
  quiet emission
    frequency spectrum 74–83
    hydrogen lines 83
    theory 508–538
    polarization 81–82
    variation with wavelength 77–80
  reversing layer of 1
  spicules of 2–3
  sporadic emission of 99–243
  supercorona of 2
  temperature of 3–7
Sunspots 9–12
  magnetic field of 10–12, 105
  radio emission of 101–102, 579–583
Supercorona of Sun 2
Synchrotron emission of e.m. waves 413–430
  from relativistic plasma 458

Temperature
  effective
    of Mars 277
    of Mercury 283
    of Moon 284
    of Saturn 269
    of Sun 74–81, 83–88, 98, 110, 155, 199, 238–240
    of Venus 278–282
  of emission, effective 23–26
  of Sun's atmosphere 3–7
Theoretical model of atmosphere of Sun 531–538
Theory of Sun's emission 509–609
  non-thermal 567–609
  thermal 509–567
Thunderstorm model for Jupiter's sporadic emission 610–612
Tropospheric absorption 63
Two-aerial polarimeter 53–55
Two-element interferometers 31–39
Type I bursts from Sun (noise storms) 135–154

  correlation with optical features 137–143
  directionality 144
  frequency spectrum 137–139
  polarization 148–154
  position of sources 144–148
  theory 606–609
  time scale 136–137
Type II bursts of Sun 154–176
  correlation with optical features 155–156
  fine structure 165–167
  frequency drift 167–176
  harmonics 157–165
  polarization 156
  theory 602–605
Type III bursts of Sun 176–194
  correlation with optical features 181–184
  frequency drifting 185–191
  polarization 180–181
  theory 589–602
  U-bursts 191–194
Type IV emission 195–200
  frequency spectrum 196–197
  motion 198
  polarization 199
  theory 583–589

Vavilov–Cherenkov effect 415–423, 432
Venus
  effective temperature of 277–282
  physical condition of 15–16
  radio emission of 247, 277–282
  theory 638–651

Wave propagation *see* E.m. wave propagation
Wide-band bursts of Sun's emission 212
Wolf number 10

X-radiation accompanying solar bursts 221–222

DATE DUE